FUNDAMENTALS OF
WEED SCIENCE

FUNDAMENTALS OF WEED SCIENCE

FIFTH EDITION

ROBERT L. ZIMDAHL

Professor Emeritus, Department of Bioagricultural Sciences and Pest Management
Colorado State University, Fort Collins, Colorado

Winner of the 2019 William Holmes McGuffey Longevity Award

This award from the Textbook & Academic Authors Association (TAA) recognises textbooks which have maintained their excellence over time. First published in 1993, this long-standing legacy title has been empowering botany and agriculture students and researchers for more than 25 years.

ACADEMIC PRESS
An imprint of Elsevier

Academic Press is an imprint of Elsevier
125 London Wall, London EC2Y 5AS, United Kingdom
525 B Street, Suite 1800, San Diego, CA 92101-4495, United States
50 Hampshire Street, 5th Floor, Cambridge, MA 02139, United States
The Boulevard, Langford Lane, Kidlington, Oxford OX5 1GB, United Kingdom

Library of Congress Cataloging-in-Publication Data
A catalog record for this book is available from the Library of Congress

British Library Cataloguing-in-Publication Data
A catalogue record for this book is available from the British Library

ISBN: 978-0-12-811143-7

For information on all Academic Press publications visit our website at
https://www.elsevier.com/books-and-journals

Working together
to grow libraries in
developing countries

www.elsevier.com • www.bookaid.org

Publisher: Andre Gerhard Wolff
Acquisition Editor: Nancy Maragioglio
Editorial Project Manager: Billie Jean Fernandez
Production Project Manager: Kiruthika Govindaraju
Designer: Mark Rogers

Typeset by TNQ Books and Journals

*This book is dedicated to the memory
of
Ann Osborn Zimdahl
and
Pamela Jeanne Zimdahl*

How little we know of what there is to know.
I wish that I were going to live a long time....
I'd like to be an old man and to really know
I wonder if you keep on learning or if there is
only a certain amount each man can understand.
I thought I knew about so many things that I
know nothing of. I wish there was more time.

Hemingway, E. 1940. For Whom the Bell Tolls Scribner. New York, NY. p.380.

Contents

8. Invasive Plants

ROBERT L. ZIMDAHL, CYNTHIA S. BROWN

9. Allelopathy

10. Methods of Weed Management

11. Weed Management in Organic Farming Systems

15. Herbicides and Soil

16. Properties and Uses of Herbicides

17. Herbicide Formulation

18. The Role and Future of Genetic Modification in Weed Science

Preface

About 50 years ago, Monsanto Company distributed a picture that hangs in my study. It shows four books[1] with several weed seedlings emerging from each. Two (Ahlgren et al. and King) were textbooks and two (Muenscher and Fernald) were plant identification books. They were the beginning of a now greatly expanded literature of weed science.

Many, but not all of the early textbooks written for undergraduate weed science courses lacked an ecological-management perspective for the rapidly developing science of weeds and their control. This book does not ignore the history of weed science and the development of chemical weed control; rather, it portrays herbicides as one management technique among many. It is undeniable that in the agricultural production system of developed countries, herbicides dominate weed control and management. Whether this creates a sustainable agricultural system is an important question. It has been and will continue to be discussed. Contributing to the discussion is part of but not the primary purpose of this book.

Whether one lives in a developed or developing country and whether one is rich or poor, male or female, educated or not, we all live in a postindustrial, information-age society. We are fortunate to live in an era of scientific achievement and technological progress perhaps unequaled in human history, which has created the good life many of us enjoy and some of the problems we experience. The achievements include:

Waking up this morning to music from your cell phone,
Preparing breakfast in your microwave as you review the news on your computer, which gives you nearly instant access to information that is orders of magnitude greater than the resources of any of the world's libraries,
Medical advances that cure what used to kill or cripple,
Immunization to prevent childhood diseases,
Elimination of smallpox and possibly polio in the near future,
Vastly improved detection and control of some diseases,
Travel at speeds and convenience unknown to our grandparents, across oceans and mountains that were once formidable barriers,
and, finally, for many, abundant food.

The problems include climate change, global warming, pollution of all forms, social inequality, environmental degradation, and soil erosion. Many citizens of developed countries know and benefit from the achievements

---

[1]The books are: Ahlgren, G.H., Klingman, G.C., Wolf, D.E., 1951. *Principles of Weed Control*. J. Wiley & Sons, New York, NY. 368 pp.; Fernald, M.L., 1970. *Gray's Manual of Botany*. Eighth ed. American Book Co.; King, L.J., 1966. *Weeds of the World: Biology and Control*. Interscience Pub., Inc., New York, NY. 526 pp.; and Muenscher, W.C., 1935. *Weeds*. The Macmillan Co., New York, NY. 577 pp.

of science and technology and are concerned about the problems that science and technology have wrought.

Despite its many benefits, science is commonly regarded with suspicion, if not fear, by the public, which enjoys and benefits from science and its technology. Scientists, including weed scientists, eagerly accepted the credit when, after World War II, advances in societal development were widely regarded as contributions of science, and in fact were. The public regarded these advances, which included herbicides and other pesticides, as desirable and benign. Now science and its technology are held responsible for many societal, environmental, and ecological problems. Herbicides are no longer regarded as benign, but as threats to humans, other creatures, and the environment. They are commonly regarded as undesirable scientific creations. The public's attitude toward science and scientists is mingled awe and fear. The practice of science is constrained because although it claims to be an end in itself, it is publicly supported and tolerated because of its utility, its practical value. It is feared because of well-known undesirable consequences (see Endnote: Agricultural Examples). Weed science is typical, and because of its close identification with herbicides, it may be regarded with more fear than some other areas of agricultural science. A few glaring errors of the agricultural community (including weed science) are illustrated in comments made by James Davidson, Emeritus Vice President for Agriculture and Natural Resources, University of Florida.[2]

> With the publication of Rachel Carson's book entitled *Silent Spring*, we, in the agricultural community, loudly and in unison stated that pesticides did not contaminate the environment—we now admit that they do. When confronted with the presence of nitrates in groundwater, we responded that it was not possible for nitrates from commercial fertilizer to reach groundwater in excess of 10 parts per million under normal productive agricultural systems—we now admit they do. When questioned about the presence of pesticides in food and food quality, we reassured the public that if the pesticide was applied in compliance with the label, agricultural products would be free of pesticides—we now admit they're not.

When criticism of herbicides and other agricultural technology was offered, disturbingly often the agricultural community responded by questioning the accuracy of the science, attacking the credibility of the scientist(s), and warning of increasing, punitive costs if the critics, who were often labeled environmentalists (an epithet) were allowed to prevail.

This book will not pursue this discussion, but it is important that students of weed science know about the issues. The public's lack of understanding or its misunderstanding of what weed scientists do will not lessen the need for what is done, but it increases the responsibility of weed scientists to be clear about the problem of weeds and proposed solutions. The responsibility is not so much to educate the public about "what we do" as it is to engage in a conversation (a dialogue, not a monologue) with the public. It is using science not just to persuade and defend a position, but to explore and discover other views. It is an engagement in public scholarship in which original, peer-reviewed intellectual work is fully integrated with the social learning of the public (Jordan et al., 2002).

This book includes herbicides and their use as an important aspect of modern weed

[2]Davidson's comments were made several years ago and cited by Kirschenmann, F, 2010. Some things are priceless. *Leopold Lett.* 22 [1]:5.

management and strives to place them in an ecological framework. (Common names of herbicides will be used throughout the text, except in some tables where they may be paired with one or more trade names.) Any book that purports to discuss current practice (and the art) of weed management would be of little consequence and limited value to students and others who wish to know about weed management if it omitted discussion of herbicides.

Many weed scientists believe agriculture to be a continuing struggle with weeds. That is, they believe that without good weed control, good, profitable agriculture is impossible and herbicides are an essential component of productive success. Weed science and other agricultural disciplines regard their role to be central to agriculture's success and continued progress. While not denying the importance of weed management to successful agriculture, its role in the larger ecological context is emphasized. The role of culture, economics, and politics in weed management is mentioned, but they are not strong themes.

This, the fifth edition, is not a complete revision of the earlier editions, but it has been changed in several significant ways, while striving to maintain an overall ecological framework.

Some references in the first edition have been omitted and 505 references have been added, 197 of which are work published after 2013. The literature review for this edition was completed in early 2017.

The chapters are arranged in a logical progression. Chapter 1 asks and answers the questions "Why study weeds and Why are they important?" The second chapter pursues discussion of the definition of weed begun in Chapter 1 and presents the characteristics and harmful aspects of weeds. It concludes with a discussion of what weeds cost. Chapter 3 classifies weeds in several

ways, and Chapter 4, unique among weed science texts, discusses the fact that not all plants that are weedy in some environments are weeds (that is, undesirable) in all places. Many plants have uses known to many and are studied by ethnobotanists. Weed reproduction and dispersal and the important topics of seed germination and dormancy are presented in Chapter 5. Chapter 6 is important because it presents the fundamental ecological base of weed science, including plant competition and the interactions of weeds and other pests. The significance of weed-crop competition is included. Chapter 7, "Population Biology of Weeds," was written by Dr. Michael Christophers, associate professor, Department of Plant Sciences, North Dakota State University, Fargo, North Dakota. It describes the role and increasing importance of population biology to weed science. Chapter 8, also unique among weed science textbooks, first appeared in the third edition. It is an extended discussion of the role and importance of invasive plant species. The coauthor is Dr. Cynthia S. Brown, professor, Department of Bioagricultural Sciences and Pest Management, Colorado State University, Fort Collins, Colorado. It is followed by a discussion of allelopathy in Chapter 9, a subject included as a minor point in many weed science texts.

Chapter 10 begins consideration of methods of weed management. For many, this is the essence of weed science, the fundamental topic: how are weeds controlled? Weed problems are an inevitable part of production agriculture and any place where we grow plants. Those who wish to control them need to ask why the weed is there, as well as how to manage or control it (Zimdahl, 1999). Key concepts of prevention, control, and management are presented followed by presentation of mechanical, nonmechanical, and cultural control techniques.

Chapter 11 discusses the challenges of weed management in organic cropping systems. Chapter 12 introduces important concepts related to biological control of weeds. Chapter 13 introduces herbicides and chemical control of weeds. This is not a "how-to" book. Herbicides are discussed in depth in this chapter and Chapter 16, but there are no recommendations about what herbicide to use in a crop. Chapters 14 and 15 are central to understanding interactions between herbicides and plants and herbicides and soil. Chapter 16, one of the longest and most difficult, classifies herbicides based on how they do what they do: their site of action and their chemical family. Herbicide formulation is covered in Chapter 17.

Chapter 18, new in this edition, deals with the role and future of genetic modification in weed science. Chapter 19 is an expansion of previously included information on the problem, study, and challenges of herbicide resistance among weeds. The influence of molecular biology on weed management is included. Chapter 20 returns to the ecological theme, but this time with information on the interaction between herbicides and the environment, including effects on water, humans, and global change. A central and intentionally unanswered question is how one balances and judges the potential harmful and beneficial aspects of herbicides. The chapter concludes with a discussion of herbicide safety. Chapter 21 is a brief presentation of the US legislative decisions required to address some of the questions raised in Chapter 19. Chapter 22 brings things together by discussing eight among many weed management systems, each of which is primarily conceptual, not completely prescriptive. A section on weed management decision aids completes the chapter. The last section, Chapter 23, presents a view of the future of weed science. It is meant to provoke thought and discussion. It is not an infallible prediction of what will be.

There is a strong, growing trend in weed science away from exclusive study of annual control techniques toward understanding weeds and the systems in which they occur. Control is important but understanding endures. Herbicides and weed control are important parts of the science and of this text, but it is hoped that understanding the principles of management and the biology and ecology of the weeds to be managed will be seen as the dominant themes. The primary objectives of the book are to introduce concepts fundamental to weed science and provide adequate citations so interested readers can pursue specific interests and learn more.

Study of weeds, weed management, and herbicides is a challenging, demanding task that requires diverse abilities. Weed science involves far more than answering the difficult question of what chemical will selectively kill weeds in a given crop. Weed science includes work on selecting methods to control weeds in a broad range of crops, on noncrop lands, in forests, and in water. Weed scientists justifiably claim repute as plant physiologists, ecologists, botanists, agronomists, organic and physical chemists, molecular biologists, and biochemists. However, lest the reader be intimidated by that list of disciplines, I hasten to add that this text will emphasize general principles (the fundamentals) of weed science and not attempt to include all applicable knowledge. It is tempting, and would not be much more difficult, to incorporate extensive, sophisticated knowledge developed by weed scientists. Although this knowledge is impressive and valuable, it is beyond the scope of an introductory text.

I have claimed, with only anecdotal data, that weeding crops consumes more human energy than anything else we do, and that much of it is done by women. I hope the book conveys some of the challenges of the world of weeds, their management, and the importance of weed problems to agriculture and society, and to meeting the demand to feed a growing world population. The aim has been to include most aspects of weed science without exhaustively exploring each. The book is designed for undergraduate weed science courses. I hope it is not too simple for sophisticated readers and that omissions of depth of coverage do not sacrifice accuracy and necessary detail. Readers should note that in nearly all cases I have used the units of measure used in the original reference rather than changing all to one measurement system.

Several colleagues provided helpful suggestions on this and earlier editions of the *Fundamentals of Weed Science*. I thank all of them, even though some comments were difficult to hear. The first edition had several errors of fact that have been corrected in subsequent editions. I thank the following colleagues for suggestions and critical review of portions of the manuscript included in this and earlier editions: Dr. Kenneth A. Barbarick, Dr. K. George Beck, Dr. Cynthia S. Brown, Dr. Sandra K. McDonald, Dr Philip Westra, Dr. Scott J. Nissen, and Mr. Steven Markovits of Colorado State University; Dr. William W. Donald US Department of Agriculture/Agricultural Research Service, University of Missouri; Dr. David L. Mortensen, The Pennsylvania State University; Dr. Robert F. Norris, University of California—Davis; Dr. Gregory L. Orr, Fort Collins, Colorado; Dr. Keith Parker, Syngenta Corp, Greensboro, North Carolina; Dr. Alan R. Putnam, Gallatin, Gateway, Montana; Dr. Albert E. Smith, Jr., University of Georgia; Dr. Malcolm D. Devine, vice president, Crop Development, Performance Plants, Saskatoon, Saskatchewan, Canada; and Dr. Steven Brunt, BASF Corp, Research Triangle Park, North Carolina. The book has been improved because of their efforts and the comments of anonymous reviewers and several colleagues on this and earlier editions.

My wife, Pamela J. Zimdahl (deceased 2012) encouraged my writing and offered comments and criticism when she thought they were appropriate. They usually were. Errors of interpretation or fact are solely my responsibility.

Robert L. Zimdahl
Fort Collins, Colorado, 2017

ENDNOTE: AGRICULTURAL EXAMPLES

- The mid-1960s controversy over the real and suspected hazards of 2,4,5-trichlorophenoxyacetic acid, a component of Agent Orange used in Operation Ranch Hand, a vegetation control program during the Vietnam war. It was the first major public debate that challenged the intellectual foundation of weed science and its dependence on herbicides (see Chapter 2).
- On December 3, 1984, a poisonous cloud of methyl isocyanate, used in the manufacture of pesticides, escaped from Union Carbide's plant in Bhopal, India, killing 14,000 people and permanently injuring 30,000.
- Regarding pesticide poisoning: No one knows for sure, but it is estimated that between one and five million cases of pesticide poisoning occur every year in the world, resulting in 20,000 deaths. Developing countries use 25% of pesticides but experience 99% of deaths (see Chapter 20).
- Debate within and outside the agricultural community over the risks and

ultimate beneficiaries of genetic modification of crops has raised legitimate economic, social, and biological concern. A major concern about weed science is the widespread occurrence of herbicide resistance.
- Air and water pollution and animal suffering have resulted from confined animal feeding operations.
- Mad cow disease, swine flu, bird flu, meat recalls, and antibiotic resistance are all of concern or have been of major societal concern in the past.
- The ecological dead zone extending into the Gulf of Mexico from the Mississippi terminus.

Literature Cited

Jordan, N., Gunsolus, J., Becker, R., White, S., 2002. Public scholarship—linking weed science with public work. Weed Sci. 50, 547–554.

Zimdahl, R.L., 1999. My view. Weed Sci. 47 (1).

1

Introduction

See them tumbling down,
Pledging their love to the ground
Lonely but free I'll be found
Drifting along with the tumbling tumbleweeds.

Cares of the past are behind
Nowhere to go but I'll find
Just where the trail will wind
Drifting along with the tumbling tumbleweeds.

I know when night has gone
That a new day starts at dawn.

I'll keep rolling along
Deep in my heart is a song
Here on the range I belong
Drifting along with the tumbling tumbleweeds. *Composed in 1932 by Bob Nolan and recorded by the Sons of the Pioneers*

Weeds have been the subject of songs and receive a lot of attention, but they have never been respected or well understood. The fact that many people earn a living and serve society by working to control and manage them is often greeted with amusement if not outright laughter. Even scientific colleagues who work in other esoteric disciplines find it hard to believe that another group of scientists could be concerned exclusively with what is perceived to be as mundane and ordinary as weeds.

Weeds have surely been with us since the advent of settled agriculture some 10,000 years ago. It has been suggested that the most common characteristic of the ancestors of our presently dominant crop plants is their weediness—their tendency to be successful, to thrive, in disturbed habitats, most notably those around human dwellings (Cox, 2006).

Bailey (1906, p.199), to whom agricultural science owes so much, spoke of the Sisyphean battle against Russian thistle in the western United States:

> What I have thus far stated is only a well-known truth in organic evolution, — that the distribution of an animal or plant upon the earth, and to a great extent the attributes of the organism itself, are the result of a struggle with other organisms. A plant which becomes a weed is only a victor in a battle with farm crops; and if the farmer is in command of the vanquished army, it speaks ill for his generalship when he is routed by a pigweed or a Russian thistle.

It is not surprising to me that students who enroll in a course about weeds often wonder why the course is recommended or, perhaps required, and what it is about. Students who enroll in chemistry or English have a reasonably good idea what the class will be about and how it fits in their curriculum. It is often not true for students of weed science. Of course, students from farms and ranches know about weeds and the problems they cause, but they do not always comprehend the complexities of weed management and the management skill required. Therefore, it is important to our pursuit that the nature of the subject be established and that the subject matter be related to students' knowledge of agricultural, biological, and general science. From the beginning, a textbook, the professor, and the student should strive to establish relationships between weed science, agriculture, and society. It is the intent of this book to introduce the fundamental concepts of weed science, show how they have changed with time, and conclude (Chapter 23) with some thoughts on the future of weed science. I risk thinking about the future in full recognition that prediction is very difficult, especially about the future.[1] However, I hasten to add a thought from Saint-Exupéry (1950, p. 155)—"As for the Future, your task is not to foresee it, but to enable it."

A brief review will lead to the conclusion that the story of agriculture is a story of the struggles that have ensued "in consequence of the sudden overturning of established conditions, and the substitution therefor of a very imperfect and one-sided system of land occupancy" (Bailey, 1906, p. 200)—what we know as modern agriculture. Agricultural history, a fascinating subject, is too large a task for this book and only small bits are included. Those interested in beginning a study of agricultural history and weed science are referred to Goodwin and Johnstone (1940) and Rasmussen (1975). The history of weed control was reviewed by Timmons (1970, republished in 2005), Appleby (2005), and Zimdahl (2010).

Formidable obstacles have been placed between humans and a continuing food supply. These include:

- Physical constraints: lack of appropriate technology, lack of good highways, inadequate or no infrastructure for transportation of supplies to farms or produce to markets. These are food distribution, not food production, problems.
- Economic constraints: lack of credit and operating funds. An overwhelming constraint to feeding people is poverty—no income or an income inadequate to purchase food.

---

[1]This saying has been attributed to Niels Bohr (It is difficult to predict, especially the future (Yale Book of Quotations, p. 92). The Yale book cites K. K. Steincke's 1948 book Goodbye and Thanks, which cites it as a pun used in the Danish parliament in the late 1930s. Others attribute it to Mark Twain who wrote about prophecy in A Connecticut Yankee in King Arthur's Court and in The Innocents Abroad, but the precise quote is not in either book. Twain may have said it, but he didn't write it.

- Environmental constraints: too much or too little water, short growing season, poor soil, highly eroded soil, uncontrolled weeds, insects, and plant diseases.
- Biological constraints: poor soil fertility, poorly adapted plant varieties, low or high soil pH, salinity, inadequate expensive fertilizer, inadequate inappropriate technology, and lack of applicable agricultural research, research scientists, and facilities.
- Political constraints: inadequate agricultural research funds and dysfunctional markets. There is no universal, or even a local, human right to food.

One of the most formidable environmental constraints has been pests. In many developing countries, between 40% and 50% of most crops are lost to insects, diseases, and inadequate storage before they reach the market. Surveys by the Food and Agriculture Organization (FAO) of the United Nations (1963, 1975) showed that in the 1960s and 1970s more than one-third of the potential annual world food harvest was destroyed by pests. In 1975, the $75 billion loss was equivalent to the value of the world's grain crop (about $65 billion) and the world's potato[2] crop (about $10 billion). This means that insects, plant diseases, nematodes, and weeds deprived humans of food worth more than the entire world crop of wheat, rye, barley, oats, corn, millet, rice, and potatoes. The data reported herein were only up to harvest and do not include damage during storage—another large sum. Currently, less complete estimates show that in spite of the abundant research and pesticide use, losses due to pests of all kinds have increased since the first FAO estimates were made.

Estimates of potential and actual losses despite the current crop protection practices are given for wheat, rice, maize, potatoes, soybeans, and cotton for 2001—03 on a regional basis (19 regions) as well as for the global total. Among crops, the total global potential loss due to pests varied from about 50% in wheat to more than 80% in cotton. The responses estimated losses of 26%—29% for soybean, wheat, and cotton, and 31%, 37%, and 40% for maize, rice, and potatoes, respectively. Overall, weeds produced the highest potential loss (34%), with animal pests and pathogens being less important (18% and 16%) (Oerke, 2006; Pinstrup-Andersen, 2001). Another study reported that weeds produced the highest potential loss (30%), with animal pests and pathogens being less important (23% and 17%). The efficacy of control of pathogens and animal pests was 32% and 39%, respectively, compared to almost 74% for weed control.[3]

Losses during production are important, but one must also acknowledge the continuing burden of food loss due to waste. Data from the Institution of Mechanical Engineers (IMechE) (Fox and Fimeche, 2013) show as much as 2 billion metric tons of food never make it to a plate. That strongly suggests that about half of all the food produced in the world is lost every year due to poor harvest practices, storage and transportation, and market and consumer waste. Fox and Fimeche estimate that 30%—50% (or 1.2—2 billion tons) of all food produced never reaches a human stomach. The waste does not reflect the fact that large amounts of land, energy, fertilizer, and water are used for no human benefit. IMechE blames the "staggering" loss on unnecessarily strict sell-by dates, buy-one-get-one-free marketing, and Western consumer demand for cosmetically perfect food, along with "poor engineering and agricultural practices," inadequate infrastructure, and poor storage facilities.

---

[2]Common and scientific names of all crops and weeds mentioned in the text are in Appendices 1 and 2.

[3]http://www.agrivi.com/yield-losses-due-to-pests/.

Per capita food loss in Europe and North America is 280–300 kg/year. In sub-Saharan Africa and South/Southeast Asia it is 120–170 kg/year. Total per capita production of food for human consumption in Europe and North America is about 900 kg/year, whereas in sub-Saharan Africa and South/Southeast Asia it is 460 kg/year. Per capita food wasted by consumers in Europe and North America is 95–115 kg/year, but only 6–11 kg/year in sub-Saharan Africa and South/Southeast Asia. Food losses in industrialized countries are too high, but in developing countries more than 40% of the food losses occur at postharvest and processing levels, while in industrialized countries, more than 40% of the food losses occur at retail and consumer levels. Food waste by consumers in industrialized countries (222 million tons) is almost as high as the total net food production in sub-Saharan Africa (230 million tons). It is a bleak tale that challenges all of agriculture.

History is filled with examples of human conflicts with pests, from biblical to modern times. Examples include:

- The desert locust (*Schistocerca gregaria* Forskal), a pest since biblical times. They unexpectedly appear and can strip a field bare in an hour. They prefer grasses, but consume a wide range of crops.
- Short-horned grasshoppers and locusts—a large family found predominantly in warmer regions. As many as 500 species are agricultural pests. Locust swarms threaten primarily grass crops on about one-third of the world's land surface (Hill, 1994).
- Late blight [*Phytophthora infestans* (Mont.) D. By.] caused the Irish potato famine of 1845–49.
- The continuing worldwide presence of Colorado potato beetles (*Leptinotarsa decemlineata* Say).
- The 1970s epidemic of Southern corn leaf blight (*Helminthosporium maydis* Nisik and Miyake).
- Western corn rootworm (*Diabrotica virgifera virgifera*).
- The spread of the mountain pine beetle (*Dendroctonus ponderosae*) in British Columbia, Canada, and the western United States has killed millions of hectares of lodgepole pine forest and released an estimated 270 million tons of carbon, converting the forest from a carbon sink to a large net carbon source.
- The *Puccinia graminis tritici* strain of wheat rust, discovered in Uganda in 1998, has spread across Africa, Asia, and the Middle East.

The battle has not ended. It has become more intense. In the industrialized (rich) world agriculture has evolved from many small farms to large industrial-scale, dominantly mono-cultural farms. The widespread use of synthetic fertilizer, chemical pesticides and the more recent introduction of genetic modification of plants have created enormous changes in the way agriculture is practiced. The most reliable estimates are from the United Nations population division, which shows the 2017 human population is 7.4 billion and is projected to be 9.7 billion by 2050 and possibly 11.2 billion by 2100. Today the world must feed almost 240,000 more people than yesterday. Population growth has increased the need and the demand for ever greater quantities of high-quality food and meat.

One must respect the prescience of Swift (1677–1745, See Williams 1937) who said:

> Hobbes clearly proves that every creature
> Lives in a state of war with nature
> ...

So, Nat'ralists observe, a Flea
Hath smaller Fleas that on him prey;
And these have smaller Fleas to bite 'em:
And so proceed ad infinitum.

De Morgan (1850), who probably had read, but did not cite, Swift's poem, expressed the ubiquity of pests several years later:

"Great fleas have little fleas upon
their backs to bite 'em,
And little fleas have lesser fleas,
and so ad infinitum,
And the great fleas themselves, in
turn, have greater fleas to go on;
While these again have greater still,
and greater still, and so on.

The subject of this book is weeds—visible, unspectacular pests, whose presence is obvious nearly everywhere, but whose effects are not. Weeds have always been with us and are included in some of our oldest literature:

Cursed is the ground for thy sake;
in sorrow shalt thou eat of it all the days of thy life;
thorns and thistles shall it bring forth to thee;
and thou shalt eat the herb of the field. *Genesis 3:17—18*

Ye shall know them by their fruits. Do men gather grapes
of thorns, or figs of thistles? *Matthew 7:16*

And thorns shall come up in her palaces, nettles and
brambles in the fortresses thereof... *Isaiah 34:13*

Weeds are also mentioned in the parables of Jesus (Matthew 13:18—23). The biblical thistles, thorns, and brambles are common weeds and have been identified as such by biblical scholars (Moldenke and Moldenke, 1952). They were and are serious threats in the continuing battle to produce enough food. The tares in the following parable (Matthew 13:25—30) are the common weed, poison ryegrass, a continuing problem in cereal culture:

The kingdom of heaven is likened unto a man which sowed good seed in his field: But while he slept, his enemy came and sowed tares among the wheat, and went his way. But when the blade was sprung up, and brought forth fruit, then appeared the tares also.

The Greek word *tares* is translated as darnel—a weed that grows in wheat. It is a grass resembling wheat or rye, but with smaller, poisonous seeds. The weed called tares in Europe today is one of several vetches native to Europe.

No agricultural enterprise or part of our environment is immune to the detrimental effects of weeds. They have interfered with human endeavors for a long time. In much of the world, including my garden, weeds are controlled by hand or with a hoe. A person with a hoe may be as close as we can come to a universal symbol for the farmer, even though most farmers in developed countries no longer weed with, or even use, hoes. For many, the hoe and the weeding done with it, symbolize the practice of agriculture. The battle to control weeds, done by people with hoes, is the farmer's primary task in much of the world.

> Bowed by the weight of centuries he leans
> Upon his hoe and gazes on the ground,
> The emptiness of ages in his face,
> And on his back the burden of the world.
> Who made him dead to rapture and despair,
> A thing that grieves not and that never hopes,
> Stolid and stunned, a brother to the ox?
> Who loosened and let down this brutal jaw?
> Whose was the hand that slanted back this brow?
> Whose breath blew out the light within this brain?
> …
> O masters, lords and rulers in all lands,
> How will the future reckon with this man?
> How answer his brute question in that hour
> When whirlwinds of rebellion shake all shores? *Excerpt from "The Man with the Hoe" Edwin Markham (1899)*

There have been four major advances in agriculture that have significantly increased food production.

First was the introduction of mineral fertilizer. Early work on plant nutrition and soil fertility proceeded directly from the pioneering studies of Justus von Liebig (see Liebig, 1942) who questioned prevailing theories of plant nutrition.

The introduction of mineral fertilizer increased food production.

The **second** major advance was rapid mechanization that began in the United States with development of Whitney's cotton gin in 1793, McCormick's reaper in 1834, and Deere's moldboard plow in 1837.

Mechanization has increased agricultural productivity.

Understanding and using genetic principles in plant and animal production was the **third** major advance for agriculture. The obscure Austrian monk, Gregor Mendel, pursued his studies quietly and in seclusion. He had no goal of pragmatic application or economic gain. The discoveries made from his research, most notably in development of plant hybrids, have had huge, generally positive, effects on our ability to produce food. The nearly simultaneous and independent rediscovery of Mendel's work by De Vries in Holland, Correns in Germany, and Tschermak in Austria in 1900 when searching the literature to confirm their own discoveries has resulted in enormous positive benefits to agriculture.

The **fourth** major advance in agriculture was the development and widespread use of fertilizers and pesticides. These heralded the chemicalization of agriculture and led to the development and growth of weed science. It is reasonable to posit that weed control has been always been part of agriculture and that weed science began coincident with herbicide development. Weed science did not develop because of herbicides although their dominance is undeniable. The preponderance of evidence is that they will continue to dominate weed science research. They are and will remain an essential part of the knowledge required to manage weeds.

Weed science is vegetation management—the employment of many techniques to manage plant populations in an area. This includes dandelions in turf, poisonous plants on rangeland, and Palmer amaranth in soybeans. Weed science might be considered a branch of applied ecology that attempts to modify the environment against natural evolutionary trends. Natural evolutionary or selection pressure tends toward the lower side of the curve in Fig. 1.1 (Shaw et al., 1960), toward what ecologists call climax vegetation, the specific composition of which will vary with latitude, altitude, and environment. A

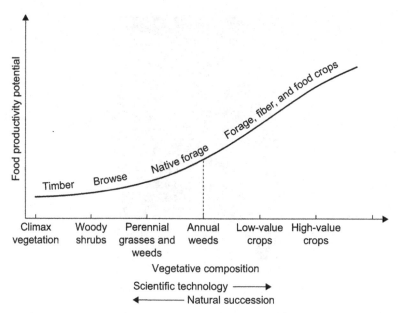

FIGURE 1.1   The food productivity potential of vegetation (Shaw et al., 1960).

climax plant community does not and cannot provide the kind or abundance of food the growing world population wants and needs. Therefore, we humans successfully modify the environment—the natural world—in many ways. We are the only species able to modify the natural world. It is no longer humans against nature. We decide and act to make the natural world what we want it to be. All other species must adapt to the environment as it is. Bronowski (1973, p. 19) reminds us that humans are not figures in the landscape—we are shapers of the landscape. All one has to do is look around to see what we have wrought. One of the most significant changes began over 10,000 years ago when humans began altering the land and the plants and animals on it to grow what was wanted. Hunting and gathering—living from what the land offered—were slowly abandoned as we learned how to dominate and subdue,[4] which many began to and still interpret as a God-given right—even a duty. Diamond (1987) suggests that the transition to settled agriculture was "the worst mistake in the history of the human race." The immediate success of settled agriculture resulted in what many regard as a continuing catastrophe. It allowed the human population to grow, perhaps beyond the limits of what the earth can support (see Malthus, 1798). In Diamond's view the mistake was choosing to increase

[4]"And God said, Let us make man in our image, after our likeness: and let them have dominion over the fish of the sea, and over the fowl of the air, and over the cattle, and over all the earth, and over every creeping animal that creepeth upon the earth. So God created man in his [own] image, in the image of God created he him; male and female created he them. And God blessed them, and God said to them, Be fruitful, and multiply, and replenish the earth, and subdue it: and have dominion over the fish of the sea, and over the fowl of the air, and over every living animal that moveth upon the earth." (Genesis 1:26–28).

TABLE 1.1 How Many People Can the Earth Support

Year	World Population (Billion)	World Gross Income (Billion $)	Number of High-income Countries	Avg. Gross Income/Cap High-income (Million $)	Billion People
2000	6.11	49,424	53	25,446	1.94
2010	6.92	65,317	67	39,981	1.63
2015	7.34	70,575	79	39,576	1.78

Data from World Bank Development Report and Data_extract-from_world_development_indicators_HI_aggregate.

the food supply (what agriculture has done so successfully) rather than limit the population. That choice has led, in his view, to exploitation of the earth, "gross social and sexual inequality, (and) the disease and despotism that curse our existence." However, it has also yielded a better, indeed a good, life for many, unfortunately not all, humans.

Population growth continues, albeit at a slower rate, and we don't know for sure if we have exceeded the earth's capacity. Grant (2000) suggested the limit may be about 1.06 billion people at the World Bank's High Income Countries standard of living. He provided a debatable, intellectually challenging, answer to the nagging question: How many people can the earth support? First one must ask: At what level? Using the World Bank Atlas from 1996, he divided the world's gross income ($US) by the average gross income of those of us who live in the 44 high income countries—a standard most would like to have. Using data from the World Bank's Development Report for 2000, 2010, and 2015 the answer was less than 2 billion (Table 1.1).

Another estimate is based on energy consumption. Holdren (1991) said the earth was consuming 13 terawatts (13 trillion watts) of human-generated energy, 75% of which was used by 1.5 billion people (about 23% of world's population) in industrialized countries, who produce about 85% of the world's economic product. Average consumption in developing countries was about 1 kW/person. Holdren projected that sometime in the 21st century demand would be an unachievable at eight times higher and asked what if demand could be held at 3 kW/person = 3× what the poor used and one-fourth of 1991 US use. Then he asked what if we could use 9 terawatts without trashing the environment? If one allows for unforeseen consequences (a 50% error margin), 6 terawatts total and 3 kW/person is reasonable. Long division = 2 billion people. These simple calculations are disturbing and challenging to all in agriculture. Bartlett's (1978) paper is relevant.

In the beginning there were no weeds. If one impartially examines the composition of natural plant communities, or the morphology of weed flowers, one can find beauty and great aesthetic appeal. The flowers of wild onion, poison hemlock, dandelion, Queen Anne's Lace, chicory, sunflower, and several of the morning glories are beautiful and worthy of artistic praise for symmetry and color. By what right do we humans call plants with beautiful flowers *weeds*? Who has the right to say some plants are unwanted? By what authority do we so easily assign the derogatory term "weed" to a plant and say it interferes with agriculture, increases costs of crop production, reduces yields, and may even detract from quality of life?

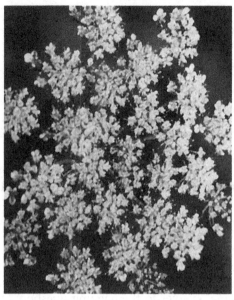

The flowers of many weeds are beautiful and have great aesthetic appeal. This is the flower of the wild carrot or Queen Anne's Lace.

Nature knows no such category as weed. In 1967 (Buchholtz), Weed Science of America defined a weed as "a plant growing where it is not desired." The definition in the 1974 Herbicide Handbook (Hilton p. xxi) of the society added, "Plants are considered weeds when they interfere with activities of man or his welfare." Desire was omitted in the 2014 Handbook (Shaner, p. 487), in which a weed was defined as "Any plant that is objectionable or interferes with the activities or welfare of man." It is important to note the continuing anthropocentric dimension of the definition. Desire is a human trait and only we have the ability to object or claim interference. Therefore a particular plant is a weed only in terms of a human attitude. Ecologists speak of weedy plants but often their use of the term is affected by preconceptions of the role of vegetation on a particular site. People say that a plant, in a certain place, is not desirable and therefore arbitrarily assign it the derogatory term "weed." Weeds are usually regarded as the lowest of the kingdom of flowering plants not because they are naturally harmful but because they are or are perceived to be harmful to us.

It is homeowners and neighbors who say that dandelions and crabgrass are unacceptable in lawns. Does grass really care what other plants live in the neighborhood? It is those who suffer from hay fever who say that ragweed or perhaps big sagebrush in the western United States are unacceptable. It is those who are allergic to poison ivy who say it is unacceptable in their environment and who want to get rid of it. Farmers say, with clear economic justification, that they want their crops to grow in a weed-free environment to maximize yield and profit. People decide what plants are weeds and when, where, and how they will be controlled.

The dandelion is considered a weed by many.

This book will discuss many aspects of weeds, their biology, and their control. It differs from other weed science texts in significant ways. The differences may not make it better, only different. For example, most, but not all, presently available weed science textbooks devote at least 50% of their content to herbicides and their use. In some it is as much as 75%. A notable exception is Aldrich and Kremer's book (1997), which does not include any major section on herbicides. Because of the undeniable success of chemical weed management, it is my view that it must be included in a complete weed science textbook. Omitting the topic will produce students who are only partially prepared for modern weed management. Therefore the book includes herbicides and their use, but only as part, albeit an important part, of the fundamentals of weed science. The book correctly claims that killing weeds with herbicides is the highly successful modern way. It is, but its very success has deterred understanding weed biology and ecology. Control has been the primary emphasis of weed science since its beginning. Several years ago the historical committee of the Weed Science Society of America identified 17 important early publications on weeds (from 1895 to 1965), 12 dealt with killing, controlling, or eradication. As you proceed, you will find that this book does not ignore these important topics, but understanding weeds is the primary emphasis.

One can establish a relationship between pesticide use and agricultural yield. Perhaps a better way to put it is that one can find a relationship between good pest management (regardless of how it is accomplished) and agricultural yield. One should not always equate good weed control with herbicide use. Good weed control depends on cultural knowledge—what a good farmer or plant grower knows. Cultural knowledge is different than the scientific knowledge that leads to herbicide development and successful use. Both kinds of knowledge—scientific to tell us what can be done and cultural to tell us what we ought to do—are essential for good weed management.

One can also postulate a relationship between the way weeds and other pests are controlled, the practice of pest management, and a nation's food supply. Fig. 1.2 shows the world's tropical and sub-tropical areas, their major crops, and the percent of the world's

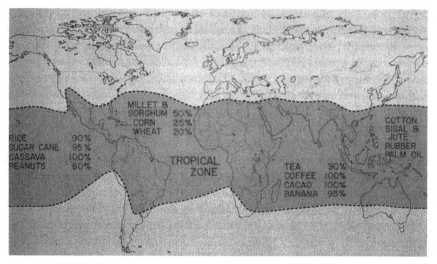

FIGURE 1.2   Crop production in the world's tropics (Holm, 1971).

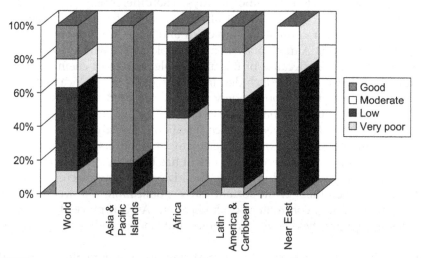

FIGURE 1.3   The level of weed control practices in the world and four regions (Labrada, 1996).

crop grown in each area. The region's ability to control weeds is shown in Fig. 1.3 with data for the world and four major areas. Each segment in Fig. 1.3 is divided into good, moderate or acceptable, low, and very poor weed management. The world's tropical and subtropical regions (Fig. 1.2) are home to about 66% of the world's people. The regions extend roughly from the Tropic of Capricorn in the south to the Tropic of some of the world's most important crops. In the industrialized, primarily temperate, world, production agriculture and the ability to manage a large array of pests has made remarkable progress since 1971 when Dr. Leroy Holm of the University of Wisconsin prepared Fig. 1.2. However, the areas identified still suffer from underdevelopment of weed science and other agricultural technology. Seventy countries in Asia, Africa, and Latin America were included in Labrada's 1996 UN/FAO survey, which, sadly, is still accurate. The countries had roughly 44% of the world's 1.4 billion acres of arable land and, of most importance to our subject, inadequate weed control technology and knowledge (Fig. 1.3).

The founder of Latin prose, Cato the Elder, reminds us in his work on agriculture that "it is thus with farming: if you do one thing late, you will be late in all your work." We are late in implementing appropriate weed management techniques in much of the world, and agriculture will not progress to its full potential without them.

The agricultural productivity of the developed world is not an accident. US agriculture and that of other advanced nations grew out of a propitious combination of scientific advancement, industrial growth, and abundant resources of soil, climate, and water. One should not regard it as just good fortune or God's benevolence that we, in the United States, can say that after the food bill is paid we have more money left over than most other folks in the world.[5] For most Americans (although, unfortunately, not all) this is true. It not only is true for US citizens, it is regarded as so common that it is treated as a right rather than as something that was created and must be maintained.

Weeds are controlled in much of the world by hand or with crude hoes. The size of a farmer's holding and yield per unit area are limited by several things, and paramount among them is the rapidity with which a family (most often its female members) can weed its crops. More human labor may be expended to weed crops than on any other single human enterprise, and most of that labor is expended by women. Weed control in the Western world and other developed areas of the world is done by sophisticated machines and by substituting chemical energy for mechanical and human energy. There is a relationship between the way farmers control weeds and the ability of a nation to feed its people. Weed science is part of that relationship. Good weed management is one of the essential ingredients to increase food production.

The early flights of the Apollo spacecrafts and subsequent space flights gave those of us bound to earth a view of the whole planet, floating in the great, black sea of space (Fig. 1.4). Many had imagined but had never seen such a picture before. My generation remembers Apollo 13. On April 11, 1970, James Lovell, Fred Haise, and John Swigert were launched toward the moon. On April 13, their oxygen system exploded. As the 1995 movie shows so well, they had only one option for survival—return to earth. We have the same option. We must inhabit this planet as one human society. There is no other place to go. There is only one land mass to live on, one ocean system, one sun, one atmosphere, one rain cycle. You might choose to regard life as having one major purpose—to sustain itself.

There are, of course, other purposes. People want more than existence (survival). We want a good life—a life that enables us to realize our full potential. We want purpose, hope, and a sense of the meaning of things. Yet in practical terms the first prerequisite is a place to live. We have inherited a place to live from those who preceded us; and if others are to live, we

---

[5]In 2014 US consumers spent 6% on food consumed at home and 11.4% overall. http://www.ers.usda.gov/data-products/food-expenditures.aspx#.UuE9EHn0Ay5. Others: Russia—29%, UK—9%, India—36%, China—34%, Peru—29%. http://wsm.wsu.edu/researcher/wsmaug11_billions.pdf.

FIGURE 1.4   National Aeronautics and Space Agency—Epic view of earth. http://www.nasa.gov/press-release/
nasa-satellite-camera-provides-epic-view-of-earth.

must pass it on, in habitable condition with a productive, secure, sustainable agricultural
foundation.

Around 1965, world food production began to lose the race with an expanding population
as Rev. T. R. Malthus (1798) predicted it would. Each year the Malthusian apocalypse he predicted is prevented, but it is a daily specter for many in the world. The world's population
now exceeds 7.4 billion and it will continue to grow, albeit at a slower rate. More than
85% of the world's people live in poor, developing countries where about 95% of the population growth will occur. As world population expands, food production is barely keeping
pace, and often slipping behind. About 10% of the world's 33 billion acres of land are arable
and while the area devoted to productive agriculture can be expanded, the cost will be great.
One must also recognize that the world may lack the social and political will to handle the
complex problems that expansion onto previously untilled land will bring. Such expansion
may be part of the solution to the world food dilemma, but an equally important solution
is use of appropriate, available technology, and development of new technology. If all the
world's people are going to enjoy higher standards of living and be able to watch their
children mature without fear of debilitating disease, malnutrition, or starvation, we must
intelligently use all present agricultural technology and continue to develop better, safer,
and equally or more effective technology. Shared technology and knowledge will permit
our neighbors in this world to farm in ways that create opportunities to realize the earth's
full agricultural and human potential.

Weed science is not a panacea for the world's agricultural problems. The problems are too complex for any simple solution and students should be suspicious of those who propose simple solutions to complex problems. The caution of H.L. Mencken (1920) is worth noting: "There is always a well-known solution to every human problem—neat, plausible, and wrong." In fact, the hope should be not to solve but to diminish, not to cure but to alleviate, and to at least anticipate the "brute question" and have some answers when "whirlwinds of rebellion strike all shores." The work of the weed scientist is fundamental to solving problems of production agriculture in our world. Weeds have achieved respect among farmers who deal with them every year in each crop. Weeds and weed scientists have achieved respect and credibility in academia and the business community. The world's weed scientists are and will continue to be in the forefront of efforts to feed the world's people.

Literature Cited

Aldrich, R.H., Kremer, R.J., 1997. Principles in Weed Management, second ed. Iowa State Univ. Press, Ames, IA. 455 pp.

Appleby, A.P., 2005. A history of weed control in the United States and Canada—a sequel. Weed Sci. 53, 762–768.

Bailey, L.H., 1906. The Survival of the Unlike: A Collection of Evolution Essays Suggested by the Study of Domestic Plants, fifth ed. Macmillan & Co., NY. 515 pp.

Bartlett, A.A., 1978. Forgotten fundamentals of the energy crisis. Am. J. Phys. 46 (9), 876–888.

Bronowski, J., 1973. The Ascent of Man. Little, Brown and Company, Boston, MS, 448 pp.

Buchholtz, K.P., 1967. Report of the terminology committee of the Weed Science Society of America. Weeds 15, 388–389.

Cato, the Elder. De Agri Cultura. 2nd C. B.C.

Cox, S., Spring 2006. Civilization's weedy roots. In: The Land Report 84, pp. 8–10.

De Morgan, A.C., 1850. A budget of paradoxes. p. 453, item 662.1. In: International Thesaurus of Quotations, 1970. Compiled by R. T. Tripp.

Diamond, J., May 1987. The worst mistake in the history of the human race. Discov. Mag. 64–66.

Food and Agriculture Organization of the United Nations, 1963. Production Yearbook.

Food and Agriculture Organization of the United Nations, 1975. Production Yearbook.

Fox, T., Fimeche, C., 2013. Global Food — Waste Not, Want Not. Institution of Mechanical Engineers, London, UK, 36 pp.

Goodwin, D.C., Johnstone, P.H., 1940. A Brief Chronology of American Agricultural History. U.S. Dep. Agric. Yearb. Agric., pp. 1184–1196

Grant, L., 2000. Too Many People — The Case for Reversing Growth. Seven Locks Press, Santa Ana, CA, 102 pp.

Hilton, J.L. (Ed.), 1974. Herbicide Handbook, third ed. Weed Science Society of America, Lawrence, KS. 430 pp.

Hill, D.S., 1994. Agricultural Entomology. Timber Press, Portland, OR, 635 pp.

Holdren, J., 1991. Population and the energy problem. Popul. Environ. 12 (3), 231–255.

Holm, L., 1971. The role of weeds in human affairs. Weed Sci. 19, 485–490.

Labrada, R., 1996. Weed management status in developing countries. In: Second Int. Weed Cont. Congress, vol. 2, pp. 579–589.

Liebig and After Liebig, 1942. American Association for Adv. of Sci, Washington, DC, 111 pp.

Malthus, T.R., 1798. An Essay on the Principle of Population as It Affects the Future Improvement of Society, with Remarks on the Speculations of Mr. Godwin, M. Condorcet, and Other Writers, 1966 ed. Macmillan and Co., Ltd. 396 pp.

Markham, E., 1899. The man with the hoe. In: The Pocket Book of Verse. 1940. Pocket Books, Inc., NY, pp. 303–304.

Mencken, H.L., 1920. Prejudices: Second Series (Chapter 4).

Moldenke, H.N., Moldenke, A.L., 1952. Plants and the Bible. Dover Publications, Inc., New York, NY. Pages 70–72, 133–134, and 153.

Oerke, E.C., 2006. Crop losses to pests. J. Agric. Sci. 144 (1), 31–43.

Pinstrup-Andersen, P., 2001. The future world food situation and the role of plant diseases. Plant Health Inst. https://doi.org/10.1094/PHI-I-2001-0425-01.

Rasmussen, W.D., 1975. Agriculture in the United States: A Documentary History, vol. 4. Random House, New York, NY.

Saint Exupéry, A., 1950. The Wisdom of the Sands. Translated by Stuart Gilbert from the French — Citadelle. Harcourt, Brace and Co., NY, 350 pp.

Shaw, W.C., Hilton, J.L., Moreland, D.E., Jansen, L.L., 1960. Herbicides in plants. In: The Nature and Fate of Chemicals Applied to Soils, Plants and Animals. USDA, ARS, 20-9, pp. 119–133.

Shaner, D.L. (Ed.), 2014. Herbicide Handbook, tenth ed. Weed Science Society of America, Lawrence, KS. 513 pp.

Timmons, F.L., 1970. A history of weed control in the United States and Canada. Weed Sci. 18, 294–307. Republished — Weed Sci. 53, 748–761.

Williams, H., 1937. The Poems of Jonathan Swift. In: On Poetry: A Rhapsody, vol. II, pp. 639–659. Specific reference on p. 651.

World Development Report, 2010. Development and climate change. The World Bank, Washington, DC, 417 pp.

Zimdahl, R.L., 2010. A History of Weed Science in the United States. Elsevier Insights, London, UK, 207 pp.

2

Weeds: The Beginning

FUNDAMENTAL CONCEPTS

- The most basic concept of weed science is embodied in the term "weed."
- Weeds are defined in many ways, but most definitions emphasize behavior that affects humans.
- Weeds share some characteristics.

- There are at least nine ways in which weeds express their undesirability.
- There are no completely accurate estimates of what weeds cost in the United States or the world. Losses due to weeds and the costs of weed control in the United States surely exceed $8 billion per year.

LEARNING OBJECTIVES

- To understand why weeds are defined as they are.
- To know the characteristics that weeds share.
- To understand how weeds cause harm.

- To appreciate how estimates of the cost of weeds are made and the magnitude of those costs.

1. THE BEGINNING

...and nothing teems
But hateful docks, rough thistles, kecksies, burs,
Losing both beauty and utility.
And as our vineyards, fallows, meads, and hedges
Defective in their natures, grow to wildness;
Even so our houses, and ourselves, and children,
Have lost, or do not learn, for want of time,
The sciences that should become our country. *Shakespeare, W., King Henry V, Act V, Scene II. The Duke of Burgundy addressing the kings of France and England.*

I will go root away
The noisome weeds, which without profit suck
The soil's fertility from wholesome flowers. *Shakespeare, W., Richard II, Act III, Scene IV. A gardener speaking to a servant in the Duke of York's garden,*

It is indisputable that farmers have always been aware of weeds in their crops, although evidence of their awareness and concern in early agriculture is nearly all anecdotal. It makes sense that farmers had to be aware of weeds although they could not do much to control them. Clark and Fletcher (1923) suggested that the "annual losses due to the occurrence of pernicious weeds upon farm lands, although acknowledged in a general way, are far greater than is realized." They thought that this was because "farmers gave little critical attention to the weeds growing among their crops." They did not deny that farmers were aware of the weeds. The original hardback book, with 71 carefully drawn color pictures of weeds and 100 seeds, cost $2. Many of the same weeds are shown in most current weed identification books. There are few (actually, very few) old books about weeds. Most are identification books as is Clark and Fletcher's. Weed science is relatively new among agricultural sciences that study pests. It cannot claim the historical lineage of entomology or plant pathology. No one disputes that weeds have been present as long as other pests, but they have not been studied as long.

Timmons (1970) reported that "available literature indicates that relatively few agricultural leaders and farmers became interested in weeds as a problem before 1200 AD or even before 1500 AD." One cannot be certain how he selected the dates, but his claim has not been disputed. We can be certain that the "critical attention" Clark and Fletcher thought was absent increased slowly primarily because the general attitude seemed to be that "weeds were a curse which must be endured, and about which little could be done except by methods which were incidental to crop production, and by laborious supplemental hand methods" (Timmons, 1970). In 1733, Jethro Tull, an English agricultural pioneer (1829), appealed for greater attention to weeds:

It is needless to go about to compute the value of the damage weeds do, since all experienced husbandmen know it to be very great, and would unanimously agree to extirpate their whole race as entirely as in England they have done the wolves, though much more innocent and less rapacious than weeds.

Farmers, however, were bound by their inability to do much about weeds except by the laborious hand methods Timmons mentioned.

Insects cause obvious human and crop problems. With a few exceptions, weeds do not cause direct harm to humans. Those that do (e.g., poison ivy and poison oak) can be avoided. Neither is common in crops or of great concern to most people. Some weeds aggravate human allergies, but many other plants are also allergenic. Leaves and stems of giant and common hogweed can cause painful blisters that may leave scars. Myrtle spurge causes dermatitis and eye irritation.

Insects and insecticides were respectively causes of and solutions to human disease problems. Weeds and herbicides were not, and less attention was paid to them. Weeds were agricultural problems, not organisms of general societal concern. Before World War II (1940—45), only a few scientists were interested in studying weeds and developing techniques to reduce crop losses caused by weeds.[1]

It is true that weeds were not of general societal concern, and in most situations still are not (lawns and golf courses are notable exceptions to this generalization). However, it is not generally recognized in weed science that weed control and the way agriculture was practiced have influenced societal structure and attitudes toward women: "cultural norms about the economic role of the sexes can be traced back to traditional farming practices" (Economist, 2011). I claim, without supporting evidence, that more human labor is spent weeding crops than is spent on any other task people do, and that most of the laborious hand-weeding is done by women. Boserup (1970) was among the first to argue that cultural norms about women's roles, and their place in society, can be traced back to traditional farming practices. Alesina et al. (2011) affirm Boserup's hypothesis. Traditional agricultural practices have had a strong influence on gender division of labor and the evolution of gender norms. Today, societies that practiced what Alesina et al. and many others call plough agriculture have significantly lower rates of female participation in "the workplace, in politics, and in entrepreneurial activities, as well as a greater prevalence of attitudes favoring gender inequality." Plough agriculture is dominated by men who work with plows, cultivators, and planting and harvesting machines to grow crops for sale. Plows and other machines require the greater upper-body strength that men have to control the machine and the animals that power it. Women's role in plough agriculture is limited. There is less need for weeding than in what Boserup calls shifting cultivation, which does not involve machines. It is labor-intensive and requires hoes and digging tools. Hoes are used for weeding, which in many cultures is done primarily, although not exclusively, by women. In societies that were and in many cases still are dominated by shifting cultivation, women are much more likely to be employed in the nonagricultural labor force. Thus, research has shown that the role of women in many societies can be traced to how agriculture was practiced and who did the weeding.

2. DEFINITION OF THE WORD "WEED"

To be fully conversant with any subject, it is mandatory to understand its basic concepts. The most basic concept of weed science is embodied in the word "weed." Each weed scientist has a clear understanding of the term, but no definition is shared by all scientists. In 1967, the Weed Science Society of America defined a weed as a plant growing where it is not desired (Buchholtz, 1967). In 1989, the Society's definition was changed to define a weed as "Any plant that is objectionable or interferes with the activities or welfare of man" (Humburg, 1989, p. 267, Vencill, 2002, p. 462). The definition was modified in 2016: A weed is a plant that causes economic losses or ecological damage, creates health problems for humans or animals, or is undesirable where it is growing.[2] The European Weed Research Society defined

---

[1]Much of the preceding paragraphs is from pages 33—34 of Zimdahl (2010a).

[2]wssa.net/wp-content/uploads/WSSA-Weed-Science-Definitions.pdf.

a weed as "any plant or vegetation, interfering with the objectives or requirements of people" (EWRS Constitution, 2008).

Although the definitions are clear, it is equally clear that there is widespread disagreement about whether a particular plant in a particular place is a weed. It is a subjective, not objective, decision. The definitions leave the burden and responsibility for specific identification and final definition with people. People determine when a particular plant is growing in a place where it is not desired or when it interferes with their activities or welfare.

The *Oxford English Dictionary* (Little et al., 1973) defines a weed as an "herbaceous plant not valued for use or beauty, growing wild and rank, and regarded as cumbering the ground or hindering the growth of superior vegetation." The human role is again clear because it is we who determine use or beauty and which plants are regarded as superior. It is important that weed scientists and vegetation managers remember the importance of definitions as determinants of their view of plants and their attitude toward them.

How one defines something has a significant role in determining one's attitude toward the thing defined. For the weed scientist and vegetation manager, the definition determines which plants are weeds and thus are undesirable and to be controlled. Like other plants, weeds lack consciousness and cannot enter the court of public opinion to claim rights. Humans can assign rights to plants and serve as their counsel to determine whether they have rights and then advocate their rights or lack thereof. Our attitude toward weedy plants do not always need to be shaped by another's definition. We do not always agree about what should be done when others define things:

> Once in a golden hour,
> I cast to earth a seed.
> Upon there came a flower,
> The people said a weed.
>
> Read my little fable:
> He that runs may read
> Most can raise the flowers now,
> For all have got the seed.
>
> And some are pretty enough,
> And some are poor indeed:
> And now again the people
> Call it but a weed. *Tennyson, A., "The Flower."*

It is clear that there is disagreement about what a weed is and what plants are weeds. Harlan and de Wet (1965) assembled several definitions (partially reproduced here) to show the diversity of definitions of the same or similar plants. The array of definitions emphasizes the care that weed scientists and vegetation managers must take in equating how something is defined, with a right or privilege to control it.

DEFINITIONS OF WEEDS BY PLANT SCIENTISTS

Robbins et al.	1942	"these obnoxious plants are known as weeds." p. 1
Muenscher	1935 and 1955	"those plants with harmful or objectionable habits or characteristics which grow where they are not wanted, usually in places where it is desired that something else should grow." p. 3
Harper	1960	"higher plants which are a nuisance." p. xi
King	1966	"the more aggressive, troublesome and undesirable elements of the world's vegetation." p. 1
Salisbury	1961	"a plant growing where we do not want it."
Klingman	1961	"a plant growing where it is not desired; or a plant out of place." p. 1

DEFINITIONS OF WEEDS BY ENTHUSIASTIC AMATEURS

Emerson	1878	"a plant whose virtues have not yet been discovered."
King	1951	"weeds have always been condemned without a fair trial."

ECOLOGICAL DEFINITIONS OF WEEDS

Bunting	1960	"weeds are pioneers of secondary succession, of which the weedy arable field is a special case." p. 12
Blatchley	1912	"a plant which contests with man for the possession of the soil."
Pritchard	1960	"those species which grasp the opportunities offered frequently become serious weeds and, by following in the wake of human activities, rapidly expand their range of distribution." p. 61.
Quammen	1998	They reproduce quickly, dispersed widely when given a chance, tolerate a fairly broad range of habitat conditions, take hold in strange places, succeed especially in disturbed ecosystems, and resist eradication once they're established. They are scrappers, generalists, opportunists. They tend to thrive in human-dominated terrain because in crucial ways they resemble *Homo sapiens*: aggressive, versatile, prolific, and ready to travel.
Salisbury	1961	"the cosmopolitan character of many weeds is perhaps a tribute both to the ubiquity of man's modification of environmental conditions and his efficiency as an agent of dispersal."

Godinho (1984) compared the definition of the French words *d'aventice* and *le mauvaise herbe* with the English "weed" and the German *unkraut*. No single definition was found for "weed" and *unkraut* because both have two meanings:

1. In the ecological sense, "weed," *unkraut*, and *d'aventice* mean a plant that grows spontaneously in an environment modified by humans.
2. In the weed science sense, "weed," *unkraut*, and *malherbe* (Italian) or *le mauvaise herbe* = unwanted plant.

In some languages, weeds are just bad (mal) plants. In Spanish, they are *mala hierba* or *malezas* and in Italian, *malherbe*. One must agree with Godinho, Fryer and Makepeace (1977), Anderson (1977), and Crafts and Robbins (1967) that neither the word "weed" nor the weedy characteristics of the plants to which the word is assigned are easy to define.

Aldo Leopold (1943) made the point well in an article that was critical of the 1926 bulletin *Weeds of Iowa*. Many of Iowa's native plants are in the bulletin, and Leopold noted that in addition to their inherent beauty, they have value as wildlife food, for nitrogen fixation, or as creators of stable plant communities. He admits that many plants that others call weeds are frequent in pastures, but argued that soil depletion, overgrazing, and needless disturbance of advanced successional stages encourage expression of their weediness and create the need for control. Leopold argues that the definition of weed is part of the problem, because not all plants that some call weeds "should be blacklisted for general persecution." Leopold's view is supported by McMichael (2000), who noted with supporting evidence that "in many rural cultures, non-crop plants (often termed weeds) represent food, fodder, and medicine." See Chapter 4 for additional supporting citations.

About 3000 of the more than 350,000 recognized plant species have been or are cultivated somewhere in the world. It is incorrect to assume that those that are not cultivated are weeds. That this is wrong is undebatable. However, when a new, unknown plant appears in a field or garden, it is difficult to maintain objectivity.

The ulterior etymology of "weed" is unknown, but an exposition of what is known was provided by King (1966). Weed has Germanic, Romance language, and Oriental roots. He concluded that weed is an "example of language as an accident of usage." He was unable to find a common word or words, in ancient languages, for the collective term "weed." The ultimate etymology of "weed" is unknown.

It is logical to assume that even if one cannot define "weed," it should be possible to identify the origin of individual species and determine certain characteristics of weeds that come from native and naturalized flora. Some plants succeed as weeds because they evolve forms adapted to disturbed environments more readily than do other species. Baker's (1965) definition, repeated here, emphasizes success in disturbed environments, a point he reiterated in a later article (Baker, 1991)"

> ... a plant is a "weed" if, in any specified geographical area, its populations grow entirely or predominantly in situations markedly disturbed by man (without, of course, being deliberately cultivated plants). Thus, for me, weeds include plants which are called *agrestals* by some writers of floras (they enter agricultural land) as well as those which are *ruderals* (and occur in waste places as well as along roadsides). It does not seem to me necessary to draw a line between these categories and accept only the agrestals as weeds (although this is advocated by some agriculturally oriented biologists) because in many cases the same species occupy both kinds of habitat. Ruderals and agrestals are faced with many similar ecological factors, and the taxa which show these distributions are, in my usage, "weedy."

If one considers weeds in the Darwinian sense of a struggle for existence, they represent one of the most successful groups of plants that have evolved simultaneously with human disruption of areas of indigenous vegetation and habitats and creation of disturbed habitats (King, 1966). Aldrich (1984) and Aldrich and Kremer (1997, p. 8) offer a definition that does not deny the validity of others, yet introduces a desirable ecological base. A weed is "a plant that originated in a natural environment and, in response to imposed or natural environments, evolved, and continues to do so, as an interfering associate with our crops and activities." This definition provides "both an origin and continuing change perspective" (Aldrich). Aldrich wants us to recognize weeds as part of a "dynamic, not static, ecosystem." His definition departs from those that regard weeds as enemies to be controlled. Its ecological base defines weeds as plants with particular, perhaps unique, adaptations that enable them to survive and prosper in disturbed environments. Navas' (1991) definition included biological and ecological aspects of plants and effects on humans. A weed was defined as "a plant that forms populations that are able to enter habitats cultivated, markedly disturbed or occupied by man, and potentially depress or displace the resident plant populations which are deliberately cultivated or are of ecological and/or aesthetic interest."

Although all do not agree on precisely what a weed is, most know they are not desirable. Those who want to manage or control weeds ought to think about their definition. When the term "weed" is borrowed from agriculture and applied to plants in natural communities, a verification of negative effects on the natural community should be a minimal expectation. Simple yield effects are not acceptable. Effects of a plant, presumed to be a weed, in a natural community should be estimated in terms of a management goals such as establishment of predisturbance conditions, preserving rare species, maximizing species diversity, or maintaining patch dynamics (Luken and Thieret, 1996). Humans are responsible for creating the negative, often deserved, image that weeds have. Weeds are detrimental and often must be controlled, but only with adequate justification for a site, its purpose, its stability, and perhaps its aesthetic appeal. They can do good things (see Chapter 4). For example, because they have the ability to grow in habitats disturbed by humans, they can be viewed as a kind of ecological Red Cross. They rush in to disturbed places to protect and restore the land. For example, The High Park forest fire west of Fort Collins, CO, burned over 87,284 acres and destroyed 259 homes in June 2012. By any measure, it was devastating: the third most damaging fire in Colorado history.[3] Figs. 2.1 and 2.2 show the burned area and moth mullein (Fig. 2.1) and Canada thistle (Fig. 2.2) growing in the destroyed forest soon after the fire ended. The ecological Red Cross (the weeds) arrived and grew to stabilize the soil and begin restoration.

3. CHARACTERISTICS OF WEEDS

Crop agriculture in developed countries is increasingly based on one plant (a monoculture) that thrives in a disturbed habitat (a cropped field) and produces an abundance of seed (e.g., wheat, rice) or vegetative growth (e.g., sugarcane, pasture). Weeds also thrive in disturbed habitats and produce an abundance of vegetative growth (e.g., johnsongrass, quack grass) or seed (e.g., wild proso millet, green foxtail), which in most cases is not useful to humans, or we choose not to use them (Manning, 2004, p. 55). Why is it

---

[3]The 2013 Black Forest fire (Black Forest, CO) destroyed 511 homes, caused two deaths, and burned 14,280 acres. The 2012 Waldo Canyon fire (San Isabel National Forest, Colorado Springs, CO) burned 18,247 acres and destroyed 346 homes.

FIGURE 2.1 Mullein acting as the ecological Red Cross after the High Park forest fire west of Fort Collins, CO, in 2012.

FIGURE 2.2 Canada thistle acting as the ecological Red Cross after the High Park forest fire west of Fort Collins, CO, in 2012.

that some plants that thrive in disturbed habitats are crops and some are weeds? What is it that makes some plants capable of growing where they are not desired? Why are they difficult to control? What are their modes of interference and survival? The most consistent trait of weedy species is not related to their morphology or taxonomic relationships. It is, as Baker (1965) noted, their ability to grow well in habitats disturbed by human activity. They are plants growing where someone does not want them, and often that is in areas that have been intentionally changed: gardens, cropped fields, and golf courses.

Not all weeds have all possible undesirable characteristics, but in addition to growing in disturbed habitats, all have some of the following 20 characteristics (see Baker, 1965):

- Rapid seedling growth and the ability to reproduce when young. Redroot pigweed can flower and produce seed when less than 8 inches tall. Crops do not do either.
- Quick maturation or only a short time in the vegetative stage. Canada thistle can produce mature seed 2 weeks after flowering. Russian thistle seeds can germinate very quickly between 28 and 110°F in late spring (Young, 1991). It would spread more, but the seed must germinate in loose soil because the coiled root unwinds as it pushes into soil and is unable to do so in dry, hard soil.
- Dual modes of reproduction. Nearly all weeds are angiosperms, which flower and reproduce by seed. Field horsetail, a member of the genus *Equisetum*, is widely regarded as a weed, but is not an angiosperm; it reproduces by spores. Many angiosperms also reproduce vegetatively (e.g., Canada thistle, field bindweed, leafy spurge, quack grass).
- Environmental plasticity. Many weeds are capable of tolerating and growing under a wide range of climatic and edaphic conditions.
- Weeds are often self-compatible. Self-pollination is not obligatory.
- If a weed is cross-pollinated, pollination is accomplished by nonspecialized flower visitors or by wind.
- Weeds resist detrimental environmental factors. Most crop seeds rot if they do not germinate shortly after planting. Weed seeds resist decay for long periods in soil. They are alive but dormant (see Chapter 5).
- Weed seeds exhibit several kinds of dormancy or dispersal in time, to escape the rigors of the environment, and germinate when conditions are most favorable for survival. Many have no special environmental requirements for germination.
- Weeds often produce seed the same size and shape as crop seed, which makes physical separation difficult and facilitates spread by humans.
- Some annual weeds produce seed as long as growing conditions permit rather than once per year.
- Each generation is capable of producing a large number of seeds per plant. Some seed is produced over a wide range of environmental conditions.
- Many weeds have specially adapted long- and short-range seed dispersal mechanisms.
- Roots of some weeds are able to penetrate and emerge from deep in soil. Whereas most roots are in the upper foot of soil, Canada thistle roots routinely penetrate 3—6 ft, and field bindweed roots have been recorded over 10 ft deep. Roots and rhizomes are capable of growing many feet per year.
- Roots and other vegetative organs of perennials are vigorous with large food reserves enabling them to withstand environmental stress and intensive cultivation.
- Perennials have brittleness in lower stem nodes or in rhizomes and roots; if severed, vegetative organs can quickly regenerate a whole plant.
- Many weeds have adaptations (e.g., spines, taste, or odor) that repel grazing.

Kudzu (*Pueraria lobata*) 1990

FIGURE 2.3   US distribution of kudzu (New York Invasive Species information: from http://www.nyis.info/index.php?action=contact).

Purple Loosestrife (*Lythrum salicaria*) 1985

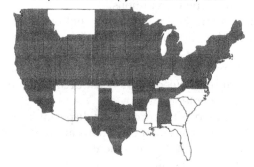

FIGURE 2.4   US distribution of purple loosestrife (Blossey, 2002). Also see: http://www.invasiveplants.net/plants/purpleloosestrife.htm (Thompson et al., 1987; Anonymous, 1990).

- Weeds have great competitive ability for nutrients, light, and water and can compete by special means (e.g., rosette formation, climbing, allelopathy).
- Weeds are ubiquitous. They exist everywhere we practice agriculture.
- They have the ability to spread to new areas and habitats (see Chapter 8 for discussion of invasive plants). This ability is illustrated by two nonindigenous species, kudzu (Fig. 2.3 and purple loosestrife Fig. 2.4). Both were introduced to the United States and both exist across large areas. Kudzu occurs primarily in southeastern United States but has invaded natural areas from Connecticut to Florida and west to Texas. Infestations have also been reported in Oregon and Washington. Purple loosestrife exists throughout the continent except Florida (Blossey, 2002).
- Weeds resist control including resistance to herbicides.

In 1941, a Swiss amateur-mountaineer and inventor, George de Mestral (died 1990), was walking in the mountains near Geneva. When he returned, he noticed that he had burdock burrs attached to his pants, jacket, and wool socks, and that they were in his dog's fur. Mr. de Mestral must have been a good observer: someone who sees what he is looking for when it is there, does not see what he is looking for when it is not there, and sees what he is not looking for when it is there. Mr. de Mestral did not go walking to collect burdock burrs on his clothing. However, he saw them and observed what he was not looking for. Microscopic examination revealed the unique hook of the burdock burr that allowed it to attach to the wool of his clothing. After his observation and after more than 7 years of work, he duplicated the grasp of the burr with nylon and in 1955 patented Velcro (from the French *velour* [velvet] and *cro* from the French *croc* or "crochet hook"). de Mestral formed Velcro S.A. in Switzerland in 1957. Subsequently, Velcro industries Ltd. was formed in 1967 with annual sales in excess of $10 million. Velcro Industries N.V. was created in 1972. Velcro is a generic term and velcro fasteners seem to be everywhere in modern society: children's clothing, airplanes, shoes, and artificial heart valves. It is worth a moment's reflection to consider that it came from a good observer's noticing the way in which one weed species disperses its seed in space.

Despite the anthropomorphic definition of "weed," weed scientists clearly know the plants that behave as undesirable weeds. The Weed Science Society of America published a Composite List of Weeds[4] in January 2010. In April 2017 it included 3488 weeds, 1412 more than the list of 2076 published in 1989 (Patterson, 1989). It seems that nearly every place must have weeds. Several books have been written to describe the weeds that exist in a place. Here are a few examples, some old, some new:

A Photo Handbook of Weeds Identification and Green Grass Lawn Care for Picture Perfect Turf (NC Weed Killer, 2014).
Aquatic and Riparian Weeds of the West (DiTomaso and Healy, 2003)
Common Weed Seedlings of the Central High Plains (Nissen and Kazarian, 2000)
Major Weeds of the Philippines (Moody et al., 1984)
Perennial Weeds: Characteristics and Identification of Selected Herbaceous Species (Anderson, 2008).
Striga *Identification and Control Handbook* (Ramaiah et al., 1983)
The Arable Weeds of Europe—With Their Seedlings and Seeds (Hanf, 1983)
Weeds of California & Other Western States (DiTomaso and Healy, 2006)
Weeds of Colorado, a Comprehensive Guide to Identification (Zimdahl, 2010b)
Weeds of Hawaii's Pastures and Natural Areas (Motooka et al., 2003)
Weeds of the Midwestern United States and Central Canada (Bryson and Defelice, 2010)
Weeds of Nebraska and the Great Plains (Stubbendieck et al., 1994)
Weeds of North America (Dickinson and Royer, 2014)

[4]http://wssa.net/weed/composite-list-of-weeds.

Weeds of the Northern U.S. and Canada: A Guide for Identification (Royer and Dickinson, 1999)
Weeds of Rice in Asia (Caton et al., 2004)
Weeds of the South (Bryson and Defelice, 2009)
Weeds of the West (Whitson et al., 1991)
XID Services of Pullman, WA sells an interactive CD for identification of 1200 Weeds of the 48 States and Canada.

An Interactive Encyclopedia of North American Weeds (Version 4.0) is available from the Weed Science Society of America and regional weed science societies. It includes 690 species, 4300 photographs, an illustrated glossary of botanical terms, and an identification key.

Larry Mitich published a series of brief, informative articles on 39 weeds: *The Intriguing World of Weeds*. They are available at wssa.net/wssa/weed/intriguing-world-of-weeds/#x.

4. HARMFUL ASPECTS OF WEEDS

Definitions of weeds usually include some aspect of trouble with crops, harm to people, animals, or the environment. Most people do not consider plants to be bad. They are assigned the descriptive, derogatory term "weed" because of something they do to us or to our environment. They interfere with our activities (plants with thorns or spines, such as puncture vine and Canada thistle) or affect our welfare (poison ivy). If they were just present and benign we would not be so concerned about them because there would be no real or perceived detrimental effects. That harmful effects occur is not questionable. It is important to understand specific effects so that appropriate action can be taken.

4.1 Plant Competition

From an agricultural perspective, weeds are of concern because they compete with crop plants for nutrients, water, and light. If they did not, those who grow crops and gardens would probably be willing to tolerate their presence. If weeds did not compete (see Chapter 6), they would not need to be managed because crop yield would not be affected by their presence. However, often complete crop failure (100% loss of marketable yield) can occur if weeds are not controlled. A summary of competitive yield losses without cost estimates is in Chapter 5 of Zimdahl (2004), with specific data on corn (Table 5.1) and soybean (Table 5.3).

4.2 Added Protection Costs

Weeds may increase protection costs because they often harbor other pests. A partial listing of diseases, insects, and nematodes that use weeds as alternate hosts is provided in Tables 2.1, 2.2, and 2.3. Weeds harbor a wide range of organisms, which increases opportunities for those organisms to persist in the environment and infest crops in succeeding years. Weeds that exist on the edges of crop fields serve as hosts when crops are not present, and as sources of infestation. Volunteer wheat is a primary host of wheat streak mosaic virus. The

TABLE 2.1   Plant Diseases Harbored by Specific Weeds

Plant Disease	Weed Host	Crop Infested	References
Blackleg	Black nightshade Common lamb's-quarter Marestail Redroot pigweed Smartweed	Potato	Dallyn and Sweet (1970)
Wilt diseases	Netseed lamb's-quarter Purslane Redroot pigweed	Potato, alfalfa	Oshima et al. (1963)
Stem canker	Netseed lamb's-quarter	Potato, beans	Oshima et al. (1963)
Soft rot	Annual sow thistle Dayflower Common lamb's-quarter	Chinese cabbage and other vegetables	Kikumoto and Sakamoto (1969)
Powdery mildew	Wild oats	Wheat, oats, Barley	Eshed and Wahl (1975)
Stripe mosaic virus	Common lamb's-quarter	Barley	Agric. Res. Serv. (1966)
Leaf curl virus	Common lamb's-quarter	Sugar beet	Agric. Res. Serv. (1966)
Cucumber mosaic virus	Black nightshade	Several	Agric. Res. Serv. (1966)
Potato virus X and Leaf roll virus	Redroot pigweed	Potato	Agric. Res. Serv. (1966)
Maize dwarf mosaic virus Maize chlorotic dwarf virus	Johnsongrass	Corn	Bendixen et al. (1979)
White rust Early blight Leaf spot Vascular wilts Cottony rot White mold Watery rot	Redroot pigweed	Potato, tomato Annual vegetables and flowers, Beans, cabbage, carrot, peanut	Commers (1967)
Leaf spot and leaf blight Stalk rot Vascular wilt Damping off Soft rot	Tall morning glory	Sugar beet, celery, pea, peanut, corn, tobacco, bean, fruits and vegetables	Commers (1967)
Stem rust Leaf spot and Leaf blight	Cocklebur	Wheat, barley, rye, celery, beets, tomato, soybean	Commers (1967)
White rust Banana Leaf spot Take-all Stem rust Rusts	Canada thistle	Crucifers, banana, wheat, rye, barley, legumes, bean, pea, fava bean	Commers (1967)

TABLE 2.2 Insects Harbored by Specific Weeds

Insect	Vector of	Weed Host	Crop Infested	References
Cabbage maggot	Blackleg	Common lamb's-quarter	Potato	Bonde (1939)
Colorado potato beetle		Black nightshade Buffalobur	Potato	Brues (1947)
Mosquito	Malaria, zika	Water lettuce	Aquatic areas	None
Beet leaf hopper	Curly top	Russian thistle	Sugar beet	Brues (1947)

TABLE 2.3 Nematodes Harbored by Specific Weeds

Nematode	Weed Host	Crop Infested	References
Criconemoides onoensis	Nutsedges Junglerice	Rice	Hollis (1972)
Ditylenchus dipsaci	9 weeds from 7 genera	Soybean, Snapbean, Pea	Edwards and Taylor (1964)
Heterodera glycines	Bitter cress Common foxglove Common pokeweed Oldfield toadflax Purslane Rocky Mountain bee plant Spotted geranium	Soybean	Riggs and Hamblen (1966)
Heterodera marioni	47 weeds from 42 genera	Pineapple	Godfrey (1935)
Heterodera schachtii	Black nightshade Lamb's-quarter Mustards Purslane Redroot pigweed Saltbush	Sugar beet	Anderson (1977)
Hoplolaimus columbus	Henbit Johnsongrass Purple nutsedge Yellow nutsedge	Soybean, cotton	Bendixen et al. (1979) Bird and Högger (1973) Högger and Bird (1974)
Meloidogyne incognita	Chickweed Johnsongrass Purple nutsedge Yellow nutsedge	Soybean, cotton	Bird and Högger (1973) Högger and Bird (1974)
Pratylenchus sp.	Johnsongrass	Corn	Bendixen et al. (1979)
Trichodorus spp.	19 weeds from 18 genera	Potatoes	Cooper and Harrison (1973)

disease is carried and transmitted by the wheat curl mite (*Aceria tosichella*) and can be transmitted up to a quarter mile from a stand of volunteer wheat. Volunteer wheat should be eliminated 3 weeks before wheat planting to eliminate the mites and prevent crop infection. This is a complex management problem in which a viral disease, an insect, and a weed host interact. Another illustration is the spread of potato blackleg disease (*Erwinia carotovora* var. *atroseptica*) and potato soft rot (*E. carotovora* var. *carotovora*) by *Erwinia* bacteria via enduring infestations of common lamb's-quarter, redroot pigweed, or black nightshade that harbor the disease organisms (Cooper and Harrison, 1973).

In addition to direct attack on crops, insects are a primary means of dispersal for many pathogenic organisms. The causal agent of aster yellows phytoplasma is vectored by the adult leafhopper (*Macrosteles quadrilineatus*) from lettuce to broadleaf plantain after lettuce emerges and during harvest. Several aphids carry pepper veinbanding mosaic virus and potato virus Y from weeds to crops (Broadbent, 1967). Fungal spores such as the conidia of *Claviceps purpurea* (the cause of ergot in rye) are transported by fungal gnats. The insects are attracted to sticky substances secreted by wounds. The fungal disease infects a wide range of grasses, including wild species. Piemiesel (1954) found that leafhoppers and curly top virus of sugar beets used weeds as breeding grounds to increase inoculum density for later crop infection.

A classic case of a weed serving as a host for a pathogen is the heteroecious stem rust fungus (*Puccinia graminis* var. *tritici*) of wheat, which uses European barberry as an alternate host. King (1966) estimated that wheat yield losses from this fungal disease were over 600 million bushels per year in the early 1960s. Over the 20 years from the 1970s to the 1990s, wheat rust caused $100 million in crop losses annually. Programs to eradicate European barberry and related species in the United States dramatically reduced stem rust epidemics. Several states in the United States joined in a program that was estimated to have saved farmers well over $30 million per year (Stakman and Harrar, 1957). It is a fungal disease that affects wheat, barley, and rye stems, leaves, and grain. The pathogen includes species of *Puccinia triticina* that cause black, brown, or yellow rust. It has caused epidemics in North and South America, Mexico, and India. Plant breeders struggled with it in the 1960s, believing they had finally beaten it into submission with new wheat strains. Weed control is an important part of rust control. It was once the most feared disease of all cereal crops. Development of resistant cultivars has reduced its incidence, but outbreaks can still occur when new pathogen races appear to which existing cultivars are not resistant. Stem rust remains an important threat to wheat in eastern Africa, and thus to the world food supply.

Russian thistle (Table 2.2) is an alternate host for the curly top virus of sugar beets and tomatoes (Young, 1991) and the beet leafhopper (*Circulifer tenellus*) (Goeden, 1968). Goeden points out that hosting a potentially damaging insect may not be a sufficient reason to control a weed. Russian thistle hosts 32 economically important insects from five different orders. These are not all harmful because some may be entomophagous enemies of harmful insects, both of which are hosted by Russian thistle. Crested wheatgrass is widely planted in the western United States for soil conservation. It and other species of *Agropyron* harbor the Russian wheat aphid (*Diuraphis noxia*), an important wheat pest that was first discovered in Texas in 1986. Now it is only an occasional pest in the United States, but it is an emerging issue in eastern Africa, where wheat production has moved into drier locations. Johnsongrass, a major weed in the southern United States, can hybridize with cultivated sorghum to produce the annual weed shattercane. Thus, a weed produces another weed (Mack, 1991).

4.3 Reduced Quality of Farm Products

Most grain growers are familiar with weed seed in grain crops and resultant decreases in quality and losses from dockage and cleaning. Weed seed in grain crops perpetuates the problem if the crop seed is replanted. A particularly bad problem is wild onion or wild garlic in wheat. Seeds and aerial bulblets of these weeds are similar in size to wheat grains and are difficult to separate. They impart an onion flavor to flour made from grain and an onion odor to milk after cows have grazed them or eaten feed containing them.

Wild oats affect the quality of bread and other wheat products and infest many acres of small grains, most notably spring wheat. Wild oats also infest barley used for feed and malting; any brewer will verify that wild oats make bad beer.

Weeds reduce the quality of seed crops. Purchasers of hybrid or certified seed expect to receive a high-quality product that will give high yields and not be infested with weed seed. This necessitates weed control in seed crops. Failure leads to high cleaning costs before sale.

Weeds cause loss of forage and reduce the carrying capacity of pastures and rangeland. Surveys in the 1990s by the Nebraska Department of Agriculture showed over 2 million acres infested with musk thistle and over 400,000 with leafy spurge. Rangeland and pasture were the dominant sites, and carrying capacity (the number of animals supported by the land) was reduced 8%–100% by musk thistle and 10%–70% by leafy spurge. In 2012, musk thistle was still a problem in central and eastern Nebraska (Roeth et al., 2012).

4.4 Reduced Quality of Animals

Many acres of western US range land are infested with poisonous larkspur, which can cause cattle death. Cattle like it and often eat it selectively. In early growth stages, within an hour, as little as 0.5% of an animal's weight ingested as larkspur can lead to toxicity; 0.7% may be fatal (Kingsbury, 1964). Locoweed and crazyweeds are also important poisonous range weeds. All ruminants are susceptible to locoweed poisoning, but only when large amounts are consumed over weeks or even months. Horses are also poisoned and symptoms appear at lower levels of intake for shorter periods of time than is true for ruminants (Kingsbury).

Halogeton grows on arid, alkaline soils and is found in many parts of the world including the western United States. It is especially toxic to sheep owing to its high oxalate content. Photosensitization or excessive sensitivity to light by cattle can be caused or aggravated by St. John's wort and mock bishop's weed (Anonymous, 1977). Common groundsel contains highly toxic alkaloids that can cause liver damage if consumed by cattle, horses, and sheep.

Weed science emphasizes the negative effects of weeds on animals grown for profit and on human food, but game animals can be affected. In western Montana, elk use of rangeland decreased as spotted knapweed increased. On native bunchgrass sites, 1575 pellet groups were found on each acre. Sites infested with spotted knapweed had only 35 pellet groups per acre (Hakim, 1975). Weeds can affect animals by providing an inadequate diet or a diet that is unpalatable because of chemical compounds in the weed. They can directly reduce the quality of animal products by affecting milk production, and fleece or hide quality. Reproductive performance is affected through toxins that may cause abortion or kill animals (Table 2.4).

TABLE 2.4 Characteristics of Some Poisonous Weeds (Evers and Link, 1972; Kingsbury, 1964)

Name	Toxic Principle	Source	Signs
Arrow grass	Hydrocyanic	Leaves	Nervousness, trembling, spasms, or convulsions
Bouncing bet	Saponin	Whole plant seeds are most toxic	Nausea, vomiting, rapid pulse, dizziness
Bracken fern	Unknown	Fronds	Fever, difficulty in breathing, salivation, congestion
Buffalobur	Solanin	Foliage and green berries	Most serious in nonruminants
Buttercup	Protoanemonin	Green shoots	Loss of condition, production drops, and milk is often red; diarrhea; nervousness; twitching; labored breathing
Chokecherry and other cherries	Glucoside-amygdalin, a cyanogenic compound	Leaves	Rapid breathing, muscle spasms, staggering, convulsions, and coma
Cocklebur	Hydroquinone	Seeds and seedlings	Nausea, depression, weakness, especially in swine
Corn cockle	A glucoside, githagin, and a saponin	Seeds	Poultry and pigs are most affected, inability to stand, rapid breathing, coma
Horsetail	Thiaminase activity, an alkaloid	Shoots	Loss of condition, excitability, staggering, rapid pulse, difficult breathing, emaciation
Indian tobacco	Alkaloids similar to nicotine	Leaves and stems	Ulcers in mouth, salivation, nausea, vomiting, nasal discharge, coma
Jimsonweed	Alkaloids	All parts	Rapid pulse and breathing, coma
Larkspur	Alkaloid	All parts	Staggering, nausea, salivation, quivering, respiratory paralysis
Nightshade	Solanine, a glycoalkaloid	Foliage and green berries	Most cases in sheep, goats, calves, pigs, and poultry; anorexia; nausea; vomiting; abdominal pain; diarrhea
Ohio buckeye	Alkaloid	Sprouts, leaves, and nuts	Uneasy or staggering gait, weakness, trembling
Water hemlock	Cicutoxin	Young leaves and roots	Convulsions
Whorled milkweed	A resinoid, galitoxin	Shoots, especially near top	Poor equilibrium, muscle tremors, depression and then nervousness, slobbering, mild convulsions

Poisonous plants may contain one or more of hundreds of toxins from nearly 20 major chemical groups including alkaloids, glucosides, saponins, resinoids, oxalates, and nitrates (Kingsbury, 1964). There is no way to determine whether a plant may be poisonous based on where it grows, when it grows, or how it changes during growth. Because poisonous plants occur in many habitats, one must learn to know the important ones in each area. There are no good antidotes after ingestion of poisonous plants by humans or animals. Signs of poisoning differ in intensity depending on the species, its stage of growth, when it is eaten, the soil in which the plant grew, the amount of other food eaten with or before the poisonous plant, and individual tolerance. Poisonous weeds can be managed once recognized. A few of the common poisonous weeds found in the United States, their toxic principle, the plant source, and some clinical signs of poisoning are shown in Table 2.4.

In addition to direct poisoning, weeds cause mechanical damage to grazing animals. Sharp spines on seed-bearing burs of puncture vine and sandbur are strong enough to penetrate tires and shoe leather. Anyone encountering a seed bur in bare feet quickly recognizes the pain and damage it could do to tender mouth tissue. Seed burs of these weeds and those of cocklebur and burdock also become entangled in sheep's wool, decreasing cleanliness, quality, and price.

It is well-known that many plants are poisonous to mammals. A few of them were mentioned earlier. Green plants dominate the terrestrial landscape. There are numerous insect and herbivore species that feed on the plants that dominate the landscape. As this section notes, weeds can and do reduce the quality of animals through their toxic principles, but the toxic principles also protect plants from predation by insects and herbivores. Plants that are relatively harmless to humans and other mammals may be and often are highly toxic to other animals, birds, fish, and especially insects (Harborne, 1988). The defensive toxicity of weeds and other plants is an important determinant of ecological relationships. The toxicity of many plants is often unknown; if it is known, it may be regarded as trivial although it is vital to ecological stability and part of the reason why plants dominate the terrestrial landscape.

4.5 Increased Production and Processing Costs

We are concerned about weeds because they do things to us or our products and increase production costs. Any weed control operation from hand hoeing to herbicide application costs money. These costs are often necessary to prevent greater crop loss or even crop failure and are regarded as necessary to gain a profit. However, if the weeds were not there, there would be no control cost. Unfortunately, the complete absence of weeds is rare and the costs of their competition and control must be included when calculating profit or loss. Costs of control are relatively easy to calculate if hourly labor, equipment, fuel, and herbicide costs are known. It has been estimated that the cost of tilling cultivated land may equal as much as 15% of a crop's value. Although tillage may be required on some soils for crop production, most is done only for weed control. There are sound agronomic reasons for tillage, including seedbed preparation, trash burial, soil aeration, promotion of water infiltration, and, of course, weed control. The ascendancy of minimum and no-tillage farming and availability of appropriate herbicides have brought many traditional tillage practices into question. Before herbicides, an experiment to investigate effects of tillage was always confounded by weeds and the need to control them. In many soils, experiments that included herbicides and tillage have shown little benefit from tillage other than weed control.

There are other, less obvious costs associated with weeds. Wild oats seed in wheat or barley, or black nightshade berries in beans lead to increased costs owing to the necessity of cleaning. Failure to remove these can lead to loss of quality, dockage losses at the point of sale, or even loss of the crop if it should heat and spoil in storage because of unripened weed seed. If a harvested crop has large amounts of weed seed in it, one can assume that

TABLE 2.5 Soybean Harvest Losses From Two Weeds (Nave and Wax, 1971)

	Percent Loss(%)		
Weed	Header	Threshing and Separating	Total
Redroot pigweed	5.35	00.73	6.08
Giant foxtail	1.55	00.81	2.36

some of the crop was lost in the field from weed competition and that additional quality was lost owing to harvest of weeds with the crop. Another cost of weeds at harvest is wear and tear on machinery. The extra bulk of weedy plants that pass through mechanical harvesting systems may cause machinery to break down more frequently and wear out sooner. These kinds of things are not usually attributed to weeds because they are not recognized as contributors to increased costs of machinery breakdown, repair, and replacement. Weeds also cost money when they remain in the field and interfere with harvest (Table 2.5).

4.6 Water Management

Weeds interfere with water management in irrigated agriculture. Water is consumed and flow is impeded by weeds growing in and along irrigation ditches. Weeds consume water intended for crops; cause water loss by seepage via root channels; transpire water; and reduce water flow in irrigation ditches, leading to evaporative water loss. Aquatic weeds may impede navigation and can ruin fisheries.

Terrestrial criteria for assessing weed competition cannot be employed in aquatic environments. There are no known appraisals of direct crop losses owing to aquatic weeds. However, Timmons (1960) reported almost 6 decades ago that man-made lakes above dams across major rivers in Africa, Asia, and Central and South America became so badly infested with weeds within 5–10 years of construction that their usefulness for power development, boat transport, and irrigation was greatly reduced. Therefore, it is reasonable to conclude that national development was impeded by weeds. Aquatic weeds quickly reduced designed flow of some irrigation canals in India by 40%–50% and in others up to 80% (Gupta, 1973). Submerged weeds retarded water flow up to 20 times, whereas floating weeds decreased it only two times (Gupta, 1976). Decreased flow reduced the possibility of irrigating distant fields and accelerated opportunities for leakage and evaporation. In addition to agricultural concerns, those who use water for recreation or enjoy the aesthetic appeal of aquatic habitats are often disturbed by weeds. Aquatic weeds are often ugly and their rotting remains are smelly, but the more important problem is that their presence and inevitable decay hasten eutrophication. Water hyacinth decreases water quality for human consumption. There is more public concern about aquatic weeds in recreational waters than in agricultural waterways.

4.7 Human Health

Those not associated with agriculture may often think of weeds, if they think of them at all, as plants that impair human health. One who has not experienced the misery of a runny nose, sneezing, and watery eyes associated with plant allergies (often called hay fever) cannot fully appreciate the animosity that people who undergo it have toward some plants. The pollen that causes hay fever often, but not always, comes from weedy plants. Ragweed and goldenrod are common causes in many parts of the United States. Sagebrush is a leading cause in the western United States.

Although allergies may be an obvious weed menace to some people, others would choose poison ivy or poison oak as the worst weeds. Swelling and very irritated, itching skin rashes after contact with the oily toxin urushiol[5] (present in both) are always bothersome and can lead to serious discomfort. The rash can be caused by contact with any portion of live plants, including roots, flowers, berries, and stems. Inhaling smoke from fire in which plants are burned can cause a whole-body reaction. Most people are quick to put poison ivy or poison oak into the category of unwanted plants after one or the other has disturbed their picnic or camping trip.

Many plants that poison when consumed are common household plants that can be especially hazardous to children. Some weedy species can lead to aberrant behavior or death when consumed by people. Examples of household plants that are poisonous when consumed include narcissus, oleander, lily of the valley, and iris.

Dead and dry weeds can be serious fire hazards. Those who live in the arid western United States know that fires spread rapidly in dry plants. Fire prevention and visibility are major reasons why weeds are controlled on roadsides, in vacant areas, and around homes in forested areas.

4.8 Decreased Land Value and Reduced Crop Choice

Perennial weeds (field bindweed, johnsongrass, or quack grass) or the annual parasitic weeds dodder, witchweed, or broomrape can lead purchasers to discount offers to buy or bankers to reduce the amount of a loan, because each recognizes a loss of productive potential. They also recognize the costs required to restore otherwise valuable land to full productivity. These weeds reduce land value and sale price because they restrict crop choice and increase production costs. Severe infestations of almost any perennial or parasitic weed will reduce yield of most crops, and dodder may completely eliminate successful growth of some crops.

4.9 Aesthetic Value

Weeds in recreation areas often must be controlled. No one wants a weedy soccer field or baseball diamond. Weeds are fire hazards around power substations and equipment, oil, or chemical storage areas. A practical need for weed control exists near traffic intersections where, in addition to being aesthetically unappealing, weeds reduce visibility and may contribute to accidents. Weeds can have serious environmental or ecological effects when they replace native vegetation (see Chapter 8 for discussion of invasive plants).

Nevertheless, Seiter (2016) and Del Tredici (2010) disagree. They advert that urban habitats characterized by high levels of disturbance, paving everywhere, and heat retention provide

---

[5]The name urushiol is derived from the Japanese word for the lacquer tree, *Toxicodendron verniciffuum*. The oxidation and polymerization of urushiol in the tree's sap in the presence of moisture forms a hard lacquer, which is used to produce traditional Chinese, Korean and Japanese lacquerware.

an environment that favors growth of stress-tolerant, early successional plants. We call such plants weeds. They are perfectly adapted to the urban environment. Seiter and Del Tredici encourage us to see weeds, one might say, through a different lens. They advocate recognition of their environmental benefits: temperature reduction, oxygen production, food and habitat for wildlife, pollution mitigation, habitat restoration, carbon dioxide absorption, erosion control and, yes, aesthetic value: pretty flowers.

5. COST OF WEEDS

There are no completely accurate estimates of the total cost of weed control and crop yield losses owing to weed competition, although several attempts have been made. One of the first estimates was reported in the 1969 United Nations Food and Agriculture Organization International Conference on Weed Control. For example, US losses owing to weeds in potatoes were estimated to be $65,000,000 in 1969 (Dallyn and Sweet, 1970).

In 1967, weeds caused an estimated 8% loss of potential US agricultural production (Irving, 1967). Cramer (1967) summarized losses attributed to pests of all kinds in the world's major crops. He calculated that 9.7% of the potential world crop yield was lost because of weeds. Parker and Fryer (1975) used Cramer's data and calculated that weeds eliminated 14.6% of the world's potential crop yield. Their estimate was a loss of 11.5% of world crop production in 1975 (Table 2.6). A comparison made in 1980 (Ahrens et al., 1981) for wheat and rice showed that losses were still about 10% despite improved control technology. Combellack (1989) estimated that the total cost of Australian weeds was $2 billion in 1986, of which $137 million was the cost of herbicides.

An estimate of crop yield losses from weeds in Canada in 1935 was $69 million (Hopkins, 1938). In 1949, the cost had risen 2.7 times to $186.2 million (McRostie, 1949), and it was $255 million in western Canada alone (Wood, 1955). By 1956, the total loss was estimated to be $468.6 million, a 150% increase over 1949 (Anderson, 1956). Friesen and Shebeski (1960) estimated that the annual loss due to weeds in Manitoba grain fields was $32.3 million in 1959. Renney and Bates (1971) estimated that losses resulting from weeds in British Columbia were $72 to $78 million per year in 1969. Their study showed that 38%–42% of yield losses owing to weeds in British Columbia were due to crop yield reduction, increased

TABLE 2.6 Estimated Food Losses Caused by Weeds in Three Classes of Crop Production[1]

Class of Crop Production	Total Cropped Area (%)	Relative Production per Unit Area	Total Food Production (%)	Loss to Weeds (%)	Loss of World Food Supply (%)	Estimated Food Loss per Year (metric tons × million)
Most highly developed	20	×1.5	30	5	1.5	37.5
Intermediate	50	×1.0	50	10	5.0	125.0
Least developed	30	×0.67	20	25	5.0	125.0
Total	100		100		11.5	287.5

[1]Estimates are not based on any statistical data. They are approximations suggested by the authors. Food losses estimated in metric tons are based on an approximate world total food production of 2.5 billion metric tons per year (Parker and Fryer, 1975).

TABLE 2.7 Estimated Average Annual Dollar Losses Owing to Weeds in Several Commodity Groups (Chandler et al., 1984)

Commodity Group	Average Annual Loss ($ × 1000)		
	United States	Western Canada	Eastern Canada
Field crops	6,408,183	616,331	69,647
Vegetables	619,072	20,972	29,956
Fruits and nuts	441,449	8,418	—
Forage seed crops	37,400	75,661	—
Hay	—	—	89,507
Total	7,506,104	722,634	189,110

insect and disease problems, dockage, harvest losses, and costs of control. If forest weeds were included, losses in yield and costs of control accounted for an additional 45%–49% of total loss. By 1984, Canadian losses were estimated to be $911.7 million per year (722.6 + 189.1) (Table 2.7) in 36 crops, nearly double what they had been in 1956.

A US soybean loss survey (Anonymous, 1971) found that weed competition caused an estimated 3.3 bushel/acre yield reduction in 28 states. Weeds were responsible for a 12% crop loss each year. Chandler (1974) summarized other estimates and concluded that weed competition in some parts of the southern United States caused as much as 20% soybean yield loss. For the entire country, 5% was regarded as an optimistically low level of loss, except on half of the most intensively farmed acreage.

Peanut farmers in the southeastern United States spent about $50 per acre for weed control. Annual losses from weeds were estimated to be $20,000,000 in Alabama, $8,000,000 in Florida, and $72,000,000 in Georgia in 1991 (Dowler, 1992). In the early 1990s, there were good herbicide choices for peanut weed control. The reasons for the large losses were a concern for farmers and weed scientists.

A US Department of Agriculture report for the 1950s (Agric. Res. Serv., 1965) estimated that annual losses owing to reduced crop yield and quality and costs of weed control in the United States were $5.1 billion. This value, which was an educated guess, became enshrined in early weed science textbooks. Although the estimate was never proven wrong, changes in the value of crops and inputs, as well as methods employed to arrive at such figures, have resulted in an increase in the dollar value of losses owing to weeds. In 1954, it was estimated that weeds caused an annual loss of more than $2 billion in 11 major US agronomic crops (Anonymous, 1962). In the 1970s, poisonous plants alone may have caused a $118 million loss to livestock producers in the US Great Plains states (Deloach, 1976). Shaw (1976) estimated that weeds caused loss of 10% of the value of food, feed, and fiber crops and ornamental plantings. The total annual loss was greater than $6 billion. He also projected that $2.7 billion was spent for cultural, ecological, and biological control, and another $2.3 billion for chemical control. The total cost of weeds was estimated to be $11 billion per year. In 1980, Shaw (1982) raised the total annual loss to $18.2 billion, with $12 billion resulting from competitive loss, $3.6 billion for chemical control, and $2.6 for other controls.

From 1975 to 1979, the annual competitive loss owing to the weeds in US agriculture for 64 crops was estimated to be $7.5 billion (Table 2.7) (Chandler et al., 1984). In a separate publication, Chandler (1985) estimated total losses of $14.1 billion, with $8 billion due to weed competition, $2.1 billion to herbicides, and $4 billion to equipment and labor.

Bridges (1992) estimated the cost of weeds in the United States for 1989–91. The report covered the entire United States except Alaska and 46 crops, including field crops, vegetables, fruits, and nuts. Research or extension weed scientists from each state estimated the percent yield loss from weeds competing in crops for which the current best management practices were employed. The same scientists also estimated losses with best management practices, but without herbicides. The loss was $4.2 billion annually just in field, nut, and fruit crops with best management strategies; 82% of the total was lost in field crops. Without herbicides, the loss rose to $19.6 billion. Total losses with best management practices were $6.2 billion. Costs of control were above $9 billion for a total loss of $15.2 billion per year. By any measure, this is a large amount of money, significantly greater than the 1984 estimate. Pimentel et al. (2000) estimated that at least $5 billion is spent annually in the United States to control nonindigenous weeds introduced to the United States that are in pastures, and another $1.5 billion is spent just on lawns, gardens, and golf courses. Control costs for nonindigenous weeds in crops were estimated to be $3 billion, and weeds caused an additional $23.4 billion in crop losses (yield not obtained) and damage to crops. Although the article (Pimentel et al.) specifically addresses nonindigenous weeds, the results can be applied to weeds in general because so many are nonindigenous.

Regional or more local estimates are often more accurate, but extrapolation to other areas, although tempting, is usually completely unwarranted. For example, leafy spurge occupies more than 150 million acres of rangeland in the northern Great Plains of the United States. Direct livestock production losses and indirect economic effects approached $110 million in 1990 (Bangsund and Leistritz, 1991). In North Dakota, losses of income by cattle producers owing to leafy spurge were $8.7 million, and the producers reduced personal spending by $14.4 million. That translates to reduced income for merchants who sell to cattlemen. In 1990, leafy spurge reduced cattle-carrying capacity about 580,000 animal-unit months or by 63,100 cows over a 7.5-month grazing season. Total annual direct grazing land losses were estimated to be $23.1 million. Indirect grazing land losses were $52.2 million and wildland losses were $2.9 million. A 40% leafy spurge infestation reduced rangeland carrying capacity 50%. Higher leafy spurge populations can reduce carrying capacity 75%. Because of only leafy spurge, North Dakota lost $87.3 million and 1000 jobs in 1980 (Leistritz et al., 1992).

A 2011 estimate concluded that weed control cost Australian farmers $1.4 billion annually. An additional $2.2 billion was lost from agricultural production. The annual cost of weeds to Australian agriculture exceeded $3.4 billion. Dependent on seasonal conditions, input, and commodity prices, the cost could be as high as $4.4 billion per year. The average annual net loss was $3.9 billion. The study suggested that the real environmental cost of weeds cannot be calculated, but it would be equal to or greater than agricultural costs.[6]

[6]http://pandora.nla.gov.au/pan/64168/20070119-0000/www.weeds.crc.org.au/documents/tech_series_8.pdf. A summary is available at http://pandora.nla.gov.au/pan/64168/20070119-0000/www.weeds.crc.org.au/documents/tech_series_8_summary.pdf.

World literature concerning domestic and international food production leaves no doubt that weeds cost money, lots of money. They are ubiquitous and their effects on yield create large losses borne by producers and consumers because production costs are inevitably reflected in food price. Globalization trends and lack of a world or country database for each crop make it unproductive to attempt more accurate estimates of world, country, region, or crop losses owing to weeds although current estimates lack precision. All estimates (by definition) are not absolutely accurate; they are created using the best information available. Because they are not the result of well-designed experiments, all data reported here cannot be regarded as absolutely true. All are estimates: educated guesses.

Weed costs are calculated in dollars associated with commodities. There are other ways to estimate costs and associated benefits of weed management. One is to examine the number of acres of crops treated for weed control. This estimates the value of weed management to farmers and is an accurate estimate of the extent of market penetration by herbicides (Table 2.8). The data do not estimate the use of other weed management techniques. Table 2.9

TABLE 2.8   Percentage of Crop Acreage Treated With Herbicides and Total Herbicide Use in the United States, 1971 and 1982 (Chandler, 1985)

Commodity	Proportion of Hectares Treated With Herbicide (%)		Herbicide Applied (million kg, by Active Ingredient)	
	1971	1982	1971	1982
ROW CROPS				
Corn	79	95	45.8	110.4
Cotton	82	97	8.9	7.8
Sorghum	46	59	5.2	6.9
Soybeans	68	93	16.6	56.8
Peanuts	92	93	2.0	2.2
Tobacco	7	71	0.1	0.7
Total	71	91	78.6	184.8
SMALL GRAIN CROPS				
Rice	95	98	3.6	6.3
Wheat	41	42	5.3	8.2
Other grain	31	45	2.5	2.7
Total	38	44	11.4	17.2
FORAGE CROPS				
Alfalfa	1	1	0.2	0.1
Other hay	a	3	a	0.3
Pasture and range	1	1	4.8	2.3
Total	1	1	4.0	1.7
Total	17	33	94.0	204.7

*a*Included in alfalfa.

TABLE 2.9 Comparison of Yield in Weeded and Unweeded Crops (Mercado, 1979)

Crop	Yield (tons/hectare)		% Increase From Weeding
	Weeded	Unweeded	
Lowland rice			
Transplanted	3.9	2.9	34
Direct-seeded	4.1	1.0	310
Upland rice	2.8	0.6	367
Corn	5.1	0.53	862
Soybean	1.15	0.48	140
Mung bean	0.75	0.57	32
Transplanted tomato	9.2	5.5	67
Direct-seeded tomato	5.1	1.5	240
Transplanted onion	10.8	0.44	2355

shows losses resulting from weeds by comparing weeded and unweeded crops in the Philippines and other Asian countries (Mercado, 1979). More recent information (A. Baltazar, personal communication[7]) confirms the scale, if not the actual cost, of the 1979 estimates. The percent increase in yield owing to weeding is an impressive statement about the value of weeding regardless of the technique by which it is done. Similar data are shown in Table 2.10 for studies performed on several crops in India, where improved methods may mean only

TABLE 2.10 Benefits From Weed Control at Various Dryland Centers in India, 1971–81

Location	Crop	Crop Yield With		Increase (%)
		Traditional Weed (kg/hectare)	Improved Control (kg/hectare)	
Varanasi	Upland rice	1700	2700	59
Dehradun	Maize	1760	4600	161
Hyderabad	Sorghum	1500	3740	149
Solapur	Pearl millet	180	950	428
Dehradun	Soybean	920	1840	100
Bangalore	Peanut	420	1910	355

Unpublished data from Friesen, G., Manitoba.

[7]Baltazar, A. Professor, Department of Agronomy, University of the Philippines—Los Baños, College, Laguna, Philippines.

better cultivation. The data should not be interpreted as a recommendation for all modern technology.

The Weed Science Society of America asked a slightly different question: What would the losses be if weeds were not controlled in corn and soybeans? The results showed that about half of both crops would be lost to uncontrolled weeds, annually costing growers about $43 billion (http://wssa.net/2016/05/wssa-calculates-billions-in-potential-economic-losses-from-uncontrolled-weeds/).

THINGS TO THINK ABOUT

1. What commonalities and differences can be found in the several definitions of the word "weed"?
2. How does the way we define something determine our attitude toward it?
3. What taxonomic, biological, morphological, and physiological traits do weeds share?
4. What is the best estimate of what weeds cost in the United States?
5. How are cost estimates obtained?
6. What are the problems with estimates of the cost of weeds?
7. Should new estimates of the cost of weeds be prepared? Why or why not?

Literature Cited

Agric. Res. Serv., 1965. A Survey of Extent and Cost of Weed Control and Specific Weed Problems. Agric. Res. Serv. Rpt. 23–1. U.S. Dept. Agric., Washington, D.C., 78 pp.

Agric. Res. Serv. USDA, 1966. Plant Pests of Importance to North American Agriculture. ARS. Handbook No. 307.

Ahrens, C., Cramer, H.H., Mogk, M., Peschel, H., 1981. Economic impact of crop losses. In: Proc. 10th Cong. of Plant Protection. Brit. Crop Prot Council, pp. 65–73.

Aldrich, R.J., 1984. Weed-Crop Ecology: Principles in Weed Management. Breton Pub., N. Scituate, MA, pp. 5–6.

Aldrich, R.J., Kremer, R.J., 1997. Principles in Weed Management, second ed. Iowa State Univ. Press, Ames, IA. 455 pp.

Alesina, A.F., Giuliano, P., Nunn, N., 2011. On the origins of gender roles: women and the plough. In: Working Paper Series. National Bureau of Economic Research, Cambridge, MA, 46 pp.

Anderson, E.G., 1956. What weeds cost us in Canada. In: Proc. California Weed Conf., vol. 8, pp. 34–45.

Anderson, W.P., 1977. Weed Science: Principles. West Pub. Co., N.Y., p. 1

Anderson, W.P., 2008. Perennial Weeds: Characteristics and Identification of Selected Herbaceous Species. Wiley, 228 pp.

Anonymous, 1962. A Survey of Extent and Cost of Weed Control and Specific Weed Problems. U.S. Dept. Agric. Agric. Res. Serv. and Fed. Ext. Serv. ARS 34-23, 65 pp.

Anonymous, 1971. Weed Losses in Soybeans. 1971 National Soybean Weed Loss Survey. Elanco Products Co., Chicago, IL.

Anonymous, 1977. Texas weed makes cattle supersensitive to sun. Chem. Eng. News 55, 44.

Anonymous, July 1990. Scourge of the south may be heading north. Natl. Geogr. 178 (1), 5.

Baker, H.G., 1965. Characteristics and modes of origin of weeds. In: Baker, H.G., Stebbins, G.L. (Eds.), Genetics of Colonizing Species, Proc. First Int. Union of Biol. Sci. Symp. on Gen. Biol. Academic Press, NY, pp. 147–172.

Baker, H.G., 1991. The continuing evolution of weeds. Econ. Bot. 45, 445–449.

Bangsund, D.A., Leistritz, F.L., 1991. Economic Impact of Leafy Spurge on Grazing Lands in the Northern Great Plains. Agric. Econ. Rpt. No. 275–5. Dept. Agric. Econ. N. Dakota State University, Fargo, ND.

Bendixen, L., Reynolds, D.A., Riedel, R.M., 1979. An annotated bibliography of weeds as reservoirs for organisms affecting crops. In: Res. Bull., vol. 1109. Ohio Agric. Res. and Ext. Center, Wooster, OH, 64 pp.

Bird, G.W., Högger, C., 1973. Nutsedge as host of plant parasitic nematodes in Georgia cotton fields. Plant Dis. Rep. 57, 402–403.

Blatchley, W.S., 1912. The Indiana Weed Book. Nature Publishing Co., Indianapolis, 191 pp.

Blossey, B., 2002. Ecology and Management of Invasive Plants.

Bonde, R., 1939. Comparative studies of the bacteria associated with potato blackleg and seed piece decay. Phytopathology 29, 831—851.

Boserup, E., 1970. Women's Role in Economic Development. G. Allen and Unwin, Ltd, London.

Bridges, D.C. (Ed.), 1992. Crop Losses due to Weeds in the United States-1992. Weed Sci Soc. of America, Champaign, IL, 403 pp.

Broadbent, L., 1967. In: Corbett, M.K., Sisler, H.D. (Eds.), Plant Virology. Univ. of Florida Press, Gainesville, 346 pp.

Brues, C.T., 1947. Insects and Human Welfare: An Account of the More Important Relations of Insects to the Health of Man to Agriculture, and to Forestry. Revised. Harvard Univ. Press, Cambridge, MA, 154 pp.

Bryson, C.T., Defelice, M.S. (Eds.), 2010. Weeds of the Midwestern United States and Central Canada. University of Georgia Press, 427 pp.

Bryson, C.T., Defelice, M.S. (Eds.), 2009. Weeds of the South. University of Georgia Press, 468 pp.

Buchholtz, K.P., 1967. Report of the terminology committee of the Weed Science Society of America. Weeds 15, 388—389.

Bunting, A.H., 1960. Some reflections on the ecology of weeds. In: Harper, J.L. (Ed.), The Biology of Weeds — A Symposium of the British Ecological Society. Blackwell Scientific Publications, Oxford, UK, pp. 11—26, 256 pp.

Caton, B.P., Mortimer, M., Hill, J.E., 2004. Weeds of Rice in Asia. Int. Rice Res. Inst., Los Baños, College, Laguna, Philippines, 116 pp.

Chandler, J.M., 1974. Economic losses due to weeds. In: Res. Rpt. 27th Ann. Mtg. Southern Weed Sci. Soc., pp. 192—214.

Chandler, J.M., 1985. Economics of weed control. In: Amer. Chem. Soc. Symposium Series 268. Chemistry of Allelopathy. American Chem. Soc., Washnigton, D.C., pp. 9—20

Chandler, J.M., Hamill, A.S., Thomas, A.G., May 1984. Crop Losses Due to Weeds in the United States and Canada. Weed Sci. Soc. of America, Champaign, IL, 22 pp.

Clark, G.H., Fletcher, J., 1923. Farm Weeds of Canada, second ed. Revised and enlarged by G.H. Clark pub in 1909. Reprinted by Canada Dept. of Agric. F.A, Acland, Ottawa, Canada.

Combellack, J.H., 1989. Resource allocations for future weed control activities. In: Proc. 42nd New Zealand Weed and Pest Cont. Conf., pp. 15—31.

Commers, I.L., 1967. An Annotated Index of Plant Diseases in Canada and Fungi Recorded on Plants in Alaska, Canada and Greenland. Pub. No. 1251 Res. Branch. Canada Dept. of Agriculture, 381 pp.

Cooper, J.I., Harrison, B.D., 1973. The role of weed hosts and the distribution and activity of vector nematodes in the ecology of tobacco rattle virus. Ann. Appl. Biol. 73, 53—66.

Crafts, A.S., Robbins, W.W., 1967. Weed Control. McGraw-Hill, N.Y., pp. 1—2

Cramer, H.H., 1967. Plant Protection and World Crop Production. English translation by J.H. Edwards. Pub. as Pflanzenschutz-Nachrichten by Bayer, A.G. Leverkusen, W. Germany. 524 pp.

Dallyn, S., Sweet, R., 1970. Weed control methods, losses and costs due to weeds and benefits of weed control in potatoes. In: FAO Int. Conf. on Weed Control, Davis, CA, pp. 210—228.

Deloach, C.J., 1976. Considerations in introducing foreign biotic agents to control native weeds of rangelands. In: 4th Int. Symp. on Biol. Cont. of Weeds. Gainesville, FL, pp. 39—50.

Del Tredici, P., 2010. Wild Urban Plants of the Northeast a Field Guide. Comstock Publishing Associates, Cornell University Press, Ithaca, NY, 374 pp.

Dickinson, R., Royer, F., 2014. Weeds of North America. University of Chicago press, 656 pp.

DiTomaso, J.D., Healy, E.A., 2003. Aquatic and riparian weeds of the west. University of California agriculture and natural resources, Oakland, CA. Publication 3421.

DiTomaso, J.D., Healy, E.A., 2006. Weeds of California and other western states (2-volume set). University of California agriculture and natural resources, Oakland, CA. Publication 3488.

Dowler, C.C., 1992. Weed survey — southern states, broadleaf crops subsection. Proc. South. Weed Sci. Soc. 45, 392—407.

Economist, July 23, 2011. Economics Focus — The Plough and the Now, p. 74.

Edwards, D.I., Taylor, D.P., 1964. Host range of an Illinois population of the stem nematode (Ditylenchus dipsaci) isolated from onion. Nematologica 9, 305—312.

Emerson, R.W., 1878. The Fortune of the Republic. Houghton, Osgood, Boston, MA, 44 pp.

Eshed, N., Wahl, I., 1975. Role of wild grasses in epidemics of powdery mildew on small grains in Israel. Phytopathology 65, 57—62.

EWRS. European Weed Res. Society, 2008. Constitution and Bye-Laws. Eur. Weed Res. Soc., 14 pp. http://www.ewrs.org/doc/ewrs_constitution_and_bye-laws_2008.pdf

Evers, R.A., Link, R.P., 1972. Poisonous Plants of the Midwest and Their Effects on Livestock. Spec. Pub. 24. Univ. IL College of Agric, 165 pp.

Friesen, G., Shebeski, L.H., 1960. Economic losses caused by weed competition in Manitoba grain fields. I. Weed species, their relative abundance and their effect on crop yields. Can. J. Plant Sci. 40, 457–467.

Fryer, J.D., Makepeace, R.J., 1977. Weed Control Handbook. Blackwell Sci. Pubs., Oxford, UK, p. 1.

Godfrey, G.H., 1935. Hitherto unreported hosts of the root-knot nematode. Plant Dis. Rep. 19, 29–31.

Godinho, I., 1984. Les de'finitions d'aventice et de mauvaise herbe. Weed Res. 24, 121–125.

Goeden, R.D., 1968. Russian thistle as an alternate hose to economically important insects. Weed Sci. 16, 102–103.

Gupta, O.P., 1973. Aquatic weed control. World Crops 25, 185–190.

Gupta, O.P., 1976. Aquatic weeds and their control in India. FAO Plant Prot. Bull. 24 (3), 76–82.

Hakim, S.E.A., 1975 (M.Sc. thesis). U. Montana, Missoula.

Hanf, M., 1983. The Arable Weeds of Europe – With Their Seedlings and Seeds. BASF Aktiengellschaft, Ludwigshafen, Germany, 494 pp.

Harbourne, J.B., 1988. Plant toxins and their effects on animals. In: Introduction to Ecological Biochemistry, third ed. Academic Press, London, UK, pp. 82–119.

Harlan, J.R., de Wet, J.M.J., 1965. Some thoughts about weeds. Econ. Bot. 19, 16–24.

Harper, J.L. (Ed.), 1960. The Biology of Weeds – A Symposium of the British Ecological Society. Blackwell Scientific Publications, Oxford, UK, 256 pp.

Högger, C.H., Bird, G.W., 1974. Weed and covercrops as overwintering hosts of plant parasitic nematodes in soybean and cotton fields in Georgia. J. Nematol. 6, 142–143.

Hollis, J.P., 1972. Competition between rice and weeds in nematode control tests. Phytopathology 62, 764.

Hopkins, E.S., 1938. The weed problem in Canada. In: Proc. Fourth Mtg. Assoc. Comm. On Weeds, East Div., pp. 11–15.

Humburg, N.E. (Ed.), 1989. Herbicide Handbook, sixth ed. Weed Sci. Soc. Am., Champaign, IL. 301 pp.

Irving, G.W., 1967. Weed control and public welfare. Weed Sci. 15, 296–299.

Kikumoto, T., Sakamoto, M., 1969. Ecological Studies on the Soft-Rot Bacteria of Vegetables. VII. The Preferential Stimulation of the Soft-Rot Bacteria in the Rhizosphere of Crop Plants and Weeds.

King, F.C., 1951. The Weed Problem, a New Approach. Faber and Faber, Ltd, London, UK, 164 pp.

King, L.J., 1966. Weeds of the World-Biology and Control. Interscience Publishers, Inc., NY.

Kingsbury, J.M., 1964. Poisonous Plants of the United States and Canada. Prentice Hall, Inc., NJ, 626 pp.

Klingman, G.C., 1961. Weed Control as a Science. J.W. wiley & Sons, NY, 421 pp.

Leistritz, F.L., Bangsund, D.A., Wallace, N.M., Leitch, J.A., 1992. Economic Impact of Leafy Spurge on Grazing Land and Wildland in North Dakota. Dept of Agric. Econ. Staff Paper Ser. AE-92005. N. Dakota State Univ., Fargo, 14 pp.

Leopold, A., 1943. What is a weed? In: Flader, S.L., Callicott, J.B. (Eds.), 1991. The River of the Mother of God and Other Essays by Aldo Leopold. Univ. Of Wisconsin Press, Madison, WI, pp. 306–309.

Little, W., Fowler, H.W., Coulson, J., 1973. The Shorter Oxford English Dictionary on Historical Principles, third ed. Rev. and ed. by C.T. Onions with etymologies revised by G.W.S. Friedrickson. Oxford, Clarendon Press. 2V. 2672 pp.

Luken, J.O., Thieret, J.W., 1996. Amur honeysuckle, its fall from grace. Bioscience 46, 18–24.

Mack, R.N., September 1991. Pathways and consequences of the introduction of non-indigenous plants in the United States. In: Rpt. to Office of Technol. Assessment.

Manning, R., 2004. Against the Grain: How Agriculture Has Hijacked Civilization. North Point Press, New York, NY, 232 pp.

McMichael, P., 2000. The power of food. Agric. Hum. Values 17, 21–33.

McRostie, G.P., 1949. Losses from weeds. Agric. Inst. Rev. 4, 87–90.

Mercado, B.L., 1979. Introduction to Weed Science. Southeast Asian Regional Center for Graduate Study and Research in Agriculture College, Laguna, Philippines, 292 pp.

Moody, K., Munroe, C.E., Lubigan, R.T., Paller Jr., E.C., 1984. Major Weeds of the Philippines. Univ. Of the Philippines at Los Baños, College, Laguna, Philippines, 328 pp.

Motooka, P., Castro, L., Nelson, D., Nagai, G., Ching, L., 2003. Weeds of Hawai'is Pastures and Natural Areas: An Indentification and Management Guide. Univ. Of Hawaii at Manoa, Honolulu, 184 pp.

Muenscher, W.C., 1935 (first ed.), 1955 (second ed.). Weeds. Macmillan Co., NY. 577 pp.

Navas, M.L., 1991. Using plant population biology in weed research: a strategy to improve weed management. Weed Res. 31, 171−179.

Nave, W.R., Wax, L.M., 1971. Effect of weeds on soybean yield and harvesting efficiency. Weed Sci. 19, 533−535.

NC Weed Killer, 2014. A Photo Handbook of Weeds Identification and Green Grass Lawn Care for Picture Perfect Turf. CreateSpace Publishing, 56 pp.

Nissen, S.J., Kazarian, D.E., 2000. Common Weed Seedlings of the Central High Plains. Colorado State Univ., 68 pp.

Oshima, N., Livingston, C.H., Harrison, M.D., 1963. Weeds are carriers of two potato pathogens in Colorado. Plant Dis. Rep. 47, 466−469.

Parker, C., Fryer, J.D., 1975. Weed control problems causing major reductions in world food supplies. FAO Plant Prod. Bull. 23, 83−95.

Patterson, D.T., 1989. Composite List of Weeds, Revised. Weed Science Society of America, Champaign, IL, 112 pp.

Piemeisel, R.L., 1954. Replacement control: changes in vegetation in relation to control of pests and diseases. Bot. Rev. 20, 1−32.

Pimentel, D., Lach, L., Zuniga, R., Morrison, D., 2000. Environmental and economic costs of non-indigenous species in the United States. Bioscience 50, 53−65.

Pritchard, T., 1960. Race formation in weedy species with especial reference to *Euphorbia cyparrissias* L. and *Hypericum perforatum* L. In: Haper, J.L. (Ed.), The Biology of Weeds − A Symposium of the British Ecological Society. Blackwell Scientific Publications, Oxford, UK, pp. 61−66, 256 pp.

Quammen, D., October 1998. Planet of weeds. Harpers Mag. 297 (1781), 57−69.

Ramaiah, K.V., Parker, C., Vasudeva, M.J., Musselman, L.J., 1983. Striga Identification and Control. ICRISAT Info. Bul. No. 15, Andhra Pradesh, India, 52 pp.

Renney, A.J., Bates, D.L., 1971. The cost of weeds. Can. Weed Comm. West. Sect. 24, 40−49.

Riggs, R.D., Hamblen, M.L., 1966. Additional hosts of *Heterodera glycines*. Plant Dis. Rep. 50, 15−16.

Robbins, W.W., Crafts, A.S., Raynor, R.N., 1942. Weed Control: A Textbook and Manual. McGraw Hill, NY, 543 pp.

Roeth, F., Melvin, S., Schluefer, I., Bernards, M., 2012. Noxious Weeds of Nebraska. Univ. of Nebraska-Lincoln Extension. EC176, 8 pp.

Royer, F., Dickinson, R., 1999. Weeds of the Northern U.S. and Canada: A Guide for Identification. University of Alberta Press, 434 pp.

Salisbury, S.E., 1961. Weeds and Aliens. Collins, London, UK, 383 pp.

Seiter, D., 2016. With future green studio. In: Spontaneous Urban Plants: Weeds in NYC. Rare Bird books, Los Angeles, CA, 248 pp.

Shaw, W.C., 1976. Weed Control Technology for Protecting Crops, Grazing Lands, Aquatic Sites, and Noncropland. U.S. Dept. Agric., Agric. Res. Serv. ARS-NRP-20280, 185 pp.

Shaw, W.C., 1982. Integrated weed management systems technology for pest management. Weed Sci. 30 (Suppl), 2−12.

Stakman, E.C., Harrar, G.J., 1957. Principles of Plant Pathology. Ronald Press, NY, 581 pp.

Stubbendieck, J., Friisoe, G.Y., Bolick, M.R., 1994. Weeds of Nebraska and the Great Plains, second ed. Nebraska Dept. Of Agriculture, Lincoln, NE. 589 pp.

Thompson, D.Q., Stuckey, R.L., Thompson, E.B., 1987. Spread, Impact, and Control of Purple Loosestrife (*Lythrum Salicaria*) in North American Wetlands. U.S. Dept. Interior, Fish and Wildlife Service, Washington, D.C.

Timmons, F.L., 1960. Control of aquatic weeds. In: FAO Int. Conf. on Weed Cont., Davis, CA, pp. 357−386.

Timmons, F.L., 1970. A history of weed control in the United States and Canada. Weed Sci. 18, 294−307. Republished Weed Sci. 53:748−761. 2005.

Tull, J., 1733. The Horse-Hoeing Husbandry: Or, an Essay Treatise on the Principles of Tillage and Vegetation, Wherein Is Taught a Method of Introducing a Sort of Vineyard Culture Into the Corn-Fields, in order to Increase Their Product and Diminish the Common Expense. William Cobbett, Publisher, London. U.K., 466 pp.

Vencill, W.K. (Ed.), 2002. Herbicide Handbook, eighth ed. Weed Sci. Soc. Am., Lawrence, KS. 493 pp.

Whitson, T.D., Burrill, L.C., Dewey, S.A., Cudney, D.W., Nelson, B.E., Lee, R.D., Parker, R., 1991. Weeds of the West. Western Soc. of Weed Sci., 630 pp.

Wood, H.E., November 10, 1955. Herbicides Used Agriculturally in Western Canada for the Control of Weeds. Mimeo, Manitoba Weeds Comm., Winnipeg, MAN.

Young, J.A., March 1991. Tumbleweed. Scientific American, pp. 82−87.

Zimdahl, R.L., 2010a. A History of Weed Science in the United States. Elsevier Inc., London, UK, 207 pp.

Zimdahl, R.L., 2004. Weed-Crop Competition a Review, second ed. Blackwell Publishing, Ames, IA. 220 pp.

Zimdahl, R.L., 2010b. Weeds of Colorado. A Comprehensive Guide to Identification. Cooperative Extension Serv., Colorado State Univ. Fort Collins, CO, 221 pp.

CHAPTER 3

Weed Classification

FUNDAMENTAL CONCEPTS

- Order in the world of weeds is recognized through systems of classification.
- Weeds can be classified in at least four ways. The most important and oldest system is based on phylogenetics or evolutionary ancestry.

LEARNING OBJECTIVES

- To learn the fundamentals of weed classification based on phylogenetics or ancestral relationships.
- To learn why and how other weed classification systems are used and their importance in weed management.
- To understand the unique habitat and role of parasitic weeds.
- To know the major groups of parasitic weeds.
- To understand the importance of a plant's scientific name.

One of the great, often unspoken hypotheses of modern science is that there is order in the world. With careful study, scientists discover and describe the order. With each discovery and consequent description, science improves understanding of how our world functions. Among those who study order in the natural world are taxonomists, who describe and classify species. All known weeds have been classified by plant taxonomists. However, although weeds are members of the plant kingdom and have been taxonomically classified, not everyone agrees about whether a particular plant is a weed, or exactly what a weed is (see Chapter 2).

There are at least 450 families of flowering plants and well over 350,000 different species. It is estimated that 3000 have been used by humans for food. Fewer than 300 have been domesticated; of these, there are about 20 that stand between humans and starvation. There are at least 100 species of great regional or local importance, but only a few species dominate the human food supply. About 18 plants have provided most of the food humans have consumed for many generations; wheat, rice, and corn provide about 60% of the human diet. In alphabetical order, other important dietary staples are: banana, barley, beans (including pigeon pea and chickpea), cassava, coconut, millet, oats, potato, peanut, rye, sorghum, soybean, sugar beet and sugarcane, and sweet potato. Nine of these are from one plant family: the Poaceae (grasses).

Twelve plant families include 68% of the 200 species that are the most important world weeds (Holm, 1978). These weeds share some characteristics including:

- Long seed life in soil
- Quick emergence
- Ability to survive and prosper under the disturbed conditions of field
- Rapid early growth
- No special environmental requirements for seed germination

They are competitive and often invasive, and react similarly to crop cultural practices. Weeds are usually defined primarily by where they are and how that makes someone feel about them. The fact that they may have shared characteristics means that we may be able to define and classify them based on what their genotype enables them to do.

Table 3.1 shows the 12 plant families that include 68% of the world's important weed problems. The Poaceae and Cyperaceae, account for 27% of the world's weed problems,

TABLE 3.1 Families of the World's Worst Weeds (Holm, 1978)

Family	Number of Species	Percent of Total[a]		
Poaceae	44			
Cyperaceae	12	27		
Asteraceae	32		43	
Polygonaceae	8			
Amaranthaceae	7			
Brassicaceae	7			68
Leguminosae	6			
Convolvulaceae	5			
Euphorbiaceae	5			
Chenopodiaceae	4			
Malvaceae	4			
Solanaceae	4			
Total	138[b]			

[a]Percentages are from Holm (1978). There are 138 species in the 12 families in the table. The percentages are based on 205 species in the 59 families included in the original table.
[b]Forty-seven other families have three species or less.

and when the Asteraceae are added, 43% of the world's most important (worst) weeds are included.[1] Nearly half of the world's worst weeds are in only three families, and any two of these include over a quarter of the world's worst weeds. The Poaceae has the most weedy species and includes the nine grasses that feed humans: wheat, rice, barley, millet, oats, rye, corn, sorghum, and sugarcane.

About two-thirds of the world's worst weeds are single-season or annual weeds. The rest are perennials in the world's temperate areas, but in the tropics they are more accurately called several-season weeds. The categories annual and perennial do not have the same meaning in tropical climates, where growth is not limited by cold weather but may be limited by low rainfall.[2] About two-thirds of the important weeds are broad-leaved or dicotyledonous species. Most of the rest are grasses, sedges, or ferns. The United States has about 70% of the world's important weeds, which are classified in different ways.

1. PHYLOGENETIC RELATIONSHIPS

Weeds are classified by taxonomists in the same way all species are: based on phylogenetic (Greek *phylo = phulon =* race or tribe plus Greek *gen =* be born of, become) relationships: their ancestry. All good identification manuals include a key to the species, and all keys are based on a classification developed over many years and, for plants, brought near their present form by the Swedish botanist Carl von Linné (Latinized form = Linnaeus, 1707–78). He established the binomial system of nomenclature (genus + species) primarily based on plants' floral characteristics, especially the presence, number, and characteristics of stamens and pistils. Before Linnaeus, all creatures were described in Latin with names that were what Bryson (2005, p. 448) calls "expansively descriptive." Bryson's example is the common weed cutleaf groundcherry, which botanists now agree is known as *Physalis angulata* L. Before Linnaeus it was known as *Physalis amno ramosissime ramis angulosis glabris foliis dentoserratis*. Students often abhor binomial nomenclature, but as difficult as it is, it is much easier than a Latin name with eight words.

Phylogenetic keys to plant species, based on ancestry and ancestral similarity, include division, subdivision, class, family, genus, and species. A brief description of a plant key for weed species follows.

1.1 Division I: Pteridophyta

Description: Fern-like, moss-like, rush-like, or aquatic plants without true flowers. Reproduce by spores.

Representative families: Salviniaceae, Equisetaceae, Polypodiaceae

---

[1] Most important and worst, in this context, means they cause the most crop yield losses.

[2] To illustrate the point — Tropical rainforests don't have winter and summer temperature extremes. Trees grow year round during rain and dry seasons. A few trees (e.g., teak) have annual growth rings, but most tropical trees do not.

1.2 Division II: Spermatophyta

Description: Plants with true flowers with stamens, pistils, or both. Reproduce by seed containing an embryo.

1.2.1 Subdivision I: Gymnospermae

Description: Ovules not in a closed ovary. Trees and shrubs with needle-shaped, linear, or scale-like, usually evergreen leaves.

Representative families: Pinaceae, Taxaceae. Almost no weedy species.

1.2.2 Subdivision II: Angiospermae

Description: Ovules borne within a closed ovary that matures into a fruit.

1.2.2.1 CLASS I: MONOCOTYLEDONEAE

Description: Stems without a central pith or annular layers but with woody fibers. Embryo with a single cotyledon. Early leaves always alternate. Flower parts in threes or sixes but never fives. Leaves mostly parallel-veined.

Representative families: Poaceae, Cyperaceae, Juncaceae, Liliaceae, Commelinaceae

1.2.2.2 CLASS II: DICOTYLEDONEAE

Description: Stems formed of bark, wood, and pith with the wood between the other two and increasing with annual growth. Leaves net-veined. Embryo with a pair of opposite cotyledons. Flower parts mostly in fours and fives.

Representative families: Polygonaceae, Chenopodiaceae, Convolvulaceae, Asteraceae, Solanaceae

All classified plants have a genus and specific name. By convention the genus is always capitalized (e.g., *Amaranthus*) and is commonly written in italics or underlined. The species name is not capitalized.

2. A NOTE ABOUT NAMES

The first question most people ask about a weed is, What is it? The expected and best answer is its name. But what name? Most plants have several names. Each has its own distinctive scientific name plus one or several common names. Common names vary among languages and among regions that share a language. For example, *Zea mays* is the plant Americans call corn, but the most of the rest of the world's people call it maize or (in Spanish) *maíz*. In England, wheat and other small grains are often known as corn. The weed *Vulpia myuros* (L.) K.C. Gmel. is called rattail fescue in the United States, but it is silvergrass in Australia. Further confusion results when common names dominate, when two different weeds share a common name. Southern sandbur and bristly starbur are different plants but have the same common name in the north and south of Brazil.

Reluctantly, but for the US reader's convenience, common names have been used throughout this book. The scientific name for all plants mentioned in the book are included in Appendix A (crop plants) and Appendix B (weeds). The scientific name is accepted throughout the world, or at least it is the name that can be used to resolve confusion that often occurs when just the common name is used.

Students resist learning scientific names because they are regarded as useless, boring, and perhaps even nonsense words designed to confuse and make learning difficult. Arguments against learning them are manifold. The first defense is that the names are difficult because they are in Latin, which, after all, is a dead language. Outside the Roman Catholic Church, few speak it, and knowing Latin certainly does not score many points with one's peers. Besides, the argument continues, common names are widely accepted and convey real meaning. Latin is difficult, but difficulty should be dismissed as an objection not worthy of one engaged in higher education. Similar to most worthy goals, obtaining an education will not be achieved without some effort. Latin is a dead language, but therein lies its advantage as a medium to naming things. A dead language does not evolve and assume new forms as daily use modifies it and introduces variation. The rules are fixed, and although the language can be manipulated, it is not malleable, as is a living language (Zimdahl, 1989).

As opposed to common names, scientific names have a universal meaning. Those who know scientific names will be able to verify a plant's identity by reference to standard texts or will immediately know the plant in question when the scientific name is used. Those who do not share the same native language can make use of unchanging Latin to share information about plants.

Scientific plant names have been derived from a vocabulary that is Latin in form and usually Latin or Greek in origin. Other peculiarities that make scientific nomenclature difficult are the frequent inclusion of personal names, Latinized location names, and words based on other languages. Taxonomists have developed and accepted rules for name creation that provide latitude for imagination and innovation but not license for their neglect (Zimdahl, 1989).

3. CLASSIFICATION METHODS

Other common and less systematic classification methods for weeds are based on life history, habitat, morphology, or plant type. Knowledge of classification is important because a plant's ancestry, length of life, the time of year during which it grows and reproduces, and its method or methods of reproduction provide clues about management methods most likely to succeed.

3.1 Type of Plant

The type of plant or general botanical group is an essential bit of knowledge but not very useful as a complete classification system. It is important that we know whether a weed is a fern or fern ally, sedge (Cyperaceae), grass (monocotyledon), or broad-leaved (dicotyledon). One should not begin to attempt to control or try to understand weedy behavior until this is known. However, when one knows the general classification, other questions about habitat or life cycle must be answered to acquire an understanding necessary to control the weed or to create a management system.

3.2 Habitat

3.2.1 Cropland

The first and most important agricultural, weedy habitat is crop land, where many annual and perennial weeds grow with disturbing regularity. Although it is essential to know the crop and whether it is agronomic or horticultural, it is not particularly useful. It tells us where the weed is but it does not tell us much about it. It is not a precise way to classify because there is so much overlap among crops. Few if any weeds occur exclusively in agronomic or horticultural crops or in just one crop. Redroot pigweed, velvetleaf, Canada thistle, and quack grass are commonly associated with agricultural crops. Others such as crabgrass, common mallow, prostrate knotweed, dandelion, and creeping wood sorrel commonly associate with horticultural crops. Each can occur in many different crops and environments.

3.2.2 Rangeland

Some weeds are almost exclusively identified with rangeland, a dry, untilled, extensive environment. Sagebrush and gray rabbitbrush are rarely weeds in corn or front lawns. Only the worst farmer or horticulturalist would attempt to grow a crop in an environment in which these weeds thrive. Range weeds include those shown in Table 3.2, and although the list is not exhaustive, it shows that rangeland weeds are commonly perennial and include many members of the Asteraceae. There are poisonous weeds such as locoweed and larkspur on rangeland and many others including thistles (of several species), dandelion, groundsel, buttercup, and vetch, but these also occur in other places.

3.2.3 Forests

In 1600, forests covered 46% of the United States (1023 million ac). In 2012, forest land was 766 million ac, about 33% of the US land area. In addition to common herbaceous annual and perennial weeds, there are others unique to the forest environment (Table 3.3). Woody

TABLE 3.2   Rangeland Weeds

Weed	Life Cycle	Family
Big sagebrush	Perennial	Asteraceae
Sand sagebrush	Perennial	Asteraceae
Fringed sagebrush	Perennial	Asteraceae
Broom snakeweed	Perennial	Asteraceae
Gray rabbitbrush	Perennial	Asteraceae
Yucca	Perennial	Liliaceae
Greasewood	Perennial	Chenopodiaceae
Halogeton	Annual	Chenopodiaceae
Mesquite	Perennial	Leguminosae
Locoweed	Perennial or annual	Leguminosae
Larkspur	Perennial	Ranunculaceae

perennials such as alder, aspen, big-leaf maple, chokecherry, cottonwood, oaks, and sumac and the herbaceous perennial bracken fern (common in the acidic soils of Pacific Northwest Douglas fir forests) are unique forest weeds. However, in other places (e.g., a windbreak, an ornamental) each of these tree species could be desirable.

Red alder was nearly eliminated by herbicides from Douglas fir forests in the 1970s. It fixes atmospheric nitrogen, and in soils deficient in nitrogen, Douglas fir will grow better with than without red alder. In the 1990s, red alder wood increased in value and some companies now plant it. Some weeds do so well they become crops! Red alder has been the target of biological control with a fungus (Dorworth, 1995).

3.2.4 *Aquatic*

Agriculture is the largest user of fresh water in the world and irrigated crops are sensitive to supply variation. Most of the world's major cities are located on a lake, ocean coast, or major river. Water, a finite resource, has been and will continue to be essential for urban and agricultural development. Aquatic weeds (Table 3.3) interfere with crop growth because they impede water flow or use water before it arrives in cropped fields. They can interfere with navigation, recreation, and power generation. Free-floating plants (e.g., water hyacinth) attract attention because their often massive infestations are so obvious. They move with wind and floods and some have stopped river or lake navigation. They float freely and never root in soil. Submersed plants (e.g., hydrilla) complete their life cycle beneath the water. Emersed aquatic weeds (e.g., common cattail) grow with their root system anchored in bottom mud and have leaves and stems that float on water or stand above it. They grow in shallow water, but all can impede flow, block boat movement, clog intakes of electric power plants and irrigation systems, and hasten eutrophication.

TABLE 3.3 Aquatic Weeds

Growth Habit	Weed	Life Cycle	Family
Free floating	Water hyacinth	Perennial	Ponterderiaceae
	Salvinia	Annual/perennial	Salviniaceae
	Water lettuce	Perennial	Araceae
	Duckweed	Annual	Lemnaceae
Submersed	*Hydrilla*	Annual/perennial	Hydrocharitaceae
	Western *Elodea*	Perennial	Hydrocharitaceae
	Pondweed	Perennial	Potamogetonaceae
	Eurasian watermilfoil	Perennial	Haloragaceae
	Coontail	Perennial	Ceratophyllaceae
Emersed	Cattail	Perennial	Typhaceae
	Alligator weed	Perennial	Amaranthaceae
	Arrowhead	Perennial	Alismataceae

3.2.5 Environmental Weeds

This category includes plants particularly obnoxious to people, such as poison ivy and poison oak, both of which cause itching and swelling when people come into contact with them. Other plants in the environmental group are goldenrod, ragweed, and big sagebrush, primary causes of hay fever–type allergies.

3.3 Life History

Another way to classify weeds is based on their life history. A plant's life history determines what in crops it might be a problem and what management methods are likely to succeed. All temperate weeds can be categorized as annual, biennial, or perennial. These groups are easy to define and observe and are useful in temperate zone agriculture. As mentioned earlier, the concept of perennation is not as useful in tropical agriculture, where temperatures extremes do not occur as they do in temperate areas.

3.3.1 Annuals

An annual is a plant that completes its life cycle from seed to seed in less than 1 year or one growing season. They produce abundant seed, grow quickly, and are usually but not always easier to control than perennials. Summer annuals germinate in spring, grow in summer, flower, and die in fall, and thus go from seed to seed in one growing season. Many common weeds such as common cocklebur, redroot and other pigweeds, crabgrass, wild buckwheat, and foxtails are annuals. The typical life cycle of an annual weed is shown in Fig. 3.1. Weed ecologists study many of the steps in this cycle. The sequence of events is qualitatively accurate but neither rates nor quantities are defined for most annual weeds. For example, it is

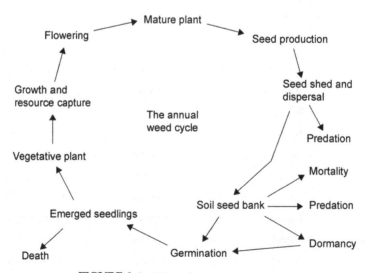

FIGURE 3.1 Life cycle of an annual weed.

known that not all seed produced by a weed survives in soil. Some die from natural causes at an unknown rate. Others experience predation by soil organisms or enter the soil seed bank where their life may be prolonged by dormancy. Quantitative understanding of the steps in a weed's life is essential to wise management.

Winter annual weeds germinate in fall or early winter, and flower and produce mature seed in spring or early summer the following year. Downy brome, shepherd's-purse, pinnate tansy mustard, and flixweed are winter annual weeds. They are particularly troublesome in winter wheat, a fall-seeded crop, and in alfalfa, a perennial.

Some parts of the world (Southern European and North African Mediterranean countries) have a winter rainy season, rarely with snow or subfreezing temperatures. This is followed by a long, dry period. Crops are planted in the fall when rains begin or just before it. The crops and their weeds begin to grow with the rain. Because the rains do not begin until late fall, annual weeds live into the next calendar year and their life cycle fits part of the definition of a winter annual. However, they are best regarded and managed as annuals because their growth is continuous and not interrupted by a cold period when plants live but do not grow.

3.3.2 *Biennials*

Biennials live more than 1 but not over 2 years. They should not be confused with winter annuals, which live during 2 calendar years but not for more than 12 months. Musk thistle, bull thistle, and common mullein are biennials. It is important to know that one is dealing with a biennial rather than a perennial. Spread of a biennial can be prevented by preventing seed production, which is not true for true perennials.

3.3.3 *Perennials*

Perennials are usually divided into two groups: simple and creeping. Simple perennials spread by seed and by vegetative reproduction. If the shoot is injured or cut off, simple perennials may regenerate a new plant vegetatively, but the normal mode of reproduction is seed. Simple perennials include dandelion, buckhorn and broadleaf plantain, and curly dock. Creeping perennials reproduce by seed and vegetatively. Vegetative reproductive organs include creeping aboveground stems (stolons), creeping belowground stems (rhizomes), tubers, aerial bulblets, and bulbs. The life cycle of a typical perennial is shown in Fig. 3.2. An excellent summary of the characteristics of 28 perennial weeds can be found in Anderson (1999). Important kinds of vegetative reproduction and examples of weeds that use each kind are (Leakey, 1981):

1. Rooting of detached shoots
   a. Turion: a dwarf vegetative shoot common among aquatic species. May replace an inflorescence (e.g., Eurasian watermilfoil)
2. Creeping stems
   a. Layers: shoots that contact soil root at nodes (e.g., annual bluegrass)
   b. Runner: a plagiotropic (tendency to grow obliquely or horizontally) shoot that may root in some shoot areas when in contact with soil (e.g., European blackberry, hedge bindweed)

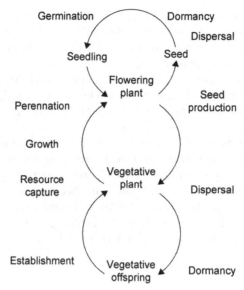

FIGURE 3.2   Life cycle of a perennial weed that produces seed and vegetative progeny (Grime, 1979).

   c. Stolons: horizontally growing stems that root at their nodes (e.g., creeping buttercup, creeping bentgrass, water hyacinth).
   d. Rhizomes: horizontal subterranean stems that send out aerial shoots (e.g., leafy spurge, quack grass, field bindweed, Johnson grass).
   e. Rhizomes and stolons (e.g., Bermuda grass)
   f. Tubers: swollen portions of underground stems (e.g., purple and yellow nutsedge)
3. Creeping roots: roots that send out new shoots (e.g., Canada thistle, field bindweed, Russian knapweed)
4. Taproot reproduction: roots that generate a new plant from root fragments (e.g., dandelion)
5. Modified shoot bases
   a. Bulbs: underground storage organs composed of swollen leaf bases or scales (e.g., wild garlic).
   b. Corms: swollen stems with dormant bulbs in the axils of scale-like leaf remnants (e.g., bulbous buttercup, tall oat grass)

3.4 Parasitic Weeds

Parasitic weeds are often placed in other sections of weed science texts. They are here because theirs is a particular and peculiar habitat. Phanerogamic parasites ("phanerogamic" comes from the Greek words *phaneros* = visible and *gamos* = marriage) include more than 3000 species distributed among 17 families. Only eight families include important parasitic weeds. The economically important species that damage crop and forest plants are all dicotyledons from five families (Table 3.4) (Sauerborn, 1991). Parasitic weeds from four families

TABLE 3.4 Important Families of Parasitic Weeds

Family	Genera	Common Name
Cuscutaceae	*Cuscuta*	Dodder
Loranthaceae/Viscaceae	*Loranthus*	
	Arceuthobium	Mistletoe
	Viscum	Mistletoe
Orobanchaceae	*Orobanche*	Broomrape
	Aeginetia	
Scrophulariaceae	*Striga*	Witchweed
	Alectra	

will be discussed briefly. Readers who want more detailed information are directed to Parker and Riches (1993).

The Cuscutaceae, which are dodders, are noxious in all US states except Alaska and are distributed throughout the world's agricultural regions. The Composite list of Weeds published by the Weed Science Society of America includes 14 species plus three synonyms.[3] A mature dodder plant, a true parasite, is a long, fine, yellow branching stem. A single stem of field dodder, one of the most important species, can grow up to 10 cm in 1 day. It is nonspecific regarding hosts. It coils, twines, and parasitizes many plants. Dodder flowers and reproduces by small, sticky seeds. Haustoria penetrate a host's cortex to the cambium and the fine stems dodder (tremble) when the wind blows. Dodder seed emerges from as deep as 4 ft in soil as a rootless, leafless seedling. The fine, yellow stem, which is 1–3 in. long, emerges as an arch, straightens, and slowly rotates in a counterclockwise direction (circumnutation) until it contacts another plant, which must be within about 1 1/4 in. Seeds have sufficient resources to search for a host for 4–9 days after which the emerged plant dies (Sauerborn, 1991). After contact and attachment, the seed (soil) connection withers and dodder plants live as obligate stem parasites. Westwood (http://wssa.net/2016/08/what-plants-sense-and-say-may-impact-the-future-of-weed-control/) adverts that dodder may tell host plants precisely how to lower their defenses so they can be more readily attacked and produce the nutrients dodder needs to thrive. It is a provocative hypothesis that may lead to new innovative management strategies. The US witchweed eradication program has been one of the world's greatest success stories in parasitic weed control (Tasker and Westwood, 2012).

Witchweed is one of three weedy, hemiparasitic species of the Scrophulariaceae in the world. It damages crop plants before they are even visible above ground. There are 35 species of *Striga*: 23 are found in Africa and at least 11 parasitize crops (Parker and Riches, 1993). Native to Asia and Africa, it is a US Federal noxious weed that was first identified in the United States, in North Carolina, in 1956 (Werth et al., 1984). Two important species are

[3]http://wssa.net/wssa/weed/composite-list-of-weeds/.

Striga hermonthica (purple or giant witchweed), which parasitizes sorghum, millet, and corn in Africa, and *Striga gesnerioides* (cowpea witchweed), the only one that parasitizes dicots. It is important to cowpeas and groundnut in East and West Africa and Asia. It is a root parasite on corn, sorghum, and other grasses in Africa, India, and the far East. Plants are normally 6–12 in. tall, but can grow up to 24 in. Leaves are linear and about 1 in. long. Flowers are less than 1/2 in. in diameter, occur in or on loose spikes, and can vary from white to yellow, red, or purple. Flowers produce a swollen seed pod that contains thousands of microscopic seeds (0.2 × 0.3 mm) per pod. Placed end to end, 1000–1500 seeds would be only 1 foot long. They survive up to 14 years in soil and one plant can produce up to 58,000 seeds.

The desert locust (*Schistocerca gregaria*) gains a great deal of publicity when it swarms in Africa. Massive efforts are made to combat it, but over the years and in any 1 year, witchweeds cause more crop losses in Africa than the desert locust. The genus has the narrowest host range of the important parasitic weeds and a narrower range of distribution than dodder. In 2014–15 it was limited to five counties in North Carolina and two in South Carolina. The successful eradication program, funded by the US Department of Agriculture (USDA), which was begun in 1957, has reduced the infestation from 450,000 ac to five North Carolina counties, 82 farms, 118 fields, and 1140.9 ac to two counties, and 15 farms, 18 fields, and only 130.3 ac in South Carolina.

Witchweeds are widely distributed in the world's tropical and subtropical regions and occur naturally in parts of Africa, Asia, and Australia. Secretions from corn (and some other grass) roots encourage germination of witchweed seed. After parasitization, corn is stunted, yellow, and wilted because of loss of nutrients and water. Many weeds, including crabgrass, serve as alternate hosts. It easily parasitizes corn because its 90- to 120-day life cycle is similar to corn's. One corn plant can support, but usually does not, up to 500 witchweed plants. Witchweed seed will not germinate in soil in the absence of a host-excreted stimulant. It may be induced to germinate with the artificial stimulant, ethylene gas. It was introduced to a corn field in North Carolina in July 1956 (Werth et al., 1984) and was recognized by a graduate student from India who had seen its effects on sorghum production in India. The USDA, via Animal Plant Health Inspection Service (APHIS) regulations, has had witchweed under quarantine in North and South Carolina since 1957 to prevent its spread throughout the United States. The quarantine has eliminated over 99% of the more than 432,000 ac that were infested within the eastern Carolinas (Iverson et al., 2011; Tasker and Westwood, 2012). South Carolina released the last acres from quarantine in 2009. Eventual eradication is predicted.

Plant parasites such as witchweed have not been controlled in susceptible crops with herbicides or weed management methods before damage occurred. The USDA/APHIS program has emphasized early detection, isolation, and quarantined areas. Crop seed coating with the benzoate herbicide pyrithiobac or the imidazolinone herbicide imazapyr offers promise for controlling witchweed in Africa (Kanampiu et al., 2003; Ransom et al., 2012). High herbicide levels can be localized on or near seed of acetolactate synthase–resistant maize. Imazapyr was optimal for seed dressings to prevent damage from witchweed, which emerged on untreated maize 6–12 weeks after planting. There was almost no emergence for 3 months on imazapyr-dressed maize seed (Kanampiu, 2001). Hand pulling escaped shoots reduced infestation and depleted the seed bank. Seed treatment gave a three- to fourfold increase over no control. The best control has been achieved with integration of improved, adapted varieties with cultural methods. Kanampiu (2008) suggests that herbicide seed treatment is a "stop-gap" measure until genetic resistance becomes available. Other herbicides offer some promise (Kabambe et al., 2008).

The most important parasite in the Loranthaceae is mistletoe. Mistletoes occur in two families: the Loranthaceae and the Viscaceae. Some taxonomists combine both families in the Loranthaceae. Dwarf mistletoe is a photosynthetic, flowering plant that parasitizes ponderosa pine in the southwestern United States. It occurs on the trunk and branches as a dense tangle of short, brown to yellow-brown stems. Seeds are dispersed by birds or by explosion of seed pods and expulsion of sticky seeds that may adhere to adjacent trees. Seeds that burst from pods can travel up to 60 miles per hour over 45 ft. The seeds are usually dispersed in August or early September in southwestern United States.

The Orobanchaceae (from Latin, *orobos* = bitter vetch and Latin, *anchein* = to strangle), or broomrapes, include over 100 species, five of which are important, obligate root holoparasites (lacking all chlorophyll) that attack carrots, broad beans, tomatoes, sunflowers, red clover, and several other important, small acreage crops in more than 58 countries (Parker and Riches, 1993; Sauerborn, 1991). The broomrapes have the broadest host range of the parasitic families. They cause major yield losses and often complete loss of some crops in many developing countries where control is not possible. They are the most important weed of cool-season food legumes (e.g., cowpea, fava bean[4]). Broomrape is found in California but is not important in most of the United States. It is important in Southern and Eastern Europe, West Asia, and North Africa. Seed of some species live in soil for up to 10 years. One plant can produce up to 200,000 small seeds; 1 g contains up to 150,000 seeds. Similar to witchweed, seed germination of *Orobanche* is stimulated by secretions from the host's root or from roots of nonhost plants. Pepper has shown some promise as a trap-crop for two *Orobanche* species (Hershenhorn et al., 1996). Germination will not occur in the absence of host-excreted chemical stimulants. Most damage from root parasites occurs before the parasite emerges and only 10%–30% of attached parasites emerge (Sauerborn, 1991).

An important aspect of parasitic weeds is the current inability to manage them with other than sophisticated chemical technology or extended fallow periods. It has been noted that as little as 100 g glyphosate/hectare (a sublethal dose) applied three times after rimsulfuron (a sulfonylurea herbicide) selectively reduced broomrape shoot numbers in potato (Haidar et al., 2005). Parasitic mechanisms and control were reviewed by Joel et al. (2013). Many of the world's people live in areas where food is scarce and agricultural technology is not modern or is absent or unaffordable. These are the places where parasitic weeds cause the greatest yield losses. Fields have been taken out of production and production area of some crops has been reduced severely owing to parasitic weeds.

Chris Parker, a retired weed scientist in the United Kingdom, prepares and distributes Haustorium, an online review of worldwide research on parasitic weeds. It is published by the International Parasitic Plant Society (http://www.parasiticplants.org) and distributed by Parker, the editor (chrisparker5@compuserve.com).

---

[4]The fava or broad bean is often called a faba bean, because the scientific name is *Vicia faba*.

THINGS TO THINK ABOUT

1. How are weed classification systems used?
2. What classification system is most likely to be used by horticulturalists, agronomists, and weed scientists?
3. Why are parasitic weeds such difficult problems and where do they exist?
4. If parasitic weeds are not important problems in most developed countries, why do we bother to study them?
5. Why bother to learn the scientific names of plants?
6. How are the scientific names of plants created?

Literature Cited

Anderson, W.P., 1999. Perennial Weeds: Characteristics and Identification of Selected Herbaceous Species. Iowa State University Press, Ames, IA, 228 pp.

Bryson, B., 2005. A Short History of Nearly Everything. Broadway Books, New York, NY, 624 pp.

Dorworth, C.C., 1995. Biological control of red alder (*Alnus rubra*) with the fungus *Nectria ditissima*. Weed Technol. 9, 243–248.

Grime, J.P., 1979. Plant Strategies and Vegetation Processes. J. Wiley and Sons, New York, p. 2.

Haidar, M.A., Sidahmed, M.M., Darwish, R., Lafta, A., 2005. Selective control of *Orobanche ramosa* in potato with rim-sulfuron and sub-lethal doses of glyphosate. Crop Prot. 24, 743–747.

Hershenhorn, J., Goldwasser, Y., Plakhine, D., Herzlinger, G., Golan, S., Russo, R., et al., 1996. Role of pepper (*Capsicum annum*) as a trap and catch crop for control of *Orobanche aegyptiaca* and *O. Cernua*. Weed Sci. 44, 948–951.

Holm, L., 1978. Some characteristics of weed problems in two worlds. Proc. West. Soc. Weed Sci. 31, 3–12.

Iverson, R.D., Westbrooks, R.G., Eplee, R.E., Tasker, A.V., 2011. Overview of the status of the witchweed (*Striga asiatica*) eradication program in the Carolinas. In: Leslie, A.R., Westbrooks, R.G. (Eds.), Invasive Plant Management Issues and Challenges in the United States: 2011 Overview. American Chemical Society, Washington, DC, pp. 51–68.

Joel, D.M., Gressel, J., Musselman, L.J. (Eds.), 2013. Parasitic Orobanche — Parasitic Mechanisms and Control Strategies. Springer, Berlin, 513 pp.

Kabambe, V.H., Kauwa, A.E., Nambuzi, S.C., 2008. Role of herbicide (metolachlor) and fertilizer application in integrated management of Striga asiatica in maize in Malawi. Afr. J. Agric. Res. 3 (2), 140–146.

Kanampiu, F.K., 2001. Herbicide seed coating: taming the Striga witchweed. Crop Prot. 20 (10), 885–895.

Kanampiu, F.L., 2008. Herbicide seed coating: taming the Striga witchweed. In: Second Session of the Open Forum on Agricultural Biotechnology in Africa. CIMMYT, Nairobi, Kenya.

Kanampiu, F.K., Kabambe, V., Massawe, C., Jasi, L., Friesen, D., Ransom, J.K., Gressel, J., 2003. Multi-site, multi-season field tests demonstrate that herbicide seed-coating herbicide-resistance maize controls *Striga* spp. and increases yields in several African countries. Crop Prot. 22, 697–706.

Leakey, R.R.B., 1981. Adaptive biology of vegetatively regenerating weeds. Adv. Appl. Biol. 6, 57–90.

Parker, C., Riches, C.R., 1993. Parasitic Weeds of the World: Biology and Control. CAB International, Wallingford, Oxon, UK, p. 332.

Ransom, J., Kanampiu, F., Gressel, J., De Groote, H., Burnet, M., Odhiambo, G., 2012. Herbicide applied to imidazolinone-resistant-maize seed as a *Striga* control option for small-scale African farmers. Weed Sci. 60, 283–289.

Sauerborn, J., 1991. Parasitic flowering plants in agricultural ecosystems of West Asia. Flora Vegetatio Mundi 9, 83–91.

Tasker, A.V., Westwood, J.H., 2012. The U.S. witchweed eradication effort turns 50: a retrospective and look-ahead on parasitic weed management. Weed Sci. 60, 267–268.

Werth, C.R., Riopel, J.L., Gillespie, N.W., 1984. Genetic uniformity in an introduced population of witchweed (*Striga asiatica*) in the United States. Weed Sci. 32, 645–648.

Zimdahl, R.L., 1989. Weeds and Words: The Etymology of the Scientific Names of Weeds and Crops. Iowa St. Univ. Press, Ames, IA, p. 125.

CHAPTER
4

Uses of Weeds - Ethnobotany

FUNDAMENTAL CONCEPTS

- Many weeds are useful for food, feed, or medicine.

- Some weeds may be useful for fuel or insulation.

- The same plant can be a weed in one place and a beneficial plant in another.

LEARNING OBJECTIVES

- To understand the many ways weeds can be used.

- To encourage thought about the importance of doing research to find uses for weeds.

There are laws in the village against weeds.
The law says a weed is wrong and shall be killed.
The weeds say life is a white and lovely thing
And the weeds come on and on in irrepressible regiments. *Sandburg, C., 1922. "Weeds." In: Smoke and Steel, Harcourt Brace and Howe, New York, p. 241.*

Ethnobotany is the study of the relationship between plants and people (Balick and Cox, 1996). It includes study of the uses of plants by humans and the relationship between humans and vegetation. It examines our dependence on plants and our effects on them. If weeds are just plants out of place and are properly regarded as useless by humans, is it possible they could also be useful? Can a single species be a weed and a useful plant?

One longs for a weed here and there, for variety;
A weed is not more than a flower in disguise,
Which is seen through at once,
if love give a man eyes. *"A Fable for Critics" (Lowell, 1848).*

Perhaps the greatest defense of weeds is contained in the last stanza of Gerard Manley Hopkins' 1918 poem "Inversnaid":

> What would the world be, once bereft,
> of wet and wildness? Let them be left.
> O let them be left; wildness and wet;
> Long live the weeds and the wilderness yet.

Ethnobotanical studies in Bolivia (Bentley et al., 2005) illustrate the multiple roles weeds have and affirm that the answer to the questions "Do we need them? Are they useful?" is indisputably *Yes!*

Some plants are weeds and farmers actively work to control them in crops. Some weeds, perhaps the same ones, are used as cattle fodder, especially in areas where cultivated land is limited and holdings are small. Bolivian farmers are familiar with the weeds and manage crops to preserve them. Some weeds are used as fodder when they grow on fallow land but are hoed out and left to rot when they grow in crops. Organic farmers (see Chapter 11) do not regard weeds with the same antipathy that modern agriculture does. Weeds are also used as construction material (cylindrical granaries), toys made by or for children, and medicine (Bentley et al., 2005). In Mabey's (2011) view, weeds provide a plethora of benefits. They stabilize soil, reduce evaporative water loss, provide shelter for insects and disease organisms (some useful and some harmful), and repair landscapes (an ecological Red Cross). He also mentions humorously, surely as a debatable aside, that weeding builds character.

It is possible, if one looks carefully, to learn something from the weeds. One of the most detested agricultural weeds is field bindweed. Yet it illustrates that beauty saves it. Its flowers are pretty. Behold, it brings glory to the morning. It is hard not to like it. It is a humble, prostrate plant that creeps along until it finally bumps into something and climbs. Although it may be humble it is also aggressive and ambitious. It grows in all directions including 10—20 ft underground. It may grow up to 10 ft radially in a single season. It thrives in soil that is dry, has low fertility, and may be unsuitable for most crops. It is not reluctant to climb on anything it bumps into; it is willing to accept help. Bindweed is patient; it waits for opportunities to succeed (Alvardo, 2016). It is true, we can learn from weeds rather than just despise them.

1. FOOD FOR HUMANS

There is interest in the possible uses of weeds, but not much. It is a sobering thought that agriculture has not found another major food crop since the soybean was rediscovered in the western world in the 20th century. Its wild relatives are native to China, where it was first domesticated more than 5000 years ago. It has been grown in China since 2838 BC, but was not grown in Europe until 1712. Soybeans were first grown in Thunderbolt, Georgia (near Savannah) in 1765 by Henry Yonge. The crop did not really develop in the United States until the early 1940s, when World War II stopped imports from China. In 2015, 85 million acres of soybeans (33% of the world crop on 20% of US crop acres) were grown and the crop was worth $12.5 billion, second only to corn in value. What would be the benefit of another food crop such as the soybean? Would it be worth a mission to the moon or a Nimitz class aircraft carrier (the United States has 10, each of which costs $4.5 billion) or a Gerald R. Ford aircraft carrier (each costs $13 billion, with the first deployment in 2019)? I suggest it would. The new crop could be a wild plant or even a weed, but few scientists are looking. It would wise to study indigenous crops grown by people in several world areas and the potential food or industrial value of weeds.

There are sources of information to assist the search. They range from an article in the *Reader's Digest* (Daniel, 1974) to books on edible native plants (Harrington, 1967; Gibbons, 1962), edible weeds (Duke, 1992; Hatfield, 1971), potential dietary uses of wild and cultivated plants (Hylton, 1974; Nyerges, 1999; Musselman and Wiggins, 2013; Kershaw, 2000), medicinal products (Moerman, 2009; Plotkin, 1993; Swerdlow, 2000), ethnobotany (Balick and Cox, 1996; Martin, 1995), new crops (Janick, 1996, 1999), and many articles in the scientific and popular literature. The national agricultural library of the US Department of Agriculture publishes the list of over 100 alternative crops especially suited for small farm diversification (see: https://www.nal.usda.gov/afsic/list-alternative-crops-and-enterprises-small-farm-diversification#toc1).

The tradition of using indigenous plants for human food is regarded by some as vegetarianism or food faddishness, but when viewed in its historical context, a long history of potentially useful food sources can be discovered. As mentioned in Chapter 3, the world has more than 300,000 species of seed-bearing plants. Perhaps as many as 30,000 of these have been used to some extent by humans as a food source. Many of these have been lost or forgotten. Fewer than 300 seed-bearing plants have become more or less domesticated, and of these about 30 provide most human food. Nine of these belong to a single plant family: the grasses. Vietmeyer (1981) calls these past foods, potential food sources. America's forgotten crops include the tepary bean, groundnut, and amaranth. His article includes jojoba, whose oil has potential as an engine lubricant, skin moisturizer, livestock feed, and leather softener, plus guayule, a potential source of natural rubber. The advantage of each of the forgotten crops Vietmeyer identified is that they grow well in arid soils where few current food crops do well. *Euphorbia lathyris* L. (caper, gopher, or myrtle spurge, or mole plant) is an erect biennial/annual native to southern Europe and introduced to eastern and western United States and Canada. It has been studied in California as a potential source of petroleum, but given the price of petroleum energy, its economic potential is low (Sachs et al., 1981).

Of 158 weed species collected from rice fields in two districts of West Bengal, India, 124 had economic importance to farmers (Vega, 1982). Young pigweeds may be eaten as salad greens and pigweed seeds can be eaten raw or parched. Several species of amaranth grow rapidly and contain abundant, high-quality protein (Hauptil and Jain, 1977). Leaves of some species contain up to 33% protein and their seeds have 16%−19% protein (Hauptil and Jain). Young leaves of shepherd's-purse are eaten as greens and dried roots can be eaten as a substitute for ginger or candied by boiling in a sugared syrup.

Instead of agonizing over dandelions in turf, why not learn to love and use them? In the late 1970s, Mayor Patrick R. Fiorello proclaimed his city, Vineland, New Jersey, to be the world's dandelion capital (a claim the city no longer makes). I have a copy of the Vineland recipe book for things ranging from dandelion Jell-O to dandelion soup (Anon., 1979). The small book is no longer available, but do not despair: others are (Wilensky, 2000;

Gail, 1990). Dandelions are harvested and sold for conversion to dandelion corn chowder, wine, or Italian dandelion casserole. Some say dandelion flowers are quite good when dipped in batter and deep-fat fried. Dandelion roots make a caffeine-free coffee substitute. Dandelions are part of the national cuisine of at least 54 countries (Gail, p. 12). Koreans can use them in kimchi (a pickled or fermented mixture of cabbage, onions, and fish with various seasonings). Germans make dandelion gravy to use on potatoes. The French use them in salads and the Italians, who call them *chigoda*, use them in many ways. The leaves are rich in vitamins A and C. More than 100,000 lb of dandelion are imported to the United States annually for use in patent medicines (Duke, 1992). The root contains a diuretic, and an old European common name is piss-a-bed. If you do not like the taste or price of your current brew, try some dandelion beer (Hatfield, 1971, p. 65).

Common purslane contains high levels of fatty acids, vitamin E at six times the level of spinach, and other nutrients. Europeans eat it in salads and it could be developed as a new vegetable crop (Anon., 1992). Omega-3 fatty acid has been linked in some studies to reduced heart disease, and purslane contains more than other green leafy vegetables (Anon.). It is well-adapted to and could be an alternative crop in arid areas. A farmer in Congerville, Illinois grows common lamb's-quarter, common purslane, and other plants that many, including weed scientists (Patterson et al., 1989), consider to be common and undesirable weeds. He sells his produce to high-end Chicago restaurants (Hale-Shelton, 2004). In fact, restaurant chefs across the country use common weeds as part of their cuisine. Wild plants have an important role in setting a restaurant's cuisine apart, making it special. Strand and DiStefano (2011) mention several weeds, each associated with a special item in a special restaurant.

Barnyard grass seeds may be eaten dry or parched and have been ground into flour. Some thistle seedlings may be eaten raw in salads if the spines are removed. Young Canada thistle roots can be peeled and the pith-like interior eaten raw or as a condiment in some cooked dishes. There are recipes for thistle-leaf tea. Seeds of wild oat can be ground into flour and one can make an artificial fly for fishing from wild oat seeds. Seeds of crabgrass, green foxtail, wild oat, and the common reed have been eaten whole.

Iroquois Indians ate burdock leaves as greens and used its dried roots in soup (Duke, 1992). Martin (1983) reported and Duke agreed, before Viagra (sildenafil) became popular, that eating raw burdock stems had been reported to stir lust and improve sexual virility. Duke reported its sale in Japan as an herb for sexual problems.

Wild mustard leaves have a hot, spicy flavor that blends well in salads with lettuce and dandelion. Wild onion has been used as a relish, to flavor cooked foods, and to cover the taste of gamey meat (Ross, 1976).

Martin (1983) reported that some Japanese eat kudzu root. It is ground to a fine powder and used as a condiment. The leaves are also eaten. Kudzu was promoted extensively by the US Department of Agriculture in the 1930s to stabilize eroding land. The Chinese have long relied on simple kudzu root extract to stop human craving for alcohol. It is sold as an over-the-counter drug in China and is 80% effective when taken for 2—4 weeks. The extract was evaluated in the United States and did not gain approval for treatment of alcoholism. Weed scientists do not welcome fields of kudzu, but other priorities may prevail. For example, sheep will eat kudzu readily, and they may be regarded as a biological control (Stewart, 2000; Corley et al., 1997). It has been studied as a technique for adsorption of heavy metals from aqueous waste streams (Brown et al., 2001), and as reinforcing material for manufacture of fiber-reinforced polypropylene composites (Luo et al., 2002).

Before hops were used, leaves of ground ivy, also called gill-over-the-ground (gill from the French *guiller* = to brew) were added to the brew to clarify and add flavor (Martin, 1983).

Alligator weed is a one of the worst weeds of waterways across the world (Holm et al., 1997). It is generally found in the warm tropics and warmer temperate regions. It forms dense mats, kills aquatic fauna, and reduces water flow and quality. Alligator weed and the related but not as vigorous weed, sessile joyweed, are used in green vegetable dishes in southeast Asia. Use of these plants as human food is not a problem, but when some Australians discovered that alligator weed was being cultivated as an herb or green vegetable, concern heightened because of the great possibility of escape and rapid spread as an invasive species.

Duckweed is one of the world's tiniest flowering plants, but it has potential in the fight against world hunger. It is rich in protein, contains high levels of all essential amino acids except one, and is nutritious and abundant. It is found in temperate and tropical regions and grows in thick green masses on surfaces of ponds and lakes. People in Thailand have eaten duckweed for generations. It can be harvested every 3–4 days and eaten in soups or stir-fried with other vegetables and meat. Duckweed could become a valuable livestock feed as well. Ten acres of duckweed could supply 60% of the nutritional needs of 100 dairy cows for 1 year. Considering that more than 100 million people each year experience severe protein/calorie malnutrition, the food potential of duckweed should be studied carefully.

N.W. Pirie of Rothamsted Experiment Station in England conducted experiments on juice pressed from a random collection of jungle plants. He was able to extract a juice with 50%–75% protein that, when coagulated, made a tasteless product that could be textured to resemble cheese.

There are other sources of information on edible weeds (See Duke, 1992; Hatfield, 1971) DeFelice (2002) describes yellow nutsedge (nutlets = tubers) as the snack food of the gods. Holm et al. (1977) classified it as one of the world's worst weeds. Both are correct. This erect, perennial, aggressive, weedy herb came from the eastern Mediterranean and has spread to all continents (except Antarctica) (DeFelice). Its tubers make it especially difficult to control, but they are quite tasty after roasting. In fact, it was imported to the United States as a potential vegetable crop in 1854 (DeFelice), which illustrates that not all imports have been good ones. Nevertheless, consumption of the roasted tubers, commonly known as chufa, can be traced to ancient Egypt. The tubers are not commonly available in the United States but are readily available in markets in West Africa.

Readers are cautioned that these examples are intended to illustrate the range of potential uses for plants and are not a recipe book or set of recommendations. Because of the danger of poisoning, allergies, digestive upset, and invasive species, specific references should be consulted before casual experiments lead to unanticipated problems.

Increased agricultural production has relied on low-cost energy and rapid genetic improvement for several decades (Boyer, 1982). These have allowed farmers to use dense plant populations adapted to high production on soil amended with purchased resources. Weeds grow well in fertilized crop fields and in some environments with limited resources. Some weeds have self-selected to do well with abundant whereas others do well with limited resources. Their genetic abilities may be important resources for plant breeders and crop producers.

A natural stand of giant ragweed in Champaign county, Illinois had an aboveground biomass similar to corn and greater than soybeans. Its seed biomass was lower than corn or soybeans but equal to the average soybean grain yield in the United States in 1975 (1610 kg/hectares [ha]). Giant ragweed is not a food crop and will not become one, but its high productivity with low inputs provides a valuable lesson for the future of food crops (Boyer, 1982) as energy and water resources decline or are directed away from agriculture.

Lewis Ziska, a scientist with the US Department of Agriculture Alternate Crop and Systems Laboratory, is engaged in virtue discovery. His research is at least partially motivated by Emerson's[1] definition of a weed as a plant whose virtues have not been discovered. He argues that wild lines of wheat, oats, and rice, which often are regarded as weeds, may be useful. They are useful because many weeds grow well when crops do not. Weeds survive extremes of temperature, droughts, low fertility, and other environmental factors that lead to crop failure. Ziska's research has shown that inclusion of wild lines in plant breeding programs may broaden genotypic or phenotypic variation and assist in selection of crops that do well with increasing temperature and atmospheric carbon dioxide levels (Ziska et al., 2014). Early differences in tiller formation may be an effective means to facilitate screening for carbon dioxide—sensitive rice genotypes (Ziska et al.). The same changes will have significant effects with respect to herbivory and on the production of secondary compounds of potential pharmacological utility (Ziska et al., 2005). Sweet wormwood, an annual native to China, is the world's primary source of artemisinin the primary worldwide treatment to counteract the effects of *Plasmodium falciparum*, the cause of the predominant form of malaria in Africa. Research by Zhu et al. (2015) has shown that projected increases in atmospheric carbon dioxide forecast global changes in artemisinin chemistry, potentially allowing a greater quantity of the drug's production. Even the much-maligned kudzu may have potential to supplement existing bioethanol feedstocks (Sage et al., 2009). Virtues remain to be discovered.

2. FEED FOR ANIMALS

In 2006, the University of Nebraska had an active research program to assist farmers in developing chicory as a crop. In 2011, farmers in western Nebraska planted 750 acres of chicory. Howlett et al. (2006) reported the same acreage in 2006. One Wyoming farmer grew more than 1000 acres in 2014. It is also grown in Colorado. Walker et al. (2011) showed that its use in livestock diets is limited. Chicory has been grown in Nebraska since the latter part of the 19th century, primarily for use as a flavoring in hot drinks such as coffee. It is now grown as a source of fructooligosaccharide, a broad term used to describe, fructans, and inulin in chicory root extract; thus, it has a role as an additive to human food. Inulin is the white milky substance found in dandelion roots. Most of the *Asteraceae* produce inulin as a storage carbohydrate. Chicory root extract is the accepted term for the fructooligosaccharides.[2] Chicory

---

[1]Emerson, R.W., March 1878. Fortune of the Republic. Part 1 — The Inheritance. A Sseries of tThree lLectures in Boston.

[2]Personal communication from Dr. Robert G. Wilson, Panhandle Research & Extension Center, University of Nebraska, Scottsbluff, NE., October 2011.

roots can also be roasted for use in coffee. Chicory derivatives are a major ingredient in Fiber One cereals and bars.

Weed seed screenings are used in many US states as animal feed, a practice with some disadvantages. A primary concern is that some seeds pass through the animal, which then effectively distributes weeds to new places.

Some rangeland plants are weeds, but cattle grazing on native blue grama and buffalo grass range achieved better gains when weeds and shrubs constituted 10%—70% of total vegetation. One should not neglect the contribution of sagebrush and other weedy range species to the diet of browse animals such as deer, elk, and antelope.

Cattle ranchers in the western United States often use kochia hay as feed. When immature, its protein content can be 17%, equal to alfalfa. However, it becomes woody as it matures, and because it is an annual it will not reseed when harvested immature for hay. An important warning is that it can accumulate high amounts of nitrates, and cattle may be intoxicated and lose weight when kochia is 90% or more of the diet.

The forage and nutritional value of many weed species is equal to that of cultivated forage crops. Marten and Anderson (1975) and Temme et al. (1979) reported that the annual broad-leaved weeds redroot pigweed, common lamb's-quarter, and common ragweed had digestible dry matter, fiber, and crude protein concentrations about equal to alfalfa when alfalfa and weeds were harvested at the same growth stage. Giant foxtail, Pennsylvania smartweed, shepherd's-purse, and yellow foxtail all had nutritional value lower than alfalfa. Dutt et al. (1982) concluded that yellow rocket reduced the feeding value of alfalfa hay but white campion and dandelion did not. Yellow rocket reduced nutritive value index, animal intake, and digestibility.

The perennial weed quack grass is a serious weed problem in the perennial crop alfalfa. It invades hay consumption and its presence in hay decreases cattle's consumption of feed. Although its nutritional value is high, its palatability is lower than alfalfa or smooth bromegrass (Marten et al., 1987). A biotype selected for broad leaves was equal or superior to smooth bromegrass and equal to alfalfa in palatability in Minnesota (Marten et al.). The research also investigated the forage value of nine perennial broad-leaved and grass weeds compared with alfalfa and smooth bromegrass. Smooth bromegrass and quack grass consistently had more neutral detergent fiber, less crude protein, and an in vitro digestibility similar to alfalfa. Jerusalem artichoke, Canada thistle, dandelion, and perennial sow thistle had crude protein and in vitro digestibility equal to or greater than alfalfa. Broad-leaved species generally had lower palatability than alfalfa or smooth bromegrass. Jerusalem artichoke, Canada thistle, curly dock, and hoary alyssum were completely rejected by grazing lambs and are therefore always weedy species in sheep pastures.

Weed forage and hay quality are correlated with plant maturity. More mature plants have lower forage quality. Marten et al. showed this to be true for nine perennial species in Minnesota, where forage quality measured as digestible dry matter, fiber, or crude protein declined with maturity. Crude protein of curly dock declined 22% from the vegetative to the mature seed stage (Bosworth et al., 1985). In hay crops, the decision to control weeds must be site specific and depends on the weeds present and their growth stage when hay is to be cut. Alfalfa stands often become weedy because of the death of alfalfa plants, not because weeds crowd them out (Sheaffer and Wyse, 1982). Control may reduce hay yields and produce hay of lower quality if all weeds are controlled, just because they are perceived to be weeds and therefore undesirable. Some weeds make good pasture and forage.

As mentioned earlier, ragweeds are palatable to grazing animals, with common ragweed being more so than giant ragweed. Equally important, seed of both species of ragweed and many other weedy species provide food for finches and other birds in the winter in the US midwestern states.

Balick and Cox (1996, p. 34–35) list 50 drugs used in human medicine that have been discovered from ethnobotanical leads. The list includes seven species that are also listed in the Weed Science Society of America (WSSA) composite list of weeds (Patterson et al., 1989) and five other plants from genera that are included in the WSSA list. One wonders how many common weeds may be sources of new pharmaceuticals.

Until the early 1960s, diagnosis of childhood leukemia was a sign of sure and fairly rapid death. Now the long-term survival of victims of childhood leukemia is 90% or greater (Swerdlow, 2000). The common weed (in the WSSA view) Madagascar or rosy periwinkle, changed sure death from childhood leukemia to probable complete recovery. The pharmaceutical potential of vinblastine (a vinca alkaloid) was one of the most significant discoveries of ethnobotany. Researchers at the Eli Lilly Company screened a collection of 400 potentially medicinal plants against cultures of p38 mouse cell leukemia and found that Madagascar periwinkle killed leukemia cells. Up to 250 kg of leaves is required to make a single 500-mg dose; therefore it was unlikely to be an effective folk medicine, but it was a healer's claim of its effects against diabetes that led to further investigation (Balick and Cox, 1996, p. 33).

Rosy periwinkle is one of more than 10,000 known plant species in Madagascar, a Texas-size island off the southeastern African coast. Many are not known elsewhere. Over the past 40 years there have been few new pharmaceutical drugs developed from plant sources. Part of the reason is the cost of finding the plants and verifying potential utility. Secondarily there is the manufacturers' legitimate concern about unpredictability and lack of scientific verification. Rosy periwinkle worked as a folk medicine for diabetes (as a tea it lowers blood sugar) but not for leukemia; that test was a random stroke of luck. Science-based development, testing, and formulation is a surer route to commercial success. Nevertheless, two-thirds of the world's people rely on the healing power of plants and the healers who use them. They could not afford modern medicines even if they were available (Swerdlow). Many plants contain bioactive chemicals with potentially beneficial effects for humans and animals. Many of them may be just weeds. We will never know until we listen to healers and study their plants.

---

3. MEDICAL USES

Plants move in one place but do not move in space while growing, as animals do. Because they are immobile and cannot escape from predators, they have evolved elaborate chemical defenses. Plants contain a myriad of mostly unknown, frequently unusual chemical compounds that may have medicinal properties (Swerdlow, 2000). Most (perhaps as many as 99% of the flowering plants) have never been tested (Bryson, 2005, p. 461). A few are known to traditional tribal healers (often called shamans). Scientists have been interested in the actual and potential use of plants in medicine for a long time. Henkel (1904) wrote about possible medicinal uses of 26 common weeds. She notes that in their fight to control weeds, farmers may also be able to "turn some of them to account." Reflecting the cultural attitudes of the time, she says "the work of handling and curing them is not excessive and can readily be done by women and children." Stepp and Moorman (2001) showed significant uses of weeds in the medicinal floras of the Highland Mayas in Chiapas, Mexico and in the medicinal flora of native North Americans. Stepp (2004) analyzed 101 plant species from which 119 contemporary pharmaceuticals are derived. He showed that 36 are weeds (see Table 2, Stepp, 2004). The frequency of the appearance of weeds as pharmaceutical products is significantly larger than would be predicted by the frequency of the appearance of weedy species in the general flora. Stepp adverts that "it's time the much maligned weeds be given more respect for their use as phytomedicines."

Plants and plant extracts have been used to treat almost every ailment known to humans, ranging from venereal disease and rheumatism to colds and bleeding. Plants with the word *officinale* (or its derivatives) were at one time included on an official drug or medicinal list. "Wort" (e.g., common St. John's *wort*), a common suffix in plant names, means "healing." A plant with "bane" added to its common name (e.g., hen*bane*) was probably once used for medicinal purposes. Readers can probably think of plants that fit in one or more of these categories. St. John's wort (Chase-devil, Klamath weed, or Tipton's weed) is a nonprescription medication (usual dose, 300 mg, three times/d) for mild depression. It can be used in lieu of the prescription drug Prozac (fluoxetine), but its efficacy varies among consumers.

Roots of yucca can be chopped and soaked in water to extract a soapy substance that western US Indians used for washing and cleaning. One must assume that this is a source of the common name soapweed. The next time you are stung by a bee, hope you are standing near a curly dock plant. Quickly rub some leaves between your hands and press them, with their juice, against the sting. Within 10–20 min the stinging sensation will be gone, but it is usually gone by then anyway. Curly dock has more vitamin C than oranges, and extracts of its yellow root have been used to treat jaundice. Plantain has similar properties. Curly dock leaves have also been boiled in vinegar to soften the fibers, and then combined with lard to make an ointment for treatment of inflammation.

A persistent human problem is the common head cold. If one boils a few ounces of sunflower seed in a quart of water, adds honey and gin, and takes the mixture three or four times daily, irritating mucous will be discharged from the nose and mouth. However, one must question whether the mixture's efficacy results from the sunflower seed extract or the gin (also a plant product made by distilling rye or other grains with juniper berries).

Yarrow and big sagebrush have been used as a tea to relieve the fever that accompanies a cold. Yarrow leaves were chewed by western pioneers to settle an upset stomach. Extracts were also used to regulate menstrual flow and to stop blood flow from a wound. Modern medicine has confirmed its efficacy (Martin, 1983).

The common European herb/weed queen of the meadow has long been used in folk medicine to treat fever and pain (Balick and Cox, 1996, p. 32). According to Balick and Cox, it is also known incorrectly as meadowsweet. In 1839 (Balick and Cox), salicylic acid was isolated from flower buds of queen of the meadow. Pure salicylic acid was used for pain relief but frequently caused stomach problems. In 1899, the Bayer Company combined acetic acid and salicylic acid to create acetylsalicylic acid, an effective pain killer and still one of the world's most widely used analgesics. The name "aspirin" was derived from "a" for acetyl and "spirin" from the genus *Spirea* (Balick and Cox, 1996, p. 32). Aspirin was first synthesized by Felix Hoffmann in Germany in 1897, patented in February 1899, and marketed the same year, as aspirin by Bayer Chemical Co. It succeeded because it had better pharmacological activity and fewer side effects than pure salicylic acid. Salicylic acid is also found in the bark of the willow tree (*Salix*). The bark was chewed by some North American Indians to relieve pain.

Dandelion has been used as a laxative, and shepherd's-purse and common St. John's wort have been used for control of diarrhea. Common burdock has been used to make a tonic and diuretic. Young shoots and the pith of young leaf stalks can be eaten raw with salt or after boiling in salt water. Water lettuce has been used to cure coughs and heal tubercular wounds. The Chinese use its leaves as an external medicine for boils; American Indians used the leaves to cure hemorrhoids (Harrington, 1967). Extracts or preparations of several weeds have been used as sedatives, including poison hemlock, jimsonweed, poppies, and, marijuana, which has achieved widespread use and popularity (although it is often illegal) for its sedative and relaxant properties. The latter weed is known to most people and experienced by many.

Some plants inspire poetry, others encourage artists, but some "evoke the dark arts" (Shyr, 2011). Some weeds are best avoided (see Kingsbury, 1964). Contact with the phototoxic chemicals in the sap of giant hogweed can cause blisters with exposure to the sun. Consuming the attractive black berries of the common garden weed deadly nightshade can affect the human nervous system. In the Middle Ages, some believed it helped witches fly. Ingesting any part of monkshood can lead to heart failure. Consuming any part of mandrake may cause severe gastroenteritis. Of course, you do not have to believe me. One can go to a real authority. Check out the effects and uses of mandrake in *Harry Potter*.

Healers used to grate the dried root of wild carrot and apply the grated material to soothe burned skin. Modern science has shown that the roots contain carotin, which heals burns when mixed with oil (Martin, 1983). Modern science has proven the utility of extracts of bouncing bet to treat jaundice and liver problems (Martin). Backpackers and campers should know bouncing bet. When torn or bruised leaves are added to cold water, a bubbly lather ensues. This source of soap has been known since the Middle Ages (about AD 1066–1450). Its unusual name comes from the white, reflexed petals that someone thought resembled the posterior view of a washerwoman (named Bet?) with her petticoats pinned up.

Drury (1992, based on Gerard, 1597) reported that sprigs of common tansy were placed in beds and bedding to discourage vermin. Tansy tea tastes terrible but has been used to treat a variety of illnesses including rheumatism and intestinal worms (Martin, 1983).

Scientists in the Philippines have studied antifertility and abortive characteristics of the common weed called sensitive plant.

4. AGRICULTURAL USES

As mentioned earlier, kochia is a good source of protein for ruminant animals. It is self-seeding, high-yielding, water-efficient plant with no serious disease or insect pests. It is a serious annual weed in many crops and common in many parts of the United States. Kochia can accumulate high levels of nitrate and will escape from cultivation and become a weed. It may cause photosensitization in cattle. In some experiments, cattle have lost weight and some have died when fed only kochia.

At least one farmer used kochia as a cover crop to suppress wild proso millet (Cramer, 1992). A thick stand of unirrigated kochia grew through the summer and was mowed before seed set. It was hard to plow because of all the biomass, but millet was suppressed effectively the next year.

Farmers in southeastern Mexico classify plants as crops or noncrops (Chacón and Gliessman, 1982). The latter are classed according to potential use and their effects on soil or crops. Chacón and Gliessman argue that local farmers understand the contribution of non-crop plants to agriculture. The authors contrast the farmers' view with the dominant view in developed countries that a weed is any plant other than the crop (see Chapter 2). In much of African agriculture, weeding is done by hand by women. It is the most labor-intensive crop production task, taking up to 280 h to weed 1 ha twice (Chikoye, 2000). In southern Nigeria, farmers have found that some dicot weeds suppress other weeds, and in time replace them (Okon and Amalu, 2003). One might say that some weeds are better than others. Siam weed (independence weed) eliminates carpet grass. Siam weed, a shrub native to South and Central America that grows up to 3 m (9.5 ft) tall, is most appropriately employed as a ground cover during a fallow period. It is not a desirable weed management technique in a crop.

Weeds have practical but often unappreciated value when used as ground cover for wild-life or to prevent soil erosion on sites that cannot be cropped or otherwise managed by humans. Weeds can conserve nitrogen in some situations. They have been introduced in many places because someone thought they would be useful (see Chapter 8 for other examples). Cogon grass was introduced into the United States in Grand Bay, Alabama, and McNeil, Mississippi (Tabor, 1952). At Grand Bay in 1912, bare root satsuma orange plants were boxed for shipping with cogon grass. The grass was discarded. The McNeil introduction was part of a search for a superior forage. Cogon grass is now a weed in many parts of the southern United States. Catclaw mimosa was introduced to Thailand from Indonesia in 1980 as a green manure cover crop in tobacco plantations and for control of ditch bank erosion (Thamsara, 1985). It was successful for both things but spread to become a weed problem. The aggressive, weedy annual, para grass was introduced into Africa from several tropical countries for fodder and pasture, and as a cover crop in banana plantations. There are many examples of plant introductions that someone thought would be useful, only to find each could become a serious weed. For further discussion of invasive species, see Chapter 8.

5. ORNAMENTAL USES

Many species of weeds have been used as ornamentals, and several species that are now weedy were first imported into the United States for ornamental purposes (see Chapter 8). One US study (Williams, 1980) documented 33 imported species that became weedy. Of these, two were imported as herbs, 12 as hay or forage crops, and 16 as ornamentals. Henbane was imported for its potential medicinal value. One was imported for use in aquaria (hydrilla), one as a fiber crop (hemp or marijuana), and one privately, just for observation (wild melon). Imported plants including Bermuda grass, jimsonweed, kochia, kudzu, musk

TABLE 4.1   Plants to Avoid in Gardening, Reclamation, and Restoration

Type	Name	Problem
Forb	Purple loosestrife	Displaces native wetland or marsh plants
Forb	Mediterranean sage	Forms monoculture and outcompetes native plants
Forb	Yellow toadflax	Displaces native vegetation
Grass	Timothy	Competes with native plants in arid areas
Shrub	European buckthorn	Competes with native vegetation in riparian areas
Shrub	Scotch broom	Problem in West Coast of United States: displaces native vegetation
Tree	Tamarisk or Salt cedar	Uses large amounts of water and displaces native vegetation
Tree	Russian olive	Seed dispersed by birds. It displaces native plants

thistle, johnsongrass, and water hyacinth have become some of our most detrimental weeds. Nevertheless weeds are often used as ornamentals despite, or in ignorance of, their weedy nature.

Several forbs, grasses, shrubs, and trees have been and still are used in gardening and landscaping, reclamation, or restoration. Some are widely acknowledged as weeds and others may become weedy because of their ability to invade and dominate. All show the ability to escape their intended habitat. A few are shown in Table 4.1. Not everyone may agree that the plants shown are weeds or could become weedy. Currently there is no civil or criminal penalty for planting any of them. The choice of what to plant is the land owner's. When an escape occurs the cost is externalized,[3] as are many of the costs of agricultural technology, and everyone pays part of the price if the species becomes weedy.

6. INSECT OR DISEASE TRAPS

A disadvantage of weeds is that they can shelter insects and disease organisms. They can also be used intentionally in agriculture as traps for insect or disease pests (Table 4.2). They do this in one of three ways (See Norris, 1982; Norris and Kogan, 2000):

1. as hosts for adult insect parasites
2. as hosts for noneconomic insects that serve as alternate hosts or food for parasites or predators
3. by increasing effectiveness of biological control organisms and thereby reducing damage to crops

[3]An externality is a cost that is not reflected in price, or more technically, a cost or benefit for which no market mechanism exists. In the accounting sense, it is a cost that a firm (a decision maker) does not have to bear, or a benefit that cannot be captured. From a self-interested view, an externality is a secondary cost or benefit that does not affect the decision maker.

TABLE 4.2 Weeds and Control of Other Pests

Cropping System	Weed Species	Pest Regulated	Reason
Beans	Goose grass Red sprangletop	Leafhoppers (*Empoasca kraemeri*)	Chemical repellency and masking
Vegetable crops	Wild carrot	Japanese beetle (*Popillia japonica*)	Increased activity of the parasitic wasp *Tiphia popilliavora*
Corn	Giant ragweed	European corn borer (*Ostrinia nubilalis*)	Provision of alternate host for the tachinid parasite *Lydella grisescens*
Cotton	Common ragweed	Boll weevil (*Anthonomus grandis*)	Provision of alternate hosts for the parasite *Eurytoma Tylodermatus*

Norris and Kogan (2000) provide an extensive review of the interactions among weeds, arthropod pests, and natural enemies in managed ecosystems (i.e., cropped fields). Their review identifies more than 90 insects that are involved in resource and habitat-driven interactions. A separate table identifies more than 50 resource and habitat-driven influences of weeds on beneficial arthropods. The extensive review also illustrates the effects of beneficial and detrimental effects of tillage and several herbicides used for weed management on arthropod populations. Norris and Kogan cite Altieri (1994, p. 40), who cited Bendixen and Horn (1981) to report that more than 70 families of arthropods are known to be potential crop pests and that the members of these families are primarily associated with weeds. Some associations may be beneficial to crops; most are not. Without much more understanding of these associations and how management of one pest may affect other pests and crops, integrated pest management programs are less likely to be successful.

Johnsongrass is an alternate host of the sorghum midge (*Contarinia sorghicola* Coquillet), an important pest of grain sorghum. Larvae develop and feed in the sorghum spikelet and prevent normal seed development. Johnsongrass maintains the first two or three generations of the insect until grain sorghum flowers are available. Time and duration of johnsongrass flowering (that can be determined by management) may affect the sorghum midge population (Holshouser and Chandler, 1996).

Showy crotalaria, a legume weed in Hawaii, is used in macadamia nut orchards to attract Southern green stinkbugs (*Nezara viridula* L.) away from macadamia nut trees. Showy crotalaria was introduced to Florida in 1921 as a green manure crop because, as a legume, it fixes nitrogen. However, the foliage and seed are toxic, especially to poultry. It is a weed in soybeans in the southern United States, where because of the toxicity of its seeds, contaminated soybean seed cannot be sold.

In parts of California, wild blackberries are grown with grapes as hosts of a noneconomic leafhopper that hosts a parasite of the grape leafhopper (*Erythroneura elegantula* Osborn). Japanese farmers graft tomato scion (shoot or bud tissue) onto the rootstock of some weedy members of the *Solanaceae* to avoid root diseases. Other examples of this kind of use can be found in Altieri (1985), Norris and Kogan (2000), and Zandstra and Motooka (1978).

Chapter 2 described how weeds serve as hosts for damaging insects and diseases. It is important to realize that not all insects or microorganisms damage other plants. If one plant harbors harmful organisms, it is only logical to assume that other plants may harbor beneficial organisms. The examples that were given verify this. Altieri and Norris and Kogan provide many other examples.

The agricultural quest for high-yielding monocultures has reduced plant diversity to the point where beneficial insects have been reduced in or eliminated from crop fields. One way to regain a desirable diversity in crop fields is to manipulate the abundance and composition of the weed flora. Weed borders, occasional weedy strips, or the presence of weeds at certain times in the crop growth cycle are possibilities. Weed scientists and farmers may even want to consider planting weeds that harbor known beneficial organisms. This could optimize plant protection and crop yield while minimizing other inputs.

Briens at Canada's agriculture ministry (Economist, 2010a; Booker et al., 2010) ground tobacco leaves and after heating in pressurized oxygen distillation produced a treacly oil. The crude oil killed larvae of *Pythium ultimum*, *Clavibacter michiganensis*, and *Streptomyces scabies*, which are, respectively, a fungus that attacks eggplant, peppers, lettuce, tomatoes, and cucumber; a bacterium that kills young plants and deforms fruit; and a bacterium that causes potato scab. Nicotine was the presumed active agent, but the oil was more effective when nicotine was removed.

7. POLLUTION CONTROL

In addition to the uses just mentioned, which most weed scientists would readily acknowledge, there are other, less well-known and perhaps esoteric but interesting and potentially valuable uses that a few creative minds have explored.

Star chickweed has been used as a vegetable and is a good source of vitamins A and C. In Elizabethan England, it was used to reduce fever (Martin, 1983). Martin reported that it has been used to predict rain. If it blooms fully, there will be no rain for at least 4 h. If blossoms shut, rain is on the way. Perhaps a look skyward would yield the same prediction.

Water hyacinth, an aquatic macrophyte, can be used for bioremediation when there is heavy metal pollution of water. It removes the heavy metals selenium, manganese, and chromium from water and may be useful in detecting them. It concentrates heavy metals up to 2000 times the level found in water. Others have studied it as a way to harvest valuable heavy metals. Limited research indicates that an acre of water hyacinth could yield 0.45 kg of silver every 4 days. Work on gold harvest has been done (Anon., 1976). Removal of heavy metals from water reduces eutrophication (Rogers and Davis, 1972; Murray, 1976). One hectare of water hyacinth, growing under optimum conditions, could absorb the average daily nitrogen and phosphorus waste of over 800 people if maximum uptake and plant growth for a whole year are assumed. The absorptive capacity would be reduced to 300–400 people if less than year-round growth was achieved. Under optimum growth conditions, 1 ha of water hyacinth may contain 2 million plants that weigh 270–400 metric tons. The plants can be dried, ground, and added to corn silage for cattle feed. The supplementary feed value is comparable to cotton seed meal or soybean oil meal. Anaerobic fermentation of the plant residue produces methane gas that can be used to produce heat or light. One pound of dried plants yields up to 6 ft^3 of methane or up to 2 million ft^3 of gas per acre of plants per year. A study in China described a cost-effective methane plant capable of producing more than 200,000 L of biogas and organic fertilizer annually from water hyacinth. A major cost would be collecting and removing the plants (EEPSEA, 2013).

There are potential problems:

1. Water hyacinth grows best in warm water in warm climates. Fortunately cold weather kills it. However, it is one of the most serious and widespread aquatic weeds in the world's warm regions. Its population can be so dense that boat travel is sometimes impossible. Its escape to become a nearly unmanageable weed is inevitable.
2. The cost of establishing and maintaining a processing facility may be too high for a tropical developing country where it grows best and where bioremediation is needed.
3. An obvious problem if water hyacinth is to be used for bioremediation is the continued need to dispose of plants to prevent eutrophication of ponds.

Uses for the plant residue have been developed. In India and Indonesia, researchers have made paper products (blotting paper and cardboard) from water hyacinth mixed with rice straw. India may have as much as 4 million ha (9.8 million acres) of water infested with water hyacinth (covered with it may be more accurate). The average yield is 50 tons/ha, which means that if all were harvested, as much as 200 million tons of nonforest raw material could be available for paper production. If only half were used and there was only a 10% conversion efficiency, 10 million tons of paper could be produced. This has not happened, but might be possible if factories were built and harvest procedures were developed. Paper making seems to be a better option than enduring the weed's bad effects or continuing to try to control it, which has been largely unsuccessful. Goswami and Saikia (1995) concluded that because water hyacinth contains relatively high amounts of hemicellulose and pentosan, it is an ideal source of raw material for producing grease-proof paper when admixed with bamboo pulp.

A joint initiative of the North-East Development Finance Corporation Limited and North Eastern Council under the ministry of Department of North-eastern Region in the state of Assam, India has changed rural Assam by successfully converting water hyacinth into a "wonder-weed" by the use of Thai technology. Rural artisans make sandals, lampshades, bags, and pen stands.[4]

ATC furniture, a leading manufacturer and exporter in Vietnam, specializes in providing handmade products from water hyacinth, seagrass, and rattan.[5] The plant's stalks are strong and make good furniture. The plant stalks (the raw material) are also used to make ropes employed to tie up elephants.[6]

The bulrush has also been identified as a cheap, effective way to remove pollutants from water (Zandstra and Motooka, 1978). Sudanese tribesmen have used it. Muddy water from the Nile river is stored in jars containing bulrush, and soon one has clean, pure water. A German company designed a municipal water treatment facility using bulrushes to absorb

[4]http://archive.tehelka.com/story_main52.asp?filename=Fw240412Should.asp.

[5]https://www.youtube.com/watch?v=3S7m8m28c8s.

[6]Thailand. http://www.deccanherald.com/content/34754/hyacinth-turns-productive-northeast.html.

pollutants such as phenols, cyanide, phosphates, and nitrates. Commercialization may not be possible, but we should be cognizant of potential uses of the plants we so easily call weeds. Germans have also experimented with Sakhalin knotgrass, which takes up cadmium and lead without self-injury. They hope it will be useful to reclaim soil treated with metal-contaminated sewage sludge so crops can be grown.

During the 1970s, there was great interest in developing systems to use plants to process sewage. Jewell (1994) reported a hydroponic or nutrient film technique originally developed in England. The technique does not require deep water or a growth-supporting medium. Most terrestrial plants can be grown in a nutrient film system. Cattails, a common weedy species, have been a good choice for the initial stages of sewage treatment in a nutrient-film system.

8. OTHER USES

Bliss (1978) included 50 common weeds, all of which share the ability to produce a dye that produces a color on wool. Some produce durable colors. Some, in her words, produce "more exciting and richer shades." Others produce colors more resistant to fading in light. None of those she identifies are rare or endangered; some are invaders, some native, others introduced, and all are weeds in some places. Plants that can be a source of dye include common ragweed, showy milkweed, dandelion, field bindweed, leafy spurge, musk thistle, redroot pigweed, and yellow sweet clover.

Ground into powder, the common water reed can be used as a home heating fuel, according to Swedish scientists (Bjork and Graneli, 1978). One kilogram of dry reeds will yield 5 kW of energy. About 10 times more energy can be obtained from the powder than is required to cultivate, harvest, grind, and transport it. Cultivation of the weed could greatly increase production per unit area and may have the added advantage of preserving and using some portions of wetlands now threatened by development. Preservation of such lands has positive environmental benefits in terms of habitat for marsh animals and waterfowl. However, harvesting the reed, a naturally growing plant, may compromise the ecological stability of the wetlands.

Scientists at the University of Arizona have compressed Russian thistle to make fireplace logs (Tumble Logs) with an energy value equal to lignite. They also investigated the biomass potential of mesquite, saltbush, and johnsongrass for energy production. Russian thistle plants have a high energy content and use little water while growing. It is primarily the stems, the skeleton of the plant, that can be pressed into fire logs. The trademarked name, Tumble Logs, is appropriate for Russian thistle, tumbleweed of the western United States.

During World War II, allied forces were denied the world's far eastern sources of natural rubber. The war could not be fought without rubber for tires, and there was abundant research in the early 1940s to develop alternate sources or substitutes for natural rubber. Gray rabbitbrush and guayule (*wy-oo-lee*) were among the plants studied. Gray rabbitbrush contains a high-grade rubber called chrysil that vulcanizes well (Ross, 1976). One-fourth of guayule's entire weight is natural rubber. It is drought tolerant, can be grown on land not suited for many other crops, and can be mechanically harvested. There was interest in guayule and other plants as sources of hydrocarbons for replacement of petroleum oil, which was scarce during World War II. Guayule, a desert shrub, produces rubber free of allergenic proteins, which makes it useful for surgical gloves (Economist, 2010b). Bell et al. (2015) used prickly lettuce, which synthesizes long-chain natural rubber polymers similar to the polymers in natural rubber (Kantor, 2015), to determine the fundamental genetic and phenotypic characteristics of rubber biosynthesis. They suggested that prickly lettuce has potential to become a temperate climate crop plant for rubber production.

Even the scourge of the homeowner's front lawn, dandelion, may be a source of rubber. Scientists at the Fraunhofer Institute for Molecular Biology and Applied Ecology in Aachen, Germany discovered a way to switch off a key enzyme, which enabled production of 500% more usable latex.[7] A Russian variety of dandelion (*Taraxacum kok-saghyz*) produces molecules of rubber in its sap and may replace the traditional source of rubber (*Hevea brasiliensis*) (Economist, 2010b).

There are several latex-bearing plants from the *Euphorbiaceae* (spurge) and *Asclepiadaceae* (milkweed) families. Many are common weeds or perhaps just plants that are not even noticed or bothersome enough to be raised to the defined category of weed.

For a brief period, the US military used "down" from mature cattail heads (part of the female flower) to fill life jackets, which had been filled with kapok, the silky fiber from the fruit of the silk-cotton tree. It had also been used to insulate clothing and as stuffing for quilts and pillows. Western Indians ate young cattail shoots, roots, stem bases, and seeds. The same "down," the pappus from female flowers, was used to make dressings for burns as well as padding and in talcum powder.

Jatropha, a native of Africa, grows all over the tropics, often in poor soil, lives up to 50 years, and is drought tolerant. Livestock will not eat it and it has no known pest problems. Oil extracted from the black seeds burns with a fifth of the carbon emissions of fossil fuels. Because it grows on marginal soils unfit for crop production, jatropha offers an advantage not enjoyed by many biofuel crops: it does not compete for crop land and therefore does not affect food production. Most biofuel crops are not "green." Those that are or may be (jatropha may be) cannot compete commercially.

If none of these ideas interests you and you like plants, buy a few acres of land and plant milkweed. It is a hardy perennial and competes well with most plants, and once established, it should thrive with care and pest control. A crop of milkweed can be grown with about three-quarters of the inputs and a quarter of the water corn requires. But why would anyone want to plant milkweed? Most people regard milkweed as a persistent, perennial, hardy weed. One must grant that it is a survivor, but also that it has never been an aggressive invader or a troublesome weed in crops. It is commonly found in pastures, roadsides, and open fields. There are at least 107 species of milkweed (Schwartz, 1987). The genus propagates by seed and by vegetative buds on the spreading underground root system. The interaction between milkweed and the Monarch butterfly is a classic in ecological studies. Monarchs feed on milkweed foliage and store its toxic alkaloids in their tissues, which make the butterfly unpalatable to birds (Morse, 1985). It has been grown as a crop in western Nebraska (Witt and Knudson, 1993; Witt and Nelson, 1992), but currently it is not. The

[7]http://www.ime.fraunhofer.de/en/presse_medien/rubber-from-dandelions.html.

grower's cost is high, the return is not certain, and the risk of loss owing to hail in western Nebraska is high. Milkweed was first grown as a crop after research by Melvin Calvin[8] suggested that its biomass could be used to produce oil. Milkweed was attractive because its seeds contained high quantities of compounds from which oil could be extracted. Standard Oil of Ohio began a research program and found that the cost of producing synthetic crude from milkweed was too high and the yield was too low to make the operation profitable. The work also revealed that the seed pappus (floss) had potential as a substitute for goose down and for use in disposable products that required absorbency (e.g., diapers). Milkweed's follicular[9] seed pod can be harvested, carded to remove seeds, and dried, and the pappus or floss can be used as a substitute for expensive goose down in jackets, sleeping bags, and other things designed to trap air and keep us warm (Lione, 1979). Milkweed seed contains an oil that may have medicinal qualities (reduce pain and restore mobility). Natural Fibers, Inc., of Ogallala, Nebraska made great strides toward commercial production of milkweed fiber for use in down comforters and pillows. It was included on pages 118−23 of the 1992 US Department of Agriculture yearbook. The Ogallala Comfort Company, which succeeded Natural Fibers, now depends on wild collection of milkweed pods (160,000 lb in 2016), mainly in Canada and Michigan, for use in bedding products. They have also developed a market in milkweed balm cream (www.milkweedbalm.com). Yes, for most people it is just a weed, but some good observers see opportunities where others see only problems.

THINGS TO THINK ABOUT

1. How many uses can you think of for a plant you thought was just a weed?
2. Are there situations in which we ought to encourage weed growth in crops? Give examples.
3. Should genetic engineering be used to create useful weeds?

Literature Cited

Alvardo, N., January/February 2016. 7 lessons bindweed is trying to teach me. Orion 6.
Anon., 1976. Roots of water hyacinth may be harvested for gold. Weed Trees Turf. 15 (2), 50.
Anon., 1979. Dandelion Recipe Book. Mayor's Special Events Office, Vineland, NJ, 12 pp.
Anon., July 1−September 30, 1992. Quarterly Report if Selected Research Projects. USDA/ARS.
Altieri, M.A., 1985. Agroecology: The Scientific Basis of Alternative Agriculture, second ed. Div. of Bio. Cont. U. of Cal., Berkeley, pp. 112−113.
Altieri, M.A., 1994. Biodiversity and Pest Management in Agroecosystems. Food Products Press, New York, NY, 185 pp.

[8]Melvin Calvin (1911−1997), Professor of Chemistry, Univ. California, Berkeley discovered the Calvin cycle (with A. Benson and J. Bassham) for which he was awarded the 1961 Nobel Prize in chemistry.

[9]Follicular = A follicle is a seed bearing organ of a flower. Milkweed''s follicle opens (dehisces) along a single seam (a suture).

Balick, M.J., Cox, P.A., 1996. Plants, People, and Culture: The Science of Ethnobotany. Scientific American Library Series, vol. 60. A division of HPHLP, New York, NY, 228 pp.

Bell, J.L., Burke, I.C., Neff, M.M., 2015. Genetic and biochemical evaluation of natural rubber from eastern Washington prickly lettuce (*Lactuca serriola* L.). J. Agric. Food Chem. 63 (2), 593—602.

Bendixen, L.E., Horn, D.J., 1981. An Annotated Bibliography of Weeds as Reservoirs for Organisms Affecting Crops. III. Insects. Agricultural Research and Development Center, Wooster, OH.

Bentley, J.W., Webb, M., Nina, S., Pérez, S., 2005. Even useful weds are pests: ethnobotany in the Bolivian Andes. Int. J. Pest Manag. 51 (3), 189—207.

Bjork, S., Graneli, W., 1978. Energy reeds and the environment. Ambio 7, 150—156.

Bliss, A., 1978. Weeds, a Guide for Dyers and Herbalists. Bliss, Juniper House, Boulder, CO, 112 pp.

Booker, C.J., Bedmutha, R., Vogel, T., Gloor, A., Xu, R., Ferrante, L., Young, K.K.-C., Scott, I.M., Conn, K.L., Berruti, F., Briens, C., 2010. Experimental investigations into the insecticidal, fungicidal, and bactericidal properties of pyrolysis bio-oil from tobacco leaves using a fluidized bed pilot plant. Ind. Eng. Chem. 49, 10074—11007.

Bosworth, S.C., Hoveland, C.J., Buchanan, G.A., 1985. Forage quality of selected cool-season weed species. Weed Sci. 34, 150—154.

Boyer, J.S., 1982. Plant productivity and environment. Science 218, 443—448.

Brown, P.A., Brown, J.M., Allen, S.J., 2001. The application of kudzu as a medium of the adsorption of heavy metals from dilute aqueous wastestreams. Bioresour. Technol. 78, 195—201.

Bryson, B., 2005. A Short History of Nearly Everything. Broadway Books, New York, NY, 624 pp.

Chacón, J.C., Gliessman, S.R., 1982. Use of the "non-weed" concept in traditional tropical agroecosystems of southeastern Mexico. Agroecosystems 8, 1—11.

Chikoye, D., 2000. Weed management in small-scale production systems in Nigeria. In: Akoroda, M.O. (Ed.), Agronomy in Nigeria. Department of Agronomy, University of Ibaden, Ibaden, Nigeria, pp. 153—156.

Corley, R.N., Woldeghebriel, A., Murphy, M.R., 1997. Evaluation of the nutritive value of kudzu (*Pueraria lobata*) as a feed for ruminants. Anim. Feed Sci. Technol. 68, 183—188.

Cramer, C., November—December 1992. Healthy harvest. New Farm Mag. 13—16.

Daniel, J., 1974. Take a weed to lunch. In: The Reader's Digest. Condensed from American Agriculturalist and the Rural New Yorker, pp. 213—215.

DeFelice, M.S., 2002. Yellow nutsedge *Cyperus esculentus* L.—snack food of the Gods. Weed Technol. 16, 901—907.

Drury, S., 1992. Plants and pest control in England circa 1400—1700: a preliminary study. Folklore 103, 103—106.

Duke, J.A., 1992. Edible Weeds. C.R.C. Press, Inc., Boca Raton, FL, 232 p.

Dutt, T.E., Harvey, R.G., Fawcett, R.S., 1982. Feed quality of hay containing perennial broadleaf weeds. Agron. J. 74, 673—676.

Economist, January 2, 2010a. Blow Out — New Sources Rubber 60.

Economist, November 20, 2010b. Smok. Them Out 93.

EEPSEA (Economic and Environment Program for Southeast Asia), 2013. Using Water Hyacinth to Clean Up Pollution: A Case Study from China. EEPSEA policy brief No. 2013—PB2. www.eepsea.net.

Gail, P.A., 1990. The Dandelion Celebration: A Guide to Unexpected Cuisine. Goosefoot Acres Press, Cleveland, OH, 156 pp.

Gerard, J., 1597. The Herball or Generall Historie of Plants, 1636 ed., p. 115

Gibbons, E., 1962. Stalking the Wild Asparagus. D. McKay Co., New York, NY, 303 pp.

Goswami, T., Saikia, C.N., 1995. Water hyacinth — a potential source of raw material for greaseproof paper. Bioresour. Technol. 50, 225—238.

Hale-Shelton, D., September 8, 2004. Weeds by any other name can actually be a feast. Summit Dly. News B-1.

Harrington, H.D., 1967. Edible Native Plants of the Rocky Mountains. Univ. of New Mexico Press, 392 pp.

Hatfield, A.W., 1971. How to Enjoy Your Weeds. Sterling Pub. Co., New York, NY, 192 pp.

Hauptil, H., Jain, S., September 1977. Amaranth and meadowfoam: two new crops. Calif. Agric. 6—7.

Henkel, A., 1904. Weeds Used in Medicine, vol. 188. U.S. Dept. of Agriculture, Farmers' Bulletin, Washington, DC, 45 pp.

Holm, L., Plucknett, D., Pancho, J., Herberger, J., 1977. The World's Worst Weeds. University of Hawaii Press, Honolulu, HI, 609 pp.

Holm, L., Doll, J., Holm, E., Pancho, J., Herberger, J., 1997. World Weeds: Natural Histories and Distribution. J. Wiley & Sons, Inc., New York, NY, 1129 p.

Holshouser, D.L., Chandler, J.M., 1996. Predicting flowering of rhizome johnsongrass (*Sorghum halepense*) populations using a temperature dependent model. Weed Sci. 44, 266–272.

Howlett, L.A., Nielsen, P.M., Wilson, R.G., 2006. Chicory production in Western Nebraska and southeastern Wyoming. Proceedings Western Society of Weed Science, p. 25.

Hylton, W.H., 1974. The Rodale Herb Book. Rodale Press, Inc., Emmaus, PA, 653 pp.

Janick, J. (Ed.), 1999. Perspectives on New Crops and New Uses. ASHS Press, Alexandria, VA, 528 pp.

Janick, J. (Ed.), 1996. Progress in New Crops. ASHS Press, Alexandria, VA, 660 pp.

Jewell, W.J., 1994. Resource-recovery wastewater treatment. Am. Sci. 82, 366–375.

Kantor, S., April 6, 2015. Study Points the Way Toward Producing Rubber from Lettuce. WSU (Washington State University) News, 2 pp.

Kershaw, L., 2000. Edible and Medicinal Plants of the Rockies. Lone Pine Publishing, Edmonton, Alberta, Canada, 270 pp.

Kingsbury, J.M., 1964. Poisonous Plants of the United States and Canada. Prentice-Hall, Englewood Cliffs, NJ, 626 pp.

Lione, A., September–October, 1979. Make a milkweed down jacket. The Mother Earth News 104–105.

Lowell, J.R., 1848. A fable for critics. In: The Complete Poetical Works of James Russell Lowell. Riverside Press, Cambridge, MA, pp. 117–148.

Luo, X., Benson, R.S., Kit, K.M., Dever, M., 2002. Kudzu fiber-reinforced polypropylene composite. J. Appl. Polym. Sci. 85, 1961–1969.

Marten, G.C., Andersen, R.N., 1975. Forage nutritive value and palatability of 12 common annual weeds. Crop Sci. 15, 821–827.

Marten, G.C., Sheaffer, C.C., Wyse, D.L., 1987. Forage nutritive value and palatability of perennial weeds. Agron. J. 79, 980–986.

Martin, L.C., 1983. Wildflower Folklore. The East Woods Press, Charlotte, NC, 256 pp.

Mabey, R., 2011. Weeds – In Defense of Nature's Most Unloved Plants. Ecco/HarperCollins Publishers, New York, 324 p.

Martin, G.J., 1995. Ethnobotany: a methods manual. Chapman & Hall, New York, 268 pp.

Moerman, D.E., 2009. Native American Medicinal Plants: An Ethnobotanical Dictionary. Timber Press, Portland, OR, 799 pp.

Morse, D.H., July 1985. Milkweeds and their visitors. Sci. Am. 112–119.

Murray, C., March 22, 1976. Weed holds promise for pollution cleanup. Chem. Eng. News 23–24.

Musselman, L.J., Wiggins, H.J., 2013. The Quick Guide to Wild Edible Plants: Easy to Pick, Easy to Prepare. The John Hopkins University Press, Baltimore, MD, 133 pp.

Norris, R.F., 1982. Interactions between weeds and other pests in the agro-ecosystem. In: Hatfield, J., Thomason, I. (Eds.), Biometerology and Pest Management. Academic Press, New York, NY, pp. 343–406.

Norris, R.F., Kogan, M., 2000. Interactions between weeds, arthropod pests, and their natural enemies in managed ecosystems. Weed Sci. 48, 94–158.

Nyerges, C., 1999. Guide to Wild Foods and Edible Plants. Chicago Review Press, Chicago, IL, 237 pp.

Okon, P.B., Amalu, U.C., 2003. Using weed to fight weed. Leisa Mag. December 21.

Patterson, D.T., et al., 1989. Composite List of Weeds. Weed Sci. Soc. of America, Champaign, IL, 112 pp.

Plotkin, M.J., 1993. Tales of a Shaman's Apprentice: An Ethnobotanist Searches for New Medicines in the Amazon Rain Forest. Penguin Books, NY, 328 pp.

Rogers, H.H., Davis, D.E., 1972. Nutrient removal by water hyacinth. Weed Sci. 20, 423–428.

Ross, R.L., 1976. Wild Edible and Medicinal Plants. Coop. Ext. Serv. Montana State Univ. Circular 1183, 7 pp.

Sachs, R.M., Low, C.B., MacDonald, J.D., Awad, A.R., Sully, M.J., July–August 1981. *Euphorbia lathyris*: a potential source of petroleum-like products. Calif. Agric. 29–32.

Sage, R.F., Coiner, H.A., Way, D.A., Runion, G.B., Prior, S.A., Torbert, H.A., Sicher, R., Ziska, L., 2009. Kudzu [*Pueraria montana* (Lour.) Merr. Variety *lobata*]: a new source of carbohydrate for bioethanol production. Biomass Bioenerg. 33, 57–61.

Schwartz, D.M., September 1987. Underachiever of the plant world. Audubon 46–61.

Sheaffer, C.C., Wyse, D.L., 1982. Common dandelion (*Taraxacum officinale*) control in alfalfa (*Medicago sativa*). Weed Sci. 30, 216–220.

Shyr, L., October 2011. Wicked beauty. In: The National Geographic. The National Geographic Society, Washington, DC, p. 24.

Stepp, J.R., 2004. The role of weeds as sources of pharmaceuticals. J. Ethnopharmacol. 92, 163–166.

Stepp, J.R., Moorman, D.E., 2001. The importance of weeds in ethnopharmacology. J. Ethnopharmacol. 75, 19–23.

Stewart, D., 2000. Kudzu – love it—or run. Smithsonian 31 (7), 65–70.

Strand, O., DiStefano, J., November 23, 2011. What weed? That's dinner. New York Times. Section D, p. 1 and 6.

Swerdlow, J.L., April 2000. Nature's Rx. Natl. Geogr. 98–117.

Tabor, P., 1952. Comments on cogon and torpedo grasses. A challenge to weed workers. Weeds 1, 374–375.

Temme, D.G., Harvey, R.G., Fawcett, R.S., Young, A.W., 1979. Effect of annual weed control on alfalfa forage quality. Agron. J. 71, 51–54.

Thamsara, S., 1985. *Mimosa pigra* L.. In: Proc. 10th Asian-Pacific Weed Sci. Soc. Conf., pp. 7–12.

Vega, M.R., September 6–10, 1982. Crop production in the total absence of weeds. In: Paper Presented at the FAO/IWSS Expert Consultation on Weed Management Strategies for the 1980's for the LDC's. FAO, Rome, Italy.

Vietmeyer, N.D., May 1981. Rediscovering America's forgotten crops. Natl. Geogr. 702–712.

Walker, J.A., Jenkins, K.H., Klopfenstein, T.J., 2011. Protein, Fiber, and Digestibility of Selected Alternative Crops for Beef Cattle. Animal Science Department. Nebraska beef cattle reports. 2 pp.

Wilensky, A.S., 2000. Dandelion: Celebrating the Magical Blossoms. Council Oaks Books, San Francisco, CA, 24 pp.

Williams, M.C., 1980. Purposefully introduced plants that have become noxious or poisonous weeds. Weed Sci. 28, 300–305.

Witt, M.D., Knudson, H.D., 1993. Milkweed cultivation for floss production. In: Janick, J., Simon, J.E. (Eds.), New Crops. J. Wiley & Sons, New York, NY, pp. 428–431.

Witt, M.D., Nelson, L.A., 1992. Milkweed as a new cultivated row crop. J. Prod. Agric. 5, 167–171.

Zandstra, H.B., Motooka, P.S., 1978. Beneficial effects of weeds in pest management—a review. PANS 24, 333–338.

Zhu, C., Zeng, Q., McMichael, A., Ebi, K.L., Ni, K., Khan, A.S., Zhu, J., Liu, G., Zhang, X., Cheng, L., Ziska, L.H., May 2015. Historical and experimental evidence for enhanced concentration of artemesinin, a global anti-malarial treatment, with recent and projected increases in atmospheric carbon dioxide. Clim. Change. https://doi.org/10.1007/s10584-015-1421-3. Published online.

Ziska, L.H., Tomecek, M.B., Gealy, D.R., 2014. Assessment of cultivated and wild, weedy rice lines to concurrent changes in CO_2 concentration and air temperatures: determining traits for enhanced seed yield with increasing atmospheric CO_2. Funct. Plant Biol. 41, 236–243.

Ziska, L.H., Emche, S.D., Johnson, E.L., Geroge, K., Reed, D.R., Sicher, R.C., 2005. Alterations in the production and concentration of selected alkaloids as a function of rising atmospheric carbon dioxide and air temperature: implications for ethno-pharmacology. Global Change Biol. 11, 1798–1807.

Weed Reproduction and Dispersal

FUNDAMENTAL CONCEPTS

- The seed bank in most agricultural soils includes millions of weed seeds per acre and is the primary source of yearly weed problems.

- In addition to plant dispersal, there are many methods for dispersal of weed seeds: human-, water-, and animal-aided, and mechanical systems.

- Continued development of understanding of the processes of seed germination and the physiological and environmental factors that affect it is essential to the development of good weed management systems.

- Seed dormancy is dispersal of seeds in time.

- Vegetative reproduction makes weed management problems difficult because vegetative reproductive organs are not susceptible to most available control techniques.

LEARNING OBJECTIVES

- To learn the size and role of the soil seed bank.

- To understand aerial seed sources and how seed is dispersed in space.

- To understand how seeds are dispersed in time via seed dormancy.

- To understand the causes, classification, and role of weed seed dormancy.

- To know the methods of vegetative reproduction and understand its role in weed management.

Indeed, as I learned, there were on the planet where the little prince lived—as on all planets—good plants and bad plants. In consequence, there were good seeds from good plants, and bad seeds from bad plants. But seeds are invisible. They sleep deep in the heart of the earth's darkness, until some one among them is seized with the desire to awaken. Then this little seed will stretch itself and begin—timidly at first to push a charming little sprig inoffensively upward toward the sun. If it is only the sprout of radish or the sprig of a rose-bush. One would let it grow wherever it might wish. But when it is a bad plant one must destroy it as soon as possible, the very first instant that one recognizes it. *de Saint Exupéry, A., 1971 (Org. 1943). The Little Prince. A Harvest/HBJ Book, New York, NY, pp. 20–21.*

> Look at the seed in the palm of a farmer's hand. It can be blown away with a puff of breath and that is the
> end of it. But it holds three lives—its own, that of the man who may feed on its increase, and that of the man
> who lives by its culture. If the seed dies, these men will not, but they may not live as they always had. They
> may be affected because the seed is dead; they may change, they may put their faith in other things. *Markham,*
> *B., 1983. West With the Night. North Point Press, San Francisco, CA, p. 132.*

Weed biology, a part of weed science, is the study of the growth, development, and reproduction of weeds. Although this is not a book about weed biology, biological knowledge is essential to understanding the fundamentals of weed science and to the development of appropriate weed management systems. It is widely agreed among weed scientists that weed biology is important; however Forcella (1997) reported that it had contributed little to weed management. Sadly, that is still true.

This chapter is divided into four sections dealing with reproduction and dispersal of weeds. The first discusses seeds and their production, the second includes dispersal of seeds in space, and the third deals with seed germination and dispersal of seeds in time or seed dormancy. The last section covers vegetative or asexual reproduction.

1. SEED PRODUCTION

Seeds are both alive and a source of life. A seed is a mature fertilized ovule (an embryo) that has stored energy reserves (sometimes missing) and a protective coat or coats. It is a small plant packaged for shipment. Survival of many flowering plants depends on production of a sufficient number of viable seeds. This is especially true for annual weeds that reproduce by seed, and therefore prevention of seed production is a key to eliminating future problems. Failure to prevent production of weed seed results in increasing numbers of seeds in soil and subsequently weeds in crops and landscapes.

The damage done to soil by the moldboard plow and how plowing usually worsened the weed problem was not discovered in the past few decades with the advent of minimum tillage and no-till farming. Faulkner (1943) questioned the very basis of agriculture: the moldboard plow. Organic material was not well incorporated into soil by the plow, but weed seeds were. Faulkner put forth the then-radical proposal that farmers should be able to farm without weeds. It was what he called a "fantastically improbable" proposition and it may still be regarded as highly improbable. He suggested that what the plow did was bury weed seeds "for future recovery every time" we plow the land. For him it was the secret of weed perpetuation, and research has shown he was right. Weed seeds were more uniformly distributed throughout the top 6 inches after moldboard and ridge-tillage than after chisel plow[1] and no-tillage (Clements et al., 1996; Table 5.1). In a 5-year Iowa study, before plowing a hay sward, weed seeds were concentrated in the upper 10 cm of the soil. After moldboard plowing, weed seeds were uniformly distributed throughout the upper 20 cm of soil (Buhler et al., 2001). Farmers thus became victims of their system of handling the

---

[1]Ridge-tillage leaves soil undisturbed from harvest to planting except for strips up to 1/3 of the row width. Planting on the ridge involves removal of the top of the ridge. A chisel plow breaks and stirs soil a foot or more beneath the surface without turning it.

TABLE 5.1 Seed Presence (%) After Three Tillage Systems and No-Tillage (Clements et al., 1996)

Soil Depth (Inches)	Moldboard Plow	Ridge-Till	Chisel Plow	No-Tillage
0–2	37	33	61	74
2–4	25	45	23	9
4–6	38	22	16	18

land, and many still are. Weed seeds near the surface are more likely to emerge and be susceptible to control than are those buried more deeply.

Johnson and Mullinix (1995) suggest that soil tillage distributes weed seed because it affects weed emergence and hence seed production. Crop cultivation, a useful weed management tactic, has been correlated with midseason emergence of Florida beggarweed in peanuts (Cardina and Hook, 1989). Mechanical control is discussed more fully in Chapter 10, Section 3.

The number of weed seeds in the plow layer of soil can be reduced by repeated tillage (Chancellor, 1985). With optimum rain, 50% of the weed seed in the plow layer of vegetable crop fields germinated within 6 weeks of cultivation (Bond and Baker, 1990). Egley and Williams (1990) increased weed emergence with frequent tillage over 4 years. Subsequently tillage had no effect on emergence, which suggested that the seed bank had been depleted. In Minnesota, wild mustard seed in soil was reduced 97% after 7 years of tillage (Warnes and Anderson, 1984). In Alabama, purple nutsedge was eradicated after 5 months of weekly or biweekly harrowing (Smith and Mayton, 1938). Therefore, it is logical to conclude that soil tillage has an important role in the availability of seed for dispersal by encouraging germination, reducing seed production, and destroying emerging seedlings.

1.1 Seed Size

Seeds produced by most weeds are small. For example, broad-leaf plantain has over 2 million seeds per pound and shepherd's-purse has nearly 5 million. Small seeds are easily dispersed by wind and water and their size precludes easy detection until they germinate and a plant emerges above the soil surface.

1.2 Seed Abundance

The number of weed seeds in arable soil is large, nay it is huge. Koch (1969) estimated the average arable soil has 30,000 to 350,000 weed seeds/m^2 (300 million to 3.5 billion/hectare (ha), or 120 million to 1.4 billion/acre).

In lowland (paddy or irrigated) rice fields in the Philippines, 804 million seeds from 12 different species (sedges dominated) were found over 1 ha 6 inches deep (Vega and Sierra, 1970). Samples of soil from Minnesota farms averaged 1600 seeds/ft^2 6 inches deep, or 70 million seeds/acre (Robinson, 1949). Other estimates range from 10.8 to 332 million seeds/ha (Klingman and Ashton, 1982).

1.3 Seed Production

In a 15-acre field that had been regularly cropped for several years, seven species were 90% of the total weed population. Good weed control reduced this up to 54% in continuous corn (Schweizer and Zimdahl, 1984b) and 26% in rotational crops in 1 year (Schweizer and Zimdahl, 1984a). Redroot pigweed populations declined 99% from 1.07 billion to 3 million seeds/ha 25 cm deep (10 inches) after 6 years of weed control in continuous corn. Common lamb's-quarter declined 94% from 153.6 billion to 8.6 million seeds/ha 25 cm deep. The total number of seeds declined 98% from 1.3 billion to 20.7 million seeds/ha 25 cm deep (Fig. 5.1). Despite this great reduction, there still were 192 weeds/ft^2 of soil if all germinated in 1 year, but that never happens. The seed bank, which was enormous at the beginning of the experiment, was still large after 6 years of good weed management. It is generally assumed that 2–10% of weed seeds in the soil seed bank emerge each year. With 192 weed seeds/ft^2 we would expect 4 to 20 plants/ft^2, a significant weed problem that must be dealt with. Emergence from weed seed banks from Ohio to Colorado and from Minnesota to Missouri showed that for 15 species found on three or more sites, average percent emergence varied from 0.6 for prostrate knotweed to 31.2 for giant foxtail. Six species had greater than 15% emergence in three or more years, four had between 5 and 8.5, and five others had less than 3.5% emergence (Forcella et al., 1992, 1997). Two percent to 10% is a reasonable average emergence percentage, but there is large variation among species.

A study in continuous corn (Schweizer and Zimdahl, 1984b) demonstrated that when atrazine was discontinued as the primary herbicide after 3 years, redroot pigweed seed numbers rose to 608 million (Fig. 5.2). Common lamb's-quarter rose to 22.8 million and the total number of seeds rose to 648.1 million/ha 10 inches deep. This contrasts with a steady decline with continued weed management (Fig. 5.2). The point is that in this system and in all cropping

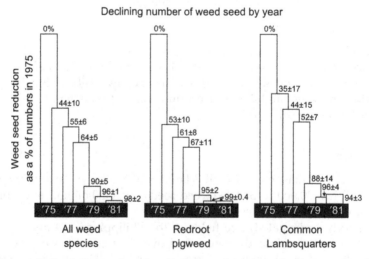

FIGURE 5.1   Percent decline in the number of seeds of all weed species, redroot pigweed, and common lamb's-quarter after 6 years of continuous corn. Standard errors shown for each weed species and year (Schweizer and Zimdahl, 1984a). *Reprinted with permission of Weed Sci. Soc. Am.*

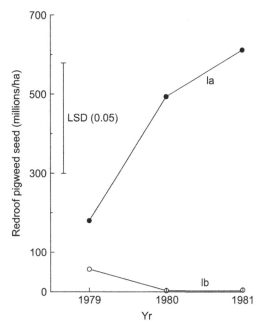

FIGURE 5.2 Number of redroot pigweed seeds present in soil each spring after conventional tillage and atrazine use in continuous corn. In weed management system Ia, 2.2 kg/hectares (ha) atrazine was applied preemergence for 6 consecutive years (Yr), beginning in 1975. In weed management system Ib, the same rate of atrazine was applied for the first 3 years and then discontinued (Schweizer and Zimdahl, 1984a). *LSD*, least significant difference. *Reprinted with permission of Weed Sci. Soc. Am.*

systems, if weeds are neglected even for just one cropping season, soil seed populations and the annual weed population rebound rapidly.

In rotational crops of barley, corn, and sugar beets, the total number of weed seeds declined 96.4% from 1.4 billion to 50 million/ha 10 inches deep after 6 years of weed management (two rotational cycles) (Schweizer and Zimdahl, 1984a). The number of redroot pigweed seeds declined over the 6-year period but the percentage of *Chenopodium* species increased because oak leaf goosefoot was more tolerant of cultivation and the herbicides used than common lamb's quarter.

After one cropping year, the decline in the number of redroot pigweed and *Chenopodium* species seed was 34% and 22%, respectively (Fig. 5.3). The next significant decline did not occur until after the fourth cropping year. After the sixth cropping year, the decline in the number of redroot pigweed and *Chenopodium* sp. seeds was 99% and 91%, respectively (Schweizer and Zimdahl). These data illustrate that weed seed populations can be reduced quickly, but continued attention is required to prevent a rapid increase when a few plants survive.

Some weeds can produce viable seed by apomixis (nonsexual reproduction; e.g., dandelion) and others are wholly self-fertile (e.g., shepherd's-purse). Weather before or during flowering is not important because with apomixis, seeds are set without pollination and with self-fertility, fertilization occurs before flowers open. These plants escape normal photoperiodic effects on flowering.

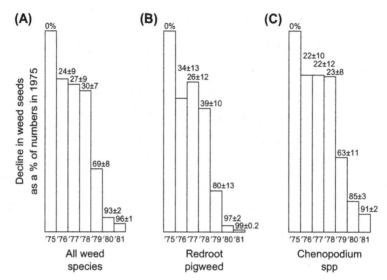

FIGURE 5.3 Percent decline in the number of weed seeds for (A) all weed species, (B) redroot pigweed, and (C) *Chenopodium* spp. when averaged over cropping sequence and weed management systems after 6 years of crop rotation. Standard errors are shown for each weed species and year (Schweizer and Zimdahl, 1984a). *Reprinted with permission of Weed Sci. Soc. Am.*

When one examines the seed-producing capacity of several weed species, it is not surprising that a few survivors rapidly increase the number of seeds in the soil bank. Data from single, undisturbed plants are shown in Table 5.2 (Stevens, 1932). The data purport to show maximum seed production and illustrate that production potential is high for many common weeds. These data have been cited in many weed science textbooks and are regarded as accurate, but there are reasons to question their accuracy. Stevens' (1932, 1957) work on 234 species was done with seed collected from diverse habitats in several areas in the United States. If the studies were redone under carefully controlled conditions, with identified seed sources and plants growing in isolation with free root growth, seed production would likely be higher.

Barnyard grass illustrates the point. Stevens (1932) reported one plant produced 7160 seeds. Barrett and Wilson (1981) reported 18,000 and Holm et al. (1977) up to 40,000. Research in California (Norris, 1992) predicts that barnyard grass growing in sugar beets averages nearly 100,000 seeds/plant and some larger plants produce more than 400,000. Reeves et al. (1981) found that wild radish produced 1030 seeds/plant with only 1 plant/m². When there were 247 wild radish plants on each square meter, seed production dropped to 67/plant. Russian thistle typically produces about 250,000 seeds (Young, 1991).

Research in irrigated row crop rotations suggests that the cropping sequence is the dominant factor that determines species composition of the soil seed bank (Ball, 1992). Herbicides and other cultural techniques vary among crops and shift seed bank composition in favor of less susceptible species. In irrigated row crops, dominant species were more prevalent near the surface after chisel as opposed to moldboard plowing (Fig. 5.4). The number of species increased more after chisel plowing and there was a greater decrease after moldboard

TABLE 5.2 Number of Seeds Produced Per Plant and Number of Seeds Per Pound for Several Common Weeds

Plant Common Name	Number of Seeds Per Plant	Number of Seeds Per Pound[a]
STEVENS (1932)		
Barnyard grass	7,160[b,c]	324,286
Black nightshade	8,460	197,391
Buckwheat, wild	11,900	64,857
Charlock	2,700	238,947
Common cocklebur	440	2,270
Toadflax	2,280	3,242,857
Dock, curly	29,500	324,286
Dodder, field	16,000[c]	585,806
Field bindweed	50	14,934
Foxtail barley	2,420	403,555
Giant ragweed	1,650	26,092
Kochia	14,600	534,118
Common lamb's-quarter	72,450	648,570
Black medic	2,350	378,333
Common mullein	223,200	5,044,444
Black mustard	13,400	267,059
Yellow nutsedge	2,420[d]	2,389,484
Wild oats	250[b]	25,913
Redroot pigweed	117,400[b]	1,194,737
Broad-leaf plantain	36,150	2,270,000
Common evening primrose	118,500	1,375,757
Prostrate knotweed	6,380	672,593
Common purslane	52,300	3,492,308
Common ragweed	3,380[b]	114,937
Sandbur	1,110[b]	67,259
Shepherd's-purse	38,500[b,c]	4,729,166
Pennsylvania smartweed	3,150	126,111
Leafy spurge	140[d]	129,714
Stink grass	82,100	6,053,333

(Continued)

TABLE 5.2   Number of Seeds Produced Per Plant and Number of Seeds Per Pound for Several Common Weeds—cont'd

Plant Common Name	Number of Seeds Per Plant	Number of Seeds Per Pound[a]
Common sunflower	7,200[b,c]	69,050
Canada thistle	680[b,c]	288,254
Witchgrass	11,400	698,462
STEVENS (1957)		
Annual bluegrass	2,050	2,270,000
Catchweed bedstraw	105	59,737
Chicory	4,600	567,500
Common chickweed	600	1,173,127
Common milkweed	600/Stem	77,080
Dandelion	12,000	709,375
Giant foxtail	4,030	238,947
Prickly sida	510	142,320
Prostrate knotweed	4,600	504,444
Redroot pigweed	229,175	1,335,294
Toothed spurge	835	97,634
Velvetleaf	4,300	51,885
Venice mallow	58,600	181,600
Wild radish	1,875	53,412

[a]*Calculated from the weight of 1000 seeds.*
[b]*Many immature seeds present.*
[c]*Many seeds shattered and lost before counting.*
[d]*Yield of one main stem.*
Adapted from Stevens, O.A., 1932. The number and weight of seeds produced by weeds. Am. J. Bot. 19, 784–794; Stevens, O.A., 1957. Weights of seeds and numbers per plant. Weeds 5, 46–55.

plowing (Ball, 1992). In a similar study, weed seed numbers dropped more under continuous corn and increased in mechanically weeded plots (Posner et al., 1995).

Forcella et al. (1992) studied weed seed bank size in eight US corn belt states and found that the total density ranged from 600 to 162,000 seeds/m^2 for three annual grasses: redroot pigweed, and common lamb's-quarter. Fifty to ninety percent of seeds were dead. Seedling emergence was inversely related to rainfall and air temperature in April and May, presumably because anoxia from high water content and high soil temperature induced secondary dormancy or killed the seeds. Forcella et al. found that viable seedlings were less than 1% of the seed bank for yellow rocket and 30% for giant foxtail.

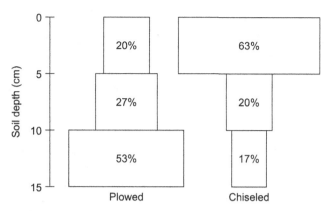

FIGURE 5.4 Influence of primary tillage on vertical distribution of weed seed 15 cm deep in soil after a dry bean crop (Ball, 1987). *Reprinted with permission of Weed Sci. Soc. Am.*

2. SEED DISPERSAL

Weed problems would be much less complicated if weed seeds just fell from plants and gravity determined their destination. One of the most obvious features of many weed seeds and seeds from many other plants is some structure that gives seed buoyancy in air or the ability to attach to something.

In 1941, a European engineer, George de Mestral (died, 1990) was walking in the Swiss mountains, near Geneva. When he returned, he noticed that he had burdock burrs attached to his pants, jacket, and wool socks, and that they were in his dog's fur. Mr. de Mestral must have been a good observer: someone who sees what he is looking for when it is there, does not see what he is looking for when it is not there, and sees what he is not looking for when it is there. Mr. de Mestral did not go walking to collect burdock burrs on his clothing. However, he saw them and observed what he was not looking for when it was there: the unique hook of the burdock burr that allowed it to attach to the wool of his clothing. After his observation and after 7 years of work, he duplicated the grasp of the burr with nylon and invented Velcro (from the French *velours* [velvet] and *cro* from the French *croc* or "crochet hook"). Velcro seems to be everywhere in modern society: children's clothing, airplanes, shoes, and artificial heart valves. It is worth a moment's reflection to consider that it came from a good observer's noticing the way in which one weed species disperses its seed in space.

2.1 Mechanical Dispersal

The long, slender awn of needle-and-thread grass facilitates movement by attaching easily to socks and other articles of clothing or to an animal's fur. The hooks on the aggregated cluster of flowers, properly called the capitulum, form the bur-like structure of cocklebur (the fruit) and hold its seeds (achenes), facilitating transport.

Burs of sandbur consist of one to several spikelets surrounded by an involucre of spiny, scabrous bristles. The burs of sandbur and the spines on the fruit of puncture vine penetrate shoe leather and tires. Bicycle riders are familiar with the hazards of puncture vine and sandbur because the spines easily penetrate bicycle tires. Dispersal in space is enabled by their sharp spines.

The seed pod of devil's claw is 2—4 inches long with a curved beak longer than the body of the pod. At maturity the pod divides into two opposite in-curved claws with an inwardly hooked, pointed tip to form an ice tong—like structure that easily attaches to livestock or equipment. Devil's claw, a stout, much branched, bushy plant, has seed pods that fall to the ground at maturity. When they dry and dehisce, the in-curved hooks can easily grab socks or legs and enlist them, involuntarily, to move their seeds.

Another form of mechanical transport of seed is illustrated by curlycup gumweed, a biennial or short-lived perennial that reproduces by seed. The flower heads are bright yellow, 0.5 to 1 inch across, and covered with a sticky resin. The sticky achenes facilitate seed transport.

The bristly capitulum of cocklebur.

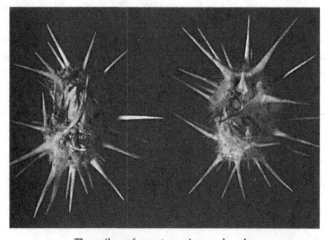

The spikes of puncture vine seed pods.

TABLE 5.3 Rate of Fall of Seeds Through Still Air (Salisbury, 1961)

Plant	Time to Fall 10 ft (Seconds)	
	Average	Range
Coltsfoot	21.3	14–45
Annual sow thistle	8.5	4.5–12
Groundsel	8.0	6.0–12
Smallflower galinsoga	3.4	2–5

2.2 Wind Dispersal

Many seeds have structural modifications that permit transport by wind. One common example, although it is not a weed, is the winged seed of maple trees. Among the weeds, examples of modification for transport by wind include the silky white pappus on dandelion achenes, the white downy pappus on Canada thistle achenes, and the tuft of silky hairs on seed of showy milkweed and prickly lettuce. It is common to see seed of one or more of these species moving with a summer breeze. The data in Table 5.3 illustrate seed dispersal by wind. Most seed is light (see number per pound in Table 5.2) and can move over great distances with light winds. On bombed sites in London after World War II, 140 different species of flowering plants were observed; those that established first, about 30% of the total, were distributed by wind (Salisbury, 1961).

Another method of wind dispersal is found in Russian thistle or tumbleweed. A mature Russian thistle plant is nearly round. When mature, it breaks off at the base, and because it is round it can tumble (the tumbling tumbleweed) or roll with the wind. The seeds, which are held by a series of twisted hairs, are released gradually because as the plant rolls and bumps, the hairs break. Other plants that roll to disperse seed include tumble pigweed and witchgrass. In the latter case, only the inflorescence breaks off and rolls.

The dandelion seed ready for dispersal.

2.3 Water Dispersal

In the western United States and other areas where irrigation is common, many seeds are dispersed by water. In Nebraska, Wilson (1980) found seeds of 77 different plant species in three main irrigation canals over two seasons. He collected a total of 30,346 seeds. Approximately 30% were viable and 26 times more were found at the end than at the beginning of canals. Most seed floated. Redroot pigweed seed was 40% of the total. He estimated that 120,000 seeds/acre per year entered fields from irrigation water. In the western United States, surface water irrigates more than 19 million acres each year and is an often-unrecognized source of weeds in irrigated fields.

It is logical to assume that because seeds are living organisms they will die quickly when submerged in water. In fact, many live a long time under water (Table 5.4). The curly dock fruit is a winged achene and the entire structure floats a long time before sinking. When seeds are deposited in water, the potential problem has not disappeared; it has moved. A Washington study (Comes et al., 1978) found 82 species in irrigation water. Twenty-four species lost viability after storage in water for 12 months or less; however, 27 endured more than 12 months, and some as long as 60. After 12 months, seed of 22% of annual monocotyledons germinated and seed of 75% of perennial monocotyledons and annual and perennial broad-leaved species germinated.

2.4 Human-Aided Dispersal

Humans have the burden of controlling weeds in crops, but we are also significant agents of distribution. A source of weed seed that was important, but which one hopes is no longer

TABLE 5.4 Germination of Weed Seeds After Storage in Fresh Water (Bruns and Rasmussen, 1953, 1957, 1958)

Species	Period of Storage (Months)	Germination (%)
Field bindweed	54	55
Canada thistle	36	About 50
	54	None
Russian knapweed	30	14
	60	None
Redroot pigweed	33	9
Quack grass	27	None
Barnyard grass	3	Less than 1%
	12	None
Halogeton	33	Less than 1%
	12	None
Hoary cress	2	5
	19	None

as important, is the farmer's grain drill. A 1965 study on the western slope of Colorado included 42 samples obtained from drills as farmers were planting. A grain probe removed a sample from the drill box in the field, and analysis of the 42 samples showed that 33% contained seed of prohibited noxious weeds and 74% contained seed of restricted noxious weeds. The farmers surveyed were planting an average of 2300 noxious weed seeds/acre. One was planting 13,000 field bindweed seeds, and another 14,000 wild oat seed/acre. A second Colorado study included 22 random samples. Fourteen percent of the drill boxes had prohibited noxious weed seed and 77% had restricted noxious weed seed. The average was 6600 noxious weed seeds planted per acre. An Iowa of study of combine run showed that 73% of oat seed had more than 20 weed seeds/lb. Sixty-three percent of all oats tested contained prohibited or restricted noxious weed seed[2]. In Minnesota, 343 drill box samples averaged 150 weed seeds/lb. One sample of red clover had 24,000 dodder seeds/lb and therefore was 10% dodder (Dunham, 1973).

The wide availability of certified seed has reduced, but not eliminated, human-aided weed dispersal when planting. In 1988, 31.3% of wheat, barley, and oat samples taken from grain drills in Utah were infested with an average of 313 weed seeds/lb of grain (Dewey and Whitesides, 1990). The worst sample had 11,118 weed seed/lb of grain. Nonnoxious weed seeds were found in 107 samples (23.8%) and noxious weeds were in 76 samples (16.9%). Wild oats, the most common noxious weed, occurred in 14% of samples at an average density of 2136 seeds/100 kg of crop seed (Dewey and Whitesides). In 1958, 52% of drill boxes were contaminated.

A cardinal rule of weed management is to buy and plant clean seed. Most farmers in the world's developed countries buy seed from a dealer and are confident that it is free of weed seed and diseases and has high germinability. Unfortunately crop seed still may be contaminated with other crop seed, weed seed, and even seed of noxious weeds. Colorado seed laboratory data from 2014 to 2015 confirm this (Table 5.5).

TABLE 5.5 Crop Seed Analysis by Colorado Seed Laboratory

Crop Seed	% Purity	% Germ	Other Crop Seed	No./lb	Weed Seed	No./lb	Noxious Weed Seed	No./lb
Smooth brome	70.8	91	Sand dropseed Intermediate wheat grass	1021 600	Tumble mustard Prickly lettuce Downy brome	1741 540 120	Field bindweed	58
Smooth brome	97.7	95	None		Common Lamb's-quarter	60	None	
Pubescent wheatgrass	98.3	94	Smooth brome	30	Downy brome	150	Wavyleaf thistle	3

[2]Personal communication. Colorado State Seed Laboratory, Colorado State Univ.

TABLE 5.6 Families of Introduced Weed Species Introduced From Europe to the United States (Fogg, 1942, 1966; Hill, 1977)

Family	Number of Species Introduced From Europe
Asteraceae	112
Poaceae	65
Brassicaceae	62
Labiatae	60
Leguminosae	54
Caryophyllaceae	37
Scrophulariaceae	30

Seed purity analysis is not common in most of the world's developing countries, where farmers keep harvested grain for planting the next year. Crop seed is often contaminated when harvested; weed seed, seed-borne pathogens, and the harvested grain may have poor germination. The best and most efficient weed management method in these situations is one that emphasizes prevention of the problem before it occurs rather than weed control in the following crop. Although the method is good, implementing it among small-scale, economically poor subsistence farmers in developing countries is difficult, if not impossible without massive government intervention; but although it is possible, it is difficult and rare.

We fail to screen or clean irrigation water and facilitate spread by mechanical means. The pattern in which the United States was populated offered a unique opportunity for spread of weeds from the east to the west coast and from the two coasts inland. Fogg (1966) pointed out the predominance of species of European origin in the United States. He found about 12.5% (1051) of the flowering plants and ferns of the central and northeastern United States and adjacent Canada in *Gray's Manual of Botany* were of foreign origin and 692 were from Europe. About 14% (1200) of the species in *Gray's Manual* are recognized as weedy, and European species dominate. Not all are weeds, but 60% of the 1200 plants were from only seven plant families (Table 5.6). The species in the seven plant families are primarily herbaceous (not woody), produce abundant seed, and are aggressive invaders or pioneering plants (see Chapter 8); that is, they have some of the traits that make weeds successful.

The United States, which is a major recipient of weeds because of immigration, has also distributed weeds to others. An example is parthenium ragweed, which was imported to India from the United States with shipments of grain during the early 1960s. It is an annual that has spread over large areas of southern India. It is especially problematic because it contains an irritating human skin toxin. The weed's common name in parts of India is AID weed, from its identification with grain distributed by the US Agency for International Development (US/AID). Weeds and their seeds have been imported to many countries in forages and feed grains.

There is a story, perhaps apocryphal, that Canada thistle was brought from Canada to the United States to feed the horses in British General John B. Burgoyne's army during the Revolutionary War. The British plan in 1777 was to divide the States by the line of the Hudson River. General Burgoyne was to proceed from Canada by way of Lake Champlain, which forms part of the boundary between northern New York and Vermont. The campaign began in January and Burgoyne was defeated on October 7 at the second battle of Bemis Heights (near Saratoga, NY). He surrendered his entire force on October 17. Burgoyne had to feed his army's horses; sadly (maybe truly), the hay contained Canada thistle. The weed is now ubiquitous in the northern United States. Canada thistle is called California thistle in Australia, a name indicative of where the Australians think it came from.

Weed seeds are also regularly transported in feed for cattle. Millers usually clean seed received for storage or processing. Screenings can contain weed seed and are routinely transported and used as cattle feed. Because of their role, seeds have good nutritional value. Their viability can be destroyed by cooking, but screenings are of low value and cooking or heating is usually not done; the seeds are fed whole. Tables 5.7 and 5.8 show examples of seed screenings that have been transported into Colorado. It is obvious that such animal feed can be important sources of weed seed; similar examples can be found for most places.

Grinding in a hammer mill does not completely destroy seed viability (Zamora and Olivare, 1994). Less than 1% of spotted knapweed, sulfur cinquefoil, timothy, and alfalfa seeds were intact after passing a 1-mm screen in a hammer mill. Of the four plants, only sulfur cinquefoil failed to germinate.

TABLE 5.7 Composition of Seed Screenings Analyzed by the Colorado State Seed Laboratory

Seed	Number/lb	Number in Average Truck Load ($\times 10^6$)
Common Lamb's-quarter	155,700	6,228
Redroot pigweed	9,225	369
Kochia	1,800	72
Russian thistle	900	36
Common sunflower	225	9
Foxtail		
Yellow	225	9
Green	1,575	63
Total noxious	2,700	108

TABLE 5.8 Composition of Seed Screenings Analyzed by the Colorado State Seed Laboratory

Item	Number/lb	Number of Seeds in Shipment ($\times 10^6$)
SAMPLE I		
Noxious weed seeds	13,511	540.4
Common weed seeds	142,650	5,706.0
Crop seeds	8,280	
SAMPLE II		
Noxious weed seeds	2,700	113.4
Common weed seeds	279,665	11,745.9
Crop seeds	30,150	1,266.3

It is common to assume that once a crop is harvested and weeds are ensiled (stored in a silo), weed seed can be forgotten about. In general, this is true. The pH of corn silage is between 4.5 and 5.8 and decreases with age. Most seed will completely lose germination after 3 weeks storage in silage. It is also true that the organic acid content of silage is 1.5–2% and silos quickly become anaerobic, both of which lead to seed death (Table 5.9). Downy brome, foxtail barley, and barnyard grass lost all viability after being ensiled for 8 weeks or undergoing rumen digestion for 24 h (Blackshaw and Róde, 1991). The same study showed that 17% of green foxtail seed was viable after 24 h of ruminant digestion. No wild oats survived rumen digestion in the first year, but 88% did in a second year of the study. This was attributed to the different diet in the two studies that changed the rumen bacterial population.

TABLE 5.9 Effect of Ensiling on Viability of Weed Seeds (Tildesley, 1937)

Weed Species	% Germination		
	Month Before	2 Weeks After	4 Weeks Later
Quack grass	99	0	0
Barnyard grass	61	0	0
Yellow foxtail	20	0	0
Wild buckwheat	64	0	0
Common lamb's-quarter	82	34	0
Cow cockle	68	0	0
Field pennycress	77	0	0
Wild mustard	93	0	0

2.5 Animal Aided

One might think that if any weed seed were fed to cattle, no problem would ensue because cattle chew things and rumen digestion is thorough. However, there is a potential problem. Beach (1909) fed a Jersey cow 6 lb of flax seed containing 212,912 weed seed/lb. The seed had 26.4% viability, which is typical for weed seed in feed. The cow voided 40 lb of feces per day and 1 oz contained about 1000 weed seeds, 4.5% of which were viable. Harmon and Keim (1934) confirmed that passage through an animal's digestive tract reduces but does not eliminate weed seed viability. Seed viability after digestion ranged from 6.4% for sheep to 9.6% for calves. Chickens destroyed all but 0.2% of viable seed.

Even after weed seeds have been voided in manure, many are alive and will germinate. Seeds left in cattle manure in the field had only 3.1% germination whereas seed in top-dressed manure hauled directly from the barn had 12.8%. Plowing under fresh manure increased seed germination to 23% (Oswald, 1908).

Ridley (1930) listed 124 weed species whose seeds were dispersed by cattle. In 36 samples of cattle manure from 20 New York dairy farms, viable seed from 13 grasses and 35 broadleaved species were found (Mt. Pleasant and Schlather, 1994). Four farms had cattle manure with no weed seed; others averaged 75 to 100 weed seeds/kg of manure. The authors concluded that manure can add seed to fields but the numbers were small compared with the soil seed bank. In contrast, in Iran, sheep manure added 10 million seeds/ha each time it was put on soil, and it was a more important source of new weed seed than the crop seed the farmer planted (182,000 seed/ha) or irrigation water (120 seed/ha) (Dastgheib, 1989). Other research in diverse locations confirms the successful passage of weed seed through cattle (Atkeson et al., 1934; Burton and Andrews, 1948; Dore and Raymond, 1942; Gardner et al., 1983).

Thill et al. (1986) studied common crupina, an introduced winter annual that invades rangelands in Idaho. They showed that its seed can be spread by cattle, deer, horses, and pheasants, but achenes were not found in sheep feces. Seed passed through the digestive tract of pheasants without loss of viability. The data support the contention that weeds are spread by game birds, wildlife, and domestic livestock.

The irrevocable conclusion is that many weeds pass through the digestive tract of several different animals without loss of viability and remain viable after storage in manure (Harmon and Keim, 1934) (Table 5.10). These data are confirmed by studies that show the effect of digestion and manure storage on germination of seed of several different species (Table 5.11).

Other data (Proctor, 1968) show that viable seed can be retained 8–12 h in the digestive tract of birds. Seed smaller than 1 mm in diameter with a hard seed coat can be retained more than 100 h. Birds distribute weed seeds. Still viable field bindweed, little mallow, and smooth sumac seeds were regurgitated from the digestive tract of killdeer (*Charadrius vociferus*) after 144, 152, and 160 h. Velvetleaf seed was intact for 77 h. Seed of many weeds can remain intact and viable in the intestinal tract of some birds long enough to be transported several thousand miles.

2.6 Machinery Dispersal

An important way in which people affect weed seed dispersal is through movement of farm machinery, especially itinerant grain combines and accompanying trucks. Dispersal

TABLE 5.10   Effect of Storage in Cow Manure on the Viability of Weed Seeds (Harmon and Keim, 1934)

Weed	% Viability Before Storage	% Viability After Storage (Months)			
		1	2	3	4
Velvetleaf	52.0	2.0	0	0	0
Field bindweed	84.0	4.0	22.0	1	0
Sweet clover	68.0	22.0	4.0	0	0
Peppergrass	34.5	0	0	0	0
Smooth dock	86.0	0	0	0	0
Smartweed	0.5	0	0	0	0
Cocklebur	60.0	0	0	0	0

TABLE 5.11   Germination Tests on Weed Seeds Before and After Passing Through Digestive Tract of Cattle and After 3 Months' Storage in Manure (Atkeson et al., 1934)

Weed Species	Germination Before Feeding (%)	Germination After 47 h Digestion (%)	Germination After 47 h Digestion + Manure Storage (%)	Decrease Owing to Manure Storage and Digestion (%)
Redroot pigweed	98	36.0	11.5	88
Common lamb's-quarter	70	58.0	22.0	69
Alfalfa	86	17.0	80.0	7
Buckhorn plantain	94	16.0	0.0	100
Curly dock	95	58.0	3.0	97
Green foxtail	21	19.5	0.0	100
Wild oats	74	10.0	0.0	100

of the seed of many weeds is aided by itinerant combine harvesters that move from field to field, often across large areas of the country. Itchgrass has grown wild in Louisiana sugarcane since the 1920s. It started to migrate when soybean farming expanded. Sugarcane has long been grown in Louisiana and cane processing machinery is a likely vector, but it rarely leaves a farm. Because soybean harvesting machinery is often itinerant, itchgrass has spread with itinerant soybean combines.

Some weed seeds are dispersed by combines because weeds are harvested with the crop and seed is dispersed by the combine as straw is spread on the field. Other weeds (e.g., wild mustard and field pennycress) shed seed before harvest. In the Northern Great Plains of the United States, wild oats, downy brome, and Canada thistle shed seed before and during harvest. Green and yellow foxtail, barnyard grass, quack grass, redroot pigweed, kochia, wild buckwheat, common lamb's-quarter, field bindweed, and Russian thistle shed seed during and after harvest. Combine harvesting facilitates seed dispersal (Donald and Nalewaja, 1991). These weeds make harvest more difficult by accumulating on the harvester's cutting bar and adding weight and green material to the combine's load and to harvested grain.

Movement and storage in combines are of concern because it has been shown that seed of slim-leaf lamb's-quarter, Venice mallow, and curly dock grew better when collected from combines that were harvesting hard red winter wheat than when the seed was harvested by hand from weedy plants in the same field (Currie and Peeper, 1988). Mechanical abrasion or scarification in the combine was the likely cause.

2.7 Mimicry

Discussing the evolutionary potential of crop pests, Gould (1991) said, "of all the crop pests, weeds boast the longest recorded history of adapting to agricultural practices." Weeds use two techniques to survive between cropping seasons: seed dormancy and crop seed mimicry. The second technique is basically hiding in crop seed to be planted the next year. It avoids all of the perils of remaining in the field, exposed to the environment and predators. By evolving to mimic the seed size, shape, or color of the crop they infest, weedy plants are passed on by humans who plant contaminated seed. Gould cites mimicry of lentil seeds by common vetch, flax seed by species of false flax, and rice by barnyard grass. In the latter case, the mimicry is in plant morphology and growth habit, not seed (Barrett, 1983). Because the plants are hard to distinguish visually, they are not removed by hand. The foregoing examples are of unrelated plants, but Gould also cites mimicry in closely related wild and domestic rice.

2.8 Other

Human activities are often sources of new weeds. It has been suggested that downy brome first entered California in packing material for glassware shipped from Europe. We also spread weeds growing in nursery stock and ornamentals. Highway construction that demands "fill" soil can easily spread weeds and their seed over wide areas.

2.9 Consequences of Weed Dispersal

Note: The following briefly introduces invasive weeds, a major consequence of weed dispersal. Invasive weeds are the subject of Chapter 8.

Although it may be interesting to know that weeds are dispersed in many ways, it is important to know that dispersal of weed seeds has real consequences. For example, data from the US Bureau of Land Management show that alien plants (some of which are weeds) are expanding their territory by 14% each year (Culotta, 1994). Leafy spurge arrived in Massachusetts in 1827. By 1909, it had arrived in the Great Plains, where there were no natural enemies. By 2005, it occupied 1.2 million acres in North Dakota (McGrath, 2005). In 2003 about 60% of the 1347-acre Devils Tower National Monument in Wyoming had been invaded by leafy spurge, annual bromes, and hound's-tongue. In 2016, leafy spurge infestation had been reduced to less than 3—5% in most experimental plots within the National Monument[3] through regular monitoring and planned management.

In 1990, it was estimated that leafy spurge costs farmers and ranchers in the Dakotas, Montana, and Wyoming $40 million each year in lost revenue and control costs, plus an estimated $89 million in secondary costs, including lost jobs (Leistritz et al., 2004; McGrath, 2005). However, a weed management program that includes herbicides and a cattle—sheep—flea—beetle biological control program significantly reduced the acres populated by leafy spurge.

Rush skeleton weed, originally from the Balkans, was first spotted near Banks, Idaho, in 1954. In 10 years it had invaded only 40 acres, but by 1994 it was on 4 million acres in Idaho alone. In 1999, it was present in nearly all of western Idaho, Oregon's Willamette valley, western California, and eastern Nebraska (Sheley and Petroff, 1999). In 2014 it had infested more than 6.2 million acres, mainly in the western United States (http://smallgrains.wsu.edu/wp-content/uploads/2014/03/Rush-Skeletonweed.pdf). It is classified as a noxious weed in nine western states[4]. It flourishes in very dry to very wet environments, and once present, it quickly dominates. It quickly displaces native indigenous plants with low species diversity and high soil erosion. As an invader, it has an added advantage of recovering quickly after fire (Kinter et al., 2007).

Weeds do not affect only crop and range land. The Selway-Bitterroot Wilderness in Idaho has prime stream habitat for salmon, but some areas of riverbank is covered with spotted knapweed. Other species do not grow with spotted knapweed, so the soil is bare between the plants. When it rains erosion increases, soil enters the water, and the quality of salmon spawning area declines (Culotta, 1994).

It was first reported in the Florida Everglades National Park in 1967. By 1995, the weedy tree melaleuca had invaded and taken over more than 488,000 acres of the Everglades and tropical wetlands of south Florida (Schmitz, 1995). Melaleuca is a native of Australia, where it is kept in check by over 400 insect species. It was expanding its range in Florida by 50 acres a day. Laroche and Ferriter (1992) showed that many melaleuca infestations have increased 50-fold over 25 years. Heavily infested sites can contain as many as 31,000 trees and saplings per acre. A mature tree can be 100 ft tall. Melaleuca control cost the state more than $2.2 million in 1994 (Laroche, 1994).

In a small book, Randall and Marinelli (1996) describe 83 foreign invaders and correctly note that they can change fundamental ecosystem processes such as the frequency of wildfires, the availability of water or nutrients, and the rate of soil erosion. Invasive weedy species such as melaleuca, rush skeleton weed, and others "change the rules of the game." Invaders that do not change basic ecosystem processes cause other problems. In forests, invading trees and vines can grow into the canopy and shade desirable species. Shrubs can dominate midstory areas and herbaceous species can colonize and dominate the forest floor. Prairies and other grasslands across the United States and in other countries are severely infested by nonnative weedy species that are also crop weeds, such as leafy spurge and yellow star thistle. Randall and Martinelli (1996) also point out that on wetlands in the northern third

---

[3]Personal communication: A.J. Wetz, Chief of Resources, Devils Tower National Monument. February 2012. Personal communication R. Ohms, Chief Resource Management, Devils Tower and I.W. Ashton, Vegetation Ecologist, National Park Service, July 2017.

[4]http://techlinenews.com/articles/2014/rush-skeletonweed-management-challenges-and-solutions.

of the United States and southern Canada, purple loosestrife formed large, dense stands that displaced native plants and changed and in many cases eliminated waterfowl habitat.

3. SEED GERMINATION: DORMANCY

Two facets of plant reproduction have been considered: seed production and seed dispersal in space. The third aspect of reproduction of special concern to weed managers is seed germination. What is really interesting is not that seeds germinate but that they *do not* germinate because they are dormant. Dormancy is dispersal in time as opposed to dispersal in space. Dormant seed can be dispersed in space without losing dormancy.

Dormancy is not well defined. To be dormant is to be sleeping or inactive. In biology, it is regarded as a state of suspended animation: alive but not actively growing. Thus, dormancy is defined as something that seeds do not do (germinate), as opposed to something they do. Scientists have described types of dormancy, but because the basic regulatory processes are unknown it is difficult to define types of dormancy or to extrapolate from one species to another (Dyer, 1995).

3.1 Causes

Chepil (1946) was among the first to report on periodic seed dormancy and germination among weeds. The phenomenon has since been documented for seed of many annual weeds (Karssen, 1982; Dyer, 1995). Dormancy is a highly developed specialization and a complex research problem. Most seed will germinate when proper environmental conditions exist; however, not all do. Soil disturbance may or may not initiate germination. Changes in soil temperature, soil water content, light, surface drying and wetting, and percent oxygen or carbon dioxide in soil can create or break dormancy. Soil microflora have a role because they affect oxygen and carbon dioxide. One microsite location, a habitable site, may provide appropriate conditions for germination whereas a nearby site may not. There is a range of special requirements for germination and other special conditions that impose dormancy.

The interaction of several factors that affect seed germination and seedling survival is illustrated well by the work of Rice (1985, 1987, 1990) on *Erodium* species in California. He examined (1985) the role of germination cueing in the dynamics of introduced broadleaf filaree and a second species (*Erodium brachycarpum*) populations exposed to local environmental variations in California grasslands. Temperature fluctuations were more important than temperature maxima for increasing germination rates. Light during germination had no effect on rate. There was a significant adaptive value for germination cueing in both species. Increased germination observed for both species exposed to temperature fluctuations supported the contention that high temperatures and temperature fluctuations were major factors that promoted the softening of hard seed. Softening (dormancy breaking) was most affected by temperature fluctuations. The persistence of

---

Continued

both species was enhanced by periodic soil disturbance by pocket gophers (Rice, 1985, 1987). Small mammal (voles and pocket gophers) herbivory prevents *Erodium* from colonizing areas of disturbed soil in grasslands that have not been grazed by sheep. Vigorous herbivory of seedlings and flowers by small mammals had a strong negative effect on *Erodium* growth. Grasslands protected from sheep grazing do not have either species of *Erodium*. Because it removes surface litter on which small mammals feed, sheep grazing is an important factor in preventing *Erodium* colonization of gopher mounds, which occurs in the absence of sheep grazing. Sheep grazing promotes the growth of *Erodium* populations.

Reproductive inequality of both species increased with increasing plant density and productivity (Rice, 1990). Seed production was controlled by rainfall. "The magnitude of reproductive inequality was dependent on the interactions of sowing density and rainfall distribution." The importance of rainfall as a determinant of population is illustrated by the observation that at low sowing density, rainfall had no effect on reproductive inequality. Rainfall's effect was seen only at the highest sowing density. "Effective population number was relatively insensitive to increases in population density because of increased inequality in reproduction at higher population densities" because of the rainfall effect.

If one wants to become famous in weed science, indeed in agricultural science, it could be accomplished by figuring out how to do one of two things. The first is to make most (perhaps all) weed seeds in soil, and those shed from plants during the previous season, dormant forever: ergo weed control would not be needed. The second is to make most or all weed seeds in soil germinate immediately. Because seed dormancy is a complex environmental/physiological/biochemical phenomenon, it is unlikely that a magic bullet solution to make or break dormancy will ever be found. However, in theory, a more complete understanding of dormancy could greatly reduce the need to control weeds that reproduce by seed. It would take time to deplete soil seed banks, but once that was done, the annual need to control weeds would decline, if not disappear. If we could make most seed in the soil seed bank germinate just before frost in the temperate zones, frost would kill most of them. In the tropical dry season, weeds could be managed with tillage. Because weeds have periodicity of germination, timing of tillage and planting can be altered when possible to encourage or discourage weed seed germination (Dyer, 1995; Gunsolus, 1990).

Weeds share many traits with what ecologists call early successional species (Roberts, 1982). Indeed, they are often the same species. For early successional species, seed germination is closely linked to soil disturbance that ensures the availability of resources for growth. Soon after soil disturbance, germination reduces the probability of competition from later successional species or crops. Early successional species and many weed seeds usually require light for germination. Exposure to light is increased by tillage. Seed germination is favored by fluctuating temperatures and low carbon dioxide concentrations, and may be affected by alternate wetting and drying cycles that tend to break seed coats. All of the conditions favorable for seed germination occur on disturbed sites, and cropped fields are disturbed sites.

Seeds of early successional species and of many weeds are dormant when shed and can quickly develop secondary dormancy. Induced dormancy (dependent upon environmental interaction) is common. Early successional species grow rapidly above and below the soil and therefore escape the surface zone of maximum environmental variability and stress. Early successional species have a high photosynthetic rate over a wide range of soil water conditions. Photosynthetic rate and environmental resource demand decline quickly with declining soil water potential, permitting survival. Weeds and successful early successional species compress environmental extremes. They are able to maintain constant leaf temperatures to ameliorate stress. They also acclimate rapidly to variable environments. Genotypic plasticity facilitates adaptation. Seed dormancy is the most important physiological trait that explains the perpetuation of annual weed species. Seed of annual weeds germinate under a narrow range of environmental conditions; they are specialists in using their opportunities.

Most weed control techniques treat symptoms rather than the problem. Weed control acts on problems either just before they appear (preemergence) or after they have appeared (postemergence). There are no reliable methods for eliminating weed problems by preventing seed dormancy or encouraging germination of dormant seed (see Kennedy, Chapter 12). Improved weed management depends on a better understanding of the interaction between seeds and the environmental conditions that cause or terminate seed dormancy. For example, late-season emergence of giant ragweed led to higher levels of embryo dormancy that prevented germination at low spring temperatures (Schutte et al., 2012). Gardarin and Colbach (2014) found that dormancy was greater in elongated than in spherical seeds, because elongated seeds remained on the soil surface longer in undisturbed habitats and immediate germination may serve to limit predation.

3.1.1 Light

Of the known causes of dormancy, light may be the most important. At least half of the annual weeds in crops have seed that require light for germination (see Table 5.12 for a few examples). This is especially so for small-seeded annual weeds. Length of day and the quality of light are also important. The light requirement is regarded as an evolutionary advantage for small-seeded plants that may not survive germination from lower in soil (Pons, 1991). Light penetrates only 1 or 2 mm in soil, so dormancy can be induced even by shallow burial. Germination of seed of mullein, curly dock, evening primrose, and buttercup is favored by light. Seed of common chickweed, common purslane, johnsongrass, kochia, lamb's-quarter, prostrate knotweed, and redroot pigweed require light for germination. However, germination of wild onion and jimsonweed is favored by darkness. Chauhan and Johnson (2010) reported the effects of burial on seven weed species. Average percent germination was 79% for seeds on the soil surface. When the same seeds were buried 0.5 cm, average germination was 3.3% (globe fringe-rush was 13.0%) and two did not germinate at all. Dormancy of crop seed has been nearly eliminated by breeding light response out of the genome; most germinate in light or dark.

The phytochrome group of photoreceptors, the primary system responsible for light interactions in plants, controls breaking dormancy by light. In a simple but accurate sense, phytochrome exists in two forms: a promoter and inhibitor. The promoting form is favored by red light and the inhibitor by far-red light. Phytochrome is also involved in flowering. The

TABLE 5.12 Effect of Light on Germination of Some Weed Seeds

| Weed | Germination % in | | Source |
	Light	Dark	
Barnyard grass	69	24	Boyd and Van Acker (2004)
Celosia	31	2	Chauhan and Johnson (2007)
Chinese sprangletop	95	19	Chauhan and Johnson (2008b)
	95	0	Benvenuti et al. (2004)
Coat buttons	79	7	Chauhan and Johnson (2008d)
Common purslane	81	2	Chauhan and Johnson (2008a)
Eclipta	83–93	0	Altom and Murray (1996) and Chauhan and Johnson (2008e)
Globe fringe rush	85	0	Chauhan and Johnson (2009a)
India crabgrass	89	0	Chauhan and Johnson (2008c)
Jungle rice	76	12	Chauhan and Johnson (2009d)
Node weed	86	0	Chauhan and Johnson (2009c)
Rice flatsedge	94	0	Chauhan and Johnson (2009a)
Siam weed	80	37	Chauhan and Johnson (2008d)
Smallflower umbrella sedge	81	0	Chauhan and Johnson (2009a)
Southern crabgrass	93	7	Chauhan and Johnson (2008c)
Winged water primrose	98	0	Chauhan and Johnson (2009b)

quantity of each form of phytochrome present at a given time is related to light. More precisely, the ratio of red to far-red light determines the response. Sunlight has abundant red light and promotes germination of imbibed seeds. Seeds do not respond to light unless they have taken up water (imbibed).

Light effects the inactive and active forms of phytochrome and thus seed germination:

$$P_R \leftrightarrow P_{FR} \rightarrow \text{Biological activity}$$

Inactive form (P_R) + red light, 600–680 nm (peak at 666 mm) $\rightarrow$ P_{FR} and germination promotion

Active form (P_{FR}) + far-red light, 700–760 nm (peak at 730) $\rightarrow$ P_R and germination inhibition

Light is needed for seed germination of many species, although it is clear that burial in soil will inhibit germination and should be used as a weed control technique. Continued seed burial is encouraged when farmers shift to minimum and no-tillage.

Unfiltered light contains a preponderance of the red wavelength that shifts phytochrome to the active (P_{fr}) form and promotes seed germination. Leaf canopies filter red light because chlorophyll absorbs it strongly. A leaf canopy shifts light transmission toward far-red and depresses germination of seeds below.

3.1.2 Immature Embryo

A second cause of dormancy is the presence of an immature embryo. Smartweed and bulrush seeds are typically shed from the plant with an immature embryo and are incapable of immediate germination. This is an example of an evolved mechanism to prevent germination at the wrong time.

3.1.3 Impermeable Seed Coat

Seeds of redroot pigweed, wild mustard, shepherd's-purse, and field pepperweed often have seed coats impermeable to water, oxygen, or both. The seeds are called "hard." It is another dormancy mechanism. The seed coat can be changed (often referred to as broken) by scarification, action of acids, or microbes. A hard seed coat presents mechanical resistance to germination because the radicle cannot penetrate it. Although water and oxygen can be absorbed, the hard seed coat prevents germination. In the laboratory, scarification or breaking of a hard seed coat can be accomplished by rubbing on sandpaper, dipping in acid, or pricking with a pin. Such techniques are obviously inappropriate for the field, but the same thing is accomplished by tillage. Anything that stirs or moves soil will inevitably move seeds and abrade seed coats. In general, dormancy increases with seed coat thickness (Gardarin and Colbach, 2014).

3.1.4 Inhibitors

Some seeds are shed with endogenous (internal) germination inhibitors (e.g., abscisic acid). Varied, complicated chemical inhibitors prevent seed germination until they are removed by leaching with water or by internal metabolic activity. There are also exogenous (external) germination inhibitors, which will be discussed in Chapter 9, on allelopathy.

3.1.5 Oxygen

Partial pressure of oxygen affects seed dormancy. Percent oxygen in soil varies from less than 1% in flooded soil to 8% or 9% in a soil with good tilth, cropped with corn. Soil carbon dioxide content may vary from 5% to 15%. One reason why most seed germinates only near the soil surface is higher oxygen concentration. Soil compaction reduces seed germination. The mechanism may be reduction in the partial pressure of oxygen.

3.1.6 Temperature

There is a minimum temperature below which no seed will germinate and a maximum temperature above which germination will not occur. The precise minimum and maximum vary among species, as does the optimum temperature for germination. In late spring, Russian thistle seed germinates readily between 28 and 110°F (Young, 1991). Wild oats will germinate at 35°F (1.7°C), which is lower than the temperature at which seed of wheat or barley germinate. The optimum temperature and pH for germination of tall morning glory is 30°C at day and 20°C night at pH 6. Temperature and pH that are higher or lower significantly reduce germination (Singh et al., 2012). Beggar-ticks germinated over a wider range (15−40°C) in alternating light and dark conditions at pH 7, but germination decreased sharply with increasing acidity or alkalinity and emergence decreased as sowing depth increased (Ramirez et al., 2012). Similarly, smooth barley density was greater under zero-tillage than conventional tillage because fewer seeds were buried (Fleet and Gill, 2012).

Temperatures of 40–60°F (4–15°C) are required for germination of seed of some winter annual weeds. Higher temperatures lead to dormancy. Redroot pigweed seed stored at a constant 68°F (20°C) will remain dormant up to 6 years. It can be induced to germinate at any time by alternating storage temperature or by partial desiccation. Germination can be induced by raising the temperature to 95°F (35°C) for a short time, rubbing the seed, and then lowering the temperature to 68°F.

3.1.7 *After-Ripening Requirement*

There is an after-ripening requirement for some seed. It is a poorly understood physiological change that is not the same as an immature embryo. A seed's embryo is fully developed but will not germinate, even if oxygen and water are absorbed at the appropriate concentration. Everything appears to be normal but the seed will not germinate until it has ripened.

3.2 Classes of Dormancy

Dormancy classification is based on observed seed behavior, not, as mentioned earlier, on a complete understanding of the physiology or biochemistry of dormancy. Two classification systems will be presented. In the first, a seed dormant when shed from the plant has primary dormancy. All other manifestations of dormancy are secondary. After primary dormancy has been lost, secondary dormancy may be induced by environmental interactions or other special conditions.

The second system of classification includes three types of dormancy (Harper, 1957):

1. **Innate dormancy** has three possible causes. It could be an inherent property of the ripened seed based on genetic control when the seed leaves the plant. There may be an after-ripening requirement, perhaps dependent on receipt of specific environmental stimuli. There could be a rudimentary or physiologically immature embryo, which is not fully developed when seed is shed (e.g., smartweed). Innate dormancy can also be caused by impermeable or mechanically resistant seed coats, i.e., hard seed. Redroot pigweed, several species of mustard, and all species of wild oats have innate dormancy. A third cause is the presence of endogenous chemical inhibitors. Some species of sumac and fireweed proliferate after forest fires, because fire creates permeability in the seed coat and rain leaches out the inhibitor. The amount of an inhibitor is often adjusted to the rainfall of an area. In its simplest form, the presence of an endogenous chemical inhibitor restricts germination to the temperature range at which survival is ensured. Innate dormancy interacts with the environment because for some species, hot, dry weather during seed maturation yields less dormancy than cool, moist conditions that are more favorable to seedling survival.
2. **Induced dormancy** occurs when a seed develops dormancy after exposure to specific environmental conditions such as dryness, high carbon dioxide concentration, or high temperature and the acquired dormancy persists after the environmental conditions change Harper (1957). Seed of winter wild oats and white mustard have induced dormancy that often develops in late spring in temperate climates and persists into fall.

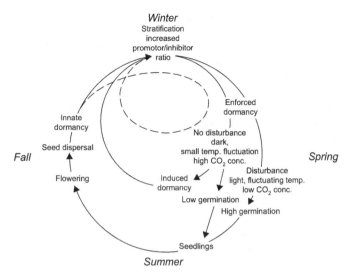

FIGURE 5.5 Schematic representation of seed germination in common ragweed, a common colonizer in old field succession and a spring annual weed. The *dashed line* represents seeds that require more than one stratification cycle to germinate and thus ensure germination and establishment across a number of seasons (Bazzaz, 1979). *conc*, concentration; *temp*, temperature.

Seed buried by tillage may not germinate when brought to the soil surface. Dormancy has been induced by environmental interaction after seed has been shed and it persists after environmental conditions change.

3. **Enforced dormancy** also depends on environmental interaction, but does not persist when conditions change. In the latter case, dormancy can be caused by lack of water or oxygen, low temperature, etc. When this external limitation is removed, the seed germinates; according to Harper (1957), the seed had enforced dormancy. There is a positive correlation between termination of dormancy and predictable environmental changes.

Wild oats exhibit all three of Harper's classes of dormancy. His system is a classification of mechanism, not of species. In general, termination of dormancy requires exposure to cool, moist conditions, the normal attributes of transition from summer to fall in temperate zones. Seeds in tropical climates have less and sometimes no seed dormancy.

Fig. 5.5 shows how common ragweed succeeds as an early successional plant and a weed. It illustrates integration of Harper's (1957) three dormancy classes (Bazzaz, 1979). Early and late successional environments are different with respect to light intensity and spectral quality. Seeds of early successional plants (many of which are common weeds) are sensitive to light, and seed germination is inhibited by light filtered through plant leaves (P_r form). This is not a problem for weeds that germinate early in the season before crop leaves shade the soil and filter light. Their germination is favored by fluctuating temperatures and low carbon dioxide concentrations in soil. They are not sensitive to soil-water fluctuations and other conditions commonly encountered in cropped fields.

3.3 Consequences of Weed Seed Dormancy

Dormancy is important because seeds survive for long times in soil and are a continuing source of infestation. It ensures survival for many years, and the aphorism that 1 year's seeding equals 7 years' weeding is reasonably accurate.

In the fall of 1879, William J. Beal (Darlington, 1951), a pioneer in the development of high-yield corn hybrids, began an unusual experiment. He buried 20 pint bottles, each containing 1000 seeds of 20 weed species, near his laboratory on the campus of Michigan State University. His aim, in an era before scientific weed control began, was to find out how long seed buried by plowing could survive and thus how long fields had to be left fallow to ensure a weed-free crop when replanted. Over 120 years later, we know the answer to Beal's question: a very long time.

Beal buried bottles upside down, uncorked, and at an angle so water and oxygen, but not light, could enter. Initially, bottles were dug up every 5 years. Since 1920 they were dug every 10 years. In the future the interval will be 20 years. Results for the last 80 years are reported in Table 5.13 (Telewski and Zeevaart [2002] reported that there is a question about the identity of the mullein seed before 1920. Seed identified as common mullein may have been moth mullein). After 120 years, only two species of mullein germinated. In 1960 (the 80th year), three species germinated, but only one (moth mullein) germinated in 1970. In 1980, seeds were planted as usual, in soil sterilized by steam. At first, nothing happened. After several weeks, the first seedling emerged; subsequently, 24 seedlings germinated. Of the survivors, 21 were common mullein, one another species of mullein and a mallow species that had not germinated since 1899 (the 20th year). The next bottle will be removed under cover of darkness, in 2020. Enough bottles remain to carry on Beal's experiment until 2100. His work shows that fallowing is not a feasible method of weed control for all species, at least in northern temperate climates.

Another experiment to investigate the consequences of seed dormancy was conducted by Duvel near Rosslyn, Virginia (Toole and Brown, 1946). In 1902, seed of 107 different species was buried 8, 22, or 42 inches deep in porous clay flowerpots covered with clay saucers. Samples removed at intervals showed no effect of depth of burial on survival but a tendency toward longer survival at 42 than at 8 inches. The results, summarized in Tables 5.14 and 5.15 show significant seed survival for 38 years. Even after 38 years, 91% of jimsonweed seed was viable and presumably capable of quickly reinfesting a cropped field.

A study in undisturbed soil in England (Lewis, 1973) showed that seeds reveal their survival potential during the first 4 years of burial. Rarely did a species that survived 4 years not survive 20 years. Seed deterioration occurred more rapidly in acid peat soil than in loam. With the exception of Timothy seed, the *Poaceae* were generally short-lived. Legumes generally persisted for the full 20 years. The weeds that survived best were common lamb's-quarter (23%), curly dock (18%), and creeping buttercup (53%).

In Mississippi, seeds of several species were buried in porous plastic bags to simulate natural conditions and avoid the clay pots of Duvel's experiment and the glass jars of Beal's study (Egley and Chandler, 1983). In contrast to Beals' northern temperate study, the primary lesson of the Mississippi experiment was that only about six of the species investigated remained viable after 5 years' burial (Table 5.16).

Seed of 41 economically important weed species of the Great Plains region of the United States were buried 20 cm deep (plowing depth) in eastern and western Nebraska in 1976

TABLE 5.13 Results of Beal's Buried Seed Study (Darlington, 1951)

Elapsed Time (Years)	Species Still Viable	Number	% Germination
120	Moth mullein[a]	23	46
	Mullein	2	4
100	Moth mullein	21	42
	Common mullein	1	2
	Common mallow	1	2
90	Moth mullein	10	20
80	Evening primrose	5	10
	Curly dock	1	2
	Moth mullein	35	70
70	Curly dock	4	8
	Evening primrose	7	14
	Moth mullein	37	74
50	Black mustard	4	8
	Marshpepper smartweed	2	4
	Evening primrose	19	38
	Curly dock	26	52
	Moth mullein	31	62
40	Common purslane	1	2
	Redroot pigweed	33	66
	Black mustard	19	38
	Curly dock	9	18
	Evening primrose	19	38
	Common ragweed	2	4
	Virginia pepperweed	1	2
	Broad-leaf plantain	5	10

[a]*Telewski and Zeevaart (2002) reported that there is a question about the identity of the mullein seed prior to 1920. Seed identified as common mullein may have been moth mullein.*

(Burnside et al., 1996). There were 11 annual grass, 14 annual broad-leaf, four biennial broad-leaf, and 12 perennial broad-leaf species. Seeds were exhumed after 1, 9, 12, and 17 years. After 1 year, germination was 57% for all annual grasses, 47% for all annual broad-leaf, 52% for biennials, and 36% for perennial broad-leaf species. Germination dropped steadily with time for each class. After 17 years, germination was 4% for annual grasses, 11% for

TABLE 5.14   Number of Weed Species Surviving (Toole and Brown, 1946)

Burial Period (Years)	Species Germinating
1	71
6	68
10	68
20	57
30	44
38	36

TABLE 5.15   Germination of Weed Seeds After
38 years (Toole and Brown, 1946)

91% of Jimsonweed

48% of mullein

38% of velvetleaf

17% of evening primrose

7% of lamb's-quarter

1% of green foxtail

1% of curly dock

TABLE 5.16   Viability of Weed Seeds After Burial (Egley and
Chandler, 1983)

	Mean Viability After Burial (years)			
	0	1.5	3.5	5.5
Species		%		
Velvetleaf	99	89	71	30
Purple moonflower	100	84	65	33
Hemp sesbania	100	77	60	18
Common cocklebur	99	27	10	1
Redroot pigweed	96	24	2	1
Common purslane	99	21	2	1
Johnsongrass	86	75	74	48

annual broad-leaf, 30% for biennials, and 8% for perennial broad-leaf species. No explanation was offered for why biennial species survived so well. A conclusion of that study was that after burial at plow depth, germination of annuals will decline rapidly but biennial species will survive well and could become problems in crops. The species with the highest survival after 17 years' burial in western Nebraska was common mullein, with 95% germination, a result consistent with the Beal study in Michigan (Darlington, 1951). Weed seed germination tended to be greater under the conditions of low rainfall and more moderate soil temperatures of western Nebraska.

Soil in Alaska is cold for much of the year. Two studies in Fairbanks, Alaska (Conn and Farris, 1987; Conn, 1990) showed that viability was higher after burial in mesh bags, at 15 rather than 2 cm deep after 21 months, and after 4–7 years. Four of 17 species had 5–10% viable seed after 4.7 years and eight species ranged from 21% to 39%. Viability of American dragonhead did not change during 4.7 years whereas viability of common hemp nettle and quack grass was zero after 2.7 and 3.7 years, respectively, and viability of two other species was less than 1%.

Taylorson (1970) found that initially nondormant seed of several weed species lost viability after burial sooner and to a greater extent than initially dormant seed. Zorner et al. (1984) found the same thing for kochia seed. Initial rates of loss were much greater in nondormant than in buried dormant populations. After 24 months of burial, the number of viable seed remaining and the depletion rates were similar for the two populations.

Woolly cup grass seeds are dormant at physiological maturity and lose dormancy during after-ripening over winter. Seeds buried below the soil surface were less dormant than those that remained on the soil surface (Franzenburg and Owen, 2002).

Studies in Michigan and Indiana showed that seed mortality of giant foxtail and velvetleaf was greatest in soil managed conventionally (using recommended rates of fertilizer and herbicide) and less in soils prepared with reduced management (nutrients from compost or organic amendments and weed control only by cultivation) (Davis et al., 2006). However, no measured soil properties were associated with seed mortality. Only management history and the soil fungal population were related to seed mortality. A 35-year study of the effect of tillage and rotation on soil weed seed banks in Ohio showed that the weed seed population in soils with a corn–oat–hay rotation differed in structure and composition from those developed under a corn or corn–soybean system (Sonoskie et al., 2006). Germinable weed seed differed in soils tilled conventionally, with no-tillage, or with minimum tillage. One assumes that such things will be true, but it is always good to have confirming data. Because such studies prove that crop sequence and tillage system affect weed seed populations and community structure, it follows that this information can be used to develop weed management systems.

An important problem in all buried seed studies is the necessity of recovering seed from soil, a complex medium. It is hard to find seed, and one must be sure that the seed found is the seed that was buried, if longevity is to be estimated. Therefore all studies use containers. Recent studies use porous mesh bags that allow transfer of air and water but do not allow other natural processes such as abrasion. Because seeds are concentrated, microbial action and seed interactions may be abnormal. It is generally thought that burial studies overestimate seed longevity. Seed dormancy is a major cause of continuing weed problems, and although a great deal is known about what causes dormancy and how to break it, no one knows how to create it, break it, or use it predictably and reliably to manage weeds in the field.

In the laboratory, it is easy to create or break dormancy with a variety of seed treatments (Anderson, 1968). These include abrasion, temperature manipulation, and chemical methods. Abrasive methods include rubbing, dehulling, dipping in sulfuric acid, and alternate wetting and drying to break the seed coat. Temperature manipulation is useful to break dormancy and is common in nature. Alternate freezing and thawing often breaks dormancy. Stratification or exposure to extremely low temperatures will break dormancy in some seed. Stratification is commonly required to break dormancy in temperate weed species but it rarely works for tropical species. It may act by decreasing the level of an endogenous inhibitor. Finally, chemical methods are used. Leaching with water may remove a chemical inhibitor and exposure to light will create chemical changes in seed. Chemicals such as potassium nitrate, gibberellic acid, cytokinins, and auxins are all used and their action is considered to be directed at overcoming the action of or inactivating an inhibitor.

In the field, breaking dormancy on demand is more difficult. Laboratory methods are obviously not suitable to field operations where seed cannot be seen. Plowing soil is a good way to break dormancy, and conversely, not disturbing soil is a good way to maintain dormancy of buried seed. Tillage exposes seed to light (see Chapter 10) and temperature changes. Field methods are nonselective and affect all seeds; therefore for some species dormancy may be promoted whereas in others it is broken. Weed management will continue to emphasize weed control until a better understanding of weed seed dormancy is obtained and methods are developed to use that knowledge for weed management.

Kremer (1993) points out that "successful weed management in agroecosystems depends on manipulating the weed seed bank in soil, the source of annual weed infestations." Despite the many successful methods for controlling weeds each year in annual crops, they inevitably appear, and Kremer correctly suggests that the source is the persistent soil weed seed bank. His work describes the many interactions of soil microorganisms and weed seeds. Fig. 5.6 shows the potential interactions. It is important to study the interactions to improve understanding of the survival of weed seeds in soil. They are also important because, as Kremer suggests, they reveal that microorganisms may be used to deplete the weed seed bank, an unexploited method of weed management.

4. VEGETATIVE OR ASEXUAL REPRODUCTION

Perennial weeds reproduce vegetatively, an unfortunate aspect of weed management. Simple and creeping perennials also reproduce by seed, but the importance of seed production varies. I suppose a good example is water hyacinth, whose pretty flowers produce seed pods with up to 300 seeds that can live 5–15 years submerged in water. Vegetative reproduction alone can double the size of an infestation in open water in 10–15 days (Leakey, 1981) to produce floating mats weighing up to 200 tons/acre. Transpired water losses from mats of water hyacinth will be three to five times the loss from an open water surface.

The reproductive organ, the depth to which it penetrates soil, and the importance of seed production for several important perennial weeds are shown in Table 5.17. Seed production is

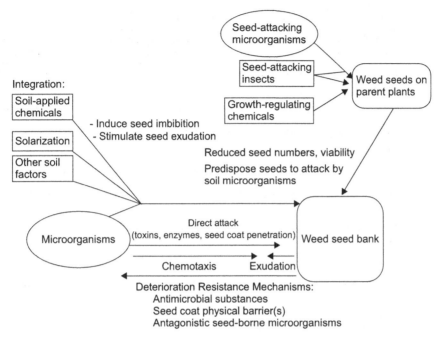

FIGURE 5.6 Relationship between microorganisms and weed seeds in soil. Several methods are indicated for depleting the weed seed bank with potential for integration with an approach including microorganisms. *Reprinted with permission of Kremer, R.J., 1993. Management of weed seed banks with microorganisms. Ecol. Appl. 3, 42–52.*

not of great importance for Canada thistle, which is dioecious. Whereas the pappus is always produced, it does not always have viable seed attached. On the other hand, seed production is important for leafy spurge, nettles, and curly dock.

Many methods of vegetative reproduction are found among weeds. Stolons or creeping aboveground stems are found in creeping bent grass, and yarrow. Rhizomes are found in Bermuda grass, quack grass, red top, hedge bindweed, and field horsetail. Bulbs and aerial bulblets are found in wild onion and wild garlic. Goldenrod has corms: thickened, vertical, underground stems that are reproductive organs. Tubers are produced by yellow and purple nutsedge and Jerusalem artichoke. Vegetative reproduction of simple perennials such as dandelion is from their tap root.

A seedling of a perennial species growing from seed has not yet assumed perennial characteristics (especially the ability to regenerate vegetatively) when it first emerges from soil and can be controlled more easily than after it assumes these characteristics. It is generally considered that quack grass assumes perennial characteristics within 6–8 weeks of emergence and johnsongrass after only 3–6 weeks. Field bindweed becomes a perennial when it has about 20 true leaves, and yellow nutsedge 4–6 weeks after it emerges from seed. These young plants can be controlled by tillage or hoeing before they assume perennial characteristics.

TABLE 5.17   Characteristics of Important Perennial Weeds (Roberts, 1982)

Species	Reproductive Parts and Overwintering State	Depth of Vegetative Reproductive Parts[a]	Importance of Seed Reproduction
Bermuda grass	Creeping rhizomes, decumbent stems spread laterally	Shallow	Moderate
Bracken fern	Rhizomes; leaves die	Deep	Reproduces by spores
Canada thistle	Creeping roots overwinter; shoots die	Deep	Occasionally produced
Coltsfoot	Rhizomes; leaves die	Very deep	Important
Common nettle	Rhizomes; short green shoots overwinter	Very shallow	Very important
Creeping bent grass	Aerial creeping stems overwinter	Aboveground	Unknown
Creeping buttercup	Procumbent stems; a few leaves overwinter	Aboveground	Very important
Curly dock	Taproots; rosette of leaves overwinter	Very shallow 7–10 cm	Very important
Dandelion	Fleshy taproot; few leaves overwinter	Shallow	Important
Field bindweed	Creeping roots overwinter; shoots die	Very deep	Important
Field horsetail	Rhizomes with tubers that overwinter	Deep	Reproduces by spores
Hedge bindweed	Rhizomes overwinter; shoots die	Deep	Rarely produced
Hoary cress	Creeping roots; small rosettes of leaves overwinter	Deep	Important
Japanese knotweed	Rhizomes, dormant underground buds; shoots die	Shallow	None produced
Leafy spurge	Creeping roots overwinter	Very deep	Very
Oxalis sp. (wood sorrel)	Bulbils, taproots, and rhizomes; leaves	Shallow	Important in some
Perennial sow thistle	Creeping roots; shoots die	Very deep	Important
Quack grass	Rhizomes with dormant underground buds; shoots overwinter	Shallow	Moderately
Red top	Rhizomes with dormant underground buds; shoots overwinter	Shallow	Very important
Roughstalk bluegrass	Short stolons; a few leaves overwinter	Aboveground	Very important
Slender speedwell	Stems creeping on surface	Aboveground	None produced
Wild onion	Offset bulbs and bulbils overwinter	Aerial or very shallow	Rarely produced
Yarrow	Stolons; terminal rosettes of leaves overwinter	Very shallow	Very

[a]Depth varies: Very shallow, 6–10 inches; shallow, 12–18 inches; deep, down to 40 inches; very deep, greater than 120 inches.

Seed production of perennials may be unimportant relative to vegetative reproduction, but it should not be neglected. In April 1990, one field bindweed seed was planted in a small planter, and on April 25, the two-leaf seedling was transplanted to a 2 × 4 × 16–ft box. The plant was harvested on October 19 by opening the box and washing all of the soil away with water. The seedling had colonized the entire box. Vertical roots (197), each about 4 ft long, grew a total of 788 ft. Horizontal root runners from the taproot (34) averaged 4 ft long and were 136 ft long. They had produced 141 new plants. The creeping roots of field bindweed can grow up to 1.5 yards in a little more than 3 months (Frazier, 1943). One little seed produced a major new weed.[5]

A similar experiment was conducted in Colorado[6] with Canada thistle. One seed was planted in a 2 × 4 × 8–ft box of soil in April 1994. In July 1995, the plant was harvested. If the height of all 142 shoots was added, the plant would have been 157 ft tall. There were 331 flowers on 60 shoots. Vegetative buds producing new shoots were found up to 4 ft below the soil surface. Total root length was estimated to be 1700 ft. Canada thistle roots have been reported to spread up to 5 yards in a single season (Bakker, 1960).

Tillage can worsen the problem after plants become perennial. Canada thistle spreads by creeping roots and pieces as small as 0.25 inches have produced new plants. Field bindweed spreads by creeping roots, and although they seldom emerge from greater than 4 ft, they can emerge from 20 ft. Pieces as small as 1 inch that contain a bud can produce a new plant. Most quack grass plants, developed from rhizomes, emerge from the top 12 inches of soil. Therefore, deep plowing may be a control method if rhizomes can be permanently buried. Permanent burial below 12 inches is highly unlikely because plowing is rarely that deep and mixing, not burial, occurs. However, most quack grass roots are 2–4 inches below the surface and shoots do not emerge from deep in soil, so it is possible. The ability of root segments to produce new plants varies with the season and is highest in spring and lowest in fall (Swan and Chancellor, 1976). Many root segments produced shoots, but regeneration of roots was largely from vertical roots.

Leafy spurge roots penetrate up to 20 ft. Over 56% of the total root weight is in the upper 6 inches of the soil profile and most leafy spurge shoots originate from buds in the top foot of soil. Shoots emerge freely from 1.5 ft deep and some emerge from as deep as 6 ft.

Vegetative buds are not killed by winter freezes. Studies in Iowa on the winter activity of Canada thistle roots showed that buds on horizontal roots continued to develop new shoots until soil was frozen 50 cm deep (Rogers, 1929). When the soil finally froze, the shoots were killed but the root bud was not. In January, when soil was still frozen, the latent buds on large roots were larger than they had been in December. By mid-January, these buds had developed thick, vigorous shoots up to 20 mm long. By February, shoots were 4–7 cm long and each had roots 10–20 cm long. When the soil thawed, root growth increased rapidly and green shoots appeared by mid-April. Rogers noted that the cycle of bud and root formation in field bindweed and skeleton leaf bur sage was similar to that described for Canada thistle.

[5]Adapted from May 1991. Agrichemical Age, May 1991, p. 16.
[6]Westra, P., 1995. Colorado State University. Personal communication.

THINGS TO THINK ABOUT

1. What is a reasonable range of the number of weed seeds likely to be found in the plow layer of a cropped field?
2. Describe the influence or lack thereof of seed dormancy on weed management.
3. How many different ways are weeds dispersed in space?
4. What are the causes of seed dormancy?
5. What are the classes of seed dormancy? How can these classes be used?
6. What are the vegetative reproductive organs weeds possess and why do they make weeds hard to control?
7. All conditions favorable for seed germination occur when soil is disturbed/plowed. Should no-tillage be recommended for all soil preparation? Why? Why not?

Literature Cited

Altom, J.V., Murray, D.S., 1996. Factors affecting eclipta (*Eclipta prostrata*) seed germination. Weed Technol. 727–731.

Anderson, R.N., 1968. Germination and Establishment of Weeds for Experimental Purposes. Weed Sci. Soc. of America, Champaign, IL, 236 pp.

Atkeson, F.W., Holbert, H.W., Warren, T.R., 1934. Effect of bovine digestion and of manure storage on the viability of weed seeds. J. Am. Soc. Agron. 26, 390–397.

Bakker, D., 1960. A comparative life-history study of *Cirsium arvense* (L.) Scop. and *Tussilago farfara* L. The most troublesome weeds in the newly reclaimed polders of the former Zuiderzee. In: Harper, J.L. (Ed.), The Biology of Weeds. Blackwell Scientific Publications, Oxford, UK, pp. 205–222.

Ball, D.A., 1987. Influence of Tillage and Herbicides on Row Crop Weed Species Composition (Ph.D. dissertation). U. Wyoming, Laramie, 157 pp.

Ball, D.A., 1992. Weed seedbank response to tillage, herbicides, and crop rotation sequences. Weed Sci. 40, 654–659.

Barrett, S.C.H., 1983. Crop mimicry in weeds. Econ. Bot. 37, 255–282.

Barrett, S.C.H., Wilson, B.F., 1981. Colonizing ability in the *Echinochloa crus-galli* complex (barnyardgrass). I. Variation in life history. Can. J. Bot. 59, 1844–1860.

Bazzaz, F.L., 1979. The physiological ecology of plant succession. Ann. Rev. Ecol. Syst. 10, 351–371.

Beach, C.L., 1909. Viability of weed seeds in feeding stuffs. Vt. Agr. Expt. Stn. Bull. 138, 11–30.

Benvenuti, S., Denelli, G., Bonetti, A., 2004. Germination ecology of (*Leptochloa chinensis*): a new weed in the Italian agro-environment. Weed Res. 44, 87–96.

Blackshaw, R.E., Rode, L.M., 1991. Effect of ensiling and rumen digestion by cattle on seed viability. Weed Sci. 39, 104–108.

Bond, W., Baker, P.J., 1990. Patterns of weed emergence following soil cultivation and its implications for weed control in vegetable crops. In: Monograph 45. Organic and Low Input Agric. Brit. Crop Prot. Council, Oxford, UK, pp. 63–68.

Boyd, N., Van Acker, R., 2004. Seed germination of common weed species as affected by oxygen concentration, light, and osmotic potential. Weed Sci. 52, 589–596.

Bruns, V.F., Rasmussen, L.W., 1953. The effects of fresh water storage on the germination of certain weed seeds. I. White top, Russian knapweed, Canada thistle, Morning glory and Poverty weed. Weeds 2, 138–147.

Bruns, V.F., Rasmussen, L.W., 1957. The effects of fresh water storage on the germination of certain weed seeds. II. White top, Russian knapweed, Canada thistle, Morning glory and Poverty weed. Weeds 5, 20–24.

Bruns, V.F., Rasmussen, L.W., 1958. The effect of fresh water storage on the germination of certain weed seeds. III. Quackgrass, Green bristlegrass, Yellow bristlegrass, Watergrass, Pigweed, and Halogeton. Weeds 6, 42–48.

Buhler, D.D., Kohler, K.A., Thompson, R.L., 2001. Weed seed bank dynamics during a five-year crop rotation. Weed Technol. 15, 170–176.

Burnside, O.C., Wilson, R.G., Weisberg, S., Hubbard, K.G., 1996. Seed longevity of 41 weed species buried 17 years in eastern and western Nebraska. Weed Sci. 44, 74–86.

Burton, G.W., Andrews, J.S., 1948. Recovery and viability of seeds of certain southern grasses and lespedeza passed through the bovine digestive tract. J. Agric. Res. 76, 95–103.

Cardina, J., Hook, J.E., 1989. Factors influencing germination and emergence of Florida beggarweed (*Desmodium tortuosum*). Weed Technol. 3, 402–407.

Chancellor, R.J., 1985. Tillage effects of annual weed germination. In: Proc. World Soybean Res. Conf. III, vol. 3, pp. 1105–1111.

Chauhan, B.S., Johnson, D.E., 2007. Effect of light, burial depth and osmotic potential on germination and emergence of *Celosia argentea* L. Indian J. Weed Sci. 39, 151–154.

Chauhan, B.S., Johnson, D.E., 2008a. Germination biology of *Portulaca oleracea*. In: Proceedings 16th Australian Weeds Conference. Queensland Weeds Society, Cairns, Queensland, Australia, pp. 183–185.

Chauhan, B.S., Johnson, D.E., 2008b. Germination ecology of Chinese sprangletop *(Leptochloa chinensis)* in the Philippines. Weed Sci. 56, 820–825.

Chauhan, B.S., Johnson, D.E., 2008c. Germination ecology of southern crabgrass (*Digitaria ciliaris*) and India crabgrass (*Digitaria longiflora*): two I and important weeds of rice in the tropics. Weed Sci. 56, 722–728.

Chauhan, B.S., Johnson, D.E., 2008d. Germination ecology is to troublesome Asteraceae species of rainfed rice: siam weed (*Chromolaena odorata*) and coat buttons (*Tridax procumbens*). Weed Sci. 56, 567–573.

Chauhan, B.S., Johnson, D.E., 2008e. Influence of environmental factors on seed germination and seedling emergence of the eclipta (*Eclipta prostrata*) in a tropical environment. Weed Sci. 56, 383–388.

Chauhan, B.S., Johnson, D.E., 2009a. Ecological studies on *Cyperus difformis*, *C. iria* and *Fimbristylis miliacea*: three troublesome annual sedge weeds of rice. Ann. Appl. Biol. 155, 103–112.

Chauhan, B.S., Johnson, D.E., 2009b. *Ludwigia hyssopifolia* emergence and growth as affected by light, burial depth and water management. Crop Prot. 28, 887–890.

Chauhan, B.S., Johnson, D.E., 2009c. Seed germination and seedling emergence of Synedrella in a tropical environment. Weed Sci. 57, 36–42.

Chauhan, B.S., Johnson, D.E., 2009d. Seed germination ecology of Junglerice (*Echinochloa colona*): a major weed of rice. Weed Sci. 57, 235–240.

Chauhan, B.S., Johnson, D.E., 2010. The role of seed ecology in improving weed management strategies in the tropics. Adv. Agron. 105, 221–262.

Chepil, W.S., 1946. Germination of weed seeds. I. Longevity, periodicity of germination and vitality of weed seeds in cultivated soil. Sci. Agric. 8, 307–346.

Clements, D.R., Benoit, D.L., Murphy, S.D., Swanton, C.J., 1996. Tillage effects on weed seed return and seedbank composition. Weed Sci. 44, 314–322.

Comes, R.D., Bruns, V.F., Kelley, A.D., 1978. Longevity of certain weeds and crop seeds in fresh water. Weed Sci. 26, 336–344.

Conn, J.S., Farris, M.L., 1987. Seed viability and dormancy of 17 weed species after 21 months in Alaska. Weed Sci. 35, 524–529.

Conn, J.S., 1990. Seed viability and dormancy of 17 weed species after burial for 4.7 years in Alaska. Weed Sci. 38, 134–138.

Culotta, E., 1994. Meeting briefs: the weed that swallowed the west. Science 265, 1178–1179.

Currie, R.S., Peeper, T.F., 1988. Combine harvesting affects weed seed germination. Weed Technol. 2, 499–504.

Darlington, H.T., 1951. The seventy-year period for Dr. Beal's seed viability experiment. Am. J. Bot. 38, 379.

Dastgheib, F., 1989. Relative importance of crop seed, manure, and irrigation water as sources of infestation. Weed Res. 29, 113–116.

Davis, A.S., Anderson, K.I., Hallet, S.G., Renner, K.A., 2006. Weed seed mortality in soils with contrasting agricultural management histories. Weed Sci. 54, 291–297.

Dewey, S.A., Whitesides, R.E., 1990. Weed seed analyses from four decades of Utah small-grain drillbox surveys. Proc. West. Soc. Weed Sci. 43, 69–70.

Donald, W.W., Nalewaja, J.D., 1991. Northern great plains. In: Donald, W.W. (Ed.), Systems of Weed Control in Wheat in North America. Monograph #6. Weed Sci. Soc. Am., Champaign, IL, pp. 90–126.

Dore, W.G., Raymond, L.C., 1942. Viable seeds in pasture soil and manure. Sci. Agric. 23, 69–79.

Dunham, R.S., 1973. The Weed Story. Inst. of Agric. Univ. of Minnesota, 86 pp.

Dyer, W.E., 1995. Exploiting weed seed dormancy and germination requirements through agronomic practices. Weed Sci. 43, 498–503.

Egley, G.H., Chandler, J.M., 1983. Longevity of weed seeds after 5.5 years in the Stoneville 50-year buried seed study. Weed Sci. 31, 264–270.

Egley, G.H., Williams, R.D., 1990. Decline of weed seeds and seedling emergence over five years as affected by soil disturbance. Weed Sci. 38, 504–510.

Faulkner, E.H., 1943. Plowman's Folly and a Second Look. Island Press, Washington, DC, 161 pp.

Fleet, B., Gill, G., 2012. Seed dormancy and seedling recruitment and smooth barley (*Hordeum murinum* spp. *glaucum*) populations in southern Australia. Weed Sci. 60, 394–400.

Fogg Jr., J.M., 1942. The silent travelers. Brooklyn Bot. Garden Rec. 31, 12–15.

Fogg Jr., J.M., 1966. The silent travelers. Plants Gardens 22, 4–7.

Forcella, F., 1997. My view. Weed Sci. 45, 327.

Forcella, F., Wilson, R.G., Renner, K.A., Dekker, J., Harvey, R.G., Alm, D.A., Buhler, D.D., Cardina, J., 1992. Weed seedbanks of the U.S. corn belt: magnitude, variation, emergence and application. Weed Sci. 40, 636–644.

Forcella, F., Wilson, R.G., Dekker, J., Kremer, R.J., Cardina, J., Anderson, R.L., Alm, D., Renner, K.A., Harvey, R.G., Clay, S., Buhler, D.D., 1997. Weed seed bank emergence across the corn belt, 1991–1994. Weed Sci. 45, 67–76.

Franzenburg, D.D., Owen, M.D.K., 2002. Effect of location and burial depth on woolly cupgrass (*Eriochloa villosa*) seed germination. Weed Technol. 16, 719–723.

Frazier, J.C., 1943. Nature and rate of development of the root system of *Convolvulus arvensis*. Bot. Gaz. 104, 417–425.

Gardarin, A., Colbach, N., 2014. How much of seed dormancy in weeds can be related to seed traits? Weed Res. 55, 14–25.

Gardner, C.J., McIvor, J.G., Jansen, A., 1983. Survival of seeds in the digestive tract and feces of cattle. In: Annual Rep. 1982–1983. CSIRO Div. of Trop. Pastures, Brisbane, Australia, pp. 120–121.

Gould, F., 1991. The evolutionary potential of crop pests. Am. Sci. 79, 496–507.

Gunsolus, J.L., 1990. Mechanical and cultural weed control in corn and soybeans. Am. J. Altern. Agric. 5, 114–119.

Harmon, G.W., Keim, F.D., 1934. The percentage and viability of weed seeds recovered in the feces of farm animals and their longevity when buried in manure. J. Am. Soc. Agron. 26, 762–767.

Harper, J.L., 1957. The ecological significance of dormancy and its importance in weed control. In: Proc. 7th Int. Conf. Plant Protection, Hamburg, pp. 415–420.

Hill, T.A., 1977. The Biology of Weeds. E. Arnold, London.

Holm, L.G., Plucknett, D.L., Pancho, J.V., Herberger, J.P., 1977. The World's Worst Weeds: Distribution and Biology. Univ. Press of Hawaii, Honolulu, pp. 32–40.

Johnson III, W.C., Mullinix Jr., B.G., 1995. Weed management in peanut using stale seedbed techniques. Weed Sci. 43, 293–297.

Karssen, C., 1982. Seasonal patterns of dormancy in weed seeds. In: Kahn, A.A. (Ed.), The Physiology and Biochemistry of Seed Development, Dormancy and Germination. Elsevier, NY, pp. 243–270.

Kinter, C.L., Mealor, B.K., Shaw, N.L., Hild, A.L., 2007. Postfire Invasion Potential of Rush Skeletonweed (*Chondrilla juncea*).

Klingman, G.C., Ashton, F.M., 1982. Weed Science: Principles and Practices, second ed. Wiley Interscience, NY.

Koch, W., 1969. Influence of Environmental Factors on the Seed Phase of Annual Weeds, Particularly from the Point of View of Weed Control. Habilitations-schrift Landw. Hochech. Univ. Hohenheim, Arbeiten der Univ. Hohenheim 50, p. 20.

Kremer, R.J., 1993. Management of weed seed banks with microorganisms. Ecol. Appl. 3, 42–52.

Laroche, F.B., Ferriter, A.P., 1992. Estimating expansion rates of melaleuca in south Florida. J. Aquat. Plant Manag. 30, 62–65.

Laroche, F.B., 1994. Melaleuca Management Plan for Florida. Exotic Pest Plant Council, 88 pp.

Leakey, R.R.B., 1981. Adaptive biology of vegetatively regenerating weeds. Adv. Appl. Biol. 6, 57–90.

Leistritz, F.L., Bangsund, D.A., Hodur, N.M., 2004. Assessing the economic impact of invasive weeds: the case of leafy spurge (*Euphorbia esula*). Weed Technol. 18, 1392–1395.

Lewis, J., 1973. Longevity of crop and weed seeds: survival after 20 years in soil. Weed Res. 13, 179–191.

McGrath, S., 2005. Attack of the alien invaders. Nat. Geogr. 207 (3), 92–117.

Mt. Pleasant, J., Schlather, K.J., 1994. Incidence of weed seed in cow (*Bos* sp.) manure and its importance as a weed source for cropland. Weed Technol. 8, 304–310.

Norris, R.F., 1992. Case history for weed competition/population ecology: barnyardgrass (*Echinochloa crus-galli*) in sugarbeets (*Beta vulgaris*). Weed Technol. 6, 220–227.

Oswald, E.J., 1908. The effect of animal digestion and fermentation of manures on the vitality of seeds. MD. Agr. Expt. Stn. Bull. 128, 265–291.

Pons, T.L., 1991. Induction of dark dormancy in seeds: its importance for the seed bank in the soil. Funct. Ecol. 5, 669–675.

Posner, J.L., Casler, M.D., Baldock, J.O., 1995. The Wisconsin integrated cropping systems trial: combining agroecology with production agronomy. Am. J. Altern. Agric. 10, 98–107.

Proctor, V.W., 1968. Long distance dispersal of seeds by retention in the digestive tract of birds. Science 160, 321—322.

Ramirez, A.H.M., Jhala, A.J., Singh, M., 2012. Germination and emergence characteristics of common beggars-tick (*Bidens alba*). Weed Sci. 60, 374—378.

Randall, J.M., Martinelli, J. (Eds.), 1996. Invasive Plants: Weeds of the Global Garden. Brooklyn Botanic Garden Handbook No. 149, 111 pp.

Reeves, T.G., Code, G.R., Piggin, C.M., 1981. Seed production and longevity, seasonal emergence, and phenology of wild radish (*Raphanus raphanistrum* L.). Aust. J. Exp. Agric. Anim. Husb. 21, 524—530.

Rice, K.J., 1985. Responses of *Erodium* to varying microsites: the role of germination cueing. Ecology 66, 1651—1657.

Rice, K.J., 1987. Interaction of disturbance patch size and herbivory in *Erodium* colonization. Ecology 68, 1113—1115.

Rice, K.J., 1990. Reproductive hierarchies in *Erodium*: effects of variation in plant density and rainfall distribution. Ecology 71, 1316—1322.

Ridley, H.N., 1930. The Dispersal of Plants Throughout the World. L. Reeve & Co., Ashford, Kent, UK, pp. 360—368.

Roberts, H.A. (Ed.), 1982. Weed Control Handbook: Principles, seventh ed. Brit. Crop Protection Council, pp. 32—33.

Robinson, H.J., 1949. Annual weeds, their viable seed population in soil and their effects on yield of oats, wheat and flax. Agron. J. 41, 515—518.

Rogers, C.F., 1929. Winter activity of the roots of perennial weeds. Science 69, 299—300.

Salisbury, E.J., 1961. Weeds and aliens. N.N. Collins, St. James Place, London, 384 pp.

Schmitz, D.C., March 13, 1995. Diversity Disappears in Florida. Newsweek, p. 14.

Schutte, B.J., Regnier, E.E., Harrison, S.K., 2012. Seed dormancy and adaptive seedling emergence timing in giant ragweed (*Ambrosia trifida*). Weed Sci. 60, 19—26.

Schweizer, E.E., Zimdahl, R.L., 1984a. Weed seed decline in irrigated soil after rotation of crops and herbicides. Weed Sci. 32, 84—89.

Schweizer, E.E., Zimdahl, R.L., 1984b. Weed seed decline in irrigated soil after six years of continuous corn (*Zea mays*) and herbicides. Weed Sci. 32, 76—83.

Sheley, R.L., Petroff, J.K., 1999. Biology and Management of Noxious Rangeland Weeds. Oregon State Univ. Press, Corvallis, OR, 438 pp.

Singh, M., Ramirez, A.H.M., Sharma, S.D., Jhala, A.J., 2012. Factors affecting the germination of tall morning glory (*Ipomoea purpurea*). Weed Sci. 60, 64—68.

Smith, E.V., Mayton, E.L., 1938. Nut grass eradication studies: II. The eradication of nut grass, *Cyperus Rotundus* L. by certain tillage treatments. J. Am. Soc. Agron. 30, 18—21.

Sonoskie, L.M., Herms, C.P., Cardina, J., 2006. Weed seedbank community composition in a 35-yr-old tillage and rotation experiment. Weed Sci. 54, 263—273.

Stevens, O.A., 1932. The number and weight of seeds produced by weeds. Am. J. Bot. 19, 784—794.

Stevens, O.A., 1957. Weights of seeds and numbers per plant. Weeds 5, 46—55.

Swan, D.G., Chancellor, R.J., 1976. Regenerative capacity of field bindweed roots. Weed Sci. 24, 306—308.

Taylorson, R.B., 1970. Changes in dormancy and viability of weed seed in soils. Weed Sci. 18, 265—269.

Telewski, F.W., Zeevaart, J.A.D., 2002. The 120-yr period for Dr. Beal's seed viability experiment. Am. J. Bot. 89 (8), 1285—1288.

Thill, D.C., Zamora, D.L., Kambitsch, D.L., 1986. The germination and viability of excreted common crupina (*Crupina vulgaris*) achenes. Weed Sci. 34, 237—241.

Tildesley, W.T., 1937. A study of some ingredients found in ensilage juice and its effect on the vitality of certain weed seeds. Sci. Agric. 17, 492—501.

Toole, E.H., Brown, E., 1946. Final results of the Duvel buried seed experiment. J. Agric. Res. 72, 201—210.

Vega, M.R., Sierra, J.N., 1970. Population of weed seeds in a lowland rice field. Philipp. Agric. 54, 1—7.

Warnes, D.D., Anderson, R.N., 1984. Decline of wild mustard (*Brassica kaber*) seeds in soil under various cultural and chemical practices. Weed Sci. 32, 214—217.

Wilson, R., 1980. Dissemination of weed seeds by surface irrigation water in Western Nebraska. Weed Sci. 28, 87—92.

Young, J.A., March 1991. Tumbleweed. Sci. Am. 82—87.

Zamora, D.L., Olivare, J.P., 1994. The viability of seeds in feed pellets. Weed Technol. 8, 148—153.

Zorner, P.S., Zimdahl, R.L., Schweizer, E.E., 1984. Effect of depth and duration of seed burial on kochia (*Kochia scoparia*). Weed Sci. 32, 602—607.

6

Weed Ecology

FUNDAMENTAL CONCEPTS

- An objective of weed ecology is to describe the adaptive mechanisms that enable weeds to do well under conditions of maximum soil disturbance.

- There is a strong human influence on the ecological relationships of weeds.

- There are good reasons for the shift toward ecologically based weed management systems.

- Ecological theory is assuming a more prominent role in the development of weed management systems.

- Species are products of natural selection that interact with their environment to obtain the resources for growth. The rate of supply and amount of resources determine growth.

- Plant competition occurs when two or more plants seek what they need and the immediate supply is below combined demand.

- Some plants possess characteristics that make them more competitive than others.

- The effect of weeds on crop yield is best described by regression analysis that yields a straight-line relationship for lower densities but a curvilinear relationship over all possible densities.

- Mathematical models are used in research studies of weed management, but are not used to manage weeds. Models may be used more in the future as they are perfected and tested against biological knowledge.

LEARNING OBJECTIVES

- To understand the importance of ecological relationships to weed management systems.

- To know the components of the weed–crop ecosystem.

- To understand weed–environment interactions.

- To know the factors affecting weed–crop associations.

- To understand the role of fundamental ecological concepts in weed management.

- To be able to define plant competition and to know for which resources plants compete.

123

- To understand the characteristics that make a plant competitive.
- To appreciate the magnitude of crop yield loss from weeds.

- To understand the current and potential role of mathematical models in crop—weed interference research and weed management recommendations.

American scientific interest from the 18th to the 20th century was dominantly focused not on theory but on the immense practical benefit to be derived from discovering the secrets of the natural world. *Jacoby (2004)*

Plant ecologists study the reciprocal arrangements between plants and the environment. The goal is to understand how climate, soil (edaphic), and biotic factors affect plant growth, development, and distribution. The primary concern of weed scientists has been learning how weed management affects weed and crop growth and development. Agriculture, a complex system, was thereby reduced to its individual parts that could be controlled. The dynamic, interdependent relationship of the system that evolves in largely unpredictable ways was ignored (Kirschenmann, 2013). For many years, ecologists emphasized only natural environmental factors in studies of reciprocal arrangements, plant distribution, and behavior. Ecologists and weed scientists now realize the importance of the role people have in agriculture's complex ecological interactions. The human role is particularly evident with weeds. Integration of ecology and weed science is increasing to the benefit of both disciplines. Without disregarding the importance of weed control, weed scientists are gradually moving toward studying weed ecology rather than simply reading, understanding, and recommending what is on an herbicide's label to accomplish weed control. In Wes Jackson's[1] words, "The descriptive world of the ecologist will meet the prescriptive world of the agriculturalist."

Ecology is the study of the interactions between individuals and their environment. Weed ecology differs only in that the organisms being studied are weeds (Booth et al., 2003). It gives special emphasis to the adaptive mechanisms that enable weeds to survive and prosper under conditions of maximum soil disturbance. Weed ecologists study the growth and adaptations that enable weeds to exploit niches in environments disturbed by people who must practice agriculture. The most successful weed management programs will be developed on a foundation of adequate ecological understanding. The questions asked by weed ecologists are those posed by Booth et al. (p. 11):

1. Are there specific characteristics or traits of weed populations?
2. Do weeds function in a certain way within communities?
3. Does weed invasion change plant community structure or function in a predictable way?
4. What types of communities are easier to invade?

High food production from annual crops requires repression of normal ecological succession. Production of food for humans from natural vegetation is presumed to be low. Fiber

[1]Jackson, W., February 2016. Personal communication. February 2106.

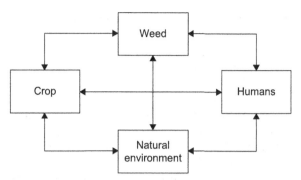

FIGURE 6.1   The weed–crop ecosystem (Aldrich, 1984).

and food crops have high food production potential, but the fields in which they are grown are unavoidably disturbed sites. Farming must often work against, rather than with, the natural order to produce high-value food and fiber crops useful to humans. Perennial crops (coconuts and apples) may create a permanent preclimax state and, although more ecologically stable, will revert to the left side of the curve shown in Fig. 6.1 if not managed to maximize production of what humans want.

Rangelands, forests, and other areas of native vegetation are relatively closed habitats that may resist but are not immune to invasion (see Chapter 8). Most agricultural weeds are not good invaders of natural sites and may not be weedy, in the ecological sense, on those sites. The dominantly monocultural cropping systems of developed world agriculture seldom use all of the moisture, nutrients, or light available in a given field and thus create open ecological niches that weeds occupy. There are fewer open niches, although there may be unused resources, in undisturbed prairies and forests.

1. HUMAN INFLUENCES ON WEED ECOLOGY

Those who grow crops try to provide conditions in which only the desired crop exists and limit the incidence and spread of weeds. The task is complicated because people have carried weed seed across the globe while traveling, in grain, in seed shipments, with armies, and when moving animals. As discussed in Chapter 5 many weeds of temperate Europe and North America migrated west from the early centers of civilization in the eastern Mediterranean region to Europe and hence to the United States with immigrants from Europe to the new world. That is a major reason many (not all) of the dominant US weeds can be traced to Europe rather than Asia. A few examples of people's major, although usually unwitting influence on weed presence follow.

When the first settlers came to the US Great Plains and some of the Pacific Northwest, they found bluebunch wheatgrass and blackseed needlegrass dominating the sides of wagon trails. When roads were cut, downy brome invaded and dominated roadsides. When chemical weed control became available, roadsides were sprayed with the triazine herbicide simazine to control downy brome. Simazine worked well, and because it persists in soil, it prevented reinvasion by downy brome and other grasses. However, it does not control sandbur, which became the dominant species on many western roadsides (Muzik, 1970).

Downy brome had arrived in the interior Pacific Northwest United States by 1889 (Mack, 1981). It was deliberately introduced at least once in the search for new grasses for over-grazed, denuded range. By 1928, it reached its current distribution (Mack) although not its current density or ubiquity on over 100 million western US acres (Devine, 1993). It may be the nation's most destructive plant. Downy brome has more than half of the 12 features Baker (1974) considered characteristic of the ideal weed (Mack). It thrived in two human-created ecosystems of the intermountain west: winter wheat and rangeland. It persisted as land was converted from one use to the other.

Russian thistle was introduced in grain imported from Russia to a farm in Bonhomme county, South Dakota about 1877 (Dewey, 1894). By the 1890s, the infestation was so extensive in North Dakota that the value of wheat production lost exceeded the taxes collected by the state (Young, 1991). Agriculture created the conditions for the weed's success and humans aided its dispersal. Without agriculture, it would have survived but remained innocuous (Young, 1988). The pioneer farmer's practice of destroying tall and midheight prairie grass to plant cereal grains created the right ecological niche to ensure Russian thistle's success. Russian thistle, which is similar to many annual weeds, competes poorly with established plants. It cannot tolerate shade or long periods of high moisture (Young). However, it thrived on dry, disturbed cereal grain land that often was fertilized.

Examination of early weed science literature reveals that the dominant weed problem in many different crops, and especially in wheat and barley in some regions, was annual broad-leaved weeds. There is also evidence of a gradual shift from annual broad-leaved weeds to annual grass weeds and broad-leaved species. A possible cause of the weed population shift could be development and widespread use of selective phenoxy acid herbicides (e.g., 2-methyl-4-chlorophenoxyacetic acid [2,4-D]). However, one also finds that wild oats have been a problem in wheat and barley for a millennium. Downy brome has been an important weed in the US Pacific Northwest since its introduction in the 1800s, well before the introduction of 2,4-D. It and jointed goat grass are adapted to the low fall and winter rainfall of the region. Their success was also encouraged when in the early 1900s, growers switched from spring to winter wheat. These and similar induced ecological changes in weed populations, although not intentional, are inevitable. Changes in herbicide identity and selectivity, weed introduction, and shifts in cultural patterns may individually or collectively be the cause. A similar change has been seen in corn, for which the widespread use of triazine herbicides eliminated many annual broad-leaved and grass weeds and created an ecological niche for invasion by annual and perennial grass weeds and, especially in the southern United States, yellow nutsedge.

Green revolution cultivars created during the 1960s helped feed the world and avoid starvation in many of the world's developing countries. The new cultivars changed the architecture (the shape) of wheat and rice plants and led to higher grain yields when appropriate production technology (fertilizer and irrigation) was provided. Green revolution cultivars also changed the harvest index (proportion of the total plant harvested as grain). Snaydon (1984) posits that the main factor leading to an increase in harvest index over several years was selection for shorter straw length, which had important effects on weeds. Short stiffer straw (stems) was less likely to lodge with higher rates of nitrogen fertilizer. This same characteristic also opened the plant canopy to more light and thereby changed the light environment for weeds. This change in plant architecture had the potential to worsen the weed problem by providing more light to stimulate seed germination and growth of seedling weeds. In fact, although weed problems may not have become worse, they changed as weeds new to an area invaded the changed habitat.

Agricultural practice has nearly always changed for reasons of convenience, to save labor, or to increase profit. It has rarely changed to reduce weed growth intentionally. The opposite has happened. Indian balsam, a Himalayan wet woodland native, is a showy plant 2 m tall, with pink flowers. It found the European climate favorable and escaped but did not become a common weed in Britain until about 1930, when use of artificial fertilizer became widespread. Until then, in attempts to maximize yield, hay was harvested right up to riverbanks where Indian balsam flourished. When fertilizer was used, yields increased and harvest of river banks was no longer necessary. The weed then flourished along previously harvested riverbanks. It has been reported that Indian balsam is invading riverbanks in the Czech Republic and moving to wet woodlands, where it crowds out the forest's less aggressive species (Anonymous, 1995). It is of concern along water courses in British Columbia.

A shift from annual to perennial weeds attributed to the extensive use of herbicides that controlled annual weeds has been documented in Japanese rice culture.[2]

US agriculture has shifted from a mix of crops on a farm to extensive monoculture. Wheat is the dominant crop in the central Great Plains of the United States and soybeans and corn dominate midwestern states. Cotton and soybeans dominate in the southern areas of the United States. Monocultural environments create ecological change that determines what weeds will succeed.

Conscious introduction, multiplication, and release of parasites and predators for biological control of pests is also ecological change. To date, this is a less important shift in ecological relationships than those mentioned previously, but weed managers need to be aware that such changes are possible.

Each agricultural practice has a potential to influence the density and survival of all species in a cropped field. The foregoing is a few examples of how human activity influences weeds. Production practices that influence weeds are shown in Table 6.1 with an estimate of the relative importance of each to species composition and weed density.

Ghersa et al. (1994) state that "in modern agriculture, social and biological systems have diverged" in their influence. Weed management practices are increasingly uncoupled from biology. They are controlled and designed by social and economic forces that are often devoid of a biological base. Management decisions have been reduced to selecting the best herbicide for the crop. It is an example of use of the precautionary principle: "When in doubt about the interaction between the weed in the crop one should control the weed.[3]" This represents the ultimate human influence on weed ecology because it is neglect of ecology. *Homo sapiens* is the only species that can and does change the environment rather than being required to adapt to it. Our technological prowess allows us to neglect ecology, perhaps in the long run, to our detriment.

[2]Itoh, K. Mat. Agric. Res. Center, Isukuba, Ibaraki, Japan, personal communication.

[3]I thank Prof. B. Maxwell of Montana State University for this insight.

TABLE 6.1   Components of Production Systems Controlled by Humans That Are Relevant to Weed Management

	Influence on[a]	
Component	Species Composition	Density
Soil tillage	9	9
Water: irrigation	9	5
Nutrient supply: fertilization	9	7
pH: liming	9	5
Date of planting	7	7
Growing period of crop	6	3
Shading period and intensity	6	8
Seed dispersal at harvest	3	5
Seed cleaning before planting	4	2
Weed control	9	9

[a]*Influence is ranked on a scale of 0 to 10, with 0 indicating no influence and 10 equaling maximum influence. Koch, W., 1988. Personal communication.*

2. THE WEED–CROP ECOSYSTEM

Herbicides have enabled neglect of the importance of weed prevention and the need to understand weed–crop ecology. Understanding weed–crop ecology will lead to more effective weed prevention, management, and control. The gradual shift toward ecologically based weed management systems is occurring for at least six reasons:

1. Weeds that have been susceptible to available herbicides have been replaced by species that are more difficult to control.
2. Herbicide resistance has developed in many weed species and for several herbicides. Multiple resistance to herbicides from chemical families with different sites of action is increasingly common.
3. There are weed problems in monocultural and organic (alternative) agriculture that cannot be solved easily with current management techniques.
4. New weed problems have appeared in reduced and minimum tillage systems.
5. Economic factors have forced consideration of alternative control methods.
6. There is increased awareness of the environmental costs of herbicides.

Aldrich (1984, p. 17) diagramed the weed–crop ecosystem (Fig. 6.1). For too long, weed scientists have focused primarily on weed–crop interactions and on protecting crops from weeds. Aldrich strongly suggests that weed management must deal with interaction of all factors rather than just two. There is a lack of knowledge about these interactions. It is not the intent of this book to discuss all ecological interactions in depth; other available books do that well (Aldrich, 1984; Aldrich and Kremer, 1997; Booth et al., 2003;

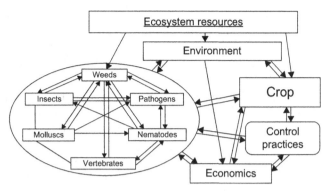

FIGURE 6.2   The interaction of weeds and other components of the agricultural production system (Norris, 1992).

Cousens and Mortimer, 1995; Harper, 1977; Radosevich and Holt, 1984; Radosevich et al., 1997, 2007). It is the intent of this book to introduce the fundamental concepts essential to development of improved weed management systems. Consulting one or more of the eight books just cited will permit those who want to pursue weed ecology in depth to do so.

Weed science has been dominated by control technology that focused on how to control (usually kill) weeds in a crop. As new weed management systems are developed, ecological knowledge will be essential and the complexity shown in Fig. 6.2 must be considered. As complex as Fig. 6.2 is, it is too simple to represent all factors that affect weed–crop relationships that should be considered as management systems are developed.

From genes to organisms (individuals), populations, and communities, relationships are the essence of life. The weed–crop system is a product of interactions; its essence is relational, as illustrated in Figs. 6.1 and 6.2. All levels of life are interdependent and no level can exist independent of another. The individual cannot survive for long independent of its population, nor can a population survive without individuals. Weed management systems directed only at weeds are founded on error, and whereas they may succeed temporarily, they are doomed to fail. They are immediately successful but not sustainable.

3. ENVIRONMENTAL INTERACTIONS

There are three important weed–environment interactions: climate, soil, and biota or living organisms. These will be discussed separately but they cannot be separated in nature, which is characterized by interactions; indeed, it depends on them.

3.1 Weeds and Climate

Factors that determine a weed's ecological interactions are light, temperature, water, wind, humidity, and their seasonal aspects: the climate. Yellow nutsedge does well in the subhumid tropics and warm, temperate regions. It does not survive well in temperate areas with prolonged frost. Purple nutsedge thrives in the humid tropics and subtropics with some excursions into subhumid temperate regions. Halogeton thrives under desert conditions of low rainfall and sometimes high alkalinity. Water hyacinth, an important aquatic weed in the tropics and subtropics, has not yet invaded temperate waters. It is sold as an ornamental to homeowners in north temperate United States but has not yet escaped and survived a winter. It was introduced to North America in 1883 and is present in warm humid areas of Asia, Australia, Africa, and New Zealand. It is found in large bodies of water in Louisiana and in Kerala, an Indian state.

Weeds are found in the environment they prefer, and weed control or weed management often may be aided by changing the environment. Irrigation and tillage are major environmental changes that lead to shifts in species composition in the affected areas. Changes in tillage practices employed for weed management can affect populations of other crop pests and, of course, weeds (Norris, 2005).

Light intensity, quality, and duration affect weed presence and survival. Photoperiodic responses govern flowering and determine the time of seed maturation. If light is too intense or days too long or short, particular plants will not flower and a species may not endure. Light and temperature response determine a species' latitudinal limits. Some weeds tolerate shade well and their ability to grow under a crop canopy is one reason they succeed. The length of the frost-free period or the time soil is frozen determine any plant's ability to survive in an environment; they determine its ecological relationships. Soil temperature is a primary determinant of seed germination and survival, especially where soil freezes. Freezing also affects winter survival of vegetative reproductive organs. Air and soil temperature are important determinants of species distribution and ecological interactions. Common chickweed survives well in cold climates because it continues to grow in winter without injury (King, 1966). When temperature is below freezing, common chickweed is often erect and continues to flower although the flowers are cleistogamous (without petals and closed). The self-pollinated seeds are fertile. Although one recognizes these relationships, humans are essentially powerless to affect them.

Kariba weed is sensitive to salinity and grows only in fresh water. The government of Kuttanad, Kerala, India wanted to develop rice production. Kuttanad is close to the sea and salt water decreases or eliminates rice production. Traditionally farmers built soil barriers across canals and rivers to prevent incursion of heavier salt water during the growing season. After harvest, the barriers were removed to allow incursion of sea water. Experts in the government planning office "knew" the farmers' practice was not going to be adequate for extensive rice production. Therefore, they built a spillway channel into the open sea and salinity was regulated with a 1400-m-long regulator channel that checked the advance of salt water. The channel and regulator worked well and the advance of salt water was halted. Invasion of kariba weed was encouraged because salt water no longer invaded the land annually and killed the weed. Kariba weed stopped rice production because engineers were good at building structures to stop the sea but knew nothing about weed ecology. The weed won.

Seasonal distribution and total supply of water determine species survival. Shortage of water at critical stages is often responsible for reproductive failure, death, or both. The world's arid areas would produce far less food if seasonal distribution and total supply of water were not modified by irrigation.

Wind can affect water supply through evaporation and an increase in the loss of transpiration. Wind also affects the microclimate within a plant canopy and the relative concentration of carbon dioxide and oxygen.

Climate will change because of increasing concentration of CO_2 and other triatomic gases that interact with radiant energy. Abundant scientific data show that the world is warming and these changes will affect weeds. It is not the purpose of this book to present the data or enter the debate. The 65-year temperature trend for Northern Colorado illustrates the change, but the implications for weed management are not clear. The data show that for every month from 1950 to 2015, average temperature has increased in Northern Colorado.[4] Similar data are available for other regions of the United States. Agriculture has always been aided and hindered by climate. Crops are vulnerable to unfavorable weather and weed management may be more difficult during rapid climate change (Patterson, 1995b). It is likely that the negative effects of all agricultural pests will increase with rapid climate change, particularly in less intensively managed production systems. Crops negatively affected (there may be positive effects on some crops) by environmental (global warming) stress will be more vulnerable to attack by insects and diseases and less competitive with weeds (Patterson).

3.2 Edaphic Factors

"Edaphic" comes from the Greek *edaphos*, meaning soil or ground. Water, aeration, temperature, pH, fertility, fertility source, the cropping system, and associated practices imposed on a soil determine what weeds survive to compete. Many weeds do well in soils too low in fertility for crop production, but others grow only in well-fertilized soil. Few weed species associate with a soil type. Most weeds can be found in soils differing widely in physical characteristics, moisture content, and pH. This adaptability explains, in part, why they are successful weeds. Some species of Asteraceae and Polygonaceae grow in soils with 1.2%—1.5% sodium chloride, and although this may not make them better weeds, it illustrates their ability to adapt to diverse environments. Kochia grows well in alkaline or saline soils but not in acidic soil. Salt grass can be a weed in turf in alkaline areas where soil pH is 8 or above. Alkali weed grows only at pH 8 and above. Other species, including common mallow and plantain, are relatively intolerant of alkaline conditions. Crabgrass, a turf weed, grows well on acid soil. Kentucky bluegrass is sensitive to acidity, and common chickweed is not common in acid soil (King, 1966).

Soil pH is an important determinant of what plants grow in an area. However, no generalizations can be made about the influence of pH on weeds. LeFevre (1956) reviewed the pH tolerance of 60 weeds and grouped them into basophile (love high pH, e.g., sow thistles, green sorrel, quack grass, and dandelion), acidophile (love acid soil, e.g., red sorrel,

[4]Data from National Climate Data Center (National Oceanic and Atmospheric Administration—NOAA)—Station 53,005. Monthly mean temperature for Northern Colorado. Personal communication Sean Poling.

corn marigold), and neutrophile (e.g., shepherd's-purse, prostrate knotweed, and common chickweed). Some nutrition is essential for plant growth but most weed species are valueless as indicators of soil reaction or fertility. Luxuriant weed growth does not indicate a potentially highly productive agricultural soil. Weed growth is determined by many factors in addition to a soil's physical and chemical properties. These include field cropping history, proximity of sources of infestation, the weed seed population present or supplied to a field, water supply, and growing season conditions. The effects of soil structure, water-holding capacity, and nutrient level are more important than soil type.

Flooding will control some weeds. Some water is required for seed germination and plant growth. Too much water changes soil ecology and can control some weeds, as it has in rice for centuries. Several species are adapted to flooding. Paddy rice is not free of weeds. No crop plants and few weeds do well in waterlogged or compact soil with poor aeration.

Warm, moist soil conditions are best for germination of weed seed and seedling growth. Seed dormancy in temperate regions is usually associated with cold, freezing conditions.

3.3 Weeds and Biota

Association of weeds and crops is determined largely by the degree of competition offered by a particular crop and weed. It is also determined by cultural operations and rotational practices associated with each crop. The factors contributing to association are discussed next.

3.3.1 Similarity of Seed Size

If a weed's seed is similar in size to a crop's seed, it can be a common, unnoticed companion when planting. It will also be more difficult to clean or separate the weed's seed from the crop's seed (see Chapter 5). Weeds have a long record of adapting to agricultural practice. A striking example of seed size mimicry is that of lentil seeds by common vetch (Gould, 1991). Lentil seed is lens shaped and vetch seed is usually more rounded. In Europe, vetch seed evolved to mimic the shape of lentil seed and made separation nearly impossible.

3.3.2 Time of Seed Germination and Formation

If a weed's seed germinates just before or within a few days after a crop's seed, the weed's chances of successful competition are enhanced. If weeds flower and set seed before the crop is harvested, that ensures its presence in the next crop. These things do not guarantee successful competition but they do not deter it. A weed whose life cycle is similar to a crop will usually be a more successful competitor than one whose life cycle is much shorter or longer than the crop with which it attempts to associate.

3.3.3 Tillage, Rotation, and Harvest Practices

Dandelions are common weeds in turf, as are several species of spurge and common chickweed. They are turf weeds because they are adapted to turf's cultural practices and withstand mowing. The perennial quack grass grows well in the perennial crop alfalfa. The association of wild oats and green foxtail with small grain crops and wild proso millet with corn is related to all of these factors. They have a similar seed size, their times of germination and ripening are nearly identical, and they easily withstand the tillage and harvest practices of the crop. It is unlikely that a weed adapted to survive in plowed fields will do equally well in no- or minimum-till fields (Gould, 1991).

Some weeds germinate after a crop is laid by (after the last tillage has occurred). These include johnsongrass, some of the foxtails, and barnyard grass. They often germinate later than broad-leaved species and elongate rapidly to compete in row crops. Downy brome competes effectively in winter wheat because it germinates in the fall after wheat has been planted, survives over the winter, and develops and sheds seed in early spring, before wheat has completed its life cycle. It is also an effective competitor during the fallow year in a wheat–fallow rotation because it is shallow-rooted, not affected by many cultural operations, and competes for water before the crop has been planted.

Other plants and animals modify the environment; grazing animals determine weed survival in pastures. Knowledge of crop competition and the relationship of weeds and biota is required to develop better control techniques and management strategies.

Plant environments, especially cropped fields, are heterogeneous. A height difference of only 5 cm between the top of a furrow and the bottom may represent a factor of 250 for the smallest weed seeds (Aldrich, 1984). When fields are irrigated, fertilized, or cultivated, we perceive uniformity, but across a large or even a small area, weed seeds experience a nonuniform environment. Nonuniform (random) seed distribution in soil is the rule, not the exception. We also know that management techniques, including herbicides, are not applied uniformly.

There are significant differences in soil temperatures determined by small amounts of litter cover or shading of soil. There are even greater influences on soil moisture and relative humidity of air just above the soil surface, determined by litter and shading. These small environmental differences explain why several different plant species occupy a single environment.

Differences in growth form are often unobserved ecological interactions. These external expressions of a plant's ability to sample its environment are illustrated in Figs. 6.3–6.8.

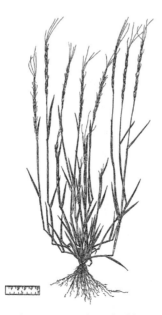

FIGURE 6.3 The upright, narrow, unbranched leaves of jointed goat grass.

FIGURE 6.4   The upright, branching, broad leaves of jimsonweed.

FIGURE 6.5   The climbing, twining growth of field bindweed.

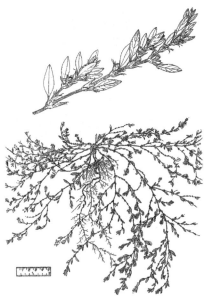

FIGURE 6.6   The prostrate growth of prostrate knotweed.

FIGURE 6.7   The taproot of redroot pigweed.

FIGURE 6.8   The fibrous roots of.downy brome.

The differences enable plants to occupy different ecological niches. Weeds create and occupy ecological niches and change the environment through their germination, growth, and death. They affect moisture, temperature, nutrient supply, and ultimately organic matter in soil. Weeds are active, not passive, participants in the agricultural environment.

4. FUNDAMENTAL ECOLOGICAL CONCEPTS

4.1 Species

Species is the fundamental biological classification. It is a subdivision of genus and each is composed of a number of individuals with a high degree of physical similarity that can generally interbreed only among themselves and show persistent differences from other species. The species *retroflexus* (redroot pigweed) of the genus *Amaranthus* is consistently different from the species *spinosus* (spiny amaranth). They do not interbreed. Species are products of natural selection and genetic manipulation that creates new gene pools. That is *what* happens, but the more important and more interesting question is *why* does it happen? Organisms are controlled in nature by the total quantity and variability of the supply of things essential for growth. All plants have a minimum requirement for various growth factors and interact with their physical and chemical environment to obtain them. Plants also have limited tolerance to various environmental components. The *why* question is usually answered in terms of rate and amount. Plant presence and growth are controlled by too little or too much of the things needed for growth and by the conditions under which they are available.

Weeds have been continually exposed to conditions that encourage speciation. The major models of speciation are allopatry and sympatry. In allopatric speciation, parent species become physically separated into daughter populations by geographic separation that restricts or eliminates gene flow between populations. This occurs because of continual movement of people and plants from continent to continent or to different regions within a continent. When weeds are introduced between continents, species development is a long-term process. Allopatric speciation is the primary mechanism for development of new species. Darwin's Galapagos Island finches are an example of allopatric speciation. Although many weeds may have originated from allopatric speciation, there are no good examples. Weeds have been imported into many places (see Chapters 5 and 8) but there has been little to no study of allopatric speciation of weeds. We assume it has happened.

In sympatric speciation, a parent species differentiates in the absence of physical restriction on gene flow. Sympatric speciation is a local, short-term process. The continual disturbance of land and changing agricultural practices provide numerous opportunities for hybridization, selection, and responses to imposed and shifting environmental conditions. Species development has not stopped and it is logical to posit that although the genus, species, and common name are the same ones that have been used for years, the weed we see can be different from what it was several years ago. Difference may be expressed in one or more of several ways. For example:

- physiological and genetic difference(s) may be revealed functionally only under certain environmental conditions,
- ecological differences are revealed by observing how it interacts with its environment, such as its drought tolerance,
- or it could be morphologically different, as shown by acclimation to mowing by growing more prostrate owing to repeated mowing.

The proposition that weeds will continue to evolve is not debatable. The plethora of studies on the development of herbicide resistance is adequate evidence that weeds evolve and do so quickly.

There are fewer examples of weeds that have evolved as a result of sympatric speciation. One is species of the genus *Passiflora* or passionflower (Harper, 1977). There are about 350 species. In 1989, the Weed Science Society of America (Patterson, 1989) listed three as weeds. In 2010, there were six species; in 2016, seven were included.[5] Nearly every species is unique. Leaf shape varies enormously, as do leaf surface characteristics. Some can be distinguished by the feeding habits of the monophagous butterflies of the genus *Heliconius*.

The ready development of ecotypes, or physiological races adapted to various climatic conditions around the world, has occurred in common chickweed and is responsible for its worldwide distribution (Fig. 6.9).

Ecotypes exist in dandelion and in members of many other genera. Their development has implications for weed management. Control techniques that work in one place may not work for the same weed in another place because, although it has the same scientific name, it is not the same weed; it is an ecotype. Ecotype development is sympatric speciation as locally adapted populations are changed. Aldrich (1984) summarized

[5]http://wssa.net//weed/composite-list-of-weeds/.

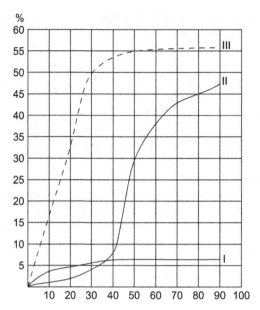

FIGURE 6.9   Average germination obtained over 100 days for seed from three ecotypes of common chickweed: I, arctic, alpine latitudes; II, oceanic latitudes; and III, maritime regions in northern latitudes (Peterson, 1936; cited in King, 1966).

examples of the development of ecotypes for a wide variety of weed species including johnsongrass (Burt, 1974; McWhorter, 1971; McWhorter and Jordan, 1976; Wedderspoon and Burt, 1974), Canada thistle (Hodgson, 1964), common ragweed (Dickerson and Sweet, 1971), yellow nutsedge (Yip, 1978), purslane (Gorske et al., 1979), annual bluegrass (Warwick and Briggs, 1978), and medusahead (Young et al., 1970).

Weeds and other plants have two survival strategies called K and r that define a population's response to disturbance. Both terms are derived from the logistic growth equation where K = environmental carrying capacity and r = the intrinsic rate of population increase. The r reproductive strategy is characterized by production of a large number of seeds (or vegetative reproductive units) and high dispersability. It is the potential rate of increase of a population for a given set of environmental conditions where there is no shortage of resources and no constraints on growth. The r strategy dominates among annual weeds and is expressed in competitive ability, seed germination, seed dormancy, and seed longevity.

Plants adopting the K reproductive strategy depend on exploitation. They produce fewer reproductive units and have relatively low dispersability and strong exploitive ability. K measures an upper size limit beyond which a population cannot go. The limit is determined by available resources and other constraints on population growth. Large-seeded annual weeds (some authors consider sunflower to be a good example) and many perennial weeds generally use K reproductive strategy. Plants with K strategy are usually not first colonizers.

Some species combine K and r strategies. Canada thistle, for example, is r for vegetative growth because it produces a large number of vegetative buds and its creeping roots disperse plants. At the same time, it is K for seed production. It is a dioecious plant and usually produces few seeds that have high dispersability but are not strongly exploitive. It is important to note that survival strategy as depicted by K and r reproduction is not equivalent to, and should not be confused with, competitive ability, which is controlled by other factors.

Undisturbed plant communities generally have a large number of a few species and a few individuals of many different species. Undisturbed communities are more complex than disturbed communities. Farmers in developed countries want crop fields to be dominated by a single species. They plant and disturb (plow, cultivate, control weeds, etc.) fields to achieve that goal. Crop dominance is favored by weed control. Weed management systems that rely on single control techniques stabilize weed populations, one hopes at a low population level, and encourage emergence of weeds that are not affected by the control technique.

4.2 The Community

The crop—weed community is important to weed management because it is the organizational level where change occurs. Change can occur within a species through mutation and ecotype development or by replacing one species with another. There are at least four reasons (Harper, 1977) why two or more species coexist. They can have:

- different nutritional requirements. For example, legumes and grass coexist in pasture and hay fields;
- different causes of mortality: for example, pastures where animals selectively graze;
- different sensitivities to environmental toxins (allelochemicals) (see Chapter 9) and human-applied toxins (herbicides); and
- a different time demand for growth factors. Many plants require the same things to grow but do not demand them at the same time. This may be the most common reason for coexistence.

Plant communities are assembled in logical, predictable ways that can be studied. Community assembly is a branch of ecology that studies how plant communities are assembled over time. Booth and Swanton (2002) propose that assembly theory should be applied because it has the potential to change the approach to weed management. The current approach emphasizes control of a species or a series of co-occurring weeds (a population) in a crop. This approach has been successful given that it is designed to manage weeds that are already present in a place. The broader approach advocated by Booth and Swanton asks why weeds occur where they do and how they interact in communities. The population approach inevitably leads to instability because as the current population is controlled successfully, as sure as night follows day, another weed or weed complex will appear. A community assembly approach leads to understanding and balancing the crop—weed community rather than destroying one weed community so another can arise. It demands knowledge of environmental (climatic extremes such as flood or drought, and variation) and weed dispersal (identity of arriving species, arrival sequence, and rate and frequency of invasion) constraints that control the species that can enter an ecological pool. Booth and Swanton cite the work of Derksen et al. (1993, 1994, 1995) and several others, not included here, to illustrate the effect of what Booth and Swanton call filters (biotic and abiotic constraints) on the trajectory of community development. Derksen et al. found that no-till, minimum-till, and conventional-tillage systems had different weed communities, and that the weed associations varied among years and sites. Tillage filtered community composition but herbicides acted as an additional, strong filter that directed composition back to the original pretillage condition. Booth and Swanton used the work of Derksen et al. to illustrate how "the multivariate approach of community analysis may bring out patterns not evident when each weed species is analyzed separately." In 2016, Swanton[6] suggested that corn and soybeans plants sense (know?) when a competitor is growing nearby. They respond via chemical changes that stunt their root systems and delay growth.

Storkey et al. (2010) used assembly theory to explain changes in weed flora in the United Kingdom, which resulted from intensification of crop management over more than 60 years. Several annual species that had been common in arable habitats have declined, whereas others continue to be present. Species that combined short stature and large seeds were shown to have "relatively greater competitive ability with low compared to high fertility." As nitrogen inputs increased, the abundance of groups that contained only common species was stable or grew, whereas groups dominated by rare or endangered species declined as fertility increased. This is a clear illustration of the benefit of community assembly as a way to understand the effects of agricultural practice on weed management.

"Understanding abundance and distribution of weed species within an agroecosystem is an important goal." Prevailing weed management practices intentionally limit plant diversity. Weeds that survive management practices are ecologically adapted and inevitably become more difficult to control. When herbicides are the dominant management practice, resistance often follows. Achieving a field free of weeds, the primary goal of weed control, has made cropped fields more vulnerable to invasion (Noka et al., 2015).

4.3 Ecological Succession

Ecological succession is a natural, continuous process. In agriculture, it occurs in continually disturbed areas from which the natural community has been removed. Agricultural ecosystems have a desired, predetermined structure and function. Management success is based on crop yield (Booth and Swanton, 2002). Environmental modification is a driving force for succession. Modern, conventional agriculture is conducted by modifying and controlling the environment via fertility, irrigation, and community structure. As mentioned previously, the ability of humans to dominate the environment rather than adapt to it stands in sharp contrast to the ability of all other species, which must adapt to their environment. There are few natural ecosystems in which species composition is independent of human involvement. Bronowski (1973, p. 19) reminds us that we are not "just figures in the landscape; we are shapers of the landscape." It is common to

[6]Http://wssa.net/2016/08/what-plants-sense-and-say-may-impact-the-future-of-weed-control/.

misinterpret Genesis 1:28: "Be fruitful and multiply, and fill the earth, and subdue it; and rule over the fish of the sea and over the birds of the sky and over every living thing that moves on the earth." We have multiplied and filled, but if we are to survive we must not dominate and subdue; we must care and protect that which we have been given. Agriculture, the essential human activity, has unavoidable environmental consequences, some of which are negative.

Creating conditions in which the crop dominates is the sine qua non of weed management in agricultural crop communities. It is often accompanied by subdominance of a few weeds (rarely only one) in cropped fields. Weed removal, their control, creates open niches and different species will move in, but usually not immediately. Therefore, weed control, especially successful control, is often a never-ending process. Must it be so? Weed management may be designed best when it achieves less than 100% control and thus is not as successful at opening niches and creating an endless process of succession. Its focus will be community management over time rather than controlling or eliminating individuals in the current population. The best weed management systems may combine techniques to gain the desired level of control but, if possible, not a completely open environment that encourages arrival of new weeds that are not controlled by current techniques, and thus which may be more difficult to control.

4.4 Interactions Between Weeds and Other Crop Pests

A fundamental although not scientifically based rule of ecology is that in the natural world it is impossible to do just one thing. Any action creates other actions and reactions. In the natural world actions interact. The rules of ecology that describe the interactions and the interconnection in the natural world can be expressed in simple, nonscientific words that make them easy to understand. The rules are (Commoner, 1971, pp. 33–46):

- Everything is connected to everything else. Humans and other species are connected and dependent on a number of other species.
- Everything must go somewhere. No matter what one does, and no matter what is used, everything must go somewhere.
- Nature knows best.
- There is no such thing as a free lunch. Every action must have a reason, a justification. This means you have to do something and know why to gain something.

Weed management and weed scientists are obliged to know and follow these simple rules. To ignore them is to court disaster and weed management failure.

Weeds live in communities. Unfortunately, they frequently choose to live in communities of the crops we want to grow or the landscape we carefully create and admire. On the other hand, perhaps it is fortunate they do. If weeds did not grow where we do not want them, weed scientists and many others would cease to have a reason to exist. Weeds are compelled to interact with crops and other pest organisms. In ecological terms, weeds are producer organisms whereas other pest organisms are consumers (Norris, 2005). Norris and Kogan (2000) reviewed the many interactions among weeds, arthropod pests, and natural enemies in managed (agricultural) ecosystems. Their review considered three mechanisms for interactions. The first is direct ecosystem energy/resource flow (trophic) interactions that occur when pests or beneficial arthropods feed directly on weeds, which may lead to allelopathic interactions. The second is alteration of the physical habitat by the presence of weeds (e.g., temperature within a plant canopy, water consumption). The final mechanism is driven by the control tactics employed to manage weeds and other pests (e.g., tillage, herbicides, other pesticides).

In addition to their productive activity, weeds support beneficial and harmful organisms. Altieri (1994, p. 195) identified more than 70 families of arthropod pests known to be potential crop pests that are primarily associated with weeds. Table 1 in Norris and Kogan's (2000) article identified more than 94 insect pests that attack 45 different crops via resource- and habitat-driven interactions, each of which is facilitated by or dependent on the presence of weeds.

A few examples, each documented by Norris and Kogan (2000), illustrate the interactions. Buffalo bur is a native host of the Colorado potato beetle (*Leptinotarsa decemlineata*). The weed's presence and that of other members of the Solanaceae in or adjacent to a potato crop can worsen damage from the potato beetle. The Russian wheat aphid (*Diuraphis noxia*), a major pest of wheat in arid areas, uses jointed goat grass and downy brome as alternate hosts. The insect can live in summer when wheat is not present; thus, the weeds enhance insect damage even when they may not be present in the crop. The tobacco budworm (*Heliothis virescens*) and the cotton bollworm (*Helicoverpa armigera* [the corn earworm]) live on several weeds whose uncontrolled presence worsens the insect problem in the crop. Similarly, several grass weeds that remain uncontrolled serve as alternate hosts for the European corn borer (*Ostrinia nubilalis*), thus potentially increasing populations that become major problems in field and sweet corn. The opposite of this situation is that weed control may worsen an insect problem by eliminating the plants on which the insects have been living and compelling migration to the crop. It is always good to remember that in the natural world, one cannot do just one thing; all things are related.

In their Table 2, Norris and Kogan (2000) also identify more than 52 beneficial insects whose resource- and habitat-driven influences provide benefits to crops. Many of these are insect predators or parasites that live on weeds. Schroeder et al. (2005) suggest three levels of interaction among polyphagous (eat many things) crop pests, including insects, pathogens, and nematodes and weeds. They acknowledge that except for work on biological control of weeds, the literature is limited on the effects of pests on weeds and the effect of pests harbored by weeds on crops. Pests hosted by susceptible weeds may have severe negative effects on insect growth and fecundity. These are of limited concern because they serve to control the weed, which then does not compete for resources. Tolerant weeds host crop pests without severe effects on an insect's growth and fecundity, but possibly with important effects on susceptible crops. This results in a larger pest population and effective crop competition. Finally, there are resistant weeds that do not host pests but compete effectively with crops. Schroeder et al. suggest that weed communities in most crops are dominated by weeds that are tolerant of or resistant to the onslaught of polyphagous pests because of constant evolutionary pressure from the pests. This suggests, in contrast to Norris (2005) and Norris and Kogan, that because reduction of existing weed populations is a dominant goal in crop management, manipulation of weed populations to benefit management of other pests is perhaps a faint hope. However, Capinera (2005) identifies the dynamic interactions between insects and weeds and the importance of weeds as a resource for insects. Weeds that are closely related to crops are especially important as reservoirs for insects that attack the crop. Wisler and Norris (2005) show the same relationships for plant pathogens and weeds.

When understood, weed populations can be manipulated to alter weed—insect interactions to benefit crops. Norris and Kogan (2000) note that the potential benefits of weed management to manage arthropod pests is much greater in perennial than in annual crops. That is because populations of weeds and insects can be observed over time and managed, which is much more difficult in annual crops that are in the field for only a few months.

5. PLANT COMPETITION

Plant competition is part of plant ecology. To compete comes from the Latin *competere*, which means to ask or sue for the same thing another does. Two reviews (Zimdahl, 1980, 2004) of the literature on weed—crop competition provide a more complete review and discussion of the topic than can or should be presented here. Three conclusions of the second review (2004) are important. First, it affirms the central hypothesis of weed science: weeds compete with crops and reduce crop yield and quality. The second major conclusion is that weed management will benefit from closer integration with plant ecology and a greater emphasis on the study and understanding of the coexistence of plants rather than continued major emphasis on weed control. In other words, that weed scientists have to change their primary questions. The questions have been "What is the identity of the problem weed and how can it be controlled?" These are good and important questions but the new questions, which are ecologically based, should be: "What is the identity of the problem weed and why is the weed where it is?" (Zimdahl, 1999). Then the management or control question logically follows. The right questions are systemic, holistic ones that accept the transformation of nature as a necessary prerequisite to food production but reject the attempt to dominate nature permanently (Zimdahl). The final conclusion of the review is that modeling (see Section XI) has become an important aspect of modern weed management systems and is likely to become more important in future weed management systems.

Imagine that you have the good fortune to receive free tickets to your favorite football team's next home game. Your football tickets are a little way up on the 50-yard line or, if you are very fortunate, in someone's private skybox. You know you will see vigorous competition as the two teams charge up and down the field competing for the ball, for scores, for glory, and perhaps, if it is a professional football game, even for money.

The next time you drive around in the spring or summer, a careful look at most agricultural fields will reveal competition just as vigorous, but not as obvious, as what you see at the football game. You will not see the plants leaping up and running around and into each other, but they will be competing vigorously and silently for environmental resources. There is no glory or financial reward for the plants. But the competition is real: it is for survival and life.

Regulating or eliminating competition between crops and weeds is why weeds are controlled. If weeds were just there and benign, we would not care as much about them. Because they cause harm to crops by competing with them, we are compelled to care and attempt to control or manage them. Many also dislike them because they are ugly in a lawn or garden. They mar the landscape.

As mentioned in Chapter 1, some of the earliest and most frequently quoted references to weeds are in the Bible:

> Cursed is the ground for thy sake; in sorrow shalt thou eat of it all the days of thy life; thorns and thistles shall it bring forth to thee; and thou shalt eat the herb of the field. *Genesis III:17–18*

> And some fell among the thorns and the thorns sprang up and choked them. *Matthew XIII:7*

The Reverend T.R. Malthus, in his 1798 essay on the principle of population, said, "The cause to which I allude is the constant tendency in all animated life to increase beyond the nourishment prepared for it." Malthus' concern was the increasing human population and consequent poverty and misery he saw in Liverpool, England. The Malthusian apocalypse (when the human population is greater than the ability of the earth to produce food) has been avoided because of the ability of the world's farmers, aided by agricultural science and technology, to increase food production. Malthus (died 1834) has been dead for nearly 200 years, but the Malthusian apocalyptic possibility, especially in the world's developing countries, still concerns many.

5.1 Plant Competition Defined

Clements et al. (1929) said that competition is a question of the reaction of a plant to the physical factors that encompass it and the effect of these upon adjacent plants. For them, competition was a purely physical process. "In the exact sense, two plants, no matter how close, do not compete with each other so long as the water content, the nutrient material, the light and heat are in excess of the needs of both." However, Swanton's research (see Footnote 5) establishes that plants can sense danger: the imminence of competition.

"Competition occurs when each of two or more organisms seek the measure they want of any particular factor or things and when the immediate supply of the factor or things is below the combined demand of the organisms" (Clements et al.). In agriculture, competition is not regarded as simply interaction without any effect on either individual. The concern is that competition in agricultural communities has negative results. The subject is discussed well by Booth et al. (2003). Clements' et al. (1929) definition makes competition different from the broader term *interference*, which includes competition and allelopathy (see Chapter 9). The dictionary defines competition as being for something in limited supply or between agents, as in a rivalry. For physiologists, competition is usually for things. For agronomists and weed scientists, competition is often for things and among individuals (Donald, 1963).

5.2 Factors Controlling the Degree of Competition

Fig. 6.10 illustrates the factors determining the degree of competition encountered by an individual plant. For weeds, density, distribution, and duration or how long weeds are present are important. For crops, density, distribution (including spacing between rows

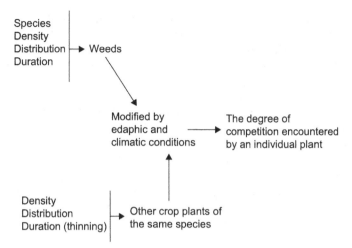

FIGURE 6.10 Schematic diagram of the competition encountered by a plant (Bleasdale, 1960).

and spacing in the row), and duration (weeding and thinning) are important. These factors, modified by soil (edaphic) and climatic conditions, determine the degree of competition encountered by each plant. The primary things foe which plants compete are nutrients, light, and water. When any one is lessened, others cannot be used as effectively. Plants may compete for heat, but it is difficult to conceptualize how they do so. However, it is well-known that accumulation of degree days enhances plant growth:

$$\text{Degree day} = \text{Daily max.} - \text{daily min. temp.} - \text{threshold temperature}/2$$

The threshold temperature differs among species. It is the temperature below which a plant does not grow. Because they do not grow well at high temperatures, there is a maximum cutoff temperature, in the range of 30°C, for many plants. Plants grow better when it is warm, but no studies have reported competition for heat, perhaps because it is not a resource that exists in a finite reservoir.

Yield reductions are generally in proportion to the amount of light, water, or nutrients that weeds use at the expense of a crop. A general rule is for every unit of weeds grown, there will be one less unit of crop grown. Inconsistent results between weed management experiments in 1 year or between years are regularly attributed to environmental (i.e., light, water, nutrient, or climatic) variation. In most cases, the data are insufficient to define cause and effect.

It is simple and neat to separate the elements of competition (nutrients, light, and water). H.L. Mencken (1880–1956) reminded us that "For every human problem, there is a solution that is simple, neat, and wrong." It is not wrong to separate the elements of competition experimentally, but it is wrong to assume that plants do so, and it is nearly impossible to separate the elements of competition in nature. deWit (1960) was among the first to point out the futility of separating the elements of competition. His work changed the approach to the study of competition. He derived mathematical expressions for competition and advocated consideration of space and what it contained rather than studies that separated the components of competition. For example, competition for light affects growth, which in turn affects a plant's ability to compete for nutrients and water. Competition will be greatest among similar species that demand the same things at the same time from the environment. Those species that best use (grow rapidly) or first capture environmental resources will succeed.

Only in recent years has research progressed to consider the spatial distribution or where weeds are in a field. Weed scientists have long been concerned with what weeds (what species) and how many (their density) are present in a field. Control has been directed toward the dominant weed or weeds. Studies of weed biology have emphasized seed production, seed dormancy and survival, and seedling growth, establishment, and survival. Results of these good studies have been translated into areas (acres or hectares) without considering the patchiness or nonuniformity of weeds in all fields. Control included the usually unstated and frequently incorrect assumption that weed distribution and density were uniform over the field. Thus, tillage for weed control and herbicides are nearly always applied uniformly over a field even though most farmers and weed scientists know that weeds are not distributed uniformly (Fig. 6.11). Because weed distribution in a field is not uniform, control practices are unnecessary in some places. Weed distribution is heterogeneous, not homogeneous (see Chapter 22 on the importance of mapping weed populations). The technology is developing for weed and crop recognition systems that permit control of weeds when they are present in parts of fields rather than whole fields. Biological knowledge is not as developed to define how the seed bank, seed dispersal, plant demography, and habitat interact to determine the stability of weed or weed seed distribution across fields and across time (Cousens and Mortimer, 1995). There is also a poor understanding of how control techniques affect weed and weed seed distribution over time. As this knowledge develops, weed managers will be able to manage weeds on less than a whole-field basis, which will lead to a reduced need for tillage and herbicides (Mortensen et al., 1998; Johnson et al., 1995). The dynamics of patches, which is defined as how inherent weed biology interacts spatially with landscape characteristics (Cousens and Mortimer), is an important area of weed management research. Weed

FIGURE 6.11   A large patch of field bindweed. *Courtesy of P. Westra, Colorado State University.*

scientists want to understand why weeds are where they are, rather than know only what species are present, and use spatial information as another tool to manage weed populations.

5.3 Competition for Nutrients

Nitrogen, phosphorus, and potassium are primary plant nutrients. One mustard plant needs twice as much nitrogen and phosphorus, four times as much potassium, and four times as much water as an oat plant. Success in gaining nutrients leads to more rapid growth and successful competition for light and water. Used to improve crop growth, fertilization may worsen the weed problem.

Table 6.2 shows the pounds of nutrients required to produce equal amounts of dry matter for three crops and five weeds that frequently compete with crops. The important point about these data is not that weeds require greater amounts of nitrogen and phosphorus than crops. Consumption of nitrogen and phosphorus by weeds and crops is similar. The point is that weeds require the same nutrients at the same time, and because of early emergence they are often more successful in obtaining them. Competition occurs when two or more organisms seek what they want or need and the supply falls below the combined demand.

Table 6.3 compares the nutrient content of weed-free corn, corn-free redroot pigweed, and corn grown with redroot pigweed (Vengris et al., 1955). When weed-free corn was set at a nutrient content of 100, in all cases except phosphorus, redroot pigweed grown alone contained more of each of the nutrients compared with corn. The more interesting data are those in the center row, which shows the nutrient content of corn infested with redroot pigweed. In every case, nutrient content was reduced. In another study (Vengris et al., 1953), corn was compared with six annual broad-leaved weeds and one annual grass (Table 6.4). Weeds contained 1.6—7.6 times more of each nutrient. In this study, application

TABLE 6.2   Kilograms of Nutrients Required to Produce Equal Amounts of Dry Matter

Plant	Nitrogen	Phosphorus
Wheat	5.5	1.2
Oats	4.9	1.7
Barley	8.4	2.6
Common lamb's-quarter	7.6	1.6
Common ragweed	6.6	1.4
Redroot pigweed	5.1	1.4
Common purslane	3.1	0.8
Mustards	9.8	2.7

TABLE 6.3 Comparison of Nutrient Content of Weed-Free Corn, Corn, and Redroot
Pigweed and Redroot Pigweed Alone (Vengris et al., 1955)

Species	Relative Nutrient Content				
	N	P₂O₅	K₂O	Ca	Mg
Weed-free corn			100		
Corn infested with redroot pigweed	58	63	46	67	77
Redroot pigweed	102	80	124	275	234

TABLE 6.4 Mineral Composition of Corn and Weeds (Vengris et al., 1953)

Species	Mean Percent Composition				
	N	P	K	Ca	Mg
Common lamb's-quarter	2.6	0.4	4.3	1.5	0.5
Common purslane	2.4	0.3	7.3	1.5	0.6
Corn	1.2	0.2	1.2	0.2	0.2
Crabgrasses	2.0	0.4	3.5	0.3	0.5
Galinsoga	2.7	0.3	4.8	2.4	0.5
Pigweeds	2.6	0.4	3.9	1.6	0.4
Ragweeds	2.4	0.3	3.1	1.4	0.3
Smartweeds	1.8	0.3	2.8	0.9	0.6

of supplemental phosphorus made several weeds more competitive. High fertility did not
reduce the detrimental effects of weeds on corn.

A similar study in Poland compared wheat, barley, sugar beets, and rape (Malicki and
Berbeciowa, 1986). Table 6.5 shows that the mineral content of most weeds is higher than
that of wheat or barley. The authors proposed that common lamb's-quarter, Canada thistle,
field bindweed, wild buckwheat, perennial sow thistle, and common chickweed are
dangerous in wheat because of their high nutrient requirement. Rapeseed (canola) responded
in the same way as the grain crops. The percentage of nutrients in roots and leaves of sugar
beets was high and few weeds exceeded it. This results in the high nutrient concentration in
the large sugar beet root.

In a crop heavily infested with weeds, it seems logical that more fertilizer should reduce
nutrient competition. If competition does not occur until the immediate supply falls below
combined demand, when supply increases, competition should decrease. Actually, although
this seems logical, it is wrong. Fertilizer usually stimulates weed growth to the crop's detriment. With low fertility, competition is primarily for nutrients; however, with high fertility,
competition is just as vigorous, and primarily for light. Yields in unweeded, fertilized plots

TABLE 6.5 Nitrogen, Phosphorus, and Potassium Content of Wheat and Barley (Grain, Straw, and Roots) and Selected Annual Weeds (Malicki and Berbeciowa, 1986)

Species	Percent Dry Matter		
	N	P	K
Barley	1.5	0.2	1.4
Canada thistle	1.6	0.3	2.0
Common chickweed	2.1	0.6	3.8
Common hemp nettle	2.0	0.4	2.3
Common lamb's-quarter	2.7	0.4	4.1
Corn speedwell	1.5	0.3	1.7
Field bindweed	2.7	0.3	2.7
Hairy vetch	3.0	0.2	1.1
Perennial sow thistle	2.3	0.5	4.0
Shepherd's-purse	1.6	0.3	2.0
Wheat	1.2	0.3	0.8
Wild buckwheat	2.7	0.4	2.5

are usually equal to those in weeded, unfertilized plots. Table 6.6 shows that increasing nitrogen reduced flax yield and tended to increase wild oat density and number of seed-bearing stems (Sexsmith and Pittman, 1963). The opposite situation is more common: nitrogen raises crop yield and then, when in excess, crop yield decreases (Table 6.7) (Okafor and DeDatta, 1976).

TABLE 6.6 Effect of Form and Timing of Nitrogen Fertilizer on Wild Oats and Flax[a] (Sexsmith and Pittman, 1963)

Fertilizer	Wild Oats Density (No./m²)	Seed-Bearing Stems (No./m²)	Flax Yield (kg/ha)
None	96 a	124 a	7.0 a
Ammonium nitrate, April 12 (early)	215 ab	254 a	4.2 ab
Ammonium sulfate, April 12 (early)	435 bc	444 b	2.4 bc
Ammonium nitrate, June 1 (seeding)	476 c	530 b	1.9 c

ha, hectare.
[a]*Means in a column followed by different letters are significantly different at P = .05.*

TABLE 6.7   Effect of Nitrogen Fertilizer on Rice Yield and Purple Nutsedge Competition (Okafor and DeDatta, 1976)

Nitrogen (kg/ha)	Purple Nutsedge (No./m²)	Rice Yield (t/ha)
0	0	1.6
0	750	1.2
60	0	4.4
60	750	2.8
120	0	4.0
120	750	2.4

ha, hectare.

Table 6.8 shows similar data on the competition of barnyard grass and barnyard grass plus the annual broad-leaved weed *Monochoria* in rice. It is apparent that increasing nitrogen fertilizer increased rice yield. It is also apparent that if barnyard grass were the only competitor, increasing nitrogen fertilizer from 0 to 60 kg/hectare (ha) would decrease the yield. Only after a further doubling of nitrogen did yield increase. Even then, yield was lower than the same amount of fertilizer with no weeds. With both weeds, neither level of nitrogen fertilizer increased yield and both yielded less than the check plot with no fertilizer and no weeds. These data are confirmed by those in Table 6.9, which show nitrogen uptake of rice and barnyard grass in two trials in Australia (Boerema, 1963).

TABLE 6.8   Weed Competition for Nitrogen in Rice (Moody, 1981)

Weed(s)	Tons/ha of Rice Grain With Nitrogen Fertilizer Applied (kg/ha)		
	0	60	120
	(tons/ha)		
None	4.5	5.3	6.6
Barnyard grass	4.4	4.0	5.5
Barnyard grass + *Monochoria*	4.1	3.1	3.5

ha, hectare.

TABLE 6.9   Nitrogen Uptake of Weeds and Rice in Two Trials (Boerema, 1963)

Species	Trial 1		Trial 2	
	Weeds Present	Weeds Absent	Weeds Present	Weeds Few
Barnyard grass	56.3	0	94.1	1.6
Rice	36.8	99.7	15.5	111.8
Total	93.1	99.7	109.6	113.6

The influence of fertility treatments for 47 years on weed types and populations was evaluated in Oklahoma (Banks et al., 1976). Plots with the lowest weed density were those that had received no fertilizer for 47 years. The highest weed density occurred on plots that received complete fertilizer (N, P, and K) and lime ($CaCO_3$). Grass weeds were most abundant with complete fertility whereas broad-leaved species declined.

Interactions of soil moisture and fertility on competition between wheat and wild buckwheat was studied in North Dakota (Fabricus and Nalewaja, 1968). Biomass of wheat growing alone increased with increasing fertility. Wheat biomass declined 30%–37% regardless of soil moisture or fertility when it grew with wild buckwheat. Wild buckwheat also reduced flax growth 47%–57% when they grew together for 90 days (Gruenhagen and Nalewaja, 1969). There was proportionately greater flax seed loss with higher fertility.

Table 6.10 shows five densities of wild oats with three levels of nitrogen. It is clear that as wild oat density increases, it is increasingly less profitable to add nitrogen. Wild oat's advantage results from its higher nitrogen use efficiency (Carlson and Hill, 1986). Increasing the rate of fertilizer application is not an economic, agronomic, or energy-efficient way to avoid or reduce crop losses owing to weed competition.

In general, weeds have a large nutrient requirement and will absorb as much as or more than crops. Nitrogen is the first nutrient to become limiting in most instances of weed–crop competition. The nitrate ion is not held strongly in soil and is highly mobile. Nitrogen depletion zones are likely to be large and similar to those for water. Therefore, rooting depth and root area of plants determine the ability to obtain resources. Relative competitiveness for nitrogen is largely determined by the soil volume occupied by roots of competing species. The amount of nitrogen taken up by plants in any combination is about equal (Table 6.9).

Movement of phosphorus and potassium is slow compared with nitrogen, and they move over shorter distances. Smaller depletion zones minimize interplant competition. Competition for phosphorus and potassium is therefore most likely to occur after plants are mature and have extensive, overlapping root development. It is reasonable to assume that competition for phosphorus will be more apparent in perennial crops. Competitiveness

TABLE 6.10   Yield of Wheat Grown in Competition With Wild Oat at Three Levels of Fertilization (Carlson and Hill, 1986)

Wild Oat Density (Plants/m²)	Wheat Yield With Preplant Nitrogen (kg/ha)			
	0	67	134	Average
0	6990	7520	7650	7390
4	6430	6660	6640	6580
8	6460	6100	6140	5230
16	5940	5200	5470	5540
32	5400	4120	3450	
Average	6240	5920	5870	

of barley cultivars with wild oats varied in response to potassium (Siddiqi et al., 1985) or phosphorus (Konesky et al., 1989) supply. There are few studies of weed–crop competition for phosphorus or potassium.

Whereas competition for nitrogen can sometimes be overcome by nitrogen fertilization, this is rarely true for phosphorus and potassium. It may be possible to prevent or delay weed invasion of perennial crops by maintaining a vigorous crop with adequate fertility.

5.4 Competition for Water

Water, or its lack, is often the primary environmental factor limiting crop production and it is probably the most critical of all plant growth requirements (King, 1966). Without irrigation, rainfall determines the geographic limit of crops. The water-use efficiency of nine weeds and nine crops is shown in Table 6.11. The point is not that weeds use a great deal more water or

TABLE 6.11   Water Use Efficiency (Dillman, 1931; Shantz and Piemeisel, 1927)

Plant	Water Use Efficiency[a]	Transpiration Coefficient[b]
WEEDS		
Common cocklebur	2.41	415
Common lamb's-quarter	1.52	383
Common purslane	3.56	317
Foxtail millet	4.63	251–274
Prostrate knotweed	1.47	678
Redroot pigweed	3.28	261–305
Russian thistle	3.18	221
Sunflower	1.73	577
Witchgrass	3.94	254
CROPS		
Alfalfa	1.15	809
Corn	3.34	361
Cotton	1.76	568
Oats	1.65	568
Smooth brome grass	1.28	791
Sorghum	3.09–3.65	268–285
Soybean	1.55	646
Sugar beets	2.65	304

[a]*Water use efficiency = efficiency of transpiration = grams dry matter produced per kilograms water transpired.*
[b]*Transpiration coefficient = milliliters water transpired per gram plant dry weight.*

TABLE 6.12 Soil Water Uptake Patterns of Common Weeds and Grain Sorghum in Summer Fallow

Weed Species	Rooting Depth (m)	Feeding Diameter (m)	Volume of Soil/Plant (m^2)	Plants to Consume Water/Hectare (No.)
Common cocklebur	2.9	8.5	17.9	704
Grain sorghum	1.7	4.3	6.5	2841
Kochia	2.2	6.7	9.5	1136
Pigweed	2.4	3.6	5.2	3853
Puncture vine	2.6	6.6	10.8	1136
Russian thistle	1.8	5.0	6.5	2149

Personal communication, adapted from Davis, R.G., Wiese, A.F., Pafford, J.L., 1965. Root moisture extraction profiles of various weeds. Weeds 13, 98–100; Davis, R.G., Johnson, W.E., Wood, F.O., 1967. Weed root profiles. Agron. J. 59, 555–556.

use water more efficiently than do crops, but rather that weeds use about the same amount as the crops with which they compete. Weeds effectively explore soil to obtain water (Table 6.12).

Comparison of rooting depth, uptake diameter, and the volume of soil from which resources can be consumed by one sorghum plant and five weeds shows why the weeds are effective competitors. Of the five weeds shown, all have a greater rooting depth and all but redroot pigweed have a larger feeding diameter and volume affected per plant than does grain sorghum. All except redroot pigweed have a greater capacity to consume water than does grain sorghum.

Classic work on the water requirements of plants was done at Akron, Colorado in the early 20th century (Briggs and Shantz, 1914; Dillman, 1931; Shantz and Piemeisel, 1927). Individual crop and weed plants were grown in separate pots and the grams of water required to produce a gram of plant dry matter were determined. Some of the data are shown in Table 6.13.[7]

Weeds compete for water, reduce water availability, and contribute to crop water stress. They require just as much water as crops, often more, and are often more successful in acquiring it. Weedy sunflowers require approximately twice as much water as corn. It takes more water to produce a potato tuber than to produce a common lamb's-quarter plant. Therefore, when common lamb's-quarter infests potato fields (as it commonly does) and water is limiting, fewer and smaller tubers will be produced. About 80 gallons of water are required to produce 1 lb of dry matter in barnyard grass, more than the 60 gallons to produce 1 lb of wheat. Crabgrass requires 83 gallons of water/lb of dry matter.

Many field, laboratory, and greenhouse studies have examined the role of water in weed–crop competition. A few illustrate water's role. One of the early studies (Wiese and Vandiver, 1970) compared growth of corn and sorghum growing in the greenhouse with three grass and five broad-leaved weeds at three soil moisture levels. Corn produced the most biomass

[7]1 Liter of water – 1000.28 mL = 1.000028 kg water.

TABLE 6.13   Water Required to Produce 1 lb Dry Matter (Shantz and
Piemeisel, 1927)

Plant	Kilograms of Water
Alfalfa	377
Barley, grain	431
Barley, whole plant	237
Bur sage	535
Common lamb's-quarter	300
Common purslane	128
Common sunflower	338
Corn	159
Mustard	1091
Potato, tuber	430
Potato, vine	150
Redroot pigweed	139
Russian thistle	1422
Sorghum	284
Wheat	227

at all moisture levels. Common cocklebur, barnyard grass, and large crabgrass normally grow well in humid regions and in irrigated crops. They were the most competitive with wet soil conditions. Kochia and Russian thistle, weeds of dry areas, were more competitive with dry soil conditions and grew poorly when soil was wet. Russian thistle produced twice as much growth in dry as in wet soil.

In field experiments in Texas (Stuart et al., 1984), water competition from smooth pigweed reduced leaf water potential and turgor pressure in cotton. Smooth pigweed was affected less by low soil water because it transpires less water and its larger root system draws water from deeper in soil. Smooth pigweed illustrates what may be called water wasting. In fact, water use is wasteful only from a human perspective or compared with another plant, a crop that uses less water. Each plant uses the water it requires. Stomata in some weeds are less sensitive to declining leaf water potential than those of crops with which they compete (Patterson, 1995a). When this is combined with a larger root system (Table 6.12) or better drought tolerance, weeds are formidable competitors for water. High water use by weeds may be ecologically advantageous to weeds in weed—crop competition especially when soil moisture is limiting (Patterson).

When soybean and velvetleaf competed in Texas, rooting depths were similar early on. After 10 weeks, soybean was able to draw water from greater soil depths and velvetleaf had little effect on soybean's water status (Munger et al., 1987). When the same species competed in Indiana, a wetter, more humid area, velvetleaf reduced soybean growth more in dry years than in wet ones (Hagood et al., 1980).

In Arkansas, soybean had higher leaf water potential than common cocklebur because of stomatal regulation of transpiration. Common cocklebur had lower stomatal resistance and higher transpiration. It is a high water user and exhausts soil water resources rapidly, to soybean's disadvantage (Geddes et al., 1979; Scott and Geddes, 1979).

Patterson (1995a) surveyed weed—crop competition studies that included water as a variable and found a slight tendency for decreased water availability to favor crops by reducing weed competition. This reasonable generalization may not always be true because it will be affected by each crop—weed combination and the cultural and environmental conditions in each crop season or over several seasons.

For example, the influence of season is shown by competition from any one of three broad-leaved weeds that reduced soybean yield more when soil moisture was adequate early followed by drought than when drought was early (Eaton et al., 1973, 1976).

Scientists in arid areas have developed fallow cropping systems. Many arid areas have sufficient rainfall to support a crop only every other year. Often wheat is grown 1 year, the land is fallowed (no crop) the next year, and rotated back to wheat in the third year. The primary purpose of this rotation is water conservation. Natural rainfall is not sufficient to grow wheat each year and extensive dryland cannot be irrigated. Therefore, minimum or no-tillage systems have been developed to conserve water. The data in Table 6.14 show the increase in water stored in the soil profile for a minimum-tillage system compared with a tilled, spring fallow system. The minimum-till system increased soil nitrate, grain protein, and wheat yield. Water is the least reliable resource for plant growth because we do not know precisely when it will arrive or how much will be received. This is the major reason why crops grown in arid areas are irrigated. Because roots grow more rapidly than shoots early in a plant's life, competition for water and nutrients usually begins before competition for light. Competition for water is determined by the relative root volume occupied by competing plants and will be greatest when roots closely intermingle and crops and weeds try to obtain water from the same volume of soil. Less competition occurs if roots of crops and weeds are concentrated in different soil areas. More competitive plants have faster-growing, large root systems that enable exploitation of a large volume of

TABLE 6.14 Conventional Tillage Versus Eco-fallow (Greb and Zimdahl, 1980)

Measurement	Treatment		
	Spring Tillage Fallow	Eco-fallow	Increase
Gain in soil water during fallow (cm)	3.9	5.4	1.5
Gain in soil nitrate during fallow (cm)	51.6	77.4	25.8
Percent gain protein	11.0	11.8	0.8
Wheat yield (bushels per acre)	34.4	41.8	7.4

soil quickly. If plants have similar root length, those with more widely spreading and less branched root systems will have a comparative advantage in competition for water.

5.5 Competition for Light

The total supply of light is the most reliable of the several environmental resources required for plant growth. In contrast to water and nutrients, light cannot be stored for later use. It must be used when received or it is lost forever (Donald, 1963). Although it varies in duration, intensity, and quality, light regulates many aspects of plant growth and development. Neighboring plants may reduce light supply by direct interception (shading). Leaves are the site of light competition. Leaves that first intercept light may reflect or absorb it, convert it to photosynthetic products or heat, or transmit it. If transmitted, the light is filtered so that it reaches lower leaves dimmer and spectrally altered. Anytime one leaf is shaded by another, there is competition for light.

Light competition is most severe when there is high fertility and adequate moisture, because plants grow vigorously and have larger foliar areas. Plants with large leaf area indices have a competitive advantage over plants with smaller leaf areas. Leaf area index, a measure of the photosynthetic surface over a given area, is correlated with potential light interception. Successful competitors do not necessarily have more foliage, but their foliage is in the most advantageous position for light interception. Thus, a plant's ability to intercept light is influenced by its angle of leaf inclination and leaf arrangement. Plants with leaves disposed horizontally to the earth's surface are more competitive for light than are those with upright leaves disposed more or less perpendicular to the earth's surface. Plants with opposite leaves are probably less competitive than are those with alternate leaves. Plants that are tall or erect have a competitive advantage for light over short, prostrate plants. A heavily shaded plant undergoes reduced photosynthesis, leading to poor growth, a smaller root system, and a reduced capacity for water or mineral uptake. The effect of shading is independent of direct competition for water or nutrients and entirely under the influence of light (Donald, 1963). Cropping practices used to manage weeds partially, such as smother crops and narrow row spacing (see Chapter 10), exploit plant responses to light (Holt, 1995). Most weeds and crops respond to shading in similar ways via morphological and physiological adaptations (Patterson, 1995a). This is not surprising because some plants evolved in disturbed habitats where shade adaptation has few selective advantages (Patterson).

Reports that crops are physiologically and genetically capable of higher productivity and photosynthetic efficiency than obtainable in the field confirm that intercepted light is a limiting factor in crop canopies (Holt, 1995). Reduced production in low light–acclimated crop plants is undesirable. Several reviews are available of responses of weeds and crops to light (Holt; Patterson, 1982, 1985, 1995a; Radosevich and Holt, 1984).

Crops and weeds differ in shade tolerance. Soybean and several of its associated weeds (e.g., eastern black nightshade, tumble pigweed, and common cocklebur) were most photosynthetically efficient under low irradiance (Regnier et al., 1988; Stoller and Myers, 1989). Many other weeds acclimate to low irradiance by plastic responses that reduce the growth-limiting effects of shading and allow restoration of high rates of photosynthesis when the plant is exposed to high irradiance (Dall'Armellina and Zimdahl, 1988; Patterson, 1979).

Bazzaz and Carlson (1982) generated photosynthetic response curves for 14 early, mid, and late successional species grown in full sunlight and 1% full sunlight. Early successional species, all common annual weeds, had the highest difference in response between sun- and shade-grown plants. The magnitude of photosynthetic flexibility decreased in plants from later successional stages. All species studied were able to change their photosynthetic output in response to light, but the change was larger for early successional annuals (Bazzaz and Carlson). These findings suggest that weeds are not only adapted to high light but are more capable of adapting to extreme variation in light, particularly deep shade. However, managing the light environment in a crop field to deter weed growth is difficult if not nearly impossible at reasonable cost, and not likely to be effective (Holt, 1995). Species that require or prefer light for germination are likely to be more prevalent in continuous no-tillage systems because a large portion of seeds remains on the surface (Chauhan et al., 2006a,b). Germination of species that require light will be impeded in no-tillage systems.

Available light is a major factor in yellow nutsedge competition with corn. More yellow nutsedge grows between corn rows than within the row because less light reaches the soil under plants. Yellow nutsedge density decreases as corn density increases (Ghafar and Watson, 1983); therefore, an acceptable yellow nutsedge management technique is to increase corn population. Increasing corn population density from 66,700 to 133,000 plants/ha reduced yellow nutsedge tuber production by 71%. Reducing corn population from 66,700 to 33,300 plants/ha increased tuber production by 41% (Ghafar and Watson). Field studies of the effect of artificial shade on yellow nutsedge concluded that rapidly developing crops (e.g., corn and potato) suppressed weed through competition for light (Keeley and Thullen, 1978). Shading greatly reduced shoot and biomass production and reduced but did not eliminate tuber production. Stoller and Woolley (1985) estimated that competition for light caused almost all soybean yield loss in competition with velvetleaf or jimsonweed, and half of the yield reduction in soybean competing with cocklebur.

Many studies have quantified the effects of light competition between weeds and crops. Cudney et al. (1991) showed that wild oats reduced light penetration and growth of wheat by growing taller. When wild oats were clipped to the height of wheat, light penetration in a mixed canopy was similar to that in monoculture wheat. Interference from wild oats planted at low densities reduced light penetration to wheat at later growth stages (Cudney et al.).

Similar height effects were observed in studies of competition between velvetleaf and soybean. Greater light interception by velvetleaf resulted from greater height and dry weight allocation to more upper branches (Akey et al., 1990). Reductions in tomato yield were greater when it grew in competition with eastern black nightshade compared with black nightshade, because eastern black nightshade is taller (McGiffen et al., 1992). These studies show that plant architecture, especially height, location of branches, and height of maximum leaf area determine competition for light and influence crop yield (Holt, 1995).

Interaction of light and water is illustrated in a study of how yield of quack grass−infested soybeans was increased by irrigation when soil moisture was limiting. Soybeans infested with quack grass yielded less than quack grass−free soybeans. Quack grass was nearly the same height or taller than soybeans at all stages of soybean development and competed for light throughout the growing period. Adequate moisture reduced quack grass competition in soybeans but did not eliminate it because quack grass continued to compete with soybeans for light (Young et al., 1983). Morphological changes in soybean resulting from expression of shade avoidance can be used to define a period of development sensitivity to low-red and far-red light, which was found to be similar to the critical period for weed control (Green-Tracewicz et al., 2012). Shade did not affect Texasweed emergence but significantly reduced its growth. In 70% and 90% shade, Texasweed height was increased 28% and 20%, respectively. The weed seemed to counter the adverse effect of shade by increasing its specific leaf area and percentage of leaf biomass (Godara et al., 2012).

Studies in India (Shetty et al., 1982) have shown that dicots are less shade-sensitive than monocots, which explains why monocots are often important tropical weeds. Broad-leaved weeds usually do not appear until after tropical crops are well established. It seems that manipulation of tropical crop canopies could suppress weeds via shading. The height of the dicot weeds celosia and coat buttons was reduced by 90% shade, but that shade level had no effect on height of southern crabgrass. Ninety percent shade reduced height of bristly starbur by 50% and purple nutsedge by 30%. The effects were most pronounced early in the growing season and similar reductions in leaf area index and plant dry matter were observed. Slender amaranth's height was not affected by shade, but as light decreased, seed production decreased. For most annuals, 90% shade reduced seed production up to 90%, and 40% shade reduced seed production by 45%. Shading reduced purple nutsedge tuber production by 89%.

Clove oil is approved for weed control in organic agriculture. Stokłosa et al. (2012) found that membrane damage from clove oil decreased with increasing light intensity. They suggested that efficacy of the oil might be affected by light intensity experienced by the plants before application.

5.6 Factors for Which Plants Generally Do Not Compete

Plants that emerge at the same time rarely compete for space even though plant density may be high. When plants emerge at different times, the first plant that occupies an area will tend to exclude all others and have a competitive advantage; in this sense, plants compete for space by occupying it first. Occupancy or competitive exclusion can be, and among plants should be, regarded as competition for the resources in a space. They compete for what a space contains. This may not be true in root crops that are planted closely, but in most cases plants compete for the light, nutrients, and water a space contains.

Booth et al. (2003) agree with this assertion but caution that plants whose roots are restricted generally have reduced shoot biomass, height, or growth. Others (Schenk et al., 1999; cited by Booth et al.) argue the controversial hypothesis that plants may be regarded as territorial because they defend the space they are in against invasion by others. That is, a plant may effectively defend its territory by preventing others from using it. Consistent with this argument, plants may do so by using or preventing the use of an area's resources by other plants.

Plants may compete for oxygen. There are no studies to document this, but it is theoretically possible. In most soils, diffusion of oxygen is rapid enough so that adequate supplies are available for all roots. Oxygen can be limiting in very wet soils. Similarly, under most circumstances, carbon dioxide concentrations are always higher than the carbon dioxide compensation point (the light intensity at which there is a balance between carbon dioxide given off by respiration and required by photosynthesis). Competition for carbon dioxide is unlikely to occur under field conditions, but crop yields can be increased by supplemental carbon dioxide (see earlier comment in this chapter on climate change). More efficient use of carbon dioxide by weeds with high photosynthetic capacities may contribute to their rapid growth and provide a competitive advantage. Therefore, a plant's competitive ability could depend on its capacity to assimilate carbon dioxide and use the photosynthate to extend foliage or increase size. Plants that fix carbon dioxide at high rates are potentially more competitive.

There is no evidence that plants compete for agents of pollination.

6. PLANT CHARACTERISTICS AND COMPETITIVENESS

In general, it is true that plants possessing one or more of the following characteristics are more competitive than are plants that lack them. This list is not in rank order and it cannot be said that a plant with a certain characteristic will always win over a plant with others. Most competitive plants have:

- rapid expansion of a tall, foliar canopy,
- horizontal leaves under overcast conditions and obliquely slanting leaves (plagiotropic) under sunny conditions,
- large leaves.
- the C_4 photosynthetic pathway and low leaf transmission of light,
- leaves forming a mosaic leaf arrangement for best light interception,
- a climbing, twining habit,
- a high allocation of dry matter to build a tall stem, and
- rapid stem extension in response to shading.

The most obvious competition among plants is what we see: foliar competition for light. Competition for nutrients and water takes place beneath soil, where it cannot be seen. The most competitive plants also share some of the following root characteristics:

- early and fast root penetration of a large soil area,
- high root density/soil volume,
- high root–shoot ratio,
- high root length per root weight,
- a high proportion of actively growing roots,
- long, abundant root hairs, and
- high uptake potential for nutrients and water.

7. RELATIONSHIP BETWEEN WEED DENSITY AND CROP YIELD

Early weed science literature assumed that the relationship shown in Fig. 6.12 described the effect of weeds on crop yield. That assumption was wrong. Fig. 6.12 says that with no weeds, crop yield will be maximized, and at some large weed density, crop yield will be zero. The real relationship is curvilinear, not linear. Such a relationship is supported by data (Fig. 6.13) showing the effect of kochia, an annual broad-leaved weed, on sugar beet root yield.

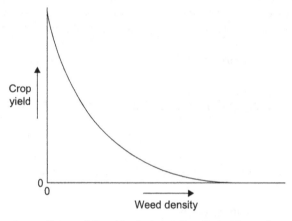

FIGURE 6.12   A schematic curvilinear relationship depicting the effect of increasing weed density on crop yield (Zimdahl, 1980).

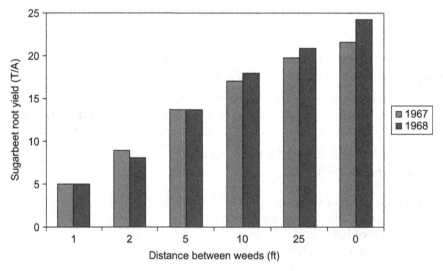

FIGURE 6.13   Effect of kochia on sugar beet yield (Weatherspoon and Schweizer, 1971). Each yield bar in each year (not between years) is significantly different from every other bar (yield) for that year.

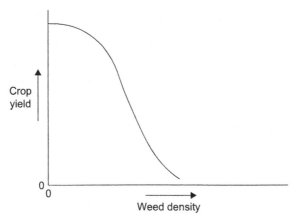

FIGURE 6.14 A schematic sigmoidal relationship depicting the effect of increasing weed density on crop yield (see Zimdahl, 1980, 2004).

Other data show the curvilinear relationship in Fig. 6.14, which, although intuitively logical, is wrong. Some of the data in Table 6.15 show that the relationship is neither linear nor curvilinear. Doubling of weed density did not double crop loss in any of these studies, and even when weed density is increased by a factor of 25, crop loss was not 100%. Therefore,

TABLE 6.15 Effect of Weed Density on Crop Yield

Crop	Weed	Weed Density	Yield Reduction From Control (%)	Source
Wheat	Wild oats	$58.5/m^2$	22.1	Bell and Nalewaja (1968)
		$134/m^2$	39.1	
Wheat	Green foxtail	$721/m^2$	20	Alex (1967)
		$1575/m^2$	35	
Rice	Barnyard grass	$1/0.09\ m^2$	57	Smith (1968)
		$5/0.09\ m^2$	80	
		$25/0.09\ m^2$	95	
Soybean	Common cocklebur	3297/ha	10	Barrentine (1974)
		6597/ha	28	
		12295/ha	43	
		25989/ha	52	
Corn	Giant foxtail	0.5/5 cm of row	4	Knake and Slife (1962)
		1/5 cm of row	7	
		3/5 cm of row	9	
		6/5 cm of row	12	
		12/5 cm of row	16	
		54/5 cm of row	24	

ha, hectare.

although the curvilinear relationship is not entirely incorrect, it is not precisely correct and can be misleading.

Smith's (1968) study of the interaction of rice and barnyard grass density showed that the appropriate relationship is neither linear nor curvilinear. The curvilinear relationship fails because it predicts that a high weed density will reduce crop yield to zero, and that does not happen. Some crop plants always survive even though they may be very small and yield is low and unprofitable. Smith's data show the interaction of crop density and how, as it increases, the effect of weed density decreases (Table 6.16).

An interpretation of the relationship between crop yield and weed density has been described by the sigmoidal curve in Fig. 6.14 (Zimdahl, 1980). At very low weed densities, there is no effect on crop yield, and as weed density increases, although there may be an effect, it is barely discernible. As weed density continues to increase, crop yield drops quickly but never goes completely to zero. Even very high weed densities do not eliminate all crop plants. This represents most weed–crop competition data and provides a picture of what happens, but *it is still not correct*. Its appeal is that it is difficult to measure the effect of a few weeds in a large area. It is not even wise to attempt to do so. There may be an effect of one weed per acre, but the weed has no measurable economic effect. However, that weed will affect nearby crop plants and produce seed, and thus can affect future crops.

There are many places in the literature of weed science that state, or the data clearly imply that, the relationship between yield loss and weed density is sigmoidal (Fig. 6.14) with little or no loss at low weed density (or nearly none). Cousens et al. (1987) state unequivocally that the data do not support this. When yields are plotted over a range of weed densities,

TABLE 6.16   Interaction of Rice and Barnyard Grass (Smith, 1968)

Rice Plants/0.09 m^2	Barnyard Grass	% Yield Reduction
3	0	0
3	1	57
3	5	80
3	25	95
10	0	0
10	1	40
10	5	66
10	25	89
31	0	0
31	1	25
31	5	59
31	25	79

there is no evidence to support a sigmoidal response. The most accurate representation of crop—weed interactions is that created by regression analysis of crop yield and weed density. This is because densities observed in the field and those used in experiments cannot represent the whole range of possible weed densities depicted in Fig. 6.14. Multiple regression models must be chosen carefully so that they reflect biological reality and not just mathematical convenience. For a more complete discussion of the role of modeling in studies of weed—crop competition, see Chapter 10 in Zimdahl (2004).

8. MAGNITUDE OF COMPETITIVE LOSS

Tables 6.17A,B, 6.18A,B, and 6.19 show the magnitude of loss in a few studies of weed competition in corn, soybeans, and small grains. This small set of data provides evidence that weeds decrease crop yield, often a great deal. The data also show that the effect of weeds is neither entirely predictable nor the effect of a particular density consistent. The data are shown as they appeared in the original publication because it makes an important point about many studies: the lack of precision of the data. There is no uniform definition of a heavy stand, a small infestation, or a natural stand (Table 6.17A) and therefore the work is not repeatable. The data also illustrate the inevitable effect of year and place. In competition studies, it is important to define precisely the number of weeds and crop plants per unit area (the density).

9. DURATION OF COMPETITION

It is obvious that a weed present for 1 day in the life of a crop will probably have no measurable effect on final yield. But what if the weed is present for 2, 20, or 200 days? The question of duration of competition has been asked in two ways. The first kind of study asks what the effect is when weeds emerge with the crop and are allowed to grow for defined periods of time. After each of these times, the crop is then kept weed-free for the rest of its growing period. These studies define the *critical duration of weed competition*. The second kind of study asks what the effect is when the crop is kept weed-free from emergence for certain periods of time and then weeds are allowed to grow for the rest of the growing season. These define what many call the *critical weed-free period*. Vega et al. (1967) was one of the

TABLE 6.17A Weed Competition in Corn (See Zimdahl 1980 for Complete Citations)

Location	Density	Yield Reduction
Illinois	Heavy stand	55%
Illinois	54 foxtail/ft of row in 4-inch band over row	25%
Iowa	Hand weeded	50% greater than unweeded
Iowa	Small infestations of foxtail	6—8 bushels/acre

TABLE 6.17 BWeed Competition in Corn (See Zimdahl 2004 for Complete Citations)

Weed Species	Density	Yield Reduction
Barnyard grass	100 m^2	18%
	200 m^2 concurrent emergence	26%–35%
	Emergence when corn had four leaves	6%
Common milkweed	11,000–45,000 plants/m^2	10%
Giant ragweed	1.7, 6.9, or 13.8 plants/m^2	13.6%–90%
Giant foxtail	10 per meter of row	13%–14%
Green foxtail	0, 29, 56, or 89/m^2	20%–56%
	129/m^2	5.8%–17.6%
Hemp dogbane	Natural stand	0%–10%
Itchgrass	2, 4, up to 14 weeks	125 kg/hectares/week of presence
	Season-long	33%
Quack grass	65–390 shoots/m^2	12%–16%
	745 shoots/m^2	37%
Palmer amaranth	0.5–8/m^2	11%–74%
Redroot pigweed	0.5 per meter of row with concurrent planting at corn's three- to five-leaf stage	5%
Wild proso millet	10/m^2	13%–22%
Yellow nutsedge	100 shoots/m^2	8% per 100 shoots
	300 tubers/m^2	17%
	700 tubers/m^2	41%

TABLE 6.18A Weed Competition in Soybeans (See Zimdahl 1980 for Complete Citations)

Location	Density	Yield Reduction
Nebraska	86 lb/acre	1 bushel/acre
Iowa	10–12 weeds/ft of row	7.5–17.1%
Illinois	54 foxtail/ft of row in 4-inch band over row	28%

first to study the effect of the duration of weed control on rice. Weeds grew for no time at all or in intervals of 10 days up to 50 days after rice was planted. They also allowed weeds to compete for 10, 20, 30, 40, or 50 days after planting and then kept the crop weed-free afterward (Table 6.20).

TABLE 6.18B Weed Competition in Soybeans (See Zimdahl 2004 for Complete Citations)

Weed Species	Density	Yield Reduction (%)
Common cocklebur	One per 1.8 m of row	7
	One per 0.9 m of row	14
	One per 0.3 m of row	30
Hemp *Sesbania*	Full season	28–41
	16 per m², full season	43
JERUSALEM ARTICHOKE		
Full season	1 tuber per meter of row	31
	2 tubers per meter of row	59
	4 tubers per meter of row	71
4 weeks after planting	4 tubers per meter of row	9
6 weeks after planting	4 tubers per meter m of row	10
8 weeks after planting	4 tubers per meter of row	38
20 weeks after planting	4 tubers per meter of row	82
Jimsonweed	0.3 per meter of row, full season	8
	1.6 per meter of row	
	2 weeks	7
	4 weeks	14
	Full season	41
Johnsongrass	Full season	59–88
Johnsongrass with early-maturing cultivar	1 week after maturity	32
	2 weeks after maturity	35
	3 weeks after maturity	36
Johnsongrass with late-maturing cultivar	1 week after maturity	27
	2 weeks after maturity	29
	3 weeks after maturity	29
Ivyleaf morning glory	1 per 15 cm of row, full season	13–36
Quack grass	Natural stand for	
	6 weeks	11
	8 weeks	23
	Full season	33
VELVETLEAF		
Mid May planting	1 per 30 cm of row, full season	27
Late June planting	1 per 30 cm of row, full season	14

TABLE 6.19   Weed Competition in Small Grains (See Zimdahl 1980, 2004)

Location	Crop	Density	Yield Reduction (Bushels Per Acre)
Montana	Spring wheat	Canada thistle/ft^2	
		3–5	4.2
		20–25	9.0
		40–45	15.3
Oregon	Winter wheat	1 fiddle-neck/ ft^2	10.0
New York	Oats	15 mustard/ ft^2	11.0
Nebraska	Sorghum	15 lb of weeds/acre	1.0

TABLE 6.20   Effect of Duration of Weed Control and Weed Competition on Rice Yield (Vega et al., 1967)

Weed Control Duration (Days After Planting)	Yield (kg/ha)	Weed Competition Duration (Days After Planting)	Yield (kg/ha)
0	46	10	2944
10	269	20	3067
20	1544	30	2752
30	2478	40	2040
40	3010	50	1088
50	2756	Not weeded	55

ha, hectare.

The data show that yield is reduced when rice is weeded only for a short time after planting. When it was weeded for 40 days, yield reached a maximum and there was no benefit from weeding an additional 10 days. In the same way, when weeds were allowed to grow up to 20 days after planting and then removed, there was no effect on yield. Therefore, rice (and many other crops) can withstand weed competition early in the growing season and do not have to be weeded immediately. Weeds in rice cannot be present more than about 30 days or yield will decrease.

Corn must be kept weed-free for 3–5 weeks after emergence or 9 weeks after seeding, depending on the location and the weeds (Table 6.21). The opposite study (Table 6.22) shows the length of early weed competition tolerated by corn. If provided with a weed-free period for 3 weeks after emergence, corn will compete effectively with weeds emerging afterward. Conversely, corn can withstand weed competition for up to 6 weeks if it is then weeded and kept weed-free.

TABLE 6.21   Weed-Free Period Required to Prevent Yield Reduction in Corn (See Zimdahl 1980, 2004)

Weed-Free Weeks Required After				
Seeding	Emergence	Competing Weeds	Location	Source
	9	Mixed annuals	Mexico City	Alemàn and Nieto (1968)
5		Mixed annuals	Vera Cruz, Mexico	Nieto (1970)
3		Giant foxtail	Illinois	Knake and Slife (1969)
After seven-leaf stage		Redroot pigweed	Ontario, Canada	Knezevic et al. (1994)
3–14 leaves		Natural stand	Ontario, Canada	Hall et al. (1992)
Six leaves		Natural stand	Ontario, Canada	Halford et al. (2001)

When barnyard grass and a mixture of redroot pigweed and Palmer amaranth were planted with alfalfa and removed by 36 days after planting, there was no effect on alfalfa yield (Fischer et al., 1988) Thereafter, yield decreased in direct proportion to the length of weed interference. When the same weeds were seeded 65 or more days after alfalfa emergence, there was no effect on alfalfa yield but weed biomass reduced hay first-cutting quality.

TABLE 6.22   Length of Early Weed Competition Tolerated Without Yield Loss in Corn: The Critical Duration (See Zimdahl 1980, 2004)

Weeks of Competition Tolerated After				
Seeding	Emergence	Competing Weeds	Location	Source
3		Mixed annuals	Vera Cruz, Mexico	Nieto (1970)
	4	Mixed annuals	Mexico City	Alemàn and Nieto (1968)
4		Mixed annuals	Chapingo, Mexico	Nieto et al. (1968)
2–4		Halberdleaf orach and Persian speedwell	England	Bunting and Ludwig (1964)
4		Green foxtail	Ontario, Canada	Sibuga and Bandeen (1978)
6		Giant foxtail	Illinois	Knake and Slife (1969)
	6	Redroot pigweed	Oregon	Williams (1971)
	2–3	Mixed annuals	New Jersey	Li (1960)
	8	Itchgrass	Zimbabwe	Thomas and Allison (1975)
4		Longspine sandbur	Colorado	Anderson (1997)
9–13 leaves		Natural stand	Ontario, Canada	Halford et al. (2001)
14 leaves		Natural stand	Ontario, Canada	Hall et al. (1992)

TABLE 6.23A   Crops With an Apparent Critical Period for Weed Competition
(See Zimdahl, 1980, 2004 for Complete Citations)

Crop	Weed-Free Weeks Required	Weeks of Weed Competition Tolerated
Corn	3–5	3–6
Potato	4–6	4–9
Rice, paddy	4–6	4–9
Soybean	2–4 after planting	4–8 after planting

These kinds of data have been used to derive the critical period for weed competition that has been defined (Table 6.23A) for a few crops and over a range of time for several crops (Table 6.23B). It is clear from Table 6.23B that weeds behave differently in different crops (e.g., compare johnsongrass in soybean and cotton or common cocklebur in bean

TABLE 6.23B   Crops With an Identified Critical Period for Weed Competition
(See Zimdahl, 2004 for Complete Citations)

Crop	Critical Period
Barley infested with wild oat	Two-node stage to maturity
Bean, snap infested with common cocklebur	Emergence to full bloom of snap bean. Note: this is too long to be a critical period
Bean, dry infested with hairy nightshade	3–9 weeks after emergence
Cotton infested with hemp *Sesbania*	≥62 days after planting
Cotton infested with johnsongrass	4–6 weeks after emergence
Cotton infested with barnyard grass	3–6 weeks after emergence
Cotton infested with Bermuda grass	4–7 weeks after emergence
Peanut infested with common cocklebur	2–12 weeks after peanut emergence
Peanut infested with bristly starbur	2–6 weeks after emergence for tolerated loss of 3%–4%
Peanut infested with horse nettle	2–6 or 8 weeks after emergence
Rice infested with bearded sprangletop	21–56 days after emergence
Soybean	9–38 days after emergence = second node (V-2) to beginning pod formation (R-3) stage
Soybean infested with giant ragweed	4–6 weeks after emergence in 1 year and 2–4 weeks in a second year
Soybean infested with johnsongrass	4–5 weeks after emergence
Tomato, transplants	24–36 days after transplanting
Watermelon infested with large crabgrass	0–6 weeks after emergence

and peanut). A critical period is not equal to the critical weed-free period mentioned previously. The critical period, defined as the time between the period after seeding when weed competition does not reduce yield and the time after which the presence of weeds does not reduce yield, has been found for several crops. It is a *time between* the early weed-free period required and the length of competition tolerated (Fig. 6.15). It is not a fixed period for a crop because it varies with season, soil, weeds, and location. The critical period is a useful measure because it gives guidance on when to weed. For example, if kept weed-free for 6 weeks, potatoes will survive the rest of the season without yield reduction, even if weeds grow. If potatoes are weeded 9 weeks after seeding, yield will not be reduced if they are subsequently kept weed-free. Therefore, for maximum benefit, potatoes should be weeded sometime between 6 and 9 weeks after seeding or yield will decrease. Critical period analyses show that preemergence weed control is not essential, nor is weed control immediately after emergence. The method of weed control dictates when it must be applied, but the lesson of critical period studies is that weed control does not have to be done in the first few weeks after crop emergence. Critical periods have practical weed management value, but Mortimer (1984) points out their limitation: all weeds are considered equally injurious and no distinction is made among the kinds of competition that can occur. Most humans would be injured in a fistfight, but they would be injured less if they got to pick their opponent (that little fellow) than if the opponent were the heavyweight boxing champion.

None of the foregoing comments on duration or magnitude of loss should be interpreted as a recommendation for stopping or limiting crop weed competition studies. They are still being done (Odero et al., 2016). However, there is sufficient evidence that weeds reduce crop yield and that there is a critical period for weeding to support the contention that the emphasis of studies should change from determining yield effects to studying the effects of components of a cropping system on competition: rotation, intercropping, seeding rate,

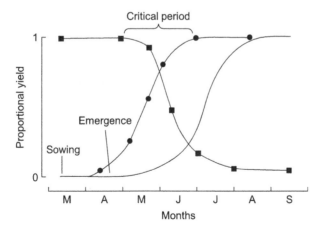

FIGURE 6.15 The "critical period of competition" illustrated for onions. Dashed lines indicate changes in crop dry weight from sowing to harvest; ■, yield response from delaying the start of continuous weed removal; ●, yield response from delaying the termination of weed removal. *Adapted by Mortimer, A.M., 1984. Population ecology and weed science. In: Dirzo, R., Sarukhan, J. (Eds.), Perspectives on Plant Population Ecology. Sinauer Assoc., Inc., Sunderland, MA, pp. 363–388.*

row spacing, and fertilizer placement (Swanton et al., 2015). For example, physiological changes attributed to a low red–far red light ratio during early weed competition result in a physiological cost to the crop plant. This can contribute to yield loss observed in weed competition studies conducted in the field (Afifi and Swanton, 2012). Because of extensive studies of the results of competition (Knezevic and Datta, 2015), weed scientists are moving toward understanding why competition occurs rather than what its effects are.

10. ECONOMIC ANALYSES

Economic analyses of weed control are available. They are not abundant because the answer is assumed: weeds reduce crop yield and have an economic effect. Farmers know that weeds reduce yield; the question they ask is not whether weeds will reduce yield, but how many weeds reduce yield by how much. Their question is, "Should I control weeds, and if so, what method(s) is best?" The farmer's definition of best usually means the method that offers the highest profit potential. The farmer knows that a few weeds are not of consequence and asks how many weeds are of consequence. The data in Table 6.24 illustrate how the answer might be provided.

The study showed, for three potential wheat yields, what the profit or loss would be for spraying, given a certain value of wheat and a defined spraying cost. For example, if a farmer has 0.5 weed/ft^2, the estimated yield loss is 5%. If the wheat yield is estimated to be 15 or 20 bushels/acre, the cost of controlling the weeds will exceed the benefit to be gained. On the other hand, if the yield will be 30 bushels, the gain will exceed the cost and the weeds should be controlled. The values in Table 6.24 are out of date but they illustrate the principle, which remains valid. A similar set of data assist with decisions on controlling wild oats in barley, wheat, or flax. These data (Table 6.25) show the potential yield loss for each crop from a wild oats density that a farmer could determine.

TABLE 6.24   Potential Profit or Loss From 2-Methyl-4-chlorophenoxyacetic Acid Application to Control Pinnate Tansy Mustard in Winter Wheat (Wiese, 1965)

| | | Potential Wheat Yield (Bushels Per Acre) | | |
| | | 15 | 20 | 30 |
Weeds Per Square Feet	Percent Estimated Yield Loss	Profit or Loss ($[a])		
0.25	2.5	−1.03	−0.87	−0.56
0.5	5	−0.56	−0.25	0.38
1	10	0.38	1.00	2.25
1	20	2.25	3.50	6.00
4	40	6.00	8.50	13.50

[a]*Profit or loss, value of yield loss if weeds are uncontrolled; wheat, $1.25/bushel; spray cost, $1.50/acre.*

TABLE 6.25 Yield Loss Caused by Wild Oats in Barley, Wheat, and Flax (Bell and Nalewaja, 1967)

Wild Oat Seedlings/m^2	Yield Reduction in Bushels Per Acre		
	Barley	Wheat	Flax
10	1.6	1.5	2.0
40	2.7	3.5	5.0
70	4.9	5.2	6.3
100	6.0	5.4	6.9
130	6.2	7.3	7.4
160	7.1	8.7	7.5

A farmer could calculate control costs and value of yield lost to determine whether control should be done. Other studies of decision models have been performed (King et al., 1986; Lybecker et al., 1984), but most decisions about what to do are still made by growers with incomplete information. Weed science needs more information on the efficacy of various weed control techniques and weed management systems in different soils and cropping systems. To make economically and agronomically sound weed management decisions, this information must be combined with information on percent emergence of the weed species in the soil seed bank, expected crop yield, weed control cost, and the farm's current economic situation.

11. MATHEMATICAL MODELS OF COMPETITION

A large number of experiments have been done to demonstrate that weeds reduce crop yield (Zimdahl, 1980, 2004). This work has demonstrated that some weeds are more detrimental to one crop than another and the effect is always modified by environmental interactions. Weed scientists do not need more experiments to establish that weeds are detrimental. In fact, the important questions in weed control and weed management cannot be answered by experiments to determine yield loss as a function of weed density.

Mortimer (1987) cited four primary issues in weed management:

- For a given crop management system, what is the likelihood of invasion by weeds?
- Given the presence of weed infestation, how rapidly will the weeds spread and what crop losses will be experienced?
- How much of any proposed control measure is required to contain the infestation or lead to total eradication?
- What are the comparative costs of different weed control measures and what risks are involved in different weed management strategies?

It is possible but not desirable to answer these questions with standard field experiments, because there is not enough time or money and there are not enough weed scientists to do so. Therefore, weed scientists are working to develop models to test experimental hypotheses and complement experimentation. Cousens et al. (1987) described four ways in which models can enhance research:

- as part of the framework to integrate available information. Critical gaps in research can be pinpointed and incompatibilities and erroneous or abnormal results may become apparent;
- mathematics is a formal, rigorous language in which theories and intuition can be expressed. Models can reduce ambiguity and describe complex systems;
- when used with an experimental program, models can increase the speed with which understanding develops. They can be used to identify critical experiments, thereby making the most economical use of resources; and
- models can be used to forecast and predict what might be observed under conditions not previously included in experiments.

Models can be empirical and describe data or a response to imposed management options. They can also be mechanistic and attempt to incorporate knowledge of processes that determine responses (Cousens et al., 1987). Much modeling effort has been expended to develop computerized decision-aid software to answer the third and fourth questions posed by Mortimer (1987). Decision-aid models are based on the knowledge that weed effects depend on population and all models attempt to predict the biological (weed density) and economic consequences of management decisions (Coble and Mortensen, 1992). Models incorporate the concept of thresholds or beginning points for weed effects. There are at least four kinds of thresholds used in decision-aid models (Coble and Mortensen):

damage: the weed population at which a negative crop yield response is detected;
economic: the weed population at which the cost of control is equal to the crop value increase from control;
period: the time or times during the crop's life when weeds are most detrimental; and
action: the point when a control measure should be initiated.

Mathematical, computer-based models are not widespread in weed science. Cousens et al. (1987) proposed that the slow development of modeling in weed science resulted from an early lack of scientists familiar with mathematical modeling and its capabilities. There was also a limited demand for model development and a high demand for problem solving. Herbicides have provided quick solutions to weed management challenges and continue to do so. Simulation models have been used primarily to predict crop yield losses from weeds. Weaver (1996) recommended linking crop–weed simulation models with biological models of population dynamics. Modeling and experimentation should proceed in tandem, not separately. Given the increasing public acceptance of environmental objections to expanding herbicide use and the difficult challenges of herbicide resistance, it may be time to move toward models that permit weed management with other than broadscale herbicide application. As models are developed, perfected, and tested against biological knowledge, they may be used more. Models are increasingly able to fulfill the basic requirements for a good weed—crop competition model (Cousens, 1985):

- Without weeds there is no yield reduction.
- At low weed densities, the effect of increasing weed density will be additive.
- Yield loss can never exceed 100%.
- At high weed densities there is a nonlinear response of crop yield to weed density.

It is beyond the scope or intent of this book to present a detailed discussion of crop—weed interference modeling in weed science (see Zimdahl, 2004 for a more complete review). Readers are directed to the references cited here and to current literature for more information.

THINGS TO THINK ABOUT

1. Why do plants compete?
2. What do plants compete for?
3. Do plants compete for space?
4. What factors determine weed—crop associations?
5. What makes a plant competitive?
6. How is the critical period of competition determined and what is it used for?
7. What is the most appropriate description of the relationship between crop yield and weed density?
8. How much yield is lost owing to weeds?
9. What must be known about crop—weed competition to make good weed management decisions?
10. How do economic analyses help make weed management decisions?
11. What is the role of mathematical models in weed science?
12. What kinds of thresholds are used in crop—weed interference models?
13. How can models aid research and weed management?

Literature Cited

Afifi, M., Swanton, C., 2012. Early physiological mechanisms of weed competition. Weed Sci. 60, 542—551.

Akey, W.C., Jurik, T.W., Dekker, J., 1990. Competition for light between velvetleaf (*Abutilon theophrasti*) and soybean (*Glycine max*). Weed Res. 30, 403—411.

Aldrich, R.J., 1984. Weed-Crop Ecology: Principles in Weed Management. Bretton Pub., N. Scituate, MA.

Aldrich, R.J., Kremer, R.J., 1997. Weed-Crop Ecology: Principles in Weed Management, second ed. Iowa State Univ. Press, Ames, IA. 455 pp.

Alemàn, F., Nieto, J.H., 1968. The critical periods of competition between weeds and corn in the high valleys of Toluca, Mexico. Weed Sci. Soc. Am. Abstr. 150.

Alex, J.F., 1967. Competition between *Setaria viridis* (green foxtail) and wheat at two fertilizer levels. Res. Rep. Can. Nat. Weed Comm. West Sect. 286—287.

Altieri, M.A., 1994. Biodiversity and Pest Management in Agroecosystems. Food Products Press, New York, NY, 185 pp.

Anderson, R.L., 1997. Longspine sandbur (*Cenchrus longispinus*) ecology and interference in irrigated corn (*Zea mays*). Weed Technol. 11, 667–671.

Anonymous, November 11, 1995. Czeching the spread of unwanted plants. Economist 82.

Banks, P.A., Santlemann, P.W., Tucker, B.B., 1976. Influence of soil fertility treatments on weed species in winter wheat. Agron. J. 68, 825–827.

Baker, H.G., 1974. The evolution of weeds. In: Johnston, R.F., Frank, P.W., Michener, C.D. (Eds.), Ann. Rev. Ecol. SystematicsAcademic Press, N.Y, pp. 1–24.

Barrentine, W.L., 1974. Common cocklebur competition in soybeans. Weed Sci. 22, 600–603.

Bazzaz, F.A., Carlson, R.W., 1982. Photosynthetic acclimation to variability in the light environment of early and late successional plants. Oecologia 54, 313–316.

Bell, A.R., Nalewaja, J.D., 1968. Competition of wild oat in wheat and barley. Weed Sci. 16, 505–508.

Bell, A.R., Nalewaja, J.D., 1967. Wild oats cost more to keep than to control. N. Dak. Farm Res. 25, 7–9.

Bleasdale, J.K.A., 1960. Studies on plant competition. In: Harper, J.L. (Ed.), The Biology of Weeds. Blackwell Sci. Publ., Oxford, Eng, pp. 133–142, 256 pp.

Boerema, E.B., 1963. Control of barnyardgrass in rice in the Murrumbridge irrigation area using 3,4-dichloropropionanilide. Aust. J. Exp. Agric. Anim. Husb. 3, 333–337.

Booth, B.D., Swanton, C.J., 2002. Assembly theory applied to weed communities. Weed Sci. 50, 2–13.

Booth, B.D., Murphy, S.D., Swanton, C.J., 2003. Weed Ecology in Natural and Agricultural Systems. CABI Pub., Wallingford, Oxon, UK, 303 pp.

Briggs, L.J., Shantz, H.L., 1914. Relative water requirement of plants. J. Agric. Res. 3, 1–63.

Bronowski, J., 1973. The Ascent of Man. Little, Brown and Company, Boston, 448 pp.

Bunting, E.S., Ludwig, J.W., 1964. Plant competition and weed control in maize. In: Proc. Seventh British Weed Cont. Conf., vol. 1, pp. 385–388.

Burt, G.W., 1974. Adaptation of johnsongrass. Weed Sci. 22, 59–63.

Capinera, J.L., 2005. Relationships between insect pests and weeds: an evolutionary perspective. Weed Sci. 53, 892–901.

Carlson, H.L., Hill, J.E., 1986. Wild oat (*Avena fatua*) competition with spring wheat: effects of nitrogen fertilization. Weed Sci. 34, 29–33.

Chauhan, B.S., Gill, G., Preston, C., 2006a. Seedling recruitment pattern and depth of recruitment of 10 weed species in minimum tillage and know-till seeding systems. Weed Sci. 54, 658–668.

Chauhan, B.S., Gill, G., Preston, C., 2006b. Tillage systems effects on weed ecology, herbicide activity and persistence: a review. Aust. J. Exp. Agric. 46, 1557–1570.

Clements, F.E., Weaver, J.E., Hanson, H.C., 1929. Plant Competition-an Analysis of Community Function. Publ. No. 398. Carnegie Inst., Washington, DC, 340 pp.

Coble, H.D., Mortensen, D.A., 1992. The threshold concept and its application to weed science. Weed Technol. 6, 191–195.

Commoner, B., 1971. The Closing Circle—Nature, Man, and Technology. Knopf, NY, 326 pp.

Cousens, R., Moss, S.R., Cussans, G.W., Wilson, B.J., 1987. Modeling weed populations in cereals. Rev. Weed Sci. 3, 93–112.

Cousens, R., 1985. A simple model relating yield loss to weed density. Am. Appl. Biol. 107, 239–252.

Cousens, R., Mortimer, M., 1995. Dynamics of Weed Populations. Cambridge Univ. Press, Cambridge, U.K., 332 pp.

Cudney, D.W., Jordan, L.S., Hall, A.E., 1991. Effect of wild oat (*Avena fatua*) infestations on light interception and growth rate of wheat (*Triticum aestivum*). Weed Sci. 39, 175–179.

Dall'Armellina, A.A., Zimdahl, R.L., 1988. Effect of light on growth and development of field bindweed (*Convolvulus arvensis*) and Russian Knapweed (*Centaurea repens*). Weed Sci. 36, 779–783. Will to 33457.

Davis, R.G., Johnson, W.E., Wood, F.O., 1967. Weed root profiles. Agron. J. 59, 555–556.

Davis, R.G., Wiese, A.F., Pafford, J.L., 1965. Root moisture extraction profiles of various weeds. Weeds 13, 98–100.

Derksen, D.A., Lafond, G.P., Thomas, A.G., Loeppky, H.A., Swanton, C.J., 1993. Impact of agronomic practices on weed communities: tillage systems. Weed Sci. 41, 409–417.

Derksen, D.A., Lafond, G.P., Thomas, A.G., Loeppky, H.A., Swanton, C.J., 1994. Impact of agronomic practices on weed communities: fallow within tillage systems. Weed Sci. 42, 184–199.

Derksen, D.A., Lafond, G.P., Thomas, A.G., Loeppky, H.A., Swanton, C.J., 1995. Impact of post-emergence herbicides on weed community diversity within conservation tillage systems. Weed Sci. 43, 311–320.

Devine, R., 1993. The cheatgrass problem. Atl. Mon. 271 (5), 40, 44–48.

Dewey, L.H., 1894. The Russian Thistle: Its History as a Weed in the United States with an Account of the Means Available for Its Eradication. Bull. 15. Div. of Botany, USDA, Washington, DC.

deWit, C.T., 1960. On Competition. Verslagen van landbouwkundige onderzoekingen, No. 66.8.

Dickerson Jr., C.T., Sweet, R.D., 1971. Common ragweed ecotypes. Weed Sci. 19, 64–66.

Dillman, A.C., 1931. The water requirements of certain crop plants and weeds in the Northern Great Plains. J. Agric. Res. 42, 187–238.

Donald, C.M., 1963. Competition among crop and pasture plants. Adv. Agron. 15, 1–118.

Eaton, B.J., Feltner, K.C., Russ, O.G., 1973. Venice mallow competition in soybeans. Weed Sci. 21, 89–94.

Eaton, B.J., Russ, O.G., Feltner, K.C., 1976. Competition of velvetleaf, prickly sida and venice mallow in soybeans. Weed Sci. 24, 224–228.

Fabricus, L.J., Nalewaja, J.D., 1968. Competition between wheat and wild buckwheat. Weed Sci. 16, 204–208.

Fischer, A.J., Dawson, J.H., Appleby, A.P., 1988. Interference of annual weeds in seedling alfalfa (*Medicago sativa*). Weed Sci. 36, 583–588.

Geddes, R.D., Scott, H.D., Oliver, L.R., 1979. Growth and water use by common cocklebur (*Xanthium pensylvanicum*) and soybeans (*Glycine max*) under field conditions. Weed Sci. 27, 206–212.

Ghafar, Z., Watson, A.K., 1983. Effect of corn (*Zea mays*) population on growth of yellow nutsedge (*Cyperus esculentus*). Weed Sci. 31, 588–591.

Ghersa, C.M., Roush, M.L., Radosevich, S.R., Cordray, S.M., 1994. Coevolution of agoecosystems and weed management. BioScience 44, 85–94.

Godara, R.K., Williams, B.J., Geaghan, J.P., 2012. Effect of shade on Texasweed (*Caperonia palustris*) emergence, growth, and reproduction. Weed Sci. 60, 593–599.

Gorske, S.F., Rhodes, A.M., Hopen, H.J., 1979. A numerical taxonomic study of *Portulaca oleracea*. Weed Sci. 27, 96–102.

Gould, F., 1991. The evolutionary potential of crop pests. Am. Sci. 79, 496–507.

Greb, B.W., Zimdahl, R.L., 1980. Ecofallow comes of age in the central great plains. J. Soil Water Cons. 35, 230–233.

Green-Tracewicz, E., Page, E.R., Swanton, C.J., 2012. Light quality and the critical period for weed control in soybean. Weed Sci. 60, 86–91.

Gruenhagen, R.D., Nalewaja, J.D., 1969. Competition between flax and wild buckwheat. Weed Sci. 17, 380–384.

Hagood Jr., E.S., Bauman, T.T., Williams Jr., J.L., Schreiber, M.M., 1980. Growth analysis of soybeans (*Glycine max*) in competition with velvetleaf (*Abutilon theophrasti*). Weed Sci. 28, 729–734.

Halford, C., Hamill, A.S., Zhang, J., Doucet, C., 2001. Critical period of weed control in no-till soybean (*Glycine max*) and corn (*Zea mays*). Weed Technol. 15, 737–744.

Hall, M.R., Swanton, C.J., Anderson, G.W., 1992. The critical period of weed control in grain corn (*Zea mays*). Weed Sci. 40, 441–447.

Harper, J.L., 1977. Population Biology of Plants. Academic Press, NY, 892 pp.

Hodgson, J.M., 1964. Variations in ecotypes of Canada thistle. Weeds 12, 167–171.

Holt, J.S., 1995. Plant responses to light: a potential tool for weed management. Weed Sci. 43, 474–482.

Jacoby, S., 2004. Freethinkers—A History of American Secularism. A Metropolitan/Owl Book. H. Holt and Co., New York, NY, p. 17.

Johnson, G.A., Mortensen, D.A., Martin, A.R., 1995. A simulation of herbicide use based on weed spatial distribution. Weed Res. 35, 197–205.

Keeley, P.E., Thullen, R.J., 1978. Light requirements of yellow nutsedge (*Cyperus esculentus*) and light interception by crops. Weed Sci. 26, 10–16.

King, L.J., 1966. Weeds of the World-Biology and Control. Interscience Pub., Inc., NY, p. 270.

King, R.P., Lybecker, D.W., Schweizer, E.E., Zimdahl, R.L., 1986. Bioeconomic modeling to simulate weed control strategies for continuous corn (*Zea mays*). Weed Sci. 34, 972–979.

Kirschenmann, F., 2013. Rethinking evolution: from competition to cooperation. Leopold Lett. 25 (4), 5. Winter.

Knake, E.L., Slife, F.W., 1962. Competition of *Setaria faberi* with corn and soybeans. Weeds 10, 26–29.

Knake, E.L., Slife, F.W., 1969. Effect of time of giant foxtail removal from corn and soybeans. Weed Sci. 17, 281–283.

Knezevic, S.Z., Datta, A., 2015. The critical period for weed control: revisiting data analysis. Weed Sci. 60, 542–551.

Knezevic, S.Z., Wiese, S.F., Swanton, C.J., 1994. Interference of redroot pigweed (*Amaranthus retroflexus*) in corn (*Zea mays*). Weed Sci. 42, 568−573.

Konesky, D.W., Siddiqi, M.Y., Glass, A.D.M., Hsiao, A.I., 1989. Wild oat and barley interactions: varietal differences in competitiveness in relation to phosphorus supply. Can. J. Bot. 67, 3366−3376.

LeFevre, P., 1956. Influence du milieu et des conditions d'exploration sur le developpement des plantes adventices. Effet particulier du pH et l'etat calcique. Ann. Agron. Paris 7, 299−347.

Li, M., 1960. An Evaluation of the Critical Period and the Effects of Weed Competition on Oats and Corn (Ph.D. dissertation). Rutgers Univ., New Brunswick, NJ.

Lybecker, D.W., King, R.P., Schweizer, E.E., Zimdahl, R.L., 1984. Economic analysis of two weed management systems for two cropping rotations. Weed Sci. 32, 90−95.

Mack, R.N., 1981. Invasion of *Bromus tectorum* L. Into western North America: an ecological chronicle. Agro-Ecosystems 7, 145−165.

Malicki, L., Berbeciowa, C., 1986. Uptake of more important mineral components by common field weeds on loess soils. Acta Agrobot. 39, 129−141.

McGiffen Jr., J.E., Masiunas, J.B., Hesketh, J.D., 1992. Competition for light between tomatoes and nightshades (*Solanum nigrum* or *S. ptycanthum*). Weed Sci. 40, 220−226.

McWhorter, C.G., 1971. Growth and development of johnsongrass ecotypes. Weed Sci. 19, 141−147.

McWhorter, C.G., Jordan, T.N., 1976. Comparative morphological development of six johnsongrass ecotypes. Weed Sci. 24, 270−275.

Moody, K., July 9 1981. Weed fertilizer interactions in rice. Int. Rice Res. Inst. Thursday Seminar.

Mortensen, D.A., Dieleman, J.A., Johnson, G.A., 1998. Weed spatial variation and weed management. In: Hatfield, J.L., Buhler, D.D., Stewart, B.A. (Eds.), Integrated Weed and Soil Management. CRC Press, Boca Raton, FL, pp. 293−309.

Mortimer, A.M., 1984. Population ecology and weed science. In: Dirzo, R., Sarukhan, J. (Eds.), Perspectives on Plant Population Ecology. Sinauer Assoc., Inc., Sunderland, MA, pp. 363−388.

Mortimer, A.M., 1987. The population ecology of weeds−implications for integrated weed management, forecasting and conservation. In: Proc. Brit. Crop Prot. Conf.−Weeds, pp. 936−944.

Munger, P.H., Chandler, J.M., Cothern, J.T., 1987. Effect of water stress on photosynthetic parameters of soybean (*Glycine max*) and velvetleaf (*Abutilon theophrasti*). Weed Sci. 35, 15−21.

Muzik, T.J., 1970. Weed Biology and Control. McGraw Hill Book Co., NY, p. 17.

Nieto, J.F., 1970. The struggle against weed in maize and Sorghum. In: FAO Int. Conf. on Weed Control. Davis, CA, pp. 79−86.

Nieto, J.F., Brando, M.A., Gonzalez, J.T., 1968. Critical periods of the crop growth cycle for competition from weeds. PANS 14, 159−166.

Noka, R., Owen, M.D.K., Swanton, C.J., 2015. Weed abundance, distribution, diversity, and community analyses. Weed Sci. 64−90 (special issue).

Norris, R.F., 2005. Ecological bases of interactions between weeds and organisms in other pest categories. Weed Sci. 53, 909−913.

Norris, R.F., 1992. Have ecological and biological studies improved weed control strategies. In: Proc. 1st Int. Weed Cont. Cong., vol. 1, pp. 7−33 (Melbourne, Australia).

Norris, R.F., Kogan, M., 2000. Interactions between weeds, arthropod pests, and their natural enemies in managed ecosystems. Weed Sci. 48, 94−158.

Odero, D.C., Duchrow, M., Havranek, N., 2016. Critical timing of fall panicum (*Panicum dichotomiflorum*) removal in sugarcane. Weed Technol. 30, 13−20.

Okafor, L.I., DeDatta, S.K., 1976. Competition between upland rice and purple nutsedge for nitrogen, moisture, and light. Weed Sci. 24, 43−46.

Patterson, D.T., 1979. The effects of shading on the growth and photosynthetic capacity of Itchgrass (*Rottboellia exaltata*). Weed Sci. 27, 549−553.

Patterson, D.T., 1982. Effects of light and temperature on weed/crop growth and competition. In: Hatfield, J.L., Thomason, I.J. (Eds.), Biometeorology in Integrated Pest Management. Academic Press, New York, pp. 407−420.

Patterson, D.T., 1985. Comparative ecophysiology of weeds and crops. In: Duke, S.O. (Ed.), Weed Physiology. CRC Press, Boca Raton, FL, pp. 101−129.

Patterson, D.T. (Ed.), 1989. Composite List of Weeds. Weed Sci. Soc. of America, Champaign, IL (Now—Lawrence, KS), 112 pp.

Patterson, D.T., 1995a. Effects of environmental stress on weed/crop interactions. Weed Sci. 43, 483–490.

Patterson, D.T., 1995b. Weeds in a changing climate. Weed Sci. 43, 685–701.

Peterson, D., 1936. Stellaria-Studien. Zur zytologie, genetik, okologie und systematik der gattung stellaria in besonders der media-gruppe. Bot. Not. 281–419.

Radosevich, S.R., Holt, J.S., 1984. Weed Ecology: Implications for Vegetation Management. Wiley Interscience, NY, 265 pp.

Radosevich, S.R., Holt, J., Ghersa, C., 1997. Weed Ecology: Implications for Vegetation Management, second ed. Wiley Interscience, New York, NY. 589 pp.

Radosevich, S.R., Holt, J., Ghersa, C., 2007. Ecology of Weeds and Invasive Plants: Relationship to Agriculture and Natural Resource Management, third ed. Wiley Interscience; J. Wiley &Sons, New York. 454 pp.

Regnier, E.E., Salvucci, M.E., Stoller, E.W., 1988. Photosynthesis and growth responses to irradiance in soy bean (Glycine max) and three broadleaf weeds. Weed Sci. 36, 487–496.

Schenk, H.J., Callaway, R.M., Mahall, B.E., 1999. Spatial root segregation: are plants territorial? Adv. Ecol. Res. 28, 146–180.

Schroeder, J., Thomas, S.H., Murray, L.W., 2005. Impacts of crop pests on weeds and weed-crop interactions. Weed Sci. 53, 918–922.

Scott, H.D., Geddes, R.D., 1979. Plant water stress of soybean (Glycine max) and common cocklebur (Xanthium pensylvanicum): a comparison under field conditions. Weed Sci. 27, 285–289.

Sexsmith, J.J., Pittman, U.J., 1963. Effect of nitrogen fertilizer on germination and stand of wild oats. Weeds 11, 99–101.

Shantz, H.L., Piemeisel, L.N., 1927. The water requirement of plants at Akron, Colorado. J. Agric. Res. 34, 1093–1190.

Shetty, S.V.R., Sivakumar, M.V.K., Ram, S.A., 1982. Effect of shading on the growth of some common weeds of the semi-arid tropics. Agron. J. 74, 1023–1029.

Sibuga, K.P., Bandeen, J.D., 1978. An evaluation of green foxtail [Setaria viridis (L.) Beauv.] and common lambsquarters (Chenopodium album) competition in corn. Weed Sci. Soc. Am. Abstr. No. 142.

Siddiqi, M.Y., Glass, A.D.M., Hsiao, A.I., Minjas, A.N., 1985. Wild oat/barley interactions: varietal differences in competitiveness in relation to K^+ supply. Ann. Bot. 56, 1–7.

Smith Jr., R.J., 1968. Weed competition in rice. Weed Sci. 16, 252–254.

Snaydon, R.W., 1984. Plant demography in an agricultural context. In: Dirzo, R., Sarvichan, J. (Eds.), Perspectives on Plant Population Ecology. Sinauer Assoc., Inc., Sunderland, MA, pp. 389–407.

Stokłosa, A., Matraszek, R., Isman, M.B., Upadhyaya, M.K., 2012. Phytotoxic activity of clove oil, its constituents, and its modification by light intensity in broccoli and common lambsquarters (chenopodium album). Weed Sci. 60, 607–611.

Stoller, E.W., Woolley, J.T., 1985. Competition for light by broadleaf weeds in soybeans (Glycine max). Weed Sci. 33, 199–202.

Stoller, E.W., Myers, R.A., 1989. Response of soybeans (Glycine max) and four broadleaf weeds to reduced irradiance. Weed Sci. 37, 570–574.

Storkey, J., Moss, S.R., Cussans, J.W., 2010. Using assembly theory to explain changes in weed flora in response to agricultural intensification. Weed Sci. 58, 39–46.

Stuart, B.L., Harrison, S.K., Abernathy, J.R., Kreig, D.R., Wendt, C.W., 1984. The response of cotton (Gossypium hirsutum) water relations to smooth pigweed (Amaranthus hybridus) competition. Weed Sci. 32, 126–132.

Swanton, C.J., Noka, R., Blackshaw, R.E., 2015. Experimental methods for crop-weed competition studies. Weed Sci. 2–11 (special issue).

Thomas, P.E.L., Allison, J.C.S., 1975. Competition between maize and Rottboellia exaltata Linn. J. Agric. Sci. 84, 305–312.

Vega, M.R., Ona, J.D., Paller Jr., E.P., 1967. Weed control in upland rice at the Univ. of the Philippines College of Agric. Philipp. Agric. 51, 397–411.

Vengris, J., Drake, M., Colby, W.G., Bart, J., 1953. Chemical composition of weeds and accompanying crop plants. Agron. J. 45, 213–218.

Vengris, J., Colby, W.G., Drake, M., 1955. Plant nutrient competition between weeds and corn. Agron. J. 47, 213–216.

Warwick, S.I., Briggs, D., 1978. The genecology of lawn weeds. 1. Population differentiation in *Poa annua* in a mosaic environment of bowling green lawn and flower beds. New Phytol. 81, 711–723.

Weatherspoon, D.M., Schweizer, E.E., 1971. Competition between sugarbeets and five densities of kochia. Weed Sci. 19, 125–128.

Weaver, S., 1996. Simulation of crop-weed competition: models and their application. Phytoprotection 77, 3–11.

Wedderspoon, I.M., Burt, G.W., 1974. Growth and development of three johnsongrass selections. Weed Sci. 22, 319–322.

Wiese, A.F., 1965. Effect of tansy mustard and 2,4-D on winter wheat. Tex. Agric. Expt. Stn. Bull. Mp-782.

Wiese, A.F., Vandiver, W., 1970. Soil moisture effects on competitive ability of weeds. Weed Sci. 18, 518–519.

Williams, C.F., 1971. Interaction of crop plant population with weed competition in corn (*Zea mays*). Bush and snap beans (*Phaseolus vulgaris*) and onion (*Allium cepa*) at differing stages of development. Diss. Abstr. 32 (4). 1944-B, 1945-B.

Wisler, G.C., Norris, R.F., 2005. Interactions between weeds and cultivated plants as related to management of plant pathogens. Weed Sci. 53, 914–917.

Yip, C.P., 1978. Yellow nutsedge ecotypes, their characteristics and responses to environment and herbicides. Diss. Abstr. Int. B 39, 1562–1563.

Young, F.L., Wyse, D.L., Jones, R.J., 1983. Effect of irrigation on quackgrass (*Agropyron repens*) interference in soybeans (*Glycine max*). Weed Sci. 31, 720–726.

Young, J.A., Evans, R.A., Kay, B.L., 1970. Phenology of reproduction of medusahead. Weed Sci. 18, 451–454.

Young, J.A., 1988. The public response to the catastrophic spread of Russian thistle (1880) and Halogeton (1945). Agric. Hist. 62, 122–130.

Young, J.A., March 1991. Tumbleweed. Sci. Am. 82–87.

Zimdahl, R.L., 1999. My view. Weed Sci. 47, 1.

Zimdahl, R.L., 1980. Weed-Crop Competition—A Review. The Int. Plant Prot. Center. Oregon State Univ., 195 pp.

Zimdahl, R.L., 2004. Weed-Crop Competition—A Review, second ed. Blackwell Pub., Ames, IA. 220 pp.

Weed Population Genetics

Michael J. Christoffers

North Dakota State University, Fargo, North Dakota

FUNDAMENTAL CONCEPTS

- A weed population consists of all individuals of a weed species in a geographic area. Genetically, a weed population is limited to individuals with potential to interbreed.

- Weeds are capable of generating diverse genetic mutations that are often inherited.

- The effects of a mutation depend on how environmental and genetic factors influence gene expression.

- Weeds use diverse sexual and asexual methods to reproduce.

- Weed populations change owing to mutation, migration, selection, or random genetic drift.

LEARNING OBJECTIVES

- To describe various chromosomal and molecular mutations possible in weeds.

- To compare and appropriately use genetic terms, including genotype, phenotype, gene, allele, mutant, wild type, polymorphism, haplotype, homozygous, heterozygous, and dominance.

- To define gene expression from different viewpoints and generalize the effects of environment and genetic factors on the expression of genes.

- To predict genotype frequencies from allele frequencies using the Hardy—Weinberg principle.

- To contrast hybridization experiments with wild populations and interpret observations from each.

- To consider the evolution of weed populations and describe how selection and other forces change allele frequencies.

1. INTRODUCTION

A weed population is composed of all individuals of the same species within a geographic area. The specific area in which a weed population exists may not be well-defined and may range from a large geographic region to a small field, depending on the context in which the population is being considered. In a genetic context, the definition of a population is refined to include only individuals with the potential to interbreed. Such a population is composed of individuals with the potential to contribute to a single gene pool, which is the collective sum of all genes from which the next generation will be derived. With this genetic definition, it may still be difficult to define individual weed populations clearly. However, the concept of the gene pool is important when considering changes in gene frequencies owing to biological factors and agronomic practices.

The genetic aspects of variation, expression, and inheritance are all important in the study of how weed populations change (i.e., population dynamics). The foundation of genetic variation is the existence of different forms or variants of genes. These gene variants are termed alleles, and allelic diversity within the gene pool of a weed population is fundamental to its ability to adapt to changing environments. The genetic makeup of each plant (that is, the specific alleles for all of its genes) is that plant's genotype. The phenotype of a plant is how it physically appears, but includes its functional metabolism. A plant's phenotype is determined by its genotype and the environment, with the environment affecting the phenotype directly or influencing the way genes are expressed. The expression of an individual allele (that is, the contribution that the allele makes to the phenotype) may also be influenced by its alternative allele on the homologous chromosome (dominance, etc.) or other genes (epistasis). The phenotype in turn may influence the relative inheritance of alleles from generation to generation, changing allelic frequencies within the gene pool and leading to adaptation of weed populations to new environments and weed control strategies.

2. GENETIC DIVERSITY

2.1 Mutation at the Chromosomal and Molecular Levels

The word "mutation" is from the Latin *mutare*, meaning "to change." Use of "mutation" to describe new genetic variations is attributed to the Dutch botanist Hugo de Vries, who is one of the recognized re-discoverers of Gregor Mendel's earlier genetic studies on garden pea. De Vries' work highlights the early role that weedy plants had in the development of genetics as a discipline. De Vries developed his mutation theory based on new variants, or mutants, of weedy redsepal evening primrose collected in 1886 in a neglected potato field in The Netherlands (Uphof, 1922).

Population genetics is primarily concerned with gametic mutations: that is, those that affect gametes. Gametes are the male and female reproductive cells that fuse during fertilization to form a zygote. Gametic mutations may originate during production of the gametes themselves or may arise within the cell line leading to the eventual production of gametes (i.e., the germ line). Gametic mutations may be passed from generation to generation and increase or decrease in frequency within populations.[1] This is why they are the primary focus of weed population genetics. In contrast, mutations that occur in somatic cells (any cell other than a germ cell) are not inherited by subsequent generations. For example, a somatic mutation occurring in the tip of a leaf would not be passed on to the next generation.

2.1.1 Chromosomal Mutations

Geneticists use the term "wild type" to denote normal, typical alleles, chromosomes, phenotypes, etc. Mutations are atypical variants, presumably derived from the wild type through a genetic mistake or alteration. Genetic studies of de Vries' redsepal evening prim-rose mutants and their progeny revealed mutations at the chromosomal level: that is, micro-scopically visible changes in chromosome number or rearrangements of chromosome structure. Wild-type plants with full chromosome sets are termed euploid, but some of de Vries' mutants were aneuploids with an extra chromosome (trisomic) (Nei and Nozawa, 2011). Another mutant was actually a polyploid with four copies of each chromosome (tetraploid) instead of the usual two copies in diploids. *Oenothera* spp. such as redsepal evening primrose are cytologically unusual owing to translocations (Cleland, 1972), which are chromosomal rearrangements in which portions of chromosomes are moved to new locations. As such, chromosome mutations are relatively common in redsepal evening primrose and have been observed in other weed species.

A classic example of phenotypic variation resulting from aneuploidy is the work done on jimsonweed by Albert Blakeslee in the early to mid-1900s (Sinnott, 1959). Wild-type, euploid jimsonweed has 12 homologous pairs of chromosomes, a total of 24 chromosomes. Blakeslee investigated jimsonweed with different seed capsule phenotypes and discovered that most were trisomic with an extra chromosome. Most of the 12 possible primary trisomics produced their own distinctive seed capsule type. Other mutants were discovered in which the extra chromosome was actually made up of a doubled half of one of the 12 chromosomes (second-ary trisomy) or sections from different chromosomes (tertiary trisomy).

Aneuploidy is rarely beneficial. Jimsonweed trisomics have a reduced growth rate compared with the wild type (Avery et al., 1959). It is also normal for chromosome mutations such as trans-locations to reduce fertility owing to the production of unbalanced gametes. Unbalanced gametes have missing or extra chromosome segments and may not produce viable offspring. However, if homologous chromosomes have the same translocation, fertility may not be affected. It is usual to find translocations when comparing the chromosomes of related species, as Yang et al. (1999) found among *Avena* spp. including wild oat. It seems that as long as a plant carries all necessary genetic information needed for health, it does not matter too much where among the chromosomes the information is located as long as gamete balance is maintained.

Alterations in chromosome number or organization may have detrimental effects on plants, or they may have little to no effect on a plant's phenotype. But do such alterations ever benefit the plant? The original derivation of polyploids from diploid ancestors is considered by many to be a type of chromosome mutation. Although it is sometimes detrimental, it is likely that many estab-lished, stable polyploid plants have an adaptive advantage over their diploid counterparts.

---

[1]A gametic mutation is not always passed on to the next generation. For example, if the mutation occurs during the final replication preceding meiosis, it may only be in one of the thousands or more gametes produced by the plant, and it will not be passed on unless involved in successful fertilization.

There is evidence that polyploidy is more common among weeds than among non-weedy plants (Bennett et al., 1998), which suggests that in some way, polyploidy helps facilitate weediness.

A plant's nuclear genome is the entire set (one haploid set) of all of its chromosomes.[2] It is designated by the letter n. Diploid weeds and other plants have two copies of their nuclear genome ($2n$), whereas polyploids have more than two genomic copies ($>2n$).[3] As a result, polyploids often have more than two copies of each gene. Additional gene copies enable the greater ability to carry different versions of each gene, and this is often beneficial for a plant when it faces a changing environment. However, more DNA per cell sometimes slows growth and development, which would be detrimental. There is also evidence that the presence of additional wild-type susceptible genes in polyploid weeds can buffer the level of herbicide resistance imparted by herbicide target-site gene mutations (Yu et al., 2013).

Additional gene copies may also be beneficial if there is an advantage for the plant to produce more of the gene product (enzyme, etc.). Polyploidy is not the only way that weeds may carry more than two copies of a gene. The gene or chromosomal segment carrying the gene may be duplicated, sometimes repeatedly, while the rest of the genome remains the same. These duplications may result in multiple copies of a gene within a genome, which is termed gene amplification.[4] Amplification of the gene for 5-enolpyruvylshikimate-3-phosphate synthase (EPSPS) has been found in several glyphosate-resistant weed species including Palmer amaranth and kochia (reviewed by Sammons and Gaines, 2014). The EPSPS enzyme is the target of glyphosate, and amplification of the wild-type, herbicide-susceptible *EPSPS* gene likely results in more EPSPS enzyme than glyphosate can inhibit at normal treatment rates.

One genetic mechanism that may result in gene duplication is unequal crossing over, an abnormal, nonreciprocal occurrence of normal, reciprocal crossing over. Normal crossing over occurs during the production of new gametes in meiosis and is a process in which sections of homologous chromosome pairs are rearranged in new combinations. This allows creation of new versions of chromosomes where the alleles for many, but not all, genes are reciprocally exchanged between the homologues. However, in unequal crossing over, because of misalignment of the homologues, one chromosome may receive the other chromosome's copy of a gene (allele) while keeping its own. The first chromosome then has a duplication and the second has a deletion. Gametes carrying these chromosomes will in turn produce progeny with a duplication or deletion, respectively. Unequal crossing over tends to produce tandem duplications in which the duplicated genes or chromosomal segments are right next to each other on the chromosome. Tandem duplications may promote further misalignment of homologous chromosomes in meiosis, increasing the chance of additional duplication via unequal crossing over. This process is suspected to have driven gene

---

[2]Plants also have extrachromosomal DNA forming the genomes of chloroplasts and mitochondria.

[3]Alternatively, $2n$ may represent all the chromosomes in the nucleus of a polyploid while x represents one genomic copy (one basic set of chromosomes). For example, a diploid plant would be $2n = 2x$ while a polyploid might be $2n = 4x$ (tetraploid) or $2n = 6x$ (hexaploid), etc. This notation is especially used for allopolyploids having dissimilar genomic copies likely originating from different progenitor species.

[4]Gene duplications are a form of gene amplification using a broad definition of the latter. A narrow definition of gene amplification limits the term to temporary synthesis of extra gene copies and would not include the duplications described here.

amplification of *EPSPS* in kochia, in which duplicated chromosomal segments carrying EPSPS are in tandem orientation (Jugulam et al., 2014).

Individual glyphosate-resistant Palmer amaranth plants have varying *EPSPS* copy numbers; some have more than 100 and the *EPSPS* genes are dispersed repeats scattered all over the genome (Gaines et al., 2011). Dispersed repeats are not likely to be the result of unequal crossing over, and the exact mechanism of gene amplification in Palmer amaranth has not been clearly identified, but the action of transposable genetic elements (transposons) is a possibility.

Barbara McClintock discovered transposons in corn (McClintock, 1950) and received the 1983 Nobel Prize in Physiology or Medicine for her work. Active transposons are able to translocate to new areas of the genome, with some transposons excising from one location and inserting into another and others leaving a copy behind at the original location. The latter are especially able to increase their genomic copy number over time.

2.1.2 Molecular Genetics

In 1944, Avery et al. identified DNA as the genetic material that Griffith (1928) had earlier called the transforming principle. Subsequent studies, including publication of DNA's structure by Watson and Crick (1953), confirmed that DNA was indeed the component of chromosomes that stored genetic information.

DNA is a double helix of two strands. Each strand is composed of a sugar-phosphate backbone onto which nitrogenous bases are attached. The four nitrogenous bases are adenine (A), cytosine (C), guanine (G), and thymine (T). They are the bases of each strand pair within the double helix. Pairing of the bases is by hydrogen bonding, with A on one strand forming two hydrogen bonds with T on the other strand, and C forming three hydrogen bonds with its pair, G, on the opposite strand. The strands are therefore considered complementary, each a complement of the other. The strands of the double helix are also antiparallel, meaning that the chemical structures of the two strands are in opposite orientations. It is therefore convenient to have a way to note the orientation of each strand, and we do so by noting strand ends as 5′ or 3′ based on the orientation of carbons within the deoxyribose sugars of the backbone.

Fig. 7.1 shows the DNA sequence of a 9−base pair (bp) portion of the acetolactate synthase (ALS) gene sequence of kochia (Foes et al., 1999). As with most but not all genes, information stored as DNA is used to produce messenger RNA (mRNA) by a process called transcription, and the mRNA is in turn used to produce a protein product by a process called translation. Although the genetic sequence of mRNA is based on DNA, these molecules differ in that mRNA is usually single stranded, has uracil (U) in place of T, and lacks intervening sequences (introns) originally present in the DNA. In the process of translation, the base sequence of mRNA is read three bases at a time, and each group of three bases is called a codon. Translation begins at a start (initiation) codon, which in eukaryotes is typically the first AUG in the mRNA (from the 5′ end). Translation then continues until one of three stop (termination) codons is reached (UAA, UAG, or UGA). With the exception of the stop codons, each codon indicates one specific amino acid, and these amino acids are chemically linked together in a polypeptide chain (protein). For example, the AUG start codon specifies the amino acid methionine. The correspondence of codons to amino acids (or translation stops) is nearly universal, so that a researcher who knows a DNA/mRNA sequence and the gene's reading frame (three-by-three grouping of bases into codons) can infer the amino

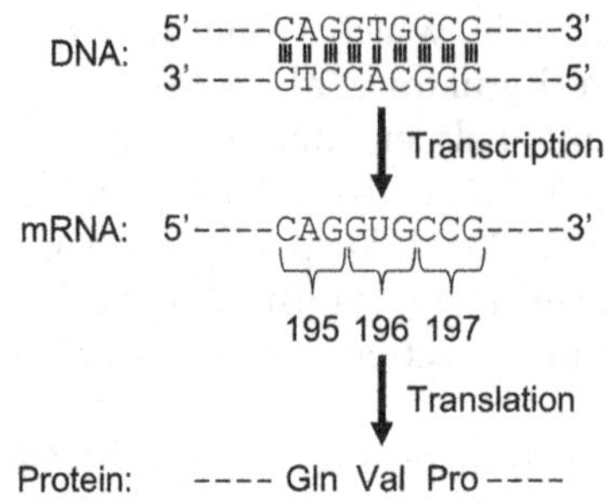

```
              5'----CAGGTGCCG----3'
DNA:              ‖‖‖‖‖‖‖‖‖
              3'----GTCCACGGC----5'

                      │ Transcription
                      ▼

mRNA:         5'----CAGGUGCCG----3'
                   └┬┘└┬┘└┬┘
                   195 196 197

                      │ Translation
                      ▼

Protein:      ---- Gln Val Pro ----
```

FIGURE 7.1 A 9−base-pair (bp) section of double-stranded DNA in which *arrows* represent hydrogen bonds and *horizontal dashes* indicate that the strands actually extend beyond the 9 bp shown here. Note the opposite orientation of strands indicated by the 5′ and 3′ ends. This sequence is part of the acetolactate synthase gene of kochia (Foes et al., 1999). The corresponding messenger RNA (mRNA) (with indicated codon position numbers) and protein sequence are shown.

acid order in the final polypeptide. Such a coding dictionary can be found in most introductory genetics textbooks.

In the ALS example (Fig. 7.1), codons CAG, GUG, and CCG specify amino acids glutamine, valine, and proline, respectively. It seems reasonable to number the amino acids of a polypeptide beginning with the initial methionine as position 1. However, many polypeptides are modified after translation and before a final functional protein is achieved, including the eventual trimming of some amino acids used for intracellular targeting such as signal peptides. The exact number of specified amino acids may also vary among genes from different species, or even among alleles within a species. Therefore, it is common to standardize amino acid position numbers to one designated, usually wild-type gene. For ALS genes of weed species, position numbers are usually standardized to the ALS gene of mouse-ear cress.[5] The three amino acids in the ALS example correspond to positions 195−197 in this numbering system.

It is not always convenient to indicate the bases for both strands when noting a DNA sequence, as in Fig. 7.1, and it is rarely necessary to do so because the sequence of one strand can be known from the other (because the strands are complementary). Thus, DNA sequences are often given only as the single-strand equivalent to the mRNA. For example, the DNA sequence in the ALS example might be given only as CAGGTGCCG, with a 5′ (left) to 3′ (right) orientation assumed. This simplified notation is used throughout the rest of this chapter.

2.1.3 Base Substitutions

Polypeptide chains fold and sometimes associate with other polypeptide chains to form proteins that perform specific functions (i.e., influence the phenotype). Wild-type ALS genes produce wild-type ALS enzyme, which is capable of performing its normal function of catalyzing the first step in the synthesis of branched chain amino acids. However, the wild-type ALS enzyme is susceptible to inhibition by sulfonylureas and imidazolinones and some other herbicides. As such, ALS is called the "target site" of these herbicides, which have ALS inhibition as their mechanism of action. However, some specific mutations within the ALS

---

[5]Arabidopsis, the genus name of mouse-ear cress dominates the literature and will be used in this chapter.

TABLE 7.1 Kochia Acetolactate Synthase Gene Sequences at Amino Acid Position 197 (Numbering System Based on *Arabidopsis*) and Corresponding Response to Chlorsulfuron Herbicide

Name	Collection Site	Codon	Amino Acid	Chlorsulfuron Response
KS-S	Kansas	CCG	Proline	Susceptible
MAN-S	Manitoba	CCG	Proline	Susceptible
ND-S	North Dakota	CCG	Proline	Susceptible
SLV-S	Colorado	CCG	Proline	Susceptible
ID#2-R	Idaho	CCG	Proline	Resistant
SD-R	South Dakota	CCG	Proline	Resistant
CO-R	Colorado	CCG	Proline	Resistant
KS-R	Kansas	ACG	Threonine	Resistant
ND-R	North Dakota	CGG	Arginine	Resistant
MAN-R	Manitoba	CTG	Leucine	Resistant
MT-R	Montana	CAG	Glutamine	Resistant
ID#5-R	Idaho	GCG	Alanine	Resistant
T7-R	Texas	GCG	Alanine	Resistant
SLV-R	Colorado	TCG	Serine	Resistant

Based on Guttieri, M.J., Eberlein, C.V., Thill, D.C., 1995. Diverse mutations in the acetolactate synthase gene confer chlorsulfuron resistance in kochia (Kochia scoparia) *biotypes. Weed Sci. 43, 175–178.*

enzyme are known to confer herbicide resistance so that the enzyme is not inhibited to the normal extent by the herbicide (Tranel and Wright, 2002).

Guttieri et al. (1995) identified several kochia accessions resistant to the sulfonylurea herbicide chlorsulfuron caused by mutations within their ALS genes at amino acid position 197 (Table 7.1). Notice that the four kochia accessions that are wild-type susceptible to chlorsulfuron also have proline at position 197, specified by the codon CCG. As in Fig. 7.1, CCG at this position is known to be the wild-type codon in plants, based on previous studies. However, three kochia accessions (ID numbers 2-R, SD-R, and CO-R) are also wild type at position 197 but have a mutant, resistant response to chlorsulfuron. A resistant ALS enzyme was confirmed in these accessions, indicating the likely presence of ALS mutation(s) outside position 197. In other words, the wild-type codon for proline at position 197 is likely within a mutant allele for resistant ALS.

An allele for herbicide resistance is often loosely referred to as a gene for herbicide resistance. For example, it might be said that many of the kochia accessions in Table 7.1 have a gene for resistance to chlorsulfuron. However, the gene in this example is not specifically for herbicide resistance. It controls a metabolic step in the synthesis of branched chain amino acids. It is the allele that is of an herbicide-resistant or herbicide-susceptible form, based on the ALS enzyme that it encodes. However, even professional geneticists often use the term "gene" for what is actually an allelic variant, with the realization that the terminology is not accurate.

For each mutant codon in Table 7.1, a single base has been changed from the wild-type sequence. A change in a single base of the DNA is a termed a base substitution, which is a type of point mutation,[6] and among kochia accessions there are several different base substitutions that have produced different ALS alleles. The single-base differences among these various alleles may also be called single nucleotide polymorphisms (SNPs). A "nucleotide" includes not only a base but also the pentose sugar (deoxyribose in DNA), and the phosphate group of the sugar-phosphate backbone. Because it is the identity of the base that is most important when considering genetic variation, the terms "nucleotide" and "base" are often used interchangeably. Polymorphism is generally used to mean the existence of different forms, and genetic polymorphism indicates genetic variation without specifically identifying which variant is wild type and which is mutant.

The base substitutions shown in Table 7.1 are nonsynonymous. This means that because of the mutation, the codon no longer specifies the same amino acid as the wild-type sequence. When a nonsynonymous mutation specifies a different amino acid, as those in Table 7.1 do, it is called a missense mutation or an amino acid substitution. Because missense mutations affect the sequence of the translated polypeptide, they may affect the function of the protein product (enzyme, etc.) and may ultimately influence the phenotype of the plant. The consequences of missense mutations in weeds are often negative or neutral but may be beneficial (e.g., when they provide herbicide resistance). Missense mutations are a common cause of herbicide target site resistance in weeds (reviewed by Powles and Yu, 2010). Although a single missense mutation is often enough to confer herbicide resistance in weeds, glyphosate resistance conferred by a proline to serine mutation at position 106 of EPSPS is increased by a second threonine to isoleucine mutation at position 102 in goosegrass (Yu et al., 2015).

The missense mutations shown in Table 7.1 were produced by single-base substitutions, but some amino acid substitutions require more than one base change within a codon. For example, an allele for ALS in wild radish has the DNA sequence TAT for tyrosine in the codon for amino acid 122 (Han et al., 2012). Compared with the wild-type sequence in which GCT designates alanine at this position, the mutant ALS allele has two substitutions corresponding to the first two bases of the codon.

Another type of nonsynonymous base substitution occurs when a codon for an amino acid is changed into a stop codon. This type of change, a nonsense mutation, is often detrimental to the function of the mutated allele's protein product, owing to truncation (shortening) of the polypeptide. Nonsense mutations represent less than 1% of nonsynonymous base substitutions in weedy sunflower (Renaut and Rieseberg, 2015).

With four different bases in DNA (A, C, G, and T), there are $4^3 = 64$ possible three-base codons. Three of these codons code for a stop codon, and $64 - 3 = 61$ codons specify amino acids. However, there are only 20 amino acids. This conflict is resolved by degeneracy of the genetic code; that is, most amino acids are specified by more than one codon. Only methionine and tryptophan are specified by a single codon, whereas other amino acids may be specified by up to six different codons. This means that some base substitutions do not change the

---

[6]Definitions of "point mutation" vary but always include single base substitutions within genes. Broader definitions may include substitutions, deletions, or insertions of one or a few bases within or outside of genes, or even larger mutations with the condition that normal recombination (crossing over) and chromosome structure is maintained.

resulting polypeptide sequence of an allele and are considered neutral. These mutations are called synonymous, same-sense, or silent mutations. Synonymous mutations can be useful in population genetic studies because they identify specific alleles while not imparting a benefit or detriment to the plant.

What causes base substitutions? These and other mutations in weeds are mostly considered to be spontaneous (i.e., they occur naturally). Cell division requires copies of plant genomes to be made for distribution to daughter cells by mitosis. However, DNA polymerase enzymes that replicate DNA are not perfect, and they sometimes make mistakes when synthesizing or replicating DNA copies. Mutations may also be incorporated when damaged DNA is repaired. However, these mutations have a useful role in populations by producing genetic diversity that may sometimes be beneficial.

Herbicide resistance in various weeds and seed shattering in weedy rice are examples in which relatively new, spontaneous mutations appear to have provided variation beneficial to weediness (reviewed by Vigueira, 2013). For weed populations treated with an herbicide, mutations conferring resistance to the herbicide are beneficial because they increase the probability that some members of the population will survive and reproduce. In some US (Thurber et al., 2010) and Japanese (Akasaka et al., 2011) weedy rice populations, the ability to disperse seeds by shattering appears to have been derived from new mutations that may have originated in cultivated rice progenitors. Seed shattering is a benefit to weedy rice because it allows seeds to remain in fields for continued infestation rather than being harvested. In situations in which herbicide resistance mutations are often, but not always, found to be base substitutions, the mutation(s) responsible for shattering in weedy rice has not been clearly classified.

Nucleotide bases may be classified by their chemical structures, with A and G as purines and C, T, and U as pyrimidines. The substitution of a purine for a purine or pyrimidine for a pyrimidine is called a transition, whereas substitution of a purine for a pyrimidine or vice versa is called a transversion. Although errors in DNA replication can cause a variety of mutations, the transition of GC bp to AT is the most common base substitution in plants (Ossowski et al., 2010). Ultraviolet light is a known cause of spontaneous GC-to-AT transitions, as is spontaneous chemical modification (deamination) of methylated C. In plants and other organisms, C bases may be methylated as a mechanism of transcriptional repression. Ossowski et al. found the mutation rate in *Arabidopsis* to be 7×10^{-9} base substitutions per site per generation, which is a reasonable estimate for plants in general.

2.1.4 Other Molecular Mutations

Not all DNA mutations at the molecular level are base substitutions. Bases may also be inserted or deleted, and together these mutations are called indels. Most of the DNA in plant genomes does not specifically code for a protein product: that is, it is not part of a gene's coding DNA sequence (CDS). The CDS of a gene is the portion from a start codon to the stop codon except for introns, which are spliced out in the processing of mRNA. Indels within non-CDS regions can sometimes affect gene expression, but within CDS regions an indel often alters the gene's reading frame (a frameshift mutation) and may drastically change the final polypeptide product. For example, deletion of a single C in the flavonoid 3'-hydroxylase (F3'H) gene of soybean results in a frameshift mutation in which all codons after the deletion have been changed (Fig. 7.2), which in soybean produces an allele for gray instead

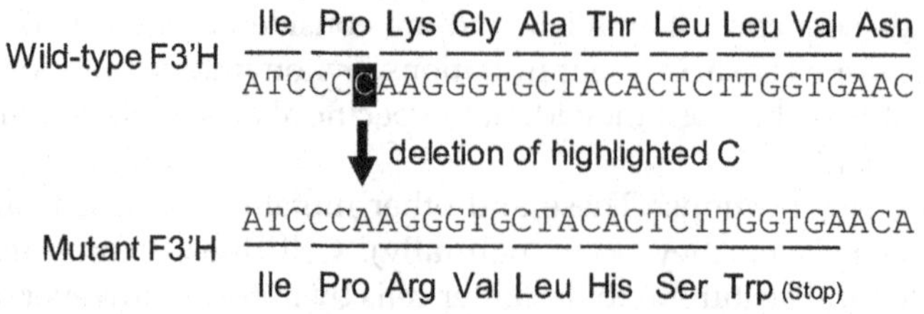

Wild-type F3'H
Ile Pro Lys Gly Ala Thr Leu Leu Val Asn
ATCCC■AAGGGTGCTACACTCTTGGTGAAC

⬇ deletion of highlighted C

Mutant F3'H
ATCCCAAGGGTGCTACACTCTTGGTGAACA
Ile Pro Arg Val Leu His Ser Trp (Stop)

FIGURE 7.2 A mutant allele for flavonoid 3′-hydroxylase (F3′H) in soybean differs from the wild-type allele by a single-base deletion of cytosine (C) that resulted in a frameshift mutation (Toda et al., 2002).

of brown pubescence (Toda et al., 2002). Note that this frameshift mutation also results in an earlier than normal stop codon.

Because codons consist of three bases, indels of a size equal to a multiple of three do not necessarily change the reading frame. For example, a three-base deletion in the *PPX2L* gene for protoporphyrinogen oxidase (PPO) confers resistance to PPO-inhibiting herbicides in water hemp (Patzoldt et al., 2006) (Fig. 7.3). Only a glycine (Gly) amino acid is deleted, and codons after the deletion are back in frame, beginning with the GAT codon for aspartic acid. Deletion of this glycine in the water hemp PPO gene is designated ΔG210.

Slippage of DNA polymerase is a proposed mechanism for the creation of short deletions, and it has been proposed as a possible cause of the *PPX2L* deletion (Gressel and Levy, 2006). Slippage of DNA polymerase may be promoted by base repeats, which is consistent with tandem repeats of GTG in wild-type *PPX2L*, and C repeated four times in wild-type F3′H. Note in Fig. 7.3 that deletion of the trinucleotide GTG is not the only route to mutant *PPX2L*. Moving one base forward, deletion of TGG would also provide the same result.

2.2 Polymorphism

Genetic polymorphism is defined as variations among chromosomes, alleles, or other DNA sequences. We know that mutations create polymorphism, and usually think of mutants as being derived from an original wild-type form. However, sometimes it is difficult to know what the original chromosome or DNA sequence was just by studying genetic variation. For example, if allele A_2 is derived from A_1 by mutation, and A_2 provides a selective advantage to plants, A_2 may eventually become the most common allele in a population or species, even though originally it was a rare mutant. It is even possible that A_2 will become

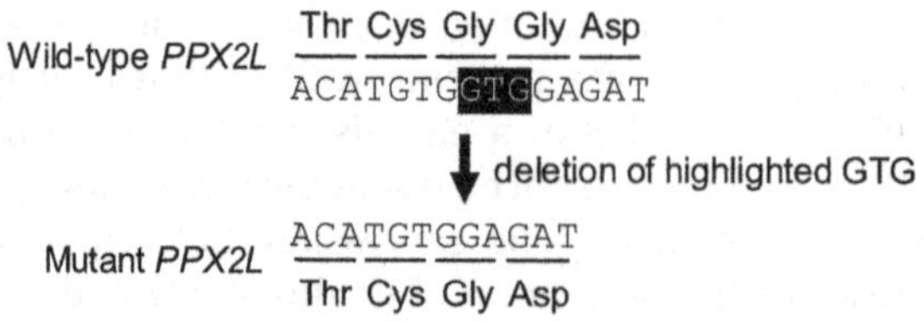

Wild-type *PPX2L*
Thr Cys Gly Gly Asp
ACATGTG■■■GAGAT

⬇ deletion of highlighted GTG

Mutant *PPX2L*
ACATGTGGAGAT
Thr Cys Gly Asp

FIGURE 7.3 A mutant allele of the water hemp *PPX2L* gene for protoporphyrinogen oxidase (PPO) differs from the wild-type allele by a three-base deletion (Patzoldt et al., 2006). The mutant allele produces a PPO enzyme that is resistant to herbicides owing to corresponding deletion of a glycine (Gly) amino acid.

fixed in the population (frequency of 1.0), whereas A_1 might be lost (frequency of 0.0). A researcher analyzing the population at this most recent stage might consider A_2 to be the normal wild-type allele, when in fact it is a mutant derived from A_1. For this reason, the wild type in population genetics is usually defined as the allele, genotype, or phenotype that predominates in natural populations (i.e., the normal, typical form). Wild-type forms are not necessarily the original; and as in the example given earlier, sometimes the original source of a mutant no longer exists in a population or species. Likewise, forms that differ from the wild type are often loosely termed mutant without direct knowledge of whether they were originally derived from the wild type. For highly selectable mutations in weed populations, such as base substitutions that confer herbicide resistance, it is usually a reasonable assumption that the resistant variant is indeed a mutant derived from the susceptible wild type.

Sometimes the frequencies of different genetic variants in a population or species are fairly similar, such that one is not confident in declaring whether a particular variant is the wild type. At the molecular level, this is especially the case with variations that may be selectively neutral, such as same sense (synonymous) SNPs, or differences among non-CDS. It is appropriate to call these variations polymorphisms without attempting to designate wild type or mutant types. Of course, clear wild-type and mutant forms may also be called polymorphisms.

Polymorphisms may be used to study the genetic diversity of weed populations even if the polymorphisms are neutral. In fact, it can be advantageous to use such markers if one wants to study the effects that breeding system and other weed species characteristics have on allele frequencies, independent of direct selection on specific alleles. In genetics, a marker is an observable characteristic that may be used to determine or score an individual's genotype. With phenotypic markers, the characteristic is observable in the phenotype (e.g., plant height and flower color). Molecular markers, on the other hand, require laboratory analysis to score the marker and infer genotype.

One of the oldest molecular marker systems is based on analysis of enzymatic proteins. Different alleles of a gene may produce enzymes that function equivalently, or nearly so, but may be slightly different in polypeptide sequence. These enzyme forms are termed allozymes. If there is an overall difference in net charge between allozymes owing to amino acid differences, the specific allozymes may often be identified in the laboratory using electrophoresis. Identification of the allozyme(s) carried by a plant reveals the underlying genetic variation that caused the allozyme difference (i.e., part of the plant's genotype).

Various DNA-based marker systems have also been developed, and the specific laboratory techniques used in these marker systems are not covered in detail. Many of the familiar DNA-based molecular marker systems are also being replaced by DNA sequencing including whole-genome sequencing to determine plant genotypes directly. However, because of the prevalence of DNA-based marker systems in the weed science literature, it is appropriate briefly to review the genetic polymorphism on which they are based.

Restriction fragment-length polymorphisms result from DNA fragment size differences after digestion of DNA with restriction enzymes, which cut DNA at specific sequences. These fragment-size polymorphisms may result from indels that affect the length of DNA between restriction enzyme recognition sites, or they may result from sequence differences at the restriction enzyme sites. Amplified fragment-length polymorphisms (AFLPs) also reveal polymorphisms in length between restriction enzyme sites, but AFLP techniques rely on polymerase chain reaction (PCR) to produce fragments observable by researchers. Randomly amplified polymorphic DNAs reveal fragment-length polymorphisms between PCR priming sites, which may be result from indels between the sites or sequence differences at the priming sites.

Tandemly repetitive sequences are useful polymorphisms for genetic studies because the number of repeats tends to be highly variable among individuals in populations. Variable number tandem repeats (VNTRs), also called minisatellites, are tandemly repeated DNA sequences of about 15–100 bp. Microsatellites, also called short tandem repeats, short sequence repeats, or simple sequence repeats, are tandem repeats of about 2–4 bp. Some researchers classify both minisatellites and microsatellites as VNTRs. In general, the number of repeats is analyzed when using these markers. Intersimple sequence repeats reveal DNA-length polymorphisms between microsatellites.

Many of these DNA marker systems do not directly reveal gene diversity because the DNA polymorphisms on which they are based may not be within an actual gene. However, the term "allele" is still used to specify individual DNA marker variants if those variants are at the same locus. A locus (plural "loci") is a physical place on a chromosome.

Various measures of population genetic diversity use results from marker analyses (Hedrick, 2011). These measures usually focus on heterozygosity as influenced by allele frequencies. Allozyme and DNA-based marker systems have revealed various levels of genetic diversity within and among weed populations (Jasieniuk and Maxwell, 2001). However, in general, weed populations tend to be diverse.

2.3 Haplotypes and Genetic Linkage

A haplotype is a set of alleles, SNPs, or other variations such as indels that tend to be inherited together because they are genetically linked (i.e., physically located in the same region of a chromosome). An allele of a gene is also sometimes termed a haplotype, especially when the SNPs that identify the allele are not known to cause phenotypic variation but are obviously linked on the same DNA molecule (chromosome). For example, consider a section of homologous chromosomes with genes A, B, C, D, and E. Alleles of gene A are A_1 and A_2, etc., and alleles for each of the homologous pair of chromosomes are shown in Fig. 7.4. This individual is homozygous for genes B and E, meaning that the alleles for these genes are the same on both homologous chromosomes (homozygous B_2 and homozygous E_1). The individual is also heterozygous for genes A, C, and D. There are two haplotypes in this example: A_1

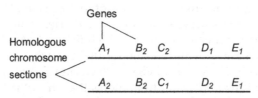

FIGURE 7.4 Homologous chromosome sections spanning genes A–E. Specific alleles are shown for each homologous chromosome.

$B_2 C_2 D_1 E_1$ and $A_2 B_2 C_1 D_2 E_1$. It is also possible for A, B, etc., to be nucleotide positions instead of alleles, in which A_1 and A_2 are SNPs or other molecular variations at position A.

Haplotypes are expected to be inherited intact unless crossing over shuffles the alleles/ SNPs or there are mutations that change the identity of the alleles/SNPs. For example, crossing over between A and B could generate haplotypes $A_2 B_2 C_2 D_1 E_1$ and $A_1 B_2 C_1 D_2$ E_1. Such DNA molecules resulting from crossing over, which is a type of recombination, are often called recombinant haplotypes, alleles, etc. However, some reserve the term "recombinant" for DNA produced through artificial genetic engineering techniques. The frequency of crossing over between genetic loci is partly related to the distance between the loci, with increased chance of crossing over correlated with increased distance. However, actual crossing-over frequency is nonuniform across most genomes, with crossing over hot spots and cold spots. Although crossing over is a beneficial mechanism by which plants and other organisms generate new genetic combinations, it can hinder population genetic studies by masking genetic history. For example, the $A_2 B_2 C_2 D_1 E_1$ haplotype resulting from the crossover in the previous example may appear to be an $A_1 B_2 C_2 D_1 E_1$ haplotype in which A_1 has mutated to A_2. Thinglum et al. (2011) suggested that crossing over may have masked the genetic history of PPO alleles carrying the ΔG210 mutation, based on an analysis of SNPs within ΔG210 alleles, which makes it difficult to determine whether the event producing ΔG210 occurred more than once.

Mutation can also generate new haplotypes, and when DNA variation surrounding putative mutations is determined not to have been significantly affected by crossing over, haplotype analysis can determine whether identical mutations are likely of independent origin. For example, Délye et al. (2004) confirmed multiple origins of identical herbicide resistance mutations among acetyl-CoA carboxylase (ACCase) allele haplotypes in black grass.

3. GENE EXPRESSION AND PHENOTYPIC DIVERSITY

Gene expression is the process by which information coded in the gene (i.e., the genotype) influences the phenotype. How specific alleles manifest as traits in the phenotype often depends on interactions with other alleles, genes, and the environment. Gene expression is an important aspect of weed population genetics because natural selection acts on phenotypes, ultimately through differential reproduction that may result from phenotypic differences. Natural selection usually does not act directly on allelic or genotypic variation; rather, it relies on expression of this genetic variation to produce a phenotype.

3.1 Gene Expression From Molecular and Classical Viewpoints

The central dogma of molecular genetics outlines the typical flow of information, viewed at the molecular level, that results in expression of a gene. Namely, that information stored as DNA is used in transcription to produce mRNA, which is used in translation to produce a protein. Control of gene expression at the molecular level is an intense area of research in terms of both understanding natural gene expression and finding ways to control expression for human benefit. Gene promoters, enhancers, transcription factors, RNA interference, chromatin structure, DNA methylation, etc., are all important in understanding the molecular processes of gene expression, but these topics are beyond what is needed for a basic understanding of weed population genetics.

In classical genetics, gene expression is studied by observing the effects that alleles have in different genetic backgrounds (i.e., in different genotypes and generations). For example, Gregor Mendel studied the expression of the unit of inheritance, which can now be defined as an allele, for short plants in pea. The short trait was observed to be true-breeding (i.e., consistent from generation to generation) in self-pollinating stock. After purposefully hybridizing short peas with true-breeding tall peas, however, the short trait disappeared among the progeny. Thus, among the hybrid progeny, there was no expression of the allele for short pea plants. After self-pollination of hybrids, some of the following generation again expressed the short trait. Notice that this classical investigation of gene expression heavily depends on observation of phenotypic variation (short and tall). Scientists studying weed populations often ask what happens to weeds of a certain phenotype (herbicide resistant,[7] etc.) under various agronomic conditions and management practices. Understanding the molecular nature of gene expression for the trait is often of secondary importance to these scientists. The most important aspect of applied population genetics is the end result: the resultant phenotype and reproductive fitness. Viewing gene expression from the end result (the phenotype) is the classical view of gene expression. Therefore, the remainder of the discussion in this section on gene expression will be mainly from the classical view.

3.2 Dominance

In the example of Mendel's short and tall peas, the expression of the short trait was masked among hybrid progeny. That is, all plants of the hybrid, first filial (F_1) generation were tall although they each carried one short allele from the short parent. The tall allele/trait was dominant and the short allele/trait was recessive.

To illustrate dominance further, consider gene A with alleles A_w and A_m. A_w is a wild-type allele and Am is a mutant allele. Homozygous wild-type and mutant plants would have A_wA_w and A_mA_m genotypes, respectively. If A_wA_w plants self-pollinate (the symbol for self-fertilization is ⊗) or cross-pollinate (symbolized by X) with other A_wA_w plants, we would expect that all of their offspring would be A_wA_w because no other alleles for the A gene are present in the parent(s) (Fig. 7.5A). In other words, there would be no segregation for gene A, i.e., all gametes from the parents(s) have the same allele. All progeny would show the A_w phenotype just like their parent(s).

If we hybridize A_w and A_m homozygotes, all F_1 progeny would be heterozygous (Fig. 7.5B). The diagrams on the right in Fig. 7.5B are Punnett squares, named after Reginald Punnett, the geneticist who developed this technique of diagramming the outcomes of fertilization. In the Punnett square for the F_1 generation, possible alleles/gametes from one parent are along the left (only A_w from the A_wA_w parent), whereas possible alleles/gametes from the other parent are along the top (only A_m from the A_mA_m parent). The rest of the Punnett square

---

[7]The state of being herbicide resistance without being exposed to herbicide is an example of what biologists might call an endophenotype—a phenotype we cannot see. The contrasting term exophenotype, describes what we can see, such as the presence or absence of herbicide symptomology after treatment.

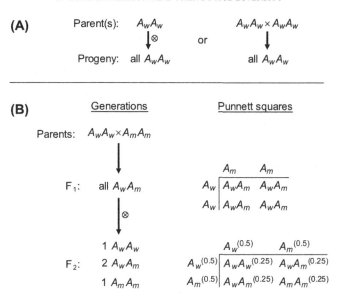

FIGURE 7.5   Progeny produced after self- or cross-pollination of homozygous A_wA_w parent(s) (A) or after cross-pollination between homozygous A_wA_w and homozygous A_mA_m parents (hybridization) (B). The top Punnett square shows progeny expected in the F_1 generation based on alleles contributed through hybridization of the parents. The bottom Punnett square shows expected F_2 progeny based on alleles contributed through self-pollination of the F_1 generation, with expected allele and genotype frequencies in superscript parentheses.

shows the outcomes of each fertilization event (the F_1 plants). Thus, what would the F_1 heterozygous plants look like? It depends on the dominance of the A_w allele versus the A_m allele. If A_w is dominant over A_m, the F_1's would have the wild-type phenotype and A_m would be considered recessive. If A_m is dominant over A_w, the F_1's would have the mutant phenotype and A_w would be considered recessive.

If these F_1 plants are self-pollinated (or pollinate among themselves), a 1:2:1 genotypic ratio would result in the F_2 generation (Fig. 7.5B). In the Punnett square for the F_2 generation, the frequencies of each allele in the F_1 parent have been added in superscript parentheses; that is, in the A_wA_m parent, the frequency of A_w is 0.5 and the frequency of A_m is 0.5. Genotype frequencies for the F_2 progeny have been added in a similar manner. In the F_2, the frequency of A_wA_w is $0.5 \times 0.5 = 0.25$ or one out of four. Likewise, the genotypic frequency of A_mA_m is 0.25, also one out of four. There are two ways for an F_2 plant to be heterozygous, namely receiving A_w from the female gamete and A_m from the male gamete, or vice versa. Thus, the overall frequency of A_wA_m heterozygotes is $(0.5 \times 0.5) + (0.5 \times 0.5) = 0.5$, or two out of four. The genotypic frequencies in the F_2 (0.25 for A_wA_w, 0.5 for A_wA_m, and 0.25 for A_mA_m) show that the F_2 genotypic ratio is indeed 1:2:1.

The phenotypic ratio in the F_2 depends on which allele is dominant. If A_w is dominant, A_wA_m heterozygotes will have the same wild-type phenotype as A_wA_w homozygotes. Their genotypic frequencies can be added to find the wild-type phenotypic frequency: $0.5 + 0.25 = 0.75$, or three out of four. Likewise, three out of four F_2 plants would have the mutant phenotype if A_m were dominant over A_w. Thus, the phenotypic ratio in the F_2 is three dominant to one recessive.

It is critical to understand that in the previous example, the 3:1 phenotypic ratio is obtained under a specific set of circumstances unlikely to be observed in most wild weed populations. Most important, the example began with exactly equal allele frequencies among the parents (0.5 for each allele), and matings between the plants were artificially controlled (forced hybridization between parents of different homozygous genotypes, followed by self-pollination or random mating of the F_1's). Phenotypic (and genotypic) ratios highly depend on allele frequency and mating system. It cannot be assumed that a trait seen among 75% of a wild population is dominant.

It is usual for heterozygotes to show an intermediate phenotype between those of homozygotes, in which neither allele is clearly dominant. This is called incomplete dominance, semidominance, or partial dominance. With incomplete dominance, the phenotype of the heterozygote is also not always at the midparent value exactly between the two homozygotes. For example, if the average height of A_1A_1 plants is about 100 cm, and the average height of A_2A_2 plants is about 50 cm, incomplete dominance for height would mean that A_1A_2 plants are likely between 50 and 100 cm tall but do not necessarily average 75 cm. The term "no dominance" is often used for situations in which the heterozygote is indeed at the midparent value.

Some examples of dominant, recessive, and incompletely dominant weed traits are given in Table 7.2. However, dominance is a measure of classical gene expression relative to other alleles, and gene expression can be influenced by environment and the specific alleles involved. To illustrate, consider an allele A_1. It is possible for this allele to be dominant over allele A_2 but recessive to a different allele (A_3). In a different environment, A_1 may actually be recessive to both A_2 and A_3. An example is that of Seefeldt et al. (1998), in which diclofop resistance in wild oat was found to be dominant at relatively low herbicide doses. As herbicide doses were increased (different environments), susceptibility became dominant.

The example of the effect of varying herbicide doses provides another opportunity to demonstrate the implications of environment on gene expression. The expression of wild-type susceptibility requires exposure to a dose of herbicide sufficient to reveal a susceptible phenotype. If no herbicide is applied, or if the dose is very low, susceptibility will not be expressed (in the classical sense of gene expression).

TABLE 7.2 Examples of Dominant, Recessive, and Incompletely Dominant Weed Traits

Weed	Trait	Expression	References
Tall morning glory	Purple flower color	Dominant (over pink)	Zufall and Rausher (2003)
Weedy rice	Seed dormancy	Dominant	Gu et al. (2003)
Wild mustard	Dicamba resistance	Dominant	Jasieniuk et al. (1995)
Prickly lettuce	Metsulfuron resistance	Incomplete dominance	Mallory-Smith et al. (1990)
Sunflower broomrape	Unpigmented plants	Incomplete dominance	Rodríguez-Ojeda et al. (2010)
Yellow star thistle	Clopyralid/picloram resistance	Recessive	Sabba et al. (2003)
Green foxtail	Trifluralin resistance	Recessive	Jasieniuk et al. (1994)

3.3 Pleiotropy

Pleiotropy occurs when a single mutation or gene/allele affects more than one phenotypic characteristic. In weedy rice, seed dormancy and red pericarp color result from a pleiotropic gene (Gu et al., 2011). Among missense base substitutions that confer herbicide resistance, several are known to reduce plant fitness compared with susceptible plants in the absence of herbicide (reviewed by Powles and Yu, 2010). A pleiotropic reduction in plant fitness caused by some herbicide resistance mutations is likely caused by reduced activity of the mutant enzyme. This phenomenon is sometimes referred to as a cost of resistance. Pleiotropy is an important concept in population genetics because it explains why some traits may always be found together among individuals of a population.

3.4 Qualitative and Quantitative Traits

Qualitative traits are those observed in distinct categories. Quantitative traits show more continuous variation among individuals. For example, consider a population in which variation for plant height is inherited (i.e., under genetic control). Plant height as a qualitative trait would mean that we can classify plants into distinct groups (e.g., tall or short). However, perhaps plant height among the plants continuously ranges from tall to short, including all heights in between. In this case, height would be quantitative.

A trait may be quantitative if it is controlled by several genes (polygenic), each with allelic variation. With increasing numbers of genes controlling a trait, there is an increasing number of gene and allele combinations, leading to continuous phenotypic variation. However, qualitative traits do not have to be polygenic. Environment may be involved in gene expression such that plants with the same genotype have different phenotypes (variable expressivity). In wild populations, nonpolygenic traits may also be quantitative because of the presence of multiple alleles for the controlling gene.

There is certainly quantitative variation for various characteristics within weed populations. However, the fundamental concepts of population genetics are most easily studied qualitatively, which is the approach in the remainder of this chapter.

4. MATING SYSTEMS

Population genetics highly depends on how alleles in a population's gene pool combine from generation to generation: that is, how plants reproduce sexually (or asexually). Such mating (or breeding) systems vary among weed species, and the effects of different mating systems can most easily be considered in idealized populations in which allele frequencies do not change. For the frequency of alleles to remain the same in a population (a theoretical concept that rarely exists), there should be no increases or decreases in allele frequency owing to mutation, and there should not be preferential migration of individuals of certain genotypes into or out of the population. Preferential selection of some alleles should not occur. Finally, the population should be large enough to buffer random changes to allele frequencies (genetic drift).

4.1 Hardy–Weinberg Principle

The Hardy–Weinberg principle is a model developed independently by Godfrey Hardy and Wilhelm Weinberg (Stern, 1943). It states that in an idealized population, allele frequencies can be used to predict genotype frequencies after one generation of random mating. Random mating is a condition in which all possible matings among individuals are equally possible. Among weeds, the mating system of kochia appears to approximate random mating (Guttieri et al., 1998). The equation for Hardy–Weinberg equilibrium is:

$$p^2 + 2pq + q^2 = 1 \qquad (7.1)$$

in which p and q are the frequencies of alternative alleles in the population. If we consider p to be the frequency of allele A_p, and q to be the frequency of allele A_q, p^2 is the frequency of A_p homozygotes, $2pq$ is the frequency of heterozygotes, and q^2 is the frequency of A_q homozygotes. We previously used a Punnett square to predict genotypic frequencies among offspring when the allele frequencies of their parents were known (0.5 for each allele if the parents were heterozygotes) (Fig. 7.5B). We can use a similar Punnett square to illustrate the genetic basis of Eq. (7.1), the Hardy–Weinberg equation (Fig. 7.6). If $p = 0.1$ and $q = 0.9$, the frequency of A_pA_p homozygotes is $0.1 \times 0.1 = 0.01$, the frequency of A_qA_q homozygotes is $0.9 \times 0.9 = 0.81$, and the frequency of heterozygotes is $2 \times 0.1 \times 0.9 = 0.18$. With continued random mating and without changes in allele frequency, the Hardy–Weinberg principle states that these genotypic frequencies will remain constant generation after generation.

Knowledge of the dominance relationship between A_p and A_q allows determination of the phenotypic frequencies from the genotypic frequencies. In the preceding example, if A_p is fully dominant to A_q, the frequency of the A_p phenotype is:

$$p^2 + 2pq = 0.01 + 0.18 = 0.19.$$

The frequency of the dominant A_p phenotype (0.19) is much less than that of the recessive phenotype ($q^2 = 0.81$). There is not a three dominant to one recessive Mendelian ratio because allele frequencies are not equal in this population.

Nonrandom mating will deviate genotype frequencies from those predicted by the Hardy–Weinberg principle. A primary example of a breeding system that alters Hardy–Weinberg frequencies is autogamy, which results in inbreeding. Autogamy is self-pollination, in which plants tend to fertilize themselves rather than hybridize with other plants. Inbreeding produced by autogamy is manifested at the population level by an increased number of homozygous

	$A_p{}^{(p)}$	$A_q{}^{(q)}$
$A_p{}^{(p)}$	$A_pA_p{}^{(p \times p)}$	$A_pA_q{}^{(p \times q)}$
$A_q{}^{(q)}$	$A_pA_q{}^{(p \times q)}$	$A_qA_q{}^{(q \times q)}$

FIGURE 7.6 Punnett square in which allele (parents) and genotype (progeny) frequencies (in superscript parentheses) are mathematically defined by population allele frequencies p and q. Note that there are two ways in which A_pA_q heterozygotes may be formed so that their overall genotypic frequency is $2(p \times q) = 2pq$. The genotypic frequencies of A_pA_p and A_qA_q homozygotes are $p \times p = p^2$ and $q \times q = q^2$, respectively.

individuals and decreased heterozygotes compared with those of random mating populations. Inbreeding within a population is measured by the inbreeding coefficient, F. When $F = 0$, there is no difference between the number of observed heterozygotes in the population and what would be expected from random mating. A completely inbred population, in which there are no heterozygotes, has an F of 1.0.

4.2 Mating Systems Among Weeds and Implications

Weeds display a variety of sexual mating or asexual reproduction systems, which suggests that no one system is best for weed reproduction (Table 7.3). Mixed mating is when plants both self-pollinate and outcross (hybridize with other plants), with little restriction on either. Mixed mating is the system that most closely approximates random mating. Other mechanisms of outcrossing, such as self-incompatibility and dioecy, may also approximate random mating unless populations are very small. Self-incompatibility is a mating system in which self-pollination is inhibited, forcing hybridization between different individuals. If a population is very small so that an individual plant represents a significant portion of the gene pool, excluding self-fertilization may be a significant deviation from random mating. Dioecy, in which there are separate male and female plants, excludes self-pollination and may not approximate random mating in small populations. Certain genotypes may also be favored by dioecy if allele frequencies are not equivalent between the male and female gene pools.

Self-pollination to the exclusion of outcrossing favors homozygosity, as discussed previously. An advantage of self-pollination is that it preserves gene combinations that may be advantageous in certain environments. However, the creation of new gene combinations is inhibited, so that adaptability may be reduced. Asexual, clonal mechanisms of reproduction, such as vegetative propagation or apomixis, also preserves gene combinations but reduces gene combination variability. Apomixis is an asexual process whereby seeds of the same genotype as the female (seed-producing) parent are produced. There is evidence that combinations of dispersed *EPSPS* gene copies in dioecious, glyphosate-resistant Palmer amaranth are sometimes maintained from generation to generation through facultative apomixis in females (Ribeiro et al., 2014).

TABLE 7.3   Examples of Sexual Mating Systems or Asexual Reproduction Among Weeds

Weed	Reproduction
Canada thistle	Dioecy and vegetative
Dandelion	Apomixis
Hydrilla	Vegetative
Kochia	Mixed mating
Wild mustard	Self-incompatibility
Wild oat	Self-pollination
Water hemp	Dioecy

Asexual reproduction and self-pollination may have an advantage in weed colonization owing to the ability of a lone plant to reproduce, especially compared with self-incompatibility and dioecy, and the reduced need for specialized pollinators (Baker, 1955, 1974; Mal and Lovett-Doust, 2005). Barrett (1992) identified asexual reproduction in aquatic, free-floating water hyacinth as a principal factor responsible for its ability to colonize. Schueller (2004) reported that island populations of hummingbird-pollinated tree tobacco have a greater capacity to self-pollinate than do mainland populations, most likely because of decreased reproductive success of plants unable to self-pollinate during initial island colonization. Similarly, Kunin (1993) observed significantly reduced seed production by self-incompatible wild mustard when plant density was reduced, most likely because of reduced opportunities for cross-pollination.

Among flowering plants, gene-encoding DNA is found not only in chromosomes of the nucleus (nuclear genes) but also within mitochondria and chloroplasts outside the nucleus (extranuclear genes). A weed's mating system has its greatest influence on the transmission of nuclear genes because these are the genes whose homozygosity or heterozygosity may be influenced by sexual mating. Extranuclear genes in flowering plants, however, are usually not transmitted through pollen but are inherited maternally (i.e., from the seed-producing "female" parent). This results in uniparental transmission of extranuclear genes from generation to generation following the maternal line, regardless of mating system. The *psb*A gene coding for the D1 protein of photosystem II (PSII) is an important extranuclear gene in weeds because mutant forms of *psb*A may confer resistance to PSII-inhibiting herbicides (Gronwald, 1994). The *psb*A gene is part of the chloroplast's genome, and extranuclear, maternal inheritance of this gene means that herbicide resistance resulting from mutant *psb*A is typically spread by seed or vegetative propagules but not pollen.

5. EVOLUTION OF WEED POPULATIONS

Evolution is change over time. Some definitions of biological evolution require the occurrence of new alleles, genes, or genetic combinations: that is, the origin of something that was not previously present. However, at the population level, changes in frequency of preexisting alleles is accurately classified as evolution. It is in the context of changing allele frequencies that this section is presented.

5.1 Fitness and Components of Fitness

Fitness is the ability of a plant to pass its alleles on to progeny. In population genetics, fitness is the main phenotypic characteristic of interest because it represents the final and ultimate measure of a plant's reproductive success. In understanding how and why weed populations evolve owing to phenotypic interactions with the environment, fitness is almost always a main player. A plant's level of fitness is determined by various genetic and environmental factors throughout its life cycle; three of the main components of fitness are survival, fecundity, and mating ability. Survival is the ability of a plant to reach reproductive maturity, and fecundity is its capacity to produce gametes. Mating ability is the likelihood that a plant's gametes will be involved in a fertilization event that results in offspring.

Directly determining weed fitness is often not a major goal of weed science research. For example, weed control measures are usually evaluated by their effectiveness in reducing weed populations or weed competition with crops. Herbicide-resistant weeds are often evaluated based on biomass after treatment compared with wild-type, susceptible plants. Although survival is indeed a component of fitness, and competitiveness or biomass may be related to reproductive ability, these measures are not complete and direct estimates of fitness. However, various studies have identified cases of weed genetic variation with clear implications for fitness. For example, the *psb*A gene codes for the D1 protein of PSII, and a missense mutation substituting glycine for serine at amino acid 264 of D1 is known to confer resistance to triazines and some other PSII-inhibiting herbicides in various weeds. However, this mutation also causes reduced fitness compared with wild-type plants grown in the absence of herbicide (reviewed by Holt and Thill, 1994). This pleiotropic reduction in fitness (fitness cost) is likely the result of reduced PSII efficiency owing to the mutated D1 protein. Triazine-resistant weeds also sometimes show reduced fitness compared with wild-type plants when treated with alternative herbicides, such as bentazon (Gadamski et al., 2000). There is also increasing evidence that other forms of herbicide resistance often, but not always, have a fitness cost (reviewed by Vila-Aiub et al., 2009).

5.2 Selection and Selection Pressure

Selection is the preferential transmission of alleles from generation to generation, so that the frequencies of some alleles in the gene pool increase and the frequencies of others decrease. With selection, there is a biological, environmental, or human cause of the preferential transmission of alleles. This differentiates selection from random genetic drift, in which the transmission of one allele may happen to be greater than another simply because of chance. Intentional plant breeding, artificial selection, selects individuals as sources of alleles (parents) for the next generation. With artificial selection, the fittest parents are not necessarily chosen for breeding. Parents may be chosen based on the usefulness of their traits to humans. With natural selection, however, population allele frequencies change without human intention. Changes in population allele frequencies owing to natural selection are the result of differences in fitness among potential parents, leading to preferential transmission of alleles. This does not mean that the activities generating selection are necessarily natural. For example, unintentional selection for herbicide resistance, with an herbicide as the selecting agent, is still considered natural selection.

Natural selection may cause allele frequencies in a gene pool to change over time as a result in reduction in the frequencies of alleles detrimental to fitness (purifying selection) or increases in the frequencies of advantageous alleles (positive selection). It is possible for alleles to be the subject of either purifying or positive selection, depending on the environment; pleiotropic effects of some herbicide resistance mutations are a good example. Délye et al. (2004) provided evidence that the high similarity of plastidic ACCase alleles among grass weeds was likely the result of purifying selection before extensive use of ACCase-inhibiting herbicides. In the absence of herbicide, there was likely selection against sequence variations that reduced weed fitness, resulting in a conserved wild-type ACCase sequence among grass weeds, especially in the important carboxyl transferase domain of the ACCase enzyme. This hypothesis is consistent with reduced ACCase enzyme activity associated with some mutant ACCase alleles, such as those with a missense mutation in which cysteine is substituted for tryptophan at position 2027 of the ACCase enzyme (Délye et al., 2005). Although ACCase alleles with cysteine 2027 may reduce weed fitness in environments where there is no treatment with ACCase-inhibiting herbicides, they also express a pleiotropic phenotype when treated with such herbicides (i.e., herbicide resistance). Thus, ACCase-inhibiting herbicides provide positive selection for alleles such as cysteine 2027 that otherwise would likely be kept at low frequency owing to purifying selection.

In a broad context, most environmental components (herbicide, soil type, climate, etc.) may be factors influencing the relative fitness of alleles so that a genotype/phenotype with a fitness advantage in one environment may be disadvantaged in another. Using simulations, Martinez-Ghersa et al. (2000) demonstrated the importance of rainfall patterns and soil type in determining the relative fitness of weeds with different seed germination phenotypes. The distinction between purifying and positive selection in different environments is also not always clear. An herbicide resistance mutation may be subject to purifying selection in the absence of herbicide or positive selection in the presence of herbicide, as in the previous example. However, positive selection for herbicide resistance may just as accurately be considered purifying selection against susceptibility.

Selection pressure is the intensity of selection as measured by changes in allele frequency per generation owing to selection (Rieger et al., 1991). The intensity of selection may be measured experimentally or predicted using population genetic parameters such as allele frequency, dominance, and coefficients of selection (Falconer, 1989). The coefficient of selection for a genotype is abbreviated s; it is a measure of the difference in fitness between itself and another genotype whose fitness is standardized to 1.0 (usually the fittest genotype or, alternatively, the wild type). For example, consider a self-pollinated weed species for which viable seed production is a reasonable measure of fitness. If genotype A_1A_1 produces 1000 seeds per plant, and genotype A_2A_2 produces 800 seeds per plant, for genotype A_2A_2:

$$s = (1000 - 800)/1000 = 0.2.$$

In other words, A_2A_2 is 20% less fit than A_1A_1 because A_2A_2 contributes 20% fewer alleles to the next generation (represented by seed). We have compared the fitness of A_2A_2 with the fittest genotype (A_1A_1) to calculate s for A_2A_2. As a result, s in this example is a measure of purifying selection against A_2A_2. Relative fitness values (often abbreviated as w) would be standardized to 1.0 for A_1A_1 and to $1.0 - s = 1.0 - 0.2 = 0.8$ for A_2A_2. Alternatively, we could have considered the fitness advantage of A_1A_1 over A_2A_2 (positive selection for A_1A_1), in which case the fitness of A_2A_2 would be standardized to 1.0 and for A_1A_1:

$$s = (1000 - 800)/800 = 0.25.$$

With natural selection considered as positive selection for A_1A_1, the relative fitness of A_1A_1 would be $1.0 + s = 1.0 + 0.25 = 1.25$. It is therefore important to consider s in the context of purifying versus positive selection, because its use as an indication of relative fitness (w) will vary ($w = 1.0 - s$ for purifying selection, or $w = 1.0 + s$ for positive selection) (see Hedrick, 2011). Also in this example, we considered a self-pollinating weed species and did not consider A_1A_2 heterozygotes. The fitness of heterozygotes would depend on the dominance of the A_1 allele compared with A_2.

5.2.1 Selection Pressure and Herbicide Resistance

Herbicide resistance is the most striking example of selection pressure in weed science. Rapid evolution of herbicide resistance in weed populations can occur after only a few years of herbicide use in a field (Maxwell and Mortimer, 1994). In part, this is because of relatively high s values associated with herbicide resistance. For example, research by P.F. Ulf-Hansen (as cited in Mortimer, 1993) estimated s to be 0.82–1.00 for resistance to chlortoluron and isoproturon in black grass (purifying selection against wild-type susceptible plants). Here, we use a simplified example of how s can be used to predict allele frequency changes from one generation to the next using population genetics theory (described by Falconer, 1989).

Consider an herbicide target-site gene (T) with two alleles in a weed population's gene pool, with T_S representing the wild-type susceptible allele and T_R representing the allele for herbicide resistance. As previously discussed, both genetic factors and environment including herbicide rate can influence gene expression, including dominance. For this example, we will assume that resistance is completely dominant and fully penetrant (meaning that the genotype always produces its expected phenotype). For simplification, we will also assume random mating and no influence of mutation, migration, or random genetic drift on allele frequencies. A final assumption is that there are distinct generations so that all individuals within the population are exposed to the selective herbicide: that is, no dormant weed seeds or other individuals that are part of the population but escape herbicide treatment. For appropriate ways to handle situations in which these assumptions may not be met, the reader is referred to Falconer (1989) or similar texts on population genetics.

Given the previous assumptions, T_S and T_R allele frequencies after a generation of herbicide selection can be predicted if we know the initial allele frequencies (p and q), and the coefficient of selection (s), by the equation:

$$q_1 = \left[q^2(1-s) + pq\right] / \left[1 - sq^2\right] \tag{7.2}$$

(Falconer, 1989). In the preceding example of herbicide resistance, p is the initial frequency of the dominant T_R allele for herbicide resistance and q is the initial frequency of the T_S allele for susceptibility, whereas q_1 is the frequency of T_S among progeny of the subsequent generation. In Eq. (7.1), p^2 and q^2 represented homozygous genotype frequencies and $2pq$ was the frequency of heterozygotes. In Eq. (7.2), q^2 in the numerator is the initial frequency of herbicide-susceptible $T_S T_S$ homozygotes. Because of herbicide selection, the allelic contribution of $T_S T_S$ homozygotes to the subsequent generation is reduced by purifying selection against susceptibility, and hence q^2 is multiplied by the quantity $1 - s$ (the relative fitness of $T_S T_S$ homozygotes). Also in the numerator of Eq. (7.2), pq represents T_S alleles from heterozygotes. The frequency of heterozygotes in Eq. (7.1) is $2pq$ (Hardy–Weinberg principle), but because only half of the gametes produced by heterozygotes are T_S, the contribution of T_S alleles from heterozygotes is $(2pq)/2 = pq$ (Eq. 7.2). Susceptibility is completely recessive in this example, so there is no reduction in the allelic contribution of heterozygotes owing to herbicide action; that is, the relative fitness of fully resistant heterozygotes is 1.0. To predict the frequency of T_S alleles in the subsequent generation (q_1), we need to divide the selection-adjusted gametic contribution of T_S alleles from the initial generation $\left[q^2(1-s) + pq\right]$ by the total selection-adjusted gametic contribution of all alleles (both T_S and T_R), which is $1 - sq^2$.

The quantity sq^2 represents the reduction in T_S alleles owing to selection against susceptible homozygotes, which is subtracted from 1.0 (the total frequency of all alleles if selection did not occur). Once q_1 is calculated, and because T_S and T_R are the only alleles in our example, the frequency of T_R alleles among progeny of the subsequent generation (after selection in the initial generation) can be calculated as $p_1 = 1 - q_1$.

We have not yet given values to p, q, and s, so that q_1 can be predicted in the example weed population described previously. As with the assumptions previously given in this example (fully dominant resistance, random mating, etc.), the values for p, q, and s would optimally be estimated by studying the population or similar populations with the same type of resistance. For illustrative purposes, let us designate the coefficient of selection for the susceptible genotype ($T_S T_S$) as $s = 0.96$ (strong purifying selection against susceptibility) and the initial frequencies of T_R and T_S alleles as $p = 0.01$ and $q = 0.99$, respectively. An initial resistance allele frequency of 0.01 is several orders of magnitude higher than what would be expected without previous herbicide exposure (see Maxwell and Mortimer, 1994), which suggests prior selection or introduction of T_R alleles into the example population. However, compared with lower allele frequencies, an initial resistance allele frequency of 0.01 better exemplifies a weed population in which one more season of selection may produce a noticeable and significant reduction in weed control. Using Eq. (7.2) and where $p = 0.01$, $q = 0.99$, and $s = 0.96$,

$$q_1 = [q^2(1-s) + pq]/[1 - sq^2] = [0.99^2(1-0.96) + (0.01)(0.99)]/[1 - (0.96)(0.99^2)]$$
$$= 0.83.$$

The new T_R resistance allele frequency is therefore $p_1 = 1 - 0.83 = 0.17$ (an increase from the initial T_R allele frequency of 0.01, which by definition indicates selection pressure). Because resistance is dominant, both homozygous $T_R T_R$ plants (initial frequency of p^2) and heterozygotes (initial frequency of $2pq$) are resistant. The initial frequency of resistant plants in this example population was therefore:

$$p^2 + 2pq = 0.01^2 + 2(0.01)(0.99) = 0.02.$$

In the subsequent generation, purifying selection against susceptibility has resulted in a new resistant plant frequency of:

$$p_1^2 + 2p_1q_1 = 0.17^2 + 2(0.17)(0.83) = 0.31.$$

The frequency of resistant plants has increased from 2% to 31% in one generation. However, this increase in resistance allele frequency and the corresponding increase in resistant plant frequency may or may not represent an increase in the total number of plants in the population, which depends on the actual number of progeny produced by each parent plant.

In natural selection, the value for s is integral to the level of selection pressure (i.e., allele frequency change) expected in a population. Typically, increased s means increased selection pressure within the population. Weed management may influence s, but it is not always easy to predict the exact effects that different weed management strategies may have on s. Returning to herbicide resistance as an example, it cannot be assumed that an increase in overall herbicide efficacy will always increase s. Because s is a measure of the difference in fitness between genotypes/phenotypes, it is intuitive that increased herbicide efficacy may increase s by reducing the fitness of susceptible plants more than resistant ones. However, if control of susceptible plants was good in the first place, so that increased herbicide efficacy improves the control of susceptible plants only slightly while providing a relatively greater increase in control of resistant plants, s may actually be reduced.

5.2.2 Minimizing and Reversing Selection

Genetic diversity within weed populations may produce differences in fitness among individual weeds, leading to selection of the fittest types. It is desirable to minimize such selection, either by keeping the fittest genotypes out of populations or by reducing coefficients of selection by minimizing phenotypic variation. Successful efforts to control weeds and stop their spread are generally helpful to minimize natural selection for adaptive traits, which are traits that positively influence fitness. Keeping weed populations small minimizes the likelihood of new mutations for adaptive traits, and reducing the spread of weeds reduces gene flow. Gene flow is the migration of gene variants (alleles) from one population to another, and gene flow may increase genetic diversity within the receiving population. Reducing gene flow minimizes the introduction of adaptive traits into new populations.

Best management practices for minimizing selection of herbicide resistance in weed populations are outlined in Norsworthy et al. (2012). These recommendations discourage repeated seasonal use of herbicides with similar mechanisms of action to reduce selection for the same herbicide resistance alleles across multiple generations. Supplementation with other biological and cultural weed control methods is also encouraged. Some genes may provide only low-level herbicide resistance by themselves, but may contribute to higher and more significant resistance when combined with other resistance genes through hybridization. Good weed control with labeled rates of herbicide minimizes selection of these minor resistance genes.

Although minimizing selection for adaptive traits such as herbicide resistance is important, is it possible to reverse selection and reduce the frequency of these traits? Herbicide resistance alleles that have a fitness cost may indeed be reduced in frequency when the corresponding herbicide is not used (see Vila-Aiub et al., 2005). However, it is not apparent that selection against herbicide resistance resulting from fitness cost alone is sufficient to restore acceptable weed control levels in most cases. Emerging genetic technologies such as gene drives[8] may one day be able to reduce resistance frequencies in weed populations through preferential inheritance of susceptibility. One such proposal would use clustered regularly interspaced short palindromic repeats—based gene editing to suppress resistance by converting these genes into a susceptible form (Esvelt et al., 2014). However, the ecological safety of such systems is still unclear.

5.2.3 Application of Genomics to Weed Population Genetics

In the previous example illustrating selection, only one gene (T) with two alleles (T_S and T_R) was considered. Of course, weeds and other plants have more than one gene in their genomes.

---

[8]Gene drives are genetic systems that promote the inheritance of specific genes, thus enhancing their spread through populations.

Based on a summary of published plant genome sequences (Michael and Jackson, 2013), an average plant has about 30,000 genes. The study of whole genomes, including the many genes that they carry, is called genomics.

Weed scientists are particularly interested in genes that contribute to weediness or are potential targets of weed control. A way to study expressed genes specifically is to analyze the transcriptome, which is the part of the genome that is transcribed (used as a template to produce mRNA by transcription). Numerous weed species have been the subject of transcriptome sequencing to investigate traits such as parasitism in dodder (Ranjan et al., 2014) and invasiveness in perennial false-brome (Fox et al., 2013). Diversity within species such as herbicide resistance in goosegrass (An et al., 2014) has also been the subject of transcriptome studies. An advantage of transcriptome analysis is that in addition to identification of sequence variation within species and populations, gene expression in different environments including after exposure to different weed control techniques can be studied. A disadvantage, however, is an inability to identify diversity that is not present among mRNAs, including introns, genes that are not being expressed, regions between genes, and overall genome organization.

Sequencing whole genomes (i.e., all DNA regardless of expression) is becoming more affordable and is leading to an increasing number of genomes available for research. The first higher plant genome to be completed was that of *Arabidopsis* (The Arabidopsis Genome Initiative, 2000). Although mouse-ear cress is often considered a weed, it is not an economically important weed and is not a good representative of plants important in most weed research. Among economically important weeds, the genomes of at least six species have been sequenced and described (Table 7.4). Not all of these genomes were sequenced because of their weediness, but all will be useful in studying the population genetics of weedy traits.

How does genomics influence our understanding of weed population genetics? Some genes are obvious targets of selection, such as genes for herbicide resistance. However, each of the many genes in a weed may have potential alleles subject to purifying or positive selection in a population. Genomics identifies potential genes for weedy traits that do not have enough clear physiological data to identify likely gene candidates. Some traits such as non−target site herbicide resistance may have many potential metabolic genes that could

TABLE 7.4   Whole-Genome Sequences of Economically Important Weeds

Weed	Primary Reason for Genomic Sequencing	Predicted Number of Protein-Coding Genes	References
Field pennycress	Potential domestication as a biofuel crop	27,390	Dorn et al. (2015)
Giant duckweed	Biofuel potential as a fast-growing aquatic plant	19,623	Wang et al. (2014)
Green foxtail	Wild ancestor of foxtail millet	24,000−29,000	Bennetzen et al. (2012)
Horseweed	Weediness including herbicide resistance	44,592	Peng et al. (2014)
Rice (weedy)	Potential to originate from cultivated rice	Not given[a]	Qiu et al. (2014)
Wild radish	Study genome evolution within mustard family	38,174	Moghe et al. (2014)

[a]*The estimated number of unique genes in cultivated rice is 39,045, not including transposons (http://rice.plantbiology.msu.edu/, as cited in McCouch et al., 2016).*

provide resistance, and genomics identifies the most likely resistance genes. Genomics is also helpful in the study of quantitative traits resulting from multiple genes. An understanding of the many genes in a weed genome may also identify new targets of future herbicides or other control methods. Genomic sequence data can be used to look for signs of recent or past hybridization with crop or other non-crop species, which may reveal the origins of genetic diversity influencing weed population dynamics. The study of variation resulting from large-scale deletions or duplications, including genome-wide alterations such as dispersed repeats from transposons, is also likely to be facilitated by genomic sequence data.

5.2.4 *Relevance and Future Prospects*

Baker (1974) proposed characteristics of an ideal weed, including the ability to germinate in many environments, seed longevity, rapid growth, the ability to produce high amounts of seed under favorable conditions while producing at least some seed under less favorable environments, and the ability to compete with other plant species (see Baker for a more extensive list). What these weedy characteristics have in common is the predicted tendency to enhance overall fitness of a weed species across the diverse environments in which weeds are typically found. These characteristics are therefore helpful in explaining why some plant species are weedy and others are not. However, as Norris (2007) notes, and as examples given in this chapter illustrate, variation in weed fitness is not observed only at the species level but within species as well. The focus of weed population genetics is this within-species variation, and advances in molecular genetic and genomic techniques are rapidly providing new ways to identify specific alleles and genotypes in weed populations. However, more research is needed to link genetic variation accurately with phenotypic diversity for characteristics that enhance weediness, and also to estimate the population genetic parameters (the coefficient of selection, dominance, etc.) of these characteristics in different environments and in different weed species. Such research will greatly enhance the weed science community's ability to evaluate weed management practices by tracking allele frequency changes within populations and to estimate the population genetic parameters of specific weed populations so that the likely success of weed management practices can be predicted.

THINGS TO THINK ABOUT

1. Do all individual weeds contribute to the gene pool each season?
2. Do all mutations affect the phenotype of a weed?
3. What is polyploidy and why might it sometimes be advantageous for weeds?
4. What role does genetic linkage have in defining haplotypes?
5. Can genetic dominance be determined by observing phenotype frequencies within a weed population?
6. What are some potential limitations of the Hardy–Weinberg principle in weed population genetics?
7. How does the mating system of a weed affect genotype frequencies?
8. How might evolution be defined within weed populations?
9. How might different herbicide rates influence selection pressure for herbicide resistance?

10. What is plant fitness and what are the biological components of fitness?
11. Are all observations of weed phenotype good measurements of weed fitness?
12. What is genomics and how might it be used to study weed population genetics?

Literature Cited

Akasaka, M., Konishi, S., Izawa, T., Ushiki, J., 2011. Histological and genetic characteristics associated with the seed-shattering habit of weedy rice (*Oryza sativa* L.) from Okayama. Jpn. Breed. Sci. 61, 168–173.

An, J., Shen, X., Ma, Q., Yang, C., Liu, S., Chen, Y., 2014. Transcriptome profiling to discover putative genes associated with paraquat resistance in goosegrass (*Eleusine indica* L.). PLoS One 9. https://doi.org/10.1371/journal.pone.0099940.

Avery, O.T., MacLeod, C.M., McCarty, M., 1944. Studies on the chemical nature of the substance inducing transformation of pneumococcal types: induction of transformation by a desoxyribonucleic acid fraction isolated from pneumococcus type III. J. Exp. Med. 79, 137–158.

Avery, A.G., Satina, S., Rietsma, J., 1959. Blakeslee: The Genus Datura. Ronald Press, New York, 289 pp.

Baker, H.G., 1955. Self-compatibility and establishment after 'long-distance' dispersal. Evolution 9, 347–349.

Baker, H.G., 1974. The evolution of weeds. Annu. Rev. Ecol. Syst. 5, 1–24.

Barrett, S.C.H., 1992. Genetics of weed invasions. In: Jain, S.K., Botsford, L.W. (Eds.), Applied Population Biology. Kluwer, Dordrecht, The Netherlands, pp. 91–119.

Bennett, M.D., Leitch, I.J., Hanson, L., 1998. DNA amounts in two samples of angiosperm weeds. Ann. Bot. 82 (Suppl. A), 121–134.

Bennetzen, J.L., Schmutz, J., Wang, H., Percifield, R., Hawkins, J., Pontaroli, A.C., et al., 2012. Reference genome sequence of the model plant *Setaria*. Nat. Biotechnol. 30, 555–561.

Cleland, R.E., 1972. Oenothera: Cytogenetics and Evolution. Academic Press, London, 370 pp.

Délye, C., Straub, C., Michel, S., Le Corre, V., 2004. Nucleotide variability at the acetyl coenzyme A carboxylase gene and the signature of herbicide selection in the grass weed *Alopecurus myosuroides* (Huds.). Mol. Biol. Evol. 21, 884–892.

Délye, C., Zhang, X.-Q., Michel, S., Matéjicek, A., Powles, S.B., 2005. Molecular bases for sensitivity to acetyl-coenzyme A carboxylase inhibitors in black-grass. Plant Physiol. 137, 794–806.

Dorn, K.M., Fankhauser, D., Wyse, D.L., Marks, M.D., 2015. A draft genome of field pennycress (*Thlaspi arvense*) provides tools for the domestication of a new winter biofuel crop. DNA Res. 22, 121–131.

Esvelt, K.M., Smidler, A.L., Catteruccia, F., Church, G.M., 2014. Concerning RNA-guided gene drives for the alteration of wild populations. eLIFE 3. https://doi.org/10.7554/eLife.03401.

Falconer, D.S., 1989. Introduction to Quantitative Genetics. Longman, London, 438 pp.

Foes, M.J., Liu, L., Vigue, G., Stoller, E.W., Wax, L.M., Tranel, P.J., 1999. A kochia (*Kochia scoparia*) biotype resistant to triazine and ALS-inhibiting herbicides. Weed Sci. 47, 20–27.

Fox, S.E., Preece, J., Kimbrel, J.A., Marchini, G.L., Sage, A., Youens-Clark, K., et al., 2013. Sequencing and de novo transcriptome assembly of *Brachypodium sylvaticum* (Poaceae). Appl. Plant Sci. 1 https://doi.org/10.3732/apps.1200011.

Gadamski, G., Ciarka, D., Gressel, J., Gawronski, S.W., 2000. Negative cross-resistance in triazine-resistant biotypes of *Echinochloa crus-galli* and *Conyza canadensis*. Weed Sci. 48, 176–180.

Gaines, T.A., Shaner, D.L., Ward, S.M., Leach, J.E., Preston, C., Westra, P., 2011. Mechanism of resistance of evolved glyphosate-resistant Palmer amaranth (*Amaranthus palmeri*). J. Agric. Food Chem. 59, 5886–5889.

Gressel, J., Levy, A.A., 2006. Agriculture: the selector of improbable mutations. Proc. Natl. Acad. Sci. U.S.A. 103, 12215–12216.

Griffith, F., 1928. The significance of pneumococcal types. J. Hyg. 27, 113–159.

Gronwald, J.W., 1994. Resistance to photosystem II inhibiting herbicides. In: Powles, S.B., Holtum, J.A.M. (Eds.), Herbicide Resistance in Plants: Biology and Biochemistry. CRC Press, Boca Raton, FL, pp. 27–60.

Gu, X.-Y., Chen, Z.-X., Foley, M.E., 2003. Inheritance of seed dormancy in weedy rice. Crop Sci. 43, 835–843.

Gu, X.-Y., Foley, M.E., Horvath, D.P., Anderson, J.V., Feng, J., Zhang, L., et al., 2011. Association between seed dormancy and pericarp color is controlled by a pleiotropic gene that regulates abscisic acid and flavonoid synthesis in weedy red rice. Genetics 189, 1515–1524.

Guttieri, M.J., Eberlein, C.V., Thill, D.C., 1995. Diverse mutations in the acetolactate synthase gene confer chlorsulfuron resistance in kochia (*Kochia scoparia*) biotypes. Weed Sci. 43, 175–178.

Guttieri, M.J., Eberlein, C.V., Souza, E.J., 1998. Inbreeding coefficients of field populations of *Kochia scoparia* using chlorsulfuron resistance as a phenotypic marker. Weed Sci. 46, 521–525.

Han, H., Yu, Q., Purba, E., Li, M., Walsh, M., Friesen, S., Powles, S.B., 2012. A novel amino acid substitution Ala-122-Tyr in ALS confers high-level and broad resistance across ALS-inhibiting herbicides. Pest Manag. Sci. 68, 1164–1170.

Hedrick, P.W., 2011. Genetics of Populations. Jones & Bartlett, Boston, 675 pp.

Holt, J.S., Thill, D.C., 1994. Growth and productivity of resistant plants. In: Powles, S.B., Holtum, J.A.M. (Eds.), Herbicide Resistance in Plants: Biology and Biochemistry. CRC Press, Boca Raton, FL, pp. 299–316.

Jasieniuk, M., Maxwell, B.D., 2001. Plant diversity: new insights from molecular biology and genomics technologies. Weed Sci. 49, 257–265.

Jasieniuk, M., Brûlé-Babel, A.L., Morrison, I.N., 1994. Inheritance of trifluralin resistance in green foxtail (*Setaria viridis*). Weed Sci. 42, 123–127.

Jasieniuk, M., Morrison, I.N., Brûlé-Babel, A.L., 1995. Inheritance of dicamba resistance in wild mustard. Weed Sci. 43, 192–195.

Jugulam, M., Niehues, K., Godar, A.S., Koo, D.-H., Danilova, T., Friebe, B., et al., 2014. Tandem amplification of a chromosomal segment harboring *5-enolpyruvylshikimate-3-phosphate synthase* locus confers glyphosate resistance in *Kochia scoparia*. Plant Physiol. 166, 1200–1207.

Kunin, W.E., 1993. Sex and the single mustard: population density and pollinator behavior effects on seed-set. Ecology 74, 2145–2160.

Mal, T.K., Lovett-Doust, J., 2005. Phenotypic plasticity in vegetative and reproductive traits in an invasive weed, *Lythrum salicaria* (Lythraceae), in response to soil moisture. Am. J. Bot. 92, 819–825.

Mallory-Smith, C.A., Thill, D.C., Dial, M.J., Zemetra, R.S., 1990. Inheritance of sulfonylurea herbicide resistance in *Lactuca* spp. Weed Technol. 4, 787–790.

Martinez-Ghersa, M.A., Ghersa, C.M., Benech-Arnold, R.L., MacDonough, R., Sanchez, R.A., 2000. Adaptive traits regulating dormancy and germination of invasive species. Plant Species Biol. 15, 127–137.

Maxwell, B.D., Mortimer, A.M., 1994. Selection for herbicide resistance. In: Powles, S.B., Holtum, J.A.M. (Eds.), Herbicide Resistance in Plants: Biology and Biochemistry. CRC Press, Boca Raton, FL, pp. 1–25.

McClintock, B., 1950. The origin and behavior of mutable loci in maize. Proc. Natl. Acad. Sci. U.S.A. 36, 344–355.

McCouch, S.R., Wright, M.H., Tung, C.-W., Maron, L.G., McNally, K.L., Fitzgerald, M., et al., 2016. Open access resources for genome-wide association mapping in rice. Nat. Commun. 7 https://doi.org/10.1038/ncomms10532.

Michael, T.P., Jackson, S., 2013. The first 50 plant genomes. Plant Genome 6. https://doi.org/10.3835/plantgenome2013.03.0001in.

Moghe, G.D., Hufnagel, D.E., Tang, H., Xiao, Y., Dworkin, I., Town, C.D., et al., 2014. Consequences of whole-genome triplication as revealed by comparative genomic analyses of the wild radish *Raphanus raphanistrum* and three other Brassicaceae species. Plant Cell 26, 1925–1937.

Mortimer, A.M., 1993. A Review of Graminicide Resistance. Monograph 1, Herbicide Resistance Action Committee–Graminicide Working Group.

Nei, M., Nozawa, M., 2011. Roles of mutation and selection in speciation: from Hugo de Vries to the modern genomic era. Genome Biol. Evol. 3, 812–829.

Norris, R.F., 2007. Weed fecundity: current status and future needs. Crop Prot. 26, 182–188.

Norsworthy, J.K., Ward, S.M., Shaw, D.R., Llewellyn, R.S., Nichols, R.L., Webster, T.M., 2012. Reducing the risks of herbicide resistance: best management practices and recommendations. Weed Sci. (Special Issue), 31–62.

Ossowski, S., Schneeberger, K., Lucas-Lledó, J.I., Warthmann, N., Clark, R.M., Shaw, R.G., Weigel, D., Lynch, M., 2010. The rate and molecular spectrum of spontaneous mutations in *Arabidopsis thaliana*. Science 327, 92–94.

Patzoldt, W.L., Hager, A.G., McCormick, J.S., Tranel, P.J., 2006. A codon deletion confers resistance to herbicides inhibiting protoporphyrinogen oxidase. Proc. Natl. Acad. Sci. U.S.A. 103, 12329–12334.

Peng, Y., Lai, Z., Lane, T., Nageswara-Rao, M., Okada, M., Jasieniuk, M., et al., 2014. De novo genome assembly of the economically important weed horseweed using integrated data from multiple sequencing platforms. Plant Physiol. 166, 1241–1254.

Powles, S.B., Yu, Q., 2010. Evolution in action: plants resistant to herbicides. Annu. Rev. Plant Biol. 61, 317–347.

Qiu, J., Zhu, J., Fu, F., Ye, C.-Y., Wang, W., Mao, L., et al., 2014. Genome re-sequencing suggested a weedy rice origin from domesticated indica-japonica hybridization: a case study from southern China. Planta 240, 1353–1363.

Ranjan, A., Ichihashi, Y., Farhi, M., Zumstein, K., Townsley, B., David-Schwartz, R., Sinha, N.R., 2014. De novo assembly and characterization of the transcriptome of the parasitic weed dodder identifies genes associated with plant parasitism. Plant Physiol. 166, 1186–1199.

Renaut, S., Rieseberg, L.H., 2015. The accumulation of deleterious mutations as a consequence of domestication and improvement in sunflowers and other Compositae crops. Mol. Biol. Evol. 32, 2273–2283.

Ribeiro, D.N., Pan, Z., Duke, S.O., Nandula, V.K., Baldwin, B.S., Shaw, D.R., et al., 2014. Involvement of facultative apomixis in inheritance of *EPSPS* gene amplification in glyphosate-resistant *Amaranthus palmeri*. Planta 239, 199–212.

Rieger, R., Michaelis, A., Green, M.M., 1991. Glossary of Genetics: Classical and Molecular. Springer-Verlag, Berlin, 553 pp.

Rodríguez-Ojeda, M.I., Velasco, L., Alonso, L.C., Fernández-Escobar, J., Pérez-Vich, B., 2010. Inheritance of the unpigmented plant trait in *Orobanche cumana*. Weed Res. 51, 151–156.

Sabba, R.P., Ray, I.M., Lownds, N., Sterling, T.M., 2003. Inheritance of resistance to clopyralid and picloram in yellow starthistle (*Centaurea solstitialis* L.) is controlled by a single nuclear recessive gene. J. Hered. 94, 523–527.

Sammons, R.D., Gaines, T.A., 2014. Glyphosate resistance: state of knowledge. Pest Manag. Sci. 70, 1367–1377.

Schueller, S.K., 2004. Self-pollination in island and mainland populations of the introduced hummingbird-pollinated plant, *Nicotiana glauca* (Solanaceae). Am. J. Bot. 91, 672–681.

Seefeldt, S.S., Hoffman, D.L., Gealy, D.R., Fuerst, E.P., 1998. Inheritance of diclofop resistance in wild oat (*Avena fatua* L.) biotypes from the Willamette Valley of Oregon. Weed Sci. 46, 170–175.

Sinnott, E.W., 1959. Albert Francis Blakeslee: 1874–1954. National Academy of Sciences, Washington, DC, 38 pp.

Stern, C., 1943. The Hardy–Weinberg law. Science 97, 137–138.

The Arabidopsis Genome Initiative, 2000. Analysis of the genome sequence of the flowering plant *Arabidopsis thaliana*. Nature 408, 796–815.

Thinglum, K.A., Riggins, C.W., Davis, A.S., Bradley, K.W., Al-Khatib, K., Tranel, P.J., 2011. Wide distribution of the waterhemp (*Amaranthus tuberculatus*) ΔG210 *PPX2* mutation, which confers resistance to PPO-inhibiting herbicides. Weed Sci. 59, 22–27.

Thurber, C.S., Reagon, M., Gross, B.L., Olsen, K.M., Jia, Y., Caicedo, A.L., 2010. Molecular evolution of shattering loci in U.S. weedy rice. Mol. Evol. 19, 3271–3284.

Toda, K., Yang, D., Yamanaka, N., Watanabe, S., Harada, K., Takahashi, R., 2002. A single base deletion in soybean 3'-hydroxylase gene is associated with gray pubescence color. Plant Mol. Biol. 50, 187–196.

Tranel, P.J., Wright, T.R., 2002. Resistance of weeds to ALS-inhibiting herbicides: what have we learned? Weed Sci. 50, 700–712.

Uphof, J.C.T., 1922. An historic spot for students of genetics. J. Hered. 13, 343–345.

Vigueira, C.C., 2013. The red queen in the corn: agricultural weeds as models of rapid adaptive evolution. Heredity 110, 303–311.

Vila-Aiub, M.M., Neve, P., Powles, S.B., 2005. Fitness costs of evolved herbicide resistance. Encycl. Pest Manag. https://doi.org/10.1081/E-EPM-120039567.

Vila-Aiub, M.M., Neve, P., Powles, S.B., 2009. Fitness costs associated with evolved herbicide resistance alleles in plants. New Phytol. 184, 751–767.

Wang, W., Haberer, G., Gundlach, H., Gläßer, C., Nussbaumer, T., Luo, M.C., et al., 2014. The *Spirodela polyrhiza* genome reveals insights into its neotenous reduction fast growth and aquatic lifestyle. Nat. Commun. 5, 3311. https://doi.org/10.1038/ncomms4311.

Watson, J.D., Crick, F.H.C., 1953. Molecular structure of nucleic acids: a structure of deoxyribose nucleic acid. Nature 171, 737–738.

Yang, Q., Hanson, L., Bennett, M.D., Leitch, I.J., 1999. Genome structure and evolution in the allohexaploid weed *Avena fatua* L. (Poaceae). Genome 42, 512–518.

Yu, Q., Ahmad-Hamdani, M.S., Han, H., Christoffers, M.J., Powles, S.B., 2013. Herbicide resistance-endowing ACCase gene mutations in hexaploid wild oat (*Avena fatua*): insights into resistance evolution in a hexaploid species. Heredity 110, 220–231.

Yu, Q., Jalaludin, A., Han, H., Chen, M., Sammons, R.D., Powles, S.B., 2015. Evolution of a double amino acid substitution in the 5-enolpyruvylshikimate-3-phosphate synthase in *Eleusine indica* conferring high-level glyphosate resistance. Plant Physiol. 167, 1440–1447.

Zufall, R.A., Rausher, M.D., 2003. The genetic basis of flower color polymorphism in the common morning glory (*Ipomoea purpurea*). J. Hered. 94, 442–448.

CHAPTER 8

Invasive Plants

Robert L. Zimdahl and Cynthia S. Brown

Colorado State University, Fort Collins, Colorado

FUNDAMENTAL CONCEPTS

- Invasive plant species can damage native plant and animal communities, alter disturbance patterns such as the frequency of wildfires, change the processing of nutrients and water in ecosystems, and interfere with human activities.

- The arrival of an invasive species is often unnoticed, whereas its effects can endure, although they may not be apparent for years or even decades after its arrival.

- Many, if not most invasive plants were intentionally introduced.

- Invasive plants are ecological, not just agricultural or horticultural problems.

- Not all nonnative plant invasions are inevitably harmful or undesirable

LEARNING OBJECTIVES

- To know what an invasive species is.

- To understand the extent of plant invasions.

- To understand the consequences of invasive plants.

- To understand why plant invasions occur.

- To know some options for managing invasive plants.

1. WHAT IS AN INVASIVE SPECIES?

As is true for many areas of study, one must first determine what it is that is to be studied. Therefore, the first question is: What is an invasive plant species? The answer to the question may be: It depends.

If an invader is simply a species that comes from somewhere else, the definition is purely geographic (Burdick, 2005), which is inadequate. Many terms can be used to indicate that a species has occurred in a new region: alien, exotic, introduced, nonnative, and nonindigenous. As we have seen with the definition of a weed, a human attitude may determine whether a newly introduced species is a threat or just a kindly new neighbor. In this case, the question of definition becomes moot.

When study of invasion biology began, the term "invasion" did not include the implication that a species had negative or positive effects in its new home (Rejmánek et al., 2002). It was used to describe the movement and spread of species on its own or with the unintentional or purposeful help of humans, into areas where it did not previously occur (Rejmánek et al.). "Invasion" and the related terms "invader" and "invasive" were not carefully defined. As concern grew over newly introduced species that were rapidly spreading, so did the use of terms to describe them. Without clear definitions as a foundation, these terms were often used to mean different things, leading to confusion and debate about which terms and definitions were most useful (e.g., Daehler, 2001; Davis and Thompson, 2000; Rejmánek et al.). These authors presented contrasting viewpoints; Colautti (2005) summarizes them and suggests alternate, neutral terminology.

Invasive species have been defined by some as species that spread in new regions (e.g., Richardson et al., 2000; Daehler, 2001; Pyšek et al., 2004) without implications for the effects they may have. Richardson et al. define the necessary terms carefully as follows. Plant introduction occurs when a plant or its propagule(s) has been transported intentionally or accidentally by human action across a major geographical barrier. They are aliens (i.e., nonnatives) and may be simply casual introductions that do not form self-replacing populations. They may become naturalized, which means they sustain populations over many life cycles without, or often despite, direct human intervention (Richardson et al.). However, these nonnative populations may not become invasive. Invasive species produce numerous offspring that reside away from the parent population. Richardson et al. suggest a scale of movement greater than 100 m to classify a species as invasive, in less than 50 years for those that spread by seed, and greater than 6 m in 3 years for taxa spreading vegetatively.

Another way to define invasive species considers their negative environmental or economic effects (e.g., Davis and Thompson, 2000; President Clinton's Executive Order 13112, 1999; Beck et al., 2008; IUCN, 2012) as well as their spread in new regions. This definition predominates in the realm of natural resource management and government policy (Colautti, 2005). Davis and Thompson use a classification scheme that includes the distance a species has moved (short or long), uniqueness of the species to the region (common or novel), and effect on the new environment (small or great). They argue that only species that are novel, have come a short or long distance, and have "great" effects should be called invaders.

Here are a couple of examples of policy definitions that include invader effects. The International Union for Conservation of Nature and Natural Resources (also known as the World Conservation Union) defines invasive species thus (IUCN, 2012):

> An *invasive species* is a species that has been introduced to an environment where it is non-native, or alien, and whose introduction causes environmental or economic damage or harm to human health.[1]

[1]http://www.iucn.org/about/work/programmes/species_work/invasive_species/.

President Clinton's Executive Order 13112 (1999) defined alien and invasive species. It was modified by President Obama's executive order of December 5, 2016: https://obamawhitehouse.archives.gov/the-press-office/2016/12/05/executive-order-safeguarding-nation-impacts-invasive-species:

> An *alien species* means, with respect to a particular ecosystem, any species including its seeds, eggs, spores or other biological material capable of propagating that species, that is not native to that ecosystem.
> An *invasive species* is an alien species whose introduction causes or is likely to cause economic or environmental harm, or harm to human, animal, or plant health.

The order also distinguishes alien and invasive from native species:

> A *native species* means, with respect to a particular ecosystem, a species that, other than as a result of an introduction, historically occurred or currently occurs in that ecosystem.

Daehler (2003) argues that uniqueness in the new range should be the criterion for determining invasiveness, not species' effects. This is because all plant invasions may be detrimental if real economic or environmental harm outweighs the benefits, and benefits and risks that are commonly context specific, subjective determinations (Daehler; Colautti, 2005). Thus, with the latter definition, the species defined as invaders will change from person to person and situation to situation, which is not useful. For example, some claim that smooth brome is invasive and causes economic and environmental harm; therefore, it should be controlled to prevent its spread. Others argue that its forage value for wildlife and domestic livestock far outweighs its harm and invasive risk, and therefore control is not needed. Whether it is a novel species and spreads on its own in many places where it has been introduced is not disputed. The discussion's focus is on whether it is harmful, and, if so, to what?

The plants with which weed scientists were historically concerned are mostly (not exclusively) herbaceous species that occur in highly artificial, species-poor habitats (cropped fields) that are environmentally homogeneous and have predictable disturbance patterns (Weber, 2003). Invasive species, what Weber suggests might be called *environmental weeds* (see subsequent discussion for further definition), occur in species-rich, natural habitats that are environmentally heterogeneous with disturbance patterns that are not primarily caused by humans. Because of the great public and environmental interest in all kinds of invasive species, the attention of many weed scientists has shifted over the past decade or two from weeds of agroecosystems to invasive plant species.

The Weed Science Society of America (WSSA) (wssa.net/wp-content/uploads/WSSA-Weed-Science-Definitions.pdf) defines an invasive weed as one that establishes, persists, and spreads widely in natural ecosystems outside the plant's native range. When in a foreign locale, invaders often lack natural enemies to curtail their growth, which enables them to overrun native plants and ecosystems. Many invaders are also classified as noxious weeds by federal, state, or local governments. A noxious weed is declared to be injurious to public health, agriculture, recreation, wildlife, or property.

Common weeds are plants that may or may not be introduced, grow where they are not wanted (in a human disturbed habitat: a cropped field, garden, landscaped area, etc.), and whose presence leads to undesirable aesthetic, economic, or environmental effects. Environmental weeds (Weber 2003, p. 1) are invasive, alien plants that spread in natural areas where they are not wanted and may have adverse effects on biodiversity, ecosystem functioning, or the economy. Not all those using the term "invasive plant" or "environmental weed" included the latter criteria in the definition, as described earlier. This is part of the reason why the answer to the question "What is an invasive species?" is "It depends." Regardless which of these definitions is used, all invasive plant species are weeds because they grow where they are not wanted, but not all weeds are invasive because they may not be nonnative, become naturalized, and spread where they are introduced. For more discussion of terminology, readers are referred to the sources mentioned previously and Sagoff (2005).

2. WHAT IS THE EXTENT OF PLANT INVASIONS?

In 2006, invasive plants dominated more than 100 million acres in the United States and were increasing at an estimated 3 million acres each year (Nat. Invasive Species Council, 2006). In 2006, they estimated noxious weeds were expanding on Bureau of Land Management land at 2300 acres/day and at twice that on western rangeland. Climatic and geographic range for nonnative plants in the United States is greater than for natives (Bradley et al., 2015). For example, tropical spiderwort was first identified as an aggressive weed in Florida in the 1990s and was established in California in 2015. Myers and Bazely (2003, pp. 18–19) compiled data for several world regions to show the percentage of all plants that have been introduced. The data clearly showed that the problem was shared by all world areas. Accurate regional data on invasive species distribution are rare (Marvin et al., 2009).

World Region	Number of Areas	Introduced Plants as Percentage of all Plants
Oceania	7	35
Canada	7	26
United States	12	16.5
Caribbean	4	25
Europe and United Kingdom	9	11.8

Invasive plants are present in all 50 United States and they will certainly be found in any country, if someone looks. As many as 100 million US acres (the area of California) are now home to one or more invasive plants and they are spreading at a rate of 14 million acres each year. Duncan and Clark (2005) estimated that the average annual rate of spread for 15 weedy species in all continental United States was 11%–17%. Oswalt and Oswalt (2011) suggested that the southern United States was an area of primary concern about the spread of nonnative invasive plants. Japanese honeysuckle and Nepalese browntop were invading forests and displacing native species throughout the southern United States.

Mullin et al. (2000) said invasive plants were one of the greatest threats to cropland, range-land, aquatic areas, and wildland. Their report included a partial list of major economic and ecologically important invasive weeds: 12 riparian, 11 aquatic or wetland, 60 range and wildland, and 14 cropland. The 2015 Federal noxious weed list includes 112 species: 19 aquatic, 5 parasitic, and 88 terrestrial (http://plants.usda.gov/java/noxious). Other sources include many more species. The Wikipedia list of invasive species in North America includes 117 plants. The Invasive Plant Atlas of the United States includes 521 herbs and forbs.

Each of the United States has an invasive species list. For example, Colorado's list includes 78 noxious weeds in three categories. When found, a plant on the A-list of 25 species must be eradicated. In 2015, Bohemian knotweed, giant knotweed, Japanese knotweed, and hairy willow herb were added to the A-list. DeMarco (2015) claimed that invasive plants are conquering the United States but occupy only 50% of their expected range. They are more widely distributed than native plants and most of the blame rests with humans.

It is important to recognize that plant invasion is not just a United States or developed world problem. China has a long history, as does the United States, of introducing plants that someone deems to be potentially beneficial or perhaps just interesting. With its fairly recent and expanding international trade, intentional introductions are now joined by unintentional introductions that may not be beneficial because they are unknown and unexpected. Diamond (2005, p. 367) notes that in Shanghai harbor, one of many international Chinese harbors, "between 1986 and 1990, examination of imported materials carried by 349 ships from 30 countries revealed as contaminants almost 200 species of foreign weeds. Weeds and poisonous grass species have spread at the expense of high-quality grass species "over as much as 90% of China's grasslands" (Diamond, p. 366).

It is clear that there is legitimate scientific concern about invasive plants and especially weed species. The WSSA's concern is illustrated by their 2008 decision to begin publishing their third journal: *Invasive Plant Science and Management*. However, this view is not univer-sal (e.g., Sagoff, 2005). Philosophical skepticism demands questioning the notion that absolutely certain knowledge about anything is possible, and ask whether claims about the extent and potential danger of invasive species are true or if they contain a bit of hyperbole. The Economist (2015a,b — March 28 and December 5) defended invaders and suggested that most campaigns against foreign plants are pointless; some are even worse than that. The article suggested that, with a few important exceptions, campaigns to erad-icate invasive species are a waste of money and effort, for reasons that are partly practical and partly philosophical. Practical reasons include that most introduced species are not successful or harmful. Second, invaders are fiendishly hard to eradicate, which is not the goal of most weed management systems: control is. Third, invasive species are oppor-tunists. They tend to invade disturbed habitats, and unfortunately the crop field is a classic example. Modern agriculture creates disturbed habitats: that is how it is practiced. The fundamental philosophical objection is that eradication is essentially impossible because it is an attempt to restore balance to nature: "a prelapsarian idyll that prevailed before human interference." However, the authors agree that we should not be passive about in-vaders. Some foreign species, particularly agricultural pests, are gaining a foothold and should be fought. Despite legitimate, watchful concerns, alien plants are rarely troublesome to native ones. In contrast to that, weed scientist agree that cheatgrass is without question the most successful invasive species in North America. It lives in every state, occupies more than 150,000 square miles of the West (Soloman, 2015), and has caused major changes to ecosystems in some places it invades (Germino et al., 2016). It is one that must be fought. Himalayan balsam, Japanese knotweed, and rhododendron in the United Kingdom, garlic mustard and kudzu in America, and rats and possums in New Zealand are all undesirable invasive species. Nevertheless, not all invaders are bad. The American honeybee (*Apis mellifera*) is an invader. Even kudzu, the vine that ate the South, was never as scary as its billing and now appears to be in retreat (Finch, 2015). Cleve Phillips (Kolb, 2015) has actually turned it into an art form.

3. WHICH SPECIES WILL BE INVASIVE?

When one suspects a species may be invasive, one must ask, how does one know? Do invaders share certain characteristics? Which species will become invasive? The WSSA definition (given earlier) provides some general characteristics but little specific guidance.

More than 50,000 nonnative, exotic species (not just plants) have been introduced to the United States over the past 200 years; about 5000 were plants but only about 14% (perhaps as many as 675 species) of all introductions are regarded as invasive (Chafe, 2005; Pimentel et al., 2000). The North American list of invasive species includes 112 plants, 23 insects, 5 aquatic arthropods, 16 mollusks, 10 reptiles, 17 bird and mammals (including domestic dogs), 11 pathogens, 20 fish, and 3 others.[2] Of the 5000 exotic plant species, intentionally introduced, beneficial species (e.g., corn, wheat, and rice, plus cattle and poultry) accounted for 98% of the productive crops grown in the US food system in 1998 (Chafe). Few of the 50,000 introductions "become capable of establishing a breeding population in the new location without further intervention by humans, and become a pest in the new location, directly threatening agriculture or local biodiversity (see footnote 2)."

In the United Kingdom, 71 of 75 nonnative crops occur at least casually or are naturalized (Williamson and Fitter, 1996). Crop plants are almost always considered to be beneficial, but as Williamson and Fitter point out, because they are strongly selected to grow where they are cultivated, they may also be adapted to grow well outside cultivation, under similar environmental conditions. Nonetheless, Williamson and Fitter found that for British animals and plants, approximately only 1 in 1000 introduced species have become pests. In January 2016, the European Union made it a punishable offense to keep, cultivate, breed, transport, sell, exchange, or intentionally or unintentionally release 14 nonnative invasive plants. Eight were plants that had been popular with gardeners.

Westbrooks (1998) claims that there are about 8000 plant species, 3% of all known plants, considered invasive. Of those, only 200—250 (less than 0.3% of known plants) are major world weeds (Holm, 1978). The most troubling weedy species are in 80 taxa (Holm et al., 1977), but the world's worst invasive species belong to only a few families and genera: Acacia, Asteraceae, Cyperaceae, Poaceae, and Mimosa (Mack et al., 2000; Holm et al.). A few plant species have invaded widely separated places on the planet, which Mack et al. equate to being "the ecological equivalent of winning repeatedly in a high-stakes lottery." "Few survive the

[2]https://en.wikipedia.org/wiki/List_of_invasive_species_in_North_America.

hazards of chronic and stochastic (random, chance) forces and only a small fraction become naturalized" (Mack et al.). Leslie and Westbrooks (2011) claim that "invasive species are everywhere" in all types of ecosystems.

A reasonable but not perfect predictor of behavior in a new place is behavior in the place of origin (the home range). Plants that behave badly in their place of origin are likely but not sure to behave badly in a new place. Firn et al. (2011) found that in general, nonnative plant species in grasslands around the world behaved similarly where they were introduced as they did where they originated.

For example, one ecotype of the common reed (*Phragmites*) is native to central Europe and can be invasive within its native range (Weber, 2003). It was assumed that the common reed had been introduced to the United States, where it dominated brackish wetlands with low salinity and high soil oxygen content relative to other wetlands (Amsberry et al., 2000). In dense stands, plants could be more than 15 ft tall. Rudrappa et al. (2009) found that North America has a native and an introduced strain of common reed coexisting. The native strain, a noninvasive weed was displaced by an invading species from the same genus (Saltonstall, 2002). It invaded low marsh habitats by first invading high marshes and then expanding through clonal growth to the lower, less favorable areas. The invasion, in the view of Amsberry et al., resulted from a variety of human-induced changes in coastal habitats and the transfer of resources through connections among clones from those in the more favorable parts of the marsh to those in less favorable locations.

Many have worked for decades to answer the question: Which species will become invasive? Early efforts were made to identify the traits of weed species, in particular the now-classic list of characteristics of weeds by Baker (1965) (see Chapter 2 of this volume).

Highly specific classifications or lists of characteristics are of little help in identifying potential plant or animal invasive organisms (Noble, 1989). Reed's (1977) study included 1200 foreign plants from 101 families that were weeds in their native range and could become serious weeds if they invaded the United States, where different "environmental and biological restraints no longer controlled their development."

Westbrooks (1998, p. 3) lists 12 characteristics of successful invasive plant species that permit them to become established in new areas and outcompete native vegetation. The list is not exhaustive, yet no single invasive plant possesses all of the characteristics Westbrooks includes:

- early maturity;
- profuse reproduction by seeds and/or vegetative structures;
- long life of seeds and vegetative parts in soil,
- seed dormancy to ensure dispersal in time,
- adaptation for dispersal by wind, water, humans, and other animals or as contaminants of seed (see the reference to seeds the size and shape of crop seeds);
- allelopathy (see Callaway and Aschehoug, 2000, Chapter 7);
- spines and thorns or other structures that cause physical injury and repel grazing animals;
- ability to parasitize other species;
- seeds that are the size and shape of crop seeds, so separation by standard cleaning techniques is ineffective;

- vegetative structures with large food storage;
- survival and seed production under adverse growing conditions; and
- high photosynthetic capacity.

Kolar and Lodge (2001) similarly found that invaders tend to reproduce vegetatively. When they do reproduce by seeds, they tend to have small seeds, short juvenile periods, and low variability in seed crops.

Many, but not all of these characteristics of invasive plants also appear in Baker's list of characteristics of weeds (Baker, 1965). More recently, others have continued the search for characteristics that permit recognition of an invader before or as it invades. In a global study, Daehler (1998) found that species with the greatest potential to become invaders are those that are primarily aquatic or semiaquatic, grasses, nitrogen fixers, climbers, and clonal trees, groupings that include a wide variety of characteristics. He found that invasive plants are often from woody plant families (Daehler). Richardson and Rejmánek (2004) studied invasive and naturalized conifer species from around the world and found that species from one genus, pines (*Pinus*), were particularly invasive. They also found that having small seeds, a short juvenile period, and short intervals between large seed crops were associated with being invasive. Grotkopp et al. (2002) evaluated 29 pine species from disturbed habitats and found that high seedling relative growth rates (i.e., the speed of plant growth), small seeds (less than 50 mg), and short intervals between generations (less than 10 years) were associated with greater invasiveness. These studies provide examples of how taxonomy (i.e., how closely related a species is to invasive species) and life-history characteristics (i.e., traits associated with growth, reproduction, and survival) may help predict invasions before they occur.

Parker et al. (2006) ranked nonindigenous species not yet naturalized in the United States by their potential to become invasive there. Their 14-character criteria fell into four categories:

1. invasive potential, based on possessing traits associated with invasiveness and reports of being invasive elsewhere;
2. geographic potential, based on the likelihood a species will spread throughout the United States;
3. damage potential, based on the likelihood a species will have a big effect owing to characteristics associated with negative effects on plants, animals, or humans; and
4. entry potential, based on being cultivated in the United States.

Many of the specific traits considered are those identified in previous studies. Their analysis indicated great potential for many weedy and pest species to become invasive in the United States.

Despite the patterns that have emerged from comparing characteristics of species known to be invasive, we are unable to predict reliably which species will be invasive using this tool alone.

Some of the worst invaders are highly adaptable generalists; that is, they have many characteristics of good invasive species. Others are not. Whereas many invaders are from plant families that have other invasive members, numerous invasive species have few or no aggressive relatives. For example, what Holm et al. (1977) call the world's worst weed, water hyacinth, is the only member of the *Eichhornia* genus that is invasive (Mack et al., 2000). That could be because of the lack of opportunity to invade offered to relatives, the characteristics of the invasion site, or a lack of the right characteristics of the potential invader for invasion (Mack et al.).

We are still left with the question: Which species will be invasive? A species' success depends not only on its traits but its interaction with the environmental conditions, both biotic and abiotic, that it encounters where it is introduced, and ultimately, on who cares.

Although new to the area, there is often an implicit assumption that invasive plants are different from and outperform co-occurring native plants. So far, the evidence does not support this (Thébaud and Simberloff, 2001; Daehler, 2003). Alien plant invaders are not more likely to be taller (Thébaud and Simberloff) or have higher growth rates, competitive ability, or fecundity (i.e., ability to reproduce) (Daehler). Daehler found that invaders' relative performance (success) depended on growing conditions, although invaders in his study had greater phenotypic plasticity than natives, which can be advantageous in variable environments (Daehler). Thus, environmental conditions have a large role in invasion success. Lemoine et al. (2016) claimed that "although decades of theory posit that introduced species require unique ecological hypotheses to explain their spread and success, recent evidence suggests that native and introduced species have similar leaf carbon capture traits chemical defenses, enemy release, allelopathy and community assembly mechanisms." Their article suggests that invasive and native species might thrive for similar reasons.

Another common assumption is that invasive species are more abundant in areas to which they are introduced than where they originate. This may be true for some species but not for all. For example, Firn et al. (2011), mentioned earlier, measured 26 grassland species that occurred at 39 sites in eight countries and found that abundances of these species were similar in introduced and native ranges. This suggests that the abundance of a species in its native territory is generally a good predictor of what its abundance will be when it arrives in a new territory. It does not support the assumption that invasive species are doing something different in their introduced range compared with their native range.

In his analysis of what makes plant communities invisible (more easily invaded), Rejmánek (1989) looked for similarities among species known to invade natural plant communities (those that experience only "natural" disturbances such as flooding) in different parts of the world. His list included 54 species from 40 families that represented all major plant growth forms. He was unable to discern characteristics shared by species that made them good invaders. He points out that it is difficult, if not impossible, to disentangle the abiotic effects of the new habitat and the biotic effects of the community residing there on the success of a newly introduced species. Instead, he suggests that factors driving plant invasions are extent and the frequency of propagule (seeds or vegetative reproduction structures) introduction and disturbance, the relative competitive ability of invaders with respect to native species, and the reduced coincidence of resource requirements between introduced and resident species (i.e., reduced competition with established occupants for resources).

Since Rejmánek's work, evidence has accumulated that particular plant communities are more easily invaded than others. For example, Zedler and Kercher (2004) point out that wetlands are especially vulnerable to invasions. Only 6% of the earth's land is wetland but 24% of the world's most invasive species are wetland species. Wetlands are what Zedler and Kercher call landscape sinks, "which accumulate debris, sediments, water, and nutrients, all of which facilitate invasions by creating canopy gaps or accelerating the growth of opportunistic plant species." That is to say, consistent with Rejmánek's (1989) hypothesis, a wetland is frequently disturbed and may have a high rate of introduction of invading propagules that have resource requirements different from those of the natives.

It is apparent that the interaction of a particular invader's characteristics with the invaded environment determines its success. Many risk assessment approaches have been developed to use these relationships in hopes of identifying which species will become invasive before they do. Pheloung et al. (1999) developed a weed risk assessment model for use in evaluating plant introductions to Australia and New Zealand. The model was based primarily on a taxon's weed status in other parts of the world, its climate and environmental preferences, and certain biological attributes (e.g., method of dispersal and seed survival). All taxa classified as serious weeds and most minor weeds were rejected (excluded) by the model, which they recommend as a screening tool. This is a semiquantitative approach, as is the ranking method developed by Parker et al. (2006) described earlier, but there are also quantitative statistical models and qualitative approaches, the latter relying on expert opinion (Hulme, 2012), that seek to achieve the same ends. Hulme concludes that none of these risk assessment approaches are accurate enough to be worth the trouble, owing to:

- the complexity of the relationships among components of invaded systems, which lead to unpredictable emergent qualities;
- the variability and uncertainty of parameters used in risk assessment models; and
- apparent overconfidence in parameter estimates that leads to underestimation of extreme events (e.g., introduction of a species that turns out to be a successful invader), which are the ones likely to have the biggest effects.

Hulme (2012) suggests that it would be better to focus resources on scenario planning in which interested parties (e.g., land managers, policy makers, and scientists) evaluate a few possible future situations that may develop in a particular habitat with a set of potential plant invaders. This stakeholder group would consider the biotic (resident and introduced species) and abiotic environments (soils, water, and climate), including human activities, which might influence the success of the hypothetical invaders and develop action plans that would be executed if one of the scenarios occurred. Resources could then be spent to identify invasions early, minimize negative effects, and manage the invasive species if it persisted.

Species distribution models are another way to combine characteristics of habitats and potential invaders for predicting plant invasions before they occur. These models can be used for invasive species risk assessments by forecasting locations most likely to be invaded. Species–environment matching and spatial modeling can be conducted using many different methods. Stohlgren et al. (2010) compared four of these methods with an ensemble model (consensus model). Ensemble models identify regions of agreement among the individual models as areas of greatest confidence and those with the most discrepancies among models as areas of least confidence about invader occurrence. Stohlgren et al. found that the ensemble model they used improved predictive power compared with modeling methods applied individually. Ensemble modeling can overcome weaknesses of individual models through combining model strengths, and may be a powerful risk assessment tool for managers (Stohlgren et al.). Barney (2014) claimed that the United States "is charging toward the largest expansion of agriculture in 10,000 years with vast acreages of primarily exotic perennial grasses planted for bioenergy." The risk is that the grasses possess many traits of invasive plants, but their invasive potential when planted over large areas has not been evaluated. Cautious integration is recommended. Smith et al. (2015) compared Australian and US weed risk assessment models for evaluation of the invasion risk of 16 potential bioenergy crops. The results of the two models were compared for evaluation of 14 agronomic crops and 10 invasive species of agronomic origin. Both models failed to distinguish weeds effectively from crops. They were unable to assess intraspecific variation accurately and were not reliable predictors of the potential invasiveness of biofuel crops.

4. WHY DO INVASIONS OCCUR?

Scientists have not reached what Simberloff (2008) calls "the Holy Grail," being able to predict the possibility or trajectory of particular invasions. Those who sell plants that may become invasive often do not know about invasion potential, at least partially because it is difficult to agree upon what it is that ensures an invader's success. Lonsdale (1999) proposed that the invasion of any environment by a new species is influenced by three things: the number of propagules entering the environment (propagule pressure), the characteristics of the new species, and the environment's susceptibility to invasion (invasibility). These causes of invasion are not that different from those Rejmánek (1989) identified 10 years earlier (see the previous section). It is safe to say that there are many reasons for plant invasions. Here we present some of the leading theories about why they occur.

4.1 Propagule Pressure

The proximal cause of any species invasion is the movement of propagules to a new location. Propagule pressure is the number of propagules (seeds, vegetative reproductive structures, or both) that are introduced. Noble (1989) points out that the absence of special long-distance dispersal mechanisms (e.g., wind-transported seed) is not a hindrance to invasion because humans are the primary vectors of transport. In contrast, short-distance dispersal mechanisms may enhance the "probability and rate of invasion." Plants that produce many reproductive propagules have enhanced invasion potential, but the characteristics of the area invaded are the critical determinant of invasion success.

Many, perhaps most invaders have been intentionally introduced and most introductions have been ornamentals (e.g., purple loosestrife) (Reichard and White, 2001). Horticulture has been what Bright (1998, p. 147) calls "a gargantuan engine of biotic mixing that has helped unleash some of the world's worst plant invasions." A few examples are shown in Tables 8.1 and 8.2. In the United States, 82% of woody species that have colonized areas outside their area of cultivation (they have become invasive) have been introduced for landscape purposes (Reichard and Hamilton, 1997). A survey of 1060 woody plant invasions found that in 624 cases, when origin could be determined, 59% came from botanical gardens (Bright).

TABLE 8.1   Plant Invaders With a Horticultural Origin (Bright, 1998, pp. 148—149)

Plant	Source	Location of Problem Invasion
Rubber vine	Madagascar	Northern Australia
Traveler's joy clematis	Northern Europe	New Zealand
Water hyacinth	South America	Southern United States, Africa, south Asia
Purple loosestrife	Europe	Northern United States
Japanese knotweed	East Asia	Europe and North America
Tamarisk/salt cedar	Central and east Asia	Most of the United States

Reichard and Campbell (1996) showed that 85% of 253 invasive woody species in the United States were introduced as ornamentals and 14% as agricultural plants. Many invaders are sold regularly in nurseries. These plants were introduced and are sold because they possess traits that are highly desirable to gardeners, landscapers, and the nursery industry (Li et al., 2004). For example, they are usually easy to establish and grow with little care, often under diverse environmental conditions. The homeowner goes to the nursery wanting a plant that will grow easily, with little care, in a bad place (shady, dry, etc., the microenvironmental factors) (Skurski et al., 2014), and there it is. What it may become in 10 years is neither asked nor revealed, if it is even known.

Many invasive species have been introduced accidentally as crop seed contaminants (e.g., leafy spurge, spotted knapweed, yellow star thistle) or simply as free riders in a shipment of unrelated things (e.g., cheatgrass). Some of the worst plant invaders have been intentionally introduced, including English ivy, Johnsongrass, kudzu, tamarisk, and water hyacinth (Westbrooks, 1998). For example, camel thorn, a native of the Turanian Desert and the Iranian Plateau, was introduced to the United States in 1915 in alfalfa seed from Turkestan and in camel dung packing around date palm offshoots (Brock, 2006). It is found in nine southwestern and western states (not in Oregon) and is common in northwestern Arizona. The common name is derived from the fact that, despite its abundant sharp thorns, camels eat it. It is unpalatable to other livestock and rapidly displaces native vegetation (Brock).

Although many plant introductions are the result of deliberate, flawed forethought (Mack et al., 2000), not all have resulted in disaster. The ornamentals camellia and azalea, originally from Asia and India, are planted widely. Both stay where they are planted. Neither escapes by vegetative fragmentation or bird dispersal of seed, and both behave as we would like them to behave (Burks, 2002). On the other hand, corallita, a perennial vine with vigorous growth and abundant pink flowers, has been praised as an ornamental (Burke and DiTomaso, 2011). However, in Florida and several tropical areas, it is classified as an invasive species owing to its vigorous growth, dominance of other species, and rapid spread.

However, the problem of invasive plant species cannot be attributed only to the desire of horticulturalists to identify and import new ornamentals. Globalization of commerce

TABLE 8.2 Purposeful Plant Introductions That Have Become Important US Weeds (Williams, 1980 and Other Miscellaneous Sources)

Weedy Species	Origin	Purpose of Introduction
Autumn olive	Asia	Wildlife attractant/erosion control
Birdsrape mustard	Eurasia	Cultivated crop
Bermuda grass	Europe	Forage crop
Bouncing bet	Europe	Ornamental
Cogon grass	Asia	Packing material
Corn cockle	Europe	Ornamental
Dalmatian toadflax	Europe	Ornamental
Hydrilla	South America	Use in aquaria
Japanese knotweed	East Asia	Ornamental
Jimsonweed	Tropics	Ornamental
Johnsongrass	Africa/Asia	Forage
Kochia	Europe	Forage/ornamental
Kudzu	East Asia	Ornamental/forage/erosion control
Lantana	Europe/Asia	Ornamental
Melaleuca	Australia	Tropical forest species
Multiflora rose	East Asia	Windbreaks/cover
Musk thistle	Europe	Ornamental
Reed canary grass	Eurasia	Forage
Salt cedar/tamarisk	Europe	Ornamental
Tansy	Europe	Herbal plant
Tree of Heaven	China	Ornamental
Tropical soda apple	Argentina	Unknown
Water hyacinth	Tropics	Ornamental
Yellow toadflax	Europe	Ornamental

leading to rapid, often unchecked movement of species is much more important. The world is becoming smaller in the sense that the speed and frequency of travel have increased. Lovelock (1979), a mathematician, was the first to argue that the earth was a single, planet-sized organism. He named it Gaia after the Greek earth goddess. His arguments were regarded as less than scientific, perhaps mystical. His proposal was largely rejected by the scientific community, which acknowledged that there has always been movement among the earth's communities, but it has been slow and controlled by natural forces. However, real, ecologically relevant geographic and climatic barriers are no longer as great as they once were. For at least three reasons, they have declined in importance as the earth has moved closer to being regarded as the unitary whole Lovelock proposed (Bright, 1998, p. 20). First, the frequency of movement has increased. Planes and ships move thousands of people daily across vast distances. Each plane and ship carries known and unknown organisms in addition to the people who bring known and unknown organisms on and in their bodies, clothing, and possessions. For centuries, natural movement across geologic barriers was slow, as was human travel. Natural movement is still slow, but the rapidity of human movement has vastly increased the speed of arrival of all kinds of organisms. Second, now movement can occur almost anywhere on almost any day. Intense biotic mixing has changed from "an occasional regional event to a chronic global occurrence" (Bright). Finally, what was impossible migration is now possible and common. Miles of salt water or desert used to be effective, impenetrable barriers to organism movement. Such barriers provided the isolation that allowed unique species and ecosystems to evolve (McNeely, 2004). With modern rapid transportation, such barriers are crossed with ease. In fact, they are not barriers to movement of any organism.

4.2 Species Change After Introduction

As observed by Hallett (2006), plants dislocated from their coevolved relationships are inevitably confronted with new relationships with which they must deal. Some succeed; many fail. In the plant kingdom, one must adapt or die. In human terms, it is similar to moving from the town in which you grew up, where everything is familiar and you knew who was friendly and who was not. Suddenly you are compelled to move to a new place where you do not know who is nice and who is not, where things are located, and perhaps most daunting, you must build new relationships. Humans often become transformed by such moves although we are not compelled to adapt as all other creatures must. Plants may also be transformed. When confronted with new relationships they undergo ecological transformation after which some succeed (adapt) and others do not. It is the transformation, in Hallett's view, that affects the ability of a plant "to become established, invasive, and naturalized in a new environment."

Many invasive species succeed only a long time after introduction. That is, some survive in a new environment but do not immediately become invasive. The time between introduction and when the population begins growing exponentially is called the lag phase (Fig. 8.1). There are a number of reasons why lag phases may occur and last for different periods of time. It may be that there are too few introduced species to be noticed initially or for them to locate mates, or the small populations may go extinct many times before conditions are favorable enough for them to persist. The spatial distribution of populations can influence the lag time because spread occurs more quickly from multiple small sources than one large one (Mack et al., 2000). If there is only one good-sized population, exponential growth may be delayed. However, once the population reaches a sufficient size and is growing exponentially, it cannot be ignored. Mack et al. provide the lag phase example of Brazilian pepper, which was introduced to Florida in the 1800s, but was not apparent until the 1960s (Schmitz et al., 1997).

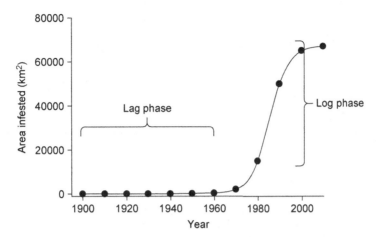

FIGURE 8.1 Invasive species often exhibit expansion patterns with a lag phase, in which the infestation appears to grow slowly, and a log phase, in which the area infested grows exponentially. *Adapted from Fig. 2 (p. 692) Mack, R.N., Simberloff, D., Lonsdale, W.M., Evans, H., Clout, M., Bazzaz, F., 2000. Biotic invasions: causes, epidemiology, global consequences and control. Ecol. Appl. 10, 689–710.*

Initiation of the exponential growth phase may be the result of various changes in the species or changes in the environment that enable a population explosion. Adaptation (genetic change as a result of natural selection that makes the population or species more successful in a habitat) to the new environment may improve the ability of a species to thrive (Mack et al., 2000) resulting in rapid population growth.

Hybridization can also stimulate the evolution of plant invasiveness and may explain the initiation of exponential growth after the lag phase, as proposed by Ellstrand and Schierenbeck (2000). Hybridization between the potential invader and "disparate source populations may serve as one stimulus for the evolution of invasiveness." They found 28 examples from 12 plant families in which invasive success was preceded by hybridization. Several of the examples were weeds. For example, they attributed the invasive success of the *texanus* subspecies of common sunflower and *Sorghum almum* to intertaxon hybridization. A few of the other invasive species that evolved after intertaxon hybridization include purple loosestrife, Scotch thistle, wild radish (after interbreeding with common radish), and a perennial rye that resulted from hybridization between common rye (a grain crop) and weedy rye.

The success of hybridization has been demonstrated with Eurasian watermilfoil populations in the United States (Moody and Les, 2002). It is present in 45 of the United States and three Canadian provinces (USDA-NRCS, 1999). Its invasive success has "resulted from hybridization between non-indigenous and native species." This resulted in *heterosis*, or increased growth and vigor in hybrid offspring, also known as hybrid vigor, which was maintained through vegetative propagation (Moody and Les). Hybridization has also had an important role in the invasion of species of tamarisk (*Tamarix* spp., also known as salt cedar). Tamarisk, a small tree or shrub from Eurasia, was intentionally introduced to North America as an ornamental and for erosion control (Lingren et al., 2010). It spreads along rivers, in flood plains and wet meadows and pastures. Genetic analysis shows that tamarisk populations in five western US states are dominated by hybrids of different tamarisk species that were geographically separated in their native ranges (Gaskin and Schaal, 2002).

Some nonnative species may become invasive only after symbionts (an organism living in a state of symbiosis) that improve their success are also introduced. Fig (*Ficus*) species are a good example of this. Over 60 species of fig have been introduced into Florida and only three have become invasive, all of them after their specific pollinator wasp was introduced (Nadel et al., 1992).

4.3 Resource Availability

Invasion of plant communities is often tied to resources that become available to invaders through different means. Resource availability can increase as a result of removal of species that would otherwise be intercepting the resources themselves, such as when a tree falls in the forest and creates a light-filled gap. Huenneke et al. (1990) showed that independent of disturbance, increased resources resulted in invasion of serpentine grasslands, which are normally resistant to invasion owing to the poor soil chemistry, by introduced annual grasses. The importance of resources is also illustrated by the work of Meekins and McCarthy (2001) who demonstrated that growth and reproduction of the nonindigenous forest herb, garlic mustard, was not dependent on disturbance. Its invasive success was determined by adequate soil moisture and available light. Similarly, more effective competition for light because of stem elongation and canopy formation enabled western elodea to invade and become established in the presence of common elodea (Barrat-Segretain and Elger, 2004).

Excess resources will be available for invaders to exploit if the resident plant community does not use them. This occurs in grasslands dominated by annual grass species with relatively shallow root systems that leave abundant water at deeper depths, which can be exploited by deep rooted invaders and improve invader success (Brown and Rice, 1998).

Davis et al. (2000) proposed that fluctuation in resource availability is a key factor that controls invasibility of plant communities. Disturbance, a major factor in invasion, usually increases resource availability. Davis and Pelsor (2001) demonstrated that changes in resource availability affect competition intensity, which affects community invasibility. Short-term increases in resource availability (a few weeks) can "temporarily reduce or suspend competition from resident vegetation," thereby increasing an environment's invasibility for as long as 12 months (Davis and Pelsor).

Goldberg (1990) argued that "effect on and response to a resource is positively correlated only to the extent that both are a function of uptake rates." Uptake may be a relatively unimportant determinant of the magnitude of the effect on and response to a resource. Research needs to be done to compare effect on and response to a required resource. Goldberg also proposes that species that make heavy demands on a resource will dominate in "communities where success is determined by size-symmetric competition," i.e., when plants in competition are of similar size, and thus light is not limiting. Good examples include cropped agricultural fields, early succession areas, and gaps in existing vegetation. In contrast, plants that tolerate and grow well with low resource levels often dominate when size-asymmetric competition occurs, as it does when seedlings germinate in mature vegetation (Goldberg). Fargione et al. (2003) showed that in constructed prairie grassland, resident species inhibited establishment of species with similar resource use patterns. The success of invaders decreased as diversity increased. This is explained by the simple mechanism of competitive inhibition of invaders that are similar to established, abundant species. These species were more likely to occur in more diverse mixtures because species composition of each community was determined by random draws from a fixed pool of 24 species.

Brown and Rice (2010) found a similar pattern in California grasslands. They showed that the early-season, shallow-rooted invader of prairie plant communities, soft brome, was least successful in mixtures of species with similar resource use patterns. The mixtures were composed of species that needed the same resources at the same time and from the same part of the soil profile as each other (termed non-complementary resource use). The late-season invader, yellow star thistle, was least successful in mixtures of species with varied resource use patterns (termed complementary resource use), some of which were similar to yellow star thistle. Invaders in this work were not as successful in plant communities containing species with resource use patterns similar to the invader. Both invaders were less successful in more diverse (greater species-rich) communities, and their success was reduced by existing vegetation independent of the resource use pattern of the invading species (Brown and Rice).

4.4 Disturbance

One of the best hypotheses about the reasons for invasion is that disturbance before or at the time of invasion is the primary cause (Mack et al., 2000). When sudden or regular changes in a particular environment occur as an invader arrives, successful invasion is more likely, often because the disturbance releases resources making them available to the invader (Gibson et al., 2011).

One of the best examples of disturbance is intentional or accidental use of fire. It has had a significant role in several biotic invasions (Mack et al., 2000). Invasions of nonnative species on the arid, temperate grasslands of Australia and North and South America were facilitated by fire (Mack et al.). D'Antonio and Vitousek (1992) provide one of the few studies that illustrate the change wrought by invading grasses (from Africa in this case) on previously forested areas of the Amazon basin. Land clearing, nutrient loss, altered microclimate, prevention of succession, and fire are significant on a local scale and are becoming significant on regional and global scales. The success of alien grasses is aided by fire and they prevent succession of native species, thereby creating an environment conducive to success of the alien at the expense of native species. Once the grasses become established, their continued success is ensured because of their rapid annual reproduction and highly flammable litter. In an ecosystem not adapted to fire, regular fire denies establishment of natives, but does not harm invading grasses; in fact, it encourages them.

The regular disturbance of grazing in systems that did not evolve with it and overgrazing often favors invaders over native species. Many of the world's currently dominant rangeland plants owe some of their success to grazing pressures (Bright, 1998 p. 41). The success of downy brome on western US rangelands was caused, in part, by continued overgrazing. Downy brome has been called the most devastating ecological problem in the western

United States (Devine, 1993; Soloman, 2015). It is a native of Eurasia and began to invade the western United States and Canada between 1889 and 1894. It appeared in most areas where it is today by 1928 (Mack, 1981). It may have been introduced intentionally as a forage, but it is more likely that it arrived in several locations as a seed contaminant (Mack). It dominates more than 100 million western US acres where its presence would not have mattered if the range's natural resilience had not been weakened by overgrazing. Some native plants are easily damaged by livestock (cattle) grazing. Some evolved without being regularly grazed by large herbivores (Devine). In contrast, downy brome evolved in Eurasia under regular grazing from camels, horses, and other animals. It is not affected detrimentally by intense grazing as long as it can set seed, which it does quickly and abundantly early in the spring. Overgrazing tends to eliminate native species and create opportunities for downy brome to thrive. Downy brome can, but often does not, enter native plant communities in the western United States and cohabit without dominating. Downy brome's invasion of the West is a clear demonstration of how successful an invader can be when "preadaptation, habitat alteration simultaneous with entry, unwitting conformation of agricultural practices to the plant's ecology and apparent susceptibility of the native flora to invasion, are all in phase" (Mack). Overgrazing and the grass fire cycle perpetuate its dominance and facilitate its aggressive invasion. Despite the weed's well-known invasive ability, ranchers do not completely despise it because it provides abundant, albeit brief, early spring grazing.

4.5 Diversity—Invasibility

Elton (1958) proposed that a community's resistance to invasion increased in direct proportion to the number of species in the community: the diversity—invasibility hypothesis. The essence of the hypothesis is that a community's species richness (the greater the number of species) indicates reduced resource availability, and therefore greater resistance to invasion, because resources are being used by the wide variety of species present in the community. All niches are occupied by one or more of the community's diverse species. It is true that there is often reduced resource availability in communities with high species richness. However, in some cases the hypothesis does not hold. All communities have resources that are not being used, and study of invasions has verified this. In fact, invasions often increase the species richness and therefore the biodiversity of a plant community (Burdick, 2005), although there are notable exceptions that will be mentioned subsequently. Biodiversity is a common, frequently undefined term in discussions of invasive species. It is the variability among living organisms of all kinds in a community, including diversity within species, among species, and of plant (and other species) communities within ecosystems (UNEP, 1992). In a comprehensive review of the ecological consequences of biodiversity, Tilman (1999) suggests that it is one of several factors that control population and ecosystem dynamics.

In addition to the work by Fargione et al. (2003) described earlier in the section on resource availability, work by Dukes (2002) supports the diversity—invasibility hypothesis. He showed that eight species grown in monoculture in containers differed widely in their ability to suppress yellow star thistle. The ability of yellow star thistle to suppress other species declined with species richness in Dukes' experiment. The work suggests that diversity can limit susceptibility to invasion and may reduce an invader's effects. Dukes' work also proposes that knowledge of the relative competitive ability of native and invasive plants can lead to effective management techniques. A monoculture of the late-season annual hayfield tarweed was the most effective competitor with yellow star thistle. No polyculture was as effective. It is reasonable to propose that hayfield tarweed's suppressive success is because its resource use patterns (requirements) are similar to those of yellow star thistle. That is, species with similar spatial and temporal resource use patterns may not do well together because they compete for the same resources, as suggested by Fargione et al. and Brown and Rice (2010). They can suppress each other, or one can suppress the other. Hayfield tarweed's success may also be attributable to rapid soil moisture depletion. Study of resource requirements and use patterns may lead to techniques for suppression of invasive species.

Second, Dukes (2002) proposes that all ecosystems are carefully structured and unless they are disturbed, there is little room for invaders because all resources are being used by the residents. Invaders have a better chance of becoming established if they are able to "make a living" employing unused or underused resources. Diverse ecosystems use resources more completely (Tilman et al., 1996) and therefore may be more resistant to invasion. There may be vacant niches to be occupied, but there are fewer such niches in diverse communities.

In contrast, Stohlgren et al. (1999) showed in the Colorado Rocky Mountains and the central grasslands of Colorado, Wyoming, South Dakota, and Minnesota, exotic species primarily invaded areas of high species richness. They argued that habitats that are good for a lot of native species (i.e., high species richness) to grow are also good places for a lot of nonnative species to grow. They concluded that sites high in herbaceous foliar cover, soil fertility, and plant diversity are highly prone to invasion. Invasibility was more a function of resource availability than species richness. Other work confirmed the susceptibility of species-rich areas to invasion (Stohlgren et al., 2003). See Ricciardi (2001) for confirmation of the hypothesis in an aquatic habitat.

Levine and D'Antonio (1999) examined the diversity−invasibility hypothesis and found it to be based on "controversial premises." As the examples presented demonstrate, experimental results have shown positive and negative effects of diversity. Diverse communities may be more susceptible to invasion owing to environmental factors rather than to diversity per se, and the scale at which investigations are conducted (neighborhood, field, watershed, etc.) can influence the patterns found (Levine and D'Antonio).

4.6 Enemy Release

Invading species may thrive because they have escaped from the biotic constraints of their previous home: the enemy release hypothesis (Mack et al., 2000; Hallett, 2006). Independent of how they reached a new place, they made the journey without their previous associates, such as other competing plant species, predators, grazers, or parasites. Such journeys are often made in the dormant or resting state as a seed or vegetative structure. The hypothesis is that the invading species does not do well where it is introduced because it necessarily possesses special invasive traits; it does well because it left its old enemies behind and has not encountered new enemies that are anywhere near as effective. Stastny et al. (2005) suggested that the competitive ability of an invader may be associated with "changes in resistance as well as tolerance to herbivory." That is, the former natural enemies (the herbivores) are not present and the new ones are not as effective. Release from herbivory may be an important key to success of highly aggressive invaders (Carpenter and Cappuccino, 2005). In a test of 39 exotic plants and 30 natives in natural areas near Ottawa, Canada, exotics experienced less herbivory than did natives. Reduced or lack of herbivory may also indicate evolution of defensive chemicals in the exotics that confer resistance to herbivory. Mitchell and Power (2003) tested 473 naturalized plant species in the United States. On average, 84% had fewer fungal and 24% fewer viral enemies than each had in its native range, which is strong support for the enemy release hypothesis. Colautti et al. (2004) argued against the simple relationship the enemy release hypothesis establishes between enemy release and the vigor, abundance, or effects of nonindigenous species. They found that many mechanisms in addition to enemy release could explain the success of invasive species. Blumenthal et al. (2009) claim that enemy release contributes most to invasion by fast-growing plant species adapted to resource-rich environments. Their work suggests that release from enemy predation and "increases in resource availability may act synergistically to favor exotic over native species."

4.7 Evolution of Increased Competitive Ability

When introduced species leave their natural enemies behind, they also leave behind the forces that select for traits that deter those enemies, such as antiherbivore compounds or structures such as thorns. The reduction in need to produce these herbivore defenses can free resources to be used in growth and resource acquisition, improving the invader's ability to compete with resident species. This is the essence of the evolution of increased competitive ability (EICA) hypothesis (Blossey and Nötzgold, 1995). Evidence to support this hypothesis includes invader traits of reduced defenses and greater ability to outcompete neighbors. Investigations testing this hypothesis have uncovered variable evidence. For some species, plants in the introduced range show reduced defenses and changes in phenotype that result in superior performance compared with plants from the original region. However, in the case of white campion, the less—well defended plants from the introduced range still outperformed those from the native range when grown in the native range (Wolfe et al., 2004). In other cases, no trade-off between defense and competitive ability can be detected (Ridenour et al., 2008). A study with hound's-tongue did not find evidence for EICA but did detect a great deal of plasticity of populations in North America and Europe when grown in both locations (Williams et al., 2008). Bossdorf et al. (2005) reviewed the evidence for EICA in molecular and field studies and found a great deal of evidence for the rapid evolution that makes EICA possible. They point out that there are many possible explanations for increased vigor of invader populations in the new habitat compared with their native ranges. They concluded that future studies should focus on other plant traits that can result in increased vigor and reproduction in addition to herbivore defense and growth (Bossdorf et al.).

4.8 Novel Weapons

In contrast, the invasive success of spotted knapweed among grass species and against native knapweeds may result from an entirely different mechanism. Its success may be due to root exudates (allelopathy) and how they affect competition for resources (Callaway and Aschehoug, 2000; Callaway and Ridenour, 2004). Success of diffuse and spotted knapweed in North America may be the result of the adaptation of its Eurasian neighbors to the allelopathic chemicals produced by the knapweeds and the lack of evolved adaptation among new North American neighbors. This is the novel weapons hypothesis (Callaway and Ridenour). The primary allelochemical in spotted knapweed [(+)-catechin] has been suggested by Thelen et al. (2005), but this now seems unlikely given the instability of the compound (Blair et al., 2005, 2006). They strongly suggested that much more research is required before the role of allelopathy in general, or the specific role of catechin in knapweeds as a novel mechanism that explains its invasive success, can be confirmed. Lorenzo et al. (2013, Chapter 1) argue that a role of allelopathy in the invasion process may include release of secondary chemical metabolites into the environment, which inhibit seedling establishment and other eco-physiological processes of native biota.

5. CONSEQUENCES OF PLANT INVASIONS

Biological invasion is a profound, global challenge to our conservation and management skills. Policy responses to the threat have been weak, uncoordinated, and often absent. Only the worst invaders (water hyacinth may be the only weed) receive serious attention, and even then, there is no systematic inquiry into the social and economic processes that launched the invasion (Bright, 1998).

Many of the negative aspects of weeds are shared by invasive plants. However, common weeds typically cause problems for human activities such as cultivated agriculture, whereas invasive plants may affect grazing lands and natural areas of all kinds, including aquatic and riparian (i.e., stream and river) systems, wetlands, and uplands. Like weeds, invasive species will compete with native and desirable species. Both common weeds and invasive plants can have negative economic effects. About $100 million was spent in the United States annually on aquatic invasive plant and weed control alone, and total costs were $120 billion on 100 million acres (Pimentel et al., 2005). The US Department of the Interior spent $100 million in 2011 on invasive species prevention, early detection, control, and management. Estimated worldwide damage from invasive species is $1.4 trillion, 5% of the global economy (PNWER, 2012). Control costs do not include the unknown costs of loss of ecosystem function, effects on human health, habitat loss among native species, and reductions in biodiversity (Li et al., 2004). In addition to similar costs associated with common weeds, such as those for control, invasive plants can result in reduced land values and loss of forage for livestock and wildlife. Sheley et al. (2015) found that savings in livestock forage on western US rangeland per animal unit month was $9.20 for each percent reduction in the spread rate of invasive weedy species over 100 years. One must conclude that annual losses are high and not decreasing with time.

Similar to some common weeds, invasive plants can pose risks to the health of humans and animals. Ragweed, a North American native, was accidentally introduced into France as a contaminant of red clover seed (Chavel and Cadet, 2011). It causes allergic reactions in some humans wherever it grows. Yellow star thistle, a native of Europe and Eurasia, has become widespread in California and other states in the western United States (Duncan and Clark, 2005). It is poisonous to horses and causes a neurological disorder called "chewing disease," which is usually lethal (Duncan and Clark). Next, we discuss the consequences of invasions by plants that are not common weeds.

5.1 Threaten Biodiversity

Invasive species are blamed for threatening biodiversity (Wilcove et al., 1998). McGrath (2005) estimates that more than 40% of all currently imperiled native US plants and animals are at risk of extinction because of invasive species. Wilcove et al. identified nonnative species as the second greatest threat to species at risk of invasion in the United States. Competition or predation from nonnative species harmed 49% of imperiled species, with plants being more affected by this threat than animals (57% and 39%, respectively). More Hawaiian plants were threatened by nonnative species than US mainland plants (99% and 30%, respectively).

Despite this grim picture, subsequent studies examining the relationship between plant invasions and extinctions have not found support for the idea that introduced plants cause the extinction of native plant species (Sax et al., 2002; Davis, 2003; Gurevitch and Padilla, 2004; Sax and Gaines, 2008). Sax et al. and Sax and Gaines demonstrate that the introduction of nonnative bird and plant species to islands around the world has resulted in greater species richness, contradicting the conclusion suggested by earlier studies of threats. Davis (2003) point out that species introductions are correlated with extinctions of natives, but causal connections have rarely been made. They argue there has not been enough work to evaluate situations carefully where invasion is implicated in extinction or where invasion is likely to lead to extinction. They reanalyzed the data of Wilcove et al. and found that most plants were affected by a combination of threats, on average 2.6 threats per species. It is difficult to identify the proportion of the threat attributable to nonnative species alone. For example, if a rare native plant species is losing its habitat to changing land use by humans, which also fosters invasion by nonnative species, teasing out the harm caused by the nonnative is complicated and may not exist.

There is no evidence that competition from plants introduced into North America has caused extinctions of any native plant species (Davis, 2003). He points out that this may be because extinctions resulting from competition take longer than the time that has passed since their introduction. Or it may be because competition is not as likely to be a cause of extinction as other factors such as herbivory or habitat loss. Nonnative species are much more likely to cause extinctions when the interactions are across trophic levels (Davis) such as if the invader is a predator or disease. The introduction of the brown tree snake into Guam is a good example. It has been blamed for the extinction of 10 bird species in only a few decades (Davis).

Although plant invasions appear unlikely to cause extinction of native plant species, they are known to affect the places they invade. Vilà et al. (2011) analyzed 199 scientific articles reporting field studies that examined the effects of invasive plants on the resident plant populations and communities and the ecosystems in which they occur. They found that effects of invasive plants were unsurprisingly highly variable, changing direction and magnitude even within single types of effect. On average, resident species abundance and diversity were reduced by the invasive plants. Hejda et al. (2009) found that all but 2 of the 13 invasive plants they studied in the Czech Republic were associated with a decrease of at least one measure of species diversity and evenness of the resident plant communities.

5.2 Change Ecosystem Function

Invasive plants may change invaded habitats by either increasing or decreasing resource availability. Norway maple is one that decreases a limiting resource: light. Native to Eurasia (Nowak and Rountree, 1990), Norway maple invades riparian areas in the northeastern and northwestern United States and contiguous areas in Canada, and greatly reduces light availability. Its own seedlings are adapted to the low light levels it creates, but the native species are not (Reinhart and Callaway, 2006). Thus, this invader facilitates its own success.

The processing of nutrients in invaded ecosystems may also be catalyzed by invasive plants. A metaanalysis conducted by Vilà et al. (2011) looked at effects on nitrogen-fixing invaders and on resident plants. These species had large effects on factors related to the processing of nitrogen, but effects on species and communities had probably already occurred by the time nutrient-cycling effects could be detected. These results are consistent with the findings of Ehrenfeld (2003) in a review of 56 invasive plant species. She found the effects on ecosystem processes such as carbon, nitrogen and water cycles to be highly variable and, in some cases, dependent on the invasion location. The overall pattern was for invasive plants to result in greater biomass, increased net primary production, nitrogen availability and fixation rates, and litter with higher decomposition rates than resident plant species (Ehrenfeld).

At least sometimes, feedback from soil microbial communities may favor invasive plants (Levine et al., 2006). For example, Klironomos (2002) showed that five of North America's worst plant invaders altered soil microbial communities in ways that increased the invading plants' growth through positive feedback.

5.3 Changes Disturbance Regimes

One of the most dramatic consequences of some invasive plants has been alteration of disturbance regimes, in particular increasing fire frequency and intensity. D'Antonio and Vitousek's classic article (1992) on grass invasions, the grass—fire cycle, and their role in global change describes how these invaders alter disturbance regimes. Invasive grasses provide fine fuels that result in more frequent, larger, and/or more intense wildfires. The invasive grasses are better able to regrow after fire than the native species, and thus increase in abundance and facilitate the recurrence of fire. This creates a positive feedback loop (Fig. 8.2). This scenario has played out in many places including North and South America (e.g., downy brome in the US Great Basin, jaragua grass in South America, buffel grass in Australia, and molasses grass in Oceania) (D'Antonio and Vitousek, 1992).

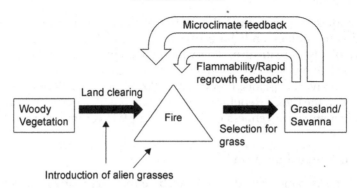

FIGURE 8.2 Grasses enhance fire owing to the flammable, dead tissue they support after each growing season; production of tissues with high surface to volume ratios, which dries quickly; fast growth rates that allow quick recovery after fire; and the creation of warmer, drier microclimates than woodlands and forests, which favor fire. These factors contribute to a positive feedback between grasses and fire. *Adapted from D'Antonio, C.M., Vitousek, P.M., 1992. Biological invasions by exotic grasses, the grass/fire cycle, and global change. Ann. Rev. Ecol. Syst. 23, 78.*

Williams and Baruch (2000) showed the effects of pasture creation and invasion by African C4 grasses (e.g., guinea grass, johnsongrass, kikuyu grass, pangola grass, para grass, signal grass) on ecosystem processes in subtropical regions of North and South America and the Caribbean. They suggest that as global warming progresses, the same grasses may negatively influence North American pasture and rangeland. The primary effects were loss of woody species and changes in the fire regime. The success of African grasses is encouraged by fire and they respond more favorably to fire disturbance than native grasses. The large amount of dead plant material left at the end of each season encourages fire and increases its intensity.

5.4 Alter Wildlife Habitat

Invasive plant species can change the habitat for wildlife in drastic, subtle ways. DiTomaso (2000) cited studies from the western and central United States reporting that areas infested with spotted knapweed had 98% lower elk use than native bunchgrass dominated land. Leafy spurge invasion was associated with 83% and 70% reduced use by bison and deer, respectively. Leslie and Spotila (2001) recount an interesting invasive plant effect on wildlife. They reported that the invasive alien bitterbush in South Africa shades the nests of Nile crocodiles. This seems to be a trivial effect, but shading lowered soil temperatures enough to change the sex ratio of crocodile hatchlings. The story illustrates the complexity of what may appear to be subtle effects of any invasive species.

5.5 Modify Trophic Interactions

Invasive plants can interact with other organisms in complex ways that can change the trophic structure of these systems. Japanese honeysuckle is a good example. It was imported to the United States as an ornamental vine more than 150 years ago. The US Department of Agriculture (USDA) promoted it as a garden and wildlife plant in the 1880s. It invades native woodlands throughout the eastern United States and is a common part of the flora in the northeastern United States (Westbrooks, 1998). It is also invasive in the United Kingdom, Australia, New Zealand, Hawaii, and many parts of the continental United States. Several cultivars have become naturalized in the United States. It grows as a ground shrub on forest floors or as a twining or trailing shrub that quickly climbs into forest canopies where light is increased in tree gaps (Weber, 2003). It forms a dense curtain on forest edges and displaces under story shrubs. It is a competent and common invader of forests. However, it has been discovered that it has actually created a new species (Cowen, 2005). Tephritid fruit flies are specifically adapted to berry-producing plants. Japanese honeysuckle has its own specially adapted fruit fly in its introduced range, but the fly did not originate or arrive with the plant. Schwarz et al. (2005) showed that the fly is a hybrid resulting from flies that live on blueberry and snowberry. Normally such a hybrid would die, but Japanese honeysuckle offered an environmental niche that was not inhabited (a residence without inhabitants, an open niche) for which there was no competition: a subtle but perhaps important change wrought by an invasive species.

Cronin and Haynes (2004) first showed that when the tall-grass prairies of North America become dominated by smooth brome (which some do not consider invasive because, in their view, its benefits outweigh its risks), the spatial and temporal dynamics of a native natural herbivore became about 50% lower within three generations. As smooth brome dominated areas, extinction rates of the natural herbivore were four to five times greater than in native plant habitats. Marler et al. (1999) showed that arbuscular mycorrhizal fungi strongly enhance the ability of spotted knapweed to invade native grasslands of western north America and compete effectively against natives such as Idaho fescue.

A few examples of plant invaders that exhibit one or more of the consequences of invasion are described subsequently. Two of these and several others are reviewed thoroughly in Duncan and Clark (2005).

6. CASE STUDIES: FOUR INVASIVE PLANTS

6.1 Tamarisk (Also Known as Salt Cedar)

The genus *Tamarisk* or salt cedar has more than 50 species. Tamarisk was introduced from Eurasia to New Jersey as an ornamental nursery plant in 1837 (Myers and Bazely, 2003, p. 25). It was introduced to the arid Western United States from central Asia in the early 1800s (Westbrooks, 1998) as an ornamental, for use in windbreaks, or to stabilize eroding stream and riverbanks. It has invaded innumerable watersheds and colonized hundreds of thousands of acres of flood plains, reservoir margins and other wetlands in western North America (Shafroth et al., 2005). It lives 50–100 years and grows to 6–26 ft tall. The branches often form thickets many feet wide. The narrow leaves are small (1.5 cm) and grayish green, and often overlap on the stems. The leaves have the appearance of an evergreen, but are deciduous. The *ramosissima* species can be distinguished from other tamarisk species by its five-toothed sepals. Tamarisk has taken over large sections of riparian ecosystems in the western United States that were once home to native cottonwoods and willows and has continued to extend its range. It is well-established on more than a million acres in the United States, especially in the southwest United States and Mexico. Its range extends to Massachusetts, Indiana, Missouri, Kansas, Colorado, Nebraska, and Oklahoma. Small populations occur in Oregon, Idaho, Montana, Wyoming, and South Dakota.

Tamarisk, a phreatophyte, is a deep-rooted plant that absorbs water from within or just above a ground water source. It is drought tolerant, can access water deeper in the soil than many native species and has invaded every major river system in the southwestern United States (Millar, 2004). It consumes up to 4 million acres per feet of water annually across 17 western US states. One mature tamarisk plant can consume as much as 200 gallons of water a day in the arid west where water is the primary limitation to agriculture and rural and urban development. Each year, tamarisk alone consumes three times more water than is used by all of the households in Los Angeles (Millar). Dense impenetrable thickets form and lower water tables owing to the plant's high water consumption.

It is also invasive in Australia and Southern Africa (Weber, 2003). Three species are present in several world areas but are not invasive; all are small deciduous trees or large shrubs. Tamarisk invades desert areas and stream banks, but grows best in damp, saline, and alkaline soil. Salt, secreted by the plant on its leaves, is washed off or drips down to increase soil salinity, of which tamarisk is more tolerant than native species. It may be the US poster plant to illustrate that invasive plants are one of the preeminent environmental problems of the 21st century. The habitat destruction, its dense thickets, and dense plant residue on the soil surface lead to displacement of native plant, animal, insect, and microbial species and to their eventual elimination from the habitat, or death.

Tamarisk worked very well for all three purposes for which it was introduced (ornamental, windbreak, and erosion control), but its invasive potential soon became the dominant feature. Mature plants can survive immersion in water for more than 1 year. The capability of a single tree to flower in its first year and to produce up to 600,000 min seeds each year enables colonization because the seeds quickly exploit suitable germinating conditions. The seeds survive about 7 weeks but are spread easily and widely by birds and small mammals (Weber, 2003). It can be removed by arduous hand labor (a weed wrench) or herbicides. However, herbicides are not always fully effective and often may not be used along waterways. Its deep root system and vegetative reproductive capability permit it to survive flooding and burning, both of which are detrimental to native plants. It comes close to being a perfect weed because it is resistant to available control measures and lacks natural enemies. Successful control has been achieved with the imidazolinone herbicide imazapyr (Duncan and McDaniel, 1998). Imazapyr applied in August or September alone at 1 lb/acre or in combination with glyphosate (0.5 lb/acre) achieved 90% or greater control. Control was less successful in dense, older stands.

The National Park Service has physically removed the plants, sprayed them with herbicides, and introduced northern tamarisk beetles (*Diorhabda carinulata*). In 2006 and 2007, the beetle was introduced in Dinosaur National Monument in Utah and Colorado along the Green and Yampa Rivers. The USDA Agricultural Research Service found that the beetles eat only tamarisk and starve when none is available. No other native North American plants have been found to be eaten by the introduced beetle. The beetle appears to be an ideal biological control. It had spread throughout the Upper Colorado River Basin by 2014 and caused extensive defoliation in some regions, but long-term consequences for tamarisk populations are not yet known (Bloodworth et al., 2016).

The value of lost ecosystems services in the western United States resulting from the presence of tamarisk was estimated to be $127 to $291 million annually as it consumed 1.4–3 billion m^3 of water/year (Zavaleta, 2000). There is considerable debate about control efforts (Shafroth et al., 2005). Water-use studies indicate that tamarisk control may result in increased water yield if, and only if, the high leaf area of tamarisk is replaced by vegetation with a lower leaf area. The composition of replacement vegetation is central to water salvage. Shafroth et al. point out that the common assumption that tamarisk invasion destroys or diminishes wildlife habitat may be false. Many wildlife taxa prefer native cottonwood, willow, and mesquite habitats to tamarisk, but it may also be adequate habitat for numerous species. It can provide nesting area for some species, but avian density and diversity decrease, and these responses are highly variable depending on geographic region and characteristics of the bird species (Shafroth et al.). It has been found that tamarisk stands supported only four species per 100 acre, compared with 154 per 100 acres of native vegetation. Tamarisk communities also tend to have fewer insects,[3] and the insects found on tamarisk are using it only as a structure because they are predators or detritivores or are generalist herbivores and pollinators (Shafroth et al.).

6.2 Kudzu

Kudzu first arrived in the United States when the Japanese Pavilion exhibited it as an ornamental vine at the US Centennial Exposition in 1876 in Philadelphia. It is a climbing, perennial vine, the planting of which was encouraged in the United States between 1935 and 1942 by the US Soil Conservation Service, and which propagated and distributed 85 million kudzu seedlings, "flinging them about the country like wedding rice" (Williams, 1994). Encouragement of its planting followed passage of the Federal Soil Conservation Act in 1935. The US Government actually paid farmers as much as $8 per acre to plant kudzu, kudzu clubs were formed, and the plant was proclaimed to be the "miracle vine" (Williams). Its foliage dies each year in cold weather but the roots survive the mild winters of the southern United States and resprout vigorously each spring (Westbrooks, 1998). When it was being promoted widely, there were some skeptics who suspected that its lack of presence in the United States was not the result of what Williams called "divine error." However, it grew well, nearly everywhere, and it helped manage the problem it was imported to solve: soil erosion. The characteristics that made it successful for erosion control also made it succeed as an invader. Soon after its introduction, many noticed that it was growing everywhere. Its 90-ft-long older stems and ground-covering mats that could be up to 6 ft thick grew over telephone poles, enveloped trees, and covered gardens, fences, and forest understories. It became the vine that ate the South. It may infest as many as 7 million acres in the southeastern United States. In 1993 the Congressional Office of Technology Assessment estimated that kudzu was costing the US economy $50 million annually in lost yields and control efforts. In 1972, the USDA finally decided it was a weed.

---

[3]For further information, interested readers are referred to http://www.columbia.edu/itc/cerc/danoff-burg/invasion_bio/inv_spp_summ/Tamarix_ramosissima.html and http://www.terrain.org/articles/27/lamberton.htm.

6.3 Water Hyacinth

At least one garden shop in my town sells small water hyacinth plants for placement in bubbling little ponds with pumps that circulate water and make pleasant sounds. They are regarded as, and are, pretty ornamentals. I suspect some people may tire of them and discard them. So far, water hyacinth has not been seen growing in natural Colorado waters. So far! Our winters are cold, although with the reality of global warming, they are not as cold for as long as they used to be. Cold temperatures kill water hyacinth, a native of South America's Amazon basin. It was introduced to the southern United States, southern Asia, and Africa in the 19th century (Bright, 1998, p. 148). It has infested Florida's and many tropical waterways ever since. It is regarded as invasive in Australia, Southern Europe, tropical and southern Africa, southeastern and western United States, and tropical Asia (Weber, 2003).

Water hyacinth is a perennial, free-floating aquatic herb. It has attractive lilac to bluish-purple, erect flowers that produce long-lived seed soon after self-pollination. The primary means of reproduction is vegetative by rhizomes and stolons. Vegetative offshoots are bound to the parent by strong stolons. Offshoots separate from the parent as a result of the action of wind and water. Stolons weaken with time, and this separates offshoots into independent plants. It rapidly colonizes large areas by forming free-floating, large mats that can completely cover lakes and rivers. Holm et al. (1977, p. 74) estimated that under good growing conditions, 25 plants could produce enough offshoots to cover a hectare in one growing season. A mat of medium-size plants may contain 2 million plants/hectare and weigh between 270 and 400 metric tons/hectare (Holm et al., 1977, pp. 73–74). The dense mats change ecological relationships, crowd out native plants, kill fish and other aquatic species, change water temperature, and lead to eutrophication (excessive nutrient concentrations). The plant can root on land with sufficient moisture, but it is primarily an aquatic, not a terrestrial agricultural problem.

Lake Victoria (also called Victoria Nyanza), the largest lake in Africa (26,828 square miles), is the world's second largest freshwater source, a biodiversity hotspot, an important regional waterway, and the primary reservoir of the Nile river, which flows out of the northern end of the lake. Parts of the lake lie within the boundaries of Kenya, Tanzania, and Uganda. The Lake Victoria basin supports one of the most dense and poorest rural populations - up to 2600 people/square mile. It is home to at least 50 million people and the regional population growth rate averages 3%/year one of the world's highest (see http://data.worldbank.org/indicator/SP.POP.GROW), which equals a doubling time of 23.3 years. In Uganda, fish account for half the nation's protein. Water hyacinth is closing down the fisheries by blocking shorelines where fish spawn, blocking access to open water, because boats cannot penetrate the large floating mats, lowering the lake's water level, and killing fish. In 1996, it was blocking 90% of the lake's shoreline (Bright, p. 90). Water hyacinth's presence in Lake Victoria has actually increased the lake's biodiversity (McNeely, 2001; Sagoff, 2005). Water hyacinth has blocked dams in Zimbabwe, often backing up enough water to burst a dam. It threatens[4] Uganda's main electric power plant by blocking the flow of water to the generators (Bright, p. 182). It has been estimated that

---

[4]See http://www.unep.org/dewa/africa/docs/en/lvicbasin_brochure.pdf and http://start.org/download/publications/lake-victoria.pdf.

water hyacinth costs seven African countries US \$20 to \$50 million/year (Joffe and Cook, 1997). World costs are much larger.

6.4 Purple Loosestrife

When Ohio's legislature attempted to restrict growth and importation of purple loosestrife, the Ohio nurserymen's association won an exemption for its hybrid cultivars because they were presumed to be sterile (Williams, 1994). Subsequently it was found that they interbred easily with wild loosestrife plants and the invasion continued. It was another example of deliberate, flawed forethought (Mack et al., 2000) and has left us with a plant that literally flaunts the power of invasive species.

Purple loosestrife is native in most of Europe, the United Kingdom, northern, tropical and temperate Asia, and southern Africa. It was introduced to the United States in the early 19th century as a contaminant in the ballast of ships from Europe and as an ornamental (Malecki et al., 1993), which in the view of some, it is. It has been valued as a medicinal plant for treatment of diarrhea, dysentery, bleeding wounds, ulcers, and sores (Stuckey, 1980). Since its introduction to the United States in the 1800s, it has spread rapidly (Thompson et al., 1987). From 1940 to 1980, its rate of spread has been about 1.5 latitude—longitude blocks/year (Westbrooks, 1998). In 1998, Westbrooks reported that it was invasive in 42 of the 50 United States. USDA[5] data for 2012 show it is in 43 of the United States (not in Arizona, Alaska, Florida, Georgia, Hawaii, Louisiana, or South Carolina) and the seven southern Canadian provinces. The distribution Westbrooks presents (p. 42) shows a few states without purple loosestrife, each surrounded by states with it. However, in 2017, it is reasonable to conclude that it is present in all US states with the possible exception of Alaska, Florida, and Hawaii. It is one of the most prevalent invasive species in the United States, covering about 400,000 acres of federal land, including wetlands, marshes, pastures and riparian meadows. Its cost in loss of forage and the cost of control is estimated to be \$45 million/year (Hall, 2000; Pimentel et al., 1999, 2005). Thompson et al. (1987) reported it was spreading to 115,000 ha each year. In 2004, Li et al. reported it was spreading to 285,000 acres each year.

It grows best in freshwater marshes, on stream banks, and on alluvial flood plains. When it invades, it forms extensive, persistent, monotypic stands in wetlands where it replaces native plants and excludes associated insects, small mammals, and other wetland inhabitants. It is noted and often prized for the beauty of its late summer inflorescence, which provides a nectar and pollen source for bees (Malecki et al., 1993). Despite its invasive characteristics, nurseries in many states continue to sell it as an ornamental. Several states include it on the state noxious weed list, which precludes its sale and importation.

Without belaboring the point made in Westbrooks (1998) and Weber (2003), these few cases illustrate that invasive plant species are major ecological problems; indeed, they can be ecological threats. Invaded ecosystems tend to be (but are not always) biologically impoverished and differ in many ways from adjacent noninvaded areas.

---

[5]See Nationalatlas.gov - profile for purple loosestrife.

7. MANAGEMENT OF INVASIVE PLANTS

Invasive plants are management challenges, as are all weeds. Action must often be taken to prevent further invasion before one can be sure of all consequences of the action. The risk of inaction is frequently regarded as greater than the risk of unanticipated consequences. A well-considered and executed management plan can minimize the negative consequences of invasive plants.

The Maui Coastal Land Trust owns a 277-acre refuge on northern shore of the Hawaiian island of Maui just outside the town of Waihe'e, known as the Waihe'e Coastal Dunes and Wetlands Refuge. The trust also has about 300 additional acres protected by conservation easements. The trust expects to acquire easements on additional land on Moloka'i. The refuge includes a 7000-ft coastal strand, a 26-acre wetland, and about 150 acres of sand dunes, which enclose the wetland and shore area. The refuge is populated by a large number of weedy species and some particularly troublesome invasive species of *Pluchea*. The species of most concern are *Pluchea carolinensis* and *Pluchea indica*, neither of which is recognized as a common weed by the WSSA nor is either mentioned in Weber (2003) or Westbrooks (1998) as important invasive species. *Pluchea camphorata* (L.) DC, or stinkweed, is recognized as a weed by WSSA. Members of the genus are often known by the common name fleabane. The genus *Pluchea* includes 40 species, all part of the Asteraceae family. Nearly all are tropical herbaceous plants or shrubs, but only a few are weedy. *P. carolinensis* is an aromatic branched shrub that grows up to 10 ft tall. It has been known as *Pluchea symphytifolia* and in older literature as *Pluchea odorata*. It is native to tropical America and was first collected on Oahu in 1931. In the Hawaiian Islands, it is common in pastures, forest, roadsides, and uncropped areas. It thrives in wet and dry areas.

P. Indica, a native of south Asia, has up to 10 branching stems but grows only 6 ft tall. The branching stems make it much more difficult to control than *P. carolinensis*. It was first collected on Oahu in 1915. *P. indica* invades wetlands whereas *P. carolinensis* does not do so as readily. *P. indica* grows well in saline soil. If *P. Carolinensis* invades wetland, the Land Trust's management strategy is to do nothing if it is in an area that will be flooded during the wet season, because it will die after about 6 days of submersion in water.

Both species probably arrived with cattle or in their fodder. They have not always been as invasive as they now seem to be. Both have significant environmental effects. They grow large, eliminate native species of plants and destroy endangered bird habitat. They displace native forage species in coastal pastures and native species in coastal marshes and wetlands. Both species are sensitive to some herbicides (e.g., 2,4-dichlorophenoxyacetic acid [2,4-D], dicamba, glyphosate, and triclopyr). Land Trust personnel use these sparingly, if at all. The best results have been achieved with mixtures of glyphosate and one of the growth regulator herbicides. The first reason for reluctant herbicide use is that the herbicides are not selective enough. That is, they kill the native as well as the invading species and return or reestablishment of the native species is a major goal of the refuge. Second, herbicides are expensive for an organization with limited funding. There are two primary control methods: hand pulling of young plants or use of a large mattock to pull the large plants out by the roots. Both are labor-intensive, arduous work. A control method that works well for both species and for their hybrid *Pluchea fosbergii* is flooding by rain water for at least 5 days. A third weed management complication is the fact that the Maui coastal land trust property includes 85 listed archaeological sites. Large-scale grubbing or grading requires several bureaucratic layers of approval. Hand tools and herbicides are acceptable because they do not lead to potential destruction of archaeological sites. The manual methods and flooding both protect native species but they are slow and might be more expensive if quick control of a large area is desired and if labor is expensive or unavailable.

The Maui coastal land also has invasive populations of the tropical weeds Brazilian peppertree, large-leaf lantana, and Java plum. Related species of the latter are used for food flavoring and to make pomanders, and the oils are used as an analgesic for toothache. All three can be invasive in the right habitat.

Perhaps the first step in developing management plans is not direct action against plants at all; it is legislative. Clout and De Poorter (2005) recommend international rather than just national action because of the increasing globalization of the world economy. They advocate an agreed-upon, effective international strategy to deal with invasive species of all kinds. To be effective, it must be combined with fundamental and applied research on all aspects of invasion biology. Appropriate economic policies that enable management must be included. This is especially important because the beneficiaries of invasive plant management are often not apparent. Everyone in an area may benefit, but when some do and others do not, deciding who should receive benefits and pay inevitable costs is not always clear and may be vigorously disputed. Agreement on how to establish the risk of the invader and how to balance that against the cost of management must be obtained. Complete condemnation of all alien species may be counterproductive because some can be tolerated and used to good advantage. They can provide essential ecological and socioeconomic services by speeding restoration (Ewel and Putz, 2004). These decisions and the threat of invasive plants have to be communicated to the public in easily comprehended ways (McNeely, 2004). Those charged with assessing risk depend on scientific research to develop and advise on biologically plausible management methods that enable fully informed regulatory decisions (Powell, 2004). Legislative approaches for preventing invasion of cacti in South Africa have been advocated (Novoa et al., 2015).

A logical first step in a national management program is prevention (Clout and De Poorter, 2005). Prevention is widely recognized as a cost-effective management strategy. Risk assessment based on scientific research and data can set priorities and formulate prevention programs. Sagoff (2000) and Stromberg et al. (2009) advocate science-based judgments rather than assumptions based on fear, not biology. Radosevich et al. (2009) point out that effective, preventive strategies are difficult to achieve because adequate descriptions of biological environmental characteristics are often lacking and predictive models of invasive plant expansion have been elusive. Smith and Sheley (2015) offer a flow model with three steps to implement prevention programs: education, early detection and eradication, and interruption of movement. Prevention is usually less costly than post-entry control (Mack et al., 2000). A preventive approach advocates changing the current policy of denying entry only to species that are known to be harmful (e.g., known harmful weeds, also known as the *black list* approach, demonstrated by the noxious weed lists maintained by every state in the United States) to one of presuming guilt until innocence can be proven (also known as the *white list* approach, e.g., the Australian Invasive Species Council, 2009) (Mack et al., 2000). This is contrary to the American system of jurisprudence (i.e., innocent until proven guilty), but it may nevertheless be wise to adopt the precautionary principle to govern imports (Clout and De Poorter), i.e., not to import species if there is uncertainty about the effects they will have. Many horticulturalists in the landscape and seed industries and plant importers object, but most environmental groups support a white list policy. If prevention is not enforced, then management devolves to the same methods available to weed managers: mechanical, cultural, biological, and chemical means. There are few other available, economically feasible choices.

South of the Colorado—Wyoming border and about 5 miles west of the I-25, the 18,771-acre Soapstone Prairie Natural Area of relatively undisturbed shortgrass and mixed-grass prairie with some wetlands and riparian areas has been preserved through the cooperative efforts of the city and county government, the Nature Conservancy, and the Legacy Land trust (see www.fcgov.com/naturalareas). The area includes the Lindenmeier Site, a registered national historic landmark, and the location of one of the oldest (about 10,000 years) known areas of human habitation (Folsom man) in the United States.

Similar to other large natural areas, Soapstone is valued for its scenic beauty and geological importance as well as its agricultural use for sheep and cattle grazing. It must be managed, and invasive, primarily weedy species are part of the management challenge. Soapstone is used for recreational purposes by citizens, which is part of the management plan.

State and local laws demand control to diminish populations of invasive noxious weeds, to maintain the rangeland's health and stop the spread of the invaders. The invasive weeds of concern within Soapstone are Canada thistle, cheatgrass, field bindweed, and dalmation toadflax. Canada thistle occurs principally in areas frequented by cattle and especially near the water tanks. Cheatgrass is found in the higher areas, whereas field bindweed and dalmatian toadflax are found in disturbed areas, especially along roadways.

Land managers use mowing where the terrain permits, and in contrast to the Maui Coastal land area, different herbicides (dicamba, imazapic, tordon, or 2,4-D) are employed for weed control. Prevention of spread is a major goal. The techniques are effective and managers claim significant population reduction in 3—5 years. Citizen concern about the weed management techniques has been minor and infrequent.

Because science cannot precisely predict either the common attributes of invaders or locales susceptible to invasion, control within a management system is the only viable option when invasive species become established. Control, regardless of technique, will always be more successful when it includes a long-term ecosystem strategy rather than a tactical, local, annual approach (Mack et al., 2000). The precise meaning of long-term is not defined, but one is sure it means many years or even decades, certainly not just a year or one crop season. The scope of the existing or anticipated invasion must be considered, and that demands a broad view that will be facilitated by discussion among weed scientists, ecologists, and conservation biologists. The latter have tended to focus on the reasons for and mechanisms of invasion and the biological and environmental effects of an invasive species, whereas the former have tended to focus on the current or potential negative economic effects of an invader and how to control it. Several people have been concerned that weed scientists and plant ecologists have been unaware of each other's activities and perhaps did not even care about what the others were doing. Greater cooperation has been advocated (D'Antonio and Jackson, 2004; Brown et al., 2008).

Hobbs and Humphries (1995) illustrate the complexity of the management challenge (Fig. 8.3). Management has often been limited to control, but they claim it must include three other components: spatial and temporal dynamics of the population, structure and dynamics of the ecosystem, and effects of human activities on all components. Control programs generally are initiated only after the problem has become obvious. Weed scientists have been persistent in their claims that aggressive, large-scale campaigns must be undertaken to prevent further spread and economic losses caused by invasive weeds. A fundamental difference in approach, which often inhibits cooperation, exists between research science and weed control and management (McPherson, 2004). Research science often strives for large-scale generalities with broad application. The weed manager however is confronted with a site and the necessity of achieving a specific objective (i.e., eliminate weeds and potentially invasive species—quickly). Weed scientists tend to use herbicides to "solve" the invasive problem. The invader is regarded, as are all weeds, as an external problem that exerts only negative effects on the natural system and on human welfare (Timmons, 1970). The thought has been that external problems must be eliminated and chemical and mechanical methods are the best way to accomplish the goal. This thinking has led to the many problems caused by herbicides because it focuses on solving

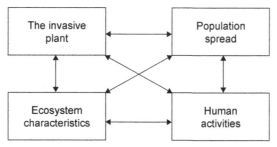

FIGURE 8.3 Components of management strategies for plant invasions. *Adapted from Hobbs, R.J., Humphries, S.E., 1995. An integrated approach to the ecology and management of plant invasions. Conserv. Biol. 9, 761—770.*

the problem (eliminating the invader) without understanding why the invader invaded. Questions such as the role of disturbance, open niches, and control practices have been regarded as less important or have been ignored in favor of development of control techniques. The control approach will inevitably lead to the same kinds of problems (herbicide resistance, control failure, public misunderstanding, and concern) that now plague weed science.

In many cases, there is little good evidence that aggressive management is the best course of action or that it is economically optimal (Eiswerth and van Kooten, 2002). For example, attempted eradication of yellow star thistle was not economically optimal, whereas strategies that attempted to control its spread were (Dukes, 2002). A framework for identifying weeds that are not yet major invaders but have the potential to become so, and then preventing invasion of new territory by eradication in their current location before they invade and become dominant has been advocated by Cunningham et al. (2004).

Ecological niche modeling may assist in determining the geographic course of an invader. Peterson et al. (2003) tested the technique and found that it effectively predicted the actual course of invasion of four North American weeds. Ecological niche modeling uses environmental characteristics of locations with known occurrences or the lack of occurrence of a species to predict areas of potential invasion. The work of Bromberg et al. (2011) and West et al. (2015) illustrate the use and benefits of niche modeling of the invasive grass downy brome for land management.

Other management techniques should not be abandoned simply because they are old. Fire is an effective technique to manage junipers in western and southwestern United States. Fire always reduces juniper canopy cover and density. It is useful when the goal is to reduce the presence and effects of junipers in an ecosystem (Ansley and Rasmussen, 2005). Vermeire and Rinella (2009) demonstrated "abrupt reductions" of 79—88% relative to nonburn areas for seedling emergence of Japanese brome, spotted knapweed, Russian knapweed, and leafy spurge.

Rana et al. (2013) show that soil pH, which has not been used as a method of weed management (see Chapter 6) can be a factor in the competitive ability and invasive potential of two species of smut grass. The relative competitive ability and aggressivity of giant smut grass was greater than Bahia grass (a desirable turf grass in Florida) across all pH levels. In contrast, the competitive ability and aggressivity of Bahia grass was greater than small smut grass at all pH levels and densities, except at pH 6.5. The work of Rana et al. demonstrates that soil pH should be considered more than it has been as a determinant of weed management success or failure.

Biological control is also available for managing invasive species, but like other techniques, it must be managed carefully. Rand et al. (2004) showed that using the flower-head weevil (*Rhinocyllus conicus*) to reduce musk thistle influenced the extent of attack on native wavy-leaf thistle. Using biological control to control an invasive species can have the undesirable result of attack on native species that help "maintain populations of the shared insect herbivore." Yager et al. (2011) advocate the biological technique of creating a vegetative barrier of woody shrubs to slow invasion of pine forests by wind-dispersed seed of cogon grass.

Li et al. (2004) propose a management technique for invasive plants that was not even conceivable a decade ago: genetic modification to create sterility. It is one of the few new choices that can now be added to those mentioned previously. The genetic solution is to create sterile cultivars of nonnative ornamentals that may be potentially invasive plants and have commercial value, which, it is proposed, would reduce or eliminate their spread. Because prohibiting imports is politically, socially, and economically (in the view of those who sell ornamentals) unfeasible, the genetic solution may be reasonable. Li et al. suggest the solution is "to neutralize the invasive characteristics of economically important nonnative species before they are planted in the landscape." Given that no one knows for sure what "invasive characteristics" are, it is nevertheless an imaginative management approach. Li et al. cite several advantages and disadvantages of the technique. Introduction of sterility can be broadly applicable to species that spread (invade) by sexual reproduction. It is inappropriate for species that spread vegetatively (e.g., Canada thistle, field bindweed). The insertion of the gene would not affect overall plant morphology; an ornamental would still be attractive. The method can be fast once the technique is perfected for a species, but insertion of a gene for sterility or parthenocarpy into many plants can be technically difficult. Not everyone can do it; special facilities are required. Clearly the technique may be useful to eliminate undesirable traits of plants that have been modified, but it will have no effect on plants that are already in the environment; it is proactive, not retroactive management. Many will object to the technique because of unease about all uses of genetic modification (potential for escape, hybridization, and dilution of the gene pool of native species). Finally, there is a hint of scientific hubris as opposed to humility in the face of nature's complexity. The method could work well. However, some plants might still produce fertile seed when pollinated by a non-sterile relative growing nearby. Even sterile plants may still possess some of the undesirable traits of an invader. Scientists always know what they are doing but may not know what they are undoing.

Myers and Bazely (2003, p. 244) criticize the genetic approach to invasive plant management because they see it as a potential time bomb. No one knows precisely what genetic and environmental traits may combine to make any plant invasive. Therefore, it may be scientific arrogance to assume that genetic modification, a relatively new management option, is well enough understood so all of its effects can be predicted. Genetic modification could increase a species invasion potential rather than decrease it. As Myers and Bazely clearly point out, "the fundamental assumption underlying the technology of genetic modification, is that genes from other organisms, introduced by bacteria to target plant species, will direct the production of (useful) proteins that are not normally synthesized by the plant." The assumption is correct. The problem is that there is little authoritative scientific evidence to support the assumption that the effects of such modification will be exclusively beneficial. They also point out that because such products (modified plants) are currently made by organizations interested in patenting and benefitting from their efforts, the questions about all effects often are not asked until after release, if they are asked at all. Then, as experience with plant introductions (kudzu, tamarisk, etc.) shows, it may be too late.

For those who would like more comprehensive treatment of invasive plant management, Myers and Bazely's (2003) review of ecological and control aspects of invasive plants is an excellent resource. Maxwell et al. (2009) and Inderjit (2009) review management strategies for invasive plants.

8. AN INVASIVE THOUGHT

As mentioned at the beginning of this chapter, some intentionally introduced exotic plants have had major benefits (e.g., corn, wheat). One must remember that not all introduced species are threats: some are beneficial, some may be weeds, and some may be ecologically detrimental. If all exotic species were considered, the benefits might exceed the costs. Observed or assumed effects "often fail to translate to ecosystem services or evidence of environmental degradation" (Hulme et al., 2013).

Kudzu, "the plant that ate the South," is an example of what Mack et al. (2000) call the result of deliberate but flawed forethought; nevertheless, it has benefits. It is a legume that fixes nitrogen and grows fast enough to cover potentially erodible soil. It makes a high-quality fodder for cattle and other livestock. Baled kudzu is actually sold. It is surely a minor use, but it is used by cooks because of its nutritious leaves and roots. However, more is available than the South's cooks can possibly use. Sagoff (2000) notes that biologists, weed scientists among them, "attribute to immigrant species some of the same characteristics that nativists and xenophobes have ascribed to immigrant humans: sexual robustness, excessive breeding, low parental involvement with the young, a preference for degraded conditions, and so on." In short, they are highly likely to be bad. If we return to the question of definitions raised earlier, none of the traits that Sagoff lists are sufficient to identify a species as exotic or as a potential invader. They are judgments based on assumptions or fear, not biology.

An important defining characteristic of the scientific enterprise is the acceptance of the fact that an experimental hypothesis can be falsified. One is reminded of the aphorism, "Seeing is believing." Many researchers have studied tamarisk and other invasive species with the preconceived, unquestioned hypothesis that accepts their vilification as aggressive, undesirable profligate consumers of water. Stromberg et al. (2009) question those assumptions. Their article cited research that challenges "the prevailing dogma regarding tamarisk's role in ecosystem function and habitat degradation." The publication questioned the "reflexively anti-exotic viewpoint," which "can provoke needlessly negative attitudes among resource managers, and young scientists, reducing scientific rigor and credibility." Stromberg and her colleagues clearly think many scientists have inverted the axiom quoted previously to "When I believe it, I will see it!" They recommend that all scientists review the literature carefully, cite sources appropriately, avoid reflexive antiexotic bias and war- and pestilence-based terminology, and acknowledge levels of scientific uncertainty.

Chew (2009) cautioned against "monstering" of introduced species. Using tamarisk as an example, he suggested that "conservation motivated scientists and their allies have focused largely on most negative outcomes and often promoted the perception that introduced species are monsters." Davis et al. (2001) urged conservationists not to judge species on their origin. Ecological effects are often assumed rather than proven, and available data do not allow prediction. Being a native is not a sign of evolutionary fitness or the species having positive effects. For example, *Dendroctonus ponderosae*, the *native* mountain pine beetle, is suspected of killing more trees than any other insect in North America.

These thoughts are not intended to deny the importance of invasive species or advocate that they should not be managed and eliminated where possible. They are costly and have negative ecological and economic consequences. Our thoughts are intended to advocate that ecologists, land managers, policy makers, and weed scientists should abandon the native—alien dichotomy and embrace more dynamic and pragmatic approaches to determine effects and appropriate management techniques. Chew (2009) cautions that scientists have created monsters, not by assembling and animating one as envisioned by Mary Shelley,[6] but by declaring that an organism once presumed tractable was flouting human intentions, and recasting it as malevolent. It is a caution worthy of thought.

Subtle changes created by a local invasive species may combine in ways that affect the earth's (Gaia's) (Lovelock, 1979) interrelated web of organic life. Weed scientists are properly concerned about the effects of any invasive plant on crop production, but its effects may go well beyond those on yield. Invasive plants could power unknown ecological changes whose consequences are equally unknown (see Tekiela and Barney, 2017). For example, increasing atmospheric carbon dioxide levels could favor the growth of an invasive species over native species. In this regard, Ziska (2006) reported that the average increase in biomass of six invasive weedy species (Canada thistle, field bindweed, leafy spurge, perennial sow thistle, spotted knapweed, and yellow star thistle) was 46% when they were grown from seed with 719 μmol/mol (the predicted atmospheric concentration of CO_2 level by the end of this century) instead of 391 μmol/mol (the 2012 atmospheric level). The largest response (73% increase in biomass) was from Canada thistle, a widespread invasive weed. Mozdzer and Megonigal (2012) showed that an introduced Eurasian genotype of common reed "expressed greater mean trait values in nearly every ecophysiological trait measured … to elevated CO_2 and nitrogen outperforming the native North American conspecific by a factor of two to three under every global change scenario."

THINGS TO THINK ABOUT

1. Are all invasive plants also weeds? Why or why not?
2. Are all weeds invasive? Why or why not?
3. What is the definition of an invasive plant?
4. What characteristics do invasive plants share?
5. Are all invasive plants of foreign origin?
6. What justification can be offered for introduction of a new plant species to a new place?
7. Describe the theories used to explain why plant invasions occur.
8. How do disturbance and invasion relate?
9. Are there examples of plants that survive in a new place and do not become invasive? Name a few.
10. What aspects of plant communities and ecosystems create susceptibility to invasion?
11. Name some examples of successful plant invasions.
12. Are all plant invasions necessarily harmful?

[6]Mary Shelley (1797—1851) wrote Frankenstein - The Modern Prometheus in 1818.

13. Is it possible for a plant with invasive potential to arrive in a new habitat and not become invasive for years or even decades?
14. What is the first step in a management plan for all weed or invasive species?
15. How can genetic modification be incorporated in an invasive species management plan?
16. Should genetic modification be used to manage invasive plants?

Literature Cited

Amsberry, L.M., Baker, A., Ewanchuk, P.J., Bertness, M.D., 2000. Clonal integration and the expansion of Phragmites australis. Ecol. Appl. 10, 1110–1118.

Ansley, R.J., Rasmussen, G.A., 2005. Managing native invasive juniper species using fire. Weed Technol. 19, 517–522.

Australian Invasive Species Council, 2009. Stopping Weed Invasions: A 'White List' Approach. https://invasives.org.au/wp-content/uploads/2014/02/fs_weedwhitelist.pdf.

Baker, H.G., 1965. Characteristics and modes of origin of weeds. In: Baker, H.G., Stebbins, G.L. (Eds.), Genetics of Colonizing Species. Proc. First Int. Union of Biol. Sci. Symp. on Gen. Biol. Academic Press, New York, pp. 147–172.

Barney, J.N., 2014. Bioenergy and invasive plants: quantifying and mitigating future risks. Invasive Plant Sci. Manage. 7, 199–209.

Barrat-Segretain, M., Elger, A., 2004. Experiments on growth interactions between two invasive macrophyte species. J. Veg. Sci. 15, 109–114.

Beck, K.G., Zimmerman, K., Schardt, J.D., Stone, J., Lukens, R.R., Reichard, S., et al., 2008. Invasive species defined in a policy context: recommendations from the federal invasive species advisory committee. Invasive Plant Sci. Manage. 1, 414–421.

Blair, A.C., Hanson, B.D., Brunk, G.R., Marrs, R.A., Westra, P., Nissen, S.J., Hufbauer, R.A., 2005. New techniques and findings in the study of a candidate allelochemical implicated in invasion success. Ecol. Lett. 8, 1039–1047.

Blair, A.C., Nissen, S.J., Brunk, G.R., Hufbauer, R.A., 2006. A lack of evidence for an ecological role of the putative allelochemical (±)-catechin in spotted knapweed invasion success. J. Chem. Ecol. 32, 2327–2331.

Bloodworth, B.R., Shafroth, P.B., Sher, A.A., Manners, R.B., Bean, D.W., Johnson, M.J., Hinojosa-Huerta, O., 2016. Tamarisk Beetle (Diorhabda spp.) in the Colorado River Basin: Synthesis of an Expert Panel Forum. Scientific and Technical Report No. 1. Colorado Mesa University, Grand Junction, CO.

Blossey, B., Notzold, R., 1995. Evolution of increased competitive ability in invasive nonindigenous plants - a hypothesis. J. Ecol. 83, 887–889.

Blumenthal, D., Mitchell, C.E., Pyšek, P., Jarošik, V., 2009. Synergy between pathogen release and resource availability in plant invasion. Proc. Natl. Acad. Sci. early edition. www.pnas.org/cgu/doi/ao.1o73/pnas.0812607106.

Bossdorf, O., Auge, H., Lafuma, L., Rogers, W.E., Siemann, E., Prati, D., 2005. Phenotypic and genetic differentiation between native and introduced plant populations. Oecologia 44, 1–11.

Bradley, B.A., Early, R., Sorte, C.J.B., 2015. Space to invade? Comparative range in filling and potential range of invasive and native plants. Global Ecol. Biogeogr. 24, 348–359.

Bright, C., 1998. Life Out of Bounds: Bioinvasion in a Borderless World. W.W. Norton & Co., New York, NY, 287 pp.

Brock, J., 2006. Ecology and management of camelthorn (Alhagi maurorum): a case study in Arizona. Proc. West. Soc. Weed Sci. 23.

Bromberg, J.E., Kumar, S., Brwon, C.S., Stohlgren, T.J., 2011. Distributional changes and range predictions of downy brome (Bromus tectorum) in Rocky Mountain National Park. Invasive Plant Sci. Manage. 4, 173–182.

Brown, C.S., Rice, K.J., 1998. Competitive Growth Characteristics of Native and Exotic Grasses. Rep. No. FHWA/CA/ESC-98/07. Calif Dept. of Transporation, 215 pp.

Brown, C.S., Rice, K.J., 2010. Effects of belowground resource use complementarity on invasion of constructed grassland plant communities. Biol. Invasions 12, 1319–1334.

Brown, C.S., Anderson, V.J., Claassen, V.P., Stannard, M.E., Wilson, L.M., Atkinson, S.Y., Bromberg, J.E., Grant III, T.A., Munis, M.D., 2008. Restoration ecology and invasive plants in the semiarid west. Invasive Plant Sci. Manage. 1, 399–413.

Burdick, A., May 2005. The truth about invasive species. Discover 35–41.

Burke, J.M., DiTomasso, A., 2011. Corallita (*Antigonon leptopus*): intentional introduction of plant with documented invasive capability. Invasive Plant Sci. Manage. 4, 265–273.

Burks, K.C., 2002. Invasive exotics: plants that smother a sense of place. Native Plants 18, 5–9.

Callaway, R.M., Aschehoug, E.T., 2000. Invasive plants versus their new and old neighbors: a mechanism for exotic invasion. Science 290, 521–523.

Callaway, R.M., Ridenour, W.M., 2004. Novel weapons: invasive success and the evolution of increased competitive ability. Front. Ecol. Environ. 2, 436–443.

Carpenter, D., Cappuccino, N., 2005. Herbivory, time since introduction and the invasiveness of exotic plants. J. Ecol. 93, 315–321.

Chafe, Z., 2005. Bioinvasions. State of the World: Redefining Global Security. W.W. Norton & Company, New York, NY, pp. 60–61.

Chavel, B., Cadet, E., 2001. Introduction and spread of an invasive species: *Ambrosia artemisiifolia* in France. Acta Bot. Gall. 158, 309–327.

Chew, M.K., 2009. The monitoring of tamarisk: how scientists made a plant into a problem. J. Hist. Biol. 42, 231–266.

Clout, M.N., De Poorter, M., 2005. International initiatives against invasive alien species. Weed Technol. 19, 523–527.

Colautti, R.I., 2005. In search of an operational lexicon for biological invasions. In: Inderjit (Ed.), Invasive Plants: Ecological and Agricultural Aspects. Birkhäuser, Verlag/Switzerland, pp. 1–15.

Colautti, R.I., Ricciardi, A., Grigorovich, I.A., MacIssac, H.J., 2004. Is invasion success explained by the enemy release hypothesis? Ecol. Lett. 7, 721–733.

Cowen, R.C., August 4, 2005. An alien invader spawns a new species. Christ. Sci. Monit. 17.

Cronin, J.T., Haynes, K.J., 2004. An invasive plant promotes unstable host-parasitoid patch dynamics. Ecology 85, 2772–2782.

Cunningham, D.C., Barry, S.C., Woldendorp, G., Burgess, M.B., 2004. A framework for prioritizing sleeper weeds for eradication. Weed Technol. 18, 1189–1193.

D'Antonio, C., Jackson, N.E., 2004. Invasive plants in natural and managed systems-Linking science and management. Weed Technol. 18, 1180–1181.

D'Antonio, C.M., Vitousek, P.M., 1992. Biological invasions by exotic grasses, the grass/fire cycle, and global change. Ann. Rev. Ecol. Syst. 23, 63–87.

Daehler, C.C., 1998. The taxonomic distribution of invasive angiosperm plants: ecological insights and comparison to agricultural weeds. Biol. Conserv. 84, 167–180.

Daehler, C.C., 2001. Two ways to be an invader, but one is more suitable for ecology. Bull. Ecol. Soc. Am. 82, 101–102.

Daehler, C.C., 2003. Performance comparisons of co-occurring native and alien invasive plants: implications for conservation and restoration. Ann. Rev. Ecol. Evol. Syst. 34, 183–211.

Davis, M.A., 2003. Biotic globalization: does competition from introduced species threaten biodiversity? Bioscience 53, 481–489.

Davis, M.A., Pelsor, M., 2001. Experimental support for a resource-based mechanistic model of invasibility. Ecol. Lett. 4, 1–8.

Davis, M.A., Thompson, K., 2000. Eight ways to be a colonizer; two ways to be an invader: a proposed nomenclature scheme for invasion ecology. Bull. Ecol. Soc. Am. 81, 226–230.

Davis, M.A., Grime, J.P., Thompson, K., 2000. Fluctuating resources in plant communities: a general theory of invasibility. J. Ecol. 88, 528–534.

Davis, M.A., et al., 2001. Don't judge species on their origins. Nature 474, 153–154.

DeMarco, E., 2015. Invasive Plants Taking Over the US. http://www.sciencemag.org/news/2015/01/invasive-plants-taking-over-us.

Devine, R., May 1993. The cheatgrass problem. Atl. Mon. 40, 44, 46, 47–48.

Diamond, J., 2005. Collapse: How Societies Choose to Fail or Succeed. Viking Press, New York, NY, 575 pp.

DiTomaso, J.M., 2000. Invasive weeds in rangelands: species, impacts, and management. Weed Sci. 48, 255–265.

Dukes, J.S., 2002. Species composition and diversity affect grassland susceptibility and response to invasion. Ecol. Appl. 12, 602–617.

Duncan, C.A., Clark, J.K., 2005. Invasive Plants of Range and Wildlands and their Environmental, Economic, and Societal Impacts. Weed Sci. Soc. America, Lawrence, KS, 222 pp.

Duncan, C.A., McDaniel, K.C., 1998. Saltcedar (*Tamarisk* spp.) management with imazapyr. Weed Technol. 12, 337–344.

Economist, December 5, 2015. Biodiversity-in Defense of Invaders, p. 18; Invasive Species — Day of the Triffids, p. 59—60.

Economist, March 28, 2015. Invasive Species -Not Weeds, p. 83.

Ehrenfeld, J.G., 2003. Effects of exotic plant invasions on soil nutrient cycling processes. Ecosystems 6, 503—523.

Eiswerth, M.E., van Kooten, G.C., 2002. Uncertainty, economics, and the spread of an invasive plant species. Am. J. Agric. Econ. 84, 1317—1322.

Ellstrand, N.C., Schierenbeck, K.A., 2000. Hybridization as a stimulus for the evolution of invasiveness in plants. Proc. Natl. Acad. Sci. 97 (13), 7043—7050.

Elton, C., 1958. The Ecology of Invasions by Plants and Animals. Methuen and Co., London, UK, 181 pp.

Ewel, J.J., Putz, F.E., 2004. A place for alien species in ecosystem restoration. Front. Ecol. Environ. 2, 354—360.

Executive order 13112, 1999. Presidential documents — invasive species, by President W. J. Clinton on Feb 3. Fed. Regist. 64 (25), 6183—6186.

Fargione, J., Brown, C.S., Tilman, D., 2003. Community assembly and invasion: an experimental test of neutral versus niche processes. Proc. Natl. Acad. Sci. 100, 8916—8920.

Finch, B., September 2015. Legend of the green monster. Smithsonian, 1921—22, 92.

Firn, J.J., Moore, L., MacDougall, A.S., Borer, E.T., Seabloom, E.W., HilleRisLambers, J., et al., 2011. Abundance of introduced species at home predicts abundance away in herbaceous communities. Ecol. Lett. 14, 274—281.

Gaskin, J.F., Schaal, B.A., 2002. Hybrid Tamarix widespread in U.S. invasion and undetected in native Asian range. Proc. U.S. Natl. Acad. Sci. 99, 11256—11259.

Germino, M.J., Chambers, J.C., Brown, C.S., 2016. Exotic Brome-Grasses in Arid and Semiarid Ecosystems of the Western US: Causes, Consequences, and Management Implications. Springer, New York, NY, 475 pp.

Gibson, D.J., Urban, J., Baer, S.G., 2011. Mowing and fertilizer effects on seedling establishment in a successional old field. J. Plant Ecol. 4, 157—168.

Goldberg, D.E., 1990. Components of resource competition in plant communities. In: Grace, J.B., Tilman, D. (Eds.), Perspectives on Plant Competition. Academic Press, San Diego, CA, pp. 27—49.

Grotkopp, E., Rejmánek, M., Rost, T.L., 2002. Toward a causal explanation of plant invasiveness: seedling growth and life-history strategies of 29 pine (*Pinus*) species. Am. Nat. 159, 396—419.

Gurevitch, J., Padilla, D.K., 2004. Are invasive species a major cause of extinctions? Trends Ecol. Evol. 19, 470—474.

Hall, M., 2000. Economic Impacts of Plants: Invasive Plants and the Nursery Industry. Undergraduate senior thesis in environmental studies. Brown Univ., Providence, RI. http://brown.edu/Research/EnvStudies,Theses/full9900/mhall/IPlants/Home.htm.

Hallett, S.G., 2006. Dislocation from coevolved relationships: a unifying theory for plant invasion and naturalization? Weed Sci. 54, 282—290.

Hejda, M., Pyšek, P., Jarošik, V., 2009. Impact of invasive plants on the species richness, diversity and composition of invaded communities. J. Ecol. 97, 393—403.

Hobbs, R.J., Humphries, S.E., 1995. An integrated approach to the ecology and management of plant invasions. Conserv. Biol. 9, 761—770.

Holm, L., 1978. Some characteristics of weed problems in two worlds. Proc. West. Soc. Weed Sci. 31, 3—12.

Holm, L., Plucknett, D., Pancho, J., Herberger, J., 1977. The World's Worst Weeds. University of Hawaii Press, Honolulu, HI, 609 pp.

Huenneke, L.F., Hamburg, S.P., Koide, R., Mooney, H.A., Vitousek, P.M., 1990. Effects of soil resources on plant invasion and community structure in Californian serpentine grassland. Ecology 71, 478—491.

Hulme, P.E., 2012. Weed risk assessment: a way forward or a waste of time? J. Appl. Ecol. 49, 10—19.

Hulme, P.E., Pysek, P., Jarošik, V., Pergl, J., Schaffner, U., Vilà, M., 2013. Bias and error in understanding a plant invasion impacts. Trends Ecol. Evol. 28 (4), 212—218.

Inderjit (Ed.), 2009. Management of Invasive Weeds. Springer Series in Invasion Ecology, vol. 5. Springer, Dordrecht, Netherlands, 363 pp.

IUCN-International Union for Conservation of Nature and Natural Resources, 2012. http://www.iucn.org/about/work/programmes/species/our_work/invasive_species/.

Joffe, S., Cook, S., 1997. Management of the Water Hyacinth and Other Aquatic Weeds. Issues for the World Bank. World Bank Technical Support Group. Commonwealth Agriculture Bureau International, Cambridge, UK, 38 pp.

Klironomos, J.N., 2002. Feedback with soil biota contributes to plant rarity and invasiveness in communities. Nature 417, 67—70.

Kolar, C.S., Lodge, D.M., 2001. Progress in invasion biology: predicting invaders. Trends Ecol. Evol. 16 (4), 199–204.

Kolb, K., November/December 2015. King of kudzu – in rural Georgia, one man turns an invasive species into art. Orion 21–27.

Lemoine, N.P., Burkepile, D.E., Parker, J.D., 2016. Quantifying differences between native and introduced species. Trends Ecol. Evol. 31 (6), 372–381.

Leslie, A.J., Spotila, J.R., 2001. Alien plant threatens Nile crocodile (*Crocodylus niloticus*) in breeding in Lake St. Lucia, South Africa. Biol. Conserv. 98, 347–355.

Leslie, A.R., Westbrooks, R.G., 2011. Invasive Plant Management Issues and Challenges in the United States: Overview. In: ACS Symposium Series, vol. 1073. American Chemical Society, Washington, DC, 272 pp.

Levine, J.M., D'Antonio, C.M., 1999. Elton revisited: a review of evidence linking diversity and invasibility. Oikos 87, 15–25.

Levine, J.M., Pachepsky, E., Kendall, B.E., Yelenik, S.G., HilleRisLambers, J., 2006. Plant-soil feedbacks and invasive spread. Ecol. Lett. 9, 1005–1014.

Li, Y., Cheng, Z., Smith, W.A., Ellis, D.R., Chen, Y., Zhneg, X., et al., 2004. Invasive ornamental plants: problems, challenges, and molecular tools to neutralize their invasiveness. Crit. Rev. Plant Sci. 25, 381–389.

Lingren, C., Pearce, C., Allison, K., 2010. The biology of invasive alien plants in Canada: 11. *Tamarix ramosissima* Ledeb., *T. chinensis* Lour. and hybrids. Can. J. Plant Sci. 90, 111–124.

Lonsdale, W.M., 1999. Global patterns of plant invasions and the concept of invasibility. Ecology 80, 1522–1536.

Lorenzo, P., Hussain, M.I., González, L., 2013. Role of allelopathy during invasion process by alien invasive plants and terrestrial ecosystems. In: Cheema, Z.A., Farooq, M., Wahid, A. (Eds.), Allelopathy Current Trends and Future Applications. Springer, Berlin, 483 pp.

Lovelock, J., 1979. Gaia: A New Look at Life on Earth. Oxford University Press, Oxford, UK, 157 pp.

Mack, R.N., 1981. Invasion of *Bromus tectorum* L. into Western North America: an ecological chronicle. Agro Ecosystems 7, 145–165.

Mack, R.N., Simberloff, D., Lonsdale, W.M., Evans, H., Clout, M., Bazzaz, F., 2000. Biotic invasions: causes, epidemiology, global consequences and control. Ecol. Appl. 10, 689–710.

Malecki, R.A., Blossey, B., Hight, S.D., Schroeder, D., Kok, L.T., Coulson, J.R., 1993. Biological control of purple loosestrife. BioScience 43, 680–686.

Marler, M.J., Zabinski, C.A., Callaway, R.M., 1999. Mycorrhizae indirectly enhance competitive effects of an invasive forb on a native bunchgrass. Ecology 80, 1180–1186.

Marvin, D.C., Bradley, B.A., Wilcove, D.S., 2009. A novel, web-based, ecosystem mapping tool using expert opinion. Nat. Areas J. 29 (3), 281–292.

Maxwell, B.D., Lehnhoff, E., Rew, L.J., 2009. The rationale for monitoring invasive plant populations as a crucial step for management. Invasive Plant Sci. Manage. 2, 1–9.

McGrath, S., 2005. Attack of the alien invaders. Nat. Geogr. 207, 92–117.

McNeely, J.A. (Ed.), 2001. The Great Reshuffling: Human Dimensions of Invasive Species. IUCN, Gland, Switzerland, 242 pp. Also available as: The great reshuffling: how alien species help feed the global economy. http://www.iucn.org/biodiversityday/introduction.html.

McNeely, J.A., July/August 2004. Strangers in our midst: the problem of invasive alien species. Environment 16–31.

McPherson, G.R., 2004. Linking science and management to mitigate impacts of nonnative plants. Weed Technol. 18, 1185–1188.

Meekins, J.F., McCarthy, B.C., 2001. Effect of environmental variation on the invasive success of a nonindigenous forest herb. Ecol. Appl. 11, 1136–1348.

Millar, H., July/August 2004. When aliens attack. Sierra 31–63.

Mitchell, C.E., Power, A.G., 2003. Release of invasive plants from fungal and viral pathogens. Nature 421, 625–627.

Moody, M.L., Les, D.H., 2002. Evidence of hybridity in invasive watermilfoil (*Myriophyllum*) populations. Proc. Natl. Acad. Sci. 99 (23), 14867–14871.

Mozdzer, T.J., Megonigal, J.P., 2012. Jack-and-Master trait responses to elevated CO_2 and N: a comparison of native and introduced *Phragmites australis*. PLoS One 7 (10), e42794. https://doi.org/10.1371/journal.pone.0042794.

Mullin, B., Anderson, L.W.J., DiTomaso, J.M., Eplee, R.E., Getsinger, K.D., February 2000. Invasive Plant Species. CAST issue paper No. 13, 17 pp.

Myers, J.H., Bazely, D.R., 2003. Ecology and Control of Introduced Plants. Cambridge University Press, Cambridge, UK, 313 pp.

Nadel, H., Frank, J.H., Knight Jr., R.J., 1992. Escapees and accomplices: the naturalization of exotic *Ficus* and their associated faunas in Florida. Fla. Entomol. 75, 29–38.

National Invasive Species Council, 2006. Meeting the Invasive Species Challenge: National Invasive Species Management Plan. U. S. Gov., Washington, DC, 89 pp.

Noble, I.R., 1989. Attributes of invaders and the invading process: terrestrial and vascular plants. In: Drake, J.A., et al. (Eds.), Biological Invasions a Global Perspective. J. Wiley & Sons, New York, NY, pp. 301–313.

Novoa, A., Kaplan, H., Kumschick, S., Wlson, J.R.U., Richardson, D.M., 2015. Soft touch or heavy hand? Legislative approaches for preventing invasions: insights from cacti in South Africa. Invasive Plant Sci. Manage. 8, 307–316.

Nowak, D.J., Rountree, A.R., 1990. History and range of Norway maple. J. Arboric. 16, 291–296.

Oswalt, S.N., Oswalt, C.M., 2011. The extent of selected nonnative invasive plants on southern forest lands. In: Proceedings of the 17th Central Hardwood Forest Conference, vol. 17, pp. 447–459.

Parker, J.D., Burkepile, D.E., Hay, M.E., 2006. Opposing effects of native and exotic herbivores on plant invasions. Science 311, 1459–1461.

Peterson, A.T., Papes, M., Kluza, D.A., 2003. Predicting the potential invasive distributions of four alien plant species in North America. Weed Sci. 51, 863–868.

Pheloung, P.C., Williams, P.A., Halloy, S.R., 1999. A weed risk assessment model for use as a biosecurity tool evaluating plant introductions. J. Environ. Manage. 57, 239–251.

Pimentel, D., Lach, L., Zuniga, R., Morrison, D., 1999. Environmental and Economic Costs of Non-indigenous Species in the United States. http://www.news.cornell.edu/releases/Jan99/species_costs.html.

Pimentel, D., Lach, L., Zuniga, R., Morrison, D., 2000. Environmental and economic costs of non-indigenous species in the United States. Bioscience 50, 53–65.

Pimentel, D., Zuniga, R., Morrison, D., 2005. Update on the environmental costs associated with alien-invasive species in the United States. Ecol. Econ. 52, 273–288.

PNWER (Pacific Northwest Economic Region), 2012. Economic Impacts of Invasive Species in the Pacific Northwest Economic Region. PNWER invasive species working group, 6 pp. http://www.aquaticnuisance.org/wordpress/wp-content/uploads/2010/06/economicimpacts_pnwer_2012.pdf.

Powell, M.R., 2004. Risk assessment for invasive plant species. Weed Technol. 18, 1305–1308.

Pyšek, P., Richardson, D.M., Rejmanek, M., Webster, G.L., Williamson, M., Kirschner, J., 2004. Alien plants in checklists and floras: towards better communication between taxonomists and ecologists. Taxon 53, 131–143.

Radosevich, S.R., Prather, T., Ghersa, C.M., Lass, L., 2009. Implementing science-based invasive plant management. In: Inderjit (Ed.), Management of Invasive Weeds, Springer Series in Invasion Ecology, vol. 5. Springer, Netherlands (Chapter 17).

Rana, N., Sellers, B.A., Ferrell, J.A., MacDonald, G.E., Silveira, M.L., Vendramini, J.M., 2013. Impact of soil pH on bahiagrass competition with giant smutgrass (*Sporobolus indicus* var. *pyramidalis*) and small smutgrass (*Sporobolus indicus* var. *indicus*). Weed Sci. 61 (1), 109–116.

Rand, T.A., Russell, F.L., Louda, S.M., 2004. Local-vs. landscape-scale indirect effects of an invasive weed on native plants. Weed Technol. 18, 1250–1254.

Reed, C.F., 1977. Economically Important Foreign Weeds: Potential Problems in the United States. U. S. Dept. Of Agriculture. Agriculture Handbook No. 498. 746 pp.

Reichard, S.H., Campbell, F., July 1996. Invited but unwanted. Am. Nurserym. 39–45.

Reichard, S.H., Hamilton, C.W., 1997. Predicting invasions of woody plants introduced into North America. Conserv. Biol. 11, 193–203.

Reichard, S.H., White, P., 2001. Horticultural invasions of invasive plant species: a North American perspective. In: McNeely, J.A. (Ed.), The Great Reshuffling: Human Dimensions of Invasive Alien Species. IUCN, Gland, Switzerland, pp. 161–170.

Reinhart, K.O., Callaway, R.M., 2006. Soil biota and invasive plants. New Phytol. 170, 445–457.

Rejmánek, M., 1989. Invasibility of plant communities. In: Drake, J.A., et al. (Eds.), Biological Invasions: A Global Perspective. J. Wiley & Sons, Ltd., New York, NY, pp. 369–388.

Rejmánek, M., Richardson, D.M., Barbour, M.G., Crawley, M.J., Hrusa, G.F., Moyle, P.B., Randall, J.M., Simberloff, D., Williamson, M., 2002. Biological invasions: politics and the discontinuity of ecological terminology. ESA Bull. 83, 131–133.

Ricciardi, A., 2001. Facilitative interactions among aquatic invaders: is an "invasional meltdown" occurring in the Great Lakes? Can. J. Fish. Aquat. Sci. 58, 2513–2525.

Richardson, D.M., Rejmánek, M., 2004. Conifers as invasive aliens: a global survey and predictive framework. Divers. Distrib. 10, 321–331.

Richardson, D.M., Pyšek, P., Rejmánek, M., Barbour, M.G., Panetta, F.D., West, C.J., 2000. Naturalization and invasion of alien plants: concepts and definitions. Divers. Distrib. 6, 93–107.

Ridenour, W.M., Vivanco, J.M., Feng, Y.L., Horiuchi, J., Callaway, R.M., 2008. No evidence for trade-offs: *Centaurea* plants from America are better competitors and defenders. Ecol. Monogr. 78, 369–386.

Rudrappa, T., Choi, Y.S., Levia, D.F., Legates, D.R., Lee, K.H., Bais, H.P., 2009. *Phragmites australis* root secreted phytotoxin undergoes photo-degradation to execute severe phytotoxicity. Plant Signal. Behav. 4 (6), 506–513.

Sagoff, M., June 23, 2000. Why exotic species are not as bad as we fear. Chron. High. Educ. 46 (42), B-7.

Sagoff, M., 2005. Do non-native species threaten the natural environment? J. Agric. Environ. Ethics 18, 215–236.

Saltonstall, K., 2002. Cryptic invasion by a non-native genotype of the common reed, Phragmites australis, into North America. Proc. Natl. Acad. Sci. 99 (4), 2445–2449.

Sax, D.F., Gaines, S.D., 2008. Species invasions and extinction: the future of native biodiversity on islands. Proc. U.S. Natl. Acad. Sci. 105, 11490–11497.

Sax, D.F., Gaines, S.D., Brown, J.H., 2002. Species invasions exceed extinctions on islands worldwide: a comparative study of plants and birds. Am. Nat. 160, 766–783.

Schmitz, D.C., Simberloff, D., Hofstetter, R.H., Haller, W., Sutton, D., 1997. The ecological impact of nonindigenous plants. In: Simberloff, D., Schmitz, D.C., Brown, T.C. (Eds.), Strangers in Paradise. Island Press, Washington, DC, p. 479.

Schwarz, B., Matta, B.M., Shakir-Botteri, N.L., McPheron, B.A., 2005. Host shift to an invasive plant triggers rapid animal hybrid speciation. Nature 436, 546–549.

Shafroth, P.B., Cleverly, J.R., Dudley, T.L., Taylor, J.P., van Riper III, C., Weeks, E.P., Stuart, J.N., 2005. Control of Tamarix in the Western United States: implications for water salvage, wildlife use, and riparian restoration. Environ. Manage. 35, 231–246.

Sheley, R.L., Sheley, J.L., Smith, B.S., 2015. Economic savings from invasive plant protection. Weed Sci. 63, 296–301.

Simberloff, D., 2008. Invasion biologists and the biofuels boom: Cassandras or colleagues? Weed Sci. 56, 86–872.

Skurski, T.C., Rew, L.J., Maxwell, B.D., 2014. Mechanisms underlying non-indigenous plant impacts: a review of recent experimental research. Invasive Plant Sci. Manage. 7, 432–444.

Smith, B.S., Sheley, R.L., 2015. Implementing strategic weed prevention programs to protect rangeland ecosystems. Invasive Plant Sci. Manage. 8, 233–242.

Smith, L.L., Tekiela, D.R., Barney, J.N., 2015. Predicting biofuel invasiveness: a relative comparison to crops and weeds. Invasive Plant Sci. Manage. 8, 323–333.

Solomon, C., 2015. Researcher finds way to fight cheatgrass, a western scourge. N.Y. Times. http://www.nytimes.com/2015/10/06/science/researcer-finds-way-to-fight-cheatgrass-a-western-scourge.html?_r=2.

Stastny, M., Schaffner, U., Elle, E., 2005. Do vigour of introduced populations and escape from specialist herbivores contribute to invasiveness? Ecology 93, 27–37.

Stohlgren, T.J., Binkley, D., Chong, G.W., Kalkhan, M.A., Schell, L.D., Ball, K.A., Otsuki, Y., Newman, G., Bashkin, M., Son, Y., 1999. Exotic plant species invade hot spots of native plant diversity. Ecol. Monogr. 69, 25–46.

Stohlgren, T.J., Barnett, D.T., Kartesz, J.T., 2003. The rich get richer: patterns of plant invasions in the United States. Research communications Front. Ecol. Environ. 1, 11–14.

Stohlgren, T.J., Ma, P., Kumar, S., Rocca, M., Morisette, J.T., Jarnevich, C.S., Benson, N., 2010. Ensemble habitat mapping of invasive plant species. Risk Assess. 30, 224–235.

Stromberg, J.C., Chew, M.K., Nagler, P.L., Glenn, E.P., 2009. Changing perceptions of change: the role of scientists in *Tamarix* and river management. Restor. Ecol. 17 (2), 177–186.

Stuckey, R.L., 1980. Distributional History of Purple Loosestrife (*Lythrum salicaria*) in North American Wetlands. Res Rept. No. 2. U.S. Fish and Wildlife Serv.

Tekiela, D.R., Barney, J.N., 2017. Invasion shadows: the accumulation and loss of ecological impacts from an invasive plant. Invasive Plant Sci. Manage. 10 (1), 1–8.

Thébaud, C., Simberloff, D., 2001. Are plants really larger in their introduced ranges? Am. Nat. 157, 231–236.

Thelen, G.C., Vivanco, J.M., Newingham, B., Good, W., Bias, H.P., Landres, P., et al., 2005. Insect herbivory stimulates allelopathic exudation by an invasive plant and the suppression of natives. Ecol. Lett. 8, 209–217.

Thompson, D., Stuckey, R., Thompson, E., 1987. Spread, Impact, and Control of Purple Loosestrife (*Lythrum salicaria*) in North American Wetlands. Fish and Wildlife Res. Rpt. No 2. U. S. Fish and Wildlife Service, Washington, DC, 55 pp.

Tilman, D., 1999. The ecological consequences of changes in biodiversity: a search for general principles. Ecology 80, 1455–1474.

Tilman, D., Wedin, D., Knops, J., 1996. Productivity and sustainability influenced by biodiversity in grassland ecosystems. Nature 379, 718–720.

Timmons, F.L., 1970. A history of weed control in the United States and Canada. Weed Sci. 18, 294–307.

UNEP (United Nations Environmental Program), 1992. The Convention on Biodiversity. http://www.biodiv.org/convention/articles.asp.

USDA-NRCS (US Department of Agriculture, Natural Resources Conservation Service), 1999. The PLANTS Database. National Plant Data Center, Baton Rouge, LA. http://plants.usda.gov/plants.

Vermeire, L.T., Rinella, M.J., 2009. Fire alters emergence of invasive plant species from soil surface-deposited seeds. Weed Sci. 57, 304–310.

Vilà, M., Espinar, J.L., Hejda, M., Hulme, P.E., Jarosik, V., Maron, J.L., Pergl, J., Schaffner, U., Sun, Y., Pysek, P., 2011. Ecological impacts of invasive alien plants: a meta-analysis of their effects on species, communities and ecosystems. Ecol. Lett. 14, 702–708.

Weber, E., 2003. Invasive Plant Species of the World: A Reference Guide to Environmental Weeds. CABI Publishing, Wallingford, Oxon, UK, 548 pp.

West, A.M., Kumar, S., Wakie, T., Brown, C.S., Stohlgren, T.J., Laituri, M., Bromberg, J., 2015. Using high-resolution future climate scenarios to forecast *Bromus tectorum* invasion in Rocky Mountain National Park. PLoS One 10 (2), e0117893. https://doi.org/10.1371/journal.pone.01178893.

Westbrooks, R.G., 1998. Invasive Plants, Changing the Landscape of America: Fact Book. Federal Interagency Committee for the Management of Noxious and Exotic Weeds, Washington, DC, 109 pp.

Wilcove, D.S., Rothstein, D., Dubow, J., Phillips, A., Losos, E., 1998. Quantifying threats to imperiled species in the United States. Bioscience 48, 607–615.

Williams, M.C., 1980. Purposefully introduced plants that have become noxious or poisonous weeds. Weed Sci. 28, 300–305.

Williams, T., September/October 1994. Invasion of the aliens. Audubon, 24–26, 28, 30, and 32.

Williams, D.G., Baruch, Z., 2000. African grass invasion in the Americas: ecosystem consequences and the role of ecophysiology. Biol. Invasions 2, 123–140.

Williams, J.L., Auge, H., Maron, J.L., 2008. Different gardens, different results: native and introduced populations exhibit contrasting phenotypes across common gardens. Oecologia 157, 239–248.

Williamson, M., Fitter, A., 1996. The varying success of invaders. Ecology 77, 1661–1666.

Wolfe, L.M., Elzinga, J.A., Biere, A., 2004. Increased susceptibility to enemies following introduction in the invasive plant *Silene latifolia*. Ecol. Lett. 7, 813–820.

Yager, L.Y., Miller, D.L., Jones, J., 2011. Woody shrubs as a barrier to invasion by cogongrass (*Imperata cylindrica*). Invasive Plant Sci. Manag. 4, 207–211.

Zavaleta, E., 2000. Valuing ecosystem services lost to *Tamarix* invasion in the United States. In: Mooney, H.A., Hobbs, R.J. (Eds.), Invasive Species in a Changing World. Island Press, Washington, DC, pp. 261–302.

Zedler, J.B., Kercher, S., 2004. Causes and consequences of invasive plants in wetlands: opportunities, opportunists, and outcomes. Crit. Rev. Plant Sci. 23, 431–452.

Ziska, L.H., 2006. Evaluation of the growth response of six invasive species to past, present and future atmospheric carbon dioxide. J. Exp. Bot. 54, 395–404.

9

Allelopathy

FUNDAMENTAL CONCEPTS

- Allelopathy is a form of plant interference that occurs when one plant, through living or decaying tissue, interferes with growth of another plant via a chemical inhibitor.

- Allelopathy may be present in many plant communities.

- Allelopathy has a potential but largely unexploited role in weed management.

LEARNING OBJECTIVES

- To know the definition of allelopathy.
- To understand the complexity of research required to discover true allelopathy.
- To understand the complexity of allelopathic chemistry.

- To understand how allelochemicals enter the environment.
- To know the application of an analogous form of Koch's postulates to allelopathy.
- To know some examples of allelopathic interference.

The Three Princes of Serendip was published in Europe in 1557 by the Venetian author Michele Tramezzino. Horace Walpole, a British statesman, read the story as a child and coined the word *serendipity* in a 1754 letter to Horace Mann, the British envoy to Florence. The story is based on an ancient Persian tale in which the characters make fortunate, unexpected, wonderful discoveries. In the story, the three princes, each vying for the hand of a princess, are assigned impossible tasks by the princess. Each failed to accomplish the assigned tasks, but wonderful, serendipitous, things happen to them as they try to do what they had been asked to do.

Serendipity is an apparent aptitude to make fortunate discoveries accidentally and make unexpected, good things happen. Serendipity may be available in weed science if the presence of allelopathic interactions (Blum, 2011, 2014) can be confirmed and used to control weeds. Organisms from microbes to mammals find food, seek mates, ward off predators, and defend themselves against disease via chemical interactions. Allelopathic interactions are chemical, and discovery of the cause and mechanism of these interactions may yield a treasure of biological and chemical approaches to control weeds. At least 25% of human medicinal products (see Chapter 4) originated in the natural world or are synthetic derivatives of naturally occurring substances. Many natural interactions are chemical interactions and some of them could influence the course of weed science.

Interference is the term assigned to adverse effects that plants exert on each other's growth. Competition is part of interference and occurs because of depletion or unavailability of one or more limiting resources. Allelopathy, another form of interference, occurs when one plant, through its living or decaying tissue, interferes with growth of another plant via a chemical inhibitor (Fig. 9.1). Allelopathy comes from the Greek *allelo* = each other, which is similar to the Greek *allelon* = one another. The second root is the Greek *patho* or *pathos* that means suffering, disease, or intense feeling. Allelopathy is therefore the influence, usually detrimental (the pathos), of one plant on another, by toxic chemical substances from living plant parts, through their release when a plant dies, or their production from decaying tissue.

There is a subset of allelochemicals known as kairomones (from Gr. *kai* = new, and *hormaein* = to set in motion, excite, stimulate) that have favorable adaptive value to organisms receiving them. A natural kairomone from water hyacinth is a powerful insect attractant for a weevil (*Neochetina eichhorniae*) and the water hyacinth mite (*Orthogalumna terebrantis*). The kairomone is liberated when water hyacinth is injured by surface wounding or by the herbicide 2,4-D. The kairomone enhances control of water hyacinth by attracting large numbers of weevils and mites to the area of the plant's wound (Messersmith and Adkins, 1995). Thus the kairomone has favorable value to the insects but not to the water hyacinth. Control of water hyacinth is enhanced when insect damage is combined with herbicide stress.

For weed management purposes, allelopathy is considered a strategy of control. Corn cockle and ryegrass seeds fail to germinate in the presence of beet seeds. If tobacco seeds germinate and grow for 6 days in petri dishes and then an extract of soil, incubated for 21 days with timothy residue is added, the root tips of tobacco blacken within 1 h while radicle elongation is unaffected. If an extract of soil, incubated with rye residue is added, the symptoms are reversed (Patrick and Koch, 1958). Residues of timothy, maize, rye, and tobacco all reduce the respiration rate of tobacco seedlings (Patrick and Koch).

Kooper (1927), a Dutch ecologist, observed the large agricultural plain of Pasuruan on the island of Java, Indonesia, where sugarcane, rice, and maize grew. After harvest, the fallowed fields developed a dense cover of weeds. Kooper observed that the postharvest floristic composition of each community was stable year after year. He found that floristic composition was determined at the earliest stages of seed germination, not by plant survival or a struggle for existence, but by differential seed germination. He showed that seeds of other species were present but could not germinate unless removed from their environment. Competition for light, nutrients, or water did not cause the consistent floristic composition. Kooper concluded that previous vegetation established a soil chemical equilibrium (an allelopathic phenomenon), determined which seeds could germinate, and subsequently, which plants dominated.

Interference = Competition + Allelopathy

FIGURE 9.1   Components of plant interference.

The word *allelopathy* was first used by Molisch (1937), an Austrian botanist. He included toxicity exerted by microorganisms and higher plants and that use has continued. The phenomenon, however, had been observed much earlier by several scientists (Putnam, 1985). A classic example of allelopathy is found in the black walnut forests of Central Asia (Stickney and Hoy, 1881). Few other plants survive under the forest plant canopy because of the presence of juglone, a quinone root toxin derived from black walnut trees (Massey, 1925). The effect of juglone couldn't be reproduced in the greenhouse because some plant metabolites, including phenolics, require ultraviolet light for their biosynthesis (Davis, 1928).

Another classic study is the work by Muller and Muller (1964) in California who observed that California chaparral often occurred near, but not intermixed with, California sagebrush. Neither species grew in the zones of contact between the respective communities; other species grew between the communities. They found volatile terpenes, particularly camphor (a monoterpene ketone) and cineole (a terpene ether) produced by the chaparral, were responsible for the no-contact zones. They concluded that plants, in this case the chaparral, are fundamentally leaky systems. Other studies are described by Rice (1974, 1979) and Thompson (1985). Zhang et al. (2012) found that volatile allelochemicals released from leaf tissue of crofton weed affected seedling growth of upland (non-paddy) rice. β tyrosine, an isomer of the common amino acid tyrosine, has been shown to contribute to the allelopathic potential (a defense function) of rice (Yan et al., 2015).

Research in China showed that the allelopathic potential of the invasive species, creeping daisy, was increased by acid rain. Simulated acid rain increased total carbon and nitrogen content, available phosphorus, decreased soil pH, and accelerated creeping daisy's litter decomposition, thereby enhancing the allelopathic potential of the weed's litter (Wang et al., 2012).

One plant does not consciously set out to affect another, rather the effect occurs as a normal, perhaps serendipitous, ecological interaction with evolutionary implications. Allelopathic species have been selected by evolutionary pressure, because they can outcompete neighbors through energy-expensive biochemical processes that produce allelochemicals. The energy expense is not a waste of resources because no species evolves successfully by wasting resources. Exploration of the phenomena will lead to better understanding of plant evolutionary strategies and, possibly, provide clues for herbicide synthesis and development.

Reviews of allelopathy are found in Putnam (1985, 1994), the proceedings of the American Chemical Society symposium on the chemistry of allelopathy (Thompson, 1985), and the Handbook of Sustainable Weed Management (Singh et al., 2006—see this Chapter, Batish et al.). Putnam (1985, 1994) lists 50 weeds alleged to interfere with one or more crops (Table 9.1). Batish et al. (2006) cited more than 25 plants with known allelopathic and therefore potential herbicidal activity. Allelopathy has also been explored with a number of crops, and there have been attempts to find crop cultivars with a competitive allelopathic edge (Putnam, 1983, 1985; Rice, 1979; Thompson, 1985; Batish et al., 2006). Residues of several crops have phytotoxic activity on other plants (Table 9.2). This effect is often incorrectly attributed to an allelochemical. All plants produce chemicals that are weakly phytotoxic (Duke, 2015).

Laboratory studies have often demonstrated allelopathy, but the evidence produced should not be regarded as conclusive of the existence of allelopathy in the environment until it is confirmed by field studies. Field studies are essential to obtain ecologically relevant data

TABLE 9.1 Some Weeds With Alleged Allelopathic Activity in
Agroecosystems (Putnam, 1983, 1994; Duke et al., 2002)[a]

Weed	Susceptible Species
Barnyardgrass	rice, wheat
Bermudagrass	barley, coffee, soybean
Bluegrass	tomato
California peppertree	cucumber, wheat
Canada thistle	several
Catnip	peas, wheat
Cogongrass	corn, cucumber, rice, sorghum, tomato
Common chickweed	barley
Common lambs quarters	cabbage, cucumber, corn, sugar beet, wheat
Common milkweed	sorghum
Common purslane	alfalfa, durum wheat, tomato
Common ragweed	several
Corn cockle	wheat
Crabgrass	cotton, trailing crown vetch,
Diffuse knapweed	ryegrass
Dock	corn, pigweed, sorghum
Field bindweed	wheat
Flaxweed	flax
Giant foxtail	corn
Giant ragweed	peas, wheat
Goosegrass	bean, corn, sorghum
Goldenrod	several
Hairy beggarticks	several
Heath	red clover
Italian ryegrass	oats, brome, lettuce, clover
Jimsonweed	several
Johnsongrass	barley, cotton, soybean, trailing crown vetch
Ladysthumb	potato, flax
Large crabgrass	several
Leafy spurge	peas, wheat

TABLE 9.1 Some Weeds With Alleged Allelopathic Activity in Agroecosystems (Putnam, 1983, 1994; Duke et al., 2002)[a]—cont'd

Weed	Susceptible Species
Mayweed	barley
Mugwort	cucumber
Mustard	several
Nutsedge, purple	barley, black mustard, broccoli, Brussels sprouts, cabbage, carrot, collards, cotton, cucumber, onion, radish, rice, sorghum, soybean, strawberries, tomato
Nutsedge, yellow	corn
Prince's feather	mustard
Prostrate spurge	several
Quackgrass	several
Redroot pigweed	soybean, wheat
Russian thistle	several
Spiny amaranth	coffee
Sunflower	barley, garden cress. jimsonweed, lettuce, redroot pigweed, tomato, wheat
Syrian sage	wheat
Velvetgrass	common barley
Velvetleaf	several
Western ragweed	several
Wild cane	wheat
Wild garlic	oats
Wild marigold	several
Wild oats	barley, flax, wheat

[a]*Complete citations for several weeds can be found in Duke et al. (2002).*

(Foy and Inderjit, 2001; Inderjit et al., 2001a). Clues from laboratory studies are not sufficient without field confirmation. For example, Norsworthy (2003) demonstrated the allelopathic potential of aqueous extracts of wild radish in controlled environment studies. The evidence indicated that aqueous extracts of wild radish or incorporated wild radish residues suppressed seed germination, radicle growth, seedling emergence, and seedling growth of "certain crops and weeds," but subsequent field confirmation is essential to establish the reality of allelopathy as an ecological phenomenon.

TABLE 9.2   Some Crops Whose Residues Have Been Reported to be
Phytotoxic (Putnam, 1994; Duke et al., 2002)[1]

Crop	Affected Species
Alfalfa	alfalfa
Apple	apple
Asparagus	tomato, asparagus. fescue spp.
Barley	white mustard
Bean	pea, wheat
Black walnut	tomato
Cabbage	mustard, lettuce, spinach, tomato
Clover, red	several
Clover, white	radish
Coffee	several
Corn	several weeds
Crambe	wheat, velvetleaf
Cucumber	several weeds
Jackbean	Brazilian satintail
Lentil	wheat
Oats	several
Pea	several
Rice	barnyardgrass, lettuce, rice
Rye	common lambs quarters
Ryegrass	several
Smooth bromegrass	several
Sorghum	fescue
Sunflower	barley, clover, garden cress, jimsonweed, lettuce, redroot pigweed, tomato, wheat
Wheat	several weeds

1. ALLELOPATHIC CHEMISTRY

Plants produce a myriad of metabolites of no known utility to their growth and development. They are often referred to as secondary plant metabolites and are defined as

[1]Complete citations for several crops can be found in Duke et al. (2002) citation.

compounds having no known essential physiological function. The idea that these compounds may injure other forms of life is not without a logical base. However, proof is questionable because most allelochemical effects occur through soil, a complex chemical matrix. Conclusive studies require extraction and isolation of the active agent from soil. Any allelopathic chemical may be chemically altered prior to or during extraction. That which is extracted, isolated, and studied may not be what the plant produced.

Secondary plant metabolites, also known as natural products, are regarded by many as "a vast repository of materials and compounds with evolved biological activity, including phytotoxicity" (Duke et al., 2002). It is proposed that some of these compounds may be useful directly as herbicides or as templates for herbicide development. According to Duke et al. (2002) they often have unique molecular target sites in plants that have not been developed or used much in agriculture or herbicide development. Several reviews of this area of research are available (Dayan et al., 1999; Duke et al., 1998, 2000a,b, 2002; Hoagland, 2001; Hoagland and Cutler, 2000). Acetic acid, the primary component of vinegar, can be used as a selective, contact herbicide in some crops (e.g., onion and sweet corn) and is approved for organic systems. Matran Pro, a 50% clove oil (eugenol, a phenylpropanoid) product, is approved for weed management in organic production systems (see Evans and Bellinder, 2006). For vinegar and clove oil success depends on the time of application and the growth stage of the crop and weeds. Both require high active ingredient application (acetic acid 34—68 gals/A, clove oil—60 gals/A) and both are expensive (up to several hundred dollars per acre) (Evans and Bellinder). Both gave fair to good control of small broadleaved species, were less effective on hairy or waxy leaves, velvetleaf, and common ragweed, and gave generally poor control of giant foxtail and other grasses.[2]

Allelochemicals vary from simple molecules such as ammonia to the quinone—juglone, the terpenes camphor and cineole, complex conjugated flavonoids such as phlorizin (isolated from apple roots), or the heterocyclic alkaloid caffeine (isolated from coffee) (Putnam, 1985; Rice, 1974; Thompson, 1985). Putnam (1985) lists several chemical groups from which allelopathic agents come: organic acids and aldehydes, aromatic acids, simple unsaturated lactones, coumarins, quinones, flavonoids, tannins, alkaloids, terpenoids and steroids, a few miscellaneous compounds such as long-chain fatty acids, alcohols, polypeptides, nucleosides, and some unknown compounds. Some of the diversity and complexity of allelopathic chemistry is shown in Table 9.3. The diversity suggests several mechanisms of action, a multiplicity of effects, and is one reason for the slow emergence of a theoretical framework. The chemistry of allelopathy is as complex as synthetic herbicide chemistry but it is a chemistry of discovery as opposed to one of synthesis.

There is little doubt that allelopathy occurs in plant communities; but there are questions about how important it is in nature and if it can be exploited in cropped fields. It has been reported for many crop and weed species (Putnam, 1983, 1985, 1994; Batish et al., 2006), but proof of its importance in nature is lacking (Foy and Inderjit, 2001). Proof will require something similar to the application of Koch's postulates (1912) proposed for plant pathology in 1883 and amended by Smith (1905).

---

[2]Curran, B., 2013. http://extension.psu.edu/pests/weeds/organic/weed-management-in-organic-crops.pdf and Curran, W.S. et al. http://extension.psu.edu/pests/weeds/organic/wssa05poster.pdf.

TABLE 9.3   Allelopathic Compounds Isolated From Plants (Putnam, 1983)

Common Name	Chemical Class	Natural Source
Acetic acid	aliphatic acid	Decomposing straw
Allylisothiocyanate	thiocyanate	Mustard
Arbutin	phenolic	Manzanita shrubs
Bialaphos	amino acid derivative	Microorganisms
Caffeine	alkaloid	Coffee
Camphor	monoterpene	*Salvia* shrubs
Cinnamic acid	aromatic acid	Guayule
Dhurrin	cyanogenic glucoside	Sorghum
Gallic acid	tannin	Spurge
Juglone	quinone	Black walnut trees
Patulin	simple lactone	*Penicillium* fungus on wheat straw
Phlorizin	Flavonoid	Apple roots
Psoralen	Furanocoumarin	*Psoralea*

The analogous postulates applied to allelopathy (Aldrich, 1984; Putnam, 1985) are:

1. Observe, describe, and quantify the degree of interference in a natural community.
2. Isolate, characterize, and synthesize the suspected toxin produced by the suspected allelopathic plant.
3. Reproduce the symptoms by application of the toxin at appropriate rates and times in nature. [Koch's (1912) postulates called for reisolation of the bacterial agent from the experimentally infected plant—an inappropriate criterion for allelopathic research.]
4. Monitor release, movement and uptake, and show they are sufficient to cause the observed effect(s).

These four steps describe difficult, expensive, complex scientific research. Rigorous proof has rarely been applied to any ecological interaction, but such proof is vital if allelopathic research is to move from description to causation. Duke (2015) added the necessity of determining whether the compounds identified are produced in sufficient quantity in time and space in soil to exert an allelopathic effect. He goes on to assert that most articles purporting to deal with allelopathy are not designed to prove it. They are designed "to discover phytotoxins that might have utility as herbicides or herbicide leads" (Duke).

In short, it is insufficient to make an observation and suspect a toxin. It is insufficient to demonstrate the toxin is produced by one plant. Specific cause and effect must be demonstrated through chemical and plant studies (Duke, 2015). It may not be necessary to prove that plant X is the source of allelochemical Y. If an allelochemical, effective as a natural herbicide, can be isolated and identified, then, in theory, it might be useful without absolute proof of its plant origin or physiological mode of action. The basic chemistry and biology would remain a scientific challenge but it might be possible to exploit the activity. Proceeding with partial knowledge is risky, but not impossible. For example, medical science still doesn't know exactly how aspirin relieves pain, and weed science doesn't know exactly how some available, useful herbicides kill a plant. However, both can be used productively and safely.

2. PRODUCTION OF ALLELOCHEMICALS

Production of allelochemicals varies with environment and associated environmental stresses. It can occur in any plant organ (Rice, 1974), but roots, seeds, and leaves are the most common sources. Source becomes important for exploitation of allelochemicals for weed control. For example, an allelochemical found in flowers or fruits would have less potential value than if it were concentrated in roots or shoots (Putnam, 1985)—a statement about availability, not allelochemical potency. For control, soil incorporation of whole plants might create proper distribution regardless of the plant part that produced the chemical. The amount is important for control purposes, and if specific effects are to be predicted in the field, total quantity and concentration must be determined (Putnam, 1985).

There is evidence that allelochemical production may be greater when plants suffer from environmental stress (Putnam, 1983, 1985; Rice, 1979). Production is influenced by light intensity, quality, and duration, with a greater quantity produced with high ultraviolet (UV) light and long days (Aldrich, 1984). Weeds, commonly, but not always, understory plants, and might be expected to produce lower quantities of allelochemicals because UV light is filtered by overshadowing crop plants. This, of course, assumes that crops provide shade and that shade effectively suppresses allelopathic activity. Quantities of allelochemicals produced are also greater under conditions of mineral deficiency, drought stress, and cool temperatures as opposed to more optimal growing conditions. In some cases, plants affected by growth regulator herbicides may increase production of allelochemicals. Because stress frequently enhances allelochemical production, it is logical to assume that stress accentuates the involvement of allelopathy in weed-crop interference and that competition for limited resources may increase allelopathic potential or sensitivity of the weed, the crop, or both. Thus, weed-crop competition and allelopathy should be regarded as intimately related components of interference in a crop ecosystem.

Allelochemicals enter the environment in a number of ways at different times, and mode and time of entry can alter their effects (Fig. 9.2). Although chemicals with allelopathic activity may be present in many species, presence does not mean that allelopathic effects will ensue. Even after a chemical has been isolated and identified, its placement and stability in the environment after plant release or its time of release may preclude expression of potential activity.

Allelochemicals enter the environment by volatilization or root exudation and may move through soil by leaching (Fig. 9.2). These entry paths are usually regarded as true allelopathy. Toxins also result from decomposition of plant residues, properly regarded as functional allelopathy, that is, environmental release of substances that are toxic as a result of transformation after their release by the plant.

Allelochemicals can be produced by weeds and affect crops, but the reverse is also true, although it has not been as widely studied (Putnam, 1994). It is probably true that some

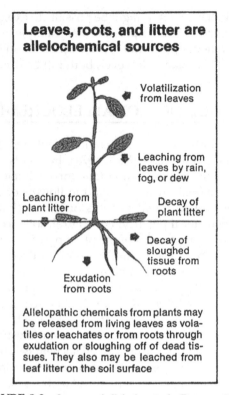

Leaves, roots, and litter are allelochemical sources

Volatilization from leaves

Leaching from leaves by rain, fog, or dew

Leaching from plant litter

Decay of plant litter

Decay of sloughed tissue from roots

Exudation from roots

Allelopathic chemicals from plants may be released from living leaves as volatiles or leachates or from roots through exudation or sloughing off of dead tissues. They also may be leached from leaf litter on the soil surface

FIGURE 9.2   Sources of allelochemicals (Putnam, 1994).

crop cultivars produce allelochemicals. Therefore, it is theoretically possible that such cultivars could be planted to take advantage of their allelochemical potential. It has been suggested that crops with allelopathic potential could be planted as rotational crops or companion plants in annual or perennial cropping systems to exert their allelopathic effect on weeds. Rye and its residues have been shown to provide good weed control in a variety of cropping systems (Barnes and Putnam, 1983). Rye residues reduced emergence of lettuce and proso millet by 58% and 35%, respectively. Rye shoot tissue inhibited lettuce seed germination 52%. It also was phytotoxic to barnyardgrass and cress (Barnes and Putnam, 1986).

3. ALLELOPATHY AND WEED-CROP ECOLOGY

Aldrich (1984) suggested allelopathy was significant for weed-crop ecology in three ways:

1. As a factor affecting changes in weed species composition;
2. As an avenue of weed interference with crop growth and yield; and
3. As a possible weed management tool.

Allelopathy should not always be implicated when other explanations do not suffice, but it should not be overlooked because of the difficulty of establishing causality.

3.1 Effects on Weed Species

Why one species succeeds another is a question that has intrigued ecologists for many years. Weed scientists are interested in the same question but perhaps too often only for the life span of an annual crop. Weed scientists accept that plants change the environment and are changed by it. It is generally agreed that many early colonizers succeed by producing large numbers of seeds, whereas late arrivals succeed through greater competitive ability. This is true in old-field succession and in annual crops. Ecologists have shown that successful plants may change the environment to their advantage through changes in soil nitrogen relationships caused by release of specific inhibitors of nitrogen fixation or nitrification (Putnam, 1985; Thompson, 1985).

3.2 Weed Interference

Weed seeds survive for long periods in soil and chemical inhibitors of microbial decay have been implicated in their longevity, but specific identification of inhibitors from weed seeds has not been accomplished. Allelochemicals have been implicated in the inability of some seeds to germinate in the presence of other seeds or in the presence of crop residues in soil. Although neither phenomenon has been exploited for weed management, there is little doubt that both occur. Eventual exploitation may depend on discovery of specific chemicals and their site(s) of action. Because of the mass and volume of plant residue compared to the mass and volume of seed (even though the number of seeds may be large), the possibility of effects from plant residues is greater than the effects from seed.

The problems with replanting the same or different crops in a field have been cited (Putnam, 1985; Rice, 1974) to show the effect of allelochemicals on crop growth. Putnam (1983) showed that the allelopathic potential of sorghum residues has been exploited for weed control in subsequent rotational crops. While there is little doubt that allelochemicals inhibit crop growth, a research challenge still exists to separate allelopathic effects from competition. Most greenhouse studies cannot be directly translated to the field because of different climatic, light, edaphic, and biological conditions, and possible effects of soil volume. Confirmation of allelopathy awaits development of appropriate experimental methods that verify its presence in field and greenhouse studies.

A fundamental assumption of biological control of weed is that damaged plants are less fit and compete poorly and therefore they will fail in the struggle for survival. That assumption, like so many in science, often is not borne out by research. When spotted knapweed control was attempted using larvae of two different root-boring insects and a parasitic fungus, its allelopathic potential increased significantly, and it had "more intense effects on native" vegetation (Thelen et al., 2005). The authors conclude that while biological control can be very effective, it can often be less effective or fail. Without an understanding of the basic ecology of the area and the plants, it is not possible to know why success or failure occurred. "An invasive species that inhibits natives via unusually deep shade might be a more appropriate target for biological control than allelopathic invaders" (Thelen et al.).

3.3 Weed Management

A living cover crop of spring-planted rye reduced early season biomass of common lambs quarters 98%, common ragweed 90%, and large crabgrass 42% compared to control plots with no rye (Barnes and Putnam, 1983). Wheat straw reduced populations of pitted morning glory and prickly sida in no-tillage culture. It was suggested that the wheat produced an allelochemical that inhibited emergence of several broadleaved species (Liebl and Worsham, 1983). Inderjit et al. (2001b) studied the allelopathic potential of wheat and perennial ryegrass. They showed in a laboratory study that root length of perennial ryegrass was suppressed by wheat and was dependent on the density of wheat seeds in a petri dish. Ryegrass shoot growth was unaffected by wheat, and ryegrass density had no effect on wheat seedling growth. The allelopathic potential of wheat straw has been demonstrated in the laboratory (Guenzi and McCalla, 1962; Guenzi et al., 1967; Hamidi et al., 2001) but not in the field.

It is reasonable to assume that many plants have allelopathic potential or some susceptibility to allelochemicals when they are present in the right amount, form, and concentration at the appropriate time. It is equally reasonable to assume that allelopathy may have no role in the interference interactions of many species. However, enough work has been done to conclude that allelopathy could be utilized for development of new weed management strategies. Trials in South Dakota showed that fields planted with sorghum had 2—4 times fewer weeds the following year than similar fields planted with soybean or corn (Kozlov, 1990). It was proposed, although not proven, that reduced weed seed germination was due to phenolic acids and cyanogenic glucosides given off by sorghum. Suppression of weeds by sorghum has been reported by Guenzi and McCalla (1966) and Hussain and Gadoon (1981). Sunflower has been reported to have an allelopathic effect against grain sorghum (Schon and Einhellig, 1982) and against other weeds (Leather, 1983). Guenzi and McCalla (1966) found allelopathic phenolic acids in oats, wheat, sorghum, and corn residues, and Lodhi et al. (1987) discussed the role of allelopathy from wheat in crop rotations. Other sources are available to describe and summarize the major findings of allelopathy research and their application for weed management (Putnam, 1983, 1985, 1994; Rice, 1974, 1979; Thompson, 1985; Batish et al., 2006 plus the reviews cited herein). A few examples follow to illustrate the research and its potential.

Walker and Jenkins (1986) were the first to demonstrate that sweet potato residues inhibited growth of sweet potato and cowpea. Decaying residues reduced uptake of calcium, magnesium, and sulfur by other plants (Walker et al., 1989). Additional studies showed that after one growing season, shoot dry weight of yellow nutsedge growing with sweet potatoes was less than 10% of the weight when yellow nutsedge was grown alone. Moreover, remaining yellow nutsedge had no effect on sweet potato growth (Harrison and Peterson, 1991). Allelochemicals were present in the tuber periderm that is continually sloughed off during root growth. Proso millet was susceptible to all extracted fractions, but other plants showed differential susceptibility, indicating that several allelochemicals may be present (Peterson and Harrison, 1991).

Plant pathogens and allelochemicals from plant pathogens and other soil microorganisms can be used as bioherbicides. This possibility has been studied for more than three decades (Hoagland, 2001). Numerous pathogens and microbial allelochemicals have been isolated and studied for their bioherbicidal potential. A good example of a microbial product is the herbicide bialaphos (active ingredient phosphinothricin), which leads to accumulation of ammonium and disruption of primary metabolism (Duke and Dayan, 2011). It is manufactured by fermentation as a metabolite of the soil microbe *Streptomyces viridochromeogenes* (Auld and McRae, 1997). The second example is the ammonium salt of phosphinothricin, glufosinate (see Chapter 16). A gene coding for the enzyme phosphinothricin acetyl transferase was isolated from the nonpathogenic bacteria *Streptomyces hygroscopicus* and cloned into several crops. The enzyme converts the herbicide glufosinate to a nonphytotoxic metabolite, and the genetically engineered crop is thus resistant to glufosinate. Another example began with a study of the root parasitic damping off fungus (*Pythium* spp.) in turf. Christians (1991, 1993) wanted to establish the fungus in the soil of a new golf course green at Iowa State University. *Pythium* was cultured in the laboratory on corn meal, a standard procedure. The culture was placed on field plots and other plots were treated with the same amount of fresh corn meal. The attempt to establish *Pythium* failed but seeded cultivars of creeping bentgrass did not germinate well on plots that had received fresh corn gluten meal, a by-product of the wet-milling process of corn grain. This was unexpected. Further study showed potential for selective control of crabgrass in Kentucky bluegrass turf. Liu et al. (1994) demonstrated that enzymatically hydrolyzed corn gluten meal was more herbicidally active than corn gluten. Corn gluten hydrolysate completely inhibited germination of crabgrass and creeping bentgrass seed and root emergence of perennial ryegrass seed. Corn gluten meal (9-0-0) is used for preemergence weed management and fertilization (Bingamen and Christians, 1995; Christians, 1993; Gough and Carlstrom, 1999). Corn gluten meal, a patented, commercially available product with ~10% nitrogen, is approved for use as a fertilizer and weed control in turf and home gardens.

Pollen can also be allelopathic. Pollen can release toxins that inhibit seed germination, seedling emergence, sporophytic growth, or sexual reproduction (Murphy, 2001). Two crops (timothy and corn) and four weeds (orange hawkweed, ragweed parthenium, yellow hawkweed, and yellow-devil hawkweed) are known to exhibit pollen allelopathy (Murphy). There may be others. Pollen allelopathy might be useful in biological weed management because the allelochemical is active in very low doses (as little as 10 grains of pollen per mm^2 on stigmas) and pollen is a small, naturally targeted distribution system. Murphy points out that pollen allelopathy has potential but is not a confirmed weed management technique. Disadvantages include weed adaptation to pollen toxicity and possible threats of toxic pollen to crop plants.

Few researchers recommend that allelopathy is a dominant way plants interact. Many argue that it is present and that nonresource competitive mechanisms should regularly be considered to account for the success of weeds and other invading species (Hierro and Callaway, 2003). Diffuse knapweed is an invasive Eurasian weed in western North America. Research and general observations suggest that diffuse knapweed produces virtual monocultures and that allelopathy may be an important component of its success. Hierro and Callaway (2003) suggest that allelopathy "may be more important in recipient communities than in origin communities because the former are more likely to be naive to the chemicals possessed by newly arrived species." They do not suggest that allelopathy is a unifying theory or a dominant way that plants interact or the only way to explain diffuse knapweed's dominance. They do suggest that such nonresource mechanisms should not be dismissed as irrelevant.

With this kind of evidence one is inclined to agree with Putnam's (1985) suggestion that not believing in allelopathy, now, is like not believing in genetic inheritance before DNA's structure was known. One area to explore might be testing for suppression of weed seed germination and seedling emergence by potential allelopathic species. Work to date has shown this to be an inconsistent effect, and, if developed, it might be useful when combined with other methods of weed management. Allelopathy is not, and will never be, a panacea for all weed problems. It is another weed management tool to be placed in the toolbox and used in combination with other techniques. It is not a technique that will finally solve all weed problems or make the hoe obsolete.

The second strategy where allelopathy may be used is weed-suppressing crops. This can be realized by discovering, incorporating, or enhancing allelopathic activity in crop plants. This technique would be most useful in crops maintained in high-density monocultures, such as turf grasses, forage grasses, or legumes. Olofsdotter (2001) notes that while allelopathy has been demonstrated with varying success, it has been much more difficult to use the principle in crop production. She suggests that if genetic mapping of quantitative traits can be linked to understanding of allelopathic mechanisms it may lead toward optimization of a plant's allelopathy and production of more competitive crops—crops with an allelopathic advantage. It may be possible with modern techniques to transfer (genetically modify) the ability of any plant to produce a weed-controlling allelochemical to a crop plant. [For example, the work on rye done by Barnes and Putnam (1983, 1986).] Much more physiological and chemical knowledge is required before this can be done successfully, but it is an enticing possibility—a crop that does more, perhaps all, of its own weed control because it has a chemical advantage.

The third area for allelopathic research and development includes the use of plant residues in cropping systems, allelopathic rotational crops, or companion plants with allelopathic potential. Many crops leave residues that are regarded as a necessary but not as a beneficial part of crop production, except as they contribute to soil fertility or tilth. Research (Putnam, 1985, 1994; Rice, 1979) indicates that plant residues have allelopathic activity, but the nature of this activity has not been explored sufficiently to permit effective use. Rotation, a neglected practice in many agricultural systems, is being studied because of its potential for weed management through competition and allelopathy. Companion cropping is a relatively new technique for agricultural systems in developing countries. Multiple cropping is common in many developing countries where allelopathy may be operational without being obvious or defined. These systems may hold valuable lessons for further agricultural development of allelopathy as a useful weed management tool.

Weed scientists need to look beyond the immediate assumption that interference is always competition and see what they may not be looking for—an allelopathic effect—an unexpected, good thing. Perhaps there are expressions of allelopathy before our eyes that we don't see because we're not looking for them. If there are compounds in nature with such great specificity, they should be examined. The patterns of herbicide development point to greater specificity, and nature may have solutions in natural products if we recognize them, learn how they work, and exploit their capabilities.

One of the first and quite potent phytotoxins found in higher plants was 1,8-cineole released by sagebrush species (Muller and Muller, 1964). Cinmethylin was developed as an herbicide for weed control in rice, cotton, soybeans, peanuts, and some vegetables, vine crops, and ornamentals. It never received US/EPA approval for use in the United States, but was used in Japan. Chemically, it is a structural analog of 1,4 cineole, which inhibits asparagine synthetase, the enzyme responsible for biosynthesis of the amino acid asparagine (Romagni et al., 2000). Cinmethylin controls many annual grasses and some broadleaf weeds and sedges. It was produced synthetically, but the thought behind it may have been derived from the known phytotoxicity of the allelopathic cineoles.

A second and clearer example of a natural herbicide is AAL-toxin a natural metabolite produced by *Alternaria alternata* f. sp. *Lycopersici*, the pathogen that causes stem canker of tomato (Abbas et al., 1995b). It disrupts sphingolipid metabolism (Abbas et al., 1995a). The phytotoxic effects were tested on 86 crop and weed species (Abbas et al., 1996). Monocots (corn and wheat) and some species of tomato were generally immune to its effects. Black nightshade, jimsonweed, and prickly sida were susceptible to low doses. Other broadleaved species were susceptible but only at higher doses. Abbas et al. (1996) proposed the differential susceptibility of species to AAL-toxin could be exploited for selective weed control. There may be other potentially valuable chemicals hidden from us because we are looking for something else. Promising observations await the good observer.

However, Duke et al. (2002) present five problems associated with natural products, including allelochemicals, that describe why there has not been more research and development:

- The most important reason is that natural products that have or potentially have phytotoxic activity are usually structurally complex and therefore expensive to manufacture.
- Secondly, they often have high mammalian toxicity (AAL-toxin is toxic to mammalian cells) (Abbas et al., 1996), which makes them undesirable from a public health standpoint.
- Many potentially beneficial natural products (phototoxins, pharmacueticals, etc.) are derived from plants found only or mainly in developing countries. These counties have charged, with adequate justification, that developed nations have exploited their resources with inadequate or no compensation. Laws have been passed in many countries to prevent exploitation of indigenous natural resources and to retain some level of ownership.
- The cost of compound identification (discovery), isolation, structural identification, and manufacture has been very high with no assurance of a return to justify the initial costs.
- Finally, many natural products have relatively short environmental half-lives. This is desirable from a nontarget species view but not from a weed management view where some environmental persistence is desirable.

Allelopathic research is now conducted primarily in Europe and Asia. Volume 40 of the Allelopathy journal was published in 2017 by the International Allelopathy Foundation (India). An annotated bibliography of allelopathy, 1700—2012 (Narwal and Willis, 2014), is available. Current trends and future applications of allelopathy were discussed in a book edited by Cheema et al. (2013). It included ecological consequences (Koocheki et al., 2013), implications of potential in agricultural systems (Jabran and Farooq, 2013), and weed management for sustainable agriculture (Narwal and Haouala, 2013). There are abundant literature citations in each chapter, but unfortunately little evidence of weed management success.

THINGS TO THINK ABOUT

1. What is the present role of allelopathy in weed management?
2. What is the potential role of allelopathy in weed management?
3. Why has so little research been done on allelopathy?
4. What are the essential ingredients of a research program to discover allelochemicals?

Literature Cited

Abbas, H.K., Duke, S.O., Paul, R.N., Riley, R.T., Tanaka, T., 1995a. AAL-toxin, a potent natural herbicide which disrupts sphingolipid metabolism of plants. Pest Manag. Sci. 43 (3), 181–187.

Abbas, H.K., Tanaka, T., Duke, S.O., Boyette, C.D., 1995b. Susceptibility of various crop and weed species to AAL-toxin, a natural herbicide. Weed Technol. 9, 125–130.

Abbas, H.K., Duke, S.O., Shier, W.T., Riley, R.T., Kraus, G.A., 1996. The chemistry and biological activities of the natural products AAL-toxin and the fumonisins. In: Singh, B.R., Tu, A.T. (Eds.), Natural Toxins 2. Structure, Mechanism of Action, and Detection, Advances in Experimental Medicine and Biology, vol. 391. Plenum Press, New York, NY, pp. 293–308.

Aldrich, R.J., 1984. Weed-Crop Ecology – Principles in Weed Management. Chapter 8 – Allelopathy in Weed Management. Breton Pub., N. Scituate, MA.

Auld, B.A., McRae, C., 1997. Emerging technologies in plant protection – bioherbicides. In: Proc. New Zealand Plant Prot. Conf., pp. 101–194.

Barnes, J.P., Putnam, A.R., 1983. Rye residues contribute to weed suppression in no-tillage cropping systems. J. Chem. Ecol. 9, 1045–1057.

Barnes, J.P., Putnam, A.R., 1986. Evidence for allelopathy by residues and aqueous extracts of rye (Secale cereale). Weed Sci. 34, 384–390.

Batish, D.R., Singh, H.P., Kohli, R.K., Dawra, G.P., 2006. Potential of allelopathy and allelochemicals for weed management. In: Singh, H.P., Batish, D.R., Kohli, R.K. (Eds.), Handbook of Sustainable Weed Management. Food Products Press, Binghamton, NY, pp. 209–256.

Bingamen, B.R., Christians, N.E., 1995. Greenhouse screening of corn gluten meal as a natural product for broadleaf and grassy weeds. HortScience 30, 1256–1259.

Blum, U., 2011. Plant-Plant Allelopathic Interactions: Phenolic Acids, Cover Crops and Weed Emergence. Springer, 200 pp.

Blum, U., 2014. Plant-Plant Allelopathic Interactions II: Laboratory Bioassays for Water-Soluble Compounds with an Emphasis on Phenolic Acids. Springer, 322 pp.

Cheema, Z.A., Farooq, M., Wahid, A. (Eds.), 2013. Allelopathy: Current Trends and Future Applications. Springer-Verlag, Berlin, 517 pp.

Christians, N.E., 1991. Preemergence Weed Control Using Corn Gluten Meal. US Patent No. 5030268, pp. 63–65.

Christians, N., October 1993. A natural product for the control of annual weeds. Golf Course Manag. 72, 74, 76.

Davis, R.F., 1928. The toxic principle of Juglans nigra as identified with synthetic juglone and its toxic effects on tomato and alfalfa plants. Am. J. Bot. 15, 620.

Dayan, F.E., Romagni, J.C., Tellez, M.R., Rimando, A.M., Duke, S.O., 1999. Managing weeds with natural products. Pestic. Outlook 10, 185–188.

Duke, S.O., 2015. Proving allelopathy in crop-weed interactions. Weed Sci. 63 (sp1), 121–132.

Duke, S.O., Canel, C., Rimando, A.M., Tellez, M.R., Duke, M.V., Paul, R.N., 2000a. Current and potential exploitation of plant glandular trichome productivity. Adv. Bot. Res. 31, 121–151.

Duke, S.O., Dayan, F.E., Rimando, A.M., 2000b. Natural products and herbicide discovery. In: Covv, A.H., Kirkwood, R.C. (Eds.), Herbicides and Their Mechanisms of Action. Academic Press, Sheffield, UK, pp. 105–133.

Duke, S.O., Dayan, F.E., 2011. Modes of action of microbially-produced phytotoxins. Toxins (Basel) 3 (8), 1038–1064.

Duke, S.O., Dayan, F.E., Rimando, A.M., 1998. Natural products as tools for weed management. Proc. Jpn. Weed Sci. Soc. (Suppl.), 1–11.

Duke, S.O., Dayan, F.E., Rimando, A.M., Schrader, K.K., Aliotta, G., Oliva, A., Romagni, J.G., 2002. Chemicals from nature for weed management. Weed Sci. 50, 138–151.

Evans, G.J., Bellinder, R., 2006. Natural products as herbicides in sweet corn and onion. Abst. No. 47 Weed Sci. Soc. Am.

Foy, C.L., Inderjit, 2001. Understanding the role of allelopathy in weed interference and declining plant diversity. Weed Technol. 15, 873–878.

Gough, R.E., Carlstrom, R., 1999. Wheat gluten meal inhibits germination and growth of broadleaf and grassy weeds. HortScience 34, 269–270.

Guenzi, W.D., McCalla, T.M., 1962. Inhibition of germination and seedling development by crop residues. Soil Sci. Soc. Am. Proc. 26, 456–458.

Guenzi, W.D., McCalla, T.M., 1966. Phenolic acids in oats, wheat, sorghum, and corn residues and their phytotoxicity. Agron. J. 58, 303–304.

Guenzi, W.D., McCalla, T.M., Norstadt, F.A., 1967. Presence and persistence of phytotoxic substances in wheat, oat, corn and sorghum residues. Agron. J. 59, 163–165.

Hamidi, B.A., Inderjit, Streibig, J.C., Olofsdotter, M., 2001. Laboratory bioassay for phytotoxicity: an example from wheat straw. Agron. J. 93, 43–48.

Harrison Jr., H.F., Peterson, J.K., 1991. Evidence that sweet potato (*Ipomoea batatas*) is allelopathic to yellow nutsedge (*Cyperus esculentus*). Weed Sci. 39, 308–312.

Hierro, J.L., Callaway, R.M., 2003. Allelopathy and exotic plant invasions. Plant Soil 256, 29–39.

Hoagland, R.E., 2001. Bioherbicides: phytotoxic natural products. In: Am. Chem. Soc. Symp. Ser., vol. 774, pp. 72–90.

Hoagland, R.E., Cutler, S.J., 2000. Plant and microbial compounds as herbicides. In: Narwal, S.S., Hoagland, R.E., Dilday, R.H., Reigosa, M.J. (Eds.), Allelopathy in Ecological Agriculture and Forestry. Kluwer Academic Pub., Amsterdam, Netherlands, pp. 73–99.

Hussain, F., Gadoon, M.A., 1981. Allelopathic effects of *Sorghum vulgare* pers. Oecologia 51, 284–288.

Inderjit, Kaur, M., Foy, C.L., 2001a. On the significance of field studies in allelopathy. Weed Technol. 15, 792–797.

Inderjit, Olofsdotter, M., Streibig, J.C., 2001b. Wheat (*Triticum aestivum*) interference with seedling growth of perennial ryegrass (*Lolium perenne*): influence of density and age. Weed Technol. 15, 807–812.

Jabran, K., Farooq, M., 2013. Implications of potential allelopathic crops and agricultural systems (Chapter 15). In: Cheema, Z.A., Farooq, M., Wahid, A. (Eds.), Allelopathy: Current Trends and Future Applications. Springer-Verlag, Berlin, pp. 349–385, 517 pp.

Koch, R., 1912. Complete works. In: 10th Int. Medical Congr. Berlin. 1890, vol. 1. Liepzig. George Thieme, pp. 650–660.

Koocheki, A., Lalegani, B., Hosseini, S.A., 2013. Ecological consequences of allelopathy (Chapter 2). In: Cheema, Z.A., Farooq, M., Wahid, A. (Eds.), Allelopathy: Current Trends and Future Applications. Springer-Verlag, Berlin, pp. 23–38, 517 pp.

Kooper, W.J.C., 1927. Sociological and ecological studies on the tropical weed-vegetation of Pasuruan (the Island of Java). Rec. Trav. Botan. Neerl. 24, 1–255.

Kozlov, A., February 1990. Weed woes. Discover 24.

Leather, G.R., 1983. Sunflowers (*Helianthus annuus*) are allelopathic to weeds. Weed Sci. 31, 37–42.

Liebl, R., Worsham, D., 1983. Inhibition of pitted morning glory (*Ipomoea lacunosa* L.) and certain other weed species by phytotoxic compounds of wheat (*Triticum aestivum* L.) straw. J. Chem. Ecol. 9, 1027–1043.

Liu, D.L.-Y., Christians, N.E., Garbutt, J.T., 1994. Herbicidal activity of hydrolyzed corn gluten meal on three species under controlled environments. J. Plant Growth Reg. 13, 221–226.

Lodhi, M.A.K., Bilal, R., Malik, K.A., 1987. Allelopathy in agroecosystems: wheat phytotoxicity and its possible role in crop rotation. J. Chem. Ecol. 13, 1881–1889.

Massey, A.B., 1925. Antagonism of the walnuts (*Juglans nigra* L. and *J. Cinerea* L.) in certain plant associations. Phytopath 16, 773–784.

Messersmith, C.G., Adkins, S., 1995. Integrating weed-feeding insects and herbicides for weed control. Weed Technol. 9, 199–208.

Molisch, H., 1937. Der einfluss einer pflanze auf die ander–Allelopathie. G. Fischer, Jena.

Muller, W.H., Muller, C.H., 1964. Volatile growth inhibitors produced by *Salvia* species. Bull. Torrey Bot. Club 91, 327–330.

Murphy, S.D., 2001. The role of pollen allelopathy in weed ecology. Weed Technol. 15, 867–872.

Narwal, S.S., Haouala, R., 2013. Role of allelopathy and weed management for sustainable agriculture (Chapter 10). In: Cheema, Z.A., Farooq, M., Wahid, A. (Eds.), Allelopathy: Current Trends and Future Applications. Springer-Verlag, Berlin, pp. 217–249, 517 pp.

Narwal, S.S., Willis, R., 2014. Annotated Bibliography of Allelopathy (1700–2012). Five Volumes. Available for purchase in PDF format — see. http://www.allelopathyjournal.org/allelopathy/.

Norsworthy, J.K., 2003. Allelopathic potential of wild radish (*Raphanus raphanistrum*). Weed Technol. 17, 307–313.

Olofsdotter, M., 2001. Getting closer to breeding for competitive ability and the role of allelopathy—an example from rice (*Oryza sativa*). Weed Technol. 15, 798–806.

Patrick, Z.A., Koch, L.W., 1958. Inhibition of respiration, germination, and growth by substances arising during the decomposition of certain plant residues in the soil. Can. J. Bot. 36, 621–647.

Peterson, J.K., Harrison Jr., H.E., 1991. Differential inhibition of seed germination by sweet potato (*Ipomoea batatas*) root periderm extracts. Weed Sci. 39, 119–123.

Putnam, A.R., 1983. Allelopathic chemicals. Chem. Eng. News 61 (19), 34–43.

Putnam, A.R., 1985. Weed allelopathy (Chapter 5). In: Duke, S.O. (Ed.), Weed Physiology, Vol. I: Reproduction and Ecophysiology. CRC Press, Boca Raton, FL, pp. 131–155.

Putnam, A.L., 1994. Phytotoxicity of plant residues. In: Unger, P.W. (Ed.), Managing Agricultural Residues. Lewis Pubs. (CRC Press), Boca Raton, FL, pp. 285–314.

Romagni, J.G., Duke, S.O., Dayan, F.E., 2000. Inhibition of plant asparagine synthetase by monoterpene cineoles. Plant Physiol. 123, 725–732.

Rice, E.L., 1974. Allelopathy. New York Academic Press, 353 pp.

Rice, E.L., 1979. Allelopathy-an update. Bot. Rev. 45, 15–109.

Schon, M.K., Einhellig, F.A., 1982. Allelopathic effects of cultivated sunflower on grain sorghum. Bot. Gaz. 143, 505–510.

Singh, H.P., Batish, D.R., Kohli, R.K., 2006. Handbook of Sustainable Weed Management. Food Products Press, Binghamton, NY, 892 pp.

Smith, E.F., 1905. Bacteria in Relation to Plant Disease, vol. 1. Carnegie Inst. of Washington, Washington, D.C.

Stickney, J.S., Hoy, P.R., 1881. Toxic action of black walnut. Trans. Wis. State Hort. Soc. 11, 166–167.

Thelen, G.C., Vivanco, J.M., Newingham, B., Good, W., Bais, H.P., Landres, P., et al., 2005. Insect herbivory simulates allelopathic exudation by an invasive plant and the suppression of natives. Ecol. Lett. 8, 209–217.

Thompson, A.L. (Ed.), 1985. The Chemistry of Allelopathy-Biochemical Interactions Among Plants. Am. Chem. Soc. Symp. Series No. 268. Am. Chem. Soc., Washington, DC, 470 pp.

Walker, D.W., Jenkins, D.D., 1986. Influence of sweet potato plant residues on growth of sweet potato vine cuttings and cowpea plants. HortScience 21, 426–428.

Walker, D.W., Hubbell, T.J., Sedberry, J.E., 1989. Influence of sweet potato crop residues on nutrient uptake of sweet potato plants. Agric. Ecosyst. Environ. 26, 45–52.

Wang, R.L., Staehlin, C., Dayan, F.E., Song, Y.Y., Su, Y.J., Zeng, R.S., 2012. Simulated acid rain accelerates litter decomposition and enhances the allelopathic potential of the invasive plant *Wedelia trilobata* (creeping daisy). Weed Sci. 60, 462–467.

Yan, J., Aboshi, T., Teraishi, M., Strickler, S.R., Spindel, J.E., Tung, C.-W., Takata, R., et al., 2015. The tyrosine aminomutase TAM1 is required for β-tyrosine biosynthesis in rice. Plant Cell 27, 1265–1278.

Zhang, F., Guo, J., Chen, F., Liu, W., Wan, F., 2012. Identification of volatile compounds released by leaves of the invasive plant croftonweed (*Ageratina adenophora*, Compositae), and their inhibition of rice seedling growth. Weed Sci. 60, 205–211.

10

Methods of Weed Management

FUNDAMENTAL CONCEPTS

- Weed prevention, control, eradication, and management use and combine technologies differently.

- Prevention is an important strategy in weed management.

- Many important agricultural weeds are escaped imports.

- Mechanical, nonmechanical, and cultural weed control techniques each have distinct advantages and disadvantages.

- Weed control using cultural techniques (changing the way agriculture is practiced) is intuitively sensible.

- No weed control method has ever been abandoned.

- Each new method of weed control introduced in large-scale agriculture has reduced the need for human and animal power.

- Other than herbicides and genetically modified crops, all weed control techniques are acceptable in alternative/organic agriculture production systems.

LEARNING OBJECTIVES

- To know the definition and relative merits of weed prevention, control, eradication, and management.

- To be familiar with weed seed laws and US Federal noxious weed laws.

- To know the practices that prevent introduction and spread of weeds.

- To consider the advantages and disadvantages of nonchemical weed control techniques.

- To know the current role and consider future weed management role of living mulches and companion cropping.

- To appreciate the role of minimum and no-tillage in weed management.

You cannot play with the animal in you without becoming wholly animal, play with falsehood without forfeiting your right to truth, play with cruelty without losing your sensitivity of mind. He who wants to keep his garden tidy doesn't reserve a plot for weeds. *Hammarskjöld, D. 1964. Markings, p. 15*

271

1. PREVENTION, CONTROL, ERADICATION, AND MANAGEMENT DEFINED

When asked, students in weed science classes often say they are taking weeds or weed control. Those who work on weeds often spend a great deal of time controlling or figuring out how to control them. However, weed science is not just about control. Weed scientists try to answer fundamental questions about weeds and weed management. For example, they want to know why weeds are problems: that is, what is the nature of weed–crop competition? Why are some weeds problems in many places and others in relatively restricted habitats? Why do different weed management strategies work differently in different cropping systems? Why are some plants so successful as weeds? Answers to these and similar questions lead to hypotheses and theories and thus greater clarity about what ought to be done to manage weeds, and why.

1.1 Prevention

The most difficult part of weed management is prevention, defined as stopping weeds from contaminating an area. It is a practical means of dealing with weeds but it takes time and careful attention to many details. Prevention is difficult to achieve. Experience has shown that it is much easier to make the case and gain support for controlling weeds. After all, if control is successful, as it usually is, results are easily observed: something good has happened. Prevention addresses a potential problem, one that does not yet exist. Results of preventive efforts are harder to observe and measure. It is difficult to demonstrate that because of weed prevention a weed problem did not appear. Science cannot prove a negative. However, it is as true for the agricultural ailment, weeds, as it is for human ailments: an ounce of prevention is worth a pound of cure. Effective preventive techniques may reduce short-term economic gain and increase long-term profit.

Preventive measures include the following:

isolating imported animals for several days,
not importing weeds or weed seeds in animal feed (buying only clean hay),
using only clean crop seed free of weed seed,
cleaning equipment between fields and especially between farms,
preventing seed production, especially by new weeds,
preventing vegetative spread of perennials,
scouting for new weeds; finding them before they become a problem,
small-patch treatment to prevent patch expansion, and
education about weeds (e.g., weed identification).

1.2 Control

Weed control includes many techniques used to limit weed infestations and minimize competition. These techniques attempt to achieve a balance between cost of control and crop yield loss, but weed control is used only after the problem exists; it is not prevention. Weed control techniques have been adopted widely because control is the easiest thing to do and is usually effective. The problem is known and can be seen, and actions can be tailored to the observed problem. Control techniques can be selected to meet short-term economic and agricultural planning goals.

1.3 Eradication

Weed eradication is the complete elimination of all live weeds, vegetative reproductive parts (e.g., rhizomes, stolons), and weed seed. It is 100% elimination (complete control). It sounds easy but it is very difficult to achieve and eradication efforts have rarely been completely successful. With modern herbicide and mechanical techniques, it is usually easy to control live plants because they can be seen and control techniques are effective. It is difficult to eliminate seed and vegetative reproductive parts in soil because they cannot be seen and do not emerge in one season. Dormancy allows emergence over years. Eradication is the best program for small populations of perennial weeds, but current technology does not make it easy.

In weed science as in medical science, prevention is better than control, but control is required because weeds and other pests arrive without notice and are present before their arrival can be prevented. Prevention and eradication require long-term thinking and planning.

1.4 Management

Weed management is a combination of the techniques of prevention, control, and eradication to manage weeds in a crop, cropping system, or environment. Weed managers recognize that a field's or area's cropping history, the grower's management objectives, available technology, financial resources, and a host of other factors must be considered before appropriate management decisions can be made. Complete weed control in a crop may be the best decision in some cases, but one should not assume that it will always be the most desirable goal. Maintenance of a weed population at some low level in a cropping system may be the most easily achieved, financially wise, and environmentally desirable goal.

2. PREVENTIVE TECHNIQUES AND WEED LAWS

People want to be and stay healthy. When we become ill, we are pleased to have competent physicians, hospitals, medical services, and health insurance. People would rather remain healthy than have to cure an illness. The same logic applies to weed management. Weed control may cure a weed invasion but does not prevent one.

A good weed management program includes vigilance (watchfulness). A good weed manager can identify weed seedlings and mature plants, and has a management program for each crop and field and appropriate follow-up plans. The good manager is ever watchful for new weeds that may become problems, and whenever possible, emphasizes prevention rather than control. Several preventive practices can be included in management programs:

- isolation of introduced livestock to prevent importation and spread of weed seeds from their digestive tract;

- use of clean farm equipment and cleaning of itinerant equipment, including combines, cultivators, and grain trucks;
- cleaning of irrigation water before it enters a field;
- mowing and other appropriate weed control practices to prevent seed production on irrigation ditch banks;
- inspection of imported nursery stock for weeds, seeds, and vegetative reproductive organs;
- inspection and cleaning of imported gravel, sand, and soil;
- special attention to fence lines, field edges, rights of way, railroads, etc., as sources of new weeds; and
- prevention of deterioration of range and pasture to stop easy entry of invaders such as downy brome (Mack, 1981).
- Seed dealers and grain handlers should clean crop seed and heat, grind, or dispose of cleanings properly.
- Fields should be surveyed regularly to identify new weeds.
- When identified, small patches of new weeds should be treated to prevent growth and further dispersal.

The first step toward weed prevention and the first step of all good weed management programs is planting clean seed. Forty or fifty years ago, it was much more necessary for farmers to be wary of possible foreign contaminants (weed seed, pathogens, debris, rat droppings, etc.) of purchased seed. Contamination was common. It was important to observe the seller's report of percent pure seed. The examples in Chapter 5 illustrate that the importance of planting clean seed has not diminished although seed sellers have taken appropriate measures to ensure quality and purity.

The first attempt to ensure clean seed was the US Federal Seed Act of 1939. It regulated transport and sale of seeds in foreign and interstate (but not intrastate) commerce. For interstate commerce, the Federal Seed Act defines noxious-weed seed as seed or bulblets of plants recognized as noxious by the law or rules and regulations of the state into which the seed is offered for sale or transported. The law specifies that it is unlawful for any person to transport or deliver for transportation in interstate commerce any agricultural seeds or mixture of seeds for planting unless each container bears a label that reveals the kinds of noxious weed seed and the rate of occurrence (percent) of each.

The law is enforced by the US Department of Agriculture (USDA), which has provided supplementary rules applicable to interstate movement of parasitic plants and noxious weed seeds. The Federal Seed Act and state laws mandate labeling of crop seed to show the kind of seed, its variety, and the state and specific locale where it was grown. Complete labels also show percent pure seed, percent weed seed, percent other crop seed, percent inert matter, percent germination of pure seed, percent hard seeds (seeds that are viable but not capable of immediate germination), and the date on which the tests were performed. Seed labels also include the name and number per pound of each noxious weed seed.

Each of the United States has a noxious weed seed law that identifies and regulates sale and movement of crop seed containing what the state law identifies as seed of a prohibited or restricted noxious weed. These laws may prohibit importation of crop seed with greater than a certain percentage of specific noxious weed seed and may require identification of each noxious weed seed. The presence of noxious weed seed in excess of 1 g/10 g of the crop seed results in exclusion from sale in most states. For large-seeded crops such as beans, the exclusion is often 1 g/100 g of crop seed. These laws may also regulate import and sale of crop seed screenings because they usually contain viable weed seed. State seed laws are designed to protect buyers (farmers and other purchasers). These laws do not mean and should not be viewed as implying equal regulation of weedy plants that may be detrimental to agriculture or the environment.

A bushel of clover seed weighed 60 lb. It was 88% clover with 35% germination. Therefore, in 1 bushel, there was 18.5 lb of live clover seed, 34.3 lb of dead clover seed, and 4.2 lb of weed seed, representing 11 different species, 2.8 lb of inert matter, and 0.2 lb of other crop seed. The purchased seed contained 7800 Canada thistle seeds per bushel, 5700 curly dock seeds per bushel, and 114,000 wild mustard seeds per bushel. This bargain seed cost $5.90 per bushel or $19.14 per 60 lb of 100% viable seed. The same variety of certified clover seed could have been purchased for $8.40 per bushel. That bushel had 99.15% purity and 95% germination or a cost of $8.84 per 60 lb of 100% viable seed. The difference in cash cost ($8.40 to $5.90 per bushel) was $2.50. The cash cost is the only thing some buyers notice. The bargain seed cost $19.14 for 100% good seed versus $8.84 for 100% good seed in the second source, a difference of $10.30 per bushel in favor of the second source (Barnes and Barnes, 1960). Purchasing bargain seed or cheap seed is rarely a good idea and can create weed problems.

Seed standards are not restricted to the United States. The regulations of the Canada Seeds Act of 1987 allow various levels of weed seed to be present depending on the crop and the level of classification desired. The standards apply to barley, buckwheat, lentil, rye, and sainfoin, and with minor variation for wheat, canola, flax, and oats. If requested, sellers must supply a certificate that states the number and kinds of weed seed present.

Most of the United States have established limits between 1% and 4% on total weed seed in crop seed. Most state laws exempt seed sold by a grower, without advertising. Prohibited noxious weed seeds are usually from perennial, biennial, or annual plants that are highly detrimental to crop yield and difficult to control. The presence of these seeds in any amount prohibits sale of crop seed for planting purposes. Restricted noxious weed seeds are those of plants that are objectionable in fields, lawns, or gardens but can be controlled by good cultural practices.

Almost 600 species are included in the prohibited and restricted noxious weed lists of the 50 United States[1]. More than 1300 are included on the USDA list of introduced, invasive, and noxious plants/weeds of the United States, six US territories and protectorates, 14 Canadian provinces, Greenland, and the self-governing territories of France: St. Pierre and Miquelon.[2] Twenty not listed by any other state are included on Hawaii's list. Weeds from 95 taxa, with

[1]https://www.ams.usda.gov/sites/default/files/media/NWS%20List%20for%202015.pdf.

[2]See plants.usda.gov/java/invasiveOne.

no tolerance, are included on the Federal noxious weed seed list. These are legal definitions that are informed by agronomic and horticultural practice.

Most states have seed laboratories that determine seed quality, among other things. One aspect of quality is the number of weed seed or other crop seed in a sample. Tables 10.1 and 10.2 illustrate how many weed seeds were sown when the purchased seeds were sown.

In January 1975, weed prevention took a major step forward when President Ford signed the Federal Noxious Weed Act of 1974 (Public Law 93-629, 88 Stat. 2148), empowered the US Secretary of Agriculture to declare plants to be noxious weeds, to control import, distribution, and interstate commerce of weeds declared to be noxious and limit the interstate spread of such plants without a permit. Previous laws regulated just seed, not plants. In 2004, President Bush signed the Noxious Weed Control and Eradication Act (Public Law 108-412). It passed, after several years of effort, as an amendment to the Federal Plant Protection Act of 2000. It was a first step and only a bit ($15 million) of the funding requested was received. It demonstrates the benefit of groups working together to pass Federal legislation and an increasing recognition of the importance of weed management.

All 50 states and the District of Columbia list some prohibited agricultural weeds in addition to those included in the Federal noxious weed law. These laws provide some protection, but in the view of many it is inadequate protection for agriculture or the environment. The Federal law included 112 weedy species in 2016. In 2010, there were 19 aquatic, 5 parasitic, and 88 terrestrial species. Surely there are more that meet the Act's definition. Many noxious weeds are agricultural problems, whereas some also exclusively invade other areas of

TABLE 10.1   Sample Seed Analysis From Colorado State Seed Testing Laboratory

Bromegrass, smooth, 61% germination (seeds/lb)	
Seeded at 4−6 lb/acre	136,000
Redroot pigweed	27,968
Japanese brome	512
Stink grass	256
Barnyard grass	64
Oldfield cinquefoil	64
	28,864
Timothy	448
Barley	64
Sweet clover	64
Sand dropseed	64
Bent grass	64
	704

TABLE 10.2 Sample Seed Analysis From Colorado State Seed Testing Laboratory

Alfalfa, Sample 1

 84% germination, 84% live

  224,000 seeds/lb   Seeded at 8–10 lb/acre

Dodder	432/lb
Mallow	180/lb
Ground-cherry	90/lb

At 10 lb/acre, 4320 dodder seeds and 2,240,000 alfalfa seeds will be sown per acre.

Alfalfa, Sample 2

 66% germination, 84% live

Russian knapweed	9/lb
Chicory	270/lb
Netseed lamb's-quarter	360/lb
Kochia	180/lb
Buckhorn plantain	117/lb
Other weeds	189/lb
Other crop	
Red clover	6930/lb

At 10 lb seed/acre, 1,478,400 alfalfa seeds, 11,250 weed seeds, and 69,300 red clover seeds will be sown/acre.

environmental concern, such as wetlands and undisturbed natural areas. Invasive species are discussed in Chapter 8.

A survey of weed and seed laws in five contiguous western states Idaho, Oregon, Utah, Washington, and Wyoming (US Congress, 1993) showed that the laws provided adequate to inadequate protection based on the likelihood of unlisted weeds causing economic or ecological problems. Many potential threatening weeds were omitted (Table 10.3). These problems have been corrected.

Federal and state laws do not include enough weedy plants and they regulate only agricultural and vegetable seed. The laws do not cover horticultural seeds including known sources of weed seed such as wildflower and native grass mixtures (US Congress, 1993).

Despite existing laws, regulations are not stringent and it is unsurprising that at least 36 weed species now resident in the United States were imported and escaped to become weeds, in some cases noxious weeds (Williams, 1980). Of the 36, 2 were imported as herbs, 12 as hay or forage crops, and 16 as ornamentals. Weeds have been imported for use as a windbreak (multiflora rose), for possible medicinal value (black henbane), for use in aquaria (*Hydrilla*), as a fiber crop, just for observation, and as a dye (dyer's woad).

TABLE 10.3 Survey of Weed and Seed Laws in Five Western States (US Congress, 1993)

State	Number of Species Listed[a]	Adequacy of Protection	Number of Potential Threats Omitted
Idaho	46	Adequate	6
Oregon	38	More than adequate	Few
Utah	35	Inadequate	11
Washington	55	More than adequate	Few
Wyoming	61	Adequate	11

[a]*See: http://www.ams.usda.gov/AMSv1.0/getfile?dDocName=STELPRD3317318 for current list.*

Bermuda grass, a valuable forage species in the southern United States and many other parts of the world and a desirable turf grass, is also an important weed in many areas and was introduced into the United States as a forage crop. In 1849, the US Cotton Office proposed and subsequently introduced a new forage grass: crabgrass (Brosten and Simmonds, 1989). More recent introductions of grassy weeds include sorghum-almum, which was promoted as a drought-resistant emergency hay/forage crop with names such as perennial Sudan grass, sorghum grass, and Columbia grass. It is a hybrid between johnsongrass and grain sorghum and was first described and cultivated in Argentina (Brosten and Simmonds). Wild proso millet is cultivated for food for mourning doves, pheasants, quail, and songbirds. It provides good winter and brood-rearing cover for upland birds. It was first recognized as a weed in the North Central United States in the early 1970s and subsequently infested several million acres in Wisconsin and Minnesota. Its habitat extends west to Colorado, the midwestern states, and Canada. It is the same species as cultivated millet and therefore is difficult to control in several crops, especially corn. A major reason it is such a good weed is that its seed germinates throughout the growing season rather than in a short period, as crop seed does. It thereby escapes control by nonresidual herbicides and single cultivations.

The latter two cases are interesting cases of failure to prevent and because of their implications for biotechnology. Hybridization of weeds and crops is uncontrolled and may be uncontrollable. Cross-pollination is inevitable when two phenologically similar, outcrossing plants share a small area (exist in an overlapping range). Research to determine the potential for gene transmittal in cropped fields from weeds to crops, or vice versa, is ongoing. There is a possibility that a crop genetically engineered for high yield or herbicide resistance will contribute to the generation of new, difficult to control weed hybrids (Brosten and Simmonds, 1989).

Two species of toadflax, introduced as ornamentals, became weeds. Jimsonweed and *Kochia* were brought to the United States as ornamentals, and *Kochia* was studied as a forage crop. The artichoke thistle escaped to become a weed in artichokes and is a recurring problem in California (Brosten and Simmonds, 1989). It is a particularly nasty noxious weed that occurs sporadically in some areas of the southern California coast. It is found on medium to heavy soils on roadsides and in permanent pastures in some high rainfall areas of southwestern Western Australia. It has been grown as an ornamental. Artichoke thistle is spiny and can grow 5—6 ft tall. It is the wild form of cultivated globe artichoke.

Water hyacinth was introduced from South America to the United States by Japanese entrepreneurs as part of a horticultural exhibit at the cotton centennial exposition in New Orleans in 1884 (Penfound and Earle, 1948). It originally came from the Orinoco River in Venezuela and single plants were given away at the cotton exposition. It has been introduced around the world primarily because of its pretty flowers. At the New Orleans exposition, people liked it so much, they took it home and put it in ponds and gardens, from which it inevitably escaped as it was discarded or water flowed out and carried the weed along. It reproduced profusely in ponds and escaped to the St. Johns river in Florida, where it became a major weed problem by clogging the waterway. Water hyacinth was brought to the Tonkin region of China (now Vietnam) in 1902 as an ornamental. It reached southern China and Hong Kong in the same year. Soon after, it was observed in Sri Lanka and then India, where the sluggish rivers of east Bengal were ideal for its growth. In the 1950s, it was discovered in Africa (Vietmeyer, 1975), and in 1958 it had infested over 1000 miles of the Nile river from Juba in the south to the Jebel Aulia Dam in northern Sudan (Heinen and Ahmad, 1964). It is a serious problem in all of these places and many others, but not in Venezuela, where its spread is controlled by natural enemies.

Cogon grass or Alang-Alang, a perennial, was introduced at Grand Bay, Alabama and McNeil, Mississippi (Tabor, 1952). Bare root orange plants were imported to Grand Bay in 1912 and the Cogon grass that lined boxes in which the plants were shipped was discarded. In McNeil, scientists were searching for better forage plants and Cogon grass escaped from farmer's fields and the experiment station and spread rapidly.

Kudzu, a nonindigenous species, was introduced to the United States at the Philadelphia Centennial Exposition in 1876 (Shurtleff and Aoyagi, 1977). It was promoted by the USDA for erosion control and forage, but it became a major weed and now grows in many areas throughout the southeastern United States and has spread to some midwestern states. It is commonly known as the weed that ate the South because it grows up and over electric poles and almost any stationary object in a field or open area.

Further evidence of distribution of the world's weeds and the necessity for vigilance to prevent introduction of new species is shown in Table 10.4 (Holm et al., 1977). Most of the important US weeds come from somewhere else, and vigilance is necessary to prevent new problems. Among 300 nonindigenous weeds in the western United States, 8 were former crops and 28 escaped from horticultural areas (US Congress, 1993).

All is not lost because weed entry is often not prevented. Most imported plants do not become weeds. In the United Kingdom, about 10% of the invaders became established but only 1% became weeds (Williamsen and Brown, 1986). In Australia, only 5% of introduced plants became naturalized and only 1% to 2% of those became weeds (Groves, 1986). Once a plant is naturalized, whether it remains insignificant or becomes a weed problem depends on the absence of damaging natural enemies and the presence of suitable soil, crops, land use, and weed management practices, and on how the plant responds to the local climate (Panetta and Mitchell, 1991). The few that become weeds can be costly. Although the chance is small, the consequences can be great. We can identify areas at risk of invasion, but weediness cannot be predicted as easily (Panetta and Mitchell).

TABLE 10.4   Origin and Distribution of Some of the World's Most Serious Weeds (Holm et al., 1977)

Weed	Origin	Distribution (Number of Countries)	Associated Crops
Purple Nutsedge	India	92	52
Bermuda grass	Africa or Indo-Malaysia	80	40
Barnyard grass	Europe and India	61	36
Jungle rice	India	60	35
Goose grass	China, India, Japan, Malaysia	60	36
Johnsongrass	Mediterranean	63	30
Cogon grass	Old World	75	35
Spiny amaranth	Tropical America	54	28
Sour paspalum	Tropical America	30	25
Tropic ageratum	Tropical America	46	36
Itchgrass	India	28	18
Carpet grass	Tropical America	27	13
Hairy beggar-ticks	Tropical America	40	31
Para grass	Tropical Africa	34	23
	Mexico, West India, tropical South America	23	13
Smallflower umbrella sedge	Old-world tropics	46	1
Rice flatsedge	Old-world tropics	22	17
Crowfoot grass	Old-world tropics	45	19
Eclipta	Asia	35	17
Globe fringerush	Tropical America	21	(Rice)
Witchweed	Europe or South America	35	2
Halogeton	Asia	Unknown	Rangeland
Russian knapweed	Asia	Unknown	>10
Quack grass	Eurasia	>80	Many

3. NONCHEMICAL METHODS OF WEED MANAGEMENT

No weed control method has ever been abandoned completely. New techniques have been added in large-scale agriculture but old ones are still used effectively, especially in small-scale and organic agriculture. Other than the use of herbicides (chemical weed control), all other methods of weed control can be and are used in alternative/organic agricultural production systems (referred to here as organic agriculture). The current chemical-, capital-, and energy-dependent agricultural production system (referred to here as modern agriculture) is dominated by chemical weed control. Sections 3.1–3.3 will describe and discuss the role of three dominant, nonchemical weed management/control techniques (i.e., mechanical, nonmechanical, and cultural).

3.1 Mechanical Control

Mechanical weed control methods have a long history. It began with people with hoes, which were replaced by horse-drawn implement with shovels and sweep implements and early harrows. They were often the primary weed control method in many crops and remain so in many of the world's developing countries. Although they have been used widely for many years, they have not been studied carefully. If mechanical methods are to become acceptable alternatives, research is required to achieve greater efficacy. This is especially true for weed management in organic agriculture, in which chemical control is forbidden and hand weeding, if labor is available, is expensive, arduous, and often ineffective because it is delayed. Mechanical control of intrarow weeds is often unsuccessful because:

- Cultivation is delayed and weeds are most susceptible to uprooting and subsequent drying in early growth stages,
- Achieving crop–weed selectivity is difficult in early crop growth stages, and
- Weed response to mechanical damage is highly dependent on weather conditions after cultivation (Kurstjens et al., 2004).

Mechanical weed control can be expensive because of the time required and the cost of equipment and fuel. Success nearly always requires several trips over the field, precise timing, and favorable subsequent weather. The farmer must have more knowledge of the weed and crop. That is, successful mechanical control requires careful planning to get the timing and tool right (Kurstjens et al., 2004). More planning is required than with what many refer to as the brute force of chemical control.

3.1.1 Tillage

When most people think of mechanical control, the first thing that comes to mind is tillage with an implement to disturb, cultivate, and mix soil. On arable land, tillage alone or combined with cropping or chemical treatment may be the most economical system of weed control. Tillage turns under crop residue, may improve soil tilth, and facilitates drainage. It controls weeds by burying them, separating shoots from roots, stimulating germination of dormant seeds and buds (to be controlled by another tillage), desiccating shoots, and exhausting carbohydrate reserves of perennial weeds.

Other reasons for tillage include breaking up compacted soil, aeration, seedbed preparation, trash incorporation, and intrarow cultivation in a crop. All of these are important, but in addition to planting preparation, the main accomplishment of most tillage in the world's developed countries is weed control. No-till and minimum-till farming have shown that tillage is not essential to grow crops and may do nothing more than control weeds. Too-frequent tillage can increase soil compaction, which is a disadvantage. Other disadvantages include exposure of soil to erosion, moisture loss, and stimulation of weed growth by encouraging germination of dormant seeds and vegetative buds. In some soils, without tillage, soil can crust and there will be poor water penetration. Decisions about the role of tillage must be made for each soil type and farming system.

Cultivating for weed control in beans.

Tillage is usually divided into primary and secondary. Primary tillage is initial soil breaking or disturbance. The depth varies from at least 6 in. (except where primitive tools are used with limited animal power) to as much as 24 in. Primary tillage implements include moldboard and chisel plows. These cut and invert soil and bury plant and other surface residue. Primary tillage is often the first step of seedbed preparation. It was made possible by Jethro Tull's (1774–1834) invention of a cast iron plow in England in 1819. That was followed by a steel-blade plowshare, introduced by John Lane in England in 1833, followed by John Deere (1804–86), who introduced the first steel moldboard plow in the United States in 1837. The moldboard plow has been characterized as the most important invention of the era. It lifted and inverted soil and greatly expanded the ability of a farmer to till and thus plant more land. Its invention came at a time when the English were never far from starvation, and it literally, saved humanity (Faulkner, 1943). Farmers had trouble then, as they still do, keeping unwanted plants from growing in their crops. Because it buried plants and debris, plowing gave the farmer time to plant and get the crop up before the weeds appeared. Agricultural scientists welcomed the plow without question for its crop production and weed control benefits. They developed what Faulkner (p. 53) called "an unquestioning reverence for the plow." Only later were the disadvantages of the plow and the intensive tillage it enabled recognized. The advantages of plowing were clear but few realized that each plowing buried weed seeds for future recovery and germination (Faulkner, p. 151).

Most disking accomplishes weed control and seedbed preparation. *Courtesy of Deere and Co., Moline, IL.*

Plowing is used to prepare land for planting and it controls weeds. *Courtesy of Deere and Co., Moline, IL.*

Secondary tillage implements, usually used subsequent to primary tillage, may be the first tillage operation. Soil is disturbed, often vigorously, but upper layers are usually not inverted. A wide selection of tools is available (Kurstjens et al., 2004). Secondary tillage is fast and inexpensive, and its tools are appropriate for large areas. The implements have been used for a long time. The first revolving disk harrow was invented in 1847. Tools available to modern farmers include the double disk, several kinds of harrows, torsion and finger weeders, field cultivators, rotary hoes, vertical row brushes, spring tooth harrows, rototillers, rod-weeders, finger weeders, and the cultipacker (combination of harrow and roller). This diverse group of implements tills but does not invert soil from a few to a maximum of 5 or 6 in. Secondary tillage implements break clods and firm soil as they remove weeds. Many regard secondary tillage implements as weed control and seedbed preparation tools.

Each type of weeding implement requires careful adjustment of angle, depth, intrarow and interrow perpendicular distance, and speed.

Primary and secondary tillage is followed, in many row crops, by interrow cultivation. Tractor-mounted cultivators or animal-drawn implements move soil between crop rows to loosen it and control weeds. In general, interrow tillage is just that: it works between crop rows. Some implements prepare interrow areas for furrow irrigation (water runs down furrows between crop rows). Implements used for interrow cultivation include a wide range of tine (long finger-like rods) and flared or straight steel shovel-like tools at the end of solid or flexible (flat, steel) shanks that travel through soil at shallow depths (1–2 in.). They break soil crusts and facilitate irrigation, but their main purpose is weed control. Intensive cultivation of peanuts is effective on weeds between rows but not in the row. A tine cultivator used early in the season to achieve cultivation perpendicular to the row direction (in the row) improves overall weed control (Johnson and Davis, 2015).

The growth of organic agriculture and its demands for nonchemical weed control combined with the application of visual sensing devices, global positioning systems, and sophisticated operational software have led to the development of improved cultivation techniques (Fennimore et al., 2016). Among these is the Robovator, a vision-based cultivator controlling weeds in row crops developed by F. Poulson Engineering in Denmark. It detects weeds by discriminating between plant size and is effective as long as weeds and crops are not the same size. It selectively removes all kinds of weeds and employs no chemical technology (see Chapter 11). Lati et al. (2016) evaluated it over for 2 years for weed control in direct-seeded broccoli and transplanted lettuce. The Robovator did not reduce crop stand or marketable yield compared to a standard cultivator. It removed 20%–40% more weeds at moderate to high weed densities and reduced hand weeding 20%–45% compared with a standard cultivator (Fennimore et al., 2016. Potential new weed tools for lettuce and spinach. http://cesantabarbara.ucanr.edu/files/198872.pdf, Lati et al., 2016).

Research (Schweizer et al., 1994; VanGessel et al., 1995) has shown that to maximize efficiency in corn, intrarow cultivation requires preceding, early-season weed control with cultivation or herbicide. Intrarow cultivators are more efficient (control more weeds) than interrow cultivators. Without herbicides, weeds in corn were always controlled better by an in-row cultivator than by the standard interrow cultivator when each operation was performed at the right time. In-row cultivators have specially designed tools (Fig. 10.1 shows some examples) that disturb soil around crop plants and uproot weeds in rows. The tools include spyders (toothed disks that move soil toward or away from crop rows) and torsion and spring hoe weeders that flex vertically and horizontally to uproot weeds in crop rows. Spinners displace weeds in crop rows. Standard interrow crop cultivators are most effective on weeds 15 cm tall or less. Interrow cultivators are most effective on weeds less than 6 cm tall (Schweizer et al., 1994). These cultivators do not work well in row crops when weed density is high and weeds are large.

Plowing and subsequent tillage do not always just prepare land for planting. Land heavily infested with perennial sod-forming grasses, a situation often encountered in developing country agriculture, cannot be prepared properly for planting. Many tillage implements give inadequate results in the crop row after the crop has emerged and begun to grow. Tillage between rows is efficient and can be done to within a few inches of crop plants. Tillage with standard between-row cultivators is not effective in the crop row except when soil is moved

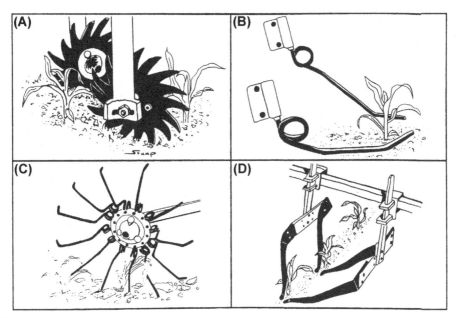

FIGURE 10.1 Types of tillage implements used for in-row cultivation. (A) A pair of spyders. (B) A pair of torsion weeders. (C) A spinner. (D) A pair of spring hoe weeders. *From Schweizer, E.E., Westra, P., Lybecker, D.W., 1994. Controlling weeds in corn (Zea mays) rows with an in-row cultivator versus decisions made by a computer model. Weed Sci. 42, 593–600. Reproduced with permission.*

into the row and weeds are buried. To maximize tillage benefits, uniform spacing of crop rows, straight rows (achieved by precision planting), gauge wheels, and depth guides are needed. Uneven stands and driver error often lead to damage from mechanical cultivation and destruction of some crop plants.

Successful weed control with tillage is determined by biological factors:

- how closely weeds resemble the crop. Weeds that share a crop's growth habit and time of emergence may be the most difficult to control with tillage, especially when they grow in crop rows. Weeds that emerge earlier or later than the crop are often easier to control;
- If a weed's seeds have a short, specific period of germination, it is easier to control them with tillage as opposed to those whose seeds germinate over a long time;
- Perennial weeds that reproduce vegetatively are particularly difficult and usually impossible to control with tillage alone.

Successful mechanical control of weeds is also determined by human factors. Gunsolus (1990) noted that science can explain why certain weed management practices work the way they do. Science develops basic principles to guide action. Human cultural knowledge is different from scientific knowledge, although each may work toward the goal of good weed management. Cultural knowledge tells one when and how to do something on a given soil and farm. Tillage is a cultural practice, and therefore, by definition it requires cultural knowledge. It requires the mind of a good farmer who knows the land, the crop, and the weeds.

Successful mechanical control requires managerial skill (cultural knowledge) that can be aided by, but cannot be acquired from, scientific knowledge. Cultural knowledge is acquired by doing and observing those who have done things well. Cultural knowledge is the art of farming whereby one knows how to select and apply scientific knowledge to solve problems. Successful mechanical control of weeds, regardless of the implement used, is always related to the timeliness of the operation. Research can determine when to do something, but knowing when to act on a particular farm is part of the cultural knowledge good farmers have.

For example, a 3-year study in Pennsylvania showed that corn yields did not differ among no-till, zone-till (surface tillage in narrow rows where corn is to be planted), strip-till (deep tillage in the row where corn is to be grown), and full tillage (chisel plowing followed by disking) (Duiker et al., 2006). The study recommended farmers use no-tillage because it saved fuel, reduced soil erosion, and improved soil and water quality. Cultural knowledge will determine whether farmers will adopt the recommended no-till practices. The scientific knowledge of what is possible will be combined with the cultural knowledge of what should be done on a piece of land.

The operative principle for use of tillage for control of perennial weeds is carbohydrate depletion. The underground reproductive system (stolons and rhizomes) of perennial weeds is a carbohydrate storehouse. When shoots grow and photosynthesize, eventually the store-house will be replenished. If shoots are cut off, the plant calls on its reserve to create new growth. When tillage is done frequently, the management assumption is that reserves will be depleted and plants will die because of exhaustion of root reserves and increased suscep-tibility to other stresses (e.g., frost or dryness). Unfortunately, root reserves are vast and outlast human patience, time, and the need to grow a crop. If tillage and destruction of foliage are delayed from a few days up to a week after emergence, the greatest depletion of root reserves occurs. With most perennial weeds, the great majority of roots and vegetative buds are in the top 6—12 in. of soil. Tillage done when a crop is growing cannot go this deep without disturbing crop roots, a disadvantage for control of perennial weeds.

Early research showed that if field bindweed was tilled 12 days after it first emerged, 16 more tillage operations at approximately 12-day intervals were required to approach eradi-cation. If it was tilled immediately after emergence, about twice as many tillage operations were needed. The efficacy and impracticality of tillage are also illustrated by a study that showed that purple nutsedge could be controlled in Alabama by disking at weekly or biweekly intervals for 5 months (Smith and Mayton, 1938). Obviously, no crop can be grown during the 5 months. Buhler et al. (1994) demonstrated that over 14 years, greater and more diverse populations of perennial weeds developed in reduced-tillage systems than on areas that were moldboard plowed. Practices used to control annual weeds and environmental factors interacted with tillage to regulate (but not eliminate) perennial weeds.

It is often thought, incorrectly, that as long as one tills, it does not depend on how or when it is done as long as the weed is there to be controlled (Schweizer and Zimdahl, 1984). Studies were established in a field where corn had been grown continuously for 6 years. Half of the plots received regular chemical weed control each year whereas the other half had herbicides for the first 3 years, then no herbicide, and only cultivation for the last 3 years. Plots that received herbicide for 3 years were also cultivated regularly in each of the 6 years. In plots with herbicide for the first 3 years and only cultivation thereafter, redroot pigweed domi-nated. At the end of the 6-year experiment, the field was divided in half; one-half was plowed in January and disked in April before normal spring planting. The other half was disked in January and again in April before normal spring planting. More redroot pigweed emerged when the field was disked in the fall than when it was plowed. Where herbicide and optimum weed management had occurred for 6 years, almost no redroot pigweed survived to produce seeds for the last 3 years of the study and tillage made no difference in the redroot pigweed population in the seventh year (Fig. 10.2).

In Michigan, Smith (2006) demonstrated the importance of tillage timing. Spring tillage led to weed communities dominated by spring annual forbs and C4[3] grasses, whereas fall tillage created communities dominated by later-emerging forbs and C3 grasses. The traits that determined species' susceptibility to tillage included the seed germination process and the plant's

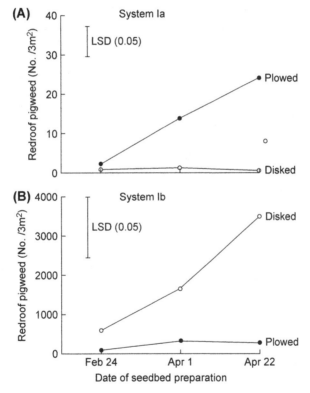

FIGURE 10.2 Population of redroot pigweed seedlings after several conventional tillage practices and atrazine use in continuous corn. In weed management system 1a, 2.2 kg/ha of atrazine was applied preemergence for 6 consecutive years. In weed management system 1b, the same rate of atrazine was applied for the first 3 years and discontinued thereafter. In the fall, one-half of each system 1a and 1b plot was plowed (*hatched line*) and the other half was disked (*solid line*). *From E.E. Schweizer, R.L. Zimdahl, Weed seed decline in irrigated soil after six years of continuous corn (Zea mays) and herbicides, Weed Sci. 32 (1984) 76–83.*

[3]All species have the more primitive C3 (carbon fixation) pathway. The C4 pathway is evolved in species in the wet and dry tropics. The first product of carbon fixation in C3 plants involves a three-carbon molecule, whereas C4 plants initially produce a four-carbon molecule that then enters the C3 cycle.

life cycle, which influence response to changes in soil resources and light availability that are related to the seasonal disturbance regime (the tillage).

Disking soil (secondary tillage) in plots that had only cultivation for 3 years enhanced germination of seeds on the soil surface by bringing them nearer the surface. Plowing (primary tillage) buried seeds. Therefore, if weed control has not been good, disking instead of plowing made the weed problem worse. If weed control has been good, the kind and time of tillage did not matter (Schweizer and Zimdahl, 1984).

Another example of the importance of tillage timing is from land to be planted to wheat in North Dakota (Donald, 1990). Moldboard plowing 18—20 cm or chisel plowing 9—15 cm deep in the fall (primary tillage) followed by a combined field cultivator-harrow in spring (secondary tillage) controlled established foxtail barley on previously untilled sites. Foxtail barley had been a problem in no-till spring wheat and other spring sown no-till crops in the northern Great Plains. Often it could be managed by changing tillage practices (e.g., rotating from no-till to primary tillage). If land was chisel plowed in spring and then harrowed, the weed was not controlled (Donald).

Research to determine the influence of the type of tillage implement and the tillage timing leads to understanding how land management and weed control may actually create weed problems. Roberts and Stokes (1965) showed that plowing distributed weed seeds throughout the plow layer. Rotary cultivation left 50% of weed seeds in the top 3 in. and 80% in the top 6 in. where they germinate best. Regardless of the type of cultivation, between 3% and 6% of the viable weed seeds in the top 10 cm of soil can be expected to produce seedlings after cultivation (Roberts, 1963; Roberts and Ricketts, 1979). Thus, one must conclude that tillage may not create more weeds to control, but it can.

Spring soil disturbance reduced seedling emergence of large crabgrass, giant foxtail, smooth pigweed, and common ragweed by 1.4—2.6 times, but emergence of eastern black nightshade and velvetleaf were unaffected by spring cultivation (Myers et al., 2005). The same study showed that the influence of soil disturbance on yellow foxtail and common lamb's-quarter varied between seasons and location. One must conclude that the type of tillage implement and tillage timing can determine the weed problem. The effect of tillage is also determined by the weeds present and the time of year tillage is done. One longs for precise generalizations, but weed management is too complex for simple rules.

In a rare study of tillage over time, Wicks (1971) grew winter wheat annually for 12 years and studied the effect of a sweep plow, one-way disk, and moldboard plowing (all primary tillage implements) on downy brome after wheat harvest. The moldboard plow eliminated the downy brome population after 12 years, compared with 94 plants/m² for sweep plowing and 24 for the one-way disk. Sweep plows do not bury seed as deeply as a moldboard plow, which buried seed that could germinate but could not emerge. Spread of downy brome was hastened by changing from spring to winter wheat because land was plowed and prepared for seeding at exactly the right time for the winter annual life cycle of downy brome (McCarty, 1982).

The same kind of evidence about the effects of timing and type of tillage is found in several farming systems. Evidence from rice culture shows that the method and timing of land preparation influenced the subsequent weed population. In fields where tractor plowing during the dry season was followed by two harrowings in the wet season, jungle rice was over 85% of the weed population, but purple nutsedge was negligible. In the same region, where two plowings and two harrowings occurred in the wet season, jungle rice was virtually nonexistent and purple nutsedge dominated (Pablico and Moody, 1984).

Annual grass weeds are likely to remain a problem with use of minimum cultivation in cereal production. Particularly when early planting was practiced (Froud-Williams et al., 1981), previously unimportant weeds became more prevalent, especially weedy species of brome in winter cereals in the United Kingdom. Working with spring-sown crops in the United States, Buhler and Oplinger (1990) showed that common lamb's-quarter density was not influenced by tillage method but redroot pigweed density was usually higher in chisel plow systems before planting soybeans. Moldboard plowing (primary tillage) followed by cultipacking (secondary tillage) always had greater densities of velvetleaf than no-till, and no-till always had more foxtail than plowing. Giant foxtail and redroot pigweed became more difficult to control when tillage was reduced, whereas velvetleaf was less of a problem.

Growers need to be aware of the effect of tillage type and timing on weed populations, and whenever possible, choose a system that contributes to weed control. That is good management. Integration of techniques follows. Reduced cultivation encourages establishment of wind-disseminated species and annual broad-leaved species decline. In corn, green foxtail density was greater in chisel plow and no-till systems than with moldboard plowing. Ridge tillage had lower green foxtail density than all other systems (Buhler, 1992). Common lamb's-quarter density was nearly $500/m^2$ after chisel plowing whereas it was only $75/m^2$ in other tillage systems. Redroot pigweed responded differently to tillage with average densities of 307 and $245/m^2$ after no-tillage and chisel plowing versus only $25/m^2$ after moldboard plowing or ridge tillage. Weed populations were affected by tillage but corn yield was not.

Many weed seeds require light to stimulate germination (see Chapter 5). Weed scientists have asked whether germination could be reduced if tillage were done at night. In Oregon's Willamette Valley, cultivating agricultural land during the day increased germination 70%—400% above the levels found after nighttime tillage (Scopel et al., 1994). The effect was attributed to light exposure during tillage. Buhler and Kohler (1994) showed that tilling soil in absolute darkness can reduce germination of some weed species up to 70%. Night tillage is most effective against small broad-leaved species such as pigweed, smartweed, ragweed, nightshade, wild mustard, and common lamb's-quarter. It reduced germination of foxtail or barnyard grass and had no effect on large-seeded, broad-leaved weeds such as velvetleaf, giant ragweed, and cocklebur. Hartmann and Nezadal (1990) reported a 7-year study showing that tillage between 1 h after sunset and 1 h before sunrise reduced weed emergence as much as 80% compared with day tillage. They recommended night tillage as a way to manipulate and control weed populations. They also advocated daytime tillage to photo-stimulate germination of dormant weeds seeds with the goal of diminishing the soil seed bank. They recommended that early primary tillage (plowing) should be carried out in full sunlight to encourage seed germination. Secondary tillage to prepare the seedbed should be done after dark to destroy emerged seedlings and discourage germination of seeds.

Do not become too enthusiastic about this idea. Although it is true that exposure to light favors germination of many weeds seeds, some are light insensitive. Light is only one of many environmental factors that affect weed seed germination. Regulating light exposure will favor management of some weeds and enhance chances for success of others. In weed management, absolute rules are hard to find.

Undisturbed in soil, most light-sensitive seeds are not photo-induced to germinate by light penetration below 1 cm. Germination stimulation comes from brief (a few seconds or less) exposure to light during soil disturbance. This observation is consistent with work by Wesson and Wareing (1969), who showed that weed seed germination depended on exposure of seeds to light during soil disturbance. Most weed seeds germinated within 2 weeks after exposure to light. They also demonstrated that stirring soil for 90 s in bright light increased weed seed germination up to 60%.

Minimum or no-tillage agriculture is practiced for economic reasons and a desire to reduce soil erosion. As emphasized previously, tillage, including minimum or no-tillage, affects the weed population. Any method of weed control that minimizes tillage is potentially beneficial to soil structure. The eco-farming data in Table 10.5 encourage minimum tillage for production of crops grown under low rainfall conditions. The point is that minimum-tillage wheat and minimum-tillage grain sorghum yield as well as production under more intensive tillage and frequently have lower production costs. Minimum-tillage, nonirrigated corn does not yield what irrigated corn does, but production costs are lower.

In vegetable fields in California, reduced tillage compared with conventional (more vigorous) tillage increased the density of shepherd's purse in the top 15 cm of soil (Fennimore and Jackson, 2003). Shepherd's purse emergence and soil seed bank densities were always lower in plots that had been organically amended (cover crops and compost). The authors suggested that organic matter additions may lead to reduced weed emergence.

The extent of use and weed control and its implications for no- or minimum-tillage have been reviewed for developing countries (Akobundu, 1982; Buckley, 1980). It has been shown that these systems rely on herbicides and may complicate soil management owing to the presence of crop residues. With an abundance of weed seed in soil, the best approach may be to use minimum or no tillage and let natural factors deplete the population of buried seed. If weed control fails after 1 year and the soil weed seed bank has been depleted (difficult to determine), the best strategy will be to plow deeply and use minimum tillage afterward (Mohler, 1993). In the first year after minimum tillage begins, no-tillage will have more seedlings than tillage, but in subsequent years, fewer weed seedlings will emerge unless dormancy is high or there is good survival of seed near the soil surface (Mohler).

TABLE 10.5   Yield and Production Costs for Different Cropping Systems in Southwest Nebraska (Klein, 1988)

Crop	Tillage	Average Yield (bushels/acre)	Production Cost ($/bushels)
Wheat	Clean fallow	37	3.88
Wheat	Stubble mulch	43	3.44
Wheat	Eco-fallow—reduced tillage	45	3.30
Sorghum	Conventional	40	3.09
Sorghum	Eco-fallow—reduced tillage	65	2.42
Corn	Conventional tillage with center-pivot irrigation	140	2.59
Corn	Eco-fallow—reduced tillage	65	2.52

Minimum and no tillage has important advantages (Phillips, 1978):

- soil erosion is reduced (a primary disadvantage of any tillage is the possibility of increased erosion);
- because of reduced erosion, land subject to erosion can be used more intensively;
- reducing tillage saves energy;
- there is less compaction with decreased travel over soil;
- because land is continually covered, soil moisture is not as limiting as it can be on bare soil;
- irrigation requirements are lower because posttillage evaporation of soil moisture is reduced; and
- less horsepower is required for land preparation and machinery costs can be reduced.

It is generally agreed that reduction or absence of tillage increases problems with perennial weeds. Tillage may increase or decrease weed seedling density (Mohler, 1993). Some studies have found more seedlings in tilled plots and others have found more without tillage. The effects of tillage vary among species, season, and locations.

Froud-Williams et al. (1981) reviewed changes in weed flora associated with reduced tillage systems. They found several studies in which perennial monocot and dicot species increased in the absence of tillage. They suggested that perennial monocot weeds with rhizomes or stolons would be the greatest threat to successful adoption of reduced tillage systems. Murphy et al. (2006) found over 6 years that tillage systems had a major effect on weed diversity and density. No-tillage promoted the highest (20 species) weed diversity and moldboard plowing promoted the lowest. Chisel plowing was intermediate. The soil seed bank declined from 41,000 seeds/m^3 of soil to 8000 seeds/m^3 over 6 years under no-tillage. Crop yield was not affected by the tillage system.

There are equally important disadvantages to reducing or eliminating tillage (Akobundu, 1982):

- average soil temperature is lower, which may delay spring planting and crop emergence;
- insect and disease problems may increase because plant residues on the soil surface provide a good environment for insects and disease pathogens (Musick and Beasley, 1978; Suryatna, 1976);
- a greater degree of managerial skill is required because:
  - fertilizer requirements and application techniques may (must be) change;
  - crop establishment may be more difficult because of surface residue;
  - irrigation systems may have to be modified;
  - weed control is essential, but as species change methods must change; and
  - there are fewer herbicides available for reduced-tillage systems.

Disadvantages have not deterred growers from learning required skills and shifting to no- or minimum-tillage. Of the 294 million acres of cropland tilled each year in the United States, conservation tillage is increasing annually. No-till increased from 10.6 to 32.9 million acres from 1972 to 1980 (Triplett, 1982). Triplett suggested that in the future, 80% of US crop acreage would be planted using some form of reduced tillage and 50% of the acreage would be no-tillage. Triplett's has been a reasonably accurate prediction. Conservation tillage, the focus of Triplett's prediction, includes any type of tillage system in which at least 30% of the soil surface is covered with crop residue after harvest to protect against erosion. These systems include no-till, strip-till, ridge-till, and mulch-till. In 1990, conservation tillage was used on 73.2 million acres, which was 26.1 of US cropland. It grew to 112.6 acres (40.7%) in 2004. No-tillage systems grew from 16.9 (6%) in 1990 to 62.4 (22.6%) in 2004.[4] By 2009, approximately 35.5% of US cropland (88 million acres) was planted to eight major crops used no-tillage (Horowitz et al., 2010)[5]. In 2012, 96 million US crop acres used no-till practices on over half of wheat (2009), corn (2010), and soybean (2012) acres (http://www.ers.usda.gov/data-products/chart-gallery/detail.aspx?chartId=49982), which improved soil organic matter and carbon sequestration. Worldwide, one sees similar trends. From 1987 to 1997, no-till grew from 15 to 110 million acres; 58% was outside the United States, with Canada, Brazil and Argentina leading the way.[6] In 1973–74, there were only 6.9 million acres worldwide, which increased to 193 million by 2003 and 386 million acres worldwide in 2013 (Friedrich et al., 2012; Kassam et al., 2015).

Fernandez-Cornejo et al. (2012) found that adoption of herbicide-tolerant crops benefitted the environment directly by increasing conservation tillage. Holland (2004) showed that conservation tillage not only influenced the quantity of herbicides used but also reduced herbicide leaching losses. He also observed that by improving soil structure, conservation tillage may reduce the risks of runoff and pollution of surface water with pesticides.

Seed burial studies (see Chapter 5) support the contention that the shift to minimum or no-tillage systems of crop production will not eliminate the need for weed management. The weeds will change as tillage systems change. Data from seed burial studies show that as tillage is reduced, biennial weeds invade cropland, partially because their seeds survive longer when buried (Burnside et al., 1996). Other annuals, adapted to no-till, will appear in cropping systems. Federal farm programs promote conservation tillage and require maintenance of plant and residue cover on the soil surface to reduce wind and water erosion. After an 18-year study of conventional tillage or no-tillage in three cropping sequences (continuous wheat, wheat–faba bean, and wheat–berseem clover), Ruisi et al. (2015) found that the tillage system did not affect the size of the total weed seed bank but altered its composition and distribution of seeds within the soil profile. Conventional tillage favored some species, primarily prostrate knotweed, whereas no-tillage favored other species (e.g., field poppy, prickly lettuce, and canary grass). Ruisi et al. recommended that no-tillage should be used only within an appropriate crop sequence. Research in Finland (Salonen et al., 2012) showed that in conventional cropping, the total abundance of weeds remained about the same from the late 1990s to the late 2000s. Overwintering species became more frequent. Increased total weed biomass was associated with organic cropping owing the lack of direct weed control methods and inadequate crop competition.

3.1.2 Mowing

Mowing to remove shoot growth prevents seed production and, if done frequently, may deplete root reserves of upright perennials. If repeated several times in a growing season,

---

[4]http://abe-research.illinois.edu/pubs/factsheets/USTillageTrends.pdf.

[5]http://www.ers.usda.gov/media/135329/eib70.pdf.

[6]http://pnwsteep.wsu.edu/directseed/conf98/world.htm.

it can be used to control upright perennials in turf. Prostrate perennials such as field bindweed and dandelion survive mowing.

Mowing followed by application of 3.3 kg/ha of glyphosate to resprouting perennial pepperweed can enhance the weed's control (Renz and Ditomaso, 1998). A similar technique has been successful for control of other perennial weeds. Renz and Ditomaso (2004) proposed that the technique was successful because mowing changed the canopy structure of perennial pepperweed and there was greater deposition of the herbicide on basal leaves with increased translocation to roots. "The delay between mowing and resprouting synchronized maximal below ground translocation rates with herbicide application timing." Brecke et al. (2005) showed similar results for a similar reason to control purple nutsedge with herbicides.

To maximize mowing's benefits, it must be done before viable seeds have been produced. Weeds should be cut in the bud stage or earlier. Table 10.6 shows the percentage of germinable seeds produced at various stages of maturity. In a real sense, waiting to mow creates the need to control.

Mowing is useful but rarely accomplishes much weed control because it is done late. It removes unsightly growth and, if done at the right time, can prevent seed production and dispersal, which are important to control annuals and biennials. Its effectiveness for control of biennial musk thistle's seed production is shown in Table 10.7 (McCarty, 1982). In contrast, Tipping (2008) found that musk thistle density declined only when dead plants that had shed seed were mowed, which at first seems counterintuitive because seeds are musk thistle's primary reproduction technique. However, seedling recruitment and survival is the most vulnerable growth stage for musk thistle. Mowing and supplemental chemical control opened the area that had been dominated by the mother plant for invasion and colonization

TABLE 10.6 Germination of Weed Seeds From Plants at Three Stages of Maturity (Gill, 1938)

Weed in Bud	Cut		
	Flowering	Medium Ripe	Ripe
Annual sow thistle	0	100	100
Canada thistle	0	0	38
Common chickweed	0	56	60
Common groundsel	0	100	100
Corn speedwell	0	69	70
Curly dock	0	88	84
Dandelion	0	0	91
Meadow barley	0	90	94
Shepherd's purse	0	82	88
Soft brome	0	18	96
Spotted cat's ear	0	0	90

TABLE 10.7   Seed Production by Musk Thistle (McCarty, 1982)

Time of Harvest	Seeds/Plant
Full bloom	26
+2 days	72
+4 days	774
Mature plant	3580

by competitors. In this case, cool-season grasses easily dominated (eliminated) the much less competitive musk thistle seedlings.[7]

Plumeless thistle, also a biennial, when mowed when most flowers were in full bud or after bloom did not reduce seed bank or plant density (Tipping, 2008), whereas mowing at full bloom reduced both, as McCarty's data would predict.

The foregoing discussion deals with mowing performed to control weeds or clean up an area. Mowing is a normal cultural operation for some crops (e.g., turfgrass, hay). However, it should not be regarded as simply a required cultural operation as it is for managing turfgrass or producing hay. It is a potential albeit underdeveloped weed management technique. Norris and Ayres (1991) showed that cutting interval (but not irrigation timing after cutting) affected yellow foxtail biomass in alfalfa and alfalfa yield. Percent yellow foxtail ground cover was greatest after a 25-day cutting interval and least after a 37-day interval (Fig. 10.3). Yellow foxtail biomass was also greatest for the short cutting interval and least for the longest interval. In the 3-year study, the 37-day cutting interval always had a higher yield than the 31- or 25-day interval (Table 10.8), thus demonstrating the utility of mowing for weed management.

Mowing is a satisfying practice. There is a pleasure in watching the plants fall away from the sickle bar mower or gather after a rotary mower passes. For me, it is one of the few farming tasks that is simple and fun. It can manage or assist in the management of some weeds. In a thorough review of mowing, Donald (2006) claimed that it is economically and environmentally wise and saves energy. Research is required to support or deny his claim. To be most beneficial, the research should focus on mowing as a desirable part of integrated weed management (Donald's view) rather than as a stand-alone weed control technique.

But is it less expensive than other control methods? Despite the pleasure that can be derived from mowing, does it make economic sense? If the goal is to clean up a site or turf maintenance, the answer is yes. If the goal is weed management, the answer is probably no. Ball (2000) reported that mowing to control broad-leaf weeds was more expensive ($10.00/acre[8]) than 2,4-dichlorophenoxyacetic acid (2,4-D) ($6.30/acre) and nearly equal to Banvel (dicamba) plus 2,4-D ($9.50/acre). A unique adaptation of mowing developed for management of aquatic weeds is discussed in Chapter 20.

[7]I am indebted to P.W. Tipping for this interpretation of his data.

[8]Based on Doane's Agricultural Report newsletter 62(21–6). 1999.

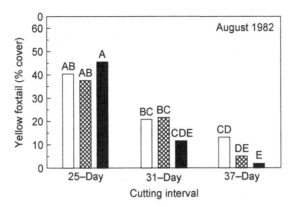

FIGURE 10.3 Percent yellow foxtail cover in relation to cutting frequency and duration of irrigation delay after cutting. Columns with different letters are different at $P = .05$ according to the least significant difference (LSD). *From R.F. Norris, D. Ayres, Cutting interval and irrigation timing in alfalfa: yellow foxtail invasion and economic analysis, Agron. J. 83 (1991) 552–558.*

TABLE 10.8 Alfalfa Dry-Matter Yield in Relation to Cutting Interval (Norris and Ayres, 1991)

	Alfalfa Yield (tons/acre) With Different Cutting Intervals (days)		
Year	25	31	37
1	10.0	12.8	14.9
2	15.0	21.7	24.0
3	11.0	16.2	20.0

3.1.3 Flooding, Salt Water, Draining, and Chaining

These techniques cause ecological change. If a normally dry area is flooded or a normally wet area is drained, ecological relationships are changed and weed species will change. The techniques are effective only when an area is immersed or drained for 3–8 weeks. Immersion, an anaerobic treatment, is not equally effective on all weeds; lowland or paddy rice fields have weeds (e.g., barnyard grass and jungle rice) that survive flooded conditions of the rice paddy as well as rice does. Flooding does not eliminate all weed problems, just some of them, and it creates an environment in which other weeds succeed. Weeds found in lowland rice are generally different from those found in upland rice. Purple nutsedge occurs in both systems. Flooding will control established perennials such as silverleaf nightshade, camel thorn, and the knapweeds in arid areas, but the expense of creating dikes and obtaining water usually make the practice economically unfeasible (Slife, 1981).

Reestablishment of natural flooding in the southwestern United States may be useful as a way to reestablish native cottonwoods. Flooding can be risky because some invasive species such as tamarisk can also be encouraged. Research by Sher et al. (2000) demonstrated that because native cottonwoods were larger and had superior competitive ability, they dominated when historical flooding regimes were restored even in the presence of an invader such as tamarisk, which responds well to disturbance.

Salty ocean water has been shown to be effective for control of mimosa vine and large crabgrass in seashore paspalum and Bermuda grass turf on the island of Guam (Weicko, 2003) but was less effective on yellow nutsedge. The turf was not flooded. Ocean water was applied as an herbicide would be at concentrations up to 55 deciSiemen/m (a measure of electrical conductivity).

Draining is an excellent control for cattails, bulrushes, and reed canary grass, which grow best in wet areas. Draining and flooding are not applicable to most agronomic or horticultural environments, but they should not be forgotten when considering weed management for appropriate sites.

Chaining has been employed on rangelands to destroy emerged vegetation. A large chain similar to a ship's anchor chain is dragged between two bulldozers and uproots sagebrush, rabbitbrush, and other range weeds. Chaining removes emerged growth and completely controls annuals, but not perennials that reproduce vegetatively. The technique is clearly not suited for cropland. Chains are also used to stop passage of weeds in irrigation channels in many countries. Removing collected weeds from the impoundment created by the chain is a labor-intensive, smelly, unpleasant operation.

3.1.4 Two Other Techniques

The Harrington Seed Destructor[9] (HSD), invented by Ray Harrington, a farmer in Western Australia, was first commercialized in 2013. Mr. Harrington recognized that when he harvested wheat, weeds in the crop had mature seeds. Modern combines separate, often smaller weed seeds, most of which are then returned to the land with the chaff and thus maintain the weed population. Harrington developed a separately powered, trailing cage mill (a grinder) to grind weed and other seeds into fine organic material and return it to the land. The seed destructor reduced the population of ryegrass seed returned to the land by 95%–99%. It is a good idea and once again shows that the man on the land, the farmer, is often the best source of innovation. The Seed Destructor has been advocated as a possible solution to the herbicide-resistant problem (Borel, 2014[10]). All seeds including those resistant to herbicides would be destroyed. The HSD, an efficient weed management technique, is also expensive. The Harrington Destructor, grain chaff collection carts, narrow windrow burning, and direct baling of straw have been combined into a harvest weed seed control system to collect weed seed during grain harvest and thereby diminish seed input to the soil seed bank, which Walsh et al. (2013) proposes as a new weed management paradigm for global agriculture.

Application of abrasive materials for weed management has great potential to increase profitability and sustainability of organic production systems. Forcella (2009), who for several years has been an innovative thinker, studied the potential of air-propelled abrasives for selective weed control. Initial work used an abrasive grit made from corn cobs expelled from a sand blaster at 517 kPa pressure. After a microsecond blast from 30 cm, common lamb's-quarter plants were killed whereas corn experience little if any damage. Research in Minnesota (F. Forcella), Illinois (S. Wortman), South Dakota (D. Humberg and S. Clay), New Zealand (A. Rahman), and Spain (M.P. Ruiz) has established the potential of air-propelled grit made from agricultural residues for weed management, especially in organic

[9]See - www.ahri.edu.au/Farmers-and-agronomists/Newsletter/.../Aut-2010-HSD.

[10]http://modernfarmer.com/2014/08/resistant-weeds-pulverize-seeds/.

agricultural systems (see Chapter 11). Forcella et al. (2010) showed that corn gluten meal, an approved organic fertilizer and preemergence herbicide, applied as a grit[11] at 500–750 kPa 30–60 cm away, sufficiently abraded one- to five-leaf seedlings of yellow foxtail to allow a competing crop (e.g., corn) to escape initial competition and suppress the weed. Forcella (2013) also found that corn cob grit did not reduce soybean yield when applied at any growth stage in the field. Application at the cotyledon stage was not recommended.

Wortman (2014, 2015) evaluated corn cob grit, corn gluten meal, green sand fertilizer, walnut shell grit, soybean meal, and bone meal fertilizer for weed control in tomato and pepper. Corn gluten meal, green sand fertilizer, walnut shell, and soybean meal provided the broadest range of postemergence weed control. One application (blast) of corn gluten meal or green sand fertilizer reduced one-leaf stage Palmer amaranth seedling biomass 95% and 100%, respectively, and green foxtail 94% and 87%, respectively.

3.2 Nonmechanical Methods

3.2.1 Hand Pulling

Hand pulling is practical and efficient, especially in gardens, but it is hard work; it is drudgery. It is very effective for annual weeds but not for perennials capable of vegetative reproduction because pulling separates shoots from roots that then produce a new shoot. An important disadvantage is that hand pulling does not get the job done when it is most needed. Most of us are too busy or too lazy to go out and weed before weeds become obvious. By the time they become obvious, easy to grab, and pull, yield reduction resulting from weed competition probably has already occurred. But if it is just your garden, that may not make much difference. Despite its obvious disadvantages, hand pulling can be effective. Persistent hand pulling after mowing or mowing plus herbicide treatments was effective in reducing isolated spotted knapweed infestations in Michigan (MacDonald et al., 2013).

Hand pulling weeds on the ground of the Imperial Palace, Kyoto, Japan.

---

[11]Grits are screened through a 20- to 40-mesh sieve. The average diameter is 0.5 mm.

3.2.2 *Use of Hand-Weeding Tools*

For the home or small-plot vegetable or ornamental plant grower, unwanted, plants out of place (weeds) are a continuing challenge. This is especially true when situations (my small garden), attitudes, or other reasons dictate that herbicides should not be used. The best weed management then is integration of cultural tactics with arduous (no fun at all), sweat-inducing manual control that may be complemented by mechanical control with an array of hoes, weed diggers, weed pullers, weed twisters, weed poppers, weed whips, weed hooks, and others. Some of these devices, e.g., pullers, do not really control weeds; they prevent seed production and dispersal. The Ergonica website (www.ergonica.com) can aid those who agonize as they try to decide what implement is most appropriate for different weed problems. One can choose a circle hoe, push-pull weeders, serrated-edge hoes, oscillating hoes, or even traditional hoes. Table 10.9 is an excerpt from the Ergonica website. It compares physical descriptions, dimensions, and user accounts of operating performance for the Angle Weeder, Weed Hound, Weed Claw, Weed Eezy, Uproot Weeder, Weed Ninja, Weed-Ho, Speedy Weedy, and others. If you encounter some really tough, deep-rooted trees or bushes, you might consider a weed wrench. Several such devices are available (http://www.pullerbear.com/compare.html, http://www.theuprooter.com/tool-features/weed-wrench-alternative/, http://www.canonbal.org/weed.html). I still use my trusty, usually dull but not rusty, old hoe. It keeps my garden fairly clean and me fairly well exercised.

TABLE 10.9   Appropriate Use of Hand-Weeding Tools (https://www.google.com/webhpsourceid=chrome-instant&ion=1&espv=2&ie=UTF-8#q=ergonica).

The Problem	Tools to Consider
Large plants, cacti and tree limbs, tall vines, potato vines, kudzu	Brush-clearing cutters, pruners, saws
Large plants and tree roots	Brush and stump removers, large plant poppers, weed wrench, root talon
Herbaceous and woody perennials, biennials, annuals, tubers (mature, late season), deep roots	Large weed poppers, Ergonica weed twister
Herbaceous perennials, biennials, annuals (mature, late season), shallow roots	Weed pullers, weed poppers, weed twisters, augers
Mature vertical grasses, ground covers	Weed whackers, trimmers, scythes, cutters, flamers
Herbaceous annuals, biennials (small seedlings)	Hoes, hand cultivators, flamers
Prostrate grasses, runners, rhizomes, vines e.g., field bindweed	Hoes, cultivators (limited effectiveness), Ergonica weed twister (crabgrass twister)
Aquatic weeds, seaweed	Pond and lake weeders, lake rakes, hand weed cutters, lake bottom forks, seaweed removers

Hand hoeing weeds in rice in the Philippines.

Hand hoeing has been used for weed control for many years. It is still the method of choice for most gardens and ornamental plantings and is used regularly in many vegetable crops, often by organic farmers. In 2004, California banned weeding of commercial crops by hand.[12] Hand hoeing controls the most persistent perennials if it is done often enough, although it may take years to achieve complete control. Although efficient and widely used, it takes a lot of time and human energy. Some data on the time required to hand weed some crops in several different places are shown in Table 10.10. If human labor is abundant and labor cost is not high, hand pulling or hoeing is an acceptable but arduous method of weed control. If human labor is not abundant and is expensive, hand methods are cost-prohibitive and inefficient.

A solar-powered machine, the Wunda Weeder (https://www.cnet.com/news/wunda-weeder-take-gardening-work-lying-down/), allows a person to pull weeds while lying prone on the carriage. Solar-powered forward and reverse movement is controlled by hand. Clearly, this novel approach is adapted only for small-scale, primarily row-crop vegetable production. A weed-picking or weed-burning robot has been invented and tested by M. Rigsby.[13] The complexity and demands of mechanical weeding (a true robotic weeder) were described by Merfield (2016).

3.2.3 *Heat*

3.2.3.1 FLAMING

Flame weeding is a primary tool in organic agriculture. It is especially useful before emergence in slow-emerging crops (e.g., carrots, parsnips) but limited to a few crops post-emergence (e.g., corn, cotton). Many plant processes are susceptible to high-temperature disruption attributed to coagulation and denaturation of protein, increased membrane permeability, and enzyme inactivation. Photosynthesis usually ceases after thermal

---

[12]Olsen, M. 2004. The end of weeding. Email from postmaster@metrofarm.com, September 30, 2004.

[13]https://hackaday.io/project/4739-weed-picking-robot.

TABLE 10.10 Time Required for Hand Weeding

Crop	Location	Hours per Hectare to Hand Weed
Soybeans	Peru	360 if 6-h day
Transplanted tomatoes	Ohio, United States	71 after herbicide; 133 after cultivation
Corn	Zimbabwe	24—48 for 6-h day
Beans	Wyoming, United States	4.4—15.5 after broadcast herbicide; 32 if no herbicide
Sugar beet	Washington, United States	2—111 after broadcast herbicide; 141 without herbicide
Vegetables	California, United States	10 after broadcast herbicide
Rice	Several	16—500 depending on location and rice culture
Wheat		101
Sorghum		50
Millet		88—298
Cotton		50—700
Jute		140
Groundnut		102—293
Cassava		115—1069

From Newsletter, Int. Weed Sci. Soc. 4 (1) (1979).

disruption of cellular membranes, followed by dehydration. Short of setting fire to an area, heat usually does not kill by combustion. The thermal death point for most plant tissue is between 45 and 55°C (113131°F) after prolonged exposure. Flame temperatures used for weed control approach 2000°F, but flamers may be used selectively when distance from the crop and speed are controlled. Datta and Knezevic's (2013) comprehensive review of flaming is available.

A flamer directs a petroleum-based fuel (usually propane) emitted under pressure. Plant size at treatment influences efficacy much more than plant density. To achieve 90% control of white mustard with one to two leaves required at least 40 kg/ha (36 lb/acre) of propane whereas plants with two to four leaves required 70 kg/ha (62 lb/acre) (Ascard, 1994). For annual dicot weeds, the dose required to control increased with growth stage. Heat-tolerant species cannot be controlled with one flaming regardless of dose (Ascard).

Weeds with unprotected meristematic areas and thin leaves such as common lamb's-quarter, common chickweed, and nettle were completely killed by 20—50 kg/ha of propane when they had less than five true leaves (Ascard, 1995). Shepherd's purse and pineapple weed have protected growing points and were killed by flaming only at very early growth stages. Annual bluegrass could not be killed with a single flaming regardless of its size or the propane rate. Plants with up to four true leaves were killed by 10—40 kg propane ha^{-1} whereas those with 4—12 leaves required 40—50 kg/ha (Ascard, 1995). Sivesind et al. (2009) showed that common lamb's-quarter was 95% controlled with 45 or more kg/ha. Comparable control of redroot pigweed required a higher dose. Shepherd's purse in the cotyledon stage required 59 kg/ha, but more than 159 kg/ha was required when plants had two to five leaves. Control of barnyard grass and yellow foxtail was poor (50% or more) for all maturity stages and propane doses. Cisneros and Zandstra's (2008) research demonstrated similar efficacy. Green foxtail with zero to two emerged leaves was killed when the flamer traveled 2, 4, or 6 km/h. A few survived 8 km/h. Barnyard grass seedlings were more tolerant. Many with up to four emerged leaves survived flaming, but fresh weight 14 days after flaming was lower. Some large crabgrass survived flaming at both growth stages (zero to two and two to four leaves). Common ragweed was most susceptible at the two- to four-leaf stage. Consistent with the work of Sivesind et al. (2009), redroot pigweed and common lamb's-quarter were killed at both growth stages (Cisneros and Zandstra).

Corn between 2 and 12 in. tall cannot withstand flaming. Before corn is 2 in. tall its meristematic region is underground and will regenerate the plant. After 12 in., the flame can be directed at the plant's base and used selectively if the weeds are shorter than the corn. Intensity and duration of exposure are important. If one held a flame on a corn plant for several minutes, the plant would die, so flamers must be kept moving. Speed affects selectivity. Flame has been used selectively in cotton and onions. When cotton stems are three-sixteenths-inch in diameter, or greater, flaming can be used. Flaming kills green shoots where tillage is impractical, such as along a railroad. Buried weed seeds or perennial plant parts are not affected. Dry seeds withstand high temperatures and long exposures because even a thin soil layer insulates and protects most seeds. Burning mature weeds destroys debris but does not prevent crop losses from competition.

Flaming controls some emerged weeds. It has no residue, a problem with many herbicides. Other than during or after rainfall, flaming is not affected by environmental conditions. It may induce erosion by eliminating vegetation that holds soil. A handheld flame cultivator was tested for control of the perennial soft rush, which occurs in established tussocks around cranberry bogs. An open flame reduced soft rush growth after 8 s, whereas infrared required 60 s to achieve similar results (Ghantous and Sandler, 2015). Heat could induce germination of dormant seeds or create conditions favorable for their germination by eliminating emerged, competing plants. This is especially true when brush is burned.

Keeping the flame 2—3 in. above the soil surface eliminates drift and improves efficacy. Keeping the speed between 2 and 4 km/h (1.2—2.4 miles/h) ensures control of most annual weeds (Cisneros and Zandstra, 2008). Flaming can achieve some insect and disease control. An additional advantage is immediate observation of results. Burning is used to eliminate vegetation along irrigation ditches. Despite its advantages and proven success, flaming is not used much except as an important (essential) control technique in organic agriculture. Flaming is not common in nonorganic agriculture because of the cost of propane and other combustible fuels and the success of other methods.

Burning is a useful weed control method. Regular fire has had a significant role in the development and stability of many ecosystems (Hatch et al., 1991). Native plants often depend on regular fires to reduce competition, remove thatch, scarify seeds (break dormancy), and cycle nutrients (Kyser and DiTomaso, 2002). In many grassland and forest communities, fire is not a hazard but a necessary part of community stability. In the absence of periodic natural or planned fires, it may be much more difficult and perhaps impossible to maintain grasslands in a natural state and prevent invasion of weedy species such as yellow star thistle (Kyser and DiTomaso). Burning has been combined successfully with the herbicide clopyralid for management of yellow star thistle in California (DiTomaso et al., 2006). The combination was most effective when burning in the first year was followed by clopyralid the second year. Thermal technologies that expose weed seeds to direct flame are most effective to reduce seed viability (White and Boyd, 2016).

3.2.3.2 SOLARIZATION

It is feasible to use solar heat (solarization) to control weeds. Weed seed germination is suppressed by high soil temperatures and seedlings are killed. Transparent and opaque polyethylene sheets raise soil temperature above the thermal death point for most seedlings and many seeds.

Solarization uses plastic sheets placed on soil, moistened to field capacity, to heat soil by trapping solar radiation just as a greenhouse does (Horowitz et al., 1983). Its effectiveness for weed control depends on a warm, moist climate and intense radiation with long days to raise soil temperature enough to kill weed seeds and seedlings. Moisture increases soil's ability to conduct heat and sensitizes seeds to high temperatures (Horowitz, 1980). Solarization also can control soil-borne diseases and increase crop growth owing to soil warming.

When different types of plastic were used for 4 weeks in Israel, the temperatures under clear plastic exceeded 45°C. Temperatures under black plastic exceeded 40°C about half the time but did not reach 45°C. UV-absorbing transparent plastic raised temperatures above 50°C. At 5 cm, temperatures increased 9°C for black and 19°C for clear plastic.

The effects of solarization on weed emergence were apparent for a short time after plastic was removed. During the first 2 months after removal, the number of emerging annuals was less than 15% of an untreated check and clear plastic was more efficient. Only clear plastic reduced weed populations 1 year after solarization (Horowitz, 1980). Table 10.11 shows some data on the sensitivity of annual weeds to solarization.

In other work, a month after solarization, field bindweed, annual sow thistle, and prostrate pigweed covered 85% of the soil surface in plots not solarized, compared with only 18% in solarized plots (Silveira and Borges, 1984). One week of solarization reduced the percentage of buried seeds of prickly sida, common cocklebur, velvetleaf, and spurred anoda in Mississippi (Egley, 1983). Solarization reduced emergence of all weeds except purple nutsedge.

TABLE 10.11 Sensitivity of Annual Weeds to Solarization (Horowitz et al., 1983)

Weed	Weeks of Solarization to Reduce Seedling Numbers to Less Than 10% of Control
Blue pimpernel	2–4
Bull mallow	>8
Fumitory	6
Heliotrope	4
Horseweed	>8
Pigweeds	2

Total weed emergence was reduced 97% by 1 week after removal of plastic and up to 77% for the season (Egley). Research in Hawaii (Miles et al., 2002) showed a different effect on purple nutsedge tubers. Five weeks of solarization with clear polyethylene film raised the mean soil temperature that was 15 cm deep by 5.8°C in spring and 7.2°C in summer; both increased the final sprouting percentage of purple nutsedge tubers from 74% to 97% in the spring and from 97% to 100% in summer. These increases may seem small (especially only 3% in summer), but because purple nutsedge is such an aggressive weed, complete or increased tuber germination is desirable for more complete control. Solarization has been combined with a green manure crop in a study of annual bluegrass survival (Peachey et al., 2001). Clear polyethylene film (0.6 mil) applied for 53 or 59 days reduced annual bluegrass seed survival 89%–100% in the upper 5 cm of soil but did not affect survival below 5 cm, and may have enhanced it. Green manure, cover crops of barley, rapeseed (canola), and Sudan grass generally increased survival of annual bluegrass seed buried 2.5–15 cm deep. Combining green manure crops and solarization did not improve annual bluegrass control over solarization alone, although solarization significantly improved the efficacy of metam sodium (a soil fumigant) control of annual bluegrass seed.

Solarization with transparent polyethylene was combined with a chicken manure mulch to study the effect on scarified and nonscarified field dodder seed (Haidar et al., 1999). Only seeds on the soil surface were consistently affected. For scarified seed, 95% germination reduction occurred after 10 days under polyethylene. Chicken manure reduced the required period of solarization for nonscarified seed from 6 to 2 weeks, but the effect of manure on total seed germination disappeared after 6 weeks. Solarization for 2–6 weeks with or without chicken manure reduced weed growth in cabbage, but manure increased yield (Haidar and Sidahmed, 2000). Solarization with clear plastic for 60 days during tomato growth killed 95% of branched broomrape seed and induced secondary dormancy in the remaining seed (Mauromicale et al., 2005). In solarized soil, no broomrape shoots emerged and no parasitic attachment to tomato roots was detected. The authors recommended solarization as a good technique for organic farming.

The major effect of high soil temperature (up to 150°F) is killing weed seedlings that germinate under the plastic. Solarization has not been employed on a large scale in field crops but is used effectively in high value vegetable crops in California's Imperial valley. Because there is no cold winter season, solarization is used for 6 weeks before crops are planted. The plastic is removed prior to planting and must be disposed of, a problem all by itself, but solarization nearly eliminates use of herbicides. Solarization has potential to improve weed management but costs, compared to other methods, preclude widespread adoption in other than high value crops.

Research in California has evaluated solarization and biosolarization as alternatives to soil fumigation. The process is known as anaerobic soil disinfestation. Plastic film covers moist soil, which is passively heated by solar radiation during warm weather. These conditions provide fumigant-like reductions in soil-borne pests. Plastic film has been combined with organic amendments including chicken compost, wheat bran, and pomaces (pulpy residue) of processing tomato and white wine grapes. Organic amendments increased soil heating to 2°C for 4 days under aerobic conditions and significantly reduced soil pH during the first week of solarization (Achmon et al., 2016; Simmons et al., 2013). Solarization and composts are both heat-based processes that select for heat-tolerant microbial communities, which tend to be competitors to plant pest organisms (Stapleton and Bañeulos, 2009; Gamliel and Stapleton, 2012).

Research by Campbell's Soup Co. in California used solarization differently (Hoekstra, 1992). Previous research showed the effectiveness of plastic mulch to heat soil and kill weeds. Campbell's studied a solar-powered lens that heated soil and killed weeds. The curved lens is an acrylic sheet made of an array of small lenses. It is cheaper and lighter than glass and not as easily damaged. Use of a lens to concentrate solar energy has two primary disadvantages:

- It does not work on cloudy days, and
- The lens must be pulled slowly over the field to focus energy sufficiently to kill seedlings. Stronger lenses capable of concentrating more energy may enable faster movement. Solar energy is concentrated by a Fresnel lens[14] and directed by manual or automated methods to kill weeds. Devices direct and control energy intensity. Global positioning systems and video cameras may be used for navigation and selection of areas and plants to be exposed. The system is covered under US patent 20090114210 A1.

Steam has been used to sterilize greenhouse and nursery soil for many years. Its use has been limited in the field, especially for weed control. Kolberg and Wiles (2002) studied steam as an alternative weed control method that does not have the disadvantages of herbicides but lacks environmental persistence. Emergence of a few common annual weeds was not affected and control was similar to glyphosate. The amount of steam applied, the speed of application, the weed species, and its growth stage at application determined effectiveness. It is a high-energy technique with potential for soil compaction.

3.2.4 Mulching

Mulching excludes light and prevents shoot growth. Wide mulches are required to subdue perennial weeds that can creep to the edge of a mulch and emerge. Mulches increase soil temperature and may promote better plant growth. Several different materials have been used as mulches: manure, paper (first used on sugarcane in Hawaii), and clear and black plastic. Mulching is common in greenhouses. Mulches are used most in high-value crops grown on small areas and in crops (e.g., sugarcane) where laying the mulch can be mechanized. Hartwig and Ammon (2002) reviewed the status and promise of cover crops and living mulches for vineyards, orchards, and some agronomic crops to evaluate their beneficial effects on soil erosion, nitrogen budgets, weed control, management of other pests, and the environment.

Shredded paper was one of the first mulches used in a crop. It has been replaced by plastic, but use of either is rare in field crops. Pellett and Heleba (1995) evaluated use of chopped paper in perennial nursery crops over two seasons. Their work showed that paper was an effective mulch that provided weed control over two seasons, especially when the paper was wetted and rolled after application. They applied 2.3 or 3.6 kg/m². The higher rate was 15 cm thick. The equivalent rate per hectare was almost 38 tons and the cost of hand application of baled paper was over $2500/ha in Wisconsin. The mulch provided good weed control for 2 years and it was possible to rototill paper into soil. A tackifier (a substance

---

[14]A Fresnel lens is a compact lens developed by Augustin-Jean Fresnel for lighthouses.

to make the paper sticky) was important to prevent paper from blowing away or piling owing to wind. The cost of the paper and its application prohibits consideration of use of paper mulch in all but high-value crops.

As the amount of wheat straw mulch increased in a wheat–corn–fallow dryland production system, weed growth decreased (Crutchfield and Wicks, 1983). Others have shown that planting no-till corn into a desiccated green wheat cover crop reduced morning glory biomass by 79% compared with a nonmulched, tilled treatment (Liebl et al., 1984). Rye mulch successfully reduced biomass of three annual broad-leaved species in three crops (Liebl et al.). Rye has been used successfully as a crop mulch in the fall and winter before corn (Almeida et al., 1984), a practice known as green manuring. Some of rye's contribution to weed control in corn can be attributed to its allelopathic activity. Rye growth was dense enough to require the use of a contact herbicide before corn planting.

Penny and Neal (2003) showed that mulching helped control mulberry weed, an invasive weed of container nurseries and landscapes in the southeastern United States. Light stimulates mulberry weed seed germination; mulches that prevent light penetration effectively prevented seed germination.

Yellow sweet clover residues left after growth ceased provided excellent weed suppression of annual and two perennial weeds (dandelion and perennial sow thistle) in Canada (Blackshaw et al., 2001). Weed suppression was similar whether yellow sweet clover was harvested as hay, grown as a green manure fallow replacement crop, incorporated in soil, or left on the surface as a mulch. Allelopathy was possible.

A mulch compost made from swine bedding residue and manure was tested to determine its effects on corn (Liebman et al., 2004) and soybean (Menalled et al., 2004) yield and weed growth in each crop. The compost consistently increased corn height but had no effect on yield compared with corn grown without swine manure compost but with nitrogen fertilizer. Similarly, the compost did not increase soybean yield, but the competitiveness of common water hemp increased. The authors concluded that if composted swine manure is to be used in corn or soybeans, other effective weed management practices must be included. In these cases, the compost/mulch provided nitrogen fertility that was equally beneficial to the crop and weeds. Meier et al. (2014) determined that when temperature was 57.2°C (137°F) or higher, composting of whole plants of water hyacinth, water lettuce, *Hydrilla*, and giant reed destroyed seed viability of each species. Little et al. (2015) reasoned supplying the proper balance of N, P, and K with manure and organic amendments "is a challenge." Their work showed that overfertilization does not benefit crop yield and high organic fertilization is likely to favor weed growth.

Black polyethylene mulch was about 1.5 times more effective (72% reduction in shoots) than clear polyethylene mulch (46% reduction) for control of yellow nutsedge in Georgia. Neither mulch was effective for control of purple nutsedge (Webster, 2005), which indicated a possible shift to purple nutsedge in mulched vegetable production systems. Anzalone et al. (2010) found that black polyethylene was superior to several biodegradable plastic mulches and several plant mulches for weed management in tomato.

A synthetic, woven black cloth available for mulching is sold commercially in rolls about 6 feet wide and can be applied by machine when trees are planted. It is easy to spread and prevents emergence of most annual weed seedlings. An additional challenge of any cloth or mulch with edges is weeds that emerge at the edges, especially perennials.

Teasdale and Mohler (2000) compared the weed control efficacy of seven mulches: corn stalks, dry plant residue of rye, crimson clover, hairy vetch, white oak and chestnut oak leaves, bark chips and landscape fabric for control of common lamb's-quarter, giant foxtail, redroot pigweed, and velvetleaf. Each was applied at six rates. The densest mulch material was corn stalks, followed by rye, crimson clover, hairy vetch, oak leaves, and, surprisingly, landscape fabric. Regardless of mulch material, the order of sensitivity was redroot pigweed, common lamb's-quarter, giant foxtail and velvetleaf. Weeds that were least well controlled were those whose seedlings were able to grow around obstructing mulch elements.

Oat straw, a flax straw mat, and a nonwoven wool mat were used for weed management in the herbs catnip and common St. John's wort (Duppong et al., 2004). All three natural mulches had fewer weeds than nonweeded plots. Oat straw was the least effective. None of the mulches affected the natural compounds in each plant.

3.2.5 Sound and Electricity

Controlling weeds with high-frequency energy and electricity has been considered since the late 19th century. Ultrahigh frequency (UHF) fields are selectively toxic to plants and seeds. The first use of sound for weed control was patented in 1895. UHF fields produce thermal and nonthermal effects, but thermal effects are the chief source of toxicity. There is a linear and positive correlation between seed water content and susceptibility to electromagnetic energy. Lower frequencies have broken seed dormancy. Commercial weed control devices using UHF fields have been developed, patented, and commercialized, but have not achieved commercial success. There has been limited use for selective vegetation control in cotton and for aquatic weeds. UHF requires a great deal of power but can be used before or after emergence. Postemergence use forces plants to conduct current and in effect boils plant liquids and ruptures cell walls. Vigneault et al. (1990) reviewed what they called electrocution for weed control. They concluded that use of electricity may have a place in high-value, specialty crops such as fine herbs when the treated area is small, no herbicides are available, and cultivation is undesirable because of the potential for root damage and risk of soil erosion. Advantages include the lack of any chemical residue and no soil disturbance.

Electric weed control (Electroherb) (Fig. 10.4) by Zasso CTO in Brazil has been used successfully in citrus, coffee, soybeans, and sugarcane. The technology has been adapted for use in gardening and urban areas and along highways. The manufacturer notes that it successfully controls weeds resistant to glyphosate (e.g., *Conyza* spp., Bengal dayflower [tropical spiderwort], signal grass, and resistant purple nutsedge). It has no preemergence activity and, perhaps surprisingly, it works well in the rain, but pools of water prevent the current from reaching roots. It has no effect on seeds in soil. The technology works well on weeds of all sizes. No plants seem to be resistant, but grasses with large roots require more power.

The usual voltage required ranges from 5000 to 11,000. Voltage is a measure of effect but not the most important one. The real concerns are power (voltage × amperage) and energy (power × time). In Ohm's law ($I = V/R$), I is the current through the conductor (in amperes), V is voltage measured across the conductor, and R is the resistance of the conductor (in ohms), a constant independent of current. The time of exposure and thus the effect is inversely proportional to speed. Another aspect of the technology is the danger of fire if used with dry straw residue. The technology should not be used with high air temperature

FIGURE 10.4   Commercial version of the electrical weed control machine developed and manufactured by the Zasso Group, Piracicaba, Brasil.

and dry plant debris on the ground. Electricity has no effect on soil integrity. The technology provides some control of deep-rooted perennial weeds.

No special protective equipment is needed for the operator. Appropriate instruction for workers is required because touching the electrodes could be fatal. Workers can move about the machine freely, but they should maintain a distance of 2 m if there is a risk of falling on the machine. Similar to all weed control techniques, it is not perfect, but it is encouraging that research continues into new techniques.

3.2.6 Light

Agriculture students are well aware of the role of light in seed germination and photosynthesis. It is not common to think of the difference in plants' light reflectivity as an aid in weed management. Research has demonstrated that different plant surfaces reflect light differently. The difference can be used to distinguish weed from crop plants and to determine whether weeds are present on a particular patch of ground. Optical sensing and optical reflectance (the ratio of red to near-infrared light: 650 versus 750 nm) can be used in weed management (Shropshire et al., 1990). Machines have been developed that use optical reflectance to determine whether a weed is present, and if one is, to turn on an herbicide spray. This reduces the amount of herbicide applied, saves money, and is environmentally good.

3.3 Cultural Weed Management

Knowledge of the unique aspects of the culture (how it is grown) of a crop, the soil, past weed problems, and the environment is an important part of nearly all weed management systems, even when these factors are not recognized. Cultural weed management techniques are especially important in crops when other weed management options are limited or unavailable. They should be included in weed management programs although they ought not to be regarded as solutions to all weed problems. Similarly, despite the outstanding success of herbicides, absolute reliance on them to solve all weed problems is economically and environmentally unfeasible (Gill et al., 1997). Gill et al. provide a complete review of nonmechanical and cultural methods of weed management.

3.3.1 Crop Competition

The techniques of cultural weed control are well-known to farmers and weed scientists. In fact, they are employed regularly but often are not conscious attempts to manage weeds. Planting a crop is a sure way to reduce weed competition because the crops compete with weeds. It is a fundamental method of weed management, but most often cultural weed control just happens rather than being a planned addition to weed management programs. Methods of cultural weed management include conscious use of crop interference, use of cropping pattern, intercropping, soil amendments, and no-tillage or minimum tillage.

Weed scientists have investigated the relative competitiveness of crop cultivars. As reported by Mohler (2001) and reviewed by Zimdahl (2004), "the role of crop genotype in weed management has received growing attention over the past 30 years." There has been attention, but the role of genotype has not been a major area of weed science research. As cited in Mohler (2001), Callaway (1992) reviewed the literature on crop varietal tolerance to weeds and Callaway and Forcella (1993) examined the prospects for breeding crops for improved weed tolerance. There are differences in crop varietal tolerance (often defined as competitive ability) to weeds. Mohler's (2001) Table 6.3 identifies 25 crops in which such differences have been found. For many crops only a few reports are included, but for the major crops (barley, beans, corn, rice, soybean, and wheat) there are many reports (e.g., 14 for soybean). However, despite many years of research and several reports, few crops have been bred to be more competitive (Caton et al., 2001). The reason is that neither weed scientists nor plant breeders know what makes a plant more competitive. No one is sure what traits to select.

Several crops exhibit genotypic differences in competitiveness (Burnside, 1972; Monks and Oliver, 1988). Weed biomass differences up to 45% have been reported among soybean genotypes (Rose et al., 1984). Wild oat competition with wheat was greater than intraspecific competition in wheat. The competitiveness of six wheat cultivars with wild oat was similar for all factors measured (Gonzalez-Ponce, 1988). The most weed-suppressive of 20 winter wheat cultivars reduced weed biomass 82% compared with the least suppressive cultivar (Wicks et al., 1986). With weed interference, the lowest-yielding varieties produced 66% and 54% of the highest-yielding varieties of wheat (Ramsel and Wicks, 1988) and rice (Smith, 1974), respectively. Other work showed that short-stemmed cultivars were more affected by taller wild oats because of light competition (Wimschneider and Bacthaler, 1979). The quest to develop integrated weed management systems has encouraged research on the competitiveness of crop cultivars. That cultivars differ in competitive ability was amply demonstrated several years ago in soybeans (McWhorter and Hartwig, 1972) (Table 10.12). Research in Denmark showed that spring barley varieties vary in weed suppression ability (Christensen, 1995). Dry matter produced by weeds growing with the most suppressive variety was 48% lower than the mean weed dry matter for all varieties, whereas it was 31% higher for the least suppressive variety. Andrew et al. (2015) reviewed the potential for development of competitive serial cultivars. They argue that although cultivars with high competitive potential have

TABLE 10.12 Yield Reduction in Selected Soybean Varieties Owing to Johnsongrass or Cocklebur Competition (McWhorter and Hartwig, 1972)

Soybean Variety	% Yield Reduction With Weed Competition From	
	Johnsongrass	Cocklebur
Davis	34	56
Lee	41	67
Semmes	23	53
Bragg	24	57
Jackson	30	67
Hardee	23	26

been identified, competitiveness has not been a priority for breeding or farmers. However, the challenge of herbicide resistance has renewed interest in cultural weed control options. Plant height, speed of development, canopy architecture, and partitioning of resources are undefined factors that affect, if not define, competitive ability. Andrew et al. advocate developing a simple protocol for assessing the competitive potential of new cultivars. It surely will not be based on a single trait, but will need to capture the combined effect of multiple traits.

It is reasonable to assume that taller, faster-growing crop cultivars are likely to be better competitors, but too little is known about what makes a cultivar competitive and whether it is a trait for which plant breeders can select and develop. Christensen's (1995) work demonstrated no correlation between varietal grain yields in pure stands and competitiveness, which suggests that breeding to optimize yield and competitive ability may be possible. Research has been done to develop crop cultivars that can be bred or managed for high levels of crop interference via high rates of resource uptake or possible allelopathic interference (see Chapter 9) with weeds (Jordan, 1993). A weed suppressive cultivar of rice had been developed in the United States and China. It is highly likely that weed suppressive cultivars of other crop species will be developed.

Alfalfa and other hay crops are often regarded as smother or cleaning crops. Land is not plowed when they are grown, which makes it hard for annuals to establish, but perennial weeds do well in perennial crops such as alfalfa. Planted in dense stands, Sudan grass and some other crops (e.g., sugarcane, vine peas, some clovers) can compete effectively against many, but not all weeds.

Crops can be favored by knowing and using the effect of row width and crop seeding rate. Khan et al. (1996) showed that spring wheat yields were as great as or greater than when earlier than normal seeding or a double seeding rate was used as a substitute for a postemergence herbicide to control foxtail species. Early and middle seeding dates favored the increase of green foxtail over yellow foxtail whereas late seeding favored yellow over green. Spring wheat competing with foxtail had a higher yield when the seeding rate was 270 kg/ha (twice the normal rate) than when it was 130 or 70 kg/ha (half the normal rate) unless the seeding was late. Yenish and Young (2004) demonstrated that the seeding rate of winter wheat in

Washington had a consistent effect on wheat yield, which was about 10% higher when the seeding rate was 60 seeds/m of row as opposed to 40 seeds/m of row when jointed goat grass was the competing weed. Tall wheat varieties competed best. Early, high seeding rates quickly increased crop density and biomass, which suppressed weed growth. Seeding wheat at higher than normal rates in Alberta, Canada improved performance of herbicides used to control wild oats (O'Donovan et al., 2006). Increasing wheat seeding rate from 75 to 150 kg/ha reduced wild oat biomass as much as 18% and the soil seed bank by 46% even when herbicides were not used. On average, wheat yield improved 19% and net economic return 16% with the higher seeding rate. In dryland organic systems in eastern Washington, increased seeding rate increased competitiveness of wheat and oats but did not affect yield. It is a practical weed management strategy.[15]

Decreased weed growth has been observed in narrow (about 8 in.) versus wide (about 30 in.) row spacing in several crops: e.g., 55% in peanuts (Buchanan and Hauser, 1980) and 37% in sorghum (Wiese et al., 1964). In most cases, yield loss was greater when soybeans were grown in 76 cm (30 in.) as opposed to 19 cm (7.5 in.) rows (Hock et al., 2006). Varying row width uses the principles of plant population biology to achieve competitive interactions that favor the crop. Experience has shown that combining narrow row production techniques with minimum tillage and the right herbicide maintains or increases crop yield, reduces soil erosion, and achieves excellent weed management. Manipulating row spacing alone is not always an effective weed management technique. Esbenshade et al. showed that row spacing had little effect on bur cucumber emergence or control in corn (2001a) and soybean (2001b). Tharp and Kells (2001) showed that corn yield was not affected by row spacing and corn population and row spacing did not influence weed emergence after glufosinate application. Common lamb's-quarter biomass was reduced as corn row width was reduced from 76 to 38 cm. In Minnesota, narrow rows (51 versus 76 cm) did not affect late-season weed density but corn grain yield increased in 2 of 3 years (Johnson and Hoverstad, 2002). Other work showed a significant reduction in weed density by careful selection of early-maturing corn hybrids planted in narrow (38-cm) versus wide (76-cm) rows (Begna et al., 2001). Combining narrow rows and high population density increased corn canopy light interception 3%–5% and decreased light available to weeds, which produced five to eight times less biomass. In contrast, Norsworthy and Oliveira (2004) suggested that increasing corn population in the row might be a more effective strategy to reduce weed competition than might decreasing row width. They found that light interception and the critical period for weed control were similar in 48- and 97-cm corn rows and the weed biomass at the end of the season was similar.

While recognizing the potential beneficial effects of row spacing and crop density, Borger et al. (2010) pursued a different aspect of light manipulation. They asked whether row orientation (east–west versus north–south) of five winter grain crops (wheat, barley, canola, lupine, and field pea) would increase shading of the interrow space and affect weed growth and crop yield. Orienting wheat and barley rows east–west reduced weed biomass 51% and 37% and increased grain yield 24% and 26%, respectively. Thus, orienting crop rows to maximize shade between rows and interception of photosynthetically active radiation by crops is a

---

[15]Manuchehri, M. Personal communication, March 2012.

useful weed control technique that has few negative effects, is environmentally friendly, and creates little if any additional cost.

An interesting study of the effect of soil amended with residue of wild radish showed that competitiveness of tomato and bell pepper with yellow nutsedge was enhanced by the weed residue compared with soil with no residue (Norsworthy and Meehan, 2005). This illustrates the previously suspected but undemonstrated potential of weed residue in weed management.

Intercropping is a common, small-scale farming system among farmers of the developing world. It has been practiced throughout the developing world where most farmers and small farms practice an integrated, subsistence farming system that has little flexibility. In Africa, corn, sorghum, and millet are commonly grown with pumpkins, cowpeas, pigeon peas, or beans. Cocoa is grown with yams or cassava. In the American tropics, corn is often grown with beans or beans and squash. In Africa and Latin America, beans or peas climb on corn while pumpkins spread over the ground. Intercropping reduces risk, especially for farmers who have limited access to modern means of weed control (Muhammed and Serkan, 2014). There are lessons in these systems.

The main reasons for mixing crops or planting in close sequence are to maximize land use and reduce risk of crop failure. Intercropping maintains soil fertility, reduces erosion, and may reduce insect problems (Altieri et al., 1983). Intercropping also gives greater yield stability over seasons and may provide yield advantages over single-crop agriculture (Altieri, 1984). The National Agricultural Library published a useful bibliography of citations on green manure and cover crops (MacLean, 1989). The positive and negative effects of *Brassica* cover cropping systems have been reviewed by Haramoto and Gallandt (2004) and Boydston and Al-Khatib (2006).

It is claimed (Altieri et al., 1983; Moody and Shetty, 1981) that one reason for intercropping is weed suppression, but other than work in Nigeria (Chikoye et al., 2001) there has been little experimental evidence to support this conclusion. Similarly, there is little evidence that intercropping requires less weed control. It is assumed that intercropping saves labor because weeding is less critical, and that planting a second crop and weeding the first one can be combined (Norman, 1973). Intercropping's effectiveness for weed control depends on the species combined, their relative proportions, and plant geometry in the field. All reports recommend additional weeding with intercropping, and weeds can often be worse than in sole crops (Moody and Shetty, 1981). Successful use of interseeded cover crops in vegetables has been limited by their tendency to suppress weeds inadequately or to suppress weeds and the crop. For example, winter rye sown in broccoli controlled weeds successfully only when sown at high density, in locations or seasons with low soil temperatures (spring), and when it was combined with other weed management methods (Brainard and Bellinder, 2004). When these conditions were not met, rye was often detrimental to weed management and reduced broccoli yield. Rye sown as a cover crop in soybean reduced total weed density and biomass compared with no cover crop. However, costs were higher and the rye cover crop system was less profitable than soybean grown without a cover crop in which weeds were controlled with conventional technology (Reddy, 2003). Similarly, five levels (0, 0.5, 1.0, 1.5, and 2.0 times the ambient level) of cereal rye residue were combined with five soybean planting densities ranging from 0 to 74 seeds/m^2 (Ryan et al., 2011). Weed biomass decreased with increasing levels of rye residue and weeds were completely suppressed at levels above $1500 \, g/m^2$ at all experimental sites. Weed biomass often, but not always, decreased with increasing soybean density. Combining rye residue and increased plant density resulted in greater weed suppression than was predicted from either method alone. Increasing the soybean planting rate (stand density) can compensate for lower cereal rye biomass. Masiunas (2006) reviewed the several aspects of using rye as a weed management tool in vegetable crops.

The weed control effects of several cover crops were compared in the moist savanna regions of Nigeria (Ekeleme et al., 2003). Weed density was negatively correlated with percent ground cover of five legume cover crops. Only one, lablab (hyacinth bean), produced adequate ground cover and adequate weed suppression in all locations independent of varying duration, distribution, and amount of rainfall. Others were successful in high rainfall regions. Variation between rainfall regions was important. The same variation is observed across the regions of the United States and Europe. No system can be developed that will work equally well in all regions. Other work with cover crops in Nigeria has been successful. For example, 12 months after planting corn, cassava, or a corn—cassava intercrop, plots with cover crops had 52%—71% less cogon grass (a hardy perennial weed that is difficult to control) and 27%—52% more corn grain yield at three locations in Nigeria (Chikoye et al., 2001). The cover crops were *Centrosema pubescens* (Centro) cowpea, hyacinth bean, egusi melon, tropical kudzu, or velvetbean. Cowpea and egusi melon are food crops; the others are green manure crops. Higher crop yield was a result of one or a combination of three things: reduced weed competition from the cover crop, a mulching effect that conserved soil moisture and prevented weed growth, and a contribution of nitrogen from the legume (cowpea, hyacinth bean, and velvetbean) cover crops. It has been demonstrated that cover crops such as hairy vetch can improve corn and soybean productivity, and when they are combined with reduced rates of environmentally benign herbicides, it minimizes the requirements for herbicides (Gallagher et al., 2003).

Weed community composition in California vineyards was affected by nine cover crops (Baumgartner et al., 2008). Weed biomass, community structure, and species diversity in the interrow (between rows) were affected. The presence of cover crops between rows had no influence on weeds in the row. Cover crops had no effect on the vine yield, growth, or nutrition. However, the authors noted that the primary purpose of cover crops between rows was to minimize soil erosion from winter rains, which they did.

Annual intercrops can enhance weed suppression and crop production compared with sole crops. Studies in Canada with wheat—canola and wheat—canola—pea intercropping demonstrated that intercropping tended to provide greater weed suppression compared with sole cropping. There was a synergism of weed suppression among the intercrops compared with any sole crop (Szumigalski and Van Acker, 2005). Studies of intercropping do not confirm that any plant grown with a crop will always provide adequate weed control. Intercropping, which is common in many agricultural systems, should be studied to discover complementary plants, to control soil erosion and prevent or reduce weed growth. It is undoubtedly true that plants that are not crops are classified by most farmers in the developed world as weeds. Other farmers classify noncrop plants in a way that judges their potential use or their effects on soil and crops. Western farmers see noncrop plants as weeds, but subsistence farmers have a different understanding of the use and value of plants that are neither crop nor weed.

A variation on intercropping is the intentional growth of spring-seeded smother plants for weed management. The intent is to eliminate the plants after the crop has become established and is a better competitor, and before the smother plants become competitive, as intercrops often do. Berseem clover, four species of medic, and yellow mustard were planted immediately after corn or soybean in a 25-cm band over the crop row. All species achieved 45% or greater ground cover within 10 weeks of seeding. Yellow mustard grew most rapidly, and it and sava medic resulted in greater weed suppression than other species. When medic was killed 30 days after planting, it reduced weed suppression but did not increase corn yield compared with a season-long presence (Buhler et al., 2001).

Research on these alternative, generally nonchemical systems of weed management is continuing as environmental concerns, sustainability questions, and debate over long-term efficacy of current weed management and crop production systems intensify. They are alternative systems, not panaceas. Weeds will adapt and change as weed management systems change, just as they have adapted to herbicides. Weeds will always be a part of agriculture.

3.3.2 *Planting Date and Population*

Early crop planting may optimize yield because yield may increase when crops have a longer growing season and photosynthesize for more days (Barrett and Witt, 1987). Early planting provides a competitive edge to adapted crop cultivars. Early-season establishment of a crop such as corn provides a competitive advantage against yellow nutsedge, a warm-season weed (Ghafar and Watson, 1983). The competitive advantage could result from the weed's light requirement for growth and from shading by the crop, which emerged first. Choice of planting date should be considered part of integrated weed management, although it may not always be appropriate. Planting date can be important, as illustrated by a 60% reduction in kochia population when proso millet was planted June 1 rather than May 15, although millet yield was not affected (Anderson, 1988). Planting date can also have a role in crop choice. Longspine sandbur emerges from late May to early June in Colorado. It flowers in late July. The seed, which are in its bur, reduces the value of hay. Foxtail millet is planted in early June, and when hay is harvested in late August (Lyon and Anderson, 1993) it may be contaminated with the bur-like seed if longspine sandbur is present. Oats, which are planted in early April and harvested in late June, will not be contaminated because longspine sandbur seed develops later.

Sunflower and safflower are oil crops in the United States Great Plains. Safflower is planted in early April and sunflower in early June. Because of its early planting, over 70% of weed seedlings emerge within 10 weeks of planting safflower. Weeds are easily controlled by tillage or herbicides and sunflower can be planted in a more weed-free field after mid-June in Colorado (Anderson, 1994). Early planting requires weed control for longer. Late planting is usually preceded by tillage that destroys emerged weeds and reduces their population. Advantages gained by later planting are often outweighed by decreased crop yield over a shorter growing season.

In Minnesota, if soybean planting was delayed until early June instead of early May, preplant tillage could be used to control early germinating weeds (Gunsolus, 1990). However, soybean yield will be reduced 10%. When corn planting was delayed from the normal time (beginning of May) until after May 25, maximum yield potential was reduced 25% (Gunsolus). The same study also showed that when either crop was young, rotary hoeing

TABLE 10.13   Effect of Row Width and Cultivation on Yield of Grain Sorghum
(Wiese et al., 1964)

Row Width (cm)	Seeding Rate (kg/ha)	Grain Yield (kg/ha)		Yield Loss (%)
		Weedy	Hand Weeded	
25	5.6	3326	4861	31
	11.2	4188	5466	23
51	5.6	3125	5152	39
	11.2	3987	4715	16
76	5.6	3237	5365	40
	11.2	3606	5029	28
102	5.6	3058	4491	32
	11.2	3203	4637	31

reduced corn population up to 10% but did not affect soybean stand. In Minnesota, a 10% loss in corn population reduced final yield by 2% but there was no similar effect on soybean yield. This small set of data illustrates agriculture's complexity. Extrapolations cannot be made between crops and certainly not between regions. Sweeping generalizations are made but rarely endure.

In a different kind of study about planting date, Khan et al. (1996) reported that crop management practices related to planting date could substitute for herbicide use to control foxtail species in wheat. Spring wheat yields in North Dakota were equal to or greater than when early seeding or a doubled seeding rate were substituted for postemergence foxtail control with an acceptable herbicide. Yield of spring wheat was greater with a high seeding rate (240 lb/acre) than with a normal (116 lb/acre) or low (62 lb/acre) one for early (late April to mid-May) or midseason (middle to late May) seeding, but not for late (early to mid-June) seeding. It is interesting to note how seeding date in this work affected certain weeds. Early and middle seeding dates favored the relative increase of green foxtail whereas the late date favored yellow foxtail. In weed management, as in ecology, one cannot do just one thing.

Planting date is often dictated by considerations other than weed management. Similarly, plant population is dictated by agronomic studies showing the population that gives the best yield. Populations are also determined by row spacings required for planting, cultivating, and harvesting machines. Increasing crop plant populations can often decrease weed density and growth. More than 60 years ago, Wiese et al. (1964) showed how row width and seeding rate interact to reduce competition from weeds in grain sorghum in Texas (Table 10.13). With 25-cm rows, yield loss from weeds was lower with the higher of two seeding rates. This relationship remained true until rows were 102 cm wide.

3.3.3 Companion Cropping and Cover Crops

The agricultural literature does not always make a clear distinction among mulching, green manure, smother crop, living mulch, catch crop, companion cropping, and cover crops

(Teasdale, 2007). Hoffman and Regnier's (2006) review describes several of the similar yet different terms. They note that using green manures as a primary nitrogen source was common agricultural practice in the early 20th century before synthetic nitrogen fertilizer became widely available.

A mulch can be other living plants, dead plants, or synthetic material (e.g., plastic). Some living crops are called companion crops (i.e., something of value is harvested), or one can be planted with no intent of harvesting anything of value. Value may be obtained through incorporating plant residue in soil or reducing soil erosion, or from nitrogen contributed by the noncrop plants. One also encounters living plant mulches, which contribute value as previously but are not to be harvested for value. These can be planted over an entire field or just between crop rows. Thus, some of what is presented subsequently could be included in the foregoing section on mulches or in the section on crop competition, just as some information from those sections could be included here. Independent of the name used, this weed management technique is most effective during the vegetative growth stage of the cover crop. New management skills are required to prevent the cover crop from becoming a weed. They consume water and nutrients, may make crop planting more difficult, may be allelopathic, and can directly compete with the crop. In a comprehensive review, Teasdale et al. (2004) reported that research has shown that most live cover crops that are effective enough to suppress weeds will also suppress the crop. "Thus, the biggest trade-offs between optimizing weed control and enhancing environmental protection occur during the cash-cropping period that follows cover cropping. The weed management skills should therefore focus on enhancing their environmental benefits to the agroecosystem rather than their contribution to weed management." The goal suggested by Teasdale et al. is to replace unmanageable weeds with a more manageable and potentially beneficial cover crop. Perfection and improvement of the cover crop/living mulch technique is a challenge to weed scientists and the "practitioners of sustainable agriculture" (Teasdale and Rosecrance, 2003). The research challenge remains of making a cover crop ecologically and agronomically beneficial, and economically profitable.

An example of the potential problem is found in the work of Chase and Mbuya (2008), who found greater interference from a living mulch than from weeds in organic broccoli production. Dorn et al. (2015) claim that although cover crops are being used more to enhance the sustainability of agro-ecosystems, their suitability for weed management "in integrated and organic conservation tillage systems is still poorly investigated." Their study demonstrated that cover crops are useful for weed management in integrated and organic conservation tillage systems.

The upper-midwest United States has no winter crops and almost no one plants a cover crop because there is no profit. Farmers are aware that their land is bare from October to May and soil erodes in spring during or just after snow melts in March and after rain in May and June. Research has shown that winter *Camelina* (an oilseed crop) and field pennycress (a weed) can be sown in autumn after crop harvest. They germinate immediately after planting and survive the winter. Both form a carpet of rosettes by November, sequester 50–75 kg/ha of nitrogen, and resume growth in spring. Both have high erucic acid levels and are used for industrial purposes. They are harvested in mid to late June, and other crops (e.g., soybean) can be planted in the stubble. They are called cash cover crops that smother spring germinating weeds.

Research indicates that the reasons why cover crops are not used more widely for weed control are that:

- They are most effective early in the season. Other management techniques may be required for acceptable season-long control (Hoffman and Regnier, 2006);
- They do not suppress all weeds effectively;
- Control is often incomplete, inconsistent, or lacking and the duration of control is short compared with other methods (Williams et al., 1998; Teasdale et al., 2005); and
- The cover crop can compete with the cash crop for nutrients, light, or water.

The importance of each of these clear disadvantages might be diminished if cover crop cultivars with improved weed-suppressive ability were developed. Hoffman and Regnier (2006) propose that overall effectiveness of the technique could be improved if the same or different cover crops were planted sequentially during the growing season. The intent would be to achieve weed management and keep the soil covered to prevent erosion.

In contrast, the review by Hartwig and Ammon (2002) describes the many benefits of cover crops and living mulches for crop production. There is no question that cover crop contributions to fertility and weed management make them a good option for alternative agricultural systems. Their effectiveness is usually improved when herbicides are included in the management system (Reddy, 2001; Teasdale, 1996; Yenish and Worsham, 1993; Yenish et al., 1996). Except in organic agriculture (see Chapter 11), herbicides should be considered a complementary technique to optimize the advantages of cover/companion cropping.

Cover crops or living mulches (Akobundu, 1980b) can be used as intercrops or companion plants to suppress weeds (Liebman, 1988; Shetty and Krantz, 1980). For many farming systems, appropriate weed control practices must consider the need to maintain soil fertility and prevent erosion. Open row crops are inimical to prevent soil erosion. Akobundu (1980a) developed integrated low- or no-tillage weed management systems, compatible with more than one crop plant in a field that reduced herbicide use, fertilizer requirements, and soil erosion. Combinations of a legume or egusi melon and sweet potato with corn showed that the companion crops or living mulches maintained corn yield, contributed to nitrogen supply, suppressed weed growth, and reduced soil erosion. Groundnut, Centro, and wild-winged bean have been used as living mulches with corn. Living mulches incorporate the attributes of organic mulch, no-tillage, and weed control. Centro and wild-winged bean grew so vigorously that a growth retardant had to be applied to bands over corn rows to gain a growth advantage for corn (Akobundu, 1980b). In unweeded no-till plots, corn grain yield was 1.6 tons/ha whereas with conventional tillage it was 2.3 tons/ha. Corn yield in unweeded, live mulch plots averaged 2.7 tons/ha. Yields were not different and live mulch plants did not reduce yield; they were complementary, not competitive. Further studies (IITA, 1980) verified these results (Table 10.14).

Clover has been grown successfully with corn and has reduced weed growth (Vrabel et al., 1980). Crimson clover and subterranean clover were the most promising cover crops in cucumbers and peppers in Georgia and contributed to effective management of diseases, nematodes, and insects (Phatak et al., 1991). Sweet corn in a living mulch of white clover had high yields in early years but lower yields later because a contact herbicide used over the corn row allowed invasion of perennial weeds that were not suppressed by white clover (Mohler, 1991). Harrowing before weed emergence reduced weed density up to a third. Over

TABLE 10.14 Effect of Weeding Frequency and Ground Cover on Weed Competition and Maize Yield (IITA, 1980)

Ground Cover	Unweeded Check[a]	
	Weed Dry Weight (ton/ha)	Grain Yield (ton/ha)
Conventional tillage	1.5 a	1.1 e
No-tillage	1.4 a	1.8 b,c,d
Maize stover	1.3 a	1.6 c,d,e
Maize and groundnut	0.3 c	1.3 d,e
Maize and wild winged bean	0.1 c	2.1 a,b,c

[a]*Values in one column followed by the same letter are not statistically different at the 95% level of probability.*

4 years, a white clover cover crop sown after preemergence weed harrowing had the highest yield for oats and wheat compared with an untreated control (Stenerud et al., 2015). Dead rye mulch decreased weed biomass but not corn yield (Mohler). A living mulch of spring-planted rye reduced early season biomass of common lamb's-quarter by 98%, large crabgrass by 42%, and common ragweed by 90% compared with unmulched controls. Barnes and Putnam (1983) also reported that the age of rye when it was killed with herbicides was important to the subsequent emergence of yellow foxtail and lettuce. Peanut and weed response to a cover crop of cereal rye, Italian ryegrass, oats, triticale, or wheat was comparable in terms of peanut yield and weed control (Lassiter et al., 2011).

Companion cropping can be a good weed control technique, but research is needed to determine how appropriate it may be in specific situations. Limited evidence supports the contention that it is effective for weed suppression, builds soil organic matter, reduces soil erosion, and improves water penetration (Andres and Clement, 1984). In some climates when spring soil moisture is limiting, cover or companion crops can deplete moisture and be detrimental to crops despite weed control advantages. Companion crops may also have to be killed before a crop is planted so they will not become competitors.

In Pennsylvania, crown vetch, a legume, was tried as a living mulch in no-tillage corn (Cardina and Hartwig, 1980; Hartwig, 1987). Crown vetch is difficult to establish, but once established it reduces soil erosion and improves fertility through reduced nutrient loss from erosion and by contributing nitrogen and weed control. It usually must be supplemented with herbicides that will not kill it. The system is amenable to rotation of corn with other crops. Work in Ohio (Hoffman et al., 1993) demonstrated use of hairy vetch for weed management (Table 10.15). Unsuppressed hairy vetch reduced weed biomass in corn by 96% in 1 year and by 58% in another. When corn was planted in late April into hairy vetch in the early bud stage of growth, corn yield was reduced by up to 76%. Hairy vetch competition was reduced or eliminated when corn was planted when vetch was in mid- or late-bloom in May or early June. Because of the shortened growing season and competition from hairy vetch, corn planted in May into untreated hairy vetch had a yield similar to corn planted in the no-cover crop, weed-free check. Use of the contact, nonresidual herbicide glyphosate to kill vetch and eliminate competition with corn was helpful with early and

TABLE 10.15   Corn Grain Yield After Planting in Hairy Vetch at Three Growth Stages (Hoffman et al., 1993)

Weed Control Treatment	Corn Grain Yield When Planted Into Hairy Vetch at Growth Stage (kg/ha)		
	Early Bud	Midbloom	Late Bloom
Untreated	130[a] a	7350 b	6520 b
Rolled with water-filled roller	40[a] a	7630 b	7510 b
Mowed with flail chopper	3000[a] a	6830[a] b	5900 b
Glyphosate 2.8 kg/ha	8020[a] a	7700 a	5630 b
Weed-free control	9770 a	8560 a	5310 b

[a]*Values are statistically different from the weed-free control in a column; lowercase letters indicate statistical differences across a row.*

midbloom planting but not with late planting because of the lack of continuing weed control. A disadvantage of hairy vetch and other cover crops is that high levels of their residue (effectively a mulch) can intercept and potentially reduce the efficacy of herbicides (Teasdale et al., 2003). A cover crop of hairy vetch was more effective than *Phacelia* and white mustard for weed management in tomato in central Italy (Campiglia et al., 2015).

In Wisconsin, spring-planted winter rye has been a successful living mulch for weed control in soybean (Ateh and Doll, 1996). A system employing just rye for weed control reduced weed shoot biomass from 60% to 90% over 3 years. Rye provided good weed control and did not reduce soybean yield when weed rye density was low and ground cover from the mulch and soil moisture were adequate for growth. Rye interference with soybean was minimal when rye was killed within 45 days after soybean planting. Rye's success as a living mulch in soybean was affirmed by Liebl et al. (1992). Control of four common annual broad-leaf weeds and giant foxtail was 90% or more and superior to corn stalk residue 5 weeks after planting.

Other successful companion crops have been low-growing plants such as cowpea and mung bean (data from India) (Shetty and Rao, 1981). Seed costs of companion plants and expected competition with the crop were offset by the value of companion plant yield, a more permanent soil cover (less erosion), reduced nitrogen fertilizer requirement, and the cost of hand weeding. The effectiveness of some other plants as cover crops to manage noxious weeds such as cogon grass in India, Malaysia, Nigeria, and Kenya has been studied (Vayssierre, 1957). The smothering effect of velvetbean on cogon grass in corn was equivalent to 1.8 kg/ha of glyphosate but less than that of imazapyr at 0.5 kg/ha in Nigeria (Udensi et al., 1999). The work suggests that planting velvetbean to manage cogon grass may be a "better alternative for farmers without the resources to purchase herbicides." Spear grass, also known as cogon grass, is difficult to control, especially in the tropics. Integration of deep ridging, deep hoe weeding (20–40 cm), and shading was more effective than the farmers' practice of hoeing and digging rhizomes. The most effective suppression of spear grass and highest corn yields were achieved when creeping varieties of cowpea were planted with maize (Vissoh et al., 2008).

Another example of a weed used to gain interspecific competition is the use of *Azolla* for weed management in lowland rice. *Azolla pinnata*, a free-floating fern, has been used in Asian rice culture because of its symbiotic relationship with *Azolla anabaena*, a nitrogen-fixing blue-green algae. This symbiotic relationship can contribute up to 100 kg of nitrogen/ha. A second use of *Azolla* is for weed control owing to the competitive effect of an *Azolla* blanket over the surface of paddy water. When *Azolla* is used, some farmers can grow rice without adding nitrogen fertilizer. Success of the technique depends on the ability of the farmer to control water supply and on the weed species present. Perennial weeds such as rushes and annuals with strong culms (e.g., barnyard grass) are not suppressed and must be controlled in other ways. Many other weeds are controlled as well.

Azolla has been successful but cannot be universally recommended because there is an increase in labor and required managerial skill. Some land must be devoted to supplying a continuing source of inoculum of *Azolla* for paddies and *Azolla* may complicate other pest problems. In fact, *Azolla* may become a weed.

An interesting twist in companion cropping is the use of genetic engineering to cause a companion crop to self-destruct. A potential problem with companion cropping is that the companion may become a competitor if it is allowed to grow too long or if it becomes too large. Herbicides or tillage may then be required to control the companion crop (Teasdale and Rosecrance, 2003; Reddy et al., 2003). Stanislaus and Cheng (2002) tried to design a cover crop that would self-destruct in response to an environmental cue. If self-destruction could be achieved, no supplemental herbicide or tillage would be required after the cover crop had completed the task of early weed control. They incorporated a heat shock–responsive promoter to direct expression of the ribonuclease barnase, which is extremely toxic to cells. The heat shock–responsive promoter effectively caused heat-regulated plant death and was sufficient to kill the transgenic plants. They concluded that although work with temperature sensitivity showed potential, temperature may not be the best factor to study. Temperature is not a completely reliable environment factor (it is not always warm). Therefore, self-destruction based on photoperiodic sensitivity is a more promising research area.

3.3.4 Crop Rotations

Crop rotation is done for economic, market, and agronomic reasons. Some weeds associate with certain crops more than with others. Barnyard grass and jungle rice are common in rice. Wild oat is common in irrigated wheat and barley but rarely in rice. Nightshades are common in potatoes, tomatoes, and beans; kochia and lamb's-quarter are frequent in sugar beets. Dandelions are common in turf but not in row crops, although without management, dandelions can increase in row crops and in pastures and long-term hay crops such as alfalfa. It is logical to assume that crop rotation will affect weed density, species, and management techniques. It has been viewed as a simple and effective method for managing weeds. However, there is little research evidence to support the effectiveness of crop rotation as a weed management technique. The few studies that have been done show that the effects of crop rotation on weed biomass and yield are minor compared with the effects of weed management in the crop.

Known associations occur because of similarity in crop and weed phenology (i.e., naturally occurring phenomena that recur periodically, e.g., flowering), adaptation to cultural practices (e.g., tillage, mowing, irrigation), similar growth habits (e.g., time to mature or to reach full height), and perhaps most important, resistance or adaptation to imposed weed control methods. When one crop is grown in the same field for many years (monoculture), some weeds, if they are present in the seed bank or simply arrive, will be favored and their populations will increase. Weed—crop associations are not accidental and can be explained. Associations can be changed by rotating crops, altering time of planting, or changing weed control methods. Annual grass weeds can be reduced in small grain crops by growing corn in the rotation and using herbicides selective in corn plus cultivation to control the grasses when corn is grown. The same herbicides and cultivation cannot be used in small grain crops. There is evidence that crop sequence does not have a great influence on species composition in a field. However, DeMol et al. (2015) argue that farmers can use crop sequence to suppress individual species. Gómez et al. (2013) postulate that diversified cropping systems are likely to have high soil microbial biomass and therefore strong potential to increase the rate of seed decay. Their work showed that in a 4-year corn—soybean—oat/alfalfa—alfalfa cropping system versus a conventional corn soybean rotation, giant foxtail seed decay was consistently greater at 2- versus 20-cm depth and was higher in the more diverse rotation. Velvetleaf seed decayed little in comparison. Gomez et al. advocated continued study of weed seed bank dynamics in different cropping systems because of the potential for determining weed management strategy.

A good rotation includes crops that reduce the population and/or deter growth of weeds that are especially troublesome in succeeding crops. Population reduction is accomplished by competition or through use of different weed control techniques in different crops. In Canada, yellow foxtail populations in flax were highest when flax followed oats, lowest after flax, and intermediate after wheat, corn, and sorghum (Kommedahl and Link, 1958). Sugar beets grown after beans in Colorado were always more weed-free than sugar beets grown after sugar beets, barley, or corn (Dotzenko et al., 1969). Beans are cultivated frequently and intensive chemical weed control is practiced. The number of weeds was highest when corn preceded sugar beets and lowest when beans preceded. Barley was intermediate (Table 10.16). In many places, barley is planted in spring before soil temperatures are ideal for germination of most annual weeds. Beans, on the other hand, are planted in late spring and tillage can be used to destroy most summer annual weeds. Brainard et al. (2008) showed that changes in weed seed bank density and composition associated with different commonly used crop rotations over a 3-year period had relatively little effect on weed management and crop yields in field crops. They suggested that effects of these rotations over long periods would be of greater significance. Their comparison of crop yield and weed seed bank size after continuous

TABLE 10.16 Effect of Preceding Crop on Number of Weeds That Germinated in 400-g Sample of Soil After 3 years of Each Sequence (Dotzenko et al.,1969)

Preceding Rotation	Number of				
	Kochia	Pigweed	Annual Grass	Lamb's-quarter	Total
Barley—beets	32	15	18	18	109
Corn—beets	67	44	48	7	166
Beans—beets	16	7	11	9	44

corn and five crop rotations showed no consistent effect of rotation across all rotations. However, rotation always affected weed populations and control. For example, integration of red clover in continuous field corn in a field with a growing seed bank led to greater emergence of summer annual weeds compared with field corn alone. However, integration of red clover in a sweet corn—pea—wheat rotation led to a 96% reduction in seed bank density of winter annuals. Cultivation alone resulted in yield losses in sweet corn (30%) and cabbage (up to 7%), but not in snap beans compared with either half or a full rate of the appropriate herbicide.

Ball and Miller (1990) showed that weed composition varied with cropping sequence among rotations of corn for 3 years, pinto beans for 3 years, or 2 years of sugar beets followed by 1 year of corn (Fig. 10.5). Hairy nightshade seed bank population increased after 3 years of pinto beans, green foxtail increased after 3 years of corn, and the sugar beet—corn sequence caused an increase in kochia. Ball and Miller attributed the differences to the herbicides used in each cropping sequence. Crop cultivation, land preparation time and method, and time of planting and harvest will favor some weeds and discourage others.

Crop rotation regularly changes the crop in each field, soil preparation practices, subsequent soil tillage, and weed control techniques. All of these affect weed populations, and although crops are not commonly rotated to control weeds, the effect of rotation as a determinant of weed problems must be recognized.

The relative dry weight of weeds in four cropping systems in the Philippines is shown in Table 10.17. Two weeds dominated but their relative magnitude in the cropping systems, on the same soil, was different. In a rice—sorghum rotation, itchgrass dominated, but with continuous sorghum, itchgrass nearly disappeared and spiny amaranth dominated. Different cropping systems affect weed populations and favor or deter some species. This is observed in vegetable crops, where intensive cultivation and weed control are regularly practiced and weed populations can be reduced (Roberts and Stokes, 1965).

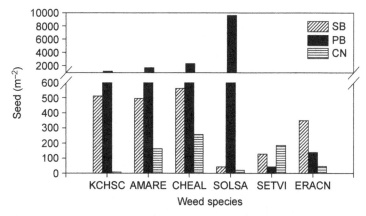

FIGURE 10.5 Influence of cropping sequence on dominant weed species in the soil seed bank 15 cm deep. *AMARE*, redroot pigweed; *CHEAL*, common lamb's-quarters; *CN*, 3 years' corn; *ERACN*, stink grass; *KCHSC*, kochia; *PB*, 3 years' pinto beans; *SB*, 2 years' sugar beets plus 1 year of corn; *SETVI*, green foxtail; *SOLSA*, hairy nightshade. *From D.A. Ball, S.D. Miller, Weed seed population response to tillage and herbicide use in three irrigated cropping sequences, Weed Sci. 38 (1990) 522—527. Reproduced with permission of Weed Sci. Soc. of America.*

TABLE 10.17   Relative Dry Weight of Weeds in Unweeded Plots in Four Cropping Systems 5 weeks After Crop Emergence (Pablico and Moody, 1984)

	(% Dry Weight)	
Cropping System	Spiny Amaranth	Itchgrass
Corn—Corn—Corn	65	21
Rice—corn	42	48
Rice—sorghum	12	83
Sorghum—sorghum—sorghum	95	3

Long-term studies to determine the effect of different cropping sequences on the population dynamics of winter wild oat (Fernandez-Quintanilla et al., 1984) showed that continuous winter cereal cropping (with or without herbicides) increased the winter wild oat soil seed bank from 26% to 80%/year. With spring barley, the soil seed bank declined 10%/year. When sunflower was a summer crop or a 12-month fallow was included in the rotation to prevent new seed production, the soil seed reserve declined 57%—80% annually. There was a great reduction in the size of the soil seed bank of winter wild oats when the cropping program was other than continuous winter cereals (Fernandez-Quintanilla et al., 1984).

Crop rotation has significant effects on the soil seed bank. A 35-year study at two locations in Ohio showed that crop rotation was a more important determinant of soil seed density than moldboard plowing, chisel plowing, or no-tillage, although the two were related (Cardina et al., 2002). Initial seed density was highest with no-tillage and declined as tillage intensity increased. The research showed how weed species composition of the soil seed bank changed in response to crop rotation and soil management and provides leads on how complex plant communities are assembled and endure. Diverse, longer rotations with "more phenologically" diverse crops can reduce the soil weed seed bank population of annual broad-leaf weeds. "Total seedbank density generally increased as tillage was reduced" (Légère et al., 2011). After 18 years, no-tillage had larger seed banks than moldboard and chisel plowing, crop rotation, and cereal monoculture, which the authors concluded confirmed the importance of annual seed production (seed rain) and seed bank management on the sustainability and success of no-tillage systems. A 10-year study in Winnipeg, Canada showed that standard (crop rotation and herbicides) weed management practices reduced weed populations below yield-loss thresholds (Gulden et al., 2011).

Rotations affect weed populations and management requirements. The effects, although qualitatively real, are not quantitatively predictable. Perhaps they never will be, because climate, crop, soil, management skill, etc., are all involved. Agriculture is a highly complicated enterprise in which all effects may never be precisely defined. Westerman et al. (2005) compared effects of a corn—soybean rotation with a corn—soybean—triticale—alfalfa—alfalfa rotation on the population dynamics of velvetleaf. They asked whether "many little hammers" were more than or equally as effective as intensive herbicide weed management. Their conclusion was that diverse rotations that employ multiple stress and mortality factors (little hammers) suppress weeds without reliance on herbicides. Heggenstaller and Liebman (2006) similarly found that corn—soybean rotations that included alfalfa and triticale "facilitated" velvetleaf suppression and reduced herbicide use.

3.3.5 Fertility Manipulation

Manipulation of soil fertility solely to manage weed populations is virtually unknown. However, as is true of most soil manipulations, fertility affects weeds. Walters (1991) suggests that most weeds can be controlled by simple manipulation of soil nutrient levels. His claim is supported by abundant anecdotal evidence, but not by peer-reviewed scientific research. Nevertheless, it should not be dismissed as idle speculation. Farmers fertilize to maximize crop yield and attain greater assurance of success and profit. They do not fertilize or withhold fertilizer to manipulate weed populations. Would research show that fertility manipulation is a viable weed management technique?

Several studies by Blackshaw et al. investigating the effects of fall versus spring fertilizer application on crops and weeds in Alberta and Saskatchewan (Blackshaw et al., 2004a, 2004b, 2005) indicate that the answer is yes. Cropping systems included continuous spring wheat, spring wheat—canola rotation, and a barley—field pea rotation. Weed infestations were created by sowing seeds as single species or mixtures at the beginning of the experiments, which ran for 4 years. In most cases, application of fertilizers (N, P, and/or S, depending on the experiment) when crops were sown in the spring (April or May) maintained or increased crop yields and had a neutral or negative effect on weed biomass relative to treatments in which fertilizer was applied the previous fall. After 4 years, spring rather than fall fertilization lowered weed seed densities in the soil seed bank 21%—24% for the wheat—canola rotation (Blackshaw et al., 2005) but had no effect on weed seed densities for the barley—field pea rotation (Blackshaw et al., 2005). In continuous wheat in which individual weed species were sown separately, spring fertilization reduced soil seed bank densities of wild oat and common lamb's-quarter but had no effect on seed density of green foxtail or wild mustard (Blackshaw et al., 2004b). With desirable or neutral, but not negative, effects on crops and weeds, spring appears to offer advantages over fall fertilization for spring sown crops on the Canadian prairies.

A study by Blackshaw and Brandt (2008) demonstrated that careful management of nitrogen fertilizer affects weed competitiveness and is species dependent. Persian darnel and Russian thistle growth and competitiveness were not influenced by nitrogen rate. On the other hand, redroot pigweed, which is highly responsive to nitrogen, increased growth as nitrogen rate increased. The competitiveness of wild oat was unaffected by nitrogen fertilizer. Their work suggests "that fertilizer management strategies that favor crops over weeds deserve greater attention when weed infestations consist of species known to be highly responsive to higher soil N levels." A similar result on the effect of nitrogen fertilizer on weed emergence and growth was reported by Sweeney et al. (2008). Spring application of nitrogen fertilizer increased inorganic soil nitrogen and weed growth. However, consistent with the results of Blackshaw and Brandt, the effect on weed emergence depended on weed species, seed source, and environmental conditions. In contrast, Wortman et al. (2011) found that soil nitrogen level influenced weed—crop interference in greenhouse studies of competition between velvetleaf and corn, but not in field research. They concluded that the difficulty of controlling nutrient dynamics in the field dictated against using corn fertilization strategy to manage weeds. Other studies led to the opposite conclusion.

Johnson et al. (2007) investigated the effects of delayed fertilizer application on corn and giant ragweed performance in Indiana, comparing a full dose of N fertilizer (200 kg N/ha) at planting with a late fertilizer treatment in which all N was applied at the five- or eight-leaf stage of corn development, and a split fertilizer treatment in which a half-dose of N was applied at corn planting and another half-dose applied at the five- or eight-leaf stage. The crop and weed were grown in mixture at fixed densities, and a weed-free corn treatment was included. Compared with at-planting N application, the late- and split-fertilization treatments increased giant ragweed late season biomass by 83% and 42%, respectively. In contrast, corn grain yield was unaffected by N fertilizer timing, and giant ragweed reduced corn yield by 19% regardless of timing. Thus, from the perspective of crop yield and weed control, at-planting N fertilization was superior. These results and those of Blackshaw et al. (2004b, 2005) indicate that the effects of fertilization are site-, crop-, and weed-specific.

Fertilizer is added to improve crop yield, but weeds are often more competitive with crops at higher nutrient levels (DiTomaso, 1995). When weed density is low, added fertilizer, particularly nitrogen, increases crop yield and makes a crop a more vigorous competitor. However, when weed density is high, it is logical to conclude that added nutrients could favor weed over crop growth. DiTomaso summarized much of the literature on this subject. Crop yield reduction when additional nitrogen fertilizer is added in the presence of weeds is illustrated by the data in Tables 6.8–6.10.

Liebman et al. (2011) studied the use of red clover and alfalfa grown as a green manure crop to supply nitrogen to a succeeding corn crop. Their work holds particular promise for reducing the need for synthetic nitrogen fertilizer and the fossil fuel energy required to manufacture it. They affirmed that both green manure crops have a positive effect on corn yield and the effect is related to their ability to supply nitrogen. Red clover supplied the equivalent of 87–184 kg N/ha, whereas alfalfa contributed 70–121 kg N/ha. Their results are particularly important to those interested in developing sustainable agricultural production systems and to organic producers.

Fertility is a major constraint to food production in sub-Saharan Africa. Research has shown (Ndufa et al., 2009) that planting monocultures of fast-growing, nitrogen-fixing legumes such as *Sesbania*, crotalaria, pigeon pea, siratro, or *Calliandra* in rotation with cereal crops is effective for improving soil fertility. Mixed-species legume fallows have the potential to produce corn yields comparable to and often greater than those obtained with 100 kg N/ha of synthetic fertilizer.

An enduring illustration of the potential of fertility manipulation as a method to change plant populations can be found in the Broadbalk and Park Grass experiments at the Rothamsted Agricultural Experiment Station in England. The Broadbalk experiment began in 1843 to investigate the relative importance of plant nutrients (N, P, K, Na, and Mg) on grain yield of winter wheat. Wheat has been grown every year since 1843. It was not designed to be a weed management experiment. Weed control was accomplished by hoeing and fallowing. Herbicides were introduced in 1964 and have been applied to most of the experimental plots (Moss et al., 2004). Weed surveys did not begin until 1991. The official title of the other long-term experiment is "The Park Grass Experiment on the Effect of Fertilizers and Liming on the Botanical Composition of Permanent Grassland and on the Yield of Hay." Both experiments were begun by Sir John B. Lawes, the son of the manor and founder of Rothamsted, as an agricultural research center, and his colleague J. H. Gilbert. In many ways, the experiments continue in their original form. The Broadbalk and Park Grass experiments are the world's longest-term ecological studies. The Broadbalk's implications for weed management were reviewed by Moss et al. (2004). Tilman et al. (1994) discussed the ecological insights of the Park Grass experiment.

Since 1843, approximately 130 weed species have been found on the Broadbalk site. About 30 are annually recorded. The influence of inorganic nitrogen fertilizer on the frequency of individual species is clear. Common chickweed was favored whereas black medic and field horsetail were not. Black grass and corn poppy populations were not affected by nitrogen. In the Park Grass study, unlimed plots amended with a complete fertilizer containing nitrogen primarily as ammonium sulfate, a pure stand of common velvet grass has developed. It was selected from the original mixture solely by fertility manipulation and lack of lime. It has had one of the heaviest hay yields of any plot, but the hay is unpalatable. With complete fertilizer and lime, plots have one of the heaviest hay yields and a diverse flora, including orchard grass and meadow foxtail. In unlimed plots amended with ammonium sulfate and no phosphorus, the vegetation is completely different from either of the previous ones. If potassium is absent, dandelions are absent because they flourish only with potassium and a pH above 5.6.

In winter wheat, downy brome was least responsive to nitrogen applied during fallow (Anderson, 1991). Nitrogen applied during winter wheat's growing season increased downy brome growth and decreased wheat yield. When crop season rainfall was only 70% of normal, applying nitrogen fertilization reduced winter wheat yield by 12%–28%.

Competition for nutrients is not independent of competition for light and water. The complexity and opportunity of fertility manipulation are well illustrated in work by Liebman (1989) and Liebman and Robichaux (1990). They demonstrated improved weed control because of differing nitrogen use efficiency of crops and weeds (Table 10.18). With no added nitrogen, total crop seed yield was identical for the long-vined Century or short-vined Alaska pea cultivars. Century's yield was 45% greater than Alaska's under these conditions. Adding nitrogen dramatically increased barley yield and reduced yield of Alaska peas. Barley could compete for the added nitrogen, but Alaska peas could not; but they do well with no added N. The seed yield of white mustard increased with nitrogen fertilization. It was much more

TABLE 10.18 Effect of Pea Cultivar and Nitrogen on Seed Yield and Final Aboveground Biomass of White Mustard in Barley–Pea Intercrop

N Treatment/Pea Cultivar	Seed Yield (g/m^2)			Dry Weight (g/m^2)
	Barley	Pea	Total	White Mustard
NO NITROGEN				
Alaska	133 (37)	230 (63)	363	189
Century	16 (5)	334 (95)	350	105
90 + 90 KG/HA				
Alaska	262 (79)	69 (21)	331	1766
Century	204 (33)	406 (67)	610	948

Numbers in parentheses = % total yield (Liebman, 1989).

competitive with short-vined Alaska than with long-vined Century peas. Results of this study were supported by greenhouse research in Canada that showed that green foxtail grown under low nitrogen required approximately six times as much of the herbicide nicosulfuron for control as plants grown under high nitrogen (10 times higher). Higher doses of four herbicides were required to achieve 50% reduction in biomass of redroot pigweed, but there was no similar effect on velvetleaf (Cathcart et al., 2004).

Further evidence of the potential role of soil fertility in weed management is from studies done in Alabama (Hoveland et al., 1976). Soils with low potassium were dominated by buckhorn plantain and curly dock. Soils with low soil phosphorus were dominated by showy crotalaria, morning glory, coffee senna, and sicklepod. The shoot and root growth of several weeds increased with added phosphorus, but the magnitude of the response varied among species. With increasing phosphorus, 17 weed species increased shoot biomass more than wheat and 19 increased shoot biomass more than canola (Blackshaw et al., 2005). The studies that have been done clearly show that manipulating soil nutrient status can change weed populations. Fertility manipulation should be regarded as a potential weed management technique.

THINGS TO THINK ABOUT

1. Why is weed prevention so difficult?
2. Why is weed eradication so difficult?
3. What are the advantages and disadvantages of each weed control method?
4. Why are perennial weeds so hard to control by mechanical methods?
5. What is the principle of carbohydrate starvation?
6. How do timing and type of tillage affect weed presence and weed control?
7. Can mowing really be used to control weeds? How?
8. Where could soil solarization be used?
9. How can living mulches and companion cropping be incorporated into modern cropping systems?
10. What role does crop rotation have in weed management?
11. What role can fertility manipulation have in weed management systems?
12. To what extent does stirring soil promote weed seed germination and thereby make weed control more difficult?
13. How do specific soil conditions (e.g., moisture, crusting, cloddiness, texture) affect weed control by mechanical means?

Literature Cited

Achmon, Y., Harrold, R.D.R., Claypool, J.T., Stapleton, J.J., VanderGheynst, J.S., Simmon, C.W., 2016. Assessment of tomato and wine processing solid wastes as soil amendments for biosolarization. Waste Manag. 48, 156–164.

Akobundu, I.O., 1980a. Live mulch: a new approach to weed control and crop production in the tropics. Proc. Br. Crop Prot. Conf. Weeds 377–380.

Akobundu, I.O., 1980b. Weed science research at the international institute of tropical agriculture and research needs in Africa. Weed Sci. 28, 439–445.

Akobundu, I.O., September 1982. The status and effectiveness of no-tillage cropping at the smallholder farmer level in the developing countries. In: Presented at the U.N./Food & Agric. Org. Expert Consultation on Weed Management Strategies for the 1980's in the Less Developed Countries. FAO, Rome, Italy, 33 pp.

Almeida, F.S., Rodrigues, B.N., Oliveira, V.F., 1984. Influence of winter crop mulches on weed infestation in maize. In: Proc. Third Eur. Weed Res. Soc. Symp. on Weed Problems in the Mediterranean Area, pp. 351–358.

Altieri, M.A., 1984. The ecological role of weeds in agroecosystems with special emphasis on crop-weed-insect interactions. In: Proc. Third Eur. Weed Res. Soc. Symp. on Weed Problems in the Mediterranean Area, pp. 359–364.

Altieri, M.A., Letourneau, D.K., Davis, J.R., 1983. Developing sustainable agroecosystems. Bioscience 33, 45–410.

Anderson, R.L., 1988. Kochia infestation levels in proso millet as affected by planing date. Res. Prog. Rpt. West. Soc. Weed Sci. 292–293.

Anderson, R.L., 1991. Timing of nitrogen application affects downy brome (*Bromus tectorum*) growth in winter wheat. Weed Technol. 5, 582–585.

Anderson, R.L., 1994. Characterizing weed community seedling emergence for a semiarid site in Colorado. Weed Technol. 8, 245–249.

Andres, L.A., Clement, S.L., 1984. Opportunities for reducing chemical inputs for weed control. In: Bexdicek, D.F., Power, J.F. (Eds.), Organic Farming: Current Technology and its Role in a Sustainable Agriculture, Am. Soc. Agron. Spec. Pub., vol. 46. Am. Soc. Agron., Madison, WI, pp. 129–140.

Andrew, I.K.S., Strokey, J., Sparkes, D.L., 2015. A review of the potential for competitive cereal cultivars as a tool in integrated weed management. Weed Res. 55, 239–248.

Anzalone, A., Cirujeda, A., Aibar, J., Pardo, G., Aragoza, C., 2010. Effect of biodegradable mulch materials on weed control in processing tomatoes. Weed Technol. 24, 369–377.

Ascard, J., 1994. Dose-response models for flame weeding in relation to plant size and density. Weed Res. 34, 377–385.

Ascard, J., 1995. Effects of flame weeding in weed species at different developmental states. Weed Res. 35, 397–411.

Ateh, C.M., Doll, J.D., 1996. Spring-planted winter rye (*Secale cereale*) as a living mulch to control weeds in soybean (*Glycine max*). Weed Technol. 10, 347–353.

Ball, D.A., Miller, S.D., 1990. Weed seed population response to tillage and herbicide use in three irrigated cropping sequences. Weed Sci. 38, 522–527.

Ball, J., 2000. How Much Does it Cost to Mow Weeds? http://www.noble.org/ag/soils/mowweeds/index.htm.

Barnes, J., Barnes, J., 1960. Weeds and Weed Seeds, Common, Noxious, Poisonous; with Commonly Used Crop Seeds. Warrex Co., Chicago, IL, 96 pp.

Barnes, J.P., Putnam, A.R., 1983. Rye residues contribute weed suppression in no-tillage cropping systems. J. Chem. Ecol. 9, 1045–1057.

Barrett, M., Witt, W.W., 1987. Alternative pest management practices. In: Helsel, Z.R. (Ed.), Energy in Plant Nutrition and Pest Control. In: Stout, B.A. (Ed.), Energy in World Agriculture, vol. 2. Elsevier Press, The Netherlands, pp. 197–234.

Baumgartner, K., Steenwerth, K.L., Veilleux, L., 2008. Cover-crop systems affect weed communities in a California vineyard. Weed Sci. 56, 596–605.

Begna, S.H., Hamilton, R.I., Dwyer, L.M., Stewart, D.W., Cloutier, C., Assemat, L., Foroutan-pour, K., Smith, D.L., 2001. Weed biomass production response to plant spacing and corn (*Zea mays*) hybrids differing in canopy architecture. Weed Technol. 15, 647–653.

Blackshaw, R.E., Beckie, H.J., Molnar, L.J., Entz, T., Moyer, J.R., 2005. Combining agronomic practices and herbicides improves weed management in wheat-canola rotations within zero tillage production systems. Weed Sci. 53, 528–535.

Blackshaw, R.E., Moyer, J.R., Doram, R.C., Boswell, A.L., 2001. Yellow sweetclover, green manure, and its residues effectively suppress weeds during fallow. Weed Sci. 49, 406–413.

Blackshaw, R.E., Brandt, R.N., Janzen, H.H., Entz, T., 2004a. Weed species response to phosphorus fertilization. Weed Sci. 52, 406–412.

Blackshaw, R.E., Molnar, L.J., Janzen, H.H., 2004b. Nitrogen fertilizer timing and application method affect weed growth and competition with spring wheat. Weed Sci. 52, 614–622.

Blackshaw, R.E., Brandt, R.N., 2008. Nitrogen fertilizer rate effects on weed competitiveness is species dependent. Weed Sci. 56, 743–747.

Borger, C.P.D., Hashem, A., Pathan, S., 2010. Manipulating crop row orientation to suppress weeds and increase crop yield. Weed Sci. 58, 174–178.

Boydston, R.A., Al-Khatib, K., 2006. Utilizing Brassica cover crops for weed suppression in annual cropping systems. In: Singh, H.P., Batish, D.R., Kohli, R.K. (Eds.), Handbook of Sustainable Weed Management. Food Products Press, Binghamton, NY, pp. 77–94.

Brainard, D.C., Bellinder, R.R., Hahn, R.R., Shah, D.A., 2008. Crop rotation, cover crop, and weed management effects on weed seedbanks and yields in snap bean, sweet corn and cabbage. Weed Sci. 56, 434–441.

Brainard, D.C., Bellinder, R.R., 2004. Weed suppression in a broccoli-winter rye intercropping system. Weed Sci. 52, 281–290.

Brecke, B.J., Stephenson IV, D.O., Unruh, J.B., 2005. Control of purple nutsedge (*Cyperus rotundus*) with herbicides and mowing. Weed Technol. 19, 809–814.

Brosten, D., Simmonds, B., May 1989. Crops gone wild. Agrichem. Age, pp. 6, 7, 26, 28.

Buchanan, G.A., Hauser, E.W., 1980. Influence of row spacing and competitiveness on yield of peanuts (*Arachis hypogaea*). Weed Sci. 28, 401–4010.

Buckley, N.G., 1980. No-tillage weed control in the tropics. In: Akobundu, I.O. (Ed.), Weeds and Their Control in the Humid and Subhumid Tropics, Int. Inst. Trop. Agric. Proc. Ser. No. 3, pp. 12–21 (Ibadan, Nigeria).

Buhler, D.D., 1992. Population dynamics and control of annual weeds in corn (*Zea mays*) as influenced by tillage systems. Weed Sci. 40, 241–248.

Buhler, D.D., Kohler, K.A., 1994. Tillage in the dark and emergence of annual weeds. Proc. North Cent. Weed Sci. Soc. 49, 142.

Buhler, D.D., Kohler, K.A., Foster, M.S., 2001. Corn, soybean, and weed responses to spring-seeded smother plants. J. Sustain. Agric. 18, 63–79.

Buhler, D.D., Oplinger, E.S., 1990. Influence of tillage systems on annual weed densities and control in solid-seeded soybean (*Glycine max*). Weed Sci. 38, 158–165.

Buhler, D.D., Stoltenberg, D.E., Becker, R.L., Gunsolus, J.L., 1994. Perennial weed populations after 14 years of variable tillage and cropping practices. Weed Sci. 42, 205–2010.

Burnside, O.C., 1972. Tolerance of soybean cultivars to weed competition and herbicides. Weed Sci. 20, 294–297.

Burnside, O.C., Wilson, R.G., Weisberg, S., Hubbard, K.G., 1996. Seed longevity of 41 weed species buried 17 years in eastern and western Nebraska. Weed Sci. 44, 74–86.

Callaway, M.B., 1992. A compendium of crop varietal tolerance to weeds. American. J. Alt. Agric. 7, 169–180.

Callaway, M.B., Forcella, F., 1993. Crop tolerance to weeds. In: Callaway, M.B., Francis, C.A. (Eds.), Crop Improvement for Sustainability. Univ. of Nebraska Press, Lincoln, NE, pp. 100–131.

Campiglia, E., Radicetti, E., Mancinelli, R., 2015. Covercrops and mulches influence weed management and weed flora composition in strip-tomato (*Solanum lycopersicum*). Weed Res. 55, 416–425.

Cardina, J., Herms, C.P., Doohan, D.J., 2002. Crop rotation and tillage systems effects on weed seedbanks. Weed Sci. 50, 448–460.

Cardina, J., Hartwig, N.L., 1980. Suppression of crownvetch for no-tillage corn. Proc. Northeast Weed Sci. Soc. 34, 53–58.

Cathcart, R.J., Chandler, K., Swanton, C.J., 2004. Fertilizer nitrogen rate and the response of weeds to herbicides. Weed Sci. 52, 291–296.

Caton, B.P., Foin, T.C., Hill, J.E., Mortimer, A.M., 2001. Measuring crop competitiveness and identifying associated traits in cultivar field trials. In: 18th Asian Pacific Weed Sci. Soc. Beijing, P.R. China, pp. 139–145.

Chase, C.A., Mbuya, O.S., 2008. Greater interference from living mulches than weeds in organic broccoli production. Weed Technol. 22, 280–295.

Chikoye, D., Ekeleme, F., Udensi, U.E., 2001. Cogongrass suppression by intercropping cover crops in corn/cassava systems. Weed Sci. 49, 658–667.

Christensen, S., 1995. Weed suppression ability of spring barley varieties. Weed Res. 35, 241–247.

Cisneros, J.J., Zandstra, B.H., 2008. Flame weeding effects on several weed species. Weed Technol. 22, 290–295.

Crutchfield, D.A., Wicks, G.A., 1983. Effect of wheat mulch level on weed control in ecofarming corn production. Weed Sci. Soc. Am. Abstr. 1.

Datta, A., Knezevic, S.Z., 2013. Flaming as an alternative weed control method for conventional and organic agronomic crop production systems: a review. Adv. Agron. 118, 399–428.

DeMol, F., von Redwitz, C., Gerowitt, B., 2015. Weed species composition of maize fields in Germany as influenced by site and crop sequence. Weed Res. 55, 574–585.

DiTomaso, J., 1995. Approaches for improving crop competitiveness through manipulation of fertilization strategies. Weed Sci. 43, 491–497.

DiTomaso, J.M., Kyser, G.B., Miller, J.R., Garcia, S., Smith, R.F., Nader, G., Connor, J.M., Orloff, S.B., 2006. Integrating prescribed burning and clopyralid for the management of yellow starthistle (*Centaurea solstitialis*). Weed Sci. 54, 757–767.

Donald, W.W., 2006. Mowing for weed management. In: Singh, H.P., Batish, D.R., Kohli, R.K. (Eds.), Handbook of Sustainable Weed Management. Food Products Press, Binghamton, NY, pp. 329–372 (Chapter 12).

Donald, W.W., 1990. Primary tillage for foxtail barley (*Hordeum jubatum*) control. Weed Technol. 4, 318–321.

Dorn, B., Jossi, W., Van der Heijden, M.G.A., 2015. Weed suppression by cover crops: comparative on-farm experiments under integrated inorganic conservation tillage. Weed Res. 55, 586–597.

Dotzenko, A.D., Ozkan, M., Storer, K.R., 1969. Influence of crop sequence, nitrogen fertilizer, and herbicides on weed seed populations in sugar beet fields. Agron. J. 61, 34–37.

Duiker, S.W., Haldemann Jr., J.F., Johnson, D.H., 2006. Tillage x maize hybrid interactions. Agron. J. 98, 436–442.

Duppong, L.M., Delate, K., Liebman, M., Horton, R., Romero, F., Kraus, G., Petrich, J., Chowdbury, P.K., 2004. The effect of natural mulches on crop performance, weed suppression and biochemical constituents of catnip and St. John's Wort. Crop Sci. 44, 861–869.

Egley, G., 1983. Weed seed and seedling reductions by soil solarization with transparent polyethylene sheets. Weed Sci. 31, 404–4010.

Ekeleme, F., Akobundu, I.O., Fadayomi, R.O., Chikoye, D., Abayomi, Y.A., 2003. Characterization of legume cover crops for weed suppression in the moist savanna of Nigeria. Weed Technol. 17, 1–13.

Esbenshade, W.R., Curran, W.S., Roth, G.W., Hartwig, N.L., Orzolek, M.D., 2001a. Effect of row spacing and herbicides on burcucumber (*Sicyos angulatus*) control in herbicide-resistant corn (*Zea mays*). Weed Technol. 15, 348–354.

Esbenshade, W.R., Curran, W.S., Roth, G.W., Hartwig, N.L., Orzolek, M.D., 2001b. Effect of tillage, row spacing, and herbicide on the emergence and control of burcucumber (*Sicyos angulatus*) in soybean (*Glycine max*). Weed Technol. 15, 229–235.

Faulkner, E.H., 1943. Plowman's Folly and a Second Look. Univ. of Oklahoma Press, Norman, OK. Reprinted by Island Press in 1987. Plowman's Folly 161 pp.

Fennimore, S.A., Slaughter, D.C., Siemens, M.C., Leon, R.G., Saber, M.N., 2016. Technology for automation of weed control in specialty crops. Weed Technol. 30, 823–837.

Fennimore, S.A., Jackson, L.E., 2003. Organic amendment and tillage effects on vegetable field weed emergence and seedbanks. Weed Technol. 17, 42–50.

Fernandez-Cornejo, J., Hallahan, C., Nehring, R., Wechsler, S., Grube, A., 2012. Conservation tillage, herbicide use, and genetically engineered crops in the United States: the case of soybeans. AgBioForum J. Agrobiotechnol. Manag. Econ. 15 (3), 231–241.

Fernandez-Quintanilla, C., Navarrete, L., Torner, C., 1984. The influence of crop rotation on the population dynamics of *Avena sterilis* (L.) ssp. *ludoviciana Dur.* In: Central Spain. Proc Third Eur. Weed Res. Society Symp. on Weed Problems of the Mediterranean Area, pp. 9–16.

Forcella, F., 2009. Potential of air-propelled abrasives for selective weed control. Weed Technol. 23, 317–320.

Forcella, F., 2013. Soybean seedlings tolerate abrasion from air-propelled grit. Weed Technol. 27, 631–635.

Forcella, F., James, T., Rahman, A., 2010. Post-emergence weed control through abrasion with an approved organic fertilizer. Renew. Agric. Food Syst. 26 (1), 31–37.

Friedrich, T., Derpsch, R., Kassam, A., 2012. Overview of the global spread of conservation agriculture. special issue J. Field Actions 6, 1–7 (On-line only).

Froud-Williams, R.J., Chancellor, R.J., Drennan, D.S.H., 1981. Potential changes in weed flora associated with reduced-cultivation systems for cereal production in temperate regions. Weed Res. 21, 99–1010.

Gallagher, R.S., Cardina, J., Loux, M., 2003. Integration of cover crops with postemergence herbicides in no-till corn and soybean. Weed Sci. 51, 995–1001.

Gamliel, A., Stapleton, J.J., 2012. Combining organic amendments and solarization for soil disinfestation. In: Gamliel, A., Katan, J. (Eds.), Soil Solarization: Theory and Practice. APS Press, Minneapolis, MN, pp. 109–120.

Ghafar, Z., Watson, A.K., 1983. Effect of corn (*Zea mays*) seeding date on the growth of yellow nutsedge (*Cyperus esculentus*). Weed Sci. 31, 572–575.

Ghantous, K.M., Sandler, H.A., 2015. Hand-held flame cultivators for spot treatment control of soft rush (*Juncus effusus*). Weed Technol. 29, 121–127.

Gill, K.S., Arshad, M.A., Moyer, J.R., 1997. Cultural control of weeds. In: Pimentel, D. (Ed.), Techniques for Reducing Pesticide Use: Economic and Environmental Benefits. J. Wiley & Sons, New York, NY, pp. 237–275.

Gill, N.T., 1938. The viability of weed seeds at various stages of maturity. Ann. Appl. Biol. 25, 447–456.

Gómez, R., Liebman, M., Munkvold, G., 2013. Weed seed decay in conventional and diversified cropping systems. Weed Res. 54, 13–25.

Gonzalez-Ponce, R., 1988. Competition between *Avena sterilis* ssp. *macrocarpa Mo.* and cultivars of wheat. Weed Res. 28, 303–307.

Groves, R.H., 1986. Plant invasions of Australia: an overview. In: Groves, R.H., Burdin, J.J. (Eds.), Ecology of Biological Invasions and Australian Perspective. Australian Acad. of Sci., Canberra, pp. 137–1410.

Gulden, R.H., Lewis, D.W., Froese, J.C., van Acker, R.C., Martens, G.B., Entz, M.H., Derksen, D.A., Bell, L.S., 2011. The effect of rotation and in-crop weed management on the germinable weed seedbank after 10 years. Weed Sci. 59, 553–561.

Gunsolus, J.L., 1990. Mechanical and cultural weed control in corn and soybeans. Am. J. Altern. Agric. 5, 114–1110.

Haidar, M.A., Sidahmed, M.M., 2000. Soil solarization and chicken manure for the control of *Orobanche crenata* and other weeds in Lebanon. Crop Prot. 19, 169–173.

Haidar, M.A., Iskandarani, N., Sidahmed, M., Baalbaki, R., 1999. Response of field dodder (*Cuscuta campestris*) seeds to soil solarization and chicken manure. Crop Prot. 18, 253–258.

Haramoto, E.R., Gallandt, E.R., 2004. Brassica cover cropping for weed management: a review. Renew. Agric. Food Syst. 19, 187–198.

Hartmann, K.M., Nezadal, W., 1990. Photocontrol of weeds without herbicides. Naturwissenschaften 77, 158–163.

Hartwig, N.L., 1987. Crownvetch and No-tillage Crop Production for Soil Erosion Control. Pennsylvania State Univ. Coop Ext. Serv, 8 pp.

Hartwig, N.L., Ammon, H.U., 2002. Cover crops and living mulches. Weed Sci. 50, 688–699.

Hatch, D.A., Bartolom, J.W., Hillyard, D.S., 1991. Testing a management strategy for restoration of California's native grasslands. In: Proc. Symposium on Natural Areas and Yosemite: Prospects for the Future. US Nat. Park Serv., Denver, CO, pp. 343–349.

Heinen, E.T., Ahmad, S.H., 1964. Water hyacinth control on the Nile river. Sudan Publ. Inf. Proc. Center, Dep. Agric., Khartoum, 56 pp.

Heggenstaller, A., Liebman, M., 2006. Demography of *Abutilon theorphrasti* and *Setaria faberii* in three crop rotation systems. Weed Res. 46, 138–151.

Hock, S.M., Knezevic, S.Z., Martin, S.R., Lindquist, J.L., 2006. Soybean row spacing and weed emergence time influence weed competitiveness and competitive indices. Weed Sci. 54, 38–46. From the Winona.

Hoekstra, B., June 13, 1992. Killer lens blasts California's weeds. New Sci. 21.

Hoffman, M.L., Regnier, E.E., 2006. Contributions to weed suppression from cover crops. In: Singh, H.P., Batish, D.R., Kohli, R.K. (Eds.), Handbook of Sustainable Weed Management. Food Products Press, Binghamton, NY, pp. 51–75.

Hoffman, M.L., Regnier, E.E., Cardina, J., 1993. Weed and corn (*Zea mays*) responses to a hairy vetch (*Vicia villosa*) cover crop. Weed Technol. 7, 594–5910.

Holland, J.M., 2004. The environmental consequences of adopting conservation tillage in Europe: reviewing the evidence. Agric. Ecosyst. Environ. 103, 1–25.

Holm, L.G., Plucknett, D.L., Pancho, J.V., Herberger, J.P., 1977. The World's Worst Weeds: Distribution and Biology. Univ. of Hawaii Press, Honolulu, HI, 609 pp.

Horowitz, J., Ebel, R., Ueda, K., 2010. "No-till" farming is a growing practice. A report from the USDA economic research service. Econ. Res. Bull. 70, 22 pp.

Horowitz, M., 1980. Weed research in Israel. Weed Sci. 31, 457–460.

Horowitz, M., Regev, Y., Herzlinger, G., 1983. Solarization for weed control. Weed Sci. 31, 170–171.

Hoveland, C.S., Buchanan, G.G., Harris, M., 1976. Response of weeds to soil phosphorus and potassium. Weed Sci. 24, 194–201.

IITA, 1980. Annual Report. Int. Inst. Trop. Agric., Ibadan, Nigeria, 185 pp.

Johnson, G.A., Hoverstad, T.R., 2002. Effect of row spacing and herbicide application timing on weed control and grain yield in corn (*Zea mays*). Weed Technol. 16, 548–553.

Johnson III, W.C., Davis, J.W., 2015. Perpendicular cultivation for improved in-row weed control in organic peanut production. Weed Technol. 29, 128–134.

Johnson, W.G., Ott, E.J., K.D Gibson, Nielsen, R.L., Bauman, T.T., 2007. Influence of nitrogen application timing on low density giant ragweed (*Ambrosia trifida*) interference in corn. Weed Technol. 21, 763–767.

Jordan, N., 1993. Prospects for weed control through crop interference. Ecol. Appl. 3, 84–91.

Kassam, A., Freidrich, T., Derpsch, R., Kienzle, J., 2015. Overview of the worldwide spread of conservation agriculture. J. Field Actions Field Actions Science Reports 8, 12 pp.

Khan, M., Donald, W.W., Prato, T., 1996. Spring wheat (*Triticum aestivum*) management can substitute for diclofop for foxtail (*Setaria* spp.) control. Weed Sci. 44, 362–372.

Klein, R.N., 1988. Economics of winter wheat production. Neb. Ecofarming Conf. 12, 4–47.

Kolberg, R.L., Wiles, L.J., 2002. Effect of steam application on cropland weeds. Weed Technol. 16, 43–49.

Kommedahl, T., Link, A.J., 1958. Ecological effects of different preceding crops on *Setaria glauca* in flax. Proc. Minn. Acad. Sci. 25–26, 91–94.

Kurstjens, D.A.G., Kropff, M.J., Perdok, U.D., 2004. Method for predicting selective uprooting by mechanical weeders from plant anchorage forces. Weed Sci. 52, 123–132.

Kyser, G.B., DiTomaso, J.M., 2002. Instability in a grassland community after the control of yellow starthistle (*Centaurea solstitalis*) with prescribed burning. Weed Sci. 50, 648–657.

Lassiter, B.R., Jordan, D.L., Wilkerson, G.G., Shew, B.B., Brandenburg, R.L., 2011. Influence of cover crops on weed management in strip tillage peanut. Weed Technol. 25, 568–573.

Lati, R.N., Siemens, M.C., Rachuy, J.S., Fennimore, S.A., 2016. Intrarow weed removal in broccoli and transplanted lettuce with an intelligent cultivator. Weed Technol. 30 (3), 655–663.

Légère, A., Stevenson, F.C., Benoit, D.L., 2011. The selective memory of weed seedbanks after 18 years of conservation tillage. Weed Sci. 59, 98–106.

Liebl, R.A., Shilling, D.G., Worsham, D.A., 1984. Suppression of broadleaf weeds by mulch in four no-till cropping systems. In: Abstr. #70 Div. Pest. Chem. 187th Mtg. Am. Chem. Soc., St. Louis, MO.

Liebl, R., Simmons, F.W., Wax, L.M., Stoller, E.W., 1992. Effect of rye (*Secale cereale*) mulch on weed control and soil moisture in soybean (*Glycine max*). Weed Technol. 6, 838–846.

Liebman, M., 1988. Ecological suppression of weeds in intercropping systems: a review. In: Altieri, M.A., Liebman, M. (Eds.), Weed Management in Agroecosystems: Ecological Approaches. CRC Press, Boca Raton, FL, pp. 197–212.

Liebman, M., 1989. Effects of nitrogen fertilizer, irrigation, and crop genotype on canopy relations and yields of an intercrop/weed mixture. Field Crops Res. 22, 83–100.

Liebman, M., Menalled, F.D., Buhler, D.D., Richard, T.L., Sundberg, D.N., Cambardella, C.A., Kohler, K.A., 2004. Impacts of composted swine manure on weed and corn nutrient uptake, growth, and seed production. Weed Sci. 52, 365–375.

Liebman, M., Robichaux, R.H., 1990. Competition by barley and pea against mustard: effects on resource acquisition, photosynthesis and yield. Agric. Ecosyst. Environ. 31, 155–172.

Liebman, M., Grael, R.L., Nettleton, D., Cambardella, C.A., 2011. Use of legume green manures as nitrogen sources for corn production. Renew. Agric. Food Syst. https://doi.org/10.1017/S1742170511000299.

Little, N.G., Mohler, C.L., Ketterings, W.M., DiTommaso, A., 2015. Effects of organic amendments on weed and crop growth. Weed Sci. 63, 710–722.

Lyon, D.J., Anderson, D.L., 1993. Crop response to follow applications of atrazine and clomazone. Weed Technol. 7, 949–953.

MacDonald, N.W., Martin, L.M., Kapolka, C.K., Botting, T.F., Brown, T.E., 2013. Hand pulling following mowing and herbicide treatments increases control of spotted knapweed (*Centaurea stoebe).* Invasive Plant Sci. Manag. 6, 470–479.

Mack, R.N., 1981. Invasion of *Bromus tectorum* L. into western North America: an ecological chronicle. Agroecosystems 7, 145–165.

MacLean, J., 1989. Green Manures and Cover Crops. Quick Bibliog. 89–58. Alternative Farming Systems Information Center, Nat. Agric. Library, Beltsville, MD.

Masiunas, J.B., 2006. Rye as a weed management tool in vegetable cropping systems. In: Singh, H.P., Batish, D.R., Kohli, R.K. (Eds.), Handbook of Sustainable Weed Management. Food Products Press, Binghamton, NY, pp. 127–158.

Mauromicale, G., Monaco, A.L., Longo, A.M.G., Restuccia, A., 2005. Soil solarization, a nonchemical method to control branched broomrape (*Orobanche ramosa*) and improve the yield of greenhouse tomato. Weed Sci. 53, 877—883.

McCarty, M.K., 1982. Musk thistle (*Carduus thoermeri*) seed production. Weed Sci. 30, 441—445.

McWhorter, C.G., Hartwig, E.E., 1972. Competition of johnsongrass and cocklebur with six soybean varieties. Weed Sci. 20, 56—510.

Meier, E.J., Waliczek, T.M., Abbott, M.L., 2014. Composting invasive plants in the Rio Grande river. Invasive Plant Sci. Manag. 7, 473—482.

Menalled, F.D., Liebman, M., Buhler, D.D., 2004. Impact of composted swine manure and tillage on common waterhemp (*Amaranthus rudis*) competition with soybean. Weed Sci. 52, 605—613.

Merfield, C.N., 2016. Robotic weeding's false dawn? Ten requirements for fully autonomous mechanical weed management. Weed Res. 56 (5), 340—344.

Miles, J.E., Kawabata, O., Nishimoto, R.K., 2002. Modeling purple nutsedge sprouting under soil solarization. Weed Sci. 50, 64—71.

Mohler, C.L., 1991. Effects of tillage and mulch on weed biomass and sweet corn yield. Weed Technol. 5, 545—552.

Mohler, C.L., 2001. Enhancing the competitive ability of crops. In: Liebman, M., Mohler, C.L., Staver, C.P. (Eds.), Ecological Management of Agricultural Weeds. Cambridge Univ. Press, New York, NY, pp. 269—320.

Mohler, C.L., 1993. A model of the effect of tillage on emergence of weed seedlings. Ecol. Appl. 3, 53—73.

Monks, D.W., Oliver, L.R., 1988. Interactions between soybean (*Glycine max*) cultivars and selected weeds. Weed Sci. 36, 770—776.

Moody, K., Shetty, S.V.R., 1981. Weed management in intercropping systems. In: Int. Crops Res. Inst. For the Semi-arid Tropics. Proc. Int. Workshop on Intercropping. Hyderabad, India, pp. 229—237.

Moss, S.R., Storkey, J., Cussans, J.W., Perryman, S.A.M., Hewitt, M.V., 2004. The Broadbalk long-term experiment at Rothamsted: what has it told us about weeds? Weed Sci. 52, 864—873.

Muhammed, D., Serkan, A., 2014. Intercropping of legumes with cereal crops in particular with the perennials to enhance forage yields and quality. In: Proceedings of UN/FAO Expert Workshop. Food and Agriculture Organization of the United Nations, Rome, pp. 221—228, 390 pp.

Murphy, S.D., Clements, D.R., Belaoussoff, S., Kevan, P.G., Swanton, C.J., 2006. Promotion of weed species diversity and reduction of weed seedbanks with conservation tillage and crop rotation. Weed Sci. 54, 69—77.

Musick, G.J., Beasley, L.E., 1978. Effect of the crop residue management system on pest problems in field corn (*Zea mays* L.) production. In: Crop Residue Management Systems. Amer. Soc. Agron. Spec. Publ. 31. Am. Soc. Agron., Madison, WI, pp. 173—186.

Myers, M.W., Curran, W.S., Van Gessel, M.J., Majek, B.A., Mortensen, D.A., Calvin, D.D., Karsten, H.D., Roth, G.W., 2005. Effect of soil disturbance on annual weed emergence in the Northeastern United States. Weed Technol. 19, 274—282.

Ndufa, J.K., Gathumbi, S.M., Kamiri, H.W., Giller, K.E., Cadish, G., 2009. Do mixed-species legume fallows provide long-term maize yield benefit compared with monoculture legume fallows? Agron. J. 101, 1352—1362.

Norman, D.W., 1973. Crop mixtures under indigenous conditions in northern part of Nigeria. In: Ofori, I.M. (Ed.), Factors of Agricultural Growth in Africa. Inst. Soc. Econ. Res., Univ. Ghana, Ghana, pp. 130—144.

Norris, R.F., Ayres, D., 1991. Cutting interval and irrigation timing in alfalfa: yellow foxtail invasion and economic analysis. Agron. J. 83, 552—558.

Norsworthy, J.K., Meehan IV, J.T., 2005. Wild radish-amended soil effects on yellow nutsedge (*Cyperus esculentus*) interference with tomato and bell pepper. Weed Sci. 53, 77—83.

Norsworthy, J.K., Oliveira, M.J., 2004. Comparison of the critical period for weed control in wide- and narrow-row corn. Weed Sci. 52, 802—807.

O'Donovan, J.T., Blackshaw, R.E., Harker, K.N., Clayton, G.W., 2006. Wheat seeding rate influences herbicide performance in wild oat (*Avena fatua* L.). Agron. J. 98, 815—822.

Pablico, P., Moody, K., 1984. Effect of different cropping patterns and weeding treatments and their residual effects on weed populations and crop yield. Philipp. Agric. 67, 70—81.

Panetta, F.D., Mitchell, N.D., 1991. Homocline analysis and the prediction of weediness. Weed Res. 31, 273—284.

Peachey, R.E., Pinkerton, J.N., Ivors, K.L., Miller, M.L., Moore, L.W., 2001. Effect of soil solarization, cover crops, and metham on field emergence and survival of buried annual bluegrass (*Poa annua*) seeds. Weed Technol. 15, 81—88.

Pellett, N.E., Heleba, D.A., 1995. Chopped newspaper for weed control in nursery crops. J. Environ. Hort. 13, 77—81.

Penfound, W.T., Earle, T.T., 1948. The biology of the water hyacinth. Ecol. Monographs 18, 447—472.

Penny, G.M., Neal, J.C., 2003. Light, temperature, seed burial, and mulch effects on mulberry weed (*Fatuoa villosa*) seed germination. Weed Technol. 17, 213–218.

Phatak, S.C., Bugg, R.L., Sumner, D.R., Gay, J.D., Brunson, K.E., Chalfant, R.B., 1991. Cover crops effect on weeds, diseases, and insects of vegetables. In: Hargrove, W.L. (Ed.), Proc. Int. Conf. on Cover Crops for Clean Water. Soil and Water Cons. Soc., Ankeny, IA, pp. 153–154.

Phillips, S.H., 1978. No-tillage, past and present. In: Phillips, R.E., Thomas, G.W., Blevins, R.L. (Eds.), No-tillage Res. Res. Rpts. and Rev. Univ. Kentucky, Lexington, pp. 1–6.

Ramsel, R.E., Wicks, G.A., 1988. Use of winter wheat (*Triticum aestivum*) cultivars and herbicides in aiding weed control in ecofallow corn (*Zea mays*) rotation. Weed Sci. 36, 394–398.

Reddy, K.M., 2001. Effects of cereal and legume cover crop residues on weeds, yield, and net return in soybean (*Glycine max*). Weed Technol. 15, 660–668.

Reddy, K.N., 2003. Impact of rye cover crop and herbicides on weeds, yield, and net return in narrow-row transgenic and conventional soybean (*Glycine max*). Weed Technol. 17, 28–35.

Reddy, K.N., Zablotowicz, R.M., Locke, M.A., Koger, C.H., 2003. Cover crop, tillage, and herbicide effects on weeds, soil properties, microbial populations, and soybean yield. Weed Sci. S987–S994.

Renz, M.J., Ditomaso, J.M., 2004. Mechanism for the enhanced effect of mowing followed by glyphosate application to resprouts of perennial pepperweed (*Lepidium latifolium*). Weed Sci. 52, 14–23.

Renz, M.J., Ditomaso, J.M., 1998. The effectiveness of mowing and herbicides to control perennial pepperweed in rangeland and roadside habitats. Proc. Calif. Weed Sci. Soc. 50, 178.

Roberts, H.A., 1963. Studies on the weeds of vegetable crops. III. Effect of different primary cultivations on the weed seeds in the soil. J. Ecol. 51, 83–95.

Roberts, H.A., Stokes, F.G., 1965. V. Final observations on an experiment with different primary cultivations. J. Appl. Ecol. 2, 307–315.

Roberts, H.A., Ricketts, M.E., 1979. Quantitative relationships between the weed flora after cultivation and the seed population in the soil. Weed Res. 19, 269–275.

Rose, S.J., Burnside, O.C., Specht, J.E., Swisher, B.A., 1984. Competition and allelopathy between soybean and weeds. Agron. J. 76, 523–528.

Ruisi, P., Frangipane, B., Amato, G., Badagliacca, G., Di Micelli, G., Plaia, A., Giambalvo, D., 2015. Weed seed bank size and composition in a long-term tillage and crop sequence experiment. Weed Res. 55, 320–328.

Ryan, M.R., Mirsky, S.B., Mortensen, D.A., Teasdale, J.R., Curran, W.S., 2011. Potential synergistic effects of cereal rye biomass and soybean planting density on weed suppression. Weed Sci. 59, 238–246.

Salonen, J., Hyvönen, T., Kaseva, J., Jalli, H., 2012. Impact of changed cropping practices on weed occurrence in spring cereals and Finland - a comparison of surveys in 1997–1999 and 2000–2009. Weed Res. 53, 110–120.

Schweizer, E.E., Westra, P., Lybecker, D.W., 1994. Controlling weeds in corn (*Zea mays*) rows with an in-row cultivator versus decisions made by a computer model. Weed Sci. 42, 593–600.

Schweizer, E.E., Zimdahl, R.L., 1984. Weed seed decline in irrigated soil after six years of continuous corn (*Zea mays*) and herbicides. Weed Sci. 32, 76–83.

Scopel, A.L., Ballare, C.L., Radosevich, S.R., 1994. Photostimulation of seed germination during soil tillage. New Phytol. 126, 145–152.

Sher, A.A., Marshall, D.L., Gilbert, S.A., 2000. Competition between native *Populus deltoides* and invasive *Tamarix ramosissima* and the implications for restablishing flooding disturbance. Conserv. Biol. 14, 1744–1754.

Shetty, S.V.R., Krantz, B.A., 1980. Weed research at ICRISAT. Weed Sci. 28, 451–453.

Shetty, S.V.R., Rao, A.N., 1981. Weed management studies in sorghum/pigeon pea and pearl millet/groundnut intercrop systems. Some observations. In: Int. Crops Res. Inst. for Semi-arid Tropics. Proc. Int. Workshop on Intercropping. Hyderabad, India, pp. 238–248.

Shropshire, G.J., Von Bargen, K., Mortensen, D.A., 1990. Optical reflectance sensor for detecting plants. Opt. Agric. 1379, 222–235.

Shurtleff, W., Aoyagi, A., 1977. The Book of Kudzu. Autumn Press, Brookline, MA, pp. 8–12.

Simmons, C.W., Guoa, H., Claypool, J.T., Marshall, M.N., Peranoa, K.M., Stapleton, J.J., VanderGheynst, J.S., 2013. Managing compost stability and amendment to soil to enhance soil heating during soil solarization. Waste Manag. 33 (5), 1090–1096.

Silveira, H.L., Borges, M.L.V., 1984. Soil solarization and weed control. In: Proc. 3rd Eur. Weed Res. Soc. Symp. on Weed Problems of the Mediterranean Area, pp. 345–349.

Sivesind, E.C., LeBlanc, M.L., Cloutier, D.C., Seguin, P., Stewart, K.A., 2009. Weed response to flame weeding at different development stages. Weed Technol. 23, 438–443.

Slife, F.W., 1981. Environmental control of weeds. In: Pimentel, D. (Ed.), Handbook of Pest Management. CRC Press, Boca Raton, FL, pp. 485–491.

Smith, E.V., Mayton, E.L., 1938. Nutgrass eradication studies. II. The eradication of nutgrass *Cyperus rotundus* L. by certain tillage treatments. J. Am. Soc. Agron. 30, 18–21.

Smith, R.G., 2006. Timing of tillage is an important filter on the assembly of weed communities. Weed Sci. 54, 705–712.

Smith, R.J., 1974. Competition of barnyardgrass with rice cultivars. Weed Sci. 22, 423–426.

Stanislaus, M.A., Cheng, C., 2002. Genetically engineered self-destruction: an alternative to herbicides for cover crop systems. Weed Sci. 50, 794–801.

Stapleton, J.J., Bañeulos, G.S., 2009. Biomass crops can be used for biological disinfestation and remediation of soils and water. Calif. Agric. 63 (1), 41–46.

Stenerud, S., Mangerud, K., Sjursen, H., Torp, T., Brandsæter, L.O., 2015. Effects of weed harrowing and undersown clover on weed growth and spring cereal yield. Weed Res. 55, 493–502.

Suryatna, E.S., 1976. Nutrient Uptake, Insects, Diseases, Labor Use and Productivity Characteristics of Selected Traditional Intercropping Patterns Which Together Affect Their Continued Use by Farmers (Ph.D. thesis). Univ. Philippipines. Los Baños, College, Laguna, Philippines, 130 pp.

Sweeney, A.E., Renner, K.A., Laboski, C., Davis, A., 2008. Effect of fertilizer nitrogen on weed emergence and growth. Weed Sci. 56, 714–721.

Szumigalski, A., Van Acker, R., 2005. Weed suppression and crop production in annual intercrops. Weed Sci. 53, 813–825.

Tabor, P., 1952. Comments on cogon and torpedo grasses. A challenge to weed workers. Weeds 1, 374–375.

Teasdale, J.R., 1996. Contribution of cover crops to weed management on sustainable agricultural systems. J. Prod. Agric. 9 (4), 475–479.

Teasdale, J.R., 2007. Cover crops and Weed Management. In: Upadhyaya, M.K., Blackshaw, R.E. (Eds.), Nonchemical Weed Management. CAB International, pp. 49–64.

Teasdale, J.R., Mohler, C.L., 2000. The quantitative relationship between weed emergence and the physical properties of mulches. Weed Sci. 48, 385–392.

Teasdale, J.R., Shelton, D.R., Sadeghi, A.M., Isensee, A.R., 2003. Influence of hairy vetch residue on atrazine and metolachlor soil solution concentration and weed emergence. Weed Sci. 51, 628–634.

Teasdale, J.R., Pillai, P., Collins, R.T., 2005. Synergism between cover crop residue and herbicide activity on emergence and early growth of weeds. Weed Sci. 53, 521–527.

Teasdale, J.R., Rosecrance, R.C., 2003. Mechanical versus her recital strategies for killing a hairy vetch cover crop and controlling weeds in minimum- tillage corn production. Am. J. Altern. Agric. 18 (2), 95–102.

Teasdale, J.R., Mangum, R.W., Radhakrishnan, J., Cavigelli, M.A., 2004. Weed seed bank dynamics in three organic farming crop rotations. Agron. J. 96, 1429–1435.

Tharp, B.E., Kells, J.J., 2001. Effect of gulfosinate-resistant corn (*Zea mays*) population and row spacing on light interception, corn yield, and common lambsquarters (*Chenopodium album*) growth. Weed Technol. 15, 413–418.

Tilman, D., Dodd, M.E., Silvertown, J., Poulton, P.R., Johnston, A.E., Crawley, M.J., 1994. The Park Grass experiment: insights from the most long-term ecological study. In: Leigh, R.A., Johnston, A.E. (Eds.), Long-term Experiments in Agricultural and Ecological Sciences. CAB International, Wallingford, UK, pp. 287–303.

Tipping, P.W., 2008. Mowing-induced changes in soil seed banks and populations of plumeless thistle (*Carduus acanthoides*) and musk thistle (*Carduus nutans*). Weed Technol. 22, 49–55.

Triplett Jr., G.B., 1982. Tillage and crop productivity. In: Recheigl Jr., M. (Ed.), CRC Handbook of Agricultural Productivity, vol. I. CRC Press Inc., Boca Raton, Florida, pp. 251–262.

Udensi, U.E., Akobundu, I.O., Ayeni, A.O., Chikoye, D., 1999. Management of cogongrass (*Imperata cylindrica*) with velvetbean (*Mucuna pruriens* var. *utilis*) and herbicides. Weed Technol. 13, 201–208.

US Congress, 1993. Office of Technology Assessment, Harmful Non-indigenous Species in the United States. OTA-f-365. US Govt. Printing Office, Washington, D.C., 391 pp.

VanGessel, M.J., Schweizer, E.E., Lybecker, D.W., Westra, P., 1995. Compatibility and efficiency of in-row cultivation for weed management in corn (*Zea mays*). Weed Technol. 9, 754–760.

Vayssierre, P., 1957. Weeds in Indo-Malaya. J. d'Agr Dulture Trop. Botanique Applique 4, 392–401.

Vietmeyer, N.D., 1975. The beautiful blue devil. Nat. Hist. Mag. 84, 64–73.

Vigneault, C., Benoit, D.L., McLaughlin, N.B., 1990. Energy aspects of weed electrocution. Rev. Weed Sci. 5, 15–25.

Vissoh, P.M., Kuyper, T.W., Gbehounou, G., Hounkonnou, D., Ahanchede, A., Röling, N.G., 2008. Improving local technologies to manage speargrass (*Imperata cylindrica*) in southern Benin. Int. J. Pest Management 54 (1), 21–29.

Vrabel, T.E., Minotti, P.L., Sweet, R.D., 1980. Seeded legumes as living mulches in sweet corn. Proc. Northeast. Weed Sci. Soc. 34, 171–175.

Walters, C., 1991. Weeds: Control without Poisons. Acres U.S.A., Metairie, LA, 352 pp.

Walsh, M., Newman, P., Powles, S., 2013. Targeting weed seeds in-crop: a new weed control paradigm for global agriculture. Weed Technol. 27, 431–436.

Webster, T.M., 2005. Mulch type affects growth and tuber production of yellow nutsedge (*Cyperus esculentus*) and purple nutsedge (*Cyperus rotundus*). Weed Sci. 53, 834–838.

Weicko, G., 2003. Ocean water as a substitute for postemergence herbicides in tropical turf. Weed Technol. 17, 788–791.

Wesson, G., Wareing, P.F., 1969. The induction of light sensitivity in weed seeds by burial. J. Exp. Bot. 20, 414–425.

Westerman, P.R., Liebman, M., Menalled, F.D., Heggenstaller, A.H., Hartzler, R.G., Dixon, P.M., 2005. Are many little hammers effective? Velvetleaf (*Abutilon theophrasti*) population dynamics in two- and four-year crop rotation systems. Weed Sci. 53, 382–392.

White, S.N., Boyd, N.S., 2016. Effect of dry heat, direct flame, and straw burning on seed germination of weed species found in lowbush blueberry fields. Weed Technol. 30, 263–270.

Wicks, G.A., 1971. Influence of soil type and depth of planting on downy brome seed. Weed Sci. 19, 82–86.

Wicks, G.A., Ramsel, R.E., Nordquist, P.T., Schmidt, J.W., Challaiah, 1986. Impact of wheat cultivars on establishment and suppression of summer annual weeds. Agron. J. 78, 59–62.

Wiese, A.F., Collier, J.F., Clark, L.E., Havelka, U.D., 1964. Effects of weeds and cultural practices on sorghum yields. Weeds 12, 209–211.

Williams, M.C., 1980. Purposefully introduced plants that have become noxious or poisonous weeds. Weed Sci. 28, 300–305.

Williams II, M.M., Mortensen, D.A., Doran, J.W., 1998. Assessment of weed and crop fitness and cover crop residues for integrated weed management. Weed Sci. 46, 595–603.

Williamsen, M.H., Brown, K.C., 1986. The analysis and modeling of British invasions. Phil. Trans. R. Soc. Lond. Ser. B 314, 505–522.

Wimschneider, W., Bacthaler, G., 1979. Untersuchungen uber die lichtkonkurrenz zwischen *Avena fatua* L. und verschiedenen sommerweizensorten. In: Proc. Eur. Weed Res. Soc. Symp. on Influence of Different Factors on the Development and Control of Weeds Mainz, Fed. Rep. of Germany, pp. 249–256.

Wortman, S.E., 2015. Air-propelled abrasives grits reduce weed abundance and increase yields in organic vegetable production. Crop Prot. 77, 157–162.

Wortman, S.E., 2014. Integrating weed and vegetable crop management with multi functional air-propelled abrasives grits. Weed Technol. 28, 243–252.

Wortman, S.E., Davis, A.S., Schutte, B.J., Lindquist, J.L., 2011. Integrating management of soil nitrogen and weeds. Weed Sci. 59, 162–170.

Yenish, J.P., Worsham, A.D., 1993. Replacing herbicides with herbage: potential use for cover crops in no-tillage. In: Bollich, P.K. (Ed.), Proceedings of the Southern Conservation Tillage Conference for Sustainable Agriculture. Monroe, LA, pp. 37–42.

Yenish, J.P., Worsham, A.D., York, A.C., 1996. Cover crops for herbicide replacement in no-tillage corn (*Zea mays*). Weed Technol. 10, 815–821.

Yenish, J.P., Young, F.L., 2004. Winter wheat competition against jointed goatgrass (*Aegilops cylindrica*) as influenced by wheat plant height, seeding rate, and seed size. Weed Sci. 52, 996–1001.

Zimdahl, R.L., 2004. Crop-Weed Competition: A Review, second ed. Blackwell Pub, Ames, IA. 220 pp.

CHAPTER

11

Weed Management in Organic Farming Systems

FUNDAMENTAL CONCEPTS

- Weed management for organic agriculture has received only minor attention from weed scientists.

- Weed management in organic agriculture is a continual process.

- Numerous weed control techniques are available to organic growers

- Although organic agriculture occupies only a small portion of US farmland and accounts for just 4% of total US food sales, it is among the fastest-growing parts of US agriculture.

LEARNING OBJECTIVES

- To know the similarities and differences among the various definitions of organic agriculture.

- To know the weed management techniques available to organic growers.

- To understand some ways in which weed management techniques can be integrated.

The choice, after all, is ours to make. If having endured much, we have at least asserted our "right to know," and if knowing, we have concluded that we are being asked to take senseless and frightening risks, then we should no longer accept the counsel of those who tell us that we must fill our world with poisonous chemicals; we should look about and see what other course is open to us. *Carson, R., 1962. Silent Spring. pp. 277–278.*

There are more things in heaven and earth, Horatio, than are dreamt of in your philosophy. *Shakespeare, Hamlet, Act 1, Scene 5.*

1. INTRODUCTION

Organic farming began in the United States in the early 1940s, when J.I. Rodale, founder of the Rodale Research Institute and organic farming and gardening magazine, became the primary source of information about what he called "nonchemical" farming. Rodale's intellectual progenitor, Sir Albert Howard, an English botanist (1873–1947), is regarded as the founder of the organic farming movement. Howard worked for 25 years as an agricultural investigator in India, first as agricultural adviser to states in Central India and Rajputana, and then as director of the Institute of Plant Industry at Indore. His most famous book, *An Agricultural Testament* (1940), is a classic organic farming text. It emphasizes the importance of maintaining humus, keeping water in the soil, and the role of mycorrhiza fungi. The other book, *Soil and Health* (1945),[1] claims "The health of soil, plant, animal and man is one and indivisible," a claim widely supported by the organic agriculture community.

All 50 United States have certified organic farms,[2] ranging from one in Alaska to 2530 in California in early 2016. When Congress passed the Organic Foods Production Act in 1990, the United States had fewer than 1 million acres of certified organic farmland. In 2006 there were 9501 certified organic farms and in 2011 there were 12,880. Organic farming has been one of the fastest-growing segments of US agriculture for over a decade. Consumer demand for organically produced goods has shown double-digit growth during most years since the 1990s.

The land area of the United States is 2264 million acres. In 2007, about 51% of the land was used for agriculture, which includes crops (408 million acres [18%]), grassland, pasture, and range (614 million [27%]) and forests (671 million acres [30%]). In 2006, 1.9 million acres of crop land and slightly less than 3 million total acres were devoted to organic agriculture. In 2011, 3,084,989 million acres of crop land were organic; total organic acres were 5,383,119 million, a large area but still only a small proportion of total US farmland. The annual percent increase in conversion to organic agriculture between 1992 and 2007 was above 40%. From 2007 to 2008, the annual increase was 14%. Any agricultural system growing at that rate deserves appropriate attention and research. Annual US organic crop sales exceed $40 billion.

Organic agriculture continues to grow, but development of weed management techniques has been neglected compared with the dominant conventional agricultural system. Weed management remains a challenge. Agricultural research scientists, especially those in publicly supported land-grant university colleges of agriculture, should be encouraged to develop weed management systems that increase productivity, profitability, and the environmental sustainability of organic farms. The increase in the number of acres of crops being grown organically has brought with it a growing need for more labor and unconventional inputs for weed control. Currently, weed control is ranked as the number one production cost by

---

[1]Both books are available from Amazon. Howard, A. 1940 (UK) 1943 (US). An Agricultural Testament. Soil and Health (originally published in 1945) is available as Howard, A. 2007. The Soil and Health: A Study of Organic Agriculture (Culture of the Land). London Ed. Faber and Faber Ltd., 352 pp. and as Howard, A. 2006. The Soil and Health: A Study of Organic Agriculture (with an introduction by W. Berry). Lexington, University Press of Kentucky, 307 pp.

[2]http://www.ers.usda.gov/data-products/organic-production.aspx#25766.

organic and many conventional growers. Over the past 10 years, development of machine-guided technologies and fundamentally herbicide-based systems for site-specific (precision) weed control, has advanced rapidly, whereas emphasis has been minimal to absent on precision weed control in organic systems.

It is reasonable to suggest that many weed scientists acknowledge, perhaps too readily, that weed management has been, is, and will continue to be a major problem in organic agricultural systems. However, organic growers manage weeds effectively. In fact, many organic growers do so with well-known techniques that have been used for many years. It is unfortunate and important to recognize that weed management in organic/alternative/low-purchased input agriculture has not benefitted from the decades of scientific research that have been devoted to development of weed management systems for modern and conventional production systems[3] that are highly dependent on herbicides. If a similar research base were available, many of the questions about managing weeds in organic agricultural systems would have been answered. Fred Kirschenmann (2010, p. 174) a third-generation North Dakota farmer, philosopher, and distinguished commentator on the challenges of conventional agriculture, said:

> If we invested research funding for ecological approaches to solving production problems that was comparable to what we spend on the engineering approach, what solutions would we find? Conversely if we convince ourselves that no alternatives to engineering external controls exist, we guarantee that ecological approaches won't be explored.

2. WHAT IS ORGANIC AGRICULTURE?

First, it is important to understand that the practice of organic agriculture is fundamentally different from conventional, energy, capital, and chemically dependent agriculture. The focus is on no use of synthetic chemicals (herbicides) for weed management. Organic agriculture is also a legal term. An organic farm is one that is certified as such according to national organic standards. To obtain certification, a farmer must document that all of the practices and inputs used conform to these standards. Some farmers practice organic or near-organic practices but eschew certification because of the certification requirements and expense. They adhere to principles and procedures required for certification, but choose not to become certified. In addition to the legal requirement, organic agriculture, a general term, has several characteristics and at least five similar names (Zimdahl, 2011, p. 200).

2.1 Agroecological

Agroecological (ecological) systems are generally labor-intensive, as are most organic systems. Agroecology is characterized by minimal, high-efficiency use of natural resources (fuel and water), no or very limited dependence on synthetic fertilizers and pesticides, use

[3]To avoid needless repetition of the descriptive terms "modern" and "conventional," the remainder of this chapter will use conventional to identify and distinguish these agricultural production systems from organic/alternative systems. For the same reason, "organic" will be used. These choices do not imply and should be interpreted as a value judgment.

of locally produced manure or compost, use of leguminous plants or trees to fertilize soil, and maintenance of a diverse cropping/livestock system that is sustainable. Advocates of agroecology have multiple objectives in addition to production. The system should achieve social (community maintenance), economic (profit), and ecological (sustainability of the environment, other creatures, soil, water, air, etc.) objectives.

2.2 Biodynamic

Biodynamic farming generally means using standards that emphasize a farm's organic unity. The foundation of biodynamic farming is the work of Steiner (1861−1925). In a series of eight lectures (Gardner, 1993; Paull, 2011), Steiner described his agriculture course over 10 days. He laid the foundation for an alternative agriculture that he envisioned would "heal the earth." He said, "In the course of this materialistic age of ours, we've lost the knowledge of what it takes to continue to care for the natural world" (Gardner). Steiner found that seeds had less vitality when farmers had grown the same crops on the same land for several years and then had to rotate crops to avoid problems. Farmers had become dependent on chemical fertilizers.[4] Biodynamic farmers are encouraged to make their own fertilizer primarily by keeping animals. It advocates an ideal, diverse, integrated farm with crops and animals (Thompson, 2010, p. 213). Biodynamic farming systems encourage the use of fermented herbal and mineral preparations as compost additives. Carpenter-Boggs et al. (2000) showed that short-term benefits from such preparations remain questionable. Biodynamic farmers emphasize the uniqueness of each farm and conclude that uniform standards are not appropriate. Farmers are encouraged "to live fully in the organic spirit, including participation in activities intended to promote the biodynamic way of life" (Thompson).

2.3 Conservation

Conservation agriculture (often called sustainable agriculture) advocates zero or minimum tillage, topsoil management to achieve preservation, crop rotations based on minimal soil disturbance, permanent soil cover, and biological (nonchemical) weed control.

2.4 Organic

Organic agriculture seeks to maintain long-term soil health. It advocates the use of crop rotation, green manure, composting, and biological pest control. Organic farmers manage pests with biological and cultural control techniques based on maintaining biological diversity (Deryckx, 2001). Northbourne (1940) is credited with coining the use of "organic" as an adjective to modify the noun "agriculture" (Darby et al., 2007). In most of the United States, a farm to be certified as organic must exclude use of synthetic fertilizers, pesticides, plant growth regulators, livestock antibiotics and hormone treatment, food additives, and genetically modified (GM) crops (Buck and Scherr, 2011). One must wonder whether and how

---

[4]Phillips, J.C., 2006. Beyond Organic: An Overview of Biodynamic Agriculture with Case Examples. Personal communication.

the standards might be modified if some promised GM crops are developed. For example, cereal varieties may be modified to survive and even prosper under drought conditions or crop cultivars may be bred or engineered to mature more quickly independent of where they are grown (Worldwatch, 2011, p. 132).

2.5 Regenerative

Regenerative agriculture is essentially agroecological farming. It is rooted in knowledge of how to manage the complex dynamic among plants, animals, water, soil, and pests required to produce crops. The name implies that the primary objective is to create a sustainable system (Buck and Scherr, 2011). Practitioners advocate crop residues and surface mulch for weed management, the use of compost and green manure, growing legumes, and the biological control of pests.

2.6 Characteristics and Objectives

The five names share three fundamental characteristics that define organic systems (Drinkwater, 2009):

- In contrast to conventional, industrial agriculture, organic agriculture has been developed by a farmer-dominated process.
- Organic farming systems use integrated weed management strategies that do not rely on a single technique.
- Farmers' goals are multidimensional. Achieving maximum yield and economic return are important, but emphasis is on the long term (several years) rather than the short term (1 year). Other goals include the development of social networks and localization of food supply, including social networks that encourage local food marketing and consumption.

It is reasonable to conclude that although there may be exceptions, organic farming systems emphasize that food is what connects humans to life; food enables life.

Kirschenmann (2010, p. 253) claims the key difference is that "organic agriculture seeks to cooperate with nature's system of production, while conventional agriculture seeks to create its own system of production alongside (and even against) nature's system." In this context, cooperation does not mean submission to nature's systems. It means knowing and understanding them while learning and benefitting from what they have to teach. It also means understanding the role of our opinions and attitudes. Several definitions of weeds are in Chapter 2, ranging from the Weed Science Society of America ("Any plant that is objectionable or interferes with the activities and welfare of man") to "I just don't like them." In the conventional agricultural sense, it may be any plant except the one planted: the crop. Conversely, organic farmers do not always regard all weeds as objectionable or interfering, but rather as useful (see Chapter 4). Ergo, there is a change in one's attitude.

Kirshenmann provides four operational techniques, which could be regarded as goals of organic farming:

- to meet basic nutrient requirements by creating self-regenerating systems to enrich soil and create a nutrient-rich growing environment;

- to achieve weed control using crop rotation systems to deprive weeds of favorable growing conditions;
- to achieve insect control through reliance on natural predators; and
- to achieve plant disease control through the use of crop rotation, careful seed selection, and general farming practices, particularly sanitation. Organic farmers cannot buy seed treated with a fungicide. Careful selection of seed suppliers is important.

Liebman and Davis (2009) propose three critical concerns or principal objectives for weed management in organic farming, which are equally applicable to conventional agricultural production systems. Their concerns are worthy goals for all weed management systems:

- limiting competitive damage caused by weeds. A primary objective is to reduce weed density to tolerable levels, which, of course means accepting less than 100% weed control, a clean field. It may mean redefining when a plant is a weed;
- minimizing the size of future weed populations by minimizing seed production and survival and survival of vegetative propagules; and
- preventing introduction of new weed species by monitoring, sanitation, and targeted eradication.

Similar objectives for weed management in organic farming systems were presented by Liebman (2001). In his view, any weed management system (organic or conventional) has to protect the crop, and that requires meeting three principal objectives:

- reducing weed density to a tolerable level. The clear implication is that 100% weed control is not required.[5] Liebman did not define a tolerable level. It is a subjective decision based on the experience and judgment of the farmer. Hundreds of studies have been done to define yield loss caused by defined weed densities (Zimdahl, 2004). These studies often define tolerable in economic terms: if the cost of control is less than the projected loss, control is appropriate. These studies confirm the central hypothesis of weed science: some weed density will compete with the crop and reduce its yield and quality. Indeed, if this were not true, weed science would lose its reason for existence. However, tolerable can be a completely subjective or economic decision.
- reducing the damage a given density of weeds causes. This can be done by:
  - minimizing resource consumption, growth, and competitive ability;
  - delaying weed emergence and, in apparent conflict, encouraging weed emergence during noncrop periods so they can be incorporated into soil and thereby contribute organic material before the next crop is planted;
  - increasing the portion of resources captured by crops; and
  - damaging but not necessarily killing weeds with one of several techniques.
- shifting the composition of weed communities to less aggressive, more easily managed species. Liebman said this can be accomplished by direct suppression of undesirable weed species through selective control or manipulation of environmental conditions (e.g., irrigation). Selective weed management requires a high level of management skill and time. It requires close attention to the efficacy of different techniques used alone

[5]The dominance of chemical weed control and herbicide resistance in conventional agriculture seem to be supportive of 100% control.

and in combination. Managers must think about what weeds are or may be present, what has worked in the past, what level of control in desired, and what method or combination of methods will best achieve the control objective.

All weed management systems share some characteristics. Organic systems are fundamentally different because of their view of the uses and purposes of nature differ (Hansen et al., 2006). They disagree about whether:

- Nature is a partner in organic production. In contrast to culture, nature is the "parts of the world which have not been touched or influenced by humans" (Liebman, 2001). Such places are few and far between, but should be protected and preserved. Bronowski (1973, p. 17) cautioned that "man is a singular creature...he is not a figure in the landscape—he is a shaper of the landscape."
- Nature has inherent biological value and there is a close relationship between the natural world and agricultural production. Nature can be uncultivated or cultivated land. It is the milieu that permits crops and animals to grow and makes farming possible. Nature is a partner in farming. This view is in sharp contrast to the modern agricultural view that nature must be tamed and controlled, or agriculture will not be possible.
- Nature is a place for relaxation of mind and body. It provides multiple sensory experiences and places for recreation, places to recreate ourselves. This view focuses on the relationship between humans and the environment. Modern agriculturalists are familiar with and sympathetic to this view but clearly separate it from their production responsibility.

Hansen et al. (2006) suggest that the question of who owns and therefore has a right to control nature is a central aspect of the debate among these different yet similar views. The distinction is drawn from within the alternative agricultural movement and is relevant to the ongoing debate about which agricultural system should be adopted: which is best. Considering all of these parameters is not an easy task. It is a task often burdened by our cultural heritage, which told us to "Be fruitful, and multiply, and replenish the earth, and subdue it: and have dominion over the fish of the sea, and over the fowl of the air, and over every living thing that moveth upon the earth" (Genesis 1:28, King James version). We have multiplied, replenished, and attempted to subdue. We must use and benefit from the earth, but first we are obligated to understand and care for the earth. Bailey (1915) told us that "a good part of agriculture is to learn how to adapt one's work to nature." That view dominates organic agriculture.

3. FEEDING THE WORLD

A common argument, if not the dominant one, against organic agriculture within weed science and the larger agricultural community is that it is fine for those who choose it, but the system is fundamentally flawed because it is not capable of feeding the world's people and furthermore, it never will be. This is an opinion. It is not a carefully constructed argument based on research and scientific data. Those who argue that organic agriculture can or cannot ever feed the world base their claim on strong opinions, but few present convincing scientific data to support their claim. Both sides complain that the other does not base their claim on replicated, peer-reviewed, scientific studies. Each laments the other's lack of scientific evidence. Science is being used to defend an opinion rather than to explore and discover a more accurate description of reality (Kirschenmann, 2010, p. 188). The discussion is not immediately relevant to decisions about weed management. Weed management is involved in

> **the pragmatic world** of production agriculture, which produces the food to feed the world,
> **the scientific world** of research and technological development, and
> **the philosophical world** that guides thinking about complex problems. Such discussions affect what weed management systems are used and those that ought to be used.

In some quarters, it is common to assert that "the only people who think organic farming can feed the world are hysterical mothers and self-righteous organic farmers" (Halweil, 2006). Halweil also asserts that many scientists and international agricultural experts believe a large shift is occurring toward organic farming that will increase the world supply. Francis and Hodges (2009) argues on behalf of all of the authors of the book Francis edited (2009), that "organic farming methods and systems hold promise to fulfill many of the food needs of the global population; indeed, some would argue that in a world of scarce resources and overwhelming pollution we have no choice in the long term but to convert to a largely organic food system." One must conclude that organic agriculture is no longer just an interesting, minor production system. It is an existing, developing agricultural production system. The American philosopher William James (1975) accurately described how it and many other new ideas evolve. They pass through three stages: "First, you know, a new theory is attacked as absurd; then it is admitted to be true, but obvious and insignificant; finally it is seen to be so important that its adversaries claim that they themselves discovered it."

Federal, state, and university research policies that favor conventional agriculture justify the position with a clear moral argument that it is their responsibility to feed the world's people, to prevent hunger and malnutrition. The justification often goes on to argue that those who favor an organic system must bear the burden of deciding who is going to starve because, as stated, no organic system can feed the world and protect the environment. The challenge to proponents of organic agriculture is to address the question: "if farmers relied on ecologically based weed management strategies and greatly reduced the use of herbicides, would they produce enough food (Mohler et al., 2001)?" That question is inevitably followed by the accusation that widespread adoption of organic, ecologically based farming will take more land, destroy wildlife habitat, and threaten biological diversity. The claim is that the only way to feed the growing world population is to maintain and improve the productive ability of conventional/industrial agriculture.

It is beyond the scope of this book to provide a complete review that addresses and tries to resolve the competing claims. A few relevant comments seem appropriate.

Each organic agricultural system can be highly productive per unit area (Bello, 2009, p. 139), but each is labor-intensive. The seasonal requirement for abundant, hard-working, low-paid labor is a disadvantage, although such labor is required in present US fruit and vegetable production. Those engaged in conventional agriculture know (perhaps everyone knows) that they do not want to do hard farm labor. It is reasonable to assume that no one wants to, or at least few citizens of the world's developed countries do. Organic modes of production may be highly productive per acre and efficient in the use of natural resources, but when labor is scarce and those who are available do not want to do farmwork, organic systems will lose in pure economic competition with conventional systems. Buck and Scherr (2011, p. 23) suggest that limiting the conventional/organic debate to whether organic agriculture can meet the entire global food demand is incorrect. If one levels the playing field and combines the (usually nonquantitative) potential achievements of organic systems to prevent environmental degradation, secure farmers' livelihood, and decrease poverty with their productive capabilities, proponents claim they will be much more competitive. Organic systems are not necessarily environmentally benign or neutral. Weeding that is done by soil cultivation uses polluting fossil fuel (unless one uses horses or mules), is likely to increase soil erosion, and disturb ground-dwelling creatures (Trewavas, 2001).

Seufert et al. (2012) compared yields of organic and conventional agricultural systems and asked whether they are comparable. They used a "comprehensive meta-analysis to examine the relative yield performance of organic and conventional farming systems globally." They concluded that "under certain conditions—that is with good management practices, particular crop types and growing conditions—organic systems can nearly match conventional yields, whereas under others it, at present, cannot." Over all crop types, plant types, and species, organic yields were 25% lower. Yields of organic fruits and oilseed crops were statistically comparable to crops grown conventionally. Organic cereals and vegetables were 26% and 33% lower, respectively. Some of the yield response can be explained by differences in the amount of available nitrogen. "Organic systems appear to be N limited, whereas conventional systems are not" because of the use of nitrogen fertilizer. A major difference Seufert at al. identify is the role of the manager. Proponents of organic agriculture agree that yields depend more on knowledge and good management practices than is true for conventional systems. Many also advocate the advisability of incorporating animals into organic systems to supply nitrogen through their manure. This is particularly true relative to weed management, which, as mentioned earlier, is not a one-time decision. In organic agriculture, weed management combines a series of practices throughout the year. Seufert et al. conclude that "there are no simple ways to determine a 'winner' for all possible farming situations." They seek what Kirschenmann (2010) advocates: Science should be the base from which to form and defend an opinion. Scientific studies should be used to explore and discover a more accurate description of reality, which should help all achieve what all want: a sustainable food supply for all.

In addition, using productivity per unit of land area as the sole criterion for rejecting organic agriculture except as a niche market for rich ideologues ignores the question of distribution (Röling, 2006). In an incisive analysis of famine, Sen (1981, pp. 162–166) concluded that whereas starvation is characterized by some people not having enough food to eat, frequently it is not accompanied by there not being enough food to eat. Lack of food production and subsequent availability can cause famine and starvation, but it is only one of many possible causes. Thus, which system can feed the world is not simply a production question, it is (or ought to be) considered as a distribution, justice, equality, and poverty question. Ignoring these aspects begs the question of whether pursuing more production is the best or the right way to "feed the world." World agriculture now produces enough food to feed everyone on earth an adequate diet. The problem is not production; it is a combination of:

- distribution: Where is the food and can it be distributed to those who need it? Is the infrastructure (ports, roads, trucks, etc.) for distribution present, and if so, is it adequate?
- access: Who has access and who does not?
- ability to buy = poverty
- equality and justice for whom?

There is no human right to food. You are not obligated to feed anyone, even those who are starving. In this view, feeding an expanding human population is not an agricultural/production problem; it is sociopolitical problem.

Making the accusation and then reaching a decision based solely on productivity per unit area "is a technocratic argument that ignores the growing international evidence that institutional factors are the real bottleneck to making the most productive use of agricultural assets" (Röling, 2006). Bindraben and Rabbinge (2005) agree that distribution is part of the problem: "Food availability per person has increased worldwide in the last 40 years by an average of 30%." At the same time, food availability to people in African countries south of the Sahara has declined an average of 12%. Bindraben and Rabbinage argue that without restoring and maintaining the soil's nutrient balance and preventing soil degradation, the production problem cannot be solved. They also claim that "currently available technologies are not sufficient to ensure sustainable development" in Africa, and perhaps everywhere. Technological innovations must be combined with favorable institutional and economic conditions, which can be, but have not been, created by national governments. Advanced technologies of the developed world (e.g., pesticides, GM crops, highly productive monoculture) may have a role in the future, but Bindraben and Rabbinage suggest that immediate progress can be achieved with production technologies used in alternative/organic agriculture. To wit:

- using specific crop sequences that diminish pest presence,
- introducing legumes to improve soil nutrient soil balance,
- planting in specially shaped holes to catch rainwater,
- using plant ground covers for fertility and water retention
  - Ground covers are an important part of organic farming practice. Their most important function is fertility. Organic growers grow, plow down, and thereby gain fertility from hairy vetch, field peas, Sudax (sorghum–Sudan grass), other crops, and weeds. Weeds plowed down before seed set are important contributors to soil fertility (see Clark, 2007).

- building ridges to retain rainwater and improve infiltration, and
- employing adapted farming practices developed through research.

Consistent with the principles of alternative agricultural systems, these techniques can be adjusted to fit local conditions and use local resources.

A study by Schaller et al. (1984) is pertinent to the last of these six points. They analyzed the US Department of Agriculture (USDA) Current Research and Information System (CRIS) database to identify research projects "relevant to organic farming." Their review included 6413 farming system research projects. The 403 projects (6.3%) with special organic relevance received 5.3% of total federal funding. Fully 87% of the projects reviewed were neutral with respect to organic systems. Lipson (1997) reviewed 30,000 active or recently completed research projects funded by various USDA agencies. Of those 4500 (15%) had key words that warranted further review to determine their relevance to organic research. Only 301 of the 30,000 (0.01%) were found to be "organic-pertinent." Of those only 34 (0.001%) were rated as "strong organic." They received a bit less than one-tenth of 1% of the USDA's annual research and education funds. The remaining 267 were classified as "weak organic," meaning that the organic context or purpose could only be inferred. In 2010, 2.4% ($32.6 million) of the $2.036 million USDA budget was allocated to organic research. The inevitable conclusion is that although great resources have been expended for research on conventional farming for many decades, little funding has been allocated to research on organic systems. However, the Organic Farming Research Foundation gave its top rating to organic programs at six US universities (Colorado State, Florida [Gainesville], Michigan State, Minnesota, Tennessee [Knoxville], and Washington State).[6] In 2012, USDA-CRIS records reported that 33 US universities had federally funded organic research projects.[7]

A more recent report from the Organic Farming Research Foundation analyzed federally funded organic research projects in Taking Stock: Analyzing and Reporting Organic Research Investments 2002—14 (https://www.ccof.org/blog/new-report-assesses-federally-funded-organic-research). The report included 189 organic research projects funded by USDA's Organic Research and Extension Initiative grants over 13 years, representing $142.2 million. About 70% studied organic crop production, 20% examined crop—livestock systems, and 10% studied livestock and poultry. Sixty-five percent emphasized soil management and 68% focused on crop pests.

The remainder of this chapter will not present scientific studies that support either view. It will describe some techniques for weed management in organic agricultural systems and provide a few examples of how these can be and have been integrated.

4. METHODS OF WEED MANAGEMENT IN ORGANIC AGRICULTURE

Weed management is much easier in conventional production systems. The large research base mentioned earlier is an important reason: it is easier to learn what to do and how

[6]Ofrf.org/education/organic-higher-education.

[7]http://cris.nifa.usda.gov/cgi-bin/starfinder/0?path=fastlink1.txt&id=anon&pass=&search=cg=*-51300-*%20AND%20gy=2004:2011&format=WEBTITLESG.

effective it will be. Other reasons why weed management with herbicides is preferred by many farmers include (Liebman et al., 2001):

- using herbicides is easy with apparently low risk to the user;
- herbicides and other pesticides are marketed aggressively. Use information is readily available when one knows where to look;
- environmental and human health costs are externalized[8]; and
- government policies favor large farms and input-intensive agriculture.

The general perception is that weed management with herbicides is much less expensive than nonchemical alternatives. Gianessi and Reigner (2007) reported that the time required to hoe a wide range of crops by hand ranged from 17 h/hectare (ha) (almond) to 408 h/ha (peanut). Earthbound Farm (http://www.ebfarm.com), one of the largest organic producers in the United States, suggests that conventional farmers spend about $50/acre on herbicides, whereas organic farmers may have to spend up to $1000/acre for hand labor in vegetables crops for which no other viable options are available. It is undeniable that hand labor is expensive, slow, and used by organic farmers. However, the claim rests on an incorrect assumption. The claim assumes that hand labor is the primary means of weed control available to organic growers, if not the only one. This is patently false. All of the control methods described in Chapter 10 are available and used by organic growers. Several old, reliable, nonchemical techniques (e.g., hoeing, opaque solar film, living mulches, mulching, stale seedbed[9]) are included in Gilman's small book (2002).

Liebman and Davis (2009) correctly claim that conventional weed management depends on herbicides, for the reasons outlined earlier. Weed management in conventional agriculture consists of deciding what informational sources (cooperative extension, manufacturers, local suppliers, or salespeople) will be used to decide when to use what herbicide at what rate. In contrast, weed management in organic systems involves considering several control options. The options include control techniques before planting, in the crop, and after the crop has been harvested. An important, frequently overlooked distinction between systems is that in the organic system, weed management is not *a* decision, it is *a series of decisions*. An organic grower may decide to use hand labor, but that is only one of several decisions that must be made during a year. Use of hand labor is never the main weed management technique. It is often a final, alternative stage of weed management, which may not be required.

The crucial role of the manager in an organic system is best characterized in a concept originated by Liebman and Gallandt (1997), who asked whether the best weed management could be obtained by combining multiple tactics, which they referred to as "many little hammers." They acknowledged that each little hammer might be individually weak, but combining them would gain the advantages of each and result in better weed management.

---

[8]An externality is a cost that is not reflected in price, or more technically, a cost or benefit for which no market mechanism exists. In the accounting sense, it is a cost that a firm (a decision maker) does not have to bear, or a benefit that cannot be captured. From a self-interested view, an externality is a secondary cost or benefit that does not affect the decision maker.

[9]Stale seedbed: preplanting shallow (1–2 in.) tillage to stimulate germination followed by another shallow cultivation to kill emerged weeds before planting.

Some claim that organic farming lacks a scientific foundation; as noted by Kirschenmann (2010, p. 174), this is correct. It does not follow that a scientific foundation is totally lacking. Organic weed management is a series of decisions about what "little hammers" are most appropriate; therefore, consideration of ecological, soil, and human health is integral. Technology is part of organic farming just as it is in conventional farming. Managerial decisions are necessary and more frequent in organic systems. Sørensen et al. (2005) showed that management alone (decision making) required 14%−19% the total labor required.

Young (2009) noted that the adjective "precision" is applied to conventional agricultural technology whereas the adjective "organic" is applied only to alternative/unconventional agricultural production systems, which by implication are presumed not to be precise. Much of the agricultural research community regards organic systems as having many characteristics, but only conventional technology is regarded as precise. Organic production systems include many common, well-known techniques such as crop rotation, cover crops, crop variety selection, employment of stale seedbeds, flaming, tine weeding and harrowing, etc. Perhaps it is only because these techniques are old that they are not regarded as precise. Perhaps this is just another opinion.

Little experimental work is available. Young[10] reviewed 111 published articles on weed control in organic cropping systems. Twenty-four were published before 2000 and all others were afterward. The work reviewed emphasized cultural and mechanical control. Of the 111 papers reviewed by Young, only seven were published in *Weed Science*, five in *Weed Technology*, and eight in *Weed Research* (18%). From 2010 to 2016, only three articles that included the key words "weed," "management," and "organic" in the title were published in *Weed Science* (DeDecker et al., 2014; Poffenbarger et al., 2015; Ryan et al., 2010b), and three in *Weed Technology* (Johnson et al., 2010; Mirsky et al., 2013; Taylor et al., 2012). Tautges et al. (2016) summarized the situation well: "A lack of information regarding weed control, relative to conventional systems, has left organic growers largely on their own when devising weed management systems for organic crops."

Organic growers acknowledge that weeds are, have been, and will continue to be a major management problem: "Organically managed fields typically have a greater abundance and diversity of weeds than conventionally managed fields" (Menalled et al., 2001; Pollnac et al., 2009). However, yield loss caused by weeds in organic fields is often not commensurate with weed abundance (Smith et al., 2010; Johnson et al., 2017). Therefore, it is logical to conclude that crops grown organically may have a greater tolerance of weeds than those grown in conventional systems (Johnson et al.) and their control may in fact be easier. Dorn et al. (2015) demonstrated that legume-based cover crops caused a 96%−100% reduction in weed dry matter in a maize- and sunflower-integrated conservation tillage production system but only 19%−87% in an organic conservation tillage system. Despite the difference, they concluded that legume-based cover crops are useful for weed suppression in organic production systems. Benaragama et al. (2016) concluded that organic management practices including increased crop rotation diversity did not increase yield or reduce yield loss owing to weeds primarily because of factors associated with soil fertility. Those who choose to grow crops organically must manage weeds. Perhaps the essential ingredient in the weed management program is crop rotation, which is not an important element of weed management in

[10]Young, S.L. Weed Control in Organic Cropping Systems (Unpublished review).

monocultural conventional agriculture.[11] Planned rotations do not eliminate weeds; they contribute (a little hammer) to overall weed management. Part of a good rotational plan is to keep the land covered at all times to prevent soil erosion.

A good weed management plan will vary with the crops being grown, the soil, the need for irrigation, available equipment, and the manager's experience and skill. Because weed management is a series of decisions, it can be regarded as beginning at any point in a growing season. One example from a vegetable producer follows:

- immediately after harvest: Disk or use a similar tool to break a crust and kill emerging weeds. If weather and market permit, plant a second crop. If a second crop is not grown, plant a cover crop that can be plowed under the next spring. Satisfactory cover crops include oats, wheat (can be harvested as grain or plowed down as a green manure, a weed management technique), pea, and vetch (both legumes). Weeds can also cover and protect soil and serve as living mulches, a technique that is most applicable in warm, wet climates. When they do, they are no longer plants out of place. They have become useful plants that protect soil, provide habitat for beneficial organisms, and contribute organic matter to soil when plowed down (Gilman, 2002).
- prepare land for planting. Beds are often made for vegetables. Organic farmers emphasize the importance of a good seedbed for crops. It is also true that weed germination is favored by a good seedbed.
- irrigate before planting to encourage germination of weeds.
- tillage: A rod- or tine-weeder running 1–2 in. deep exposes seedling weeds to desiccation on the soil surface. Shallow tillage is preferred, so weed seeds that are deeper are not brought to the surface. It is important to wait a few days before planting a crop to be sure the seedlings have died (Gilman, 2002).
- tillage: The most important tool for many organic farming systems is the rotary hoe or rotary harrow. Many crops (e.g., corn, small grains, spinach with four to six leaves) can be rotary hoed after the crop has a root system. Careful rotary hoeing does not harm the crop and significantly reduces the weed population. A rod weeder is not appropriate for shallowly planted (0.25 in.) crops (e.g., lettuce).
- plant the crop. Knowing the optimum plant spacing in the row and between rows is part of weed management.
- flaming. Use of a flame directly over and a few inches above the crop row after planting when seedling weeds have emerged, but the crop has not, can achieve nearly 90% control. Some crops (e.g., corn, onion) have foliar emergence before the growing point emerges. They can be flamed after emergence.

  A few notes on flaming:
  - Its success is a function of the crop, soil temperature, and planting depth.
  - Some crops take longer to emerge; seed germination is a function of soil temperature, which is related to planting depth.
  - Flaming is most effective in crops that emerge slower than the dominant weeds.

[11]I acknowledge Mr. L.O. Grant (deceased 2014) of Fort Collins, Colorado for his guidance and advice on weed management in organic agriculture.

- Thus, crop emergence and the weeds in the field must be known. One must know which weeds are likely to emerge before the crop.
- The decision of how and when to flame requires careful and continual observation. Weed management requires a plan that varies with the crop being grown, the soil, irrigation, equipment, and most important, the manager's experience and skill

White and Boyd (2016) showed that germination of seeds of witchgrass, spreading dogbane, and meadow salsify was reduced more than 90% after 1-s exposure to direct flame. Burning straw did not consistently reduce germination of hair fescue or winter bent grass.

- In-row cultivation.
- Hand-hoeing if required.

Several chemical products are allowed in organic agriculture.[12] None are synthetic organic chemicals. Household vinegar is usually 5% acetic acid. At 10%, 15%, or 20% acetic acid, several available products control many seedling, annual broad-leaf weeds. Control is improved with high application volume (40, 80, or 100 gallons/acre). It is a contact, nontranslocated, nonselective herbicide with no soil residual activity. Because it is not translocated, it is not effective on perennial weeds. Vinegar is expensive and can injure some crops (e.g., bell pepper and broccoli) (Evans et al., 2011). Other approved products include herbicidal soaps, iron-based herbicides, salt-based herbicides, phytotoxic oils (clove, peppermint, pine, and citronella), corn gluten meal, and combination products (including ingredients from multiple categories) (Dayan et al., 2009).

Turner et al. (2007) and Tautges et al. (2016) affirm that controlling perennial weeds (e.g., Canada thistle and field bindweed) is the most difficult weed management problem in organic culture. Growing perennial forage crops (alfalfa and grass hay) can reduce perennial weeds but will not eliminate them.

In Pennsylvania, organic maize (corn) sustained crop yields equal to conventional methods despite higher weed density (Ryan et al., 2010a). A primary reason was prior incorporation of weeds. If mechanical cultivation had been (or was able to be) more effective, organic would have yielded more. The yield was improved by greater soil resource availability. This finding is consistent with the organic farmer's constant goals of achieving the highest soil health, providing relative freedom from weeds and soil-borne pathogens, and making a profit. Many conventional farmers have similar goals, but achieving a profit takes precedence.[13] The effectiveness of interrow cultivation of corn and soybeans during the first two rotations of a corn—soybean—spelt organic grain cropping system was 73% for soybeans and 91% for corn. Control in corn was more effective because more soil was moved into the corn row. Perennial weeds were not controlled as well as those emerging from seed (Mohler et al., 2016). Interrow cultivation was an effective control in snap beans because they are a short-season crop and compete well. Weed management was more challenging in sweet corn because of its longer season and the lack of effective weed competition. Interrow cultivation

[12]The complete list can be found at http://www.ecfr.gov/cgi-bin/text-idx?rgn=div6&node=7:3.1.1.9.32.7#se7.3. 205_1601.

[13]Achieving a profit: making money is a logical, understandable goal. Surviving: being able to farm again tomorrow is a goal all farmers have. Organic farmers do not always regard maximizing profit as the only or primary goal.

was effective, but three separate cultivations were required to obtain yields similar to the use of herbicides. However, net profit was increased because of the organic price premium (Johnson et al., 2010). No differences were found between the effects of the systems on weed seed persistence or microbial abundance. Viable seed half-life for smooth pigweed and common lamb's-quarter was statistically equal (Ullrich et al., 2011).

Weed scientists have not paid a great deal of attention to techniques that affect survival of weed seeds in soil. Menallad et al. (2006) ask whether seed predation could become a "component of management strategies that reduce reliance on herbicide application and cultivation." Such studies are increasing but are not abundant, although the researchers' review includes 166 literature citations. In their view, studies of seed predation are evidence of a desirable shift in weed science research toward study of and more reliance on knowledge of ecological processes rather than herbicides. Future research will focus on identifying management practices that promote survival of microorganisms that are seed predators. Desirable practices could include several things consistent with organic farming: disturbance reduction, conservation of noncrop habitats in close proximity to cropped areas, and delaying postharvest tillage. In Finland Salonen et al. (2012), research showed that in conventional cropping, total abundance of weeds remained about the same from the late 1990s to the late 2000s. Overwintering species became more frequent. Increased total weed biomass was associated with organic cropping owing to the lack of direct weed control methods and inadequate crop competition. Williams et al. (2009) showed that crop habitat influenced seed predation of velvetleaf and giant foxtail. Predation was greater during the 4 years of a corn–soybean–small grain plus alfalfa rotation than it was during a 2-year corn–soybean rotation. Menalled et al. (2005) demonstrated that adding composted swine manure, a good organic practice, posed a small but significant risk of increasing soil weed seed abundance.

5. A DIFFERENT VIEW

A different, and in the view of many, radical approach to agriculture may have particular applicability to organic agriculture. The fundamental objective has been to develop high-yielding, seed-bearing, perennial polycultures. These crops will be most applicable to large acreage, full-season crops (e.g., corn, sorghum, wheat). The guiding principle has been: Nature knows best.

Wes Jackson founded the Land Institute in Salina, Kansas in 1976. The Institute staff has studied agricultural production for 40 years from a perspective different from that found in most land-grant colleges of agriculture. Jackson decided that modern chemical-, energy-, and capital-intensive agriculture, although highly productive of food and fiber, was inevitably unsustainable and doomed to fail. Many agree. What was needed was a new model for agriculture, one based on the land ethic of Aldo Leopold (1966):

> A thing is right when it tends to preserve the integrity, stability, and beauty of the biotic community. It is wrong when it tends otherwise.

The model Jackson chose to achieve the goal that Leopold established was the prairie. Many well-established, respected agricultural people scoffed. The comments were similar to these: Prairies do not produce edible food or fiber; they produce prairies. Prairies might be good for some cattle grazing but not for producing food crops. Another common response from the conventional agricultural community was the reasonable botanical claim that perennials put energy into perennial structures (roots, rhizomes, stolons, and tubers), not into seed production. Perennation and high seed yield do not go together. Besides, the doubters said, perennial plants that grow on prairies do not produce seeds that humans will eat. They are not palatable; they taste bad.

Ah, but prairies are highly sustainable. Without human disturbance, they have survived and remained productive for centuries. Prairies do not have weed, disease, insect, or fertility problems. If they do, they recover on their own. Prairies sponsor their own pest control and fertility and do not need to be cultivated (plowed) or irrigated. A prairie does all that a sustainable agricultural system ought to do. They do what modern agricultural systems fail to do. A prairie is place to begin to try to learn what nature has to teach.

Scientists at The Land Institute consult nature, which is regarded as the source and measure of the membership of humans in the natural world. Similar to the goal of many organic producers, The Institute's goal is to develop an agriculture that preserves rather than destroys soil and thereby ensures the life of rural agricultural communities. They have demonstrated the scientific feasibility of natural systems agriculture, which they claim can and ought to be adopted worldwide. Prairies are obviously not the proper model for all world agriculture, but for those who look, models can be found in local ecosystems. The Land Institute's research goals are compatible with the goals of organic agriculture.

Perennial crops will keep soil covered and protected, reduce or eliminate regular soil tillage, reduce erosion, sequester more carbon, conserve soil organic matter and nutrients, and reduce reliance on annual fertilizer application. Because they eliminate the need to plant each year, perennials usually diminish competition from annual weeds, but perennial weeds will be effective competitors (Glover et al., 2010). With one exception, perennial, grain-producing crop species currently exist only in the development stage, and much genetic research and breeding remain to be done before perennial grain crops can be deployed in agriculture (Batello et al., 2014). The exception is Kernza, a selection of a breeding program that began with intermediate wheatgrass at the Rodale Institute in 1988 and was successfully continued at the Land Institute and the University of Minnesota.

Research at the Land Institute has also demonstrated that continuous improvement in perennial sorghum can be accomplished with traditional plant breeding techniques, provided that selection for more domesticated phenotypes, adaptation, and higher grain yield and grain weight are practiced simultaneously (Bontz, 2011; Nabukalu and Cox, 2016).

As is true of any production system, there are disadvantages. Because perennial crops reduce the opportunity for and may eliminate crop rotation, the habitat for some pests may improve. Research on perennial upland (nonpaddy) rice production began at the International Rice Research Institute in 1988 and continues at the Yunnan Academy of Agricultural Sciences in China, which is close to releasing perennial rice varieties (Bird, 2015). Perennial crop research proceeds with several other annual crops: wheat, rye, corn, millet, barley, and oats (Bozzini, 2014). The eventual release of perennial crops to the world's farmers will change weed management in organic and conventional agriculture. Teasdale and Mirsky (2015) claim that "weed control is a major constraint to adoption of reduced tillage practices for organic grain production." Timing of planting can determine the dominant weed species. Perennial grain crops will not eliminate the need for weed management, but the species to be managed and techniques will change.

Picasso et al. (2008, 2011) proposed that cropping systems that rely on renewable energy and resources and are based on ecological principles are much more likely to be stable (sustainable) into the future than are conventional systems. The compelling rational for the Land Institute's program is that in diversity there is strength and that strength is lacking in the dominant, monocultural, conventional production systems. Non-agricultural, natural ecosystems communities with higher diversity have higher productivity and sustain themselves. Picasso et al. studied the "diversity effects … of seven perennial forage species and two-to six-species mixtures" versus monoculture. The work showed that biomass productivity increased with seeded species richness in all environments. They concluded that "choosing a single well-adapted species for maximizing productivity may not be the best alternative. High levels of species diversity should be included in the design of productive and ecologically sound agricultural systems." Their conclusion is consistent with the stated goal of most organic production systems.

THINGS TO THINK ABOUT

1. Why is weed management in organic farming systems so challenging?
2. What is the essential difference between weed management in organic/alternative farming systems and conventional agriculture?
3. Can organic farming systems feed the world? Why? Why not?
4. Describe the weed management techniques available to organic agriculture.
5. Why is weed management in organic agriculture a continual process?
6. How do the several definitions of organic/alternative agriculture differ?

Literature Cited

Bailey, L.H., 1915. The Holy Earth. Sowers Printing Company, Lebanon, PA, 117 pp.
Batello, C., Wade, L., Cox, S., Pogna, N., Bozzini, A., Choptiany, J. (Eds.), 2014. Perennial Crops for Food Security. Proceedings of UN/FAO Expert Workshop. Food and Agriculture Organization of the United Nations, Rome, 390 pp.
Bello, W., 2009. The Food Wars. Verso, London, UK, 176 pp.
Benaragama, D., Shirtliffe, S.J., Johnson, E.N., Duddu, H.S.N., Syrovy, L.D., 2016. Does yield loss due to weed competition differ between organic and conventional cropping systems? Weed Res. 56 (4), 274–283.
Bindraben, P., Rabbinge, R., 2005. Development perspectives for agriculture in Africa—technology on the shelf is inadequate. In: North-South Discussion Papers. Wagenigen, The Netherlands. 7 pp.
Bird, W., 2015. Perennial rice: in search of a greener, hardier staple crop. In: Yale environment 360. Yale School of Forestry and Environmental Studies, 5 pp. See: http://e360.yale.edu/feature/perennial_rice_in_search_of_a_greener_hardier_staple_crop/2853/.
Bontz, S., Fall 2011. Overcoming negative relationships. In: Land Report 101. The Land Institute, Salina, KS, pp. 7–15.
Bozzini, A., 2014. Present situation concerning the introduction of perennial habit into most important annual crops. In: Proceedings of UN/FAO Expert Workshop. Food and Agriculture Organization of the United Nations, Rome, pp. 376–379, 390 pp.
Bronowski, J., 1973. The Ascent of Man. Little, Brown and Company, Boston, MA, 448 pp.
Buck, L.B., Scherr, S.J., 2011. State of the World—Innovations that Nourish the Planet. W.W. Norton & Company, Inc., New York, NY, pp. 15–26, 237 pp.

Carpenter-Boggs, L., Reganold, J.P., Kennedy, A.C., 2000. Biodynamic preparations: short- term effects on crop, soils, and weed populations. Am. J. Altern. Agric. 15 (3), 110–118.

Clark, A. (Ed.), 2007. Managing Cover Crops Profitably Handbook Series Book, third ed., vol. 9. Sustainable Agriculture Network, Beltsville, MD. 244 pp.

Darby, H., Dawson, J., Delate, K., Goldstein, W., Heckman, J., Leval, K., Seiter, S., Summer 2007. Letters to the editor in response to 'going organic' – spring 2007. Crops Soils 14.

Dayan, F.E., Cantrell, C.L., Duke, S.O., 2009. Natural products in crop protection. Bioorg. Med. Chem. 17, 4002–4034.

DeDecker, J.J., Masiunas, J.B., Davis, A.S., Flint, C.G., 2014. Weed management practice selection among midwest U.S. organic growers. Weed Sci. 62 (3), 520–531.

Deryckx, W., 2001. My view. Weed Sci. 49, 1.

Dorn, B., Jossi, W., Van der Heijden, M.G.A., 2015. Weed suppression by cover crops: comparative on-farm experiments under integrated and organic conservation tillage. Weed Res. 55, 586–597.

Drinkwater, L.E., 2009. Ecological knowledge: foundation for sustainable organic agriculture. Monograph 54. In: Francis, C.F. (Ed.), Organic Farming: The Ecological System. American Society of Agronomy, Madison, WI, pp. 19–47.

Evans, G.J., Bellinder, R.R., Hahn, R.R., 2011. Integration of vinegar for in-row weed control in transplanted bell pepper and broccoli. Weed Technol. 25, 459–465.

Monograph 54. In: Francis, C.F. (Ed.), 2009. Organic Farming: The Ecological System. American Society of Agronomy, Madison, WI, 353 pp.

Francis, C., Hodges, L., 2009. Human ecology in future organic farming systems. Monograph 54. In: Francis, C.F. (Ed.), Organic Farming: The Ecological System. American Society of Agronomy, Madison, WI, pp. 301–323.

Gardner, M., 1993. Steiner, R. (1924b). Report to members of the Anthroposophical Society after the agriculture course, Dornach, Switzerland, June 20, 1924. In: Gardner, M. (Ed.), Spiritual Foundations for the Renewal of Agriculture by Rudolf Steiner. Bio-Dynamic Farming and Gardening Association, Kimberton, PA, pp. 1–12.

Gianessi, L.P., Reigner, N.P., 2007. The value of herbicides in U.S. crop production. Weed Technol. 21, 559–566.

Gilman, S., 2002. Organic Weed Management. Chelsea Green Pubs., White River Junction, VT, 62 pp.

Glover, J., Reganold, D.J.P., Bell, L.W., Borevitz, J., Brummer, E.C., Buckler, E.S., Cox, C.M., et al., 2010. Increasing food and ecosystem security through perennial grain breeding. Science 328 (5986), 1638–1639.

Halweil, B., May/June 2006. Can organic farming the us all? Worldwatch Mag. 19 (3), 3 pp.

Hansen, L., Now, E., Højring, K., 2006. Nature and nature values in organic agriculture. An analysis of contested concepts and values among different actors in organic farming. J. Agric. Environ. Ethics 19, 147–168.

Howard, A., 1940. An Agricultural Testament. New York, London, 253 pp.

James, W., 1975. Pragmatism. Harvard Univ. Press, Cambridge, MA, p. 95. Originally published in 1907.

Johnson, H.J., Colquhoun, J.D., Bussan, A.J., Rittmeyer, R.A., 2010. Feasibility of organic weed management in sweet corn and snap beans for processing. Weed Technol. 24, 544–550.

Johnson, S.P., Miller, Z.J., Lehnoff, E.A., Miller, P.R., Menalled, F.D., 2017. Cropping systems modify the impacts of biotic plant-soil feedbacks on wheat (*Triticum aestivum* L.) growth and competitive ability. Plants Soils.

Kirschenmann, F., 2010. Cultivating and Ecological Conscience: Essays from a Farmer Philosopher. Counterpoint, Berkeley, CA, 403 pp.

Leopold, A., 1966. A Sand County Almanac – With Essays on Conservation from Round River. Ballantine Books, New York, NY, p. 262.

Liebman, M., 2001. Weed management: a need for ecological approaches. In: Liebman, M., Mohler, C.L., Staver, C.P. (Eds.), Ecological Management of Agricultural Weeds. Cambridge University Press, Cambridge, UK, pp. 1–39.

Liebman, M., Davis, A.S., 2009. Managing weeds in organic farming systems: an ecological approach. Monograph 54. In: Francis, C.F. (Ed.), Organic Farming: The Ecological System. American Society of Agronomy, Madison, WI, pp. 173–196.

Liebman, M., Gallandt, E.R., 1997. Many little hammers: ecological approaches for management of crop-weed interactions. In: Jackson, L.E. (Ed.), Ecology in Agriculture. Academic Press, San Diego, CA, pp. 291–343.

Liebman, M., Mohler, C.L., Staver, C.P., 2001. Weed management: the broader context. In: Liebman, M., Mohler, C.L., Staver, C.P. (Eds.), Ecological Management of Agricultural Weeds. Cambridge University Press, Cambridge, UK, pp. 494–518.

Lipson, M., 1997. Searching for the "O-word" – Analyzing the USDA Current Research Information System for Pertinence to Organic Farming. Organic Farming Research Foundation, Santa Cruz, CA, 84 pp.

Menallad, F.D., Liebman, M., Renner, K.A., 2006. The ecology of seed predation in herbaceous crop systems. In: Singh, H.P., Batish, D.R., Kohli, R.K. (Eds.), Handbook of Sustainable Weed Management. Food Products Press, Binghamton, NY, pp. 297–327.

Menalled, F.D., Gross, K.L., Hammond, M., 2001. Weed aboveground and seed bank community responses to agricultural management systems. Ecol. Appl. 11, 1586–1601.

Menalled, F.D., Kohler, K.A., Buhler, D.D., Liebman, M., 2005. Effects of composted swine manure on weed seedbank. Agric. Ecosyst. Environ. 111, 63–68.

Mirsky, S.B., Ryan, M.R., Teasdale, J.R., Curran, W.S., Reberg-Horton, C.S., Spargo, J.T., et al., 2013. Overcoming weed management challenges in cover crop–based organic rotational no-till soybean production in the Eastern United States. Weed Technol. 27 (1), 193–203.

Mohler, C.L., Liebman, M., Staver, C.P., 2001. Weed management: the need for ecological approaches. In: Mohler, C.L., Liebmnan, M., Staver, C.P. (Eds.), Ecological Management of Agricultural Weeds. Cambridge University Press, Cambridge, UK, p. 139.

Mohler, C.L., Marschner, C.A., Caldwell, B.A., DiTomasso, A., 2016. Weed mortality caused by row-crop cultivation in organic corn-soybean-spelt cropping systems. Weed Technol. 30 (3), 648–654.

Nabukalu, P., Cox, T.S., 2016. Response to selection in the initial stages of a perennial sorghum breeding program. Euphytica 209 (1), 103–111.

Northbourne, C.J., 5th Lord., 2003. Look to the Land. 2nd Rev., Spec. ed. Sophia Perennis, Hillsdale, NY, 114 pp. First ed. 1940. J.M. Dent & Sons.

Paull, J., 2011. Attending the first organic agriculture course: Rudolf Steiner's Agriculture Course at Koberwitz, 1924. Eur. J. Soc. Sci. 21 (1), 64–70.

Picasso, V.D., Brummer, E.C., Liebman, M., Dixon, P.M., Wilsey, B.J., 2008. Crop species diversity affects productivity and weed suppression in perennial polycultures under two management strategies. Crop Sci. 48 (1), 331–342.

Picasso, V.D., Brummer, E.C., Liebman, M., Dixon, P.M., Wilsey, B.J., 2011. Diverse perennial crop mixtures sustain higher productivity over time based on ecological complementarity. Renew. Agric. Food Syst. 26 (4), 317–327.

Poffenbarger, H.J., Mirsky, S.B., Teasdale, J.R., Spargo, J.T., Cavigelli, M.A., Kramer, M., 2015. Nitrogen competition between corn and weeds in soils under organic and conventional management. Weed Sci. 63 (2), 461–476.

Pollnac, F.W., Maxwell, B.D., Menalled, F.D., 2009. Using species-area curves to examine weed communities in organic and conventional spring wheat systems. Weed Sci. 57, 241–247.

Röling, N., March 23, 2006. Organic Agriculture and World Food Security. Zie en Onzin van Biologische Landbouw – An academic debate, Wageningen, Netherlands.

Ryan, M.R., Mortensen, D.A., Bastiaans, L., Teasdale, J.R., Mirsky, S.B., Curran, W.S., Seidel, R., Wilson, D.O., Hepperly, P.R., 2010a. Elucidating the apparent maize tolerance to weed competition in long-term organically managed systems. Weed Res. 50, 25–36.

Ryan, M.R., Smith, R.G., Mirsky, S.B., Mortensen, D.A., Seidel, R., 2010b. Management filters and species traits: weed community assembly in long-term organic and conventional systems. Weed Sci. 58 (3), 265–277.

Salonen, J., Hyvönen, T., Kaseva, J., Jalli, H., 2012. Impact of changed cropping practices on weed occurrence in spring cereals and Finland – a comparison of surveys in 1997–1999 and 2000–2009. Weed Res. 53, 110–120.

Schaller, F.W., Thompson, H.E., Smith, C.M., 1984. Conventional and Organic Related Farming Systems Research: An Assessment of USDA and State Research Priorities. Iowa State University, Ames, IA, 74 pp.

Sen, A., 1981. Poverty and Famines: An Essay on Entitlement and Deprivation. Clarendon Press, Oxford, UK, 257 pp.

Seufert, V., Ramankutty, N., Foley, J.A., 2012. Comparing the yields of organic and conventional agriculture [Research letter] Nature 485, 229–232.

Smith, R.G., Mortensen, D.A., Ryan, M.R., 2010. A new hypothesis for the functional role of diversity in mediating resource pools and weed-crop competition in agroecosystems. Weed Res. 50, 37–48.

Sørensen, C.G., Madsen, N.A., Jacobsen, B.H., 2005. Organic farming scenarios: operational analysis and costs of implementing innovative technologies. Biosyst. Eng. 91, 127–137.

Tautges, N.E., Goldberger, J.R., Burke, I.C., 2016. A survey of weed management in organic small grains and forage systems in the northwest U.S. Weed Sci. https://doi.org/10.1614/WS-D-15-00186.1. On-line.

Taylor, E.C., Renner, K.A., Sprague, C.L., 2012. Organic weed management in field crops with a propane flamer and rotary hoe. Weed Technol. 26 (4), 793–799.

Teasdale, J.R., Mirsky, S.B., 2015. Tillage and planting date effects on weed dormancy, emergence, and early growth in organic corn. Weed Sci. 63, 477–490.

Thompson, P.B., 2010. The Agrarian Vision: Sustainability and Environmental Ethics. University Press of Kentucky, 323 pp.

Trewavas, A., 2001. Urban myths of organic farming. Nature 410, 409–410.

Turner, R.J., Davies, G., Moore, H., Grundy, A.C., Mead, A., 2007. Organic weed management: a review of the current UK farmer perspective. Crop Prot. 26, 377–382.

Ullrich, S.D., Buyer, J.S., Cavigelli, M.A., Seidel, R., Teasdale, J.R., 2011. Weed seed persistence and microbial abundance in long-term organic and convention and cropping systems. Weed Sci. 59, 202–209.

White, S.N., Boyd, N.S., 2016. Effect of dry heat, direct flame, and straw burning on seed germination of weed species found in lowbush blueberry. Weed Technol. 30 (1), 263–270.

Williams, C.L., Liebman, M., Westerman, P.R., Borza, J., Sundberg, D., Danielson, B., 2009. Over-winter predation of *Abutilon theophrasti* and *Setaria faberi* seeds in arable land. Weed Res. 49, 439–447.

Worldwatch Institute, 2011. State of the World – Innovations that Nourish the Planet. W.W. Norton & Company, Inc., New York, NY, 237 pp.

Young, S.L., 2009. My view. Weed Sci. 57, 449–450.

Zimdahl, R.L., 2004. Weed-Crop Competition: A Review, second ed. Blackwell Publishing, Ames, IA. 220 pp.

Zimdahl, R.L., 2011. Agriculture's Ethical Horizon, second ed. Elsevier, London, UK. 274 pp.

Biological Weed Control

FUNDAMENTAL CONCEPTS

- Biological control is the use of living organisms (e.g., parasites, predators, or pathogens) to reduce another organism's population to a lower average density than would occur in their absence.

- Most biological control organisms have not escaped to become pests.

- Biological weed control alone cannot solve all weed problems and is best regarded as a technique to be used in integrated weed management systems.

LEARNING OBJECTIVES

- To know the advantages and disadvantages of biological weed control.

- To understand the importance of specificity in the development of biological control strategies.

- To know the different kinds of organisms that have been used for biological control of weeds.

- To know the ways that biological weed control can be used.

- To appreciate the opportunities for integration of biological and other weed control methods.

1. GENERAL

Plant distribution is determined by edaphic, climatic, and biotic factors. On a given site, soil type and climate can be studied but cannot be controlled by humans. The biotic environment can be manipulated. If manipulation is through stable interactions, biological control may be possible. For an extended discussion of biological control, interested readers are referred to Hajek (2004) or one of several other authoritative books.

Biological control has been successful in a few cases, but it will never be the solution to every weed problem. Where it is successful it is employed as one weed management practice among many. Primarily because of well-known problems with chemical weed control, biological control may be important, but it has not become more so relative to other management techniques. In view of the current ease of use, cost, and success of available control and management techniques, one cannot be sanguine about future widespread use of biological control, especially in agronomic and horticultural crops. Hinz et al. (2014) describe five successful, classical, biological weed control programs in the United States (St. John's wort, field bindweed, purple loosestrife, toadflaxes, and tropical soda apple). They suggest that biological control is "one of the more cost-effective and the only truly sustained means for the management of exotic invasive plants in the United States." It "has significantly advanced as a scientific discipline since the 1950s" and has "a long history of safe and effective implementation around the world." But in their view, it is "being held to a higher ecological standard than other control tactics." None of the control agents introduced for management of those five weeds would be approved for use today. Three would be rejected because they could attack economically important plants, and two because of their potential attack on threatened or endangered plants. Thus, one must conclude that biological control programs face significant scientific and regulatory challenges.

However, it is wrong to assume that nothing is happening. Sources of specific information include Turner (1992), Julien (1992), Coombs et al. (2004), and Rees et al. (1996). Coombs et al. (2004) include 25 weeds for which biological control agents have been released and 15 other weeds for which studies were in progress in 2004, but for which no releases had been made. Tu et al. (2001) listed 63 weeds with released or available biocontrol agents, 11 with available native biocontrol agents, and 27 with agents being studied. A more complete story is available in the massive compilation by Winston et al. (2014). The first edition (Julien, 1992) of the catalog included 499 releases of exotic agents between the late 19th century and 1980. The second edition (1987) recorded 100 new releases by 1985, the third (1992) recorded 132 releases by 1990, and the fourth (1988) recorded 222 more. The fifth edition includes 2042 entries from 132 countries and 551 biocontrol agents targeting 224 weeds from 55 plant families. Clearly much as happened in the world of biological weed control.

Charudattan began studying biological control of weeds in 1971. The idea of "deliberately using pathogens to control weeds was novel, untested, and met with skepticism and resistance." Charu's reflection on and review of the field (2010) noted spectacular success and his view that some opportunities were missed. Weed "biocontrol research remained largely preoccupied with agent or product development and deployment while great strides were made … in phytopathology to understand the genetic-molecular basis of virulence, host range, host specificity, host response to infection, cell death, and pathogen population structure." Given future cooperation in these areas, Charudattan is optimistic about the future development of weed control pathogens.

1.1 Definition

Biological control, a term first used by H.S. Smith (DeBach, 1964), of any pest organism is the use of living organism(s) (e.g., a parasite, predator, or pathogen) to reduce another organism's population to a lower average density than would occur in its absence. Biological control is usually thought of as requiring intentional introduction of a parasite, predator, or pathogen to achieve control, but it is also a natural phenomenon. Scientists can introduce an agent or discover the control potential of a natural biological control agent and use either to achieve human ends. The aim is to maintain the offending organism's population at a lower average density, not to eradicate but to reduce it to a level that has little or no negative environmental effects, and enhances or improves production of desirable plants.

1.2 Advantages

In its classical or idealized form, biological weed control can be permanent weed management because once an organism is released, it may be self-perpetuating and control will continue without further human intervention (Table 12.1). This is true when some fungal species are released in an inundative approach to control a weed. Some have called this the myco-herbicide approach. Subsequently, if theory becomes reality, the weed does not have to be actively managed by people, the biological organism provides control without further human intervention. This ideal biocontrol is certainly not always achieved, and in fact rarely is.

Self-perpetuation is an advantage that other weed control techniques do not have. There are no chemical environmental residues from biological control other than the organism, which some consider a potential problem because it is foreign or unnatural in the environment in which it is released. In the classical, idealized version of biological control, this does not happen because extensive research before release establishes (one hopes) that the

TABLE 12.1 Summary of Advantages and Disadvantages of Biological Weed Control (Wapshere et al., 1989)

Advantages	Disadvantages
1. Reasonably permanent	1. Control is slow
2. Self-perpetuating	2. No guarantee of results
3. No additional inputs required once agent is established successfully	3. Establishment may fail for many reasons
4. No harmful human side effects	4. There may be unknown ecological effects; mutation to an undesirable form is possible
5. Attack is limited to target weed and a few close relatives	5. If target is related to a crop, the number of potential biocontrol agents is low
6. Risks are known and evaluated before release	6. Some risks may not be known and cannot be evaluated before release
7. Control often depends on host density	7. Does not work well in short-term cropping cycles; works best in stable environments
8. Self-dispersing spread to suitable host habitats	8. Restriction of spread to area of initial dispersal is impossible
9. Costs are nonrenewing	9. Initial investment of time, money, and personnel can be high
10. High benefit–cost ratio for successful programs	10. Eradication is not possible; must maintain host population at low level to maintain control agent

organism is environmentally benign. In theory, there is no environmental pollution. Of course, there is no chemical pollution, but there may be environmental or biological pollution from a particular biocontrol organism. There is no environmental or mammalian toxicity as there may be after chemical use. Because they most often invade environmentally sensitive areas where all kinds of pollution are to be avoided (aquatic and stream-side areas), biocontrols may be the best option for management of invasive species (see Chapter 7) (Myers and Bazely, 2003, p. 12). In ideal cases, initial costs are nonrecurring, and usually once an organism is established, no further inputs are needed. Development costs may be lower than for herbicides (Auld, 1991). Although all of these advantages do not accrue to all organisms developed for biological weed control, they are cited commonly to justify research and greater employment of biological control.

1.3 Disadvantages

There are some situations in which biological control is not appropriate. If a plant is a weed in one place and valued in another in the same general geographic region, biological control is inappropriate (Table 12.1). Once introduced, spread of a biological control organism cannot be controlled. The control organism is unable to distinguish plants humans may regard as valuable from weedy relatives. For example, artichoke thistle (also called cardoon) is a weed on some California rangeland. It is closely related to cultivated artichoke. Introduction of a biocontrol agent to control the weedy artichoke thistle is discouraged by artichoke growers because it is assumed that the biocontrol agent would lack specificity. A Eurasian weevil was introduced to North America to control the invasive and generally weedy musk thistle but it attacks native, nonpest thistles (Mack et al., 2000). There are ornamental species of delphinium related to weedy larkspurs, which make the weeds questionable targets for biological control. Other weedy species may be related to valuable native plants. Controversy regarding problems related to how closely related potential target species are to native or desired species focuses on two issues (Mack et al.). The first asks whether there is sufficient administrative infrastructure to monitor and detect nontarget effects. The second, a scientific question, addresses the likelihood that an introduced biocontrol agent will evolve to attack other possible hosts. Scientists who study biocontrol usually have "very limited knowledge of the factors that limit effectiveness of control organisms and much of that knowledge is subjective" (McEvoy, 2002). Safety concerns are best addressed by research on host specificity because it is the criterion that provides the best assurance that a biocontrol organism will suppress the host without harming other species (McEvoy). An absolute demand for specificity of a biocontrol agent means development must be research intensive; it may often require a large budget and several years of research. Research must address the uncertainty about organism movement, evolution, indirect effects, and the probability, severity, and consequences of nontarget effects (McEvoy). These are not easy, inexpensive tasks. Hobbs and Humphries (1995) proposed that it is highly unlikely that biological weed management will ever become an important weed management technique except for a very few species. The accuracy of their prediction has been repeatedly verified.

Biological control is inherently slow and results are not guaranteed. In many crops, but not in noncrop or natural areas, weeds must be controlled during a brief, critical period (Zimdahl, 2004, pp. 109–130), often only days or weeks, to prevent yield reduction. In addition, because eradication is an inappropriate, indeed usually impossible goal for biological control, weeds that should be eradicated on some sites (e.g., larkspur on rangeland) may be managed better with other techniques. Some species are geographically local, minor weeds, and development of a biological control would be expensive and financially unwise because of the small infested area. Cropland weeds exist in an ecologically unstable habitat that is often a poor environment for the successful introduction, survival, and population growth of biocontrol organisms. Cropland weeds also exist in a weed complex, rarely as a single species. Because biocontrol is necessarily directed at a single species, it is an inappropriate choice for managing the weed complex found in most crops. Projects are often constrained by the expense of finding a natural enemy in the native habitat. Locating the natural or native habitat is a difficult research task; and even if it is found, aggressive natural enemies may not be abundant, if they even exist.

Because science can never know all possible ramifications of any technological intervention, other cautions should be considered. Release of a biological control organism can induce competitive suppression or extinction of native biological control organisms and other desirable organisms. A corollary is that other harmful or beneficial species may increase in abundance. Such events could lead to loss of biological diversity, loss of existing biocontrol, release of species from competitive regulation, disruption of plant community structure, suppression of organisms essential to ecological integrity, and disruption of food chains and nutrient cycling (Lockwood, 1993).

Biological control is slow, often less effective, and commonly less certain than herbicides or mechanical control. Biocontrol, particularly in disturbed cropping situations, will not control many different weeds as other techniques do. It will not eradicate weed problems, but most other techniques also will not. It is an intervention technique that may have unanticipated effects, as herbicides do. An argument in favor of other control techniques (tillage, flame, rotation, and herbicides) is that they are reversible. That is the method can be stopped, whereas once a biological control agent is released, it cannot be controlled or recalled. This is clearly a weak argument relative to herbicides, which have often desirable residual effects. It seems that biological control is simultaneously praised and chastised because it is permanent.

One example of an unanticipated effect is shown by the work of Callaway et al. (1999). Their 2-year study demonstrated that the widely used knapweed root moth (*Agapeta zoegana*) had no significant effect on the biomass of spotted knapweed. The counterintuitive result of their work was that herbivory by the moth may lead to increased negative effects of spotted knapweed on neighboring native plant species. In this case, use of the biocontrol to weaken the invader (spotted knapweed) so natives could gain a competitive advantage led to the opposite result.

1.4 Use Considerations

Research to find and introduce agents for biological control of weeds has revealed that it is usually (but not always) easier to control an introduced weed species that was freed of natural predators once it was introduced. Second, research has shown that it is best to introduce predators that have been freed of their natural predators during introduction to the weed's new habitat. These are not absolute requirements, but they mean that development of successful biological control can be achieved only by careful research that addresses at least the following seven questions about the organism proposed for biological control of any weed and an additional regulatory question.

1.4.1 Does the Weed to Be Controlled Have a Native Habitat?

The common agricultural weeds redroot pigweed, groundsel, common lamb's-quarter, and common chickweed are distributed worldwide. Their origin, where they came from, is unknown. When the native habitat is unknown, one cannot go to it to find a predator. Some suggest that many weeds are homeless, having evolved from diverse parentage under various kinds of human-created agricultural pressure (Ghersa et al., 1994). Their home habitat cannot be known.

1.4.2 Will an Insect or Disease Control the Weed?

The question is, can an effective natural enemy be found? The quest is difficult and without assured success because many plants do not have effective natural enemies.

1.4.3 Will It Thrive and Reproduce in a New Habitat?

If an effective natural enemy is found, whether it is an insect or a disease, one must determine whether it will thrive in the weed's habitat. Will it be free of its old predators and not be subject to new ones in its new habitat? If it does not find its new neighborhood(s) to be congenial, it will fail. This is a reason some potentially good biological control agents are abandoned: they meet many perhaps stronger or better-adapted new enemies in their new home. Will the introduced agent have fecundity and the ability to reproduce in the new habitat? Furthermore, will it be able to occupy all niches the weed (the host) infests successfully?

1.4.4 Does It Have the Same Genetic Composition?

Another way to ask the question is: Is the target weed's genetic composition in its present home identical to its now distant relatives in its old home? Has migration to a new habitat changed the weed to be controlled in any significant way that may render the proposed organism ineffective? Although this is an interesting question, its relevance has not been determined.

1.4.5 Can It Be Reared in Captivity?

If it cannot, it will be necessary to import large quantities of the organism, which, of course, may not be possible.

1.4.6 Will It Have Adequate Searching Capacity?

After the organism is released in its new environment, will it be able to search for the weed to be controlled and be self-dispersing in the right places?

1.4.7 Will It Be Specific?

The most important criterion and the absolute rule for successful biological control is that if an insect or disease is able to clear all of these hurdles, it must be specific. Specificity means that it will attack and control one plant (the weed) and no others. This is the acid test for biocontrol agents.

1.4.8 Will It Obtain Regulatory Approval?

Increased concern about all kinds of invasive species have effectively increased the regulatory hurdles. Greater specificity of risk and benefit at the habitat level must be considered. The primary goal of specificity mentioned previously remains important and a defined post-release monitoring program is required (Hinz et al., 2014).

The answer to each of these questions is important and organisms proposed for biocontrol of weeds often fail because one or more of the answers is negative. Mistakes have been made when the entire complexity was not understood or ignored. In fact, the history of biological control has demonstrated that it is easy to make mistakes when a biocontrol agent is introduced. Each introduction creates a new combination of organism and environment. Both must be understood, and often they are not (US Congress, 1993). The vacant niche hypothesis has been used to rationalize or prevent introductions. The hypothesis is that some ecological roles (i.e., a population-regulating organism) are not filled in a place where biological control is desired; thus the niche is empty and can be filled. However, few species fit the narrow ecological vacancy identified by those who wish to control weeds, and it is virtually impossible to predetermine the role a species will have after release (US Congress).

There are several examples of poor understanding. The mongoose (*Marathi mangus*) was imported to Hawaii to control rats, which made sugarcane harvest unpleasant and reduced yield. A mongoose will kill a rat when it meets one. However, rats are nocturnal and the mongoose hunts during the day. They rarely meet. The mongoose eats bird eggs, had no natural enemies in Hawaii, and became and remains a pest.

Problems can arise when an introduced species moves beyond the area intended. The cactus moth (*Cactoblastis cactorum*) was introduced to the West Indies to control prickly pear cactus, a native of tropical America, a task it did well. It moved north to Florida, where it threatens indigenous, nonweedy prickly pear cacti in Florida and neighboring states (Kass, 1990); several Florida *Opuntia* species are rare and endangered and occur nowhere else in the United States (Stilling and Moon, 2001; US Congress, 1993).

The seven-spotted lady beetle (*Coccinella septempunctata*), an aphid predator, has dispersed throughout much of the United States. It appears to be outcompeting the native nine-spotted lady beetle (*Coccinella novemnotata*) and has displaced that species in alfalfa. Finally, the US Environmental Protection Agency and the Oregon Department of Environmental Quality funded a large project to eradicate weeds in Devil's Lake on the northern Oregon coast. About 30,000 weed-eating carp (*Ctenopharyngodon idella*), a successful aquatic weed control agent, were introduced into the lake to control Eurasian water milfoil. The liquefied fecal waste from the fish created new, unprecedented algal blooms and new weed crops. Six years after the project was initiated, there was no significant reduction in the total amount of aquatic vegetation and only 4000 carp survived. Intensive real estate development in the lake's pristine watershed, clear-cut logging, and recreation proceeded without inhibition and were all major contributors to the lake's pollution and eutrophication (Larson, 1996).

The southern United States has a love—hate relationship with kudzu, a native of China and India. Another Asian import, the bean plataspid (*Megacopta cribraria*), is eating kudzu. No one knows for sure how the beetle arrived in the United States. It just appeared in 2009 in Atlanta, Georgia. If people think kudzu is a weed, which many do, they will inevitably favor the beetle's spread. But if people participate in or really think kudzu festivals, kudzu jelly, and kudzu queens are neat and ought to be preserved, the beetle is probably not a good idea.

Thus, the controversy. Besides kudzu has been around since 1876, when the US Department of Agriculture introduced it for erosion control at the Philadelphia Centennial Exposition. Why, it is practically a native, and no matter what is done, it cannot be eradicated.

We humans are residents of the world, not its rulers. We can assume temporary custodianship; we can care for it. But to care properly, we must learn to understand nature's purposes and our assumed and proper role in aiding or changing them. Similar to other technology, biological weed control can lead us toward harmony with nature or away from it. Scientists must determine and citizens and users must understand the place of biological weed control in nature's scheme.

It would be a tragedy if a biocontrol agent were released to control a particular weed and it was discovered after the weed's population was reduced that the biocontrol organism had a natural appetite for rosebushes. Only a few of the more than 100 organisms released for biocontrol of weeds worldwide have become pests subsequent to their release, but as these examples show, it has happened. Biological control research is difficult and crucial to success. Plants in the weed's host range that are tested to ensure specificity include (Strobel, 1991):

- those related to the target weed,
- those not adequately exposed to the agent for ecological or geographic reasons,
- those whose natural enemies are unknown or about which little is known,
- those with secondary chemicals or morphological structures similar to those of the target weed, and
- those attacked by close relatives of the agent.

About 40% of the successful instances of pest biocontrol have involved an unrelated natural enemy. These were new associations between a host and biocontrol agent, and the host entirely lacked natural resistance to the new enemy (Pimentel, 1963, 1991). Finding and introducing a biocontrol species and ensuring that it clears all of the hurdles (eight points just mentioned) is a difficult, challenging task. A major risk is misunderstanding the nature of host specificity. Not enough is known about how natural enemies find and control weed hosts. Why do they do it?

In addition to the fundamental biological questions, there are questions that those who develop to sell must ask (Auld, 1991). These include concern about the size and stability of the market and what competing management techniques there are. Manufacturers must also be concerned about the possibility of patenting their product (the organism, not the procedure) to protect their investment and create a reasonable guarantee of profit. Finally, they must ask what is known or not about the organism and how much it will cost to develop it into a biological control agent (Auld).

Given the advantages and disadvantages of biological weed control, there are, and will continue to be, conflicting interests. A plant that is a pest in one place may be beneficial in another, or at least it may be liked by some (e.g., kudzu, prickly pear cacti). The spread of an organism once it has been released cannot be controlled. Future and present values must be considered, as must minority and majority interests, neighbors (local, national, and international), direct and indirect effects on other species, and the environment (Huffaker, 1964). A few examples illustrate the complexity (Huffaker).

Prickly pear, mentioned earlier, is one of the best examples of successful biological control of a weed by an insect. The first prickly pear was introduced to Australia from Brazil by

Captain Arthur Phillip in 1788. Myers and Bazely (2003) suggest that this was done to develop a source of red dye for the red coats of the British army. The cochineal bugs that feed on the cactus were the source of the dye. It spread widely in Australia as an ornamental and a hedgerow plant. It was also valued as a source of fruit and forage for cattle during drought. Some species are still used as hedgerow fences and for fruit in North Africa. By 1916, prickly pear had invaded more than 60 million acres in Australia and was estimated to be spreading at 1 million acres a year. It was an environmental disaster.

In 1924, exploration in Argentina found a moth borer (*C. cactorum*) that attacked a variety of cactus (*Opuntia*) species in Argentina. The moth is native to northern Argentina and parts of Peru and Paraguay. Moth eggs were collected in Argentina, sent to Australia by ship, reared, and released from 1925 to 1929 (Myers and Bazely, 2003). Within 3 years of the introduction of the moth borer, the prickly pear area was transformed as if by magic from a wilderness of 60 million acres of prickly pear to prosperous agricultural land. No one in Australia objected. The moth was also introduced in South Africa.

The cactus moth was introduced into the Caribbean islands in the 1960s, and by 1986 it had found its way to Florida (Habeck and Bennett, 1990). By 2003, the cactus moth was established in Florida and Georgia, and along the Atlantic coast almost as far north as Charleston, South Carolina (Hight et al., 2002; Bloem, 2003). By the end of 2003, it had spread as far west as Pensacola, Florida, near the Alabama state line. In 2004 it was found in Alabama and northern South Carolina. Now it has been identified in Arizona, California, Nevada, New Mexico, and Texas.

Although it is the best example of successful biological control of a weed, it was not welcome in other places. In Hawaii, there were vigorous objections to introduction of the same moth borer. Cattlemen objected because the tree cactus was useful as feed and as a source of otherwise unavailable water on some ranges. The introduction program was opposed on the US mainland for similar reasons.

Mexico, a center of cactus diversity, has more than 100 species, some of which are used and valued as fruit and vegetable crops. There are more than 3 million hectares (ha) in wild areas. Cactus is cultivated for fodder on 150,000 ha, for production of the pear on 60,000 ha, and for leaf on 10,500 ha. The plant is worth more than $100 million/year to the Mexican economy. It is an important food crop for the rural poor, a major component of arid and semiarid ecosystems, an important desert plant, wildlife food, a guardian of ecological balance, and an integral part of the culture. It had not been documented in Mexico in 2004, although there had been interceptions (Simonson et al., 2005). It was found along the Gulf Coast in 2004, but the populations were eradicated (Anonymous, 2009). The moth was detected on Isla Mujeres on Jul. 31, 2006; in Isla Contoy National Park on May 4, 2007, eradication followed. As of 2009, after three biological cycles no new infestations had been found and it was officially declared eradicated.[1] The cactus is such an important part of Mexican cultural and social history that it appears on the national flag (Myers and Bazely, 2003). The prickly pear cactus biocontrol story is at once one of great success and an illustration of how little is known about what might happen.

In California, control of yellow star thistle involves cattlemen, beekeepers, fruit growers, and seed crop growers. The weed invades and damages grazing land, grain, and seed crops. Cattlemen, those primarily affected, want to get rid of it. However, the thistle is a key plant in

---

[1]https://www.iaea.org/technicalcooperation/documents/Success-Stories/mex5029.pdf.

maintenance of the bee industry for pollination of fruit and seed crops in California. The fruit and seed crop industry dominated the early debate. Five insects were introduced and are established in California, Idaho, Oregon, and Washington. A rust fungus (*Puccinia jaceae* var. *solstitialis*) originally from Turkey was released in California in 2003 for control of yellow star thistle (Rees et al., 1996; Coombs et al., 2004).

Wood-boring insects are important for control of mesquite because trees infested with wood borers are easier to burn, a primary control technique (Ueckert and Wright, 1973). Defoliating mesquite with the herbicide 2,4,5-trichlorophenoxyacetic acid caused the wood boring insects to die and resulted in trees that were more difficult to burn. Control techniques can conflict even when each is designed to accomplish the same end.

Readers interested in information on specific weeds and biocontrol agents are directed to the truly monumental worldwide compilation of Winston et al. (2014). It includes 2042 entries from 130 countries with information on 551 biocontrol agents, which target 224 weeds. The book records 319 releases since 1988. Given that the first edition published in 1982 recorded 499 releases between the late 19th century and 1980, it is clear that there is a great deal of worldwide research on biological control of weeds. The fifth edition includes information on bioherbicides (pathogens used similarly to chemical herbicides) for 18 different weeds.

In 2013, the US Department of Agriculture's (USDA's) Animal, Plant Health Inspection Service (APHIS) report[2] listed 121 exotic biological control agents approved and released for management of 34 weeds in the United States and 94 agents released for management of 22 weeds in Canada. In most cases, the same agents were approved in both countries. The decade of first release and approval in the continental United States is shown in Table 12.2. Clearly, releases have declined in the 21st century. The five weeds approved

TABLE 12.2   Approval and Release of Exotic Biological Control Agents in United States, by Decade

Decade	Number of Agents Released
1920	2
1930	0
1940	2
1950	5
1960	16
1970	19
1980	35
1990	40
2000	5
2010 (to 2016)	0

[2]https://www.aphis.usda.gov/plant_health/permits/tag/downloads/Weeds%20and%20Biological%20Control%20
Agents%20Released%20in%20US%20and%20Canada.pdf.

for biological control in the United States in the 21st century are *Arundo*, Russian knapweed, mile-a-minute weed, salt cedar (tamarisk), and yellow star thistle. There have been three 21st century releases in Canada: Russian knapweed, rush skeleton weed, and false cleavers. Since the 1920s there have been 10 accidental introductions in the United States and five in Canada for the same weed and organism.

2. METHODS OF APPLICATION

There are four methods—scientific hypotheses used to apply biological control agents (Wapshere et al., 1989; Turner, 1992; Hajek, 2004).

2.1 Classical

This method has been limited to weeds that are not closely related to crop plants and that belong to sharply defined genera or families that are taxonomically well separated from other families (Wapshere et al., 1989). Classical biological control is the intentional introduction, usually from a geographically distant place, of a host-specific, exotic, natural enemy (or several) adapted to introduced (exotic) or native weeds. The goal is permanent, long-term control. The great majority of weeds and nearly all of the worst weeds have been introduced to agricultural habitats. How they were introduced may or may not be known; it does not affect the control's success. The theory is that when a weed is introduced to a new region, it is highly likely that it was freed of natural enemies that regulated its population in its native place. Frequently natural regulation in the weed's native place was by an innocuous species. That is the proposed biological control agent and the target weed are not pests in its native habitat because the population of the target weed is regulated by the proposed agent whose population is regulated by its natural enemies.

This is an ecological approach. Introduced weeds often occur on undisturbed rangelands or in infrequently disturbed habitats (e.g., a pasture or perennial crop). Classical control works best in habitats with minimal disturbance by humans. It is the most used and most successful long-term method.

The required procedure includes consideration of the eight questions enumerated earlier: weed identification, identification of the native habitat, searching for and importation of a natural enemy, research on rearing, specificity, regulation, and ultimately (one hopes) release. The method is appropriate only with highly specific natural enemies. Arthropods and fungal pests are first choice because they are likely be specific. Vertebrate animals are usually nonspecific feeders and not suitable for importation (Turner, 1992). The targets for classical control have almost always been economically important weeds for which no other control methods have been successful and whose range has expanded to areas where it is not cost effective to control them with available methods (e.g., puncture vine, Russian thistle, diffuse and spotted knapweeds).

2.2 Augmentation

Augmentation involves artificially manipulating a natural enemy's population to achieve control. Whole habitats rather than just the weed's habitat can be modified with this technique. Ecological appropriateness, effectiveness, and the organism's virulence are still important, but they can be changed by the population or the stocking rate of the control agent. Safety and specificity are less important for the same reason. A good example is use of selectively polyphagous grazing animals. Fences or shepherds are required to manage grazing animals and expenses are high but control is possible. Use of goats in the western United States to manage leafy spurge and other noxious weed infestations is an example of broad-spectrum biological control. There are two techniques to augment populations of biocontrol agents: inundation and inoculation.

2.2.1 *Inundation*

When large numbers of control agents are raised and released, their abundance is augmented because the target area is inundated with them. When achieved, control is short-term (a growing season or shorter). Therefore, to achieve control it may be necessary to repeat releases throughout a crop season. The controlling and target organisms are usually natives, but the controlling organism may not be. Inundative control employs ecological knowledge but is essentially technological and short-term. The method eliminates costly international searches for a weed's native habitat and an organism suitable for import. It augments the inherent phytotoxicity of organisms by abruptly increasing their population. Specificity must be guaranteed. The best agents must be amenable to large-scale captive rearing and rapid reproduction. This requirement alone has inhibited inundation. A stable but easily changed resting or spore stage is helpful. Organisms used for inundation have been pathogens or nematodes rather than arthropods, which do not satisfy the criteria mentioned earlier. A cochineal scale is redistributed each year in some areas to control prickly pear, a natural process that has been going on a long time. The conscious use of inundative techniques is relatively recent. It can be regarded as a natural, evolutionary process.

2.2.2 *Innoculation*

Innoculation is the intentional release of a known biological control organism. Its success is based on the expectation that the organism has fecundity adequate to produce a population quickly that will continue to grow rapidly and be maintained at a level high enough to achieve control for extended periods. Permanent, long-term control is the objective.

2.3 Conservation

If the number of identified native parasites, predators, and diseases of the desired species could be conserved or protected and thereby allowed to increase naturally in their present environment, they ought to be effective and might give permanent, less expensive control. No imported natural enemies are released. Conservation and protection of the known control organism(s) require environmental modification to maintain the required habitat. It is also likely that current production practices, especially the use of herbicides that may be toxic to the control organism or to destroy or diminish its habitat, may have to change. Environmental and production modification must protect and ideally enhance the population of a known natural enemy. This process rests on the assumption that if the population of a potential control organism could be enhanced, its control potential would be maximized. If it is confirmed that the target organism can be managed biologically, in theory, conservation will allow a resident control organism to fulfill its control potential. It is the same principle involved in importation, but the approach is different. For example, the insect *Aroga websteri* eats foliage of big sagebrush. It has not been exploited for biological control, but presumably it could be increased in its natural habitat.

3. BIOLOGICAL CONTROL AGENTS

Biological control of weeds began after the technique was used to control insects. It began in the United States in Hawaii in 1902, when eight fruit- and flower-feeding insects were introduced from Mexico to control large-leaf lantana, a perennial shrub native to Central America. Lantana is used throughout the world as an ornamental and has escaped to become a weed (Goeden, 1988; Huffaker, 1964). Many early biocontrol efforts emphasized insects that bored in roots, stems, or seed. Boring is advantageous because it creates avenues for secondary infection by bacteria and fungi, and boring insects are usually host specific. Early efforts also emphasized agents that destroyed flowers, in contrast to those that fed only on foliage. Experience has shown that leaf eaters may be just as safe and equally effective. Now many organisms other than insects are used for biological control of weeds (Andres, 1966; Goeden et al., 1974; Holloway, 1964). A summary of 73 biological control agents approved for 26 species and several other potential agents is available for weeds in the western United States (Rees et al., 1996). Coombs et al. (2004) provided a broader summary of biological control agents for invasive species in the United States.

3.1 Classical—Inoculative Biological Control

3.1.1 Insects

Classical biological control has been used for many years. The earliest record was the release of the cochineal insect *Dactylopius ceylonicus* from Brazil to northern India in 1795 to control prickly pear cactus (Goeden, 1988). Actually the insect was not identified correctly and was believed to be a species that produced carmine dye (Goeden). It readily transferred to its natural host plant and was subsequently introduced in southern India from 1836 to 1838, where it successfully controlled prickly pear cactus. Shortly before 1865, the insect was transferred to Sri Lanka and accomplished the same thing. This was the first successful transfer of a natural enemy between countries for biological weed control (Goeden).

An example of classical biological control of prickly pear cactus was introduction of *Dactylopius opuntiae* in 1951 to Santa Cruz island off the coast of southern California. It is perhaps the best example of successful biological control of a native US weed with introduced insects (Goeden and Ricker, 1980). Over many years, the insect has given partial to complete control of prickly pear (Goeden and Ricker, 1980; Goeden et al., 1967).

A second example of weed control by an insect is the use of the French chrysomelid leaf beetle *Chrysolina quadrigemina* for control of St. John's wort. After the beetle's introduction to California in 1946, St. John's wort was removed from the state's noxious weed list (Coombs et al., 2004). The beetle's success is due to its great specificity and the synchronization of its requirements with St. John's wort's growth. It has been successful in the western United States and has been introduced to British Columbia, where it has adapted to the colder winters (Peschken, 1972). Adult beetles strip the plant at flowering in spring and early summer, and larvae feed in fall and winter (Huffaker and Kennett, 1959). The beetle's effectiveness and that of a related species (*Chrysolina hyperici*) is limited by fall rainfall. Biological control of St. John's wort has been aided because the two *Chrysolina* beetles have been joined by a root-boring insect, a gall midge, and a moth (Coombs et al.). St. John's wort is widely distributed in Australia, Canada, New Zealand, and South Africa. In the United States, it is especially prominent in California, Montana, Oregon, and Washington. Its spread has been associated with sheep movement. St. John's wort is susceptible to herbicides, but their cost and expanse and the inaccessibility of infested rangeland are disadvantages.

Biological control has been successful in Northern California, Washington, and Oregon against the poisonous, biennial weed of rangeland, tansy ragwort. Two insect species were imported from Europe (Pemberton and Turner, 1990). A cinnabar moth (*Tyria jacobaeae*) attacks leafy and flowering shoots, and larvae of the ragwort flea beetle (*Longitarsus jacobaeae*) attack the roots. These have reduced the weed to less than 1% of its density before their introduction (Turner, 1992). They have been joined by a seed head fly, which is not particularly effective but is the only well-established biocontrol agent east of the Cascade mountains in the US Pacific Northwest.

Control of puncture vine in California and Colorado is one of the few victories over an annual weed (Turner, 1992). Two weevils, the seed feeder *Microlarinus lareynii* and the stem and crown feeder *Microlarinus lypriformis,* were introduced from Italy beginning in 1961 (Maddox, 1976).

The first release of *M. lareynii* for puncture vine control in the United States was in 1961 in Nevada and California. Subsequent releases were made in Arizona, California, Colorado, Utah, and Washington. The weevil became established in Arizona, California, and Nevada, but failed in Colorado,[3] Washington, and Utah. The weevils work best where the climate is warm. They do not overwinter well in cold climates (Turner, 1992). An Oregon company[4] sells a packet of 250 adult weevils for $90.00 (shipment begins Jul. 1 each year), the recommended release per acre for moderate infestations. It is suggested that using the weevils will prevent injuries to bare feet and flat tires on bicycles caused by the tough seed pods. The company, IRV Goatheads, encourages combining biological control with burning, pulling, and herbicides; the efficacy of the latter was confirmed by Marston (2005). A combination of control techniques is advocated for managing several weeds where biological control has achieved some success.

Research has been conducted on several insects for control of leafy spurge. At least 11 different insects have been released in the United States with success varying from minor to spectacular. The leafy spurge hawk moth (*Hyles euphorbiae*), imported from Austria, Hungary, and India, eats leaves and flowers during the caterpillar stage (Harris et al., 1985), but has had only minor success. A stem- and root-boring beetle (*Oberea erythrocephala*) imported from Hungary and Italy was established in Montana and North Dakota (Leininger, 1988). The beetles puncture stems and lay eggs. Larvae bore into roots where they mature

[3]Personal experience with flat bicycle tires confirms that the weevils failed to establish in Colorado.

[4]IRV Goatheads - See www.goatheads.com, Umatilla, OR.

and exist on carbohydrate root reserves. There is evidence that the beetle prefers some biotypes over others (Coombs et al., 2004). Six species of chrysomelid flea beetles of the *Aphthona* genus (*abdominalis, cyparissae, czwalinae, flava, lacertosa*, and *nigriscutis*) were imported to the United States from Europe. Adult *Aphthona* beetles live up to 3 months and feed on leaves. Adults females lay an average of 250 eggs on stems. Larvae bore into stems and cause extensive damage by feeding on primary and secondary roots and root hairs. Control by *A. nigriscutis*, first released in Canada in 1983, has been spectacular (Coombs et al.). Two clearwing moths, two gall midges, and a hawk moth have been moderately successful for some leafy spurge biotypes. Reports of spectacular biological control success have been met with skepticism by weed scientists in the western United States. Leafy spurge is well-known as an aggressive, difficult to control, invasive perennial that is present on too many acres.

Another chrysomelid beetle was imported from Argentina to Florida in 1965 and successfully controlled aquatic alligator weed (Coulson, 1977). Alligator weed was introduced to the United States about 1894 from South America in ship ballast and had infested nearly 70,000 acres in the southern states by the 1960s. Impressive control has been achieved, but the insect's success is influenced by temperature, rate of water flow, other plants, water nutrition, and plant vigor. The weed's population has been reduced wherever the beetle has been introduced. A stem borer and alligator weed thrips have also been successful (Coombs et al., 2004).

A weevil from southern Germany (*Rhinocyllus conicus* Froelich) was introduced to Canada in 1968 and to West Virginia in 1969 for musk thistle control. The adult weevils are dark brown with small yellow spots on their back and are only three-sixteenths to one-quarter inch long. After feeding and mating on thistles, females lay eggs on the bracts of developing flowers in late spring. The larvae hatch, bore into the base of the flower receptacle, and prevent development of some or all seed. It takes a large number of larvae to destroy seed production completely. Because musk thistle is a biennial, a key to its control is preventing seed production. Plants produce seed for 7–9 weeks and the average plant produces 4000 seeds. Egg laying is favored by hot, humid weather and late flowers may not be affected. *Rhinocyllus* has given up to 90% control on some pasture sites where plant competition provided additional stress, but it has not been a complete control for musk thistle. Unfortunately, the weevil may be a bad case of biological control and an exception to the statement that no biological control has ever escaped to become a pest. It attacks native thistles and can move to other species including the endangered Sacramento thistle in New Mexico. *Rhinocyllus* may have a fatal flaw for a good biological control: it is not host specific.

Purple loosestrife has been managed biologically with insects. There are at least 120 species of phytophagous insects associated with purple loosestrife in Europe (Malecki et al., 1993).

In 1997, three insects from Europe had been approved by the USDA/APHIS for use as biological control agents. These plant-eating insects include a root-mining weevil (*Hylobius transversovittatus*), which attacks the main storage tissue and two leaf-feeding beetles (*Galerucella calmariensis* and *Galerucella pusilla*). Two flower-feeding beetles (*Nanophyes*) that feed on various parts of purple loosestrife plants were still under investigation in 1997. *Galerucella* and *Hylobius* have been released experimentally in natural areas in 16 northern states from Oregon to New York. They are capable of defoliating entire plants. Although these beetles have been observed occasionally feeding on native plant species, their potential effect on nontarget species is considered to be low. Malecki et al. predict that once these species establish in the field, the combination of defoliation by the chrysomelid beetles and root destruction by the weevils will lead to long-term negative effects on stands of purple loosestrife.

A major effort is under way to find biological control insects for melaleuca and Brazilian peppertree. Both are major invasive species in Florida. Melaleuca, also known as punk or paperbark tea tree, was brought to Florida from Australia around 1900 to assist in drying out swampy land and as a garden plant. It was first reported in Everglades National Park in 1967. By 1993 it was estimated to cover 488,000 acres in South Florida. Once widely planted in Florida, but no longer, it formed dense thickets and displaced native vegetation on wet pine flatwoods, saw grass marshes, and cypress swamps in the southern part of the state. Medium-size tree average 50–70 ft tall and can grow to 100 ft. Its invasive ability raised serious environmental issues in Florida's Everglades because of its ecological damage and harm to the surrounding economy. Over 200 insects that feed on melaleuca have been found in Australia, its natural habitat. Scientists from the Australian Biological Control Laboratory assisted in solving the problem by releasing insects that are natural predators of melaleuca in Australia.[5] However, similar to many weedy species, melaleuca has at least one redeeming quality. There is a market for melaleuca mulch, one of the most termite-resistant mulches available.

Brazilian peppertree is also an aggressive nonindigenous plant, native to Argentina, Paraguay, and Brazil. It has replaced large natural plant communities in Florida (US Congress, 1993; Langeland, 1990) and is an important threat to biodiversity because it disrupts native plant and animal communities. Introduction to Florida as a cultivated ornamental was sometime between 1842 and 1849. The genus name *Schinus* is a Greek word for mastic tree, a plant with resinous sap, which the pepper tree resembles. In 2004 (Coombs et al., 2004) no biological controls had been released in the United States for Brazilian peppertree. Over 200 insects have been identified that feed on Brazilian peppertrees in the tree's native land, but their specificity and reproductive ability in the United States have not been confirmed.

University of Florida scientists identified several insects from exploratory surveys in Argentina, Brazil, and Paraguay as potential biological control agents for Brazilian peppertree (McKay et al., 2009). Five insects (a thrip, sawfly, leaf roller, stem-boring weevil, and a leaf galling psyllid) were selected for further study because they visibly damaged the plant in its native range. They were collected only from Brazilian peppertree or a few closely related species during field surveys (Cuda et al., 2013).

A major concern in the western United States is salt cedar or tamarisk, an import from Eurasia. Four species dominate: *Tamarix parviflora*, *Tamarix ramosissima*, *Tamarix chinensis*, and *Tamarix gallica*. Each is a deciduous shrub that grows 3–30 ft tall (average, 9–21 ft). Mature salt cedar plants are capable of producing 2.5×10^8 tiny wind-dispersed seeds per year (Stevens, 1989). Seeds are short-lived (less than 2 months in summer), have no dormancy requirements, and germinate in less than 24 h. Salt cedar is tolerant of saline soil (up to 30,000 parts per million salt). It also tolerates fire, drought, flooding, and cold temperatures (Coombs et al., 2004). Each species is a successful invader and all are survivors able to adapt

---

[5]Germplasm Resources Information Network. US Department of Agriculture. January 27, 2009. http://www.ars-grin.gov/cgi-bin/npgs/html/genus.pl?7400.

to many environments. The main concern is their ability to adapt to riparian areas in the western United States, in which they rapidly exclude native vegetation and use prodigious quantities of ground water. A single large plant can absorb 200 gallons of water a day. Large stands in riparian areas average of 4—5.5 acre feet of water per year (1 acre foot = 325,000 gallons).

More than 350 species of insects feed on salt cedar in central Asia and southern Europe. In its native range, 115 insect species and four mites are known to attack salt cedar. However, when it was brought to the United States in the 1830s, almost all of its natural enemies were left behind. Research found at least a dozen insect species that might be useful in fighting salt cedar, but none are currently available. Two possibilities are salt cedar leaf beetle (*Diorhabda elongata*), and manna scale (*Trabutina mannipara*). Larvae and adults of the salt cedar leaf beetle feed on foliage and have been released in several areas of the western United States. *D. elongata* beetles, collected from Possidi, in Northern Greece, where the climate is similar to the Texas High Plains, were released in early spring, 2004. They established on salt cedar and survived two winters. Additional releases were made in Borger in 2006 with beetles from Uzbekistan. Studies have shown that beetles collected from different locations can have varying efficacy and establishment potential. The beetles do not destroy the stems, which have to be removed manually. Initial action is slow, but by the third year of beetle infestation, large areas can be defoliated. Complete control has not been achieved.

Hajek (2004) reviewed research on the European corn borer (*Ostrinia nubilalis*). It is a difficult insect to control because the larvae live in protected locations on the roots of corn plants. Several species of the egg parasitoid *Trichogramma* that attack moth eggs have been tested using inundative control techniques. The releases were rarely considered economically feasible in the United States, although they were more successful in Québec and Europe.

3.2 Inundative or Augmentive

3.2.1 Fungi

An endemic anthracnose disease was used in the early 2000s to control Northern joint vetch, a grassy weed in rice and soybeans in southeastern United States. Application of a dry, powdered formulation of the fungus *Colletotrichum gloeosporioides* (Penz) Sacc. f. sp. *aeschymonene* as a mycoherbicide[6] (Collego - released in 1982) was effective. Collego was reregistered in Canada under the name LockDown and has been annually produced and sold directly to farmers (Bailey, 2014). Daniel et al. (1973) introduced the concept of mycoherbicide (Wilson, 1969; TeBeest, 1991). It was possible to spray the formulated fungal spores on rice infested with northern joint vetch (Daniel et al.). After a 4- to 7-day incubation period, northern joint vetch died in 5 weeks. The fungus is specific and can be produced in large quantities in artificial cultures, and the cultures are infective in the field. Two isolates of the fungus were combined for effective control of northern joint vetch and winged water primrose in rice (Boyette et al., 1979). Neither product is currently available. Although mycoherbicides worked, it was clear that a phenoxyacid herbicide did the same job in 2 weeks at less cost

---

[6]Mycoherbicide, biopesticide, and bioherbicide are not well-defined terms. They are not precise synonyms although they are often used as such.

and faster. The herbicide was the logical choice. The fungus (and the herbicide) had to be sprayed annually and neither was used if the weed was not present. Introduction did not permanently increase the organism's population. It was augmentive, but consistent with the definition, it was short-term and temporary.

The active ingredient in all mycoherbicides is applied in inundative doses. They frequently fail because of the pathogen's requirement for extended periods of dew or rain (Auld and McRae, 1997). The first mycoherbicide derived from fungi was successful for control of strangler vine in citrus orchards in Florida after application of live chlamydospores of *Phytophthora palmivora* (Butl.) Butl. It was first registered in 1981 and marketed as Devine by Abbott laboratories (subsequently by Valent Corp). Live chlamydospores germinated 6—10 h after application to a wet soil surface. The fungus initiated a root infection that killed strangler vine in 2—10 weeks, depending on the vines size and vigor when Devine was applied. Complete control was not obtained in 1 year, but the fungus persisted and was effective for up to 5 years, which was a sales disadvantage. Drift to susceptible plants including cucumber, squash, watermelon, rhododendron, begonia, and snapdragon is a problem. In addition to its persistence and effect on other species, the formulation rapidly lost viability after preparation. It had to be treated like fresh milk, and even with refrigerated storage it could not be stored for use another year.

Mycoherbicides are preparations of living inoculum of a fungal plant pathogen. They are formulated and applied similarly to chemical herbicides. They have been available since the early 1960s in the United States and China (Auld and McRae, 1997). Although few commercially successful products have been developed, there has been international interest in them (e.g., Li et al., 2003 in China). Bioherbicides were reviewed by Boyetchko (2001). Examples of living organisms used as biopesticides include inundative application of lady beetles (worldwide more than 5000 species of *Coccinellids* are called lady beetles) to control aphids (about 4400 species of 10 families are known), the predatory mite *Neoseiulus cucumeris* to control thrips (816 genera in 11 families), and beneficial nematodes to control fungal gnats (six families) (Hajek, 2004, p. 63).

Another mycoherbicide, BioMal, manufactured by Philom Bios of Saskatchewan, is registered in Canada for control of common mallow in several crops, but its market success has been limited. The active agent, *C. gloeosporioides* f.sp. *malvae*, was discovered by fortuitous observation of blight on seedlings of round-leaved mallow, a serious weed pest in prairie agriculture (Vincent et al., 2007). There are good reasons why more such products are not commercially available (Auld, 1995; Watson, 1989). The most important may be that herbicides have been so successful for control of each targeted weed and for the commonly encountered weed complexes. Specificity is the essence of success for biological control agents, but it may lead to commercial failure because weeds usually exist in complex communities. Removal of one weed with a specific biocontrol agent creates a situation in which others flourish once they are released from competition. Equally important is that each product targeted a specific weed, which inevitably made its market small. Other reasons include the difficulty of mass producing and formulating the infective agent so it could be applied. Low pathogen virulence is a common problem. Chemical herbicides can be applied under a range of environmental conditions with reasonable expectation of success. If mycoherbicides are applied with unfavorable moisture or temperature, failure is common. Success normally requires a long dew period that is difficult to obtain in dry climates, comprehensive understanding of the pathogen, the biology and population dynamics of the target weed, the optimum requirements for disease initiation, and the interactions within the host—pathogen system (Watson, 1989). Chemical herbicides are similar to the brute force required to win in Japanese sumo wrestling, whereas bioherbicides more closely resemble the finesse of successful judo wrestling.

Fusarium oxysporum f. sp. *cannabis* could provide safe, efficient control of marijuana (McCain, 1978). In inoculation studies and in nature, only marijuana was infected. All marijuana types tested were susceptible except that cultivars grown only for hemp were resistant. Inoculum for field use can be grown efficiently on mixtures of barley straw combined with alfalfa or soybean oil meal. Inoculum spread at 10 kg/ha resulted in 50% mortality of seeded marijuana. Three-quarters of subsequent marijuana plantings died. The fungus caused disease over a wide temperature range, and once a field is infested, marijuana cannot be grown for many years. There is no known danger from the fungus to humans, animals, or other plants. A potential disadvantage is that marijuana is often grown in seclusion, not in readily accessible places.

A potentially more important application of *F. oxysporum* is control of witchweed, one of the world's worst parasitic weeds. It is considered by many to be the greatest constraint to food production in Africa, particularly in the sub-Saharan region. (For additional information on witchweed, see Chapter 3.) *Fusarium* species from West Africa, grown on sorghum straw, have successfully prevented all emergence of witchweed and increased sorghum dry weight as much as 400%. In growth chambers, the fungus inhibited germination and attachment of witchweed to sorghum roots (Ciotola et al., 1995).

The most extensively studied group of plant pathogens is the fungal genus *Colletotrichum*. *Colletotrichum coccodes* isolated from eastern black nightshade (Anderson and Walker, 1985) did not kill velvetleaf, but another isolate did (Wymore et al., 1987). Other strains of the fungus kill tomatoes and potatoes, but the identified strain is harmless to all crops tested. It causes disease of velvetleaf over a wide range of dew periods and temperatures, but is most effective after a 24-h dew period at 75°F (Wymore et al.).

Peng et al. (2004) showed that mycelial suspensions of *Pyricularia setariae* had strong specificity for control of green foxtail with no significant pathogenicity on more than 25 other species including wheat, barley, and oats. When the fungus was applied with 10^5 spores/mL, green foxtail fresh weight was reduced 34%. If 10^7 spores were used, fresh weight was reduced 87%. The efficacy was comparable to the herbicide sethoxydim. The 80% of green foxtail that was resistant to the herbicide was also controlled.

Several strains of the rust fungus *Puccinia chondrillina* have been tested for control of rush skeleton weed (Lee, 1986) to find one for importation. It was successful in Australia and dry Mediterranean areas. It was the first exotic plant pathogen successfully used for weed control in North America (Coombs et al., 2004). A strain of *P. chondrilla* was released successfully in California in 1976 and spread to Oregon in 2 years (Lee). The rust has controlled skeleton weed successfully and is specific. Rush skeleton weed is also affected by a root moth, a gall midge, and a gall mite (Coombs et al.).

Puccinia punctiformis, a rust fungus, is an obligate parasite specific to Canada thistle (Cummins, 1978) and infection can lead to death. Infection reduces flowering and vegetative reproduction (Thomas et al., 1995). However, Canada thistle has been difficult to control everywhere it exists, and although the rust fungus is present, it has not always been effective.

When sprayed in an aqueous-phase, crop oil emulsion Conidia of the fungal pathogen *Myrothecium verrucaria* controlled red ivyleaf, smallflower and tall morning glory in the three- to five-leaf growth stage (Millhollen et al., 2003). Conidia continued to be effective after autoclaving, which indicated that the action was not caused by fungal infection. Chemical analysis showed the presence of several macrocyclic trichothecenes (potent protein synthesis inhibitors), some of which are known phytotoxins (Lee et al., 1999).

Chandromohan et al. (2002) found that a mixture of fungal pathogens isolated from fungi native to Florida controlled six annual and one perennial grass as well as any one of the pathogens alone. It was possible to manage all seven weeds in the field with an emulsion mixture of the pathogens. The weeds have been difficult to control because available herbicides were ineffective and their growth habits enable them to resist other control practices.

Despite early enthusiasm (TeBeest and Templeton, 1985), no new mycoherbicides have entered the market and old ones have disappeared. Most commercial formulations have been based on fungal species. However, few have been successful (Harding and Raizada, 2015). This could be because of a lack of interest on the part of industrial concerns, which have regarded mycoherbicides as economically unattractive. It is true the market is small, profits are surely low, and the risks related to efficacy and user acceptance are high. In addition, chemical herbicides are often less expensive and success is more assured. However, a more important reason may be legal challenges primarily from environmental groups that have stopped or significantly delayed issuance of permits. The Plant Protection and Quarantine division of the APHIS of the USDA has not issued a permit for field release of any weed biological control agents since 2012 (Cuda, 2016).[7] Therefore, despite their environmental attractiveness, the lack of significant regulatory permits has essentially eliminated mycoherbicides as a weed management option.

3.2.2 Bacteria

Phytopathogenic bacteria have not been considered to have good potential as biological agents because, despite their known activity, they do not penetrate plants well. This deficiency has been overcome by combining bacteria with surfactants or with a cultural operation that injures plants, such as mowing. Spray application of *Pseudomonas syringae* in an aqueous buffer with a surfactant produced severe disease in several members of the Asteraceae including Canada thistle (Johnson et al., 1996). Spray application without surfactant failed to produce disease in any plants. A stem-mining weevil (*Hadroplontus litura*), a relatively weak biological control agent, provided some Canada thistle suppression when it was combined with sunflower, a desirable competitive crop (Burns et al., 2013). *Xanthomonas campestris* pv. *poannua* controlled several annual bluegrass biotypes in Bermuda grass golf greens when it was sprayed during mowing, but not when it was applied without mowing. Prior mowing injured the grass and allowed the bacteria to enter and cause lethal systemic wilt (Johnson et al.). This technique might have led to further development of bacterial herbicides, but apparently it did not. They are not obligate biological agents, but because they do not persist they may escape the disadvantage of lack of specificity. They must be applied annually. They have an advantage over fungi because a dew period (wet period) is not required to activate them.

---

[7]Personal communication, Professor J.P. Cuda, Department of Entomology and Nematology, University of Florida, Gainesville. February 2016.

Kremer and Kennedy (1996) propose that deleterious rhizobacteria have been largely overlooked as potential biological control agents. They are able to colonize root surfaces of weed seedlings and suppress plant growth. de Luna et al. (2005) reported that enhanced growth of bacteria due to rhizosphere effects depends on soil moisture, temperature, and substrate availability and favor the use of bacteria for biological control. Kennedy has consistently challenged the prevailing view of the lack of efficacy of bacteria as weed control agents. Several research articles have demonstrated the efficacy of *Pseudomonas fluorescens*-D7 for control of downy brome (Kennedy et al., 1991; Johnson et al., 1993; Gurusiddaiah et al., 1994; Ibekwe et al., 2010). A critical aspect of using a naturally occurring soil bacterium is survival of a sufficient population of the organism under field conditions. Stubbs et al. (2014) showed that after 63 days, the level of *P. fluorescens* was adequate to suppress downy brome under field or range conditions. de Luna et al. (2011) developed procedures to isolate and screen individual organisms to determine their potential for seed decay and weed biocontrol, particularly for wild oat. One must conclude that the potential for biological control of weed seeds by bacteria is greatly underestimated.

3.2.3 A Summary

Wilson (1969) described principles for control of weeds with phytopathogens that are still applicable. The first is that host resistance is the primary deterrent to success and may often restrict disease to insignificant levels. Weeds usually have several, rarely fatal, disease lesions on their foliage. Natural weed populations resist insects and diseases because of climate and soil variability and the regular presence of natural, but not fatal, enemies. Disease susceptibility is the exception rather than the rule. Disease epidemics result from importation of new diseases or more virulent strains rather than just the presence of known pathogens. These principles, although generally true, may fail in specific cases. Weed scientists have isolated, cultured, and redistributed local pathogens such as the previously mentioned anthracnose disease to achieve weed control. In 1972, more than 40 years ago, Zettler and Freeman suggested that continued research on biological control for terrestrial and aquatic weeds offered promise. Several others have echoed their view. Biocontrol has been an active research area, but commercial success has been elusive. A review of biological control with plant pathogens reported four projects with bacteria, 42 with fungi, 3 with nematodes, and 6 with viruses (Charudattan and Walker, 1982). Coombs et al. (2004) identified 25 target weeds (21 perennials) and one or more biological control agents active on each. They also listed biological control research projects for 15 weeds (10 perennials). The book by TeBeest (1991) has many more, but widespread use for weed management and commercial success have been limited. Harding and Raizada (2015) discussed 20 North American Biocontrol agents, only four were available.

3.3 Broad-Spectrum

3.3.1 Fish

The white amur or grass carp (*C. idella* Valenciennes) is an herbivorous fish native to the Amur River, which forms most of the boundary between northeast China and southeast Russia. Its range extends into southern China. It can consume 3–5 lb/day of aquatic plants (especially *Hydrilla*). Adults may weigh 70–100 lb. It does not spawn in warm water, so it is possible to control its population (Van Zon, 1984). The grass carp breeds, but only in large rivers or canals with high water volume and velocity. It feeds on grass and other terrestrial vegetation. Scientists have discovered a way to ensure production of sterile offspring. Researchers have tried to cross the white amur and the big-head carp to produce a voracious weed-eating hybrid. There are 240,000 miles of irrigation canals, ditches, and drains in the 17 western United States and many have aquatic weeds. A theoretical advantage of plant-eating fish is that they may be harvested for food, and if sterile, their population should be controllable. A risk of one becoming an aggressive, invasive species that escapes and threatens biological diversity and other species is always a legitimate concern.

The white amur or grass carp (an herbivorous fish) eats aquatic vegetation.

Authorized and unauthorized stocking of grass carp has occurred for biological control of vegetation. The first importation to the United States was in 1963 to aquaculture facilities in Auburn, Alabama, and Stuttgart, Arkansas. The Auburn stock came from Taiwan, and the Arkansas stock was imported from Malaysia (Courtenay et al., 1984). The first (unplanned) release into open waters took place in Stuttgart, Arkansas, when fish escaped the Fish Farming Experimental Station[8] (Courtenay et al.). In 2012, grass carp has been found is all of the United States except Alaska, Maine, Montana, Rhode Island, and Vermont.[9]

Resistance to introduction of the grass carp or its hybrids centers on their potential to cause problems similar to those that occurred after introduction of the common carp. These include degradation of water quality owing to bottom feeding that disturbs sediments and muddies the water, and crowding out desirable fish because of the carp's rapid population growth in the absence of natural enemies. A single female grass carp may produce up to a million eggs; therefore, research has emphasized sterility in released populations.

Selectivity remains a central concern with a fish, as it is with any other organism introduced for biological control. Will the grass carp prefer and eat selectively the weeds those who introduce it wish to control? Will the fish consider a species desirable that a Department of Natural Resources or biological control scientist considers undesirable? Considering all aspects, use of grass carp as a control for aquatic vegetation is best suited when total elimination of the macrophyte community is desired, because overconsumption of vegetation will

[8]http://nas.er.usgs.gov/queries/factsheet.aspx?SpeciesID=514.

[9]Nas.er.usgs.gov/viewer/omap.aspx?speciesID=514.

surely have negative effects. In other words, they are effective, but do not put too much reliance on their selectivity.

3.3.2 Aquatic Mammals

The sea manatee (*Trichechus manatus*), a fully aquatic, herbivorous marine mammal, eats cattails, water hyacinth, and other aquatic vegetation and can weigh over 1 ton. It is not discriminatory in its diet and eats many kinds of aquatic vegetation. Manatees are large gray aquatic mammals with bodies that taper to a flat, paddle-shaped tail. The average adult is 10 ft long and weighs 800−1200 lb. The head and face are wrinkled with whiskers on the snout. The West Indian manatee is related to the West African manatee, the Amazonian manatee, the Dugong, and Steller's sea cow, which was hunted to extinction in 1768. The manatee's closest relatives are the elephant and the hyrax (a small, gopher-sized mammal). It is believed that they can live 60 years or more. They reproduce in fresh and salt water, breathe oxygen from air, and have no natural enemies. They often lie just below the surface of the water where they may be hit by boat propellers. In fact, most fatalities are from human-related causes: collisions with watercraft; being crushed and/or drowned in canal locks and flood control structures; ingestion of fish hooks, litter, and monofilament line; and entanglement in crab trap lines. Ultimately, loss of habitat is the most serious threat facing manatees in the United States. There were about 6000 manatees in February 2016.[10] All three species are now listed by the World Conservation Union as vulnerable to extinction. One must conclude that even if they could be developed as effective aquatic control agents, it is unlikely to happen, perhaps because so few people care about large, ugly sea mammals or could be encouraged to do so. Manatees have cleared up to half a mile of canal and banks of a major aquatic waterway in Florida in 3 weeks. However, they are not good biological controls (Etheridge et al., 1985). Estimates of consumption of *Hydrilla* in Kings Bay, Florida showed that it would take 10 times as many manatees just to consume the standing biomass of *Hydrilla*. The natural manatee population (116) was not small for the area, but it was inadequate to control *Hydrilla* without even considering the prospect of increased growth of *Hydrilla* during the winter season.

The sea manatee eats cattails, water hyacinth, and other aquatic vegetation.

[10]www.savethemanatee.org/manfacts.htm.

It is said that sailors may have seen sea manatees with their fishtail and thought they were mermaids. If you see one, you may think the sailors had a little too much grog. Grog originally referred to a drink made with water and rum, which British Vice Admiral Edward Vernon introduced in 1740. Vernon wore a coat of grogram cloth and was nicknamed Old Grogram or Old Grog. The manatee really looks a bit like former President Grover Cleveland because it is fat, has whiskers, and thick, wrinkled skin.

3.3.3 Vertebrates

Sheep and goats graze plants that cattle will not eat, such as leafy spurge. Kleppel and LaBarge (2011) found that sheep grazing purple loosestrife prevented flowering and reduced its cover by 40%, and species richness increased 20%. Goats relish shrubby species and eat more than sheep, but both are difficult to contain. It takes special attention to fencing to keep them in a place, or it takes a careful herder (usually human and canine); goats love to roam. Sheep and goat-proof fencing are expensive, as is herding. Goats can be used as a follow-up to mechanical treatments and have killed root sprouts of Gambel oak. They prefer oak over other plants and do not compete with cattle for forage. They eat, brush, leaves, and twigs and almost anything that is organic. Goats have been used successfully to control salt cedar, but 17 months later the control was not as good as that obtained with the herbicide imazapyr (Richards and Whitesides, 2006). A combination of grazing 1 year followed by imazapyr the following year was more successful. Do not expect a few goats to eat the entire population of a weed. Similar to herbicides, concentration matters. Overstocking of goats can lead to intensive grazing of the target weeds and brush, which will shift grazing to grasses and desirable species. Therefore, they must be removed when they have eaten 90%–95% of the weedy foliage or they begin to compete with cattle. Aggressive grazing can denude an area and destroy wildlife habitat.

But not all is bad. Goats control brush, weeds including several poisonous plants without disturbing existing grass or the soil. They would rather eat brush and weeds (70% of their diet) than the grasses cattle prefer. They are browsers, not grazers as cattle are, and eat up to 25% of their body weight each day. Cattle eat as they move whereas goats move to eat. Goats also fertilize as they go and the action of their small hooves literally helps to plant the seeds of the next grazing crop. Those who manage goats know that young goats learn their eating habits from their mothers. Some mothers prefer grass. Therefore, if you want a brush-eating goat, make sure it was raised by a brush-eating mother.[11]

Goats have been called the next hip thing in eco-friendly weed management (Rosner, 2003). They are useful for weed management in urban areas and on rangeland. Each summer the Lawrence Berkeley National Laboratory (the Berkeley Lab) in California hires a local

[11]See - http://www.goatseatweeds.com/.

company, Goats R Us,[12] to release dozens of goats around the hillside terrain of their 202-acre property to graze and reduce the incidence of brush fires. Goats have proven to be an environmentally friendly, effective weed control technique. Each goat eats about 10 lb of vegetation per day.

Their omnivorous diet restricts their use in fields with agricultural crops. Other than the energy to get to a site, no petroleum energy is required for goats to do their work, no chemicals are required, and soil is not disturbed. Some people think they are cute, most people seem to like them, but some might complain about a barnyard odor.

Geese, ducks, and chickens have been used to weed strawberries, raspberries, and some vegetables. They will selectively remove grasses and small broad-leaved weeds without crop damage. Chickens and geese can selectively control nutsedge in several crops. They are not selective in grass crops. Experiments have shown advantages for geese for weed control during establishment of tree seedlings (Wurtz, 1995). Geese feed almost exclusively on grasses and broad-leaved weeds whereas chickens are omnivorous and eat weed seedlings, seeds, insects, and soil invertebrates (Clark and Gage, 1996). Chickens did not affect weed abundance or crop productivity (Clark and Gage). Geese used for weed management in potatoes were more effective than chickens because they reduced weed abundance and improved yield. A problem with geese is that they are picky and do not eat all weeds. Species unpalatable to geese such as curly dock and daisy fleabane increased in abundance (Clark and Gage). An unsuspected benefit of weeding by geese was a reduction in damage to apples by the plum curculio (*Conotrachelus nenuphar* Herbst). Clark and Gage attributed this to reduced humidity at the soil surface caused by weed removal, which reduced plum curculio activity.

A novel, unusual biocontrol approach has been tried in Hungary (Economist, 2016). Rabbits were trained to add common milkweed to their diet. They have not yet been released owing to reluctance to add to a large population and the unknown and possibly undesirable effects of introducing a species with no dietary preferences.

4. INTEGRATION OF TECHNIQUES

Successful, sustainable weed management systems employ a combination of techniques rather than relying on one. Biological control is easy to combine with other methods because ideally, once established, it can be self-perpetuating. To be successful, an integrated system requires thorough knowledge of the ecological relationships within the weed–crop system. Knowledge of a farmer's production goals and farming system is necessary but not sufficient. When the goal is sustainable weed management rather than annual control, thorough ecological understanding is required (Wapshere et al., 1989). Successful, sustainable weed management means that a weed population will be reduced and maintained at a noneconomic level. When annual control is the primary aim, rougher techniques can be employed that require less biological knowledge and management skill.

[12]http://www.goatsrus.com. Also see - https://www.youtube.com/watch?v=c-zO8ieiukw.

There are several examples of cropping systems in which weed presence actually facilitates biological control of some pest organisms (Table 4.2 includes a few examples). Therefore it is reasonable to conclude that in some cases complete weed control is an undesirable goal.

Biological and chemical control have been combined to manage common groundsel (Frantzen et al., 2002). Biological control was based on stimulating epidemic infections of the rust fungus *Puccinia lagenophorae* to reduce the weed's competitiveness at the population level. Common groundsel was controlled well by the photosynthetic inhibiting herbicide monolinuron and the fungus was not needed. The fungus was not compatible with metoxuron and biological control was contraindicated. However when pendimethalin was the herbicide of choice, use of the fungus against common groundsel was complementary to pendimethalin's action on other weeds.

The fungus *Cochliobolus lunatus* is endemic on barnyard grass (Scheepens, 1987). Studies in the Netherlands have demonstrated successful (not 100%) control of barnyard grass in the field. It has potential as a biological control agent but does not have sufficient activity alone to kill barnyard grass. It has been successful when combined with sublethal doses (a dose that will not control the weed) of the herbicide atrazine. Under appropriate conditions, the fungus produces leaf necrosis and kills seedlings with fewer than two leaves. Plants with more than two leaves recover, although their growth is slowed. It can be used successfully in beans, barley, corn, oats, rye, tomatoes, and wheat. Combining the fungus with sublethal doses of atrazine enhances control over that achieved with the fungus or atrazine alone (Table 12.3). This is especially true as the weed gets older.

The success of the *Chrysolina* beetle for control of St. John's wort has already been mentioned. A successful, integrated system for pastures was developed in Australia (Campbell, 1979). It combined the beetle (biological control agent), an herbicide, and use of plant competition through reseeding in areas where the weed's population had been reduced. On arable land, a combination of mechanical and cultural control was integrated with biological control. Land is plowed in summer to expose and dry roots, then cultivated in late summer to continue drying and to prepare for seeding an improved pasture mixture. Adequate fertilization is required to guarantee the seeding's success. On nonarable land, five techniques are combined. In addition to the beetle, heavy grazing by sheep or cattle is used to remove plants that shade the weed. This is followed by applying 2,4-dichlorophenoxyacetic acid, planting the proper pasture mixture with adequate fertilization to take full advantage of plant

TABLE 12.3 Effect of *Cochliobolus lunatus* and Atrazine on Barnyard Grass in a Growth Chamber (Scheepens, 1987)

Treatment	% Necrosis After 9 Days		
	22-Day-Old Plants	30-Day-Old Plants	47-Day-Old Plants
Untreated	0	0	0
Cochliobolus lunatus	60 ± 21	60 ± 18	15 ± 9
Atrazine at 40 g/acre	60 ± 19	60 ± 19	3 ± 3
C. lunatus + atrazine at 40 g/acre	100	100	75 ± 13

competition, and well-managed, light grazing and additional fertility to maximize the crop's advantage and competitive pressure on the weed. These methods seem so obvious that one is inclined to say, "Of course, that is what should be done." If these or similar methods are tried in other environments and cropping systems, they might fail unless the ecological relationships have been analyzed and understood. Ecological understanding leads to selection of the best combination of techniques to manage the weed population rather than the best method to obtain a quick kill, but without long-term reduction of the weed's population. Chapter 21 has additional examples of integrated weed management systems.

Zorner et al. (1993) concluded that the commercial success of bioherbicides "depends upon devoting major efforts toward developing appropriate fermentation, stabilization and delivery technology." Their conclusion is equally applicable to biological weed management. It is undeniably true that biological control works and research is proceeding, albeit slowly. It is equally true that only a few weeds have been effectively controlled and the recommended major efforts have not followed. Its promise is real but unfulfilled.

THINGS TO THINK ABOUT

1. What applications are there for biological control?
2. Why has biological control not been used more widely?
3. What are some good examples of successful biological control of a weed?
4. What is a bio- or mycoherbicide and how are they used?
5. Where are vertebrate animals best used for biological weed control?
6. How can biological control be integrated with other methods?
7. What are the economic advantages of biological control?
8. Compare and contrast the advantages and disadvantages of biological weed control.
9. Describe and discuss legislative and regulatory influences on biological control techniques.

Literature Cited

Anderson, R.N., Walker, H.L., 1985. *Colletotrichum coccodes*: a pathogen of eastern black nightshade (*Solanum ptycanthum*). Weed Sci. 33, 902–905.

Andres, L.A., 1966. The role of biological agents in the control of weeds. In: Symposium on Pest Control by Chemical, Biological, Genetic, and Physical Means, pp. 75–82. ARS Pub. No. 33-110.

Anonymous, 2009. USDA APHIS. *C. cactorum* Program. Technical Working Group Report, New Orleans, LA.

Auld, B.A., 1991. Economic aspects of biological weed control with plant pathogens. In: TeBeest, D.O. (Ed.), Microbial Control of Weeds. Chapman and Hall, NY, pp. 262–273.

Auld, B.A., 1995. Constraints in the development of bioherbicides. Weed Technol. 9, 638–652.

Auld, B.A., McRae, C., 1997. Emerging technologies in plant protection – bioherbicides. In: Proc. 50th New Zealand Plant Prot. Conf., pp. 191–194.

Bailey, K.L., 2014. The bioherbicide approach to weed control using plant pathogens. In: Abrol, D.P. (Ed.), Integrated Pest Management. Elsevier, Inc., pp. 245–265

Bloem, K.A., 2003. Overview of the cactus moth problem. In: Presented at the Cactus Moth *Cactoblastis cactorum* Planning Meeting, December 9–10, 2003 in Miami, Florida.

Boyetchko, S.M., 2001. Biological herbicides in the future. In: Ivany, J.A. (Ed.), Weed Management in Transition, Topics in Canadian Weed Sci., vol. 2. Canadian Weed Sci. Soc. Sainte-Anne-de-Bellevue, Quebec, Canada, pp. 29—45.

Boyette, C.D., Templeton, G.E., Smith, R.J., 1979. Control of winged waterprimrose (*Jussica decurrens*) and northern jointvetch (*Aeschynomene virginica*) with fungal pathogens. Weed Sci. 27, 49—51.

Burns, E.E., Prischmann-Voldseth, D.A., Gramig, G.G., 2013. Integrated management of Canada Thistle (*Cirsium arvense*) with insect biological control and plant competition under variable soil nutrients. Invasive Plant Sci. Manag. 6, 512—520.

Callaway, R.M., DeLuca, T.H., Belliveau, W.M., 1999. Biological-control herbivores may increase competitive ability of the noxious weed *Centaurea maculosa*. Ecology 80, 1196—1201.

Campbell, M.H., 1979. St. John's Wort. New South Wales Dept. of Agriculture, Agdex 642. New South Wales, Australia, 16 pp.

Chandromohan, S., Charudattan, R., Snoda, R.M., Singh, M., 2002. Field evaluation of a fungal pathogen mixture for the control of seven weedy grasses. Weed Sci. 50, 204—213.

Charudattan, R., 2010. A reflection on my research in weed biological control: using what we have learned for future applications. Weed Technol. 24, 208—217.

Charudattan, R., Walker, H.L. (Eds.), 1982. Biological Control of Weeds with Plant Pathogens. J. Wiley and Sons, Inc., NY, 293 pp.

Ciotola, M., Watson, A.K., Hallett, S.G., 1995. Discovery of an isolate of *Fusarium oxysporum* with potential to control *Striga hermonthica* in Africa. Weed Res. 35, 303—309.

Clark, M.S., Gage, S.H., 1996. Effects of free-range chickens and geese on insect pests and weeds in an agroecosystem. Am. J. Altern. Agric. 11, 39—47.

Coombs, E.M., Clark, J.K., Piper, G.L., Cofrancesco Jr., A.F., 2004. Biological Control of Invasive Plants in the United States. Oregon State Univ. Press, Corvallis, OR, 467 pp.

Coulson, J.R., 1977. Biological Control of Alligatorweed, 1959—1972. A Review and Evaluation. U.S. Dep. Agric. Tech. Bull., 1547, p. 98.

Courtenay Jr., W.R., et al., 1984. Distribution of exotic fishes in the continental United States. In: Courtenay Jr., W.R., Stauffer Jr., J.R. (Eds.), Distribution, Biology and Management of Exotic Fishes. Johns Hopkins University press, Baltimore, MD, pp. 41—77.

Cuda, J.P., 2016. Novel approaches for reversible field releases of candidate weed biological control agents: putting the genie back into the bottle. In: Shroder, J.F., Sivanpillai, R. (Eds.), Biological and Environmental Hazards, Risk, and Disasters. Elsevier, New York, pp. 137—151 (Chapter 7).

Cuda, J.P., Medal, J.C., Overholt, W.A., Vitorino, M.D., Habeck, D.H., 2013. Classical Biological Control of Brazilian Peppertree (*Schinus terebinthifolia*) in Florida. IFAS Extension, University of Florida.

Cummins, G.B., 1978. Rust Fungi on Legumes and Composites in N. America. Univ. Arizona Press, Tucson, 138 pp.

Daniel, J.T., Templeton, G.E., Smith Jr., R.J., Fox, W.T., 1973. Biological control of northern jointvetch in rice with an endemic fungal disease. Weed Sci. 21, 303—307.

de Luna, L.Z., Stubbs, T.L., Kennedy, A.C., Kremer, R.J., 2005. Deleterious bacteria in the rhizosphere. In: Roots and Soil Management: Interaction Between Roots and the Soil. Agronomy Monograph No. 48. American Society of Agronomy, Madison, WI, pp. P233—P261 (Chapter 13).

de Luna, L.Z., Kennedy, A.C., Hanson, J.C., Paulitz, T.C., Gallagher, R.S., Fuerst, E.P., 2011. Microbiota on wild oat (*Avena fatua* L.) seed and their caryopsis decay potential. Plant Health Prog. https://doi.org/10.1094/PHP-2011-0210-01-RS.

DeBach, P., 1964. The scope of biological control. In: DeBach, P. (Ed.), Biological Control of Insect Pests and Weeds. Reinhold Pub. Corp., NY, pp. 1—20.

Economist, August 20, 2016. Now Try This, p. 68.

Etheridge, K., Rathbun, G.B., Powell, J.A., Kochman, H.L., 1985. Consumption of aquatic plants by the West Indian manatee. J. Aquat. Plant Manag. 23, 21—25.

Frantzen, J., Rossi, F., Müller-Schärer, H., 2002. Integration of biological control of common groundsel (*Senecio vulgaris*) and chemical control. Weed Sci. 50, 787—793.

Ghersa, C.M., Roush, M.L., Radosevich, S.R., Cordray, S.M., 1994. Coevolution of agroecosystems and weed management. Bioscience 44, 85—94.

Goeden, R.D., 1988. A capsule history of biological control of weeds. Biocontrol News Inf. 9, 55—61.

Goeden, R.D., Ricker, D.W., 1980. Santa Cruz island-revisited. Sequential photography records the causation, rates of progress, and lasting benefits of successful biological weed control. In: Proc. V Int. Symp. Biol. Contr. Weeds, Brisbane, Australia, pp. 355–365.

Goeden, R.D., Fleschner, C.A., Ricker, D.W., 1967. Biological control of prickly pear cacti on Santa Cruz Island, California. Hilgardia 38 (16), 579–606.

Goeden, R.D., Andres, A., Freeman, T.E., Harris, P., Pienkowski, R.L., Walker, C.R., 1974. Present status of projects on biological control of weeds with insects and plant pathogens in the United States and Canada. Weed Sci. 22, 490–495.

Gurusiddaiah, S., Gealy, D.R., Kennedy, A.C., Ogg Jr., A.G., 1994. Isolation and characterization of metabolites from (*Pseudomonas fluorescens*-D7) for control of downy brome (*Bromus tectorum*). Weed Sci. 42, 492–501.

Habeck, D.H., Bennett, F.D., 1990. *Cactoblastis cactorum* Berg (Lepidoptera: Pyralidae, a Phycitine New to Florida. Entomology Circular No. 333. Florida Department of Agriculture and Consumer Services, Tallahassee, FL.

Hajek, A., 2004. Natural Enemies – An Introduction to Biological Control. Cambridge University Press, Cambridge, UK, 378 pp.

Harding, D.P., Raizada, M.N., 2015. Controlling weeds with fungi, bacteria and viruses: a review. Front. Plant Sci. 6, 659. https://doi.org/10.3389/fpls.2015.00659.

Harris, P., Dunn, P.H., Schroeder, D., Vormos, R., 1985. Biological control of leafy spurge in North America. In: Watson, A.K. (Ed.), Leafy Spurge. Monograph No. 3. Weed Sci. Soc. America, Champaign, IL, pp. 79–82.

Hight, S.D., Carpenter, J.E., Bloem, K.A., Bloem, S., Pemberton, R.W., Stiling, P., 2002. Expanding geographical range of *Cactoblastis cactorum* (Lepidoptera: Pyralidae) in North America. Fla. Entomol. 85 (3), 527–529.

Hinz, H.L., Schwarzländer, M., Gassman, A., Bourchier, R.S., 2014. Successes we may not have had: a retrospective analysis of selected weed biological control agents in the United States. Invasive Plant Sci. Manag. 7, 565–579.

Hobbs, R.J., Humphries, S.E., 1995. An integrated approach to the ecology and management of plant invasions. Conserv. Biol. 9, 761–770.

Holloway, J.K., 1964. Projects in biological control of weeds. In: DeBach, P. (Ed.), Biological Control of Insect Pests and Weeds. Reinhold Pub. Corp., N.Y. (Chapter 23).

Huffaker, C.B., 1964. Fundamentals of biological weed control. In: DeBach, P. (Ed.), Biological Control of Insect Pests and Weeds. Reinhold Pub. Corp., N.Y (Chapter 22).

Huffaker, C.B., Kennett, C.E., 1959. A ten-year study of vegetation change associated with biological control of Klamath weed. J. Range Manag. 12, 69–82.

Ibekwe, A.M., Kennedy, A.C., Stubbs, T.L., 2010. An assessment of environmental conditions for control of downy brome by (*Pseudomonas fluorescens*-D7). Int. J. Environ. Technol. Manag. 12 (1), 27–46.

Johnson, B.N., Kennedy, A.C., Ogg Jr., A.G., 1993. Suppression of downy brome growth by rhizobacterium in controlled environments. Soil Sci. Soc. Am. J. 57, 73–77.

Johnson, D.R., Wyse, D.L., Jones, K.J., 1996. Controlling weeds with phytopathogenic bacteria. Weed Technol. 10, 621–624.

Julien, M.H. (Ed.), 1992. Biological Control of Weeds: A World Catalogue of Agents and Their Target Weeds, third ed. CAB. Int., UK.

Kass, H., 1990. Once a savior (*Cactoblastis cactorum*), moth is now a scourge. Plant Conserv. 5, 3.

Kennedy, A.C., Elliott, L.F., Young, F.L., Douglas, A.L., 1991. Rhizobacteria suppressive to the weed downy brome. Soil Sci. Soc. Am. J. 55, 722–727.

Kleppel, G.S., LaBarge, E., 2011. Using sheep to control purple loosestrife (*Lythrum salicaria*). Invasive Plant Sci. Manag. 4, 50–57.

Kremer, R.J., Kennedy, A.C., 1996. Rhizobacteria as biocontrol agents of weeds. Weed Technol. 10, 601–609.

Langeland, K., 1990. Exotic Woody Plant Control. Circular 868. Florida Coop. Ext Serv. Univ of Fl, Gainesville, 16 pp.

Larson, D.W., 1996. Curing the incurable. Am. Sci. 84, 7–9.

Lee, G.A., 1986. Integrated control of rush skeletonweed (*Chondrilla juncea*) in the western U.S. Weed Sci. 34 (Suppl.), 2–6.

Lee, M.-G., Li, S., Jarvis, B.B., Pestka, J.J., 1999. Effects of satratoxins and other macrocyclic trichothecenes and viability of EL-4 thymoma cells. J. Toxicol. Environ. Health A 57, 459–474.

Leininger, W.C., 1988. Non-chemical Alternatives for Managing Selected Plant Species in the Western United States. Colo. State Univ. Ext. Ser. Pub. No. XCM-118. 40 pp.

Li, Y., Sun, Z., Zhuang, X., Xu, L., Chen, S., Li, M., 2003. Research progress on microbial herbicides. Crop Prot. 22, 247–252.

Lockwood, J.A., 1993. Environmental issues involved in biological control of rangeland grasshoppers (Orthoptera: Acrididae) with exotic agents. Environ. Entomol. 22, 503–518.

Mack, R.N., Simberloff, D., Lonsdale, W.M., Evans, H., Clout, M., Bazzaz, F., 2000. Biotic invasions: causes, epidemiology, global consequences and control. Ecol. Appl. 10, 689–710.

Maddox, D.M., 1976. History of weevils on puncturevine in and near the United States. Weed Sci. 24, 414–416.

Malecki, R.A., Blossey, B., Hight, S.D., Schroeder, D., Kok, L.T., Coulson, J.R., 1993. Biological control of purple loosestrife. BioScience 43, 680–686.

Marston, B., October 17, 2005. Heard around the West – California. High Ctry. News 24.

McCain, A.H., 1978. The feasibility of using *Fusarium* wilt to control marijuana. Phytopath. News 12, 129.

McEvoy, P.B., 2002. The promise and peril of biological weed control (abst.). In: The 7th Annual Janet Meakin Poor Research Symposium Invasive Plants – Global Issues, Local Challenges. Chicago Botanic Garden, Glencoe, IL, pp. 14–15.

McKay, F., Oleiro, M., Walsh, G.C., Gandolfo, D., Cuda, J.P., Wheeler, G.S., 2009. Natural enemies of Brazilian Peppertree (*Sapindales: Anacardiaceae*) from Argentina: their possible use for biological control in the USA. Fla. Entomol. 92 (2), 292–303.

Millhollen, R.W., Berner, D.K., Paxson, L.K., Jarvis, B.B., Bean, G.W., 2003. *Myrothecium verrucaria* for control of annual morningglories in sugarcane. Weed Technol. 17, 276–283.

Myers, J.H., Bazely, D., 2003. Ecology and Control of Introduced Plants. Cambridge Univ. Press, Cambridge, UK, 313 pp.

Pemberton, R.W., Turner, C.E., 1990. Biological control of *Senecio jacobaeae* in northern California, an enduring success. Entomophaga 35, 71–77.

Peng, G., Byer, K.N., Bailey, K.I., 2004. *Pyricularia setariae*: a potential bioherbicide agent for control of green foxtail (*Setaria viridis*). Weed Sci. 52, 105–114.

Peschken, D.P., 1972. *Chrysolina quadrigemina* (Coleoptera: Chrysomelidae) introduced from California to British Columbia against the weed *Hypericum perforatum*: comparison of behavior, physiology, and colour in association with post-colonization adaptation. Can. J. Entomol. 104, 1689–1698.

Pimentel, D., 1963. Introducing parasites and predators to control native pests. Can. J. Entomol. 95, 785–792.

Pimentel, D., 1991. Diversification of biological control strategies in agriculture. Crop Prot. 10, 243–253.

Rees, N.E., Quimby Jr., P.C., Piper, G.L., Coombs, E.M., Turner, C.E., Spencer, N.R., Knutson, L.V. (Eds.), 1996. Biological Control of Weeds in the West. Montana State Univ., Bozeman, MT.

Richards, R., Whitesides, R.E., 2006. Biological Control of Saltcedar by Grazing with Goats Compared to Herbicide Treatments. Weed Sci. Soc. Am. Abst. No. 129.

Rosner, H., 2003. Getting Her Goat: Goats Are the Hip New Thing in Eco-friendly Weed Management. http:www://grist.org/news/maindish/2003/09/02/getting/.

Scheepens, P.C., 1987. Joint action of *Cochliobolus lunatus* and atrazine on *Echinochloa crus-galli* (L.) Beauv. Weed Res. 27, 43–47.

Simonson, S.E., Stohlgren, T.J., Tyler, L., Gregg, W.P., Muir, R., Garrett, L.J., 2005. Preliminary Assessment of the Potential Impacts and Risks of the Invasive Cactus Moth, *Cactoblastis cactorum* Berg, in the U.S. and Mexico. Final Report to the International Atomic Energy Agency, IAEA, Vienna, Austria.

Stevens, L.E., 1989. The status of ecological research on tamarisk (Tamaricaceae: *Tamarix ramosissima*) in Arizona. In: Kunzman, M.R., Johnson, R.R., Bennett, P.S. (Eds.), Tamarisk Control in Southwestern United States. Cooperative National Park Resources Study Unit Special Report Number 9, Tucson, AZ, pp. 99–105.

Stilling, P., Moon, D.C., 2001. Protecting rare Florida cacti from attack by the exotic cactus moth *Cactoblastis Cactorum* (Lepidoptera: Pyralidae). Fla. Entomol. 84 (4), 506–509.

Strobel, G., July 1991. Biological control of weeds. Sci. Am. 72–78.

Stubbs, T.L., Kennedy, A.C., Skipper, H.D., 2014. Survival of rifampicin-resistant *Pseudomonas fluorescens* strain in nine mollisols. Appl. Environ. Soil Sci. 7 pp. Article ID 306348. https://doi.org/10.1155/2014/306348.

TeBeest, D.O. (Ed.), 1991. Microbial Control of Weeds. Chapman and Hall, NY, 284 pp.

TeBeest, D.O., Templeton, G.E., 1985. Mycoherbicides: progress in the biological control of weeds. Plant Dis. 69, 6–10.

Thomas, R.F., Tworkoski, T.J., French, R.C., Leather, G.R., 1995. *Puccinia punctiformis* affects growth and reproduction of Canada thistle (*Cirsium arvense*). Weed Technol. 8, 488–493.

Tu, M., Hurd, C., Randall, J.M., 2001. Weed Control Methods Handbook: Tools & Techniques for Use in Natural Areas. The Nature Conservancy, Davis, CA, p. 219.

Turner, C.E., 1992. Beyond Pesticides: Biological Approaches to Weed Management in California. Div. of Agric. and Nat. Res. Univ. of Calif., Albany, CA, pp. 32–67.

Ueckert, D.N., Wright, H.A., 1973. Wood boring insect infestations in relation to mesquite control practices. J. Range Manag. 27, 383–386.

US Congress, September 1993. Office of Technology Assessment, Harmful Non-indigenous Species in the United States, OTA-F-565. U.S. Govt. Printing Office, Washington, D.C.

Van Zon, J.C.J., 1984. Economic weed control with grass carp. Trop. Pest Manag. 30, 179–185.

Vincent, C., Goettel, M.S., Lazarovits, G., 2007. Biological Control: A Global Perspective. CABI, London, UK, 466 pp.

Wapshere, A.J., Delfosse, E.S., Cullen, J.M., 1989. Recent developments in biological control of weeds. Crop Prot. 8, 227–250.

Watson, A.K., 1989. Current Advances in Bioherbicide Research. http://eap.mcgill.ca/_private/vl_head.htm.

Wilson, C.L., 1969. Use of plant pathogens in weed control. Ann. Rev. Phytopathol. 7, 411–434.

Winston, R.L., Schwarzländer, M., Hinz, H.L., Day, M.D., Cock, M.J.W., Julien, M.H. (Eds.), 2014. Biological Control of Weeds: A World Catalogue of Agents and Their Target Weeds, fifth ed. USDA Forest Service, Forest Health Technology Enterprise Team, Morgantown, WVA. FHTET-2014-04. 838 pp.

Wurtz, T.L., 1995. Domestic geese: biological weed control in an agricultural setting. Ecol. Appl. 5, 570–578.

Wymore, L.A., Watson, A.K., Gotlieb, A.R., 1987. Interaction between *Colletotrichum coccodes* and thidiazuron for control of velvetleaf (*Abutilon theophrasti* L.). Weed Sci. 35, 377–382.

Zettler, F.W., Freeman, T.E., 1972. Plant pathogens as biocontrols of aquatic weeds. Ann. Rev. Phytopathol. 10, 455–470.

Zimdahl, R.L., 2004. Weed-Crop Competition – A Review, second ed. Blackwell Publishing, Ames, IA. 220 pp.

Zorner, P.S., Evans, S.L., Savage, S.D., 1993. Perspectives on providing a realistic technical foundation for the commercialization of bioherbicides. In: Duke, S.O., Menn, J.J., Plimmer, J.R. (Eds.), Pest Control with Enhanced Environmental Safety, Amer. Chem. Soc Symp. Series 524, pp. 79–86.

13

Introduction to Chemical Weed Control

FUNDAMENTAL CONCEPTS

- Herbicides created a major change in the way agriculture is practiced by substituting chemical energy for human and animal energy.

- Herbicides have several advantages and disadvantages all of which should be considered prior to use.

- Herbicides can be classified in several useful ways; but no way integrates all.

- Classifications based on chemical structure and mechanism (site/mode) of action are common.

LEARNING OBJECTIVES

- To understand the history of chemical weed control.

- To know and understand the advantages and disadvantages of herbicides.

- To understand the different ways of classifying herbicides and the use of each classification system.

Herbicides greatly expanded the opportunities and range of methods for vegetation management/weed control. The word comes from the Latin *herba* (plant) and *caedere* (to kill). The etymology confirms that herbicides kill plants. The definition accepted by the Weed Science Society of America (WSSA) has changed. The third edition of the WSSA's Herbicide Handbook (Hilton, 1974) defined an herbicide as "A chemical used for killing plants or severely interrupting their normal growth processes." The definition was modified in the fifth edition (Beste, 1983), "A chemical used to control, suppress, or kill plants, or to severely interrupt their normal growth processes." The sixth edition (Humberg, 1989) definition is "A chemical substance or cultured biological organism used to kill or suppress the growth of plants." The

10th edition (Shaner) retained that definition. The WSSA definition is similar to dictionary definitions, which do not agree. A few examples:

> Webster's Third New International Dictionary (1965): An agent (as a chemical) used to destroy or inhibit plant growth; specifically: a selective weed killer that is not injurious to crop plants.
> The Unabridged Random House Dictionary of the English language (1966): A substance or preparation for killing plants esp. weeds.
> The 1972 College edition of Webster's New World Dictionary of the American language: Any chemical substance used to destroy plants, especially weeds, or to check their growth.
> Second College edition of the American Heritage Dictionary (1995): A substance used to destroy plants, especially weeds.

Weed scientists know that herbicides disrupt the physiology of a plant over a long enough period of time to kill them or severely reduce their growth. Pesticides are chemicals used to control pests. Herbicides, a subcategory, are pesticides used to control/manage plants. They are different from other pesticides because their sphere of influence extends beyond their ability to kill or control plants. Herbicides change the chemical environment of plants, which can be more easily manipulated than the climatic, edaphic, or biotic environment.

Herbicides reduce or eliminate labor and machine requirements and modify crop-production techniques. When used appropriately they are production tools that can increase efficiency, reduce horsepower, and perhaps reduce energy requirements. Herbicides do not eliminate energy requirements because they are petroleum based and machines are required for application.

Understanding the nature, properties, effects, and uses of herbicides is essential if one is to be conversant with modern weed management. Weed management is not accomplished exclusively by herbicides, but they dominate in the developed world. Whether one likes them or deplores them, they cannot be ignored. To ignore them is to be unaware of the opportunities and problems of modern weed management. Ignoring or dismissing herbicides may lead to an inability to solve weed problems in many agricultural systems and may delay development of better weed-management systems.

1. HISTORY OF CHEMICAL WEED CONTROL

1.1 The Blood, Sweat, and Tears Era

Agriculture can be thought of as having had three eras. During the blood, sweat, and tears era, famine and fatigue were common and inadequate food supply was common. Most people were farmers and many farms were small and operated at a subsistence level. Life was, for most people, in the words of the British philosopher Thomas Hobbes (1588–1679),

> wherein men live without other security, than their own strength, and their own invention shall furnish them...In such conditions there is...no knowledge of the face of the earth; no account of time; no arts; no letters, no society; and which is worst of all, continual fear and danger of violent death; and the life of man, solitary, poor, nasty, brutish, and short.

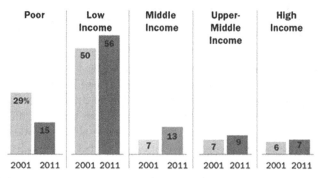

FIGURE 13.1 Global poverty 2001 to 2010 (http://www.pewglobal.org/2015/07/08/a-global-middle-class-is-more-promise-than-reality/).

Hobbes's (1651) dismal view still characterizes the lives of too many people. At the end of April 2017 there were more than 7.5 billion people in the world. In 2010, the UN Food and Agriculture Organization (UN/FAO) estimated that 925 million, 13%, were hungry. Thus, in 2010 almost one in seven of the people on this earth was hungry—truly a dismal statistic. Things were better five years later, but only marginally so according to the Pew trust (see Fig. 13.1) (Kochar, 2015). In 2012, 12.7%—almost 900 million people—lived at or below US$1.90/day. In 2016, 767 million people lived in extreme poverty (less than $1.90/day) (Economist, 2016). The good news is that extreme poverty is down 37% from 1990 to 44% from 1981. It is generally agreed that these, our fellow human beings, who survive on less than US$2/day (Nielsen, 2005, p. 170) live a Hobbesian existence, which is morally objectionable, indeed it is reprehensible. The data are a challenge to the agricultural enterprise whose primary moral responsibility is to produce food to feed a world population, projected to be 9 billion or more in 2050. Without adequate weed control, that challenge cannot be met.

1.2 The Mechanical Era

The mechanical era of agriculture began with invention of labor-saving machines, with farming becoming mechanized. In 1793, Eli Whitney invented the first workable cotton gin. Cyrus McCormick invented the reaper in 1834 and began to manufacture it in 1840. John Deere perfected the steel moldboard plow in 1837. In 1830, four farmers in the United States supported five nonfarmers. In 1910, a farmer fed himself (in 1910 farmers were men) and six others. By 1930, one farmer was able to support 10 nonfarmers. One farmer supported 40 nonfarmers in 1965. Today one United States farmer supports more than 100 nonfarmers in the United States and several non-US people through food exports.

1.3 The Chemical Era

The third era of agriculture, the chemical era, boosted production. It really began when nitrogen fertilizer became readily available and boosted production of newly available hybrid corn. In the early 1930s when these things began, a quarter of the American population lived on farms. When nitrogen fertilizer was combined with hybrid corn varieties, first experimented with by Henry A. Wallace[1] in 1913, yields went up rapidly. Wallace's early work led to his founding the Hi-Bred Corn Company in 1926, which later became Pioneer Hi-Bred Corn Company (purchased by DuPont in 1999). Hybrid corn was popularized by Roswell Garst, an Iowa farmer. In 1933, corn sold for 10 cents a bushel and a fraction of 1% of Iowa land was planted with hybrid seed. In 1943, 99.5% of Iowa corn was hybrid. In 1933 corn yield was 24.1 bushels per acre, about what it was during the Civil War. In 1943, it was 31, but by 1981 it had grown to 109 bushels per acre (Hyde, 2002). Iowa corn yields were 173 bu/A in 2005 and 189 bu/A in 2015; some have achieved 500 bu/A.

After 1945, when pesticides were developed and became widely available, yields continued to increase. In 1992, about 1% of US citizens were farmers (about 2.8 million) and each farmer fed 128 others (94.3 Americans and 33.7 people in other countries) (Krebs, 1992). Now, in modern agriculture's chemical era, less than 1% of US citizens farm and produce more than their grandfathers and great grandfathers ever dreamed possible. Between 2007 and 2012 the number of farms declined in 34 states, while production, average farm size, and age (58+) of the average farmer increased. The average US farm was 147 acres in 1900 and 434 in 2012.[2] Ninety-eight percent of US farms are family, family partnerships, family corporations, or individual ownership. About three-quarters of all US farms gross only $50,000 a year and account for just 4% of product sales. Farms with sales of more than $250,000, 9% of the 2.1 million US farms, produce 80% of all agricultural sales. Large farms with sales of $1 million or more are 4% of all farms but account for 66% of sales.

These changes are not unique to American agriculture. At every census since 1841, the percentage of people working in agriculture and fishing has declined in the United Kingdom (UK). In 1841, 22% of people worked in this industry and by 2014 fewer than 1.5% did. The number of farms decreased 6%, and those working on farms decreased 19% from 2000 to 2010 even though agriculture occupies 70% of the UK's land area. Production from each British agricultural worker increased at about twice the rate of increase for the rest of the economy (Malcolm, 1993). In Germany the number of farms and the number of farm workers decreased 25% between 2000 and 2010. Less than 1% of the population farms.

Increases in crop production and labor productivity in each agricultural era were caused by extensive farm mechanization, the use of agricultural chemicals, improved education of farmers, improved crop varieties, and improved farming practices. Developed country agriculture is now in the era of extensive and intensive use of chemical fertilizers and pesticides and is fully engaged in agriculture's next era—the era of biotechnology.

The chemical era of agriculture developed rapidly after 1945, but it did not begin then. In 1000 BCE, the Greek poet Homer wrote of pest-averting sulfur. Theophrastus, regarded as the father of modern botany (372?-287? BCE), reported that trees, especially young trees, could be killed by pouring oil, presumably olive oil, over their roots. The Greek philosopher Democritus (460?-370? BCE) suggested that forests could be cleared by sprinkling tree roots with the juice of hemlock in which lupine flowers had been soaked. In the first century BCE, the

---

[1]Wallace was the 11th US Secretary of Agriculture 1933–40, the 33rd Vice President of the United States 1941–45, and US Sec. of Commerce 1945–46.

[2]http://www.agcensus.usda.gov/Publications/2012/Preliminary_Report/Highlights.pdf.

Roman philosopher Cato advocated the use of amurca, the watery residue left after the oil is drained from crushed olives, for weed control (Smith and Secoy, 1975).

Historians tells us of the sack[3] of Carthage by the Romans in 146 BCE. The Romans put salt on the fields to prevent crop growth. Later, salt was used as an herbicide in England. Chemicals have been used as herbicides in agriculture for a long time, but their use was sporadic, frequently ineffective, and lacked any scientific basis (Smith and Secoy, 1975, 1976).

In 1755, mercurous chloride ($HgCl_2$) was used as a fungicide and seed treatment. In 1763, nicotine was used for aphid control. As early as 1803, copper sulfate was used as a foliar spray for diseases. Copper sulfate (blue vitriol) was first used for weed control in 1821. In 1855, sulfuric acid was used in Germany for selective weed control in cereals and onions. The US Army Corps of Engineers used sodium arsenite in 1902 to control water hyacinth in Louisiana.

Desperate potato growers in Ireland tried many things to control the Colorado potato beetle (*Leptinotarsa decemlineata*). All attempts failed until sometime in 1868, when a French farmer threw some leftover green paint on his potato plants. We do not know if the story is true, and, if it is true, we cannot know why he did it, but it worked. The green pigment in the paint was Paris green, a combination of arsenic and copper, commonly used in paint, fabric, and wallpaper.

Subsequently farmers diluted it with flour and dusted it on their potatoes or mixed it in water and sprayed it. Paris green (copper acetoarsenite) was a godsend to farmers. In Mann's (2011) view, chemists saw it as something with which they could tinker. If arsenic killed potato beetles, what other insects would it kill? If Paris green worked for potato beetles, would other chemicals work for other agricultural problems? The answer, of course, was yes. A French scientist found that a solution of copper sulfate and lime killed the causal organism (*Phytophthora infestans*) of late blight. Thus, spraying potatoes with Paris green and copper sulfate relieved (solved?) the beetle and the blight problem. The modern pesticide industry had begun.

Bordeaux mixture (copper sulfate, lime, and water), also green, was applied to grapevines for the control of downy mildew in the late 19th century. Actually it was not first applied to control anything. An unknown French grower put it on the vines on the edge of his vineyard to discourage passersby from stealing his grapes. Who wants a grape that has an unnatural green coating? In another apocryphal story, a passerby, a careful observer, noticed that Bordeaux mixture also turned yellow charlock leaves black. That led Bonnett, in France in 1896, to show that a solution of copper sulfate would selectively kill yellow charlock plants growing with cereals. In 1911, Rabaté demonstrated that dilute sulfuric acid could be used for the same purpose. The discovery that salts of heavy metals might be used for selective weed control led, in the early part of the 20th century, to weed control research with heavy metal salts by the Frenchmen Bonnett, Martin, and Duclos and the German Schultz (cited in Crafts and Robbins, 1962, p. 173). Nearly concurrently in the United States, Bolley (1908) studied iron sulfate, copper sulfate, copper nitrate, and sodium arsenite for selective control of broadleaved weeds in cereal grains. Bolley, a plant pathologist at the North Dakota State Agricultural College, is widely acknowledged as the first in the United States to report on selective

---

[3]*To sack* is not a verb that originated with the US National Football League.

use of salts of heavy metals as herbicides in cereals. The postemergence action was caustic or burning with little, if any, translocation.

Succeeding work in Europe observed the selective herbicidal effects of metallic salt solutions or acids in cereal crops (Zimdahl, 2010). The important early workers were Rabaté in France (1911, 1934), Morettini in Italy (1915), and Korsmo in Norway (1932).

Use of inorganic herbicides developed rapidly in Europe and England but not in the United States. In fact, weed control in cereal grains is still more widespread in Europe and England than in the United States. Some of the reasons for slow development in the United States included lack of adequate equipment and frequent failure because the heavy metal salts were dependent on foliar uptake that did not readily occur in the low humidity of the primary grain-growing areas—the western states. The heavy metal salts worked well only with adequate rainfall and high relative humidity. There were other agronomic practices such as increased use of fertilizer, improved tillage, and new varieties that increased crop yield in the United States without weed control. US farmers also could always move on to the endless frontier and were not as interested, as they would be later, in yield-enhancing technology.

Carbon bisulfide was first used in agriculture in 1854 as an insecticide in France. It was applied as a soil fumigant in Colorado to control *Phylloxera*, a root-borne disease of grapes. In 1906, it was introduced as a soil fumigant for control of Canada thistle and field bindweed. It smells like rotten eggs and may have reached its peak usage in Idaho in 1936 when over 300,000 gallons were used.

Petroleum oils were introduced for weed control along irrigation ditches and in carrots in 1914. They are still used in some areas for weed control. Field bindweed was controlled successfully in France in 1923 with sodium chlorate, which is now used as a soil sterilant in combination with organic herbicides. Arsenic trichloride was introduced as a product called kill morning glory in the 1920s. Sulfuric acid was used for weed control in Britain in the 1930s. It was and still is a very good herbicide, but is very corrosive to equipment and harmful to people.

The first synthetic organic chemical for selective weed control in cereals was 2-(1-methylpropyl)-4,6-dinitrophenol (Dinoseb), introduced in France in 1932 (King, 1966, p. 285). It was used for many years for selective control of some broadleaved weeds and grasses in large-seeded crops such as beans. It is included in the sixth edition of the Herbicide Handbook (Anonymous, 1989) but not in the seventh (Ahrens, 1994) or subsequent editions, although dinoterb, a close chemical relative, which is not sold in the United States is. Both are included in the 2016 WSSA list of approved herbicides.[4] Dithiocarbamates were patented as fungicides in 1934. In 1940, ammonium sulfamate was introduced for control of woody plants.

Historians of weed science will note 1941 as an important year. Pokorny (1941) first synthesized 2,4-dichlorophenoxyacetic acid (2,4-D). It was reported to have no activity as a fungicide or insecticide. Accounts vary about when the first work on growth-regulator herbicides was done (Akamine, 1948). Zimmerman and Hitchcock (1942) of the Boyce-Thompson Institute (formerly in Yonkers, NY, now at Cornell University) first described the substituted phenoxy acids (2,4-D is one) as growth regulators (auxins), but did not report

[4]http://wssa.net/wssa/weed/herbicides/.

herbicidal activity. They also worked with other compounds that eventually became herbicides. They were the first to demonstrate that these molecules had physiological activity in cell elongation, morphogenesis, root development, and parthenocarpy (King, 1966). A Chicago carnation grower's question, "What is the effect of illuminating gas[5] on carnations?" led to the eventual discovery of plant growth-regulating substances by Boyce—Thompson scientists (King).

E. J. Kraus, Head of the University of Chicago Botany Department, had studied plant growth regulation for several years. He supervised the doctoral programs of J. W. Mitchell and C. L. Hamner who in the early 1940s were working as plant physiologists with the US Department of Agriculture Plant Industry Station at Beltsville, MD. Kraus thought these new, potential plant growth regulators that often distorted plant growth when used at higher than growth-regulating doses, and even killed plants, might be used beneficially to selectively kill plants. He saw potential use as "chemical plant killers" (herbicides) and advocated purposeful application in toxic doses for plant control. Because of World War II (WWII) and the potential for biological warfare against an enemy's crops (e.g., German potatoes), much of this work was done under contract from the US Army (Peterson, 1967; Troyer, 2001). Similar work for similar reasons was done in Great Britain (Kirby, 1980). The chemicals were not used for biological warfare during WWII. A much more complete chronology and history of development of the hormone herbicides is available in Kirby, Troyer, and Zimdahl (2010).

Hamner and Tukey (1944a,b) reported the first field trials with 2,4-D for successful selective control of broadleaved weeds. They also worked with 2,4,5-T as a brush killer. At nearly the same time, Slade et al. (1945), in England, discovered that naphthaleneacetic acid at 25 lbs/acre would selectively remove charlock from oats with little injury to oats. Slade et al. also discovered the broadleaved herbicidal properties of the sodium salt of MCPA (later called Methoxone, King, 1966), a compound closely related to 2,4-D. Slade et al. confirmed the selective activity of 2,4-D. Marth and Mitchell (1944), former students of E. J. Kraus, first reported the use of 2,4-D for killing dandelions and other broadleaved weeds selectively in Kentucky bluegrass. Marth and Mitchell attribute the quest for selective activity of these compounds to Kraus. These discoveries were the beginning of modern chemical weed control. All previous herbicides were just a prelude to the rapid development that occurred following discovery of the selective activity of the phenoxyacetic acid herbicides. The first US patent (No. 2390941) for 2,4-D as an herbicide was obtained by F. D. Jones of the American Chemical Paint Co. in 1945 (King). There had been an earlier patent (No. 2322761) in 1943 of 2,4-D as a growth-regulating substance (King, 1966). It is interesting to note that Jones patented only its activity (the fact that it killed plants) but made no claim about selective action (the fact that it killed some, but not all, plants) (King).

The effectiveness of monuron, a substituted urea, for control of annual and perennial grasses was reported by Bucha and Todd (1951). This was the first of many new selective chemical groups with herbicidal activity. The first triazine herbicide appeared in 1956 and the first acylanilide in 1953 (Zimdahl, 2010) followed by CDAA, the first alphachloroacetamide in 1956 (Hamm, 1974).

The great era of herbicide development came at a time when world agriculture was involved in a revolution of labor reduction, increased mechanization, and new methods to

---

[5]Acetylene gas.

improve crop quality and produce higher yields at reduced cost. Herbicide development built on and contributed to changing agriculture. Farmers were ready for improved methods of selective weed control. The rapid acceptance of technological developments that changed the practice of agriculture have been characterized in terms of economic, social, political, and philosophical attitudes by Perkins and Holochuck (1993). Farmers wanted to improve their operation in competition with other farmers and were willing to adopt new technology that enabled them to improve their *economic* competitiveness. New technology was *socially* acceptable because as independent entrepreneurs many technological innovations could be used to gain advantage independent of neighbors. *Politically*, farmers welcomed technical assistance that came from public laboratories (land-grant universities) and government price-support systems that allowed farm operations to remain private. Fiercely independent farmers welcomed opportunities to do what they wanted on their farms. They welcomed technology developed at no apparent cost to them that could be adopted without interference from anyone. Finally, *philosophically*, farmers perceived that a major part of their task was controlling nature—bending it to human will. Although this was a never-ending challenge, success was apparent when technology that increased production was readily available. Herbicides fit well in each category.

It is true that no weed control method has ever been abandoned, new ones have been added, and the relative importance of methods has changed. The need for cultivation, hoeing, etc. has not disappeared. These methods persist in small-scale agriculture (I hoe my garden). Older methods have become less important in developed world agriculture because of the rising costs of labor and narrower profit margins (Table 13.1).

The amount of herbicides used increased steadily after WWII. Table 13.2 shows sales growth of all pesticides since 1960. One must assume the amount used also increased. Herbicide sales dominate. In 1997, 1 billion pounds of pesticides were used in the United States and over 47% (461.4 million pounds) was herbicides Gianessi and Silvers (2000) (Table 13.2). Just 10 herbicides accounted for 75% of sales (Gianessi and Marcelli, 2000). US farmers routinely apply herbicides to more than 85% of US crop acres (Gianessi and Sankula, 2003). A study of 40 crops showed treatment of 220 million acres at a cost of $6.6 billion (Gianessi and Sankula). In 2001, the global market for nonagricultural pesticides was more than US$7 billion per year and was growing about 4% a year. The global market just for turf

TABLE 13.1 The Evolution of Weed-Control Methods in the United States (Alder et al., 1977)

	% Control by Year in the United States			
Year	Human Energy	Animal Energy	Mechanical Energy (Tractor)	Chemical Energy
1920	40	60		
1947	20	10	70	
1975	5	TR[a]	40	55
1990	<1	TR	24	75
2012 (estimated)	<1	TR	10	90

[a]TR, *trace*.

TABLE 13.2 World Sales of Crop Protection Products 1960–2001 (Gianessi and Silvers, 2000; Hopkins, 1994, and Other Sources)

	1960	1970	1980	1990	1997	2001
Pesticide			(Million US Dollars)			
Herbicides	160	918	4,756	12,600	14,700	14,118
Insecticides	288	945	3,944	7,840	9,100	8,763
Fungicides	320	702	2,204	5,600	5,400	6,027
Other	32	135	696	1,960	1,700	2,848
Total	800	2700	11,600	28,000	30,900	31,756

pesticides is ~US$850 million per year; about half is used on golf courses. Each year US lawn care firms apply about $440 million of pesticides.

The global herbicide market was estimated to be $13.5 billion from 1990 to 1993, and one-third ($4.5 billion) was the US market. Japan was the next largest with $1.5 billion in sales. When the entire European market was considered, it was second largest, with France ($1.25 billion) the largest single country (Hopkins, 1994). In 2007, world expenditures on all herbicides were US$15,512 million; 47% was herbicides. The United States spent $5856 million for 531 million pounds of active ingredient. This was lower than purchases in 2000, but they have returned to the levels last seen in the early 1970s (US/EPA, 2004). Of these amounts, 78% is used in agriculture, with the rest nearly evenly divided between industrial/commercial/government (12%) and home and garden use (10%).

The National Agricultural Statistics service of the US Department of Agriculture regularly surveys selected states and selected crops to determine the extent of fertilizer and pesticide use. The 1997 report (1996 data) shows that herbicides were used on a major portion of the acreage of each crop surveyed. The specific figures for the five crops surveyed in 1997 and in 2010[6] follow.

	% of Acres Treated With Herbicides	
Crop	1997	2010
Corn	97	98
Cotton	92	99
Potato	87	94
Soybean	97	97
% of acres treated with glyphosate		87

(Continued)

[6]http://usda.mannlib.cornell.edu/reports/n...emical_usage_field_crops_summary_09/03.97, http://www.nass.usda.gov/Surveys/Guide_to_NASS_Surveys/Chemical_Use.

	% of Acres Treated With Herbicides	
Crop	1997	2010
Tobacco		75
Onions		88
Peanut		98
Tomato		9
Wheat, durum		99
Wheat, winter		45

Crops surveyed in each year show similar use and each illustrates the dominance of herbicides for weed control. The soybean data show the dominance of glyphosate (Roundup) resistant soybean, which is also true for corn and cotton. The wheat data may illustrate the low profitability of the crop and the lack of weed problems for which herbicide solutions exist. Tomatoes, grown for fresh market and processing, are weak competitors with weeds. Several herbicides are available, but growers use several alternative weed management techniques: plastic mulch, crop rotation, and cultivation before and after the crop.

In the world in 1990, about 45% of total pesticide sales were herbicides (similar to the United States), insecticides were 28%, and fungicides ~20% (Hopkins, 1994). Over 85% of herbicides are used in agriculture. The worldwide market is becoming increasingly concentrated in the hands of a few multinational corporations. Nearly half the companies in pesticide discovery (but not in development and marketing) in 1994 were Japanese (Hopkins, 1994). The number of companies marketing herbicides in the United States has steadily shrunk from 46 in 1970 to 10 in 2012 (Appleby, 2005; personal communication April 2012). Three are based in the United States and the others are based in Europe and Japan but operate in the United States.

While the number of companies engaged in herbicide discovery, development, and sales has steadily declined (Appleby, 2005), the number of available herbicides has steadily increased. The number of herbicides in the first (1967) through the ninth (2007) edition of the WSSA's Herbicide Handbook increased as did the number of different chemical families in which herbicidal activity had been discovered (Table 13.3). Similarly the number of WSSA-approved herbicides increased from 304 in 1995 to 357 in 2004, but decreased slightly to 325 in 2011 (Anonymous, 1995, 2004, 2011).

Because of their significant advantages and extensive research base, herbicides dominate modern weed control. Timmons (1970) reported 75 herbicides marketed between 1950 and 1969. Appleby (2005) included 184 herbicides marketed between 1970 and 2005, an increase of 2.4 times. Although the herbicide chemical industry has undergone extensive consolidation, as have many other manufacturing industries, it has not diminished discovery and development of new herbicides in older chemical families. Worldwide sales have continued to increase. World exports of pesticides of all kinds totaled US$15.9 billion in 2004, a new high in sales for the global chemical industry (Jordan, 2006), which increased to $31.9 billion

TABLE 13.3   The Number of Herbicides and Chemical Families in Each of the 10 Editions of the Herbicide Handbook of the Weed Science Society of America

	Year of Publication and Edition									
	1	**2**	**3**	**4**	**5**	**6**	**7**	**8**	**9**	**10**
	1967	**1970**	**1974**	**1979**	**1983**	**1989**	**1994 + 1998 supp.**	**2002**	**2007**	**2014[7]**
Herbicides	97	115	125	137	130	145	163	211	217	232
Chemical families	27	27	32	37	35	43	63	75	67	67

in 2015.[8] Use of all kinds of pesticides has risen from nearly 0.5 kg/ha in 1960 to 2 kg/ha in 2004. The increase was attributed mainly to the increased use of herbicides on genetically modified crops in China (Jordan, 2006) and in the world (Benbrook, 2012).

2. ADVANTAGES OF HERBICIDES

2.1 Energy Use

Agriculture collects and stores the electromagnetic energy of the sun as food energy (chemical energy) in plant and animal products. Food is available because plants do this so well and few other life forms can do it in a way that makes food for people. Farming invests about three calories of fuel energy (which some call cultural energy) in soil tillage, fertilizer, pesticides, irrigation, and harvest to help plants convert sunlight into one calorie of stored chemical energy in food (Lovins et al., 1984). When the energy costs for processing, distribution, and preparation are included, ~9.8 calories of energy are expended for every calorie of food energy produced in the United States (Lovins et al., 1984)—ergo, an energetically inefficient enterprise. However, many argue that it is not the task of agriculture to produce energy; the task is to produce food and that is done very well. Herbicides contribute to abundant production. All pesticides account for only 3% of the energy used in agricultural production. On-farm energy consumption represents only a little over 3% of total US energy consumption and only 18% of the energy used in entire the US food system from farm to table (Lovins et al.).

A technology that does not create reproducible gains in value will not succeed. Weed management is an essential part of agriculture. Although the data on energy use for weed management are old, no recent studies have been done. Its energy efficiency is important, but not studied because it must be done and there are few desirable alternatives to the dominant herbicide/chemical-based system. The energy efficiency of Table 13.4 shows energy relationships for weed control in corn in Minnesota (Nalewaja, 1974), and Table 13.5 shows

[7]The herbicides in the 9th and 10th editions are divided into 16 mechanism-of-action groups. Twenty herbicides of diverse chemical structure whose mechanism of action is unknown or uncertain are not included in the group of 67.

[8]http://www.worldstopexports.com/top-pesticides-exporters/.

TABLE 13.4   Energy Relationships in Weed Control in Six Experiments on Corn in Minnesota (Nalewaja, 1974)[a]

Method of Weed Control	Energy for Weed Control (MJ/ha)	Corn Yield kg/ha	Corn Yield MJ/ha	Net Profit From Weed Control $/ha	Net Profit From Weed Control MJ/ha	Energy Output/Input	Man-Hour/ha
None	0	3387	56,528	—	—	0	
Cultivation	579	5080	84,387	151	27,550	48/1	1.41
Herbicide	391	5645	93,763	194	37,113	95/1	0.12
Hand Labor	337	5770	—	163	39,251	116/1	148.15

[a]The land was plowed, disked, and prepared for planting corn in the conventional manner. Calculations are based on an average of 2.5 cultivations using 2.8 L of gasoline/ha and spraying using 0.8 L gasoline/ha. Atrazine was applied at 3.4 kg/ha.

TABLE 13.5   Energy and Cost for Weed Control in Cotton (Barrett and Witt, 1987)

Weed Control Income Method	Energy for Weed Control (MJ/ha)	Cotton Yield Lint (kg/ha)	Cotton Yield Seed (kg/ha)	Cotton Yield Energy (MJ/ha)	Ratio of Energy in Crop to Weed Control Input	Values ($/ha) Cost of Weed Control	Values ($/ha) Income From Lint and Seed	Values ($/ha) Income From Cost of Weeding
Four herbicides, no cultivation	2093	619	856	28,310	13/1	64.47	901.11	836.64
Three herbicides, two cultivations	1898	545	754	24,935	13/1	56.96	860.91	803.95
No herbicides, five cultivations	1220	177	244	8,082	7/1	24.69	257.43	232.7
No herbicides, five cultivations, 185 man-hour/ha hand hoeing	1641	592	819	27,085	17/1	645.06	862.02	216.97

Adapted by Barrett, M., Witt, W.W., 1987. Alternative pest management practices. In: Helsel, Z.R. (Ed.), Energy in Plant Nutrition and Pest Control. Vol. 2 in B.A. Stout ed. Energy in World Agriculture, Elsevier Press, NY, pp. 195–234 from Nalewaja, J.D., 1974. Energy requirements for various weed control practices. In: Proc. N. Central Weed Control Conf., vol. 29, pp. 19–23 and Dowler, C.C., Hauser, E.W., 1975. Weed control systems in cotton in Tifton loamy sand soil. Weed Sci. 23, 40–42. Hoeing cost was estimated to be $91 for a 40 h week.

similar data for cotton in Georgia (Dowler and Hauser, 1975). The weed density in corn was low but it was high in cotton. In both cases, cost-benefit analysis favored herbicides over other methods. Hand labor gave the greatest energy output-input ratio. The data do not consider the energy to house and feed workers or the fact that such work is seasonal (Barrett and Witt, 1987). Soil was plowed, disked, and prepared for planting corn or cotton in conventional ways and all other cultural practices were uniform. With no weed control, there was, of course, no profit due to weed control and a low yield and crop value. Cultivation and herbicides were not very different but hand labor produced a net loss because of its high cost and poor weed control.

TABLE 13.6 Weed-Control Energy Relationships for Six Corn Experiments in Minnesota (Nalewaja, 1974)[a]

| Method of Weed Control | Energy Input for Weed Control (kcal/A) | Yield of Corn/A | | Net Profit Due to Weed Control (kcal/A) |
		Bushels	Kilocalories	
None	0	54	5,443,200	—
Cultivation	56,005	81	8,164,800	2,665,595
Herbicide	37,920	90	9,072,000	3,590,880
Hand Labor	32,655	92	9,273,600	3,797,745

[a]The land was plowed, disked, and prepared for planting corn in the conventional manner.

Herbicide energy efficiency is reinforced by data from the same study showing the energy relationships for methods of weed control. Herbicides require more petroleum energy than hand labor but less than cultivation. Herbicides compare favorably to other methods in net energy profit due to weed control because yield was nearly as high as that achieved with hand labor (Table 13.6).

The total energy required for herbicides for corn, a crop that requires a great deal of energy, is relatively small (Table 13.7) (Pimentel and Pimentel, 1979). About 3% of the total

TABLE 13.7 Average Energy Input to US Corn Production System in 1975 (Pimentel and Pimentel, 1979)

Input	% of Total (kcal/ha)
Labor	0.09
Machinery	8.5
Diesel Fuel	19.6
Nitrogen	28.8
Phosphorus	3.3
Potassium	2.0
Lime	0.5
Seed	8.0
Irrigation	11.9
Insecticide	1.3
Herbicide	3.1
Drying	6.5
Electricity	5.8
Transportation	0.5
Kcal output/Kcal input	2.93

TABLE 13.8 Energy and Yield Comparisons of Corn Production Systems (Pimentel and Pimentel, 1979)

Corn Production System	Total		
	kcal Output/Input	Yield (kg/ha)	T/ha
United States	2.93	5394	2.4
Philippines w/animal power	5.06	941	0.42
Mexico			
w/oxen	4.34	941	0.42
w/manpower	10.74	1944	0.87
Nigeria w/human labor	6.41	1004	0.45

energy input for the US corn production system is directly related to herbicides that are used on 98% of US corn acreage (Gianessi and Sankula, 2003). The major energy consumers in US corn production are nitrogen fertilizer, diesel fuel, and irrigation. The US corn production system has one of the lowest energy efficiencies among the world's crop production systems (Table 13.8). Available data verify that US agricultural energy efficiency is low, but yields are high, which one must assume is good for growers and consumers.

2.2 Time/Profit

Broadcast herbicide use in corn is the least time-consuming weeding strategy, whereas preemergence rotary hoeing followed by two cultivations required the most time (Lague and Khelifi, 2001). Others (Swanton et al., 1996) have shown that for the Canadian province of Ontario energy use per hectare decreased by 19.7% for corn and by 46.3% for soybean production systems from 1975 to 1991. The reasons were the increased use of no-till production systems and herbicides.

An important criterion for a grower is profit-return on investment in technology. Becker (1983, cited in Barrett and Witt, 1987) attributed corn and soybean yield increases after herbicide use to improved weed control and earlier planting when herbicides were available. Combining these two factors reduced cost of production and increased profit about 10% (Becker) (Table 13.9). Abernathy (1981) calculated the additional land required to maintain production of seven major US crops without herbicides. He used estimates from several sources to determine likely losses and their value. All aspects of loss were considered including additional cultivation required to control weeds when herbicides were not used. Abernathy proposed a net loss greater than $23 billion and the need for an additional 28.2 million acres to maintain production. Gianessi and Sankula (2003) calculated a net loss of $21 billion for 40 crops if herbicides were not used on the 220 million acres on which the crops were grown. Three crops (corn, cotton, and soybean) accounted for $7.9 billion of the loss (37%). These data show that the great majority of land devoted to the major US crops is treated annually with a herbicide. They have been adopted rapidly and nearly completely since their

TABLE 13.9 Cost-Benefit Assessment of Herbicide Use in Corn and Soybean (Barrett and Witt, 1987)

	Corn (Cost in $/ha)		Soybean (Cost in $/ha)	
	Cultural Weed Control	Herbicide	Cultural Weed Control	Herbicide
Yield (kg/ha)	7212	8179	2554	2084
Herbicide + application cost	0	43.51	0	50.32
Savings in tillage cost	0	9.70	0	7.68
Total herbicide cost and tillage savings	0	33.80	0	44.64
Total crop production costs	799.75	833.59	587.65	632.30
Cost per kg produced	0.11	0.10	0.23	0.21

Data also available in Becker, R., 1983. Selling the benefits of weed control. In: Proc. 35th Ann. Fertilizer and Agric. Chem. Dealers Conf., Bul. CE-185—4e. Coop. Ext. Serv., Iowa St. Univ., Ames, IA.

introduction in the late 1950s. Gianessi and Reigner (2007) proposed that US crop production would decline by about 20% if herbicides were not used. Herbicides are, have been, and are projected to continue to be a major part of modern agriculture. Because they are less expensive and more efficient, they eliminated much tillage and its disadvantages (time, erosion, crop damage) and "they replaced the use of millions of workers to pull and hoe weeds by hand" (Gianessi and Reigner). Herbicides thus contributed to "Creative Destruction,"[9] a feature of capitalism whereby jobs are created by technological advance, while others, and their associated businesses, are eliminated. What happened to the "millions of workers" who lost their jobs has not been an agricultural or societal concern. Progress has a price.

Pimentel et al. (1978) estimated costs and losses with herbicide use and with alternative methods of weed control but reached a different conclusion. The primary reason for the difference is that Pimentel et al. assumed that with careful management, little additional crop loss (only $341 million) would occur by switching from herbicides to alternative weed-management techniques. More-intensive weed management is proposed frequently as a necessary part of alternative (i.e., nonherbicide) weed management.

No one knows who is correct in the debate, which seems to have ceased. Clements et al. (1995) provide a clue. They confirm the proposition put forth earlier (see Table 13.7) that energy for weed management represents a small proportion of on-farm energy use for food production. Clements et al. propose that a "large portion of energy allocated to weed control could be conserved in alternative weed management systems by elimination strategies for reduction of tillage and/or herbicide use." They showed that potential energy savings from reduction or elimination of tillage was greater than for elimination of herbicides. They also suggested that there would be a potentially high-energy requirement for tillage

[9]The concept of creative destruction was popularized by J. Schumpeter in Capitalism, Socialism and Democracy (1942). It refers to the inevitably linked processes of accumulation (creation) and annihilation (destruction) of wealth under capitalism. It was first described in The Communist Manifesto (Marx and Engels, 1848). Creative destruction describes how economic development under capitalism is derived from the destruction of a prior economic order.

if herbicides were eliminated, particularly when numerous interrow cultivations were required for weed control. Most alternative methods of weed management are more energy efficient than those based on herbicides (Clements et al.). Energy savings are being achieved by alternative herbicide-use techniques such as reducing the total area of application, using band application instead of broadcast spraying, and by choosing herbicides that require less energy to produce (e.g., trifluralin or atrazine as opposed to paraquat or bentazon) (Clements et al.). The latter is probably not a primary criterion for most farmers—efficacy is.

As claimed before, many in agriculture argue that the purpose of agriculture is to produce food, not energy. Others argue that the US system is so dependent on petroleum energy that it is not sustainable. Many believe that modern weed control with herbicides is essential to maintain the present, highly productive US agricultural system and is justified because herbicides represent only a small part of the total energy input.

2.3 Labor Requirement

Herbicides are advantageous when labor is expensive. Gianessi and Sankula (2003) claim that growers used to paying 10 cents an hour for labor suddenly found it necessary to pay 50 cents in the early 1950s, and $1 in the 1960s, and much more now. Herbicides reduced or eliminated labor costs. Gianessi and Sankula say that a 1957 onion experiment showed $8/acre for herbicide application substituted for 55 h of labor at a cost of $41/acre. They claimed the costs of weed control for organic vegetable growers in California can be as high as $1000/acre compared to $50 for herbicides for the same acres and the same, or better, weed control. More current data are not available.

Herbicides can control weeds in crop rows where most mechanical methods are ineffective.

Herbicides are not only beneficial and profitable where labor is scarce or expensive. They may be advantageous where labor is plentiful and cheap. Herbicides control weeds in crop rows where cultivation is not possible. They can be used in places where other methods don't work. Preemergence herbicides provide early season control when weed competition may cause the greatest yield reduction and when other methods are less efficient or impossible to use (e.g., it is impossible to mechanically cultivate when soil is wet).

2.4 Tillage

Cultivation can injure crop roots and foliage. Selective herbicides reduce the need for tillage and control of weeds in crop rows where tillage cannot reach. Herbicides reduce destruction of soil structure by decreasing the need for tillage and the number of trips over the field with heavy equipment.

Herbicides permit selective weed control in orchards. Proper herbicide selection maintains plant cover and reduces or eliminates the need for tillage that may encourage soil erosion. Erosion in orchards and in other perennial crops can be prevented by maintenance of a sod cover with selective herbicides. Tillage to eliminate weeds is not required at all or as often when herbicides are used. Many perennial species cannot be controlled effectively with hand labor, and herbicides are often the only reasonable option. Erosion of cropland declined from about 3.8 billion tons in 1938 to 1 billion tons in 1997 (Gianessi and Sankula, 2003). A billion tons is still way too much, but herbicides help reduce the need for tillage that can lead to soil erosion.

Other studies[10] show that conservation due to US government policies reduced total soil erosion between 1982 and 1992 by 32% and the sheet and rill erosion rate from an average of 4.1 tons/acre/year in 1982 to 3.1 in 1992 while the wind erosion rate fell from an average of 3.3 tons/acre/year to 2.4 tons. Soil erosion is imposing substantial social costs. It is reasonable to postulate that because food production is dependent on soil, erosion—a natural, manageable process—is second only to population growth as the biggest environmental problem the world faces. The United States is losing soil 10 times faster—and China and India are losing soil 30–40 times faster—than the natural replenishment rate. The economic effect of soil erosion in the United States costs the nation about $37.6 billion each year in productivity losses. Damage from soil erosion worldwide is estimated to be $400 billion/year. As a result of erosion over the past 40 years, 30% of the world's arable land has become unproductive (see Pimentel et al., 1987, 1995). Too many people regard soil as just dirt. Without soil, agriculture as we know it will be impossible. In 2007, 99 million acres (28% of all cropland) were eroding above soil loss tolerance (T) rates.[11] This compares to 169 million acres (40% of cropland) in 1982. Soil erosion has been reduced, but further reduction below the T rate is required.

Herbicides save labor and energy by reducing the need for hand labor and mechanical tillage. They can reduce fertilizer and irrigation requirements by eliminating competing weeds. They reduce harvest costs by eliminating interfering weeds and can reduce grain drying costs because green, weedy plant material is absent. Other methods of weed control will, of course, also accomplish these things but not as efficiently and often not as cheaply.

It should be noted that the previous advantages and following disadvantages are often presented quite differently. For example, cost is presented herein as a disadvantage—herbicides are expensive. Others regard the cost of herbicides as an advantage, because with analysis, they are usually less expensive than tillage with its potential soil erosion.

---

[10]http://www.ncbi.nlm.nih.gov/pubmed/11281426.

[11]The US Dept. of Agriculture T rate is 5 tons/acre/year, well above the natural soil formation rate of $^{1}/_{2}$ ton/acre/year.

One must be aware of the quality of the evidence presented in support of a conclusion and the bias of the evidence. In the current jargon, one must ask and know, where the speaker is coming from.

3. DISADVANTAGES OF HERBICIDES

3.1 Cost

It is often suggested that herbicides reduce crop-production costs. Many disagree and suggest herbicides are a net cost because they are expensive, the equipment for applying them is an added cost, and, of most importance, there are large externalized societal costs.[12] The debate continues, and its elements vary with different crop-production systems.

The cost of manufacturing, developing, and introducing a new herbicide to market has steadily increased. Crop Life America estimated that each new product cost $286 million and takes 11 years of research and development. The crop protection industry spent $2.6 billion on new innovations, in 2014,[13] but no new site-of-action herbicides appeared. Development costs have become so high that crops that used to be regarded as major markets are now minor due to financial investments required and the increasing possibility that initial costs may not be recovered in sales (Ivany, 2001). Simultaneously the availability of older herbicides is decreasing as more stringent environmental and toxicological requirements result in voluntary removal from the market. But, in spite of regulatory restraints, the number of available herbicides has continued to rise (see Table 13.3).

3.2 Mammalian Toxicity

A major concern about herbicides is mammalian toxicity. All have some toxicity to humans and other plant and animal species. Some are no more toxic in terms of their LD_{50} than many common chemicals (e.g., aspirin, mothballs, gasoline, and table salt). Many people are concerned about herbicide toxicity because all must eat and therefore there is no choice about potentially toxic residues in food, especially when one does not know they are present. For example, in 1996, there were 441 definite/probable cases of pesticide intoxication in California and 271 positive cases. Of these, only 3 and 22, respectively, were attributed to herbicides. Most (65%) were due to insecticides (http://www.cdpr.ca.gov/docs/whs/1996pisp.htm). (Note: data can be obtained for any state when that state's two-letter initial is substituted in the above URL.) No one knows for sure, but it is estimated that between 1 and 5 million cases of pesticide poisoning occur every year in the world, resulting in 20,000 deaths. Developing countries use 25% of pesticides but experience 99% of the deaths (Goldman, 2004). The World Health Organization data show that about 3 million people a year suffer from severe pesticide poisoning (Jordan, 2006).

---

[12]An externality is a cost that is not reflected in price, or more technically, a cost or benefit for which no market mechanism exists. In the accounting sense, it is a cost that a firm (a decision maker) does not have to bear, or a benefit that cannot be captured. From a self-interested view, an externality is a secondary cost or benefit that does not affect the decision maker.

[13]http://www.croplifeamerica.org/cost-of-crop-protection-innovation-increases-to-286-million-per-product/.

3.3 Environmental Persistence

Some herbicides persist in the environment. None persist forever, but all have a measurable environmental life. In some cases, but not all, an herbicide can persist in the soil from one crop season to the next, or longer. This restricts rotational possibilities and may injure succeeding crops. Therefore, herbicides can be hazards to plants planted after they are used. Plants that are not targets may be affected by drift or inappropriate application. Although weed scientists and farmers are well aware of the problems caused by excessive persistence, they still occur. Greenland (2003) showed that vegetable crops could be injured by flumetsulam (cabbage and squash) or nicosulfuron (cabbage and onion), especially when double or higher rates were used. As has been known for many years, warm summer temperatures and adequate soil moisture enhance microbial degradation of herbicides and should reduce injury.

3.4 Weed Resistance to Herbicides

This is a significant problem of increasing concern to weed scientists, manufacturers, and growers. It is discussed in Chapter 19.

3.5 Monoculture

Agriculture in the world's industrialized/developed nations is characterized by monoculture—large land areas devoted to a single crop. This is ideal for use of selective herbicides, and many have criticized herbicides because they encourage monoculture and discourage diversity. Unquestioned expansion of herbicide technology into developing countries is not always wise because of their existing agricultural plant diversity. There is ecological strength in diversity and it should not be inhibited or reduced by extensive use of herbicides for weed control, especially where the long-term agricultural and social consequences of extensive monoculture, if not unknown, have not been thoroughly examined.

3.6 Other

Herbicides are often inconsistent in their control efficacy because they are affected by environmental conditions and results of these interactions are not always predictable. Herbicide use in many crops may, intentionally or accidentally, eliminate all plants except the crop and that may lead to excessive soil erosion.

Precision is required when herbicides are used. One must think carefully about what herbicide to use, when to use it, how much to use, and how surplus chemical will be handled. They cannot be used casually; intelligence is required for use and for disposal of surplus chemicals and empty containers.

Finally, because herbicides are so good at what they do, they may actually create problems after their use. Herbicides control certain weeds while leaving a crop unscathed. They are designed to be selective. Natural plant communities are usually polycultures (this is not a universally true generalization). Diversity—that is, multiple species assemblies—is the norm. When all plants are eliminated save the crop, other plants (weeds) will move into the environment created and they may be more difficult to control than those eliminated.

For example, Florida pusley was a common weed in peanut production before herbicides were introduced (Johnson and Mullinix, 1995). When herbicides became integral to peanut production, Florida pusley was controlled, but previously minor weeds—Florida beggar-weed, Texas panicum, and yellow nutsedge—increased. When herbicides were discontinued, even after several years of use, Florida pusley again dominated (Johnson and Mullinix). A second example of replacement is from a rice-corn-soybean rotation in Peru. Weeds prior to regular herbicide use in each crop were 60% grass, 25% sedges, and 15% broadleaved. The grasses were large crabgrass and goosegrass. After 2 years and six crops, the weeds were 80% grass (85% a species of itchgrass), 13% broadleaved, and 7% a species of dayflower (Mt. Pleasant and McCollum, 1987). A disadvantage of selective herbicides is their ability to control some weeds, which can lead to creation of open niches in which other weeds succeed.

Herbicides, like any technology, have advantages and disadvantages that must be weighed carefully to consider intended and unintended consequences prior to use.

4. CLASSIFICATION OF HERBICIDES

An adequate classification system should be more than an index and should, insofar as possible, integrate all dimensions of the objects being classified. Although there are several methods of herbicide classification, no one method is completely adequate. This is because of the great diversity of uses, sites of action, and chemical families. Not many years ago, it was possible to classify herbicides on the basis of chemical structure. That is no longer possible because diversity of structures and sites of action have increased. In spite of the inadequacy of all systems of classification, all are used because each has some utility. To become familiar with chemical weed control, one must understand some of the jargon, and much of it is found in the language used to classify herbicides. The objectives of the next section are to understand why herbicides are grouped as they are and to enable use of the several systems of classification of herbicides. Understanding systems of classification will permit explanation of field observations in terms users will understand.

4.1 Crop of Use

One often hears that a particular herbicide is a corn herbicide or a turf herbicide. This is useful information because one immediately knows where it should and, by implication, where it should not be used. Frequently such statements represent only the narrow geographic or crop perspective of the speaker, and therefore crop of use cannot be a complete classification system. To illustrate its inappropriateness, one need only consider herbicides that were approved for use in 2015 in soybeans. There are 29 individual herbicides and four combinations approved for soil application to control broadleaf and grass weeds. Thirty-three different herbicides and 10 combinations were approved for postemergence application. Several others were approved for application in the spring or fall to control weeds that emerge before or after planting. Both groups include representatives of several chemical families. (For specific information, see http://www.ianrpubs.unl.edu/sendIt/ec130.pdf.) These herbicides have different sites of action and are applied at two different times relative to soybean growth. The same case can be made for several crops.

Similarly, describing 2,4-D as a turf herbicide is accurate but not reflective of its many other uses. Classification by crop is essential knowledge but includes such a diversity of other factors that it is impossible to integrate the subject. Individual, US Environmental Protection Agency approved, manufacturer prepared herbicide labels and state weed-control recommendations are excellent sources for information on the different crops and sites recommended for a herbicide. If one does not know the crops in which a particular herbicide can be used, or conversely, what crops it cannot be used on, one is not conversant with modern weed management and probably should not be applying herbicides.

4.2 Observed Effect

A second system of classification is based on effects observed after emergence. Some herbicides including bipyridiliums, dinitrophenols, and petroleum oils have a burning effect. This describes what one sees but not how the herbicide actually works. Other herbicides cause chlorosis (amitrole, clomazone) or gradual chlorosis, which is characteristic of photosynthetic inhibitors. Other herbicides affect hormone function, which results in abnormal growth. Growth abnormalities are so imprecisely defined, and so many herbicides affect growth, that the category merely serves to distinguish these effects from chlorosis or burning but does not describe what happens. Therein is the problem with observed effects as a system of classification.

4.3 Site of Uptake

A third, frequently used, system of classification is based on site of uptake and distinguishes between foliar and soil-applied herbicides. An herbicide that acts after contact with plant foliage falls in the foliar-active group. Other herbicides can be foliar active *and* soil active with the distinction often based on rate of application. The diphenyl ether herbicides, most of the phenoxy acids, the arsenicals, selective oils, and bipyridilium herbicides act primarily via foliage. Phenoxy acid herbicides and arsenicals translocate readily whereas bipyridiliums and selective oils do not. The sulfonylureas, imidazolinones, triazines, chloroacetamides, thiocarbamates, and dinitroanilines are taken up by roots. An adequate classification cannot be created for any group as large as herbicides by dividing the group in two.

4.4 Contact Versus Systemic Activity

Many herbicides are defined by noting they have contact as opposed to systemic activity. Translocation from point of application to site of action is synonymous with systemic activity. Some herbicides move only upward (acropetally) while others move acropetally and basipetally (down). This system, like many others, is useful because it reveals how an herbicide is likely to behave, but it does not tell us how it does what it does, nor does it mesh well with any other category.

4.5 Selectivity

Knowledge of selectivity is essential for wise use because it reveals the plants affected and unaffected. The first herbicides, iron and copper salts and dilute sulfuric acid, were selective because of differential wetting. Droplets of water solutions or suspensions of the herbicide bounced off or ran down upright cereal leaves and tended to stay on broad leaves. Selective herbicides kill or stunt weeds in a crop without harming the crop beyond the point of economic recovery. Nonselective herbicides kill all plants when applied at the right rate. No herbicide belongs rigidly to either group because selectivity is a function of rate and many other factors, including:

- plant age and stage of growth
- plant morphology
- absorption
- translocation
- type of treatment (e.g., broadcast vs. band or specific site application)
- time and method of application
- herbicide formulation
- environmental conditions

Because selectivity is a function of the combined action of these variables, it is not a precise system of classification. It is essential knowledge but does not integrate the subject.

4.6 Time of Application

Almost all herbicides must be applied at a particular time to maximize control and selectivity. Therefore, knowledge of when to apply to obtain the desired goal is essential to wise use. Unfortunately, some herbicides can be applied successfully at different times, and this system, like preceding systems, does not integrate the subject even though it is essential knowledge for wise use. There are three times when herbicides are applied and each can be specified relative to the weed or the crop. The first is prior to planting (preplanting). Sometimes application is immediately before planting or as early as several weeks prior to planting. Often preplanting applications include soil incorporation or mixing into soil. Incorporation can be combined with any time of application but is most common prior to planting. Use of incorporation is a function of the herbicide and control goal. A second application time is preemergence to the crop, the weed, or both. It is after planting but prior to emergence of the crop or weed. Postemergence applications are after the crop, weed, or both have emerged. Postemergence herbicides are often applied to foliage but can be applied to soil. The exact time for postemergence application varies with the crop, the herbicide, and the weed.

4.7 Chemical Structure

There is no simple relationship between an herbicide's chemical structure (its chemical family) and its biochemical behavior (its site of action). Classification based on structural formulas has been used, but with the ever-increasing number of new structures, chemical structure no longer integrates the subject.

Herbicides as the primary tools for weed control in the United States is shown by the fact that the number of herbicides included in the Herbicide Handbook of the Weed Science of WSSA has grown from 97 representing 27 site-of-action groups in 1967 (Anonymous, 1967, first ed.) to 232 from 67 chemical structural groups divided into 24 site-of-action groups in 2014 (Shaner, 2014, tenth ed.). Milne (2004) listed 442 herbicides approved for use somewhere in the world.

The structural formula in two dimensions bears little relationship to three-dimensional shape, physical properties, electronic disposition, and steric factors, all of which determine biological behavior. It is interesting that the number of herbicides increased steadily from 1967 to 2014. It is also important to know that no new site of action has been introduced in the past 20 years (Duke, 2011). Variations of chemical structure that alter activity have been developed.

To illustrate, there are two chemicals with nearly identical structures, and similar sites of action,[14] but quite different outcomes. One is testosterone, the dominant male hormone, which differs from progesterone, the dominant female hormone, by two carbon and two hydrogen atoms. These very similar structures illustrate the lack of desirability of classifying herbicides and many other things based only on structure.

4.8 Site of Action

Why do herbicides affect growth, or kill, some plants and not others? What is their mechanism/site/mode of action? Knowing an herbicide's site of action may not lead directly to better weed control, but it gives a firmer knowledge base from which to derive conclusions based on field observations. This book does not emphasize detailed knowledge of chemistry or biochemistry, and each edition has removed more chemical structures because as important as they are, they are not essential to understanding the fundamentals of weed science. In addition, structures are readily available in several places, most notably in the all editions of the WSSA's Herbicide Handbook.

This book assumes that readers know the difference between photosynthesis and respiration. Knowledge of the details of the light reactions of photosynthesis or the tricarboxylic acid cycle is not required. Determination of site of action is a complex study of chemistry, biochemistry, and plant physiology. It is defined as the entire chain of events from first contact to final effect and that detail is beyond the scope of this book.

Site of action is a system of classification. However, if one knows only site of action and nothing about the other, albeit incomplete, systems of classification, knowledge of herbicides is incomplete.

The discussion in Chapter 16 on the properties and uses of herbicides uses the several systems of classification mentioned. The primary system is based on site of action with some essential discussion of chemical structure.

---

[14]The weed science literature uses site of action, mode of action, and mechanism of action as synonyms. Apparently the user chooses.

THINGS TO THINK ABOUT

1. When did chemical weed control begin? How long is its history?
2. How did herbicides change the practice of agriculture?
3. Do herbicides affect energy use in American agriculture? If so, what is the effect?
4. What do herbicides do that other weed control techniques can't do?
5. What are the advantages and disadvantages of herbicides?
6. What are the attributes of a good classification system for herbicides?
7. Why are there so many ways to classify herbicides?
8. What is wrong with classifying herbicides based on crop of use or time of application?
9. Can all herbicides be classed as contact or systemic?
10. 10.What is the most important determinant of selectivity?
11. What are the problems with classifying herbicides only by chemical structure or site of action?

Literature Cited

Abernathy, J.R., 1981. Estimated crop losses due to weeds with nonchemical management. In: Pimentel, D. (Ed.), Handbook of Pest Management in Agriculture, vol. 1. CRC Press, Boca Raton, FL, pp. 159–167.

Ahrens, W.H., 1994. Herbicide Handbook, seventh ed. Weed Sci. Soc. Am., Champaign, IL. 352 pp.

Akamine, E.K., 1948. Plant growth regulators as selective herbicides. Hawaii Agric. Exp. Stn. Circ. 26, 1–43.

Alder, E.F., Wright, W.L., Klingman, G.C., 1977. Development of the American Herbicide Industry (Chapter 3). In: Plimmer, J.R. (Ed.), Pesticide Chemistry in the 20th Century. Amer. Chem. Soc. Symp. Ser. 37.

Anonymous, 1995. Common and chemical names of herbicides approved by the Weed Science Society of America. Weed Sci. 43, 328–336.

Anonymous, 2004. Common and chemical names of herbicides approved by the Weed Science Society of America. Weed Sci. 52, 1054–1060.

Anonymous, 2011. Common and chemical names of herbicides approved by the Weed Science Society of America. Weed Sci. 58, 611–618.

Anonymous, 1967. Herbicide Handbook, first ed. Weed Sci. Soc. Am., Champaign, IL. 293 pp.

Anonymous, 1989. Herbicide Handbook, sixth ed. Weed Sci. Soc. Am., Champaign, IL. 301 pp.

Appleby, A.P., 2005. A history of weed control in the United States and Canada—a sequel. Weed Sci. 53, 762–768.

Barrett, M., Witt, W.W., 1987. Alternative pest management practices. In: Helsel, Z.R. (Ed.), Energy in Plant Nutrition and Pest Control. In: Stout, B.A. (Ed.), Energy in World Agriculture, vol. 2. Elsevier Press, NY, pp. 195–234.

Becker, R., 1983. Selling the benefits of weed control. In: Proc. 35th Ann. Fertilizer and Agric. Chem. Dealers Conf., Bul. CE-185–4e. Coop. Ext. Serv., Iowa St. Univ., Ames, IA.

Benbrook, C.M., 2012. Impacts of genetically engineered crops on pesticide use in the U.S. – the first sixteen years. Environ. Sci. Eur. 24, 24. https://doi.org/10.1186/2190-4715-24-24.

Beste, C.E. (Ed.), 1983. Herbicide Handbook, fifth ed. Weed Science Society of America, Lawrence, KS. 515 pp.

Bolley, H.L., 1908. Weeds and methods of eradication and weed control by means of chemical sprays. N. Dak. Agric. Coll. Exp. Stn. Bul. 80, 511–574.

Bucha, H.C., Todd, C.W., 1951. 3-(p-chlorophenyl)-1,1-dimethylurea—a new herbicide. Science 114, 493–494.

Clements, D.R., Wiese, S.F., Brown, R., Stonehouse, D.P., Hume, D.J., Swanton, C.J., 1995. Energy analysis of tillage and herbicide inputs in alternative weed management systems. Agric. Ecosyst. Environ. 52, 119–128.

Crafts, A.S., Robbins, W.W., 1962. Weed Control: A Textbook and Manual, third ed. McGraw-Hill, New York, NY. 660 pp.

Dowler, C.C., Hauser, E.W., 1975. Weed control systems in cotton in Tifton loamy sand soil. Weed Sci. 23, 40–42.

Duke, S.O., 2011. Comparing conventional and biotechnology-based weed management. J. Agric. Food Chem. 59, 5793–5798.

Economist, 2016. How the Other Tenth Lives, p. 69.

Gianessi, L.P., Marcelli, M.B., 2000. Pesticide Use in U.S. Crop Production: 1997. Nat. Center for Food and Agricultural Policy, Washington, D.C.

Gianessi, L.P., Reigner, N.P., 2007. The value of herbicides in U.S. crop production. Weed Technol. 21, 559–566.

Gianessi, L.P., Sankula, S., 2003. The Value of Herbicides in U.S. Crop Production – Executive Summary. Nat. Center for Food and Agricultural Policy, Washington, D.C.

Gianessi, L.P., Silvers, C.S., 2000. Trends in Crop Pesticide Use: Comparing 1992 and 1997. Nat. Center for Food and Agricultural Policy, Washington, D.C., 165 pp.

Goldman, L., 2004. Childhood Pesticide Poisoning: Information for Advocacy and Action. United Nations Environment Programme. Châtelaine, Switzerland, 37 pp.

Greenland, R.G., 2003. Injury to vegetable crops from herbicides applied in previous years. Weed Technol. 17, 73–78.

Hamm, P.C., 1974. Discovery, development, and current status of the chloroacetamide herbicides. Weed Sci. 22, 541–545.

Hamner, C.L., Tukey, H.B., 1944a. The herbicidal action of 2,4-dichlorophenoxy acetic and 2,4,5-trichlorophenoxyacetic acid on bindweed. Science 100, 154–155.

Hamner, C.L., Tukey, H.B., 1944b. Selective herbicidal action of midsummer and fall applications of 2,4-dichlorophenoxyacetic acid. Bot. Gaz. 106, 232–245.

Hilton, J.L. (Ed.), 1974. Herbicide Handbook, third ed. Weed Science Society of America, Lawrence, KS. 430 pp.

Hobbes, T., 1651. Leviathan (Part 1, Chapter 13).

Hopkins, W.L., 1994. Global Herbicide Directory, first ed. Ag. Chem. Information Services, Indianapolis, IN. 181 pp.

Humberg, N.E. (Ed.), 1989. Herbicide Handbook, sixth ed. Weed Science Society of America, Lawrence, KS. 301 pp.

Hyde, J., 2002. Four Iowans Who Fed the World. http://hoover.archives.gov/programs/4Iowans/Hyde-Culver.html.

Ivany, J.A., 2001. Introduction. In: Ivany, J.A. (Ed.), Weed Management in Transition, Topics in Canadian Weed Management, vol. 2. Canadian Weed Sci. Soc. Sainte-Anne-de-Bellevue, Quebec, Canada, pp. 3–4.

Johnson III, W.C., Mullinix Jr., B.C., 1995. Weed management in peanut using stale seedbed techniques. Weed Sci. 43, 293–297.

Jordan, L.H., 2006. Pesticide trade shows new market trends. In: Assadourian, E. (Ed.), Vital Signs: The Trends That Are Shaping Our Future. W.W. Norton & Co., New York, NY, pp. 28–29.

King, L.J., 1966. Weeds of the World: Biology and Control. Interscience Pub., Inc., New York, NY, 526 pp.

Kirby, C., 1980. The Hormone Weedkillers: A Short History of Their Discovery and Development. Brit. Crop Prot. Council Pub., Croydon, UK, 55 pp.

Kochar, R., 2015. A Global Middle Class Is More Promise Than Reality. http://www.pewglobal.org/2015/07/08/a-global-middle-class-is-more-promise-than-reality/.

Korsmo, E., 1932. Undersok elser. 1916–1923. In: Over ugressets skadevirkninger og dets bekjempelse. I. Aker brucket. Johnson and Nielsens Boktrykkeri, Oslo, Norway.

Krebs, A.V., 1992. The Corporate Reapers: The Book of Agribusiness. Essential Books, Washington, D.C., p. 16

Lague, C., Khelifi, M., 2001. Energy use and time requirements for different weeding strategies in grain corn. Can. Biosyst. Eng. 43, 213–221.

Lovins, A., Lovins, L.H., Bender, M., 1984. Energy and agriculture. In: Jackson, W. (Ed.), Meeting the Expectations of the Land. North Point Press, San Francisco, CA, pp. 68–86.

Malcolm, J., 1993. The farmer's need for agrochemicals. In: Gareth Jones, J. (Ed.), Agriculture and the Environment. E. Horwood Pub., London, UK, pp. 3–9.

Mann, C., 2011. The eyes have it. Smithsonian 42 (7), 86–89, 92–94, 96–98, 100–102, 1–4–106.

Marth, P.C., Mitchell, J.W., 1944. 2,4-dichlorophenoxyacetic acid as a differential herbicide. Bot. Gaz. 106, 224–232.

Milne, G.W.A. (Ed.), 2004. Pesticides: An International Guide to 1800 Pest Control Chemicals, second ed. Ashgate Publication LTD., Aldershot, Hampshire, U.K. 609 pp.

Morettini, A., 1915. L'impegio dell'acido sulfurico per combattere le erbe infeste nel frumento. Staz. Sper. Agr. Ital. 48, 693–716.

Mt. Pleasant, J., McCollum, R., 1987. Effect of weed control practices on weed population dynamics in intensive-managed continuous cropping system in the Peruvian Amazon. Agron. Absts. 41.

Nalewaja, J.D., 1974. Energy requirements for various weed control practices. In: Proc. N. Central Weed Control Conf., vol. 29, pp. 19–23.

Nielsen, R., 2005. The Little Green Handbook: Seven Trends Shaping the Future of Our Planet. Picador Press, New York, NY, p. 365.

Perkins, J.H., Holochuck, N.C., 1993. Pesticides: historical changes demand ethical choices. In: Pimentel, D., Lehman, H. (Eds.), The Pesticide Question: Environment, Economics, and Ethics. Chapman & Hall, New York, NY, pp. 390–417, 441 pp.

Peterson, G.E., 1967. The discovery and development of 2,4-D. Agric. Hist. 41, 243–253.

Pimentel, D., Krummel, J., Gallahah, D., Hough, J., Merrill, A., Schreiner, I., et al., 1978. Benefits and costs of pesticides in U.S. food production. Bioscience 28, 772–783.

Pimentel, D., Pimentel, M., 1979. Food, Energy, and Society. E. Arnold, London, UK, pp. 65–69.

Pimentel, D., Allen, J., Beers, A., Guinand, L., Linder, R., McLaughlin, P., et al., 1987. World agriculture and soil erosion. Bioscience 37 (4), 277–283.

Pimentel, D., Harvey, C., Resosudarmo, P., Sinclair, K., Kurz, D., McNair, M., et al., 1995. Environmental and economic costs of soil erosion and conservation benefits. Science 267 (5201), 1117–1123.

Pokorny, R., 1941. Some chlorophenoxyacetic acids. J. Am. Chem. Soc. 63, 1768.

Rabaté, E., 1911. Destruction des revenelles par l'acid sulfurique. J d'agr. Prat (N.S. 21) 75, 497–509.

Rabaté, E., 1934. La destruction des mauvaises herbes, third ed. Paris.

Shaner, D.L. (Ed.), 2014. Herbicide Handbook, tenth ed. Weed Science Society of America, Lawrence, KS. 513 pp.

Slade, R.E., Templeman, W.G., Sexton, W.A., 1945. Plant growth substances as selective weed killers. Nature (Lond.) 155, 497–498.

Smith, A.E., Secoy, D.M., 1975. Forerunners of pesticides in classical Greece and Rome. J. Agric. Food Chem. 23, 1050–1055.

Smith, A.E., Secoy, D.M., 1976. Early chemical control of weeds in Europe. Weed Sci. 24, 594–597.

Swanton, C.J., Murphy, S.D., Hume, D.J., Clements, D.R., 1996. Recent improvements in the energy efficiency of agriculture: case studies from Ontario, Canada. Agric. Syst. 52, 399–418.

Timmons, F.L., 1970. A history of weed control in the United States and Canada. Weed Sci. 18, 294–307. Republished — Weed Sci 53,748–761, 2005.

Troyer, J.R., 2001. In the beginning: the multiple discovery of the first hormone herbicides. Weed Sci. 49, 290–297.

US/EPA, 2004. Pesticide Industry Sales and Usage: 2000 and 2001 Market Estimates. http://www.epa.gov/oppbead1/pestsales/index.htm.

Zimdahl, R.L., 2010. A History of Weed Science in the United States. Elsevier Insights, London, UK, 207 pp.

Zimmerman, P.W., Hitchcock, A.E., 1942. Substituted phenoxy and benzoic acid growth substances and the relation of structure to physiological activity. Contrib. Boyce Thompson Inst. 12, 321–343.

Herbicides and Plants

FUNDAMENTAL CONCEPTS

- Several environmental, chemical, and physiological factors affect herbicide activity and selectivity.
- The most important determinants of herbicide selectivity are rate of absorption and the amount absorbed, translocated, and metabolized by weed and crop plants.

- Several plant and environmental factors interact to determine selectivity.
- Sprayer calibration is an important aspect of successful herbicide application.
- Forward speed, pressure, and nozzle tip orifice size are the primary things that can be adjusted to change a sprayer's calibration.

LEARNING OBJECTIVES

- To understand the difference between herbicide drift and volatility and the importance of each.
- To know techniques to control drift and volatility.
- To understand the fundamental importance of sprayer calibration.
- To know the external factors that influence spray retention and herbicide absorption.

- To know the effect of moisture, temperature, and light on herbicide action.
- To know the relative advantages and disadvantages of foliar- and soil-applied herbicides.
- To understand the difference between shoot and root absorption of herbicides.
- To understand the role of absorption, translocation, and metabolism as determinants of selectivity.

1. FACTORS AFFECTING HERBICIDE PERFORMANCE

This discussion of factors affecting herbicide performance in plants assumes that users have an applicator appropriate to the task and that it has been calibrated to apply the correct volume and the proper amount of active ingredient per acre. The discussion also assumes that the correct herbicide has been selected and that it will be applied at the right time. If these things are not assured, they will negatively affect herbicide performance and environmental quality. Because human errors and their results are not precisely predictable (we can't plan our accidents), the discussion herein assumes human error has been avoided. The discussion will include factors affecting performance from the time an herbicide molecule leaves the applicator (usually this means the nozzle tip) until it hits a plant target and acts.

2. GENERAL

2.1 Sprayer Calibration

It is important to understand the equipment required to apply herbicides properly. Although size and reliability of equipment have changed, they remain basically the same (McWhorter and Gebhardt, 1987). Nearly all herbicides are applied with hydraulic sprayers that have the same four basic components: a tank, pressure regulator, pump, and spray nozzles. It has been and continues to be the most widely used method of herbicide application. Its accuracy and reliability have improved significantly. Most herbicides are applied as an aqueous mixture that uses simple nozzles to break the pressurized liquid stream into droplets. When fuel, herbicide, and the farmer's time were inexpensive and environmental contamination was a minor concern, simple applicators were used. These conditions have changed and more effort is now being expended to improve the precision of herbicide application.

It is common to apply herbicides to an entire field or a band over the crop row. Not all of the area sprayed may have weeds but it is all sprayed, thus the herbicide may be unavoidably applied where there are no weeds. Weed scientists know weeds can dominate a field, often exist in patches (see Fig. 14.1), and only rarely as a uniform stand. The logical conclusion is that applying herbicide to the entire field is not necessary. Nevertheless spraying the entire area is more efficient for weed managers because it is easier and the added cost is small.

FIGURE 14.1   Patch distribution of field bindweed. *Courtesy of Dr. Philip Westra, Colorado State University.*

The technology for detecting and spraying just the weeds is available and improving, but it is expensive. Research on optical detection systems that permit spraying only when a weed is detected is proceeding. Felton (1990) demonstrated it is theoretically possible to apply one herbicide to one species and another herbicide to a second species in one pass across a field, if the weeds can be detected. They can be when the leaf tissue of each differs in reflectance. Microprocessors turn the sprayer on only when weeds are sensed. An optical detection sprayer will not spray when no weeds are detected. This and similar systems reduce total herbicide use and cost, don't waste herbicide, reduce environmental presence, and the likelihood of off-target movement and nontarget effects. Biller (1998) found 30%–70% less herbicide used with 100% control of weeds in corn. Girmaa et al. (2004) achieved effective control of cheat grass and ryegrass in winter wheat with an optical spectral system. Paap (2014) demonstrated detection of weeds in cotton and sugarcane with 85%–95% accuracy. When morphological and foliar reflectance characteristics of different species are incorporated, control of specific weeds will be possible. Determining the reflective characteristics for the range of weeds one will encounter in a range of crops in fields with different cropping histories remains a challenge. These advances combined with global positioning system technology will allow an applicator to know and the machine to remember where species are and achieve very precise herbicide application. But reduced herbicide use, cost, and environmental presence may not be sufficient to justify the expense of the technology.

Several years ago there was great interest in controlled droplet applicators (CDAs), but it has waned. CDA technology produces droplets over a narrow, predictable size range. They worked well for low-volume applications and were quite effective for drift reduction (see Section 2.2.1) by reducing spray volume and the amount of herbicide applied. A CDA sprayer produces droplets of a relatively uniform size that are not so small that they will drift in wind. Adequate ground coverage is achieved with small drops while eliminating large droplets that tend to bounce off leaves. CDA sprayers used low pressure to produce an even droplet spectrum that achieved even and effective application with a much smaller amount of water.

Herbicides can be, but usually are not, applied as granules by applicators that can be calibrated. Granule application can often be combined easily with crop planting. Because of its exclusive foliar activity, glyphosate led to the development of wiper applicators, which have been abandoned. Wipers could be nylon ropes or material similar to a paint roller that act as wicks but do not drip herbicide on nontarget species. Weeds emerging above a crop canopy received a lethal dose of glyphosate when wiped. Shag carpet–covered rollers were replaced by rope wicks. Both technologies are now rarely used and are primarily of historical interest.

Each kind of herbicide applicator can be calibrated with the same basic technique. The applicator is driven over a known area and output is measured, or output is measured for a certain time with the applicator stationary. Special devices are available to assist with calibration by direct reading during spraying or while stationary. No technique is difficult or complex but each takes time.

Even with sophisticated, specialized knowledge of herbicide chemistry, site of action, application timing, rate of application, selectivity, and activity, herbicides may fail to control weeds, achieve desired crop selectivity, and may leave undesirable environmental residues.

A major reason for failure is not lack of knowledge about how the herbicide acts but rather that herbicides are frequently not applied properly. A study in Nebraska (Reichenberger, 1980), found that two of every three pesticide applicators made application errors due to inaccurate calibration, incorrect mixing, worn equipment, or failure to read and understand the product label. These mistakes caused over- and underapplication and cost farmers between \$2 and \$12 per acre in added chemical expense, potential crop damage, and lost weed control. When results were extrapolated to the entire United States, it was equal to a billion dollar application blunder each year. Other studies of farmers' sprayers have shown similar problems (Ozkan, 1987). One must hope that these kinds of mistakes have diminished since the 1980s.

It is not totally inaccurate to say that a major problem with agricultural chemicals is the people who apply them. In spite of all the specialized research and technology required to develop and market an herbicide, the end result is often dependent on decisions made by a user, just prior to use. These decisions are frequently wrong. The reason more accidents haven't occurred is that herbicides developed to be reasonably idiotproof are not completely so; all mistakes are not tolerable.

Because of application blunders and concern for human and environmental safety, government regulation of herbicides has increased. No legislative body can enforce a law against stupidity but all can pass laws that make penalties for stupidity greater and encourage use of reasonable intelligence. Such laws become more likely when reasonable intelligence is not the norm.

The metallic salts, the first selective herbicides, were applied at 100+ lbs/A in at least 100 gals of water/A. Invention of the compressed air sprayer in the early 1900s improved application (Gebhardt and McWhorter, 1987) but didn't reduce the amount of herbicide required. Early weed sprayers applied high volumes (100 gals/acre) from wooden tanks. Later, sprayers capable of applying lower volumes had steel or plastic tanks. As mentioned, the first sprayers and modern sprayers have basically the same parts: a tank, pressure regulator, pump, and nozzles. Today 90% of all herbicides are applied with low-pressure ground sprayers drawn by a tractor or self-propelled (Felton, 1990). Herbicides are also sprayed by airplane.

Spraying may be followed by soil incorporation to reduce or control volatility, put the herbicide in position to maximize plant uptake, and promote control of emerging seedlings or root uptake. Failure to incorporate well was a frequent reason for poor herbicide performance. Power rototillers are the best incorporation implements but are not used on most farms. Disking is probably the most common incorporation technique and works best if done twice, with the second pass at right angles to the first. A single disking produces zones of high herbicide concentration and other areas with virtually no herbicide because of the tendency of the disk to ridge soil. Few herbicides used in the 21st century require incorporation.

Herbicides can be applied by injection into water flowing in furrows or ditches and through sprinklers. This technique, called herbigation, is effective for herbicides taken up by plant roots from soil but is not effective for all herbicides. It is common in irrigated agriculture.

2.2 Reaching the Target Plant

2.2.1 *Drift*

Spray drift is movement of airborne, liquid spray particles immediately after application. It is often unseen, and while often unavoidable, it can be minimized. Drift increases with wind speed and height above ground when drops are released. It increases with spray droplet size. Ideally, uniform drops between 500 microns (moderate rain) and 1 mm (1000 microns, heavy rain) in diameter are desired. Drops of this size minimize but do not eliminate drift, especially if spraying is done when wind speed is above 5 mph. It is not uncommon, especially in arid environments, for water to evaporate within 200–300 feet of the point of delivery, so only the herbicide and associated organic solvents remain to drift. Table 14.1 shows spray droplet size, droplet lifetime, and the potential effect on drift (Brooks, 1947; Hartley and Graham-Bryce, 1980). For comparison, the diameter of number 2 pencil lead is about 2000 microns, a paper clip 850, a toothbrush bristle 300, and a human hair about 100 microns. A comprehensive review of spray drift and its mitigation was published in 2005 (Felsot, 2005). US spray drift research was summarized in an assessment prepared by the Spray Drift Task Force (Felsot). It was created in 1990 by 38 agricultural chemical companies to generate data to fulfill US Environmental Protection Agency (EPA) spray drift data requirements. An evaluation tool to estimate the environmental exposure from spray drift at time of application is available.[1]

The particular shape of a nozzle tip (fan, cone, stream) creates a spray pattern as it breaks a liquid stream into small droplets. Hydraulic nozzles produce a range of droplet sizes. Droplet size is a function of orifice size, operating pressure, and surface tension of the solution or suspension. A smaller nozzle orifice, higher pressure, and lower surface tension produces

TABLE 14.1 The Effect of Spray Droplet Size on Evaporation and Drift

Droplet Diameter (microns)	Type of Drop	Precipitation (in./hr)	Drops (No./in.²)	Evaporating Water		Time to Fall 10 feet (sec)	Distance Traveled While Falling 10 feet in a 3 mph Wind
				Drop Life (sec)	Lifetime Fall Distance (in.)		
5	Dry fog	0.04	9,220,000	0.04	<1	3,960	3 miles
20	Wet fog	–	144,000	0.7	<1	–	–
100	Misty rain	0.04	1,150	16	96	10	409 feet
200	Light rain	–	144	65	1,512	–	–
500	Mod. rain	3.9	9	400	>1,500	1.5	7 feet
1000	Heavy rain	39	1	1,620	>>1,5000	1.0	4.7 feet

Adapted from Bode, L.E., Wolf, R.E.. Techniques for Applying Postemergence Herbicides. Univ. Illinois, Urbana, IL. 5 pp. Undated.

[1]http://www.agdrift.com/AgDRIFt2/DownloadAgDrift2_0.htm.

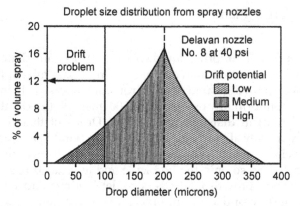

FIGURE 14.2   Normal distribution of spray drop size from a hydraulic nozzle.

more small drops. All hydraulic nozzles produce a normal (Fig. 14.2) distribution of spray drop sizes. As drop size decreases and pressure increases, a greater percentage of small droplets is produced and drift potential increases.

The influence of wind on droplets is illustrated in Table 14.1. Small droplets can drift a long way in a light breeze. Large drops decrease drift. Spraying in strong wind should be avoided, but when large areas must be sprayed with herbicides that require application at particular growth stages or before crop emergence, timing often takes precedence. To maximize benefit, herbicides must be applied at the proper time. However, if other considerations are regarded as more important than drift avoidance, problems may ensue when the applicator's or a neighbor's crop is injured or the environment is contaminated by improper application.

Sprayer boom height is normally fixed. However, as illustrated (Table 14.1), release height influences drift potential simply by allowing drops to remain suspended longer.

Because drift can occur, several techniques have been developed to control it. The first, and simplest, is to reduce spray pressure and create fewer small drops. Increasing drop size from 20 to 200 microns decreases coverage 200 times and increases drop lifetime from 0.7 to 65 s (Table 14.1). Small drops attain a horizontal trajectory quickly, and water may evaporate before the drop contacts a plant. After water evaporation, pesticides can become airborne aerosols that fall out with rain or sprinkler irrigation, but one cannot be sure where they will fall. Droplets larger than 150 microns normally resist evaporation long enough to reach the intended or another target.

Low-volume applicators [ultra-low volume (ULV) or controlled drop (CDA)] use a rotary atomizer and reduce water requirements and equipment weight. Drop size is usually controlled between 150 and 300 microns and total volume can be as low as 0.5 gal/A. CDA applicators, although available for several years, have not achieved a high level of commercial acceptance due to high cost, frequent performance failures, and the widespread acceptance and availability of hydraulic nozzles.

Nozzles that incorporate air or facilitate use of a foaming adjuvant are available. They produce coarse droplets, but up to 5% loss is still possible within 1000 feet of the point of application. Foam adjuvants increase spray volume two to three times, may act as wetting agents, and thereby increase phytotoxicity. Water-soluble thickening agents (thixotropic agents) increase average droplet size. These are water-imbibing polymers that create a particulated gel. The smallest droplet size is predetermined by the polymer's particle size. There is usually no phytotoxic benefit, but drift is reduced. Chapter 17 mentions use of now largely abandoned, invert emulsions to increase spray solution viscosity and reduce drift.

A recirculating sprayer was developed (McWhorter, 1970). The hypothesis was that environmental contamination could be reduced without loss of weed control if spray that did not strike a target plant was captured, recirculated, and reused. This was achieved by spraying horizontally above the crop's foliar canopy. The system was useful only with foliar applied, postemergence herbicides, which could be applied to weeds growing above the crop canopy, by which the weed's negative effects on yield had been expressed. The recirculating sprayer eliminated vertical spraying. The sprayer successfully applied glyphosate to control weeds above the canopy of cotton and soybeans in the southern United States, but in spite of the fact that it was a unique, innovative idea, it was never a commercial success.

Wax bars impregnated with 2,4-D (McWhorter, 1966) were once used to control weeds primarily in turf. They were not satisfactory because 2,4-D was hard to impregnate uniformly, bars tended to self-destruct, and wax melting was not uniform because temperature was not uniform or hot enough, so herbicide application was not uniform and successful. They are no longer manufactured.

The development of glyphosate led to renewed interest in the question, why spray? As mentioned before, wiper technology was developed (Derting, 1987) first with shag carpet and eventually with rope wicks. When the ropes were moved horizontally above the crop, weeds growing above the crop contacted the herbicide in the rope and, through control of solution concentration, sufficient herbicide was applied to kill them without affecting the crop; but, once again, weeds growing above the crop had already reduced yield. This eliminated drift, but the system is no longer used; it is another idea now only of historical interest.

Drift is not just a problem of historical interest. In the late 1980s, clomazone drift affected many nontarget plants after application to soybeans in the midwestern United States. The EPA was concerned about drift from sulfonylurea herbicides that can damage flowers, seeds, and fruits of grapes, alfalfa, cherries, and asparagus. The state of Washington prohibited use of sulfonylurea herbicides within 14 miles of nontarget crops because of drift potential and required 24 h notification of intent to spray (Anonymous, 1996). In 2016, in spite of state and federal regulations, dicamba, an EPA-approved herbicide for 12 crops, drifted to soybean and cotton causing extensive damage.

New resistant soybean and cotton cropping systems based on the synthetic auxin herbicides give farmers new options for managing Palmer amaranth and other broadleaf weeds resistant to glyphosate. But scientists with the Weed Science Society of America recommend precautions. Auxin herbicides are known to drift and cause harm to sensitive, off-target plants. The EPA has issued 2–5 year time-limited registrations for the auxin herbicides dicamba and 2,4-D, respectively. The product labels have details on management of drift and other risks and should be carefully followed to reduce off-site movement. If drift occurs, the registrations could be revoked.

Bish and Bradley (2017) suggested that there is much to be learned based on their survey of more than 2,300 commercial and noncommercial applicators. They found less than half were familiar with volatility and temperature inversions that can influence off-target movement. EPA says drift of herbicide damaged crops in 25 states (Lipton, 2017).

2.2.2 Volatility

Volatility measures the tendency of a chemical to vaporize, that is, to move from the liquid to the gaseous state. Drift is movement as a liquid. Volatility is movement as a gas. It is a function of an herbicide's vapor pressure and the ambient temperature. Volatilized herbicides may cause damage in another place or reduce effectiveness at the point of application. An historical example of volatilization is found with some esters of phenoxyacetic acids (e.g., 2,4-D). Fig. 14.3 shows the effect on germination of pea seeds after exposure to

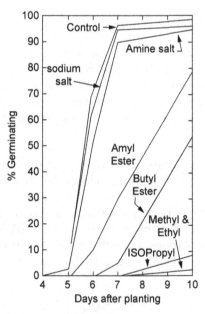

FIGURE 14.3   Percent germination of pea seeds after exposure to 2,4-D formulations (Mullison and Hummer, 1949).

volatile 2,4-D (Mullison and Hummer, 1949). Because of the experiment's design, only volatility could have caused the observed effects. Methyl and ethyl (1- and 2-carbon) esters are more volatile than the 5-carbon amyl ester, and all esters are more volatile than amine or sodium salts. Because of the high volatility of short-chain esters of 2,4–D, it is no longer possible to purchase and use them. Volatility is not limited to phenoxyacetic acids. Some carbamothioates and dinitroanilines are volatile and can be lost from the area of application if not incorporated in soil. Volatility problems are not going to go away, and intelligent herbicide use demands continuing attention to the risk. Methods to study and measure volatility have been summarized by Mueller (2015).

3. FOLIAR ACTIVE HERBICIDES

3.1 Spray Retention

If foliar active herbicides do not drift or volatilize, their performance is affected by retention on plant surfaces. A foliar herbicide must remain on leaves long enough for absorption to occur. Plants differ in their ability to retain water on leaf surfaces (Blackman et al., 1958) (Table 14.2). Barley has upright leaves disposed nearly perpendicular to the soil surface and liquid droplets run off easily. Pea leaves have a waxy cuticle that makes it difficult for liquids to remain on the surface. Flax is an upright plant with small vertical leaves, whereas sunflowers and white mustard are large with broad leaves disposed parallel to the ground that retain liquid droplets more easily than grass leaves.

Propanil was used for selective control of green foxtail in hard red spring and durum wheat. It illustrates the influence of spray retention on herbicide activity (Eberlein and Behrens, 1984 – Table 14.3). The data show that wheat has a slightly higher concentration in terms of mg of propanil retained per plant, but green foxtail absorbs and retains more. Propanil is selective because of rapid metabolism rather than differential retention (Yih et al., 1968a,b).

3.1.1 Leaf Properties

The ability of an herbicide to control weeds selectively can depend on morphology (shape) and chemical variations between plant surfaces (see Harr and Guggenheim, 1995, for detailed

TABLE 14.2 Spray Retention by Different Species (Blackman et al., 1958)

Species	Number of Leaves	Height (cm)	mL Water Retained/g Shoot Weight
White mustard	2	5–7	2.5
Sunflower	2	6	2.0
Flax	2	5	1.1
Pea	2	5–7	0.4
Barley	3	15–20	0.3

TABLE 14.3   Retention of Propanil by Wheat and Green Foxtail Plants at the Three-Leaf Stage of Growth (Eberlein and Behrens, 1984)

Species	Propanil Retained	
	mg/plant	mg/g Fresh Weight
Green foxtail	0.69	19.09
Wheat	1.20	2.49

descriptions of leaf surfaces of major crop plants). Large, broad leaves disposed parallel to the soil surface are easier to hit with spray solutions applied by most sprayers. Herbicide molecules are more likely to contact and remain on the broad leaves of dicots than on grass leaves, which are usually disposed perpendicular to the soil surface.

Velvetleaf, crabgrass, and some species of mallow have hairy leaf surfaces that prevent direct, quick contact of spray droplets with the leaf surface. But when a hairy surface becomes saturated, herbicide entry may be promoted because hairiness delays evaporation.

Because leaves are one of the principal entry points for herbicides, their structure and function are important. The primary leaf tissues are epidermis, mesophyll, and vascular. The epidermis is present on upper and lower leaf surfaces and consists of a single layer of interlocked cells with no chloroplasts. It is covered by the cuticle that is often layered with waxes, a varnish-like layer or film that retards movement of water in and out of leaves.

All leaves have cuticles, a formidable barrier to herbicide entry, yet herbicides do enter. Surface-active agents (surfactants) are used in some formulations to assist entry, and they often determine the degree of herbicidal activity obtained because of their effect on leaf surface penetration. Water is not compatible with many plant surfaces, especially those with thick or very waxy cuticles. Surfactants aid penetration because they lower surface tension of liquids, increasing their tendency to spread and wetting of leaf surfaces.

It is incorrect to assume that plants with thick, waxy cuticles absorb less herbicide or absorb the same amount more slowly than plants with thin cuticles. The reason is that cuticle hydration and composition are more important factors in herbicide absorption than cuticle thickness. Plant leaves growing in shade generally have thinner cuticles than those growing in full sun, and young leaves have thinner cuticles than old ones. Thinner cuticles are one, but not the only reason, young plants are more susceptible to herbicides than old plants.

Stomata appear to be obvious entry points, but most herbicides enter plants through leaf surfaces. Liquid spray droplets or volatile gases can enter stomata, but even after stomatal entry, herbicides must penetrate the thin cuticle in substomatal chambers. Stomata vary in number, location, and size among different plant species, and while they can be located on upper and lower surfaces, most agricultural plants have the majority of stomata on lower surfaces. There is as much as tenfold variation among species in stomatal number.

Another problem with entry through stomatal openings is the surface tension of spray solutions. It is possible, but not very likely, for a droplet of a liquid with high surface tension to bridge a stomatal opening and not enter it. Surface tension is a more important determinant of the tendency to spread than it is of stomatal entry.

Often stomata are not open during the day when herbicides are most commonly applied; they close during the heat of day and open during cool mornings and evenings. To achieve easy stomatal penetration, an herbicide spray must have low surface tension and high wetting power, a difficult combination.

3.1.2 Other Factors

The location of growing points or plant meristematic areas can determine herbicide selectivity. In grasses, growing points are usually at the base of plants and are protected from foliar herbicides by surrounding leaves. In some plants, growing points are actually below the soil surface and not exposed to direct contact by foliar-applied herbicides. In contrast, broadleaved plants usually have terminal, exposed growing points that may be more readily targeted by an herbicide.

Selectivity can be obtained through herbicide placement. An herbicide can be applied to plant foliage, only to soil, only to soil and weeds between crop rows, or only over the crop row. A nonselective herbicide can be used selectively by controlling where it is applied. Selective placement can also be obtained using granular herbicides that have little or no foliar activity because granules do not adhere to foliage.

The stage of plant growth at application is an important determinant of herbicide activity. It is a good generalization that seedling plants are more easily controlled than mature plants (see 2,4-D, Chapter 16).

3.1.3 Characteristics of Spray Solution

Composition of the spray solution is a very important aspect of selectivity and activity. A spray solution with little or no surfactant may have high surface tension and just bubble up on a cuticular surface as water does on a newly waxed car. In this case, there is less opportunity for absorption because the contact area between the applied herbicide and the plant surface is limited. On the other hand, a surfactant decreases surface tension, spreads out water droplets, increases surface coverage, and wets the surface thereby promoting penetration. Frequently, nonphytotoxic crop oils are included in spray mixtures to promote herbicide penetration and activity. Diesel fuel was, but is no longer, used as an adjuvant in a water-based spray system for control of plants on rangeland. It promotes penetration of leaves but is phytotoxic.

Another factor, often not controlled, that influences herbicide activity is drop size. For a given amount of herbicide per unit area, activity usually increases as droplet size decreases (McKinlay et al., 1972) (Table 14.4). For a fixed droplet size, effective dosage can be increased equally well by increasing herbicide concentration in each drop or by increasing the number of drops per unit area. If a spray solution has 0.86 g of 2,4-D/liter, three 400-micron drops/cm^2 will apply 64 times more spray volume and active ingredient/cm^2 than three 100-micron droplets (McKinlay et al., 1972). Drop size is difficult to control in most hydraulic sprayers, but is fixed, in a narrow range, in CDAs.

---

[2]http://marketplace.unl.edu/extension/extpubs/ec130.html. The information is on p. 159 of the 2015 Guide for Weed Management in Nebraska with insecticide and fungicide information.

TABLE 14.4   Dry Weight of Oat Seedlings Selectively Exposed to Diallate
(Appleby and Furtick, 1965)

Plant Part Exposed	Dry Weight (mg)
Coleoptile	0
Root	205.2
Coleoptile and Root	0
Untreated control	303.8

3.2 Environmental Factors

The influence of environmental factors on herbicide phytotoxicity is almost always related to differential absorption, translocation, or metabolism. These can be affected by morphological characteristics imposed on plants by the environment. Altered plant susceptibility to herbicides can often be traced to environmental stress that alters a plant's ability to absorb or metabolize herbicides.

3.2.1 *Moisture*

If an herbicide molecule doesn't drift or volatilize, reaches its target, and is retained on the plant surface, its activity can be affected by environmental factors. Herbicide users want to know the likely effects of weather (rain, dry, wet, hot, cold, dry) on herbicide performance. If the sun comes out or it rains immediately after application or even during application, as opposed to application on a gray, cloudy day, does that affect an herbicide's activity and selectivity? Phenoxy acids formulated as esters are more lipid(fat) soluble than water soluble. Therefore, on a warm day, leaf cuticles may be more fluid and more readily penetrated by lipid-soluble compounds such as esters. Warm days aid penetration and activity. It is a good generalization that the warmer it is, the better herbicide activity will be. For noncontact or soil-active herbicides, temperature at time of application is less important. Temperature influences a plant's metabolic rate and physiological activity. If a plant is rapidly metabolizing and photosynthesizing, it will translocate herbicides rapidly enhancing their activity.

What if an herbicide is applied and it rains soon after? With many herbicides, penetration occurs within an hour (phenoxy acids are rain fast after 1 h), so rain several hours after application does not affect activity. On the other hand, some herbicides are not rain fast for up to 6 h after application. Atrazine, a soil-applied herbicide, is rarely applied to plant foliage because it penetrates poorly and rain will wash some off even if it rains as many as 7 days after application. University of Nebraska weed scientists annually publish a guide to weed management, which includes a table showing how long it takes after application for more than 100 herbicides to become rain fast.[2] The time ranges from 0.5 to 8 h. The best recommendation for foliar herbicides is that they should be applied on warm, sunny days with little chance of rain within 24 h after application. Product labels should be consulted when questions arise. Activity of a soil-applied herbicide may be enhanced by a light rain, which moves it into soil, shortly after application.

In general, high temperatures and low humidity are detrimental to cuticular absorption. Plants growing under these conditions may produce thicker, less penetrable cuticles; others naturally have thin, poorly hydrated cuticles that are not easily penetrated. Sprays dry rapidly, and water stress may cause stomatal closure. High relative humidity reduces water stress, delays drying, and favors open stoma. Warmth (temperature less than 100°F) usually promotes herbicide penetration and action. Rain and hard winds before treatment may weather (break and crack) cuticle. More spray may be trapped and taken up by weathered leaves.

3.2.2 Temperature

Herbicides are most effective when temperatures during application favor uniform plant germination and growth. High temperatures during application generally increase herbicide action by favoring more rapid uptake, but the effect may be offset by rapid drying on leaf surfaces.

3.2.3 Light

Light, an important, uncontrollable environmental factor is essential for photosynthesis, but photosynthetic inhibitors do not have to be applied during the day. Many photosynthetic inhibitors are taken up by roots and can be applied at any time. Good light conditions may open stomates, increase photosynthetic rate, and increase transport of photosynthate and herbicide.

4. PHYSIOLOGY OF HERBICIDES IN PLANTS

4.1 Foliar Absorption

If all of the preceding factors that decrease or eliminate activity are avoided and an herbicide resides on a plant surface long enough, it must be absorbed for activity to follow. Very few herbicides are true contact materials that solubilize cuticles and membranes and enter plants without absorption through the cuticle to achieve activity. Most herbicides must enter plants and reach an appropriate site of action before toxicity can be expressed. After absorption, successful action requires translocation in the plant and avoidance of detoxification (loss of activity) prior to an attack at the molecular level on some process vital to plant growth.

No general description of the entire process of herbicide action is applicable to all herbicides any more than such a description can be provided for all antibiotics or pharmaceuticals. How does aspirin work?[3] Most people have no idea. However, the fact that we don't know does not mean that it can't be use intelligently.

---

[3]The active ingredient in aspirin is acetylsalicylic acid. It inhibits production of prostaglandins, which increase perception of pain, fever, and inflammation that accompany injury.

Absorption of an herbicide can be regarded as passage through a series of barriers any one of which may limit or prohibit action. With crops and weeds, functioning of such barriers can be the basis of selectivity. Modern herbicide formulations have been created with full knowledge of absorptive barriers, and while selectivity is most often explained by metabolism, absorption must still occur (see Section 4.4).

The terms *symplast* and *apoplast* are helpful when thinking about uptake and distribution. The essence of the concept of symplast is that all living cells of an organized, multicellular plant form a functionally integrated unit. The apoplast is the continuous nonliving cell wall structure that surrounds and contains the symplast. The xylem, nonliving tissue that conducts water and solutes from roots to shoots, is part of the apoplast. The apoplast varies in composition from the highly lipid cuticle to aqueous pectin and cellulose cell walls. It is interposed between the symplast and the environment. All herbicides that enter plants do so via the apoplast and bring about death by action on the symplast. There are several barriers to apoplastic penetration. The role of each barrier varies with each herbicide-plant-environment combination. These barriers include stomata, cuticle, epidermis, and cell walls.

4.1.1 Stomatal Penetration

As discussed in Section 3.1.1, stomatal presence, exposure, and distribution vary between species and between plants of the same species grown in different environments. Stoma are an obvious port of entry but are not very important because stomatal openings vary under field conditions and the maximum opening may be different than the time of application. Rapid drying of solutions (Table 14.1) also allows little time for stomatal penetration. Cuticular penetration occurs regardless of stomatal presence or aperture size when herbicides are properly formulated and applied.

4.1.2 Cuticular Penetration

The cuticle is a waxy layer on the leaf surface, the thickness and composition of which varies among species. The composition and thickness of cuticle varies when plants of the same species are grown in different environments. Apart from root absorption and stomatal entry, cuticular penetration is the way most foliar herbicides enter plants. Cuticular entry is possible when stomates are closed and occurs under a range of environmental conditions. There are aqueous and a lipid routes of entry through the cuticle. Both are available for simultaneous entry. The relative rate of entry depends on the molecule entering and the environment.

Cuticles are somewhat open, sponge-like structures made up of a lipid frame with interspersed pectin (water-soluble) strands and possibly open pores. Pores can fill in a water-saturated atmosphere to provide an accessible water diffusion continuum. Herbicides concentrate as the applied spray dries and may gather in depressions, commonly over anticlinal (sloping downward) walls, prior to absorption. Cuticular penetration is by diffusion through a water or lipid continuum. When a plant is under stress, pores fill with air, which acts as a barrier to water penetration, but lipoidal routes are still available.

4.1.3 *Fate of Foliar Herbicides*

There are five possible fates of all herbicides applied to plant foliage:

1. Volatilization from foliar surfaces and loss to the atmosphere.
2. Retention on leaves in a viscous liquid or crystalline form,
3. Penetration of the cuticle and retention there in lipid solution.
4. Adsorption by the cuticle.
5. Penetration of the cuticle.

While the first four fates are theoretically possible, for all practical purposes they can be neglected because manufacturers are aware of these potential fates and develop formulations that eliminate the possibility.

Penetration of the cuticle is intended. That should be followed by penetration of the aqueous portion of the apoplast (epidermal cell walls) and migration via anticlinal walls to the vascular system. If an herbicide is not phloem mobile, it will remain in the apoplast and move with the transpiration stream to acropetal (upper) leaves. Some herbicides that move this way cannot cross the plasmalemma barrier and translocate only acropetally in xylem. Many others cross easily to phloem. Because xylem translocation is much more rapid than phloem, the herbicide may appear to be translocated only or dominantly in xylem even though phloem translocation occurs.

Finally, after penetrating the cuticle, via the aqueous phase of the apoplast molecules are absorbed into the living cellular system (symplast) and translocated in phloem out of leaves in the assimilate stream. These molecules can become systemic and move throughout the plant to sites of high metabolic activity (e.g., meristematic regions). Many herbicides follow this route.

4.1.4 *Advantages and Disadvantages of Foliar Herbicides*

There are obvious advantages to foliar herbicides. Foliage is a readily available site of entry. There is often a high efficiency of foliar absorption, and treatments can be designed and scaled to control specific, observable weed problems. There are equally important disadvantages. Application timing is often critical because the herbicide may be most effective when applied at a certain stage of plant growth (e.g., herbicides active only post-emergence). Some herbicides are not absorbed well by foliage but are readily absorbed by roots. Wetting plant surfaces is difficult and weather conditions at the time of application affect performance. Herbicides control small plants better than large ones, but small plants do not have many leaves and contact and absorption may be inefficient. It often takes several days or even weeks for some plants (e.g., perennials) to grow enough foliage so good absorption and activity can be obtained.

4.2 Absorption From Soil

4.2.1 *General*

Some herbicides are directed at soil without any intention of foliar entry. Most foliar herbicides are applied as broadcast sprays and much (perhaps as much as 90%) of the herbicide hits soil because it misses plant foliage. Because many herbicides are applied when plants are young, most of the soil surface is exposed. Thus, soil becomes an unavoidable target and repository for much of what is applied. Herbicide fate in soil becomes a significant determinant of performance and environmental effects (see Chapter 15).

4.2.2 *Advantages and Disadvantages of Soil-Applied Herbicides*

Application timing of soil-applied herbicides may be convenient and economical because it can be combined with other operations. The effectiveness of preplant or preemergence soil-applied herbicides is not dependent on stage of plant growth or physiological condition at time of application.

Soil-applied herbicides have important disadvantages. There is a tremendous dilution by soil and soil water following application. Therefore the amount available to plants is low. There is fixation by soil colloids (adsorption) that reduces the amount of herbicide available for plant absorption. Foliar herbicides are affected by weather conditions at application whereas soil applied herbicides are more affected by weather subsequent to application, especially dry conditions. There is often dependence on rainfall, irrigation, or soil incorporation for distribution and action. Persistent residues that may injure subsequent crops can occur after use of some soil-applied herbicides.

4.2.3 *Root Absorption*

It is generally conceded that herbicides enter roots via root hairs and the symplastic system via the same pathway that inorganic ions (plant nutrients) follow. Passive and active uptake occurs, but most is passive with absorbed water and subsequent movement in the apoplast. Active uptake involves respiration energy, oxygen, entry into cells, and movement in the symplast. There is accumulation of herbicides at points of activity in the symplast. It is where selectivity is expressed. Most phenylureas, sulfonylureas, triazines, and uracils are absorbed by roots and move upward apoplastically. Root absorption is highly dependent on an herbicide's lipophilicity (lipid solubility).

4.2.4 *Influence of Soil pH*

For weak aromatic acids such as dicamba and 2,4-D, phytotoxicity increases as soil pH increases and reaches a maximum at pH 6.5 (Corbin et al., 1971). The same is true for weak bases such as prometon and amitrole. Soil pH between 4.3 and 7.5 had no effect on phytotoxicity of picloram, weak aromatic acids, and the nonionic herbicides dichlobenil and diuron. The conclusion is that no generalizations can be made about effects of soil pH on herbicide absorption (Corbin et al., 1971). There is an influence and the effect of pH cannot be ignored, but there is no basis for predicting what it will be in every case.

Many soil-applied herbicides, including the triazines (atrazine and simazine), the asymmetrical triazine (metribuzin), the phenylurea (linuron), and several of the sulfonylureas show increased activity when soil pH is above 7.5. Often on areas of exposed calcareous soil more plant injury occurs and selectivity is reduced. This is because there is less herbicide adsorbed at high pH and more is biologically available.

When soil pH was raised from 5 to 7, soil microflora and degradation rate of EPTC increased and phytotoxicity was shortened 2–3 weeks. A similar increase in rate of degradation of EPTC was found when manure was added (Lode and Skuterud, 1983). Therefore,

EPTC and, presumably, other herbicides are less effective on soils with high, effective microbiological activity and high pH. These examples further illustrate the point that phytotoxicity is affected by soil pH, but no generalizations can be made.

4.3 Shoot Versus Root Absorption

Different plants absorb herbicides at different sites. Grasses vary in seedling morphology, location of the mesocotyl, and depth of seed germination. Selectivity of triallate between wheat and wild oats is due to differences in location of the site of herbicide uptake (Appleby and Furtick, 1965). In wild oats, the mesocotyl elongates into herbicide-treated soil where the herbicide is absorbed after seed germination. Wheat and barley have a short mesocotyl that does not elongate into the herbicide-treated zone. Depth of the herbicide zone in soil can be controlled by incorporation depth, and positional selectivity can be obtained when these herbicides are used to control wild oats in wheat or barley (Table 14.4). Root exposure has an effect but wild oats survive. Coleoptile exposure results in plant death because of absorption by mesocotyls of emerging seedlings.

Parker (1966) confirmed these results and demonstrated preferential root or shoot absorption by sorghum with five herbicides (Table 14.5). Dichlobenil and trifluralin are dependent on root absorption whereas EPTC and diallate (no longer available) depend on absorption by shoots of emerging seedlings. Triallate was equally effective on sorghum when absorbed through roots or shoots.

Studies with yellow nutsedge have shown that most tuber (often called nutlets) sprouts come from below 2 in. because tubers in the top 2 in. of soil often winter kill. When tubers sprout they develop a crown meristem about 1.5 to 2 in. below the soil surface; roots and new rhizomes arise from this crown. Many herbicides that have activity on nutsedge are soil incorporated at least 2 in. deep to maximize absorption.

The area near or above the shoot of green foxtail is the primary area for herbicide uptake. Herbicide placement in the top 1–2 in. of soil is essential for good control. Corn and sorghum have a different site of uptake because of deeper planting. This provides an opportunity to achieve selectivity through herbicide placement.

TABLE 14.5 Herbicide Dose Required to Cause 50% Reduction in Root or Shoot Dry Weight of Sorghum (Parker, 1966)

| Herbicide | Equally Effective Concentration (ppm) | | Exposure Ratio |
	Root	Shoot	
Diallate	>8	2.5	1:0.33
Dichlobenil	0.055	1.25	1:23
EPTC	>16	0.8	1:0.05
Triallate	>4	4	1:1
Trifluralin	0.065	2.7	1:42

4.4 Absorption as a Determinant of Selectivity

Selectivity is a function of three factors: absorption, translocation, and metabolism. In some cases, differential absorption explains why an herbicide affects one plant and not another.

Peas are tolerant to 2,4-D and tomatoes are susceptible because peas absorb 2,4-D for only 24 h after exposure while tomatoes absorb greater quantities over 7 days (Fang, 1958). Wheat and corn absorb 2,4-D more slowly than beans and the low rate of absorption by monocots is a factor in selectivity.

For many herbicides, absorption is not a major barrier to activity. Studies of herbicide selectivity frequently find metabolic ability or rate of application to be the defining difference between species. However, not all differences in selectivity are due to metabolism. Differences in activity of glufosinate on tolerant barley and sensitive green foxtail were explained by differences in foliar absorption and translocation but not by metabolism (Mersey et al., 1990). Improved control of common milkweed and poor activity on hemp dogbane were attributed to improved foliar absorption of glyphosate by milkweed when surfactants were used (Wyrill and Burnside, 1977).

A major determinant of herbicide selectivity is a plant's growth stage when a herbicide is applied. Some plants show maximum susceptibility in early seedling stages and greatly reduced susceptibility after fruiting. Much of this can be traced to absorption. Fig. 14.4 shows the growth stages for wheat, oats, barley, and rye. Each is susceptible to growth-regulator herbicides when they are applied during stages one to three. A growth regulator herbicide applied between stages three and nine has little effect. Application during stage 10, increases

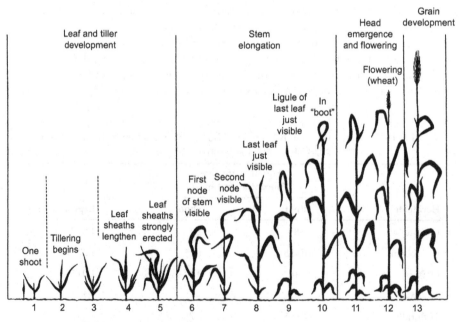

FIGURE 14.4 Growth stages in wheat, oats, barley, and rye.

susceptibility, but it is not as great as it is in stages one through three. A primary reason is greater absorption by young seedling plants and direct access to floral structures in stage 10. Susceptibility of small grains follows a consistent general pattern during various stages of growth. Thus, growth-regulator herbicides should not be applied to small-grain crops before tillers are formed. After tillering, susceptibility decreases and application is safe.

4.5 Translocation

To be effective, herbicides must move to sites of action; they must be translocated—moved elsewhere. Translocation takes place through phloem and xylem, the transport tissues in plants. It is common to find a direct correlation between foliar absorption and phloem transport and root absorption and xylem transport.

It is helpful to think of phloem translocation as movement from source to sink. It is movement of photosynthate from regions of high carbohydrate synthesis (the source) to regions of high use (the sink). Sinks are sites of high metabolic activity where herbicides express their toxicity.

Herbicide movement in plants is determined frequently by patterns of photosynthate distribution and by the relative activity of sources and sinks. For example, movement from cotyledons and young leaves is predominantly to roots. From lower leaves of mature plants there is either no movement or movement to roots. From older leaves there is transport to roots and shoot tips and meristematic areas. From upper, mature leaves transport is to shoot tips, flowers, and fruits.

Herbicides that enter phloem can pass from it to xylem and become systemic. The reverse is rare but occurs. Herbicides that move symplastically and migrate to xylem can move up or down, whereas those that move only apoplastically (root uptake dominates) translocate only acropetally in the transpiration stream. Table 14.6 shows the primary translocation pathway for several herbicides (Ashton and Crafts, 1981). Rate of translocation for one herbicide varies between species and with different environmental conditions for one species. Many patterns are possible and no absolute generalizations can be made.

4.6 Translocation as a Determinant of Selectivity

A few old experiments illustrate the role of translocation. 2,4,5-T is more mobile than 2,4-D in burcucumber (Slife et al., 1962), but burcucumber is resistant to 2,4-D and susceptible to 2,4,5-T. Translocation of 2,4-D is initially slow and there is no movement after 24 h. Slow continual movement occurs in domestic cucumbers over 8 days. Burcucumber avoids 2,4-D injury by immobilizing it, whereas 2,4,5-T is translocated to sites of action.

Bean leaves absorb 2,4-D and it seldom moves elsewhere. It is strongly absorbed by roots but moves in stems at low concentrations but not into leaves after root absorption (Crafts, 1966). Bean roots absorb 2,4-D but translocate little into stems and none into leaves. After foliar absorption there is no apoplastic movement and no retransport. Barley leaves, on the other hand, absorb 2,4-D and translocate it symplastically but not apoplastically. Barley roots absorb it but transport very little. Thus, translocation may partially explain bean's susceptibility to 2,4-D and barley's low susceptibility.

TABLE 14.6 Mobility and Primary Translocation Pathways of Some Herbicides in Plants

Free Mobility			Limited Mobility			
Apoplast	Symplast	Both	Apoplast	Symplast	Both	Little Mobility
Chloroacetamides	Glyphosate	Amitrole,	Chloroxuron diquat	Phenoxy acids[a]	Endothall naptalam	Bensulide, DCPA,
Desmedipham		dicamba	Fluridone[a]			dinitroanilines,
Diphenamid[a]		DSMA			Nitriles	diphenyl ethers
Methazole		MSMA			Phenoxy	
Napropamide[a]		Picloram			acids	
Norflurazon[a]		Imidazolinones			Propanil	
Phenmedipham		Sulfonylureas				
Pronamide						
Thiolcarbamates						
Triazines						
Uracils						
Ureas[b]						

[a]*Translocation rate varies widely between species.*
[b]*Except chloroxuron, limited apoplast.*
Adapted from Ashton, F.M., Crafts, A.S., 1981. Mode of Action of Herbicides, second ed. J. Wiley & Sons, NY. p. 34.

Two-year-old white ash trees treated for 4 weeks with 10 ppm picloram in nutrient culture were only slightly injured, but young red maple, treated in the same way, died in 2 weeks (Mitchell and Stephenson, 1973). Rate of root uptake, acropetal translocation, and leaf accumulation was lower in red maple and would explain what happened except that red maple died. Foliar penetration was similar in both species, therefore absorption could not explain selectivity. Picloram was metabolized at equal rates in both, therefore metabolism did not explain selectivity. Tolerance of white ash was not related to lower rates of uptake or faster metabolism. Red maple's high susceptibility was due to blockage of xylem by undifferentiated callus (hardened tissue) growth caused by picloram activity. Death was caused by lack of normal translocation and subsequent desiccation of leaves and stems. Picloram's activity prevented necessary translocation.

Glasgow and Dicks (1980) asked why beans and peas differed in their tolerance to dimefuron (a substituted urea). Beans were susceptible when it was applied to roots and peas were tolerant. Dimefuron was translocated from roots to shoots in beans but not in peas. Beans are therefore tolerant to preemergence field applications because there is root absorption but no shoot absorption. When only roots are exposed, absorption is low and poor translocation explains selectivity.

4.7 Metabolism

Once an herbicide is absorbed, it is susceptible to metabolism and loss of biological activity. The faster an herbicide is metabolized, the less will be available for translocation and activity at the site of toxic action. An example of plant metabolism is conversion of simazine to hydroxysimazine, a derivative with no herbicidal properties (Fig. 14.5).

Many metabolic reactions occur in plants. The most important are oxidation, reduction, hydrolysis, and conjugation (Hatzios and Penner, 1982). Plant metabolic reactions have been separated into three phases (Hatzios and Penner; Shimabukuro et al., 1981). Phase one includes nonsynthetic, generally destructive processes such as oxidation, reduction, and hydrolysis. Phase one reactions add OH, NH_2, SH, or COOH functional groups that usually change phytotoxicity, increase polarity, and lead to a predisposition for further metabolism. Reactions can be enzymatic or nonenzymatic. An example of the latter is photochemical reduction (detoxification) of bipyridyllium herbicides.

Phase two reactions are conjugations (L. *conjugare* - to unite) that result in synthesis of a new molecule. Conjugation yields metabolites with reduced or no phytotoxicity, higher water solubility, and reduced plant mobility. Conjugation occurs with glutathione, amino acids, and glucose and other sugars. Phase three metabolism is unique to plants because they cannot excrete metabolites as animals can. Conjugated metabolites must be compartmentalized in plant cells or somehow removed from further metabolic activity. Herbicides become more water soluble as they are metabolized from phase one to two, and they remain water soluble or become insoluble in phase three. Phytotoxicity is reduced by each phase, and herbicides metabolized in phase three are no longer toxic. Table 14.7 shows some reactions and the herbicide groups affected. A complete discussion of these reactions is beyond the scope of this text (see Ashton and Crafts, 1981; Corbett et al., 1984; Hatzios and Penner, 1982).

4.8 Metabolism as a Determinant of Selectivity

Herbicide activity and selectivity are often directly attributed to differences in plant metabolism. For example, black currant is susceptible to 2,4-D and decarboxylates only 2% of applied 2,4-D. Red currant is tolerant and decarboxylates 50% of applied 2,4-D in the same time (Luckwill and Lloyd-Jones, 1960). The different rate of metabolism accounts for observed selectivity. Catchweed bedstraw is selectively controlled by mecoprop (MCPP) but not by MCPA. There is no difference in absorption or translocation. Ten days after treatment, MCPP was not metabolized at all and MCPA was completely metabolized

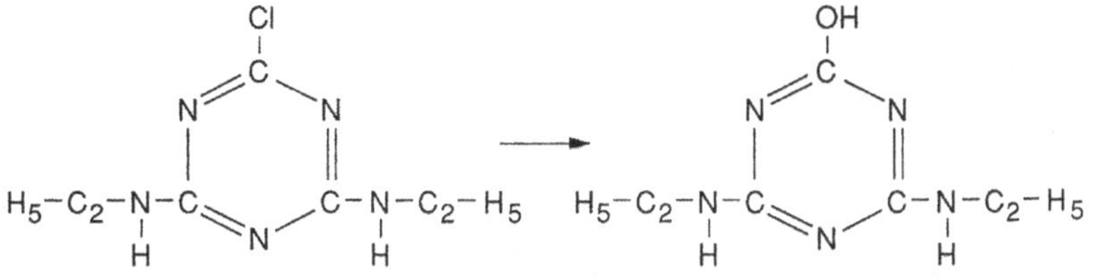

FIGURE 14.5 Conversion of simazine to hydroxysimazine.

TABLE 14.7 Plant Metabolic Reactions and the Herbicide Chemical Groups Affected

Chemical Reaction	Affected Chemical Groups
Hydroxylation	Triazines, phenoxy acids, imidazolinones
Oxidation	Phenoxy acids
Decarboxylation	Benzoic acids, picolinic acids
Deamination	Ureas, dinitroanilines
Dethioation	Carbamothioates
Dealkylation	Dinitroanilines, triazines
Hydrolysis	Carbamates, sulfonylureas, imidazolinones
Conjugation with plant constituents e.g., Glucosidation	Benzoic acids, Imidazolinones

(Leafe, 1962). Rapid metabolism of MCPA explains its lack of effect, and no metabolism of MCPP leads to death.

Broadleaf plantain, common chickweed, and strawberry are resistant to the phenoxyacids, especially 2,4-D. Dandelion, cucumber, soybean, pea, common lambs quarters, and wild buckwheat are moderately sensitive, and sunflower, mustards, and cotton are very sensitive to the same herbicides (Hatzios and Penner, 1982). These differences are explained by differences in rate of metabolism among the plants.

A portion of atrazine's selectivity can be explained by differential metabolism (Negi et al., 1964). Data in Table 14.8 show the amount of atrazine remaining 10 days after preemergence application to eight different plants. Nonsusceptible species have a low concentration because they metabolize atrazine to a nontoxic form. Species intermediate in susceptibility have a

TABLE 14.8 The Amount of Atrazine in Shoots of Eight Plant Species 10 days After Preemergence Application (Negi et al., 1964)

Species	Susceptibility	ppm Atrazine
Johnsongrass	none	19
Grain sorghum	none	8
Corn	none	10
Cotton	intermediate	222
Peanuts	intermediate	97
Oats	high	376
Soybean	high	322
Bean	high	227

higher concentration than nonsusceptible species but a lower concentration than susceptible ones. Susceptible species, especially oats and soybeans, have the highest concentration and are, in part, susceptible because of their inability to metabolize atrazine. But beans and cotton differ by only 5 ppm, hardly a significant amount. The reason cotton is intermediate in its susceptibility is not related solely to metabolism. Atrazine accumulation in lysigenous (oil-bearing) glands of cotton is an isolating, protective mechanism. Higher concentrations exist, but due to isolation a lower active concentration is present. This research illustrates the complexity of explaining selectivity.

Metabolism is the basis for differential atrazine tolerance among warm-season forage grasses (Weimer et al., 1988). Big blue stem and switchgrass are not very susceptible to atrazine, and yellow indiangrass and side oats grama are susceptible in the seedling stage. Atrazine metabolism in big bluestem and switchgrass occurred primarily by glutathione conjugation. Conjugation occurred faster than N-dealkylation of atrazine in yellow indian-grass and side oats grama. Differential tolerance to atrazine among these four grasses is due to the metabolic route by which atrazine is detoxified and the rate and type of metabolism that dominated in susceptible and resistant species.

Propanil has been used for selective control of green foxtail in hard red spring and durum wheat. Green foxtail moved into niches created when broadleaved species were controlled by growth-regulator herbicides. Other research has shown that propanil was selective in rice because it is rapidly metabolized (Yih et al., 1968a,b). Green foxtail retained more spray solution than wheat, but less propanil (Table 14.3), and retention is important. Both plants had rapid absorption during the first 12 h after treatment. Green foxtail absorbed about 10% more, but differences in absorption after 48 h did not account for selectivity. Because over 95% of applied propanil remained in leaves, translocation was not a major factor in selectivity.

Retention was important, but the most important determinant of propanil selectivity was metabolism (Table 14.9). Propanil was metabolized by wheat, not by green foxtail. Only 34% of the amount applied remained active 72 h after application to wheat, whereas over 90% remained in green foxtail. Propanil is selective in wheat and rice through rapid metabolism; wheat has the added advantage that it does not retain the amount of spray solution that green foxtail does.

Many studies include absorption, translocation, and metabolism because it is generally recognized that all must be considered if selectivity is to be understood. Results of two

TABLE 14.9   Rate of Propanil Metabolism by Wheat and Green Foxtail (Eberlein and Behrens, 1984)

Hours After Application	Percent Applied Propanil Remaining	
	Wheat	Green Foxtail
24	69.8	93.7
48	42.8	93.5
72	34.2	93.5

studies are summarized following to illustrate their scope and complexity. Wilcut et al. (1989) studied selectivity of the sulfonylurea herbicide chlorimuron among soybean, peanut, and four broadleaved weeds. Absorption was similar in five species after 72 h but lower in Florida beggarweed. There was slight symplastic and apoplastic translocation in all species. Peanut showed more tolerance with age because of reduced absorption by older plants and faster metabolism. Neither absorption nor translocation differences explained differential selectivity among the two crops and four weeds. Further experiments showed tolerance was directly correlated with the amount of unmetabolized chlorimuron. Rate of metabolism was greatest in soybean and lowest in common cocklebur. After 24 h, nearly two times as much unmetabolized chlorimuron was found in the four weeds compared to the two crops. After 72 h, 16.7% of the applied chlorimuron was present in soybean and 25.6% in peanut. Prickly sida, classified as intermediate in susceptibility, retained 29.6% of applied chlorimuron that was not metabolized after 72 h. Sicklepod and common cocklebur, susceptible species, had 39.9% and 60.6%, respectively. Florida beggarweed is susceptible to chlorimuron even though only 16.6% of the herbicide remained after 72 h. This was equal to soybeans and should have made the weed tolerant if metabolism was the only factor. Florida beggarweed actually had over five times as much chlorimuron compared to soybean when chlorimuron was calculated as amount per gram dry plant weight. Its susceptibility occurred in spite of the fact that it absorbed less than half as much as soybean and seemed to metabolize it rapidly. Chlorimuron concentration remained very high in Florida beggarweed even though the total amount was low (Wilcut et al.).

Field violet was controlled by terbacil, but only when it is applied to emerging seedlings with fewer than three leaves. Established plants with 12 leaves were not controlled by terbacil applied to control weeds in strawberries. Doohan et al. (1992) demonstrated that field violet with 12 leaves absorbed less terbacil per gram fresh weight than three-leaf plants. Young plants translocated twice as much terbacil to foliage after root uptake. Metabolism studies showed 79% of terbacil was still intact after 96 h in three-leaf plants, whereas in resistant 12-leaf plants only 40% of terbacil remained. Young plants were susceptible because although they absorbed less they translocated twice as much of what was absorbed and metabolized it more slowly than twelve-leaf plants. A similar explanation was offered for the selectivity of fluroxypyr among four species. More fluroxypyr was recovered in susceptible wild buckwheat and field bindweed (about 70%) than in tolerant Canada thistle and common lambs quarters (about 30%) 120 h after application (MacDonald et al., 1994). Fifteen and 10% of applied fluroxypyr was translocated in Canada thistle and common lambs quarters, respectively, whereas 40% was translocated in the two susceptible species. Selectivity was due to limited translocation in tolerant species and more rapid metabolism (MacDonald et al.).

One of the most striking features of herbicides is selectivity—the ability to kill or affect the growth of one plant without affecting another. Factors that affect selectivity include:

• Distribution as affected by drift, volatility, soil incorporation, and selective placement;
• Retention by plants as affected by leaf morphology, herbicide formulation, and the herbicide's chemical and physical properties;
• Absorption by plants as affected by site of uptake (root vs. shoot), cuticle, weather, soil, and the herbicide's chemical and physical properties;

- Immobilization versus translocation in plants as affected by plant age, weather, soil, and the specific herbicide and its formulation;
- Metabolism or molecular change of herbicides in plants and soil as affected by the herbicide and soil microorganisms;
- Plant age;
- Weather;
- Physiological factors including translocation and inactivation without molecular change.

In general, for maximum effectiveness the ideal herbicide should have:

- Ability to enter plants at various sites;
- Ability to enter plants without local damage;
- Activity or ability to affect plant growth that is not confined to a particular stage of plant development or plant size;
- Ability to translocate in plants to site(s) of action;
- Metabolism or degradation to inactivity in target plants should be slow enough to permit full expression of activity;
- Moderate soil adsorption to decrease leaching;
- Reasonable stability in soil except for foliar active contact herbicides where soil persistence is of no consequence to action but may have environmental consequences;
- A wide weed-control spectrum or specific activity against target weeds.

There are no ideal herbicides. Some come close, but none meet all of these criteria. Herbicide selectivity means that all plants do not respond in the same way to all herbicides. Their use in agronomic and horticultural crops, lawn and turf, forestry, aquatic sites, or for general vegetation control is dependent on selective activity. Herbicide selectivity is dependent on morphological and metabolic differences between weed and crop. For most herbicides, selective action occurs over a relatively wide dose range. This gives users some assurance of selectivity and avoids catastrophe if small errors in sprayer calibration or application are made. The selective action and effectiveness of herbicides depends on differences in their toxicity at the cellular level. Selective action also depends on all the factors that influence the amount of herbicide that reaches sites of toxic action in cells.

For herbicides, dose is the most important determinant of selectivity. All herbicides have a recommended dose for particular tasks and applicators need to know and apply the correct dose for each crop-weed situation. The EPA-approved herbicide label is the primary source of information on use rates.

THINGS TO THINK ABOUT

1. Can drift and volatility be eliminated?
2. How can drift and volatility be controlled?
3. Are drift and volatility current problems? Why? What are appropriate solutions?
4. How does plant morphology affect selectivity?
5. How can spray solutions be modified to affect selectivity?
6. How does weather affect performance of foliar-applied herbicides?

7. How does weather affect performance of soil-applied herbicides?
8. What can happen to foliar-applied herbicides after they contact plants?
9. Foliar- and soil-applied herbicides each has advantages and disadvantages. Compare and contrast them.
10. How do absorption, translocation, and metabolism interact to determine selectivity?
11. What are the phases of herbicide degradation in plants? What is the significance of each?
12. What factors determine an herbicide's selectivity?
13. What characteristics should an herbicide have to maximize activity?

Literature Cited

Anonymous, February 1996. Herbicide drift could be potent problem. Success. Farming 94 (2), 13.

Appleby, A.P., Furtick, W.R., 1965. A technique for controlled exposure of emerging grass seedlings to soil-active herbicides. Weeds 13, 172–173.

Ashton, F.M., Crafts, A.S., 1981. Mode of Action of Herbicides, second ed. J. Wiley & Sons, NY, p. 34.

Biller, R.H., 1998. Reduced input of herbicides by use of optoelectronic sensors. J. Agric. Eng. Res. 71 (4), 357–362.

Bish, M.D., Bradley, K.D., 2017. Survey of Missouri pesticide applicator practices, knowledge, and perceptions. Weed Technol. 31 (2), 165–177.

Blackman, G.E., Bruce, R.S., Holly, K., 1958. Studies in the principles of phytotoxicity. V. Interrelationships between specific differences in spray retention and selective toxicity. J. Exp. Bot. 9, 175–205.

Brooks, F.A., 1947. The drifting of poisonous dusts applied by airplane and land rigs. Agric. Eng. 28, 233–239.

Corbett, J.R., Wright, K., Baillie, A.C., 1984. The Biochemical Mode of Action of Pesticides, second ed. Academic Press, London, UK. 382 pp.

Corbin, F.T., Upchurch, R.P., Selman, F.L., 1971. Influence of pH on the phytotoxicity of herbicides in soil. Weed Sci. 19, 233–239.

Crafts, A.S., 1966. Comparative movement of labeled tracers in beans and barley. In: Proc. Symp. Int. Atomic Energy Agency. Isotopes in Weed Control. Vienna, pp. 212–214.

Derting, C.W., 1987. Wiper application. In: McWhorter, C.G., Gebhardt, M.R. (Eds.), Methods of Applying Herbicides. Weed Sci. Soc. of America Monograph No. 4, Champaign, IL (Chapter 14).

Doohan, D.J., Monaco, T.J., Sheets, T.J., 1992. Effect of field violet (*Viola arvensis*) growth stage on uptake, translocation, and metabolism of terbacil. Weed Sci. 40, 180–183.

Eberlein, C.V., Behrens, R., 1984. Propanil selectivity for green foxtail *(Setaria viridis)* in wheat *(Triticum aestivum)*. Weed Sci. 32, 13–16.

Fang, S.C., 1958. Absorption, translocation, and metabolism of 2,4-D-1-C^{14} in peas and tomato plants. Weeds 6, 179–186.

Felsot, A.S., 2005. Evaluation and Mitigation of Spray Drift, Proc. International Workshop on Crop Protection Chemistry in Latin America. Harmonized Approaches for Environmental Assessment and Regulation, San Jose, Costa Rica. http://feql.wsu.edu/esrp531/Fall05/FelsotCostaRicaDrift.pdf.

Felton, W.L., 1990. Use of weed detection for fallow weed control. In: Proceedings of Central Great Plains conservation Tillage Symp, pp. 241–244.

Gebhardt, M.R., McWhorter, C.G., 1987. Introduction to herbicide application technology. In: McWhorter, C.G., Gebhardt, M.R. (Eds.), Methods of Applying Herbicides. Monograph 4. Weed Sci. Soc. of America, Champaign, IL (Chapter 1).

Girmaa, K., Mosalia, J., Raun, W.R., Freemana, K.W., Martina, K.L., Solieb, J.B., Stoneb, M.L., 2004. Identification of optical spectral signatures for detecting cheat and ryegrass in winter wheat. Crop Sci. 45 (2), 477–485.

Glasgow, J.L., Dicks, J.W., 1980. The basis of field tolerance of field bean and pea to dimefuron. Weed Res. 20, 17–23.

Harr, J., Guggenheim, R., 1995. The Leaf Surface of Major Crops. Friedrich Reinhardt Verlag, Basel, Switzerland, 98 pp.

Hartley, B.S., Graham-Bryce, I.J., 1980. Physical Principles of Pesticide Behavior, vol. 2. Academic Press, NY, p. 93. Appendix 5.

Hatzios, K.K., Penner, D., 1982. Metabolism of Herbicides in Higher Plants. Burgess Pub. Co., Minneapolis, MN, pp. 6−7.

Leafe, E.L., 1962. Metabolism and selectivity of plant growth regulator herbicides. Nature 193, 485−486.

Lipton, E., November 2, 2017. N.Y. Times, pp. B1−B3.

Lode, O., Skuterud, R., 1983. EPTC persistence and phytotoxicity influenced by pH and manure. Weed Res. 23, 19−25.

Luckwill, L.C., Lloyd-Jones, C.P., 1960. Metabolism of plant growth regulators. I. 2,4- dichlorophenoxyacetic acid in leaves of red and black currant. Ann. Appl. Biol. 48, 613−625.

MacDonald, R.J., Swanton, C.J., Hall, J.C., 1994. Basis for selective action of fluroxypyr. Weed Res. 34, 333−334.

McKinlay, K.S., Brandt, S.A., Morse, P., Ashford, R., 1972. Droplet size and phytotoxicity of herbicides. Weed Sci. 20, 450−452.

McWhorter, C.G., 1966. Sesbania control in soybeans with 2,4-D wax bars. Weeds 14, 152−155.

McWhorter, C.G., 1970. A recirculating spray system for postemergence weed control in row crops. Weed Sci. 18, 285−287.

McWhorter, C.G., Gebhardt, M.R. (Eds.), 1987. Methods of Applying Herbicides. Monograph 4. Weed Sci. Soc. of America, Champaign, IL, 358 pp.

Mersey, B.G., Hall, J.C., Anderson, D.M., Swanton, C.J., 1990. Factors affecting the herbicidal activity of glufosinate-ammonium: absorption, translocation, and metabolism in barley and green foxtail. Pestic. Biochem. Physiol. 37, 90−98.

Mitchell, J.F., Stephenson, G.R., 1973. The selective action of picloram in red maple and white ash. Weed Res. 13, 169−178.

Mullison, W.R., Hummer, R.W., 1949. Some effects of the vapor of 2,4-Dichlorophenoxyacetic acid derivatives on various field crops and vegetable seeds. Bot. Gaz. 111, 77−85.

Mueller, T.C., 2015. Methods to measure herbicide volatility. Weed Sci. (Special Issue) 63, 116−120.

Negi, N.S., Funderburk Jr., H.H., Davis, D.E., 1964. Metabolism of atrazine by susceptible and resistant plants. Weeds 12, 53−57.

Ozkan, H.E., 1987. Sprayer performance evaluation with microcomputers. Appl. Eng. Agric. 3, 36−41.

Paap, A.J., 2014. Development of an Optical Sensor for Realtime Weed Detection Using Laser Based Spectroscopy. Doctoral dissertation. Edith Cowan University, Perth, Western Australia, 163 pp.

Parker, C., 1966. The importance of shoot entry in the action of herbicides applied to the soil. Weeds 14, 117−121.

Reichenberger, L.R., April 1980. Chemical application: the billion-dollar blunder. Success. Farming, 5 pp.

Shimabukuro, R.H., Lamoureux, G.L., Frear, D.S., 1981. Pesticide metabolism in plants: principles and mechanisms. In: Matsumura, F. (Ed.), Biological Degradation of Pesticides. Plenum Press, NY.

Slife, F.W., Key, J.L., Yamaguchi, S., Crafts, A.S., 1962. Penetration, translocation, and metabolism of 2,4-D and 2,4,5-T in wild and cultivated cucumber plants. Weeds 10, 29−35.

Wilcut, J.W., Wehtje, G.R., Patterson, M.G., Cole, T.A., Hicks, T.V., 1989. Absorption, translocation, and metabolism of foliar applied chlorimuron in soybeans (*Glycine max*), peanuts (*Arachis hypogaea*) and selected weeds. Weed Sci. 37, 175−180.

Weimer, M.R., Swisher, B.A., Vogel, K.P., 1988. Metabolism as a basis for differential atrazine tolerance in warm-season forage grasses. Weed Sci. 36, 436−440.

Wyrill, J.B., Burnside, O.C., 1977. Glyphosate toxicity to common milkweed and hemp dogbane as influenced by surfactant. Weed Sci. 25, 275−287.

Yih, R.Y., McRae, D.H., Wilson, H.F., 1968a. Mechanism of selective action of 3′,4′- Dichloropropionanilide. Plant Physiol. 43, 1291−1296.

Yih, R.Y., McRae, D.H., Wilson, H.F., 1968b. Metabolism of 3′,4′-Dichloropropionanilide: 3-4-Dichloroaniline-lignin complex in rice plants. Science 161, 376−377.

15

Herbicides and Soil

FUNDAMENTAL CONCEPTS

- There are three important concerns about herbicides in soil:
  1. A concentration equilibrium among the soil's gaseous, liquid, and solid phases,
  2. Susceptibility to degradation, and
  3. Possible effects on soil flora and fauna.
- Soil is a living medium with a vast adsorptive surface that plays a major role in

determining an herbicide's activity and environmental fate.

- Several physical and chemical factors interact to determine herbicide activity and fate in soil.
- Herbicides are degraded by soil microorganisms, nonenzymatic and photochemical processes.
- Most herbicides do not accumulate in the environment.

LEARNING OBJECTIVES

- To understand the effect of soil colloidal surfaces on herbicide activity and environmental fate.
- To know the physical and chemical factors that affect herbicide activity and performance in soil.
- To understand the importance of adsorption to an herbicide's fate in soil.
- To know the relationship between herbicide adsorption, leaching, volatility, and degradation.

- To understand the role of soil microorganisms in herbicide degradation.
- To understand the role of chemical (nonenzymatic) and photodegradation of herbicides.
- To understand the role of herbicides that persist in soil and their effect on weed management.

All natural resources, except only subterranean minerals, are soil or derivatives of soil. Farms, ranges, crops and livestock, forests, irrigation water, and even water power resolve themselves into questions of soil. Soil is therefore the basic natural resource. It follows that the destruction of soil is the most fundamental kind of economic loss the human race can suffer. With enough time and money, a neglected farm can be put back on its feet—if the soil is still there. With enough patience and scientific knowledge, an overgrazed range can be restored—if the soil is still there. By expensive replanting and with a generation or two of waiting, a ruined forest can again be made productive—if the soil is still there. With infinitely expensive works, a ruined watershed may again fill our ditches or turn our mills—if the soil is still there. But if the soil is gone, the loss is absolute and irrevocable. *Leopold (1921).*

Independent of method of application, some of any applied herbicide reaches soil. Even with careful, controlled application soil contact of some portion of any application is inevitable. Foliar applications may be washed off to soil. Other herbicides are applied directly to soil. Understanding the interactions of herbicides and soil is an essential aspect of proper use and environmental stewardship. Soil is not an instrument of crop production similar to a tractors, fertilizers, or pesticides because it can be destroyed and not recovered. It is a complex, living, fragile, medium that must be understood and protected. It is agriculture's living, dominant medium for plant growth with a myriad of biological and chemical activities. The thin mantle of soil on the earth's surface is properly regarded as humankind's most essential and least appreciated resource. But this important, essential resource is in danger because of expanding cities, deforestation, unsustainable land use and management practices, pollution, overgrazing, and climate change. Leopold is right, "destruction of soil is the most fundamental kind of economic loss the human race can suffer." Without soil there will be no agriculture.

The earth's diameter is about 8000 miles. The thin soil mantle is about 7×10^{-5} of the earth's total thickness, which equals 3 to 4 ft in the world's temperate zones. Presently, that thin mantle is not the only place (hydroponics) where our food grows, but it is indisputably the most important one. It is what Logan (1995) calls "The Ecstatic Skin of the Earth." Every day we walk on and enjoy the benefits of the most complex and crucial part of the Earth's ecosystem that is the home of a quarter of the planet's biodiversity. A handful of soil is the home of 1000 different microbial animals. "A single gram has billions of cells, thousands of species, and far more information that the human genome" (Buckley, D. Cited in Cordova, 2017). It is "the root of our existence" (Montgomery, 2007). It is definitely not just dirt. The physiocrats of the 18th century saw the land as the primary guarantor of wealth. Adam Smith included land with labor and capital as one of the three factors of production. Malthus saw its scarcity as ensuring catastrophe in the face of increasing population growth (Anon., 2015). For several years developed countries were able to ignore the land because technology increased production. The value of land relative to GDP fell. Now scientists and citizens recognize the importance of the land and that it must be protected.

There are three major concerns about herbicides in soil. The first is the reciprocal equilibrium of exchange and distribution of any material in liquid, solid, and gaseous phases. The second is its susceptibility to and rate of degradation. The third involves possible influences of herbicides on soil, soil fertility, and soil microorganisms. After soil application, there is no immediate, direct contact between an herbicide and plant roots or emerging shoots. The physical processes of diffusion in and mass flow of water bring herbicides to plant roots. These processes are necessarily weather dependent (especially rainfall or irrigation), and the dose that creates a biological response is a function of weather, soil properties, and rate of application. Some control of these factors is possible, especially in irrigated agriculture. An essential property of a successful soil-applied herbicide is activity over a fairly wide range of environmental conditions with reproducible reliability.

1. SOIL

Soil contains heterogeneous organic and inorganic compounds. It is a dynamic system in which components are constantly displaced mechanically and chemically and biochemically transformed. There are gaseous, liquid, solid, and living phases. The solid phase, what we see, is present in a finely distributed form that creates large, internal surface areas. This is of great importance to the soil behavior of herbicides. Table 15.1 shows how surface area increases with decreasing particle diameter. The fine, colloidal clay minerals with their large surface area determine herbicide behavior primarily because of the properties of their surfaces rather than because of their chemical composition. Fig. 15.1 (Dubach, 1971) shows the basic structure of the clay minerals kaolinite and montmorillonite. The thin molecular layers are held together by chemical attraction. In kaolinite, there is one silicon tetrahedral layer and one aluminum octahedral layer in a fixed lattice. Attraction between the two

TABLE 15.1   Increase of Soil Surface Area With Decreasing Particle Diameter

Size Fraction	Particle Diameter (mm)	Approximate Surface Area (cm^2 per gram)
Stones	>200	—
Coarse gravel	200–20	—
Fine gravel	20–2	—
Coarse sand	2–0.2	21
Fine sand	0.2–0.02	210
Silt	0.02–0.002	2,100
Clay	<0.002	23,000

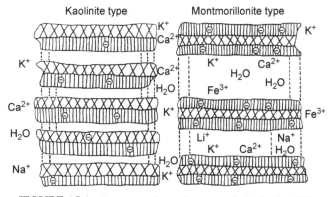

FIGURE 15.1   Structure of two clay minerals (Dubach, 1971).

layers is so strong that water molecules and chemical ions cannot penetrate the nonexpanding, fixed layers. In montmorillonite, there are two silicon tetrahedral layers and one aluminum octahedral layer in an expanding lattice structure. Individual layers are weakly held together in a lattice structure capable of expanding. Molecules and ions can penetrate between layers. Internal and external surfaces are available for chemical activity in expanding lattice clays whereas only external surfaces are available in nonexpanding (fixed) lattice clays.

The molecular lattice of clay colloids interacts with positively charged ions and molecules, from the soil solution, on the clay's predominantly negatively charged surfaces. These molecules and ions are exchangeable between surfaces and the soil solution. Most do not become permanently fixed to clay surfaces. The sum of negative charges, the cation exchange capacity, varies between clays (Table 15.2). Soils also contain negatively charged organic colloids that have large internal and external surfaces, and an exchange capacity equal to or greater than that of expanding lattice clays (Table 15.2).

Herbicides in soil are subject to electrostatic attractive forces from soil colloidal surfaces and are adsorbed on those surfaces. When adsorbed, it is difficult, often impossible, for them to be taken up by plants or microorganisms, and they are at least partially protected from attack by microbes and nonenzymatic (chemical) reactions. If an herbicide is sorbed by action at one moiety (part), an exposed or nonadsorbed portion can be susceptible to microbial or chemical attack, but the entire herbicide molecule cannot be absorbed by plants while it is adsorbed (bound) to a soil surface.

There is a well-established, negative correlation between an herbicide's soil activity and the soil's clay and organic matter content. Soil pH and soil water content are also important. In the field, rainfall, temperature, clay and organic matter content (sorptive capacity) are important determinants of activity, but each alone and all collectively can be affected by temporary, significant, alterations of their interaction (e.g., flooding, drought). Great differences in herbicide activity are determined by whether application is on dry soil and rain falls afterwards, or whether soil is moist and there is or is not precipitation or irrigation afterwards. Interaction of factors is important. Most soil interactions are well understood, and good generalizations based on laboratory and field experiments permit accurate prediction of field behavior. Such predictions of herbicide activity are qualitatively accurate although they are not always quantitatively precise.

TABLE 15.2   Comparison of Cation Exchange Capacities and Surface Area
of Three Clay Minerals and Soil Organic Matter

Exchange Surface	Exchange Capacity (cmols (+)/kg)	Surface Area (m²/g)
Organic matter	100–300	500–800
Montmorillonite	100	600–800
Illite	0	65–100
Kaolinite	10	7–30

2. FACTORS AFFECTING SOIL-APPLIED HERBICIDES

2.1 Physical Factors

Five physical factors affect herbicides in soil: placement, volatility (evaporation), adsorption, leaching, and soil moisture. Table 15.3 shows the points of entry for any pesticide, the active environmental processes, and the interacting processes that affect a pesticide's movement and environmental fate.

2.1.1 Placement

Some herbicides are taken up more readily by roots than by shoots and vice versa (see Chapter 14). Therefore an herbicide can be placed in or on soil to enhance or reduce uptake and to contact specific weeds or avoid crops. This seems obvious and easy but is difficult because control of movement after application is impossible.

2.1.2 Time of Application

Herbicides can be applied before planting with or without soil incorporation, before crop or weed emergence (preemergence), or after emergence of the crop or the weeds

TABLE 15.3 Pesticide Entry Into the Environment and Processes Affecting Movement and Loss

Environmental Zone	Environmental Processes		Processes Affecting Pesticide Fate	
			Addition	Loss/Immobilization
Atmosphere	Evaporation	Precipitation	Application Drift	Photodegradation Condensation
Above ground	Transportation Evaporation	Precipitation Irrigation	Foliar application Volatilization Drift Condensation	Plant absorption Photodegradation Drift Wash-off
On the soil surface	Irrigation/rainfall	leading to runoff		
Unsaturated soil to root depth	Evaporation	Leaching Root uptake	Surface - subsurface appl. Seed appl. Wash-off Transport	Degradation (chem/bio.) Adsorption Root absorption Transport
Unsaturated soil (below root depth = Vadose zone)	Movement Upward Lateral Downward	Transport		Degradation (chem./bio.) Adsorption Root absorption Transport
Saturated zone (groundwater)	Movement Upward Lateral Downward		Transport	Degradation (chem./bio.) Adsorption

(postemergence). The specific (best) times of application are always specified on the product label. Time of application can determine residual activity and soil persistence. Late summer or fall and early spring application normally yields good phytotoxic activity and decreases the possibility of leaching due to lower soil temperatures, reduced evapotranspiration, and a higher probability of rainfall. Application when soil is dry may lead to no or reduced activity and extend soil life.

2.1.3 Volatility

Volatility or evaporation affects location. It changes a molecule's physical state from liquid to gas but does not result in chemical change or molecular degradation. It is movement in the gaseous state, which is distinctly different from drift, which is movement as a liquid (see Chapter 14 and Table 15.3). All herbicides have a vapor pressure, although for many it is negligible. The vapor pressure of mothballs, gasoline, and ether is high, and their presence is easy to detect because of their scent. Vapor pressure—the tendency to volatilize—increases with temperature and is measured in millimeters (mm) of mercury (Hg) at a specific temperature, usually 25°C. Volatilization of herbicides with low phytotoxicity does not create an obvious hazard to plants, but it can, and in the view of many it usually does, affect environmental quality. Volatilization of herbicides that are toxic to other plants, other species, or that may affect the environment is undesirable and should not be tolerated.

Volatility can occur from soil or plant surfaces. Herbicides that volatilize from the soil surface move through the atmosphere, the easiest and most available route. Herbicides that volatilize in soil move laterally and toward the surface. Incorporation in soil decreases atmospheric volatility and is required for some dinitroaniline and carbamothioate herbicides.

Application of herbicides to a dry soil surface followed by surface wetting and high temperature and low humidity can move herbicides to the soil surface and increase volatility. Some desirable lateral and upward movement of volatile herbicides occurs after incorporation in soil.

Techniques to reduce or eliminate volatility after application are available. Mueller (2015) described methods to measure volatility. Formulation of phenoxyacetic acids as long-chain esters or complex ester chains with an ether linkage reduces volatility (see Chapter 14). Soil incorporation increases adsorption, thereby reducing volatility and perhaps enhancing activity through placement near plant roots. The relative volatility of some herbicides is shown in Table 15.4. They are divided into high, medium, and low volatility according to their vapor pressure (in mm Hg @25°C). Herbicides with high volatility have low vapor pressure (10^{-2}–10^{-4} mm Hg) and a high tendency to change state from liquid to gas at normal atmospheric pressure. High volatility is a caution but not an automatic hazard. Most herbicides have low volatility (10^{-7} mm Hg or less).

2.1.4 Adsorption

Adsorption is a process of accumulation at an interface and is contrasted with absorption, or passage through an interface. In soil, clay and organic matter solid surfaces are interfaces between soil's gaseous and liquid phases. Through cation exchange and physical attraction, herbicides can be concentrated at adsorptive surfaces and removed from the soil solution, from which plant uptake occurs. Adsorption is one of the most important mechanisms that reduces herbicide concentration in soil solution. Few herbicides

TABLE 15.4 Relative Volatility of Some Herbicides

Volatility	Herbicides
High Vapor pressure 10^{-2} to 10^{-4} mm Hg	butylate, EPTC, clomazone, trifluralin, short-chain esters of phenoxy acids
Medium Vapor pressure 10^{-5} to 10^{-6} mm Hg	alachlor, benefin, butachlor, clopyralid, DCPA, dicamba, ethalfluralin, linuron, napropamide, oxyflourfen, pendimethalin, pronamide, long-chain esters of phenoxy acids
Low Vapor pressure $>10^{-7}$ mm Hg	acetochlor, atrazine, amitrole, bentazon, bromacil, cyanazine, diclofop, bipyridilliums, ethofumesate, fluazifop, fluometuron, glyphosate, hexazinone, imidazolinones, oryzalin, picloram, sethoxydim, most sulfonylureas

completely escape adsorptive interactions. Manufacturers develop application rates to compensate for adsorption and to keep enough herbicide desorbed (in solution) for activity. The organic arsenicals, dipyridiliums, and glyphosate are adsorbed quickly and extensively and therefore have no soil residual activity.

Adsorption affects movement and availability in soil and rate of degradation. It regulates degradation by soil microorganisms and chemical reactions. The adsorption-desorption equilibrium determines the amount adsorbed and the amount in solution available for plant absorption. The equilibrium is the ratio of adsorbed herbicide to solution concentration and can be expressed mathematically given specific herbicide-soil combinations.

There are two adsorptive factors to consider, strength and extent of binding. It is not true that the most extensively bound herbicide will be the most strongly bound; both must be determined. Table 15.5 compares strength of adsorption for several common herbicides. The n-octanol water partition coefficient, (K_{ow}), is used in hydrogeology to predict and model the migration of dissolved hydrophobic organic compounds in soil and groundwater and is applicable to an herbicide's behavior in soil. It is a coefficient

TABLE 15.5 Adsorption Strength for Several Herbicides

Adsorption Strength	Herbicides
Very strong $K_{oc} > 2000$	bromoxynil, DCPA, diclofop, difenzoquat, ethalfluralin, fenoxaprop-P, fluazifop, glyphosate, paraquat, oxyflourfen, pendimethalin, prodiamine, triallate, trifluralin,
Strong K_{oc} 500–1999	bensulide, butachlor, cycloate, desmedipham, dithiopyr, ethalfluralin, fluridone, napropamide, norflurazon, oryzalin, oxadiazon, quizalofop-P
Moderate K_{oc} 100–499	aciflurofen, alachlor, amitrole, atrazine, benoxacor, bensulfuron-methyl, butachlor, clomazone, chlorimuron-ethyl, dichlobenil, diuron, EPTC, flumetsulam, glufosinate, soxaben, MCPA, linuron, oryzalin, quizalofop-P, siduron, simazine, vernolate
Weak $K_{oc} < 99$	bentazon, bromacil, chlorsulfuron, clopyralid, dicamba, haloxyfop, hexazinone, imazamox, imazapyr, imzethapyr, isoxaben, metribuzin, metsulfuron, nicosulfuron, picloram, primisulfuron, sethoxydim, sulfometuron-methyl, tebuthiuron, terbacil, tribenuron-methyl, 2,4-D

representing the ratio of the solubility of an herbicide's concentration in a known volume of n-octanol (a nonpolar solvent) to its concentration in a known volume of water (a polar solvent) after octanol and water have reached equilibrium. The higher the K_{ow}, the more nonpolar the compound. Log K_{ow} is generally used as a relative indicator of the tendency of an herbicide to adsorb to soil. It is generally inversely related to aqueous solubility and directly proportional to molecular weight. The hydrophobicity (water-fearing) of an herbicide indicates how easily it might be taken up as a groundwater pollutant, that is, not adsorbed.

The groups in Table 15.5 vary from very strong to weakly adsorbed. They were created based on the each herbicide's K_{oc}, the more common measure for herbicides. It is the "distribution coefficient" (K_d) normalized to total organic carbon content. K_{oc} values are used to predict the mobility of organic soil contaminants. Higher K_{oc} values correlate to less-mobile herbicides while a low K_{oc} predicts greater mobility. K_{oc} is the herbicide's K_d (distribution coefficient) divided by the weight fraction of organic carbon in a soil.

$$K_{oc} = k_d/\text{weight fraction of organic carbon in soil}$$

K_d, usually expressed in L/Kg or ml/g, is the ratio of sorbed to dissolved herbicide at equilibrium in a soil-water slurry.

$$K_d = \text{herbicide sorbed (mg/kg)/herbicide in solution (μm/L)}$$

These standard measures are available for most herbicides (Shaner, 2014). Hydrophobic herbicides tend to be more active, generally have longer half-lives, and therefore may pose increased risk of adverse environmental effects.

Bipyridilium herbicides are susceptible to cation exchange. Because they are cations they are adsorbed tightly and extensively by negatively charged surfaces. Imidazolinones and sulfonylureas are both acidic herbicides and are not adsorbed extensively or tightly. Their sorptive interactions are governed by soil pH and sorption increases as soil pH decreases. With acidic pHs, soil adsorption is greater because the molecules are negatively charged. As pH becomes more basic, the molecules are neutral and not sorbed extensively.

Clay and organic soils usually require higher herbicide rates (amount per acre) for equal activity compared to sandy soil. High levels of organic matter and clay adsorb herbicides, and sorbed residues persist longer than in sandy soils or those low in clay and organic matter. For example, 1 pint of a 4 lb ai/Gal formulation might be recommended/acre for coarse (sandy), 1.5 pints for medium (loams), and 3 pints for clay soil. The rate should increase with the organic content of soil. Rate dependence on soil texture and organic matter is especially important when preemergent herbicides are used to manage weeds.

Specific label instructions should always be consulted prior to use.[1] No matter how such products are used, the use rate by soil type is very important for achieving good weed control within the constraints of good crop safety.

---

[1]Labels for a wide range of agricultural chemicals can be found at Crop Data Management Systems' website (www.cdms.net).

There should be no doubt that soil clay and organic matter content are important determinants of efficacy. However, one must be aware that specific rate recommendations are required for:

- Herbicides applied at very low rates (e.g., sulfonylureas and imidazolinones);
- Contact herbicides (e.g., glyphosate); and
- Those applied to GM crops, which may have diminished the need for rate adjustment based on soil organic and clay content.

2.1.5 Leaching

Leaching is movement of a herbicide with water, usually, but not always, downward. It is of environmental concern because of the possibility of offsite movement and groundwater contamination. It can determine an herbicide's effectiveness by moving it into or out of the zone of action.

Leaching is analogous to a chromatographic process in which soil is the stationary phase and water the moving phase. Given that one knows the necessary soil and herbicide properties, leaching can be predicted mathematically. It is inversely related to percent organic matter and clay, and therefore to adsorption. The greater the adsorption of an herbicide and the adsorptive capacity of a soil, the less leaching will occur.

The extent of leaching is determined by:

- Adsorptive interactions between herbicide and soil.
- Water solubility—the greater an herbicide's water solubility, the greater its leaching potential.
- Soil pH—because they are weakly charged, sorption and leaching of imidazolinone and sulfonylurea herbicides are governed by soil pH. Adsorption increases as pH decreases and at low pH because more of the herbicides will be sorbed, leaching is reduced.
- pKa—is a measure of alkalinity. It is a property of the herbicide, not the soil. The higher the pKa, the greater the leachability. At the pKa, one-half of a molecule is neutral and one-half is ionized. For example, small amounts of acidic herbicides (phenoxyacetic acids, dicamba, picloram) are adsorbed on clay colloids when the pH equals the pKa, and molecular and anionic species occur in relatively equal amounts. For acidic herbicides, when the pH is above the pKa, anionic species dominate and adsorption will be lower. When soil pH is below the pKa, molecular species dominate and adsorption can increase. Basic molecules (e.g., triazines) have a high pKa and adsorption is greatest at low pH.
- The amount of water moving through the profile. The more water moving due to rainfall or irrigation, the more likely it is that leaching will occur.
- In theory, leaching will be greater at higher temperature, but this is very difficult to measure in the field. Sorption is an exothermic reaction, so with increasing temperature sorption will decrease and leaching can increase.

If an herbicide is very water soluble (greater than 3000 ppm), it is much more likely to leach. But water solubility is not the sole determinant of leaching. A 4.5 in. rain or equivalent irrigation weighs a bit less than 1 million lbs. That's enough water to leach most herbicides out of the soil profile if their water solubility is greater than 1 ppm, but

TABLE 15.6   Relative Mobility of Herbicides in Soil[a]

Mobility class				
5	4	3	2	1
Very leachable				**Essentially immobile**
Bromacil	Carfentrazone	Acetochlor	Bentazon	Benefin
Clopyralid	Dichlobenil	Acifluorfen	Bromacil	Chlorosulfuron
Dicamba	Isoxflutole	Alachlor	Bromoxynil	Diclofop
Dithiopyr	Pronamide	Atrazine	Chlorimuron	DCPA
Fenoxaprop-ethyl	Siduron	Diuron	Dimethenamid	Diquat
Fluazifop	Simazine	Ethofumesate	EPTC	Hexazinone
Haloxyfop	Tralkoxydim	Linuron	Imazethapyr	Glyphosate
Isoxaben	Triallate	Primisulfuron	MCPA	Gramoxone
Nicosulfuron	Tribenuron	Prometon	Mecoprop	Imazamox
Oxyfluorfen		Pyrazon	Metoachlor	Imazapyr
Pendimethalin		Pramitol	Metribuzin	Quizalofop
Trifluralin			Metsulfuron	Rimsulfuron
			Picloram	Sethoxydim
			Terbacil	
			2,4-D amine and ester	

[a]*Table compiled from Helling (1971) and published K_{oc} values from Shaner (2014).*

that doesn't happen. Even after 10, 15, or 20 in. of water as rain or irrigation, some herbicide remains in upper soil layers in spite of the fact that the water is capable of dissolving much more than was applied. The reason is adsorption. Table 15.6 shows the relative mobility of herbicides in soils. The table is approximately the inverse of Table 15.5 because adsorption and leaching are inversely related—the greater the adsorption, the lower the amount leached. Herbicides with weak adsorption are in mobility class 5, and those with very strong adsorption are in mobility class 1. There is no question that herbicides or their transformation products are present in ground and surface water in the United States.[2] A US Geological Survey (USGS) study found pesticides or their transformation products in groundwater in more than 43 states. The 2008 study found trace amounts of degradation products of EPTC, DCPA, terbacil, and three nonherbicide degradation products in public water systems between 2000 and 2005. Overall, the detection rate was less than 3% for all water samples. Thus, there is reason to be concerned about herbicide leaching and degradation, but neither appears to be a major problem. The USGS National

[2]US Geological Survey (http://water.wr.usgs.gov/pnsp/gw/gw_4.html, USGS fact sheet FS-244-95) and www.epa.gov/safewater EPA 815.R.08—013, June 2008.

Water-Quality Assessment project began in 1992 and continues to monitor the occurrence and behavior of pesticides in streams and groundwater of the United States and their potential for adversely affecting drinking water and aquatic ecosystems.

In general, leaching is movement downward, but it can occur laterally and upward. Upward movement occurs when, after application, there is movement of water upward by capillary action owing to a high rate of water evaporation from the soil surface. Herbicides can move up with evaporating water.

2.1.6 Interactions With Soil Moisture

If soil is wet and air is dry, plants transpire more. Roots absorb water from soil to replace transpired water, and herbicides in soil, if not tightly adsorbed, can move to roots by mass flow. More herbicide will be absorbed and phytoactivity will increase, although the increase, while theoretically possible, will, in all likelihood, not be readily observable. Sometimes dry air and wind cause rapid foliar water loss. When not enough water is taken up by roots, plants wilt. Stomata then close, water movement in plants slows, and herbicide uptake decreases. Although, once again, while this is theoretically possible, herbicide uptake normally occurs soon after application, whereas weather changes occur over a longer time scale. Soil drying can increase soil adsorption and decrease root uptake.

Rainfall or irrigation is essential to move herbicides into the top soil layers where most weed seeds germinate. Some rain (perhaps an inch) may be essential to activate herbicides such as the triazines, which are taken up by roots. No rain or irrigation for 10—14 days after application can cause weed control failures. Heavy rains, on the other hand, may move herbicides below the zone of activity. Excess rainfall can leach herbicides through a zone of action unless they are adsorbed.

The effect of soil moisture cannot be generalized for all herbicides, crops, or application times. Pendimethalin controlled itchgrass in upland rice irrespective of soil moisture after preemergence application (Pathak et al., 1989). When bentazon or 2,4-D were applied postemergence, they controlled purple nutsedge well but only when soil moisture was above the demands of plant evapotranspiration (Pathak et al.). Fig. 15.2 integrates the several environmental fates of an herbicide and clearly shows that little of the amount applied remains available for weed control after 2—3 weeks.

2.2 Chemical Factors

2.2.1 Microbial or Enzymatic Degradation

The most enduring lesson of my brief study of soil microbiology is that things in soil do not just rot or disintegrate. They are decomposed by active chemical processes. It is the large, (very large) heterogeneous microorganism population of soil that mediates most pesticide decomposition. If one scoops up a handful of forest soil (any forest), that handful will contain 10 billion bacteria, a million yeast cells, perhaps 200,000 mold fungi, 10,000 protozoans, and assorted other creatures known as cryptozoa (Bryson, 2005, p. 460). A single handful of well-managed agricultural soil contains hundreds of different species (not just hundreds of creatures) and billions of microorganisms all together in a complex ecological web. The handful will contain algae, fungi, actinomycetes, and bacteria. For soil degradation of herbicides, bacteria and fungi are most important.

FIGURE 15.2 A general description of the fate of any organic chemical in soil. *Source unknown.*

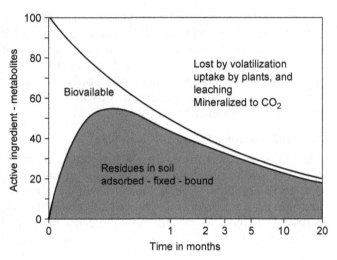

Herbicides in the soil solution can be adsorbed by soil colloids or be degraded by microorganisms. Many herbicides provide a carbon source from which microorganisms derive energy. Some herbicides (perhaps most) are degraded as an incidental process as microorganisms degrade other soil organic material. Herbicide degradation is enhanced by warm, moist, aerobic conditions that favor microbial growth. Under similar temperature and moisture conditions, herbicide degradation occurs more rapidly in soils that are rich in organic material and have high microbial activity. In general, with high soil adsorptive capacity, herbicides persist longer and are less available for microbial activity/degradation. Soil adsorption and microbial action as influenced by environmental conditions determine rate of degradation.

Microbial degradation proceeds by many pathways, none of which are unique to herbicide degradation. These include:

Dehalogenation—loss of a halogen atom.
Dealkylation—loss of a methyl or methylene group.
Decarboxylation—loss of carboxylic acid group.
Oxidation—structural change by addition of oxygen.
Hydrolysis—attack by water.
Hydroxylation—addition of an OH group.
Ether cleavage—breaking the R-O-R linkage.
Conjugation—usually with sugars or amino acids, sometimes with proteins.
Ring cleavage—Breaking ring integrity.

In a few cases, decomposition leads to activity, but in most cases it results in loss of phytotoxic activity.

2.2.2 *Chemical or Nonenzymatic Degradation*

Several herbicides are degraded nonenzymatically by chemical reactions not mediated by soil microorganisms. Some triazines (e.g., atrazine) are degraded by hydroxylation and removal of chlorine at the 6-position that converts them to the nonphytotoxic hydroxy derivative through a purely chemical, nonenzymatic process. In many cases, herbicides are decomposed by nonenzymatic *and* enzymatic processes that work in concert at different points on a single molecule.

The sulfonylurea herbicides are degraded by simple hydrolysis if soil pH is acidic. As pH approaches neutrality, or under basic conditions, enzymatic degradation by microorganisms dominates. Degradation in acidic soils is more rapid because of the high rate of acid hydrolysis. Imazaquin (an imidazolinone) persists longer and is more active at low pH (5.1 or lower) (Marsh and Lloyd, 1996). Studies have been done to determine the influence of soil pH on herbicide degradation and activity. For most herbicides, rate of degradation is slower as soil pH rises. Methods related to study of herbicide degradation have been described by Mueller and Senseman (2015). Their paper includes citation of several review articles and reports that approximately 170 studies are published annually.

2.2.3 Photodegradation

Photodegradation, the effect of radiation on internal chemical bonds, is a form of chemical degradation. It is well established that many herbicides, particularly heterocyclic molecules (with carbon and nitrogen in a ring), and nitrogen-containing compounds are susceptible to photodecomposition. Photochemical reactions have been reported for many herbicides, including phenoxy acids, dinitroanilines, propanil, benzoic acids, and others. Absorption of electromagnetic radiation at wavelengths between 290 and 450 nm affects the excitation states of electrons and leads to bond rupture, which can energize several common reactions including oxidation, reduction, hydrolysis, substitution, and isomerization. While there is no question that photodecomposition occurs, its importance as a determinant of activity or selectivity under field conditions is minor and may not even occur. Photooxidations are important environmental reactions because of the abundance of oxygen in air, soil, and water. Reactions can occur in a matter of hours and can affect any herbicide during its time in air or on an exposed surface. Photooxidations are important especially for herbicides that remain in the atmosphere or move back into the air after application. Once a compound is incorporated in soil, the importance of photodecomposition is negligible.

3. SOIL PERSISTENCE OF HERBICIDES

Table 15.7 shows the rate range, crop of use, and expected soil life for several herbicides. Soil persistence is agriculturally important because residual herbicides control weeds over time but may also injure crops. Soil residues can contaminate crops and water and affect nontarget species. Residues may kill or cause temporary or permanent effects on soil microorganisms. It would have enormous agricultural and environmental consequences if an herbicide was released only to find that it caused serious depression of activity or even death of nitrifying bacteria in soil. This has not happened because questions about possible effects on soil microorganisms are asked repetitively by manufacturers. It is generally, but not universally, accepted that registered herbicides do not cause permanent damage to the soil microflora when they are used according to label directions.

TABLE 15.7 Soil Persistence of Phytotoxic Activity and Use of Some Herbicides[a]

Herbicide Name		Major crop(s)	lb, oz, or g ai/A	Soil Persistence			
Common	Trade			1 week	>1 month	2 month	<3 month
Alachlor	Lasso	Soybean, corn, bean	2–3			X	
Amitrole	Amitrol	Noncropland	1.8–8		X		
Atrazine	Aatrex	Corn	1–3				X
Benefin	Balan	Lettuce, peanut, tobacco, turf	1–1.5				X
Bromacil	Hyvar	Noncropland, citrus	1.5–12				X
Chlorimuron	Classic	Soybean, noncrop	0.13–1.3 oz			X	
Chlorsulfuron	Glean	Wheat	1/6–1/2 oz		X		
Clomazone	Command	Soybean, pepper, pea, pumpkin	0.5–1.5		X		
Cyanazine	Bladex	Corn	0.8–4.8		X		
DCPA	Dacthal	Turf	10.5–15				X
Dithiopyr	Dimension	Rice, turf	0.05–0.5		X		
Dicamba	Banvel	Small grains	0.63–0.125				X
Dichlobenil	Casoron	Aquatics, ornamentals	2.5–8			X	
Diclofop	Hoelon	Wheat, barley, pea, lentil	0.5–1	X		X	
Diclosulam	Strongarm	Peanut, soybean	10 g		X	X	
Diuron	Karmex	Cotton, alfalfa, orchards	0.5–3				X
EPTC	Eptam	Potato, beans, alfalfa	2–8.5		X		
Ethalfluralin	Sonalan	Soybean, sunflower, peanut	0.5–1.2			X	
Flumetsulam	Broadstrike	Soybean, corn	52–78 g			X	
Fluridone	Sonar	Aquatic	0.06–0.09 mg		X		
Fosamine	Krenite	noncrop- brush	8–48	X			
Glyphosate	Round-up	Contact-nonselective	0.2–2	X			
Hexazinone	Velpar	Alfalfa, pineapple, sugarcane	0.5–1.7				X

Common name	Trade name	Crops	Rate				
Imazapyr	Arsenal	noncropland	0.5–1.5	X			
Imazaquin	Scepter	Soybean	0.09–0.125		X		
Imazethapyr	Pursuit	Alfalfa, soybean, peanut	0.05–0.09	X			
MCPA	Several	Wheat, rice	0.25–1.5			X	
Metribuzin	Sencor	Soybean, potato, alfalfa	0.25–1		X		
Metsulfuron	Ally	Wheat, barley	0.06 oz			X	
Napropamide	Devrinol	Artichoke, asparagus, mint	4+	X			
Nicosulfuron	Accent	Corn	35–70 g			X	
Oxyflurofen	Goal	Conifers	0.5–2			X	
Paraquat	Paraquat	Desiccant, mintillage	0.5–1				X
Pelargonic acid	Scythe	Green vegetation	3%–7% v/v solution				No
Pendimethalin	Prowl	Corn, sorghum, soybean, cotton, potato, tobacco, peanut	0.5–2	X			
Picloram	Tordon	Brush, rangeland	0.25–1.5	X			
Primisulfuron	Beacon	Corn	0.02–0.05			X	
Prometryn	Caparol	Cotton	0.5–2.5		X		
Propachlor	Ramrod	Corn	2.5–6				X
Pyrithiobac	Staple	Cotton	35–105 g		X		
Pyrazon	Pyramin	Sugarbeet	3–7.5		X		
Quinclorac	Facet	Rice	0.25–0.5	I yr			
Sethoxydim	Poast	Soybean, peanut, alfalfa	0.1–0.5				X
Simazine	Princep	Corn, orchards	2–4	X			
Sulfometuron	Oust	Conifers, hardwoods, noncrop	53–420 g			X	
2,4-D	Several	Corn, turf, small grain	0.25–2				X
Terbacil	Sinbar	Peppermint	0.8–1.6	X			
Triallate	Avadex	Small grains	1–1.5			X	
Trifluralin	Treflan	Cotton, soybean, alfalfa	0.75–2	X			

[a] Persistence of any herbicide varies with rate, climatic, and soil conditions. For additional information, consult Shaner (2014).

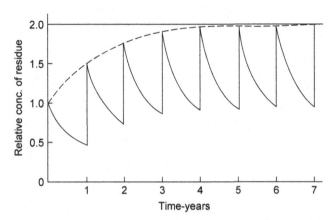

FIGURE 15.3   Residue pattern for a single, annual application and a half-life of 1 year (Hamaker, 1976).

Weed scientists, agriculturalists, and manufacturers must also know whether herbicides will accumulate in soil at a rate faster than their rate of dissipation. Under normal agricultural use patterns the answer is "no" (Fig. 15.3, Hamaker, 1976). The assumptions used to make this conclusion are that 1 lb of herbicide is applied annually at about the same time and the time it takes for half of it to degrade (Its half-life = $t_{1/2}$) is 1 year. After 1 year, 0.5 lb would remain and another pound would be applied. Continuing this sequence produces the sawtooth pattern in Fig. 15.3. Residues will never exceed twice the annual rate of application, and, therefore, except in very unusual circumstances (very cold, very dry, half-life greater than 1 year) herbicides do not accumulate in soil with repeated annual use. All herbicides in use today have soil half-lives much shorter than 1 year, and with some exceptions (see Chapter 19 − Herbicide Resistance), most are not used repetitively on the same field over several years. Therefore, the example is an extreme one, but illustrates the point well.

In most cases initial rate of degradation is independent of herbicide concentration in soil. If high rates degraded more slowly or if an herbicide was not degraded, residues would accumulate. Residues could accumulate if there were several applications in one growing season. While this has been unusual, it is now common for an herbicide to be applied more than once to a crop when it has been genetically modified to be resistant to the herbicide (see Chapter 19).

When an herbicide is applied to soil, adsorption determines availability to plants and leachability. Absorption by and translocation in plants lead to activity. Lability with respect to degradation in soil and susceptibility to physical processes in soil that do not degrade herbicides, but move them to different sites, must be understood to assure proper use and environmental care. Herbicide persistence in soil can be a problem, but it is essential for performance of some herbicides and for control of some weeds. Phytotoxicity that disappears too rapidly is not always good because weeds do not all emerge at once. Continued research will direct future use and take advantage of persistence when necessary and manage it where possible.

THINGS TO THINK ABOUT

1. How does the physical structure of soil affect herbicides?
2. How does a soil's chemical properties affect herbicides?
3. How does adsorption determine herbicide activity?
4. Describe possible interactions between adsorption and leaching.
5. What is the interaction between adsorption and volatility?
6. How do soil microorganisms affect herbicides in soil?
7. What are the nonenzymatic reactions that affect herbicides in soil?
8. What role does photodegradation of herbicides play?
9. What factors determine how long an herbicide persists in soil?
10. Do herbicides accumulate in the environment? Why? Why not?

Literature Cited

Note: Some of these references are cited in the text. Others are provided as a source of additional information.

Anon., April 4, 2015. The paradox of soil. Economist 21–23.

Bryson, B., 2005. A Short History of Nearly Everything. Broadway Books, New York, NY, 624 pp.

Cordova, M., January 17, 2017. Mocrobiome experts to speak at world economic forum. Cornell Chron.

Dubach, P., 1971. Dynamics of Herbicides in the Soil. CIBA-GEIGY, Ltd., Basle, Switzerland, 20 pp.

Hamaker, J.W., 1976. Mathematical prediction of cumulative levels of pesticides in soil. In: Gould, R.F. (Ed.), Organic Pesticides in the Environment, Adv. in Chem. Series 60. Amer. Chem. Soc., Washington, D.C., pp. 122–131

Helling, C.S., 1971. Pesticide mobility in soils. II. Application of soil thin-layer chromatography. Soil Sci. Soc. Am. Proc. 35, 737–743.

Leopold, A., 1921. Biographical Materials: Family Correspondence, 1918–1948. Letter January 18. http://digital.library.wisc.edu/1711.dl/AldoLeopold.ALBMCorr1819 or http://digital.library.wisc.edu/1711.dl/AldoLeopold.

Logan, W.B., 1995. Dirt – The Ecstatic Skin of the Earth. Riverhead Books, NY, 202 pp.

Marsh, B.H., Lloyd, R.W., 1996. Soil pH effect on imazaquin persistence in soil. Weed Technol. 10, 337–340.

Montgomery, D.R., 2007. Dirt - the Erosion of Civilizations. University of California Press, Berkeley, CA, 285 pp.

Mueller, T.C., 2015. Methods to measure volatility. Weed Sci. – Spec. Issue 63, 116–120.

Mueller, T.C., Senseman, S.A., 2015. Methods related to herbicide dissipation or degradation under field or laboratory conditions. Weed Sci. – Spec. Issue 133–139.

Pathak, A.K., Sankaran, S., De Datta, S.K., 1989. Effect of herbicide and moisture level on *Rottboellia cochinchinensis* and *Cyperus rotundus* in upland rice. Trop. Pest Man. 35, 311–315.

Shaner, D.L. (Ed.), 2014. Herbicide Handbook, tenth ed. Weed Sci. Soc. Am. Lawrence, KS. 513 pp.

Further Reading

Anon., 1972. Degradation of synthetic organic molecules in the biosphere: natural, pesticidal and various other man-made compounds. In: Proc. of a Conference. Nat. Acad. of Sci., Washington, DC, 350 pp.

Cheng, H.H. (Ed.), 1990. Pesticides in the Soil Environment: Processes, Impacts, and Modeling. Soil Sci. Soc. America, Series 2, Madison, WI, 520pp.

Garner, W.Y., Honeycutt, R.C., Higg, H.N., 1986. Evaluation of Pesticides in Groundwater. Amer. Chem. Soc. Symp. Series 315, Washington, DC, 573 pp.

Goring, C.A.I., Hamaker, J.W., 1972. In: Organic Chemicals in the Soil Environment, vols. I & II. M. Dekker, Inc., NY.

Gould, R.F., 1966. Organic pesticides in the environment. In: Advances in Chemistry Series. No. 60. Amer. Chem. Soc, Washington, DC, 309 pp.

Grover, R., 1988. Environmental Chemistry of Herbicides, vol. I. CRC Press, Boca Raton, FL, 207 pp.

Guenzi, W.D. (Ed.), 1974. Pesticides in Soil and Water. Soil Sci. Soc. America, Madison, WI, 562 pp.

Hance, R.J. (Ed.), 1980. Interactions between Herbicides and the Soil. Academic Press, NY, 349 pp.

Kearney, P.C., Kaufman, D.D. (Eds.), 1976, Herbicides: Chemistry, Degradation, and Mode of Action, second ed., vols. I–III. M. Dekker, Inc., NY.

Linn, D.M. (Ed.), 1993. Sorption and Degradation of Pesticides and Organic Chemicals in Soil. Soil Sci. Soc. America Spec. Pub. No. 32, Madison, WI, 260 pp.

Meine, C., Knight, R.L. (Eds.), 1999. The Essential Aldo Leopold: Quotations and Commentaries. University of Wisconsin Press, Madison, WI, 362 pp.

Saltzman, S., Yaron, B. (Eds.), 1986. Pesticides in Soil. Van Nostrand Reinhold Co., NY, 379 pp.

Sawhney, B.L., Brown, K. (Eds.), 1989. Reactions and Movement of Organic Chemicals in Soil. Soil Sci. Soc. America Spec. Pub. No. 22, Madison, WI, 474 pp.

16

Properties and Uses of Herbicides

FUNDAMENTAL CONCEPTS

- There are many ways to classify herbicides, but because of their chemical complexity, no single way integrates the subject.
- There are several sites of herbicide action in plants.
- The specific molecular site of action is known for most herbicides.

- Herbicide site of action and chemical structure can be integrated in a classification system.
- There is great diversity of chemical structures, with 27 major sites of herbicide action.

LEARNING OBJECTIVES

- To understand the underlying rationale for a classification that integrates herbicide site of action and chemical structure.
- To be familiar with the major groups of herbicides and major sites of action.

- To be able to use and expand the classification system with new herbicides.
- To understand the complexity of herbicide chemistry.

1. INTRODUCTION

The well-developed classification system used herein is based on the site of herbicide action in plants. The September 11, 2016 revision is available on the Weed Science Society of America (WSSA) website (http://wssa.net/wp-content/uploads/WSSA-Herbicide-MOA-20160911.pdf) and on pages 11−15 of the 10th edition of the Herbicide Handbook of the WSSA (Shaner, 2014). Each is a summary of herbicide sites of action according to the WSSA and the Herbicide Resistance Action Committee (HRAC) classification. The classification necessarily includes comments on chemical structure. The Herbicide Handbook includes time of use, observed effect, site of absorption, and selectivity. These details are included herein when essential to understanding, but they are not emphasized. This chapter and the aforementioned references are not and should not be used as use recommendations.

An herbicide's site of action is the precise biochemical (e.g., inhibition of a specific enzyme) or biophysical (e.g., inhibition of electron flow, binding to a protein, or interference with cell division) lesion that creates the herbicide's initial phytotoxic effect (other effects, some obvious and some not, may follow). This book uses the term *site of action* rather than the frequently encountered *mechanism* or *mode of action* because the former is more common.

In the 1900s, it was common to speak of herbicides as members of a structurally related, chemical family. The family defined performance characteristics and site of action. This is still possible for some families (e.g., triazines and sulfonylureas), but herbicide chemistry is now so diverse that such generalizations are no longer useful. Many herbicides have a primary site of action and several secondary sites. It is generally true that all herbicides have multiple actions. The primary site of action is known for most herbicides and is emphasized herein.

Without detailed information on the relationship between chemical structure and activity, it is not possible to predict either the site of action from examination of an herbicide's chemical structure or make reliable recommendations for herbicide use and resistance management. Structure and activity are related, but the relationship is often not clear until the necessary research has been conducted and published (interested readers are referred to Baker and Percival, 1991; Corbett, 1974; Devine et al., 1993; Duke, 1985; Fedtke, 1982; Hassall, 1990; Kearney and Kaufman, 1976, 1988, and Moreland et al., 1982). Few herbicides developed specifically to target a new site of action (site-directed research) have achieved commercial success.

The site of action of most herbicides has been determined after activity and selectivity have been determined through mass screening.[1] Specific activity and selectivity are commonly determined after patenting and before marketing. Study of herbicide site of action has advanced rapidly and is known for most herbicides. However, around the turn of the century discovery of activity and selectivity of new chemical structures slowed markedly, primarily but not solely due to the success of Roundup Ready and similar genetically modified technology. Discovery and release of new herbicides has continued, but no new sites of action have been discovered for at least 20 years (Duke, 2012). The reasons include company consolidation, rational management decisions, availability of generic formulations, the dominance of genetically modified crops, and the possibility that the best target sites are known (Duke), although the latter is debatable.

Herbicides as the primary tools for weed control in the United States and most developed countries is shown by the fact that the number of herbicides included in the Herbicide Handbook of the WSSA has grown from 97 in the first edition (Hull, 1967) to 232 in the 10th edition (Shaner, 2014). The total pounds of herbicides used on 21 selected crops, which

---

[1]Mass screening — Use of selected plant species to ascertain if a candidate herbicide has activity (kills or affects plant growth) and selectivity (affects some, but not all, species).

account for more than 70% of pesticide use, tripled from 191 million pounds in 1960 to 632 million pounds in 1981.[2] This declined to 516 million in 2008 due to more efficient active ingredients, integrated pest management, and the widespread planting of herbicide resistant crops.

It is neither feasible nor wise in a book of this kind to describe all herbicides or all sites of action. The classification scheme described below is intended to be a framework into which other herbicides can be integrated. This chapter will not attempt to describe structure-activity relationships or quantitative structure-activity relationships. Both are active research areas that are beyond the scope and intent of this book. The chemical structure of most herbicide groups has not been included, not because it is unimportant but because structures are readily available in other easy-to-find sources (e.g., several editions of the Herbicide Handbook of the WSSA, Vencill, 2002, Senseman, 2007, Shaner, 2014). More importantly, they are not essential to understanding how herbicides do what they do and how they should be used. This chapter describes most major herbicide activity and structural groups, but it is not a complete description of all groups. Books that provide a complete description of action of many herbicides are listed in the literature cited at the end of the chapter. Some aspects of the development of herbicides after 1945 can be found in Zimdahl (2010).

The chapter's intent is to acquaint students with the primary plant functions affected, the diversity of herbicide sites of action, and an herbicide classification system. Readers should note that supporting evidence for claims in this chapter about activity, selectivity, and rate of use can be found in the aforementioned Herbicide Handbooks and other sources, which, to avoid repetitive citations, are not cited throughout the chapter.

For clarity and simplicity, the classification is divided into 10 functional groups (Sections 2–11, e.g., major effect on amino acid, photosynthesis, growth regulation, etc.) plus an unknown group (XII). The system is based on the classification of WSSA and HRAC mentioned before. Earlier classifications can be found in Devine et al. (1993), Retzinger and Mallory-Smith (1997), Mallory-Smith and Retzinger (2003), and the University of Nebraska (2015). This book's primary division is the plant function affected followed by the specific site of action (e.g., lipid synthesis, photosystem II – site A, auxins, etc.). The 1997 and 2003 classifications were, as their titles indicate, done "to help maintain the usefulness of herbicides as a tool of crop production, and to delay the selection of herbicide-resistant weeds." Anyone who attempts to classify a large number of things (plants, chemicals, etc.) must decide whether they will be "lumpers" or "splitters." Lumpers tend to put similar things together in large groups, whereas splitters divide into smaller, more numerous groups based on small but real differences. My tendency is to be a lumper.

The 10 groups and several subdivisions plus an unknown category follow. In creating these groups technical language has been minimized, but some terms are necessary. The more important terms are defined in the glossary. Clarity, not confusion, is the goal. There is no attempt to mention all herbicides in each group. The classification outline is in Table 16.1.

[2]http://www.ers.usda.gov/webdocs/publications/eib124/46736_eib124_summary.pdf.

TABLE 16.1   The Herbicide Classification System

1. Introduction
2. Inhibitors of Lipid Synthesis
   a. Inhibitors of acetyl CoA carboxylase = ACC inhibitors
      i. Arylphenoxy-propionates (Fops)
      ii. Cyclohexanediones — Dims
   b. Inhibitors of Lipid Synthesis — Not ACC Inhibition
      i. Thiocarbamothioates
      ii. Benzofuran
      iii. Phosphorodithioates
      iv. Isoxazoline
3. Inhibitors of Amino Acid Synthesis
   a. Inhibition of branched-chain amino acid synthesis, specifically inhibition of acetolactate synthase (ALS) and acetohydroxy acid synthase (AHAS)
      i. Sulfonylureas
      ii. Imidazolinones
      iii. Pyrimidinylthiobenzoate
      iv. Sulfonylamino-carbonyltriazolinone
      v. Triazolopyrimidines — sulfonanilides
   b. Inhibition of aromatic amino acid synthesis, specifically inhibition of 5-enolpyruvyl-shikimate-3-phosphate synthase (EPSPS)
4. Seedling Growth Inhibition
   a. Inhibition of Mitosis/Cell Division
      i. Acetamides
      ii. Chloroacetamides
      iii. Pyridines
      iv. Benzamides
   b. Inhibition of Microtubule Assembly
      i. Dinitroanilines
      ii. DCPA
      iii. Carbanilates
5. Growth Regulators
   a. Synthetic Auxins
      i. Benzoic or Arylaliphatic Acids
      ii. Phenoxyacetic acids
      iii. Pyridine carboxylic Acids = picolinic acids
      iv. Pyrimidine carboxylic acid
      v. Quinolinecarboxylic acid
   b. Auxin transport inhibition
6. Photosynthesis Inhibitors — Photosystem II
   a. Inhibitors of photosystem II — Site A
      i. Phenyl-carbamates
      ii. Pyridazinone
      iii. Triazine
      iv. Triazinone
      v. Uracil
   b. Inhibitors of photosynthesis at photosystem II site A with different binding behavior
   c. Inhibition of Photosynthesis at Photosystem II Site B
      i. Benzothiadiazole
      ii. Nitriles
      iii. Phenylpyridazine

TABLE 16.1   The Herbicide Classification System—cont'd

7. Photosynthesis Inhibitors — Photosystem I — Electron Diverters
8. Cell Membrane Disruptors
   a. Inhibition of protoporphyrinogen
       oxidase (Protox = PPO)
      i. Diphenylethers
      ii. Oxadiazole
      iii. Triazones
      iv. Other
   b. Cell wall synthesis inhibition
9. Inhibitors of carotenoid biosynthesis Inhibition = Inhibitors of Pigment Production
   a. Inhibitors of 4-hydroxyphenyl-pyruvatedioxygenase (4-HPPD)
   b. Inhibitors of phytoene desaturase (PDS)
   c. Inhibitors of 1-deoxy-D-xylose 5-phosphate synthetase (DOXP synthase)
10. Nitrogen Metabolism Inhibition
   a. Glutamine synthetase inhibition
11. Inhibitors of Respiration
   a. Uncouplers of Oxidative Phosphorylation
12. Unknown and Miscellaneous
13. Summary

2. INHIBITORS OF LIPID SYNTHESIS

Salt and acid herbicides used prior to World War II (WWII) were contact chemicals that destroyed plant structure by acting on membranes. Their exact mechanism of action has never been determined. Many, still available, contact herbicides act by modifying membrane structure (Ashton and Crafts, 1981) through effects on lipid biosynthesis or production of toxic radicals. Lipids include fatty acids, neutral fats, and steroids. Plant surfaces are covered with a complex mixture of lipids, often in crystalline form, which are generally referred to as plant waxes. They form the cuticle or noncellular outer skin of plants and are integral components of intracellular plant membranes. Five chemical families that include two primary sites of action are in the general category of inhibition of lipid synthesis:

1. Inhibition of acetyl-CoA carboxylase (ACC) by aryloxyphenoxy-propionates and cyclohexanediones [WSSA class [1(A)]] and
2. Inhibition of lipid synthesis, especially long-chain fatty acids, but not ACC inhibition.

2.1 Inhibitors of Acetyl-CoA Carboxylase = ACC Inhibitors

ACC is a biotin-dependent enzyme that catalyzes the irreversible carboxylation of acetyl-CoA to produce malonyl-CoA through its two catalytic activities, biotin carboxylase and carboxyltransferase. ACC is a multisubunit enzyme in the chloroplasts of most plants and algae. The most important function of ACC is to provide the malonyl-CoA substrate for the biosynthesis of fatty acids.

TABLE 16.2   A Summary of Information About Aryloxyphenoxypropionate Herbicides (The Fops)

Common Name	Trade Name[3]	Primary Uses[4]
Clodinafop	Discover	Rice
Cyhalofop	Clincher	Rice
Diclofop	Hoe-Grass	Wheat, barley, lentils flax, and sugar beet
Fenoxaprop	Baghera	Soybean, wheat, and turf
Fluazifop	Fusilade	Cotton, soybeans, stone fruits (e.g., cherry), coffee, and several horticultural crops
Haloxyfop	Galent	Soybean, sunflower, rape, potato, bean, flax, and peanut
Quizalofop	Assure	Soybean

TABLE 16.3   A Summary of Information About Cyclohexanedione Herbicides (The Dims)

Common Name	Trade Name[5]	Primary Uses[6]
Clethodim	Select	Cotton and soybeans
Sethoxydim	Poast	Soybean, peanut, alfalfa, sugar beet, sunflower, and cotton
Tralkoxydim	Achieve	Wheat and barley

Aryloxyphenoxypropionates and cyclohexanediones are commonly called "fops" and "dims," respectively. Each is used postemergence for selective control of annual and perennial grasses in some dicotyledonous crops and in some cereal crops. They are often referred to as graminicides (Gronwald, 1991b). Herbicides in these groups are readily absorbed and translocated to meristems where they are toxic to grasses and have similar selectivity. Observed symptoms in susceptible plants are also similar because both inhibit the enzyme ACC, which catalyzes the first step in fatty acid synthesis. Tables 16.2 and 16.3 list representatives of these two chemically different but mechanistically similar families.

2.1.1 Aryloxyphenoxy-propionate (Fops) (Table 16.2)

Aryloxyphenoxy-propionates (fops) are foliar graminicides that selectively remove annual grasses from grass crops such as wheat and barley and in many broadleaved crops. Diclofop is a phenoxy derivative as are all fops. Selectivity is due to differential rates of metabolism to inactive products in susceptible and tolerant species. Control of wild oats and other grasses is

[3]Each herbicide may have several trade names.

[4]All aryloxyphenoxypropionates control only annual and some perennial grasses.

[5]See footnote 3.

[6]Cyclohexanediones control only annual and some perennial grasses.

growth-stage dependent, with the best control obtained when grasses have two to four leaves. Clodinafop-propargyl, a broad-spectrum graminicide, controls many annual and perennial species in wheat. Cyhalofop-butyl controls grass weeds in upland (dry) and lowland (paddy/water seeded) rice.

Fluazifop-P is a postemergence grass herbicide that is selective in broadleaved crops. It and other members of this group control young (three to six leaves), actively growing grasses best. If fenaoxprop-P is applied alone to wheat it kills it. Selectivity is achieved when it is applied in combination with phenoxy acid herbicides or one of the sulfonylureas that is selective in wheat. All fops are selective in many important broadleaved crops.

2.1.2 Cyclohexanedione – Dims (Table 16.3)

Among the cyclohexanediones, sethoxydim and clethodim, applied postemergence, selectively control nearly all annual and perennial grasses in all broadleaved crops. There are seven dims and nine registered[7] fops included in the 10th edition of the Weed Science Society's Herbicide Handbook (Shaner, 2014). Among herbicides they are structurally unique because they are based on a hexane rather than a benzene ring. Rate of application varies with the grass species to be controlled; higher rates are needed for larger plants and perennials. Combination with cultivation often improves control of perennials.

2.2 Inhibitors of Lipid Synthesis — Not ACC Inhibition

2.2.1 Thiocarbamothioate – [WSSA Class 8(N)] (Table 16.4)

Thiocarbamates = carbamothiates inhibit growth of emerging seedlings when applied to soil prior to weed emergence. Each herbicide in this group also may secondarily inhibit

TABLE 16.4 A Summary of Information About Thiocarbamate Herbicides

Common Name	Trade Name[8]	Primary Uses
Butylate	Sutan	Preplant incorporated application controls several annual grasses, yellow and purple nutsedge, and a few broadleaved species in corn
Cycloate	Ro-Neet	Preplant incorporated application controls annual grasses and some annual broadleaved species in sugar beet, table beet, and spinach
EPTC	Eptam	Preplant incorporated application controls weeds in several crops including alfalfa, bean, flax, potato, sugar beet, sunflower, citrus, pea, walnut, almond, and tomato
	Eradicane	EPTC formulated with a safener for preplant incorporated weed control in corn.
Triallate	Avadex	Pre- or postplant incorporated control

[7]The process of gaining approval (registration) from the US Environmental Protection Agency (US/EPA) to sell an herbicide or other pesticide as governed by the amended Federal Insecticide, Fungicide, and Rodenticide act.

[8]See footnote 3.

photosynthesis and carotenoid biosynthesis. Rate of application and plant species determine the dominant mechanism of action. The primary site of action is inhibition of fatty acids, particularly long-chain acids, and thereby, production of lipids.

Cuticles, the varnish-like outer layers of leaves, usually have a waxy, lipid layer underneath. Cuticles protect plants against water loss, injury from wind, physical abrasion, frost, radiation, pathogens, and chemical entry. Loss of one or more of these functions due to the inability to synthesize cuticular lipids may lead to death. Interference with the integrity of internal plant membranes also leads to death. The primary mechanism of action of these herbicides is inhibition of fatty acid biosynthesis and lipid and fatty acid elongation, although present evidence is inconclusive. Secondarily, they may play a role in gibberellin biosynthesis (Wilkinson, 1983, 1986).

The primary visual symptoms of thiocarbamate injury are shoot inhibition (aberrant morphology), abnormal growth and emergence of leaves in grasses, often seen as the leaf's inability to emerge or unfurl, and formation of foliar loops as leaves fail to emerge properly. Many thiocarbamates are volatile and must be incorporated to prevent loss. Soil persistence is short and residue carryover problems have not occurred. In general, they are much more effective against annual grass weeds than against annual broadleaved weeds.

Triallate is selective preemergence in small grains. It is not effective postemergence, except in some cases with granular application. Incorporation throughout the top 4–5 in. (10–13 cm) of soil can injure the crop, whereas incorporation in the top 1–3 in. (2.5–7.5 cm) does not. This selectivity is because of the different growth habits of the mesocotyl of small grains and wild oats. In small grains, the mesocotyl, the primary area for absorption of triallate and other carbamothioates and the apical meristem remain near the seed during emergence. In wild oats, these regions are pushed toward the surface and more herbicide is absorbed.

EPTC was released in 1954. In the 1980s it was discovered that after use, its activity could completely disappear within a matter of days. This enhanced degradation in soil occurs after repeated use on the same site. Microorganisms adapt to EPTC, and other carbamothioates, and degrade them quickly. The problem can be avoided by identifying soils where enhanced degradation is likely, by rotating crops, using different herbicides, diverse weed management techniques, or, if the same crop must be grown, by rotating herbicides.

It is used in corn to control grassy weeds not controlled well by other herbicides. If used alone, it injures corn. A safener that permits higher rates and crop tolerance is required. The mechanism of action is enhanced metabolism. EPTC is not an active herbicide until it is converted in plants to EPTC-sulfoxide, which interferes with vital plant processes. Resistant plants detoxify sulfoxides by converting them to a glutathione derivative via conjugation. The safener increases levels of the necessary enzyme and of glutathione. It is also possible that there is a direct competition between the safener and EPTC for sites of action. The safener expanded the selectivity range of EPTC in corn and allowed application through irrigation systems and when impregnated on dry fertilizer. EPTC with the safener is marketed as Eradicane.

Other thiocarbamates illustrate the diversity of uses found in this group. EPTC can be used in sugar beets, but cycloate controls as many annual grass and some broadleaf weeds with less crop injury. Butylate is the only thiocarbamate that is selective in corn without a safener. It has a relatively short (about 2 weeks) soil persistence. Thiobencarb controls some grasses, rushes, sedges, and some broadleaf weeds in rice. Molinate is also used in rice with incorporation by flooding to control annual grass weeds.

2.2.2 Benzofuran

Ethofumesate is used pre- or postemergence in grass seed crops and in sod to control a range of annual broadleaf and grass weeds. It is also approved for use in sugar beets with mechanical or irrigation incorporation. Its specific mechanism appears to be lipid inhibition, but it is generally regarded as unknown.

2.2.3 Phosphorodithioate

Bensulide is an organophosphate or phosphorodithioate. Although its site of action is not well understood, it is usually grouped with inhibitors of lipid synthesis. It is known that it inhibits root elongation and cell division (mitosis). Bensulide illustrates why classification by chemical family, although common, is not sufficient to integrate all herbicides; site of action is required. Bensulide is not as volatile as most thiocarbamates. It is used preplanting with incorporation or preemergence in vegetable crops, e.g., squash and pumpkins (preplant incorporated only), broccoli, cabbage, lettuce, onions, and peppers. Irrigation soon after application is required for maximum activity.

2.2.4 Isoxazoline

Pyroxasulfone reduces the synthesis of very long-chain fatty acids and controls grasses and small-seeded broad leaves in wheat and soybeans.

3. INHIBITORS OF AMINO ACID SYNTHESIS

When a cell divides, the information necessary to form new cells is carried in genes by DNA and is subsequently expressed in structural and enzymatic proteins. Information in plant cells flows from nucleic acids (via mRNA) to proteins but not in the opposite direction. Any disruption of this information flow leads to growth inhibition. Protein synthesis is necessarily preceded by amino acid synthesis. Three sites for amino acid biosynthesis that include three different enzyme systems are important sites of herbicide action (Duke, 1990). They are:

1. Inhibition of branched-chain amino acid synthesis, specifically inhibition of acetolactate synthase (ALS) [WSSA class 2[(B)]] and acetohydroxy acid synthase (AHAS) [WSSA class 2[(B)]].
2. Inhibition of aromatic amino acid synthesis, specifically inhibition of 5-enolpyruvyl shikimate-3-phosphate synthase (EPSPS) (WSSA class 9[(g)]).

Plants synthesize all essential amino acids and, in theory, blocking biosynthesis of any one will lead to death. The three enzymes above are firmly established as primary sites of action of five herbicide chemical families.

3.1 Inhibition of Branched-Chain Amino Acid Synthesis, Specifically Inhibition of Acetolactate Synthase [WSSA Class 2$^{(B)}$] and Acetohydroxy Acid Synthase [WSSA Class 2$^{(B)}$]

3.1.1 Sulfonylurea (Table 16.5)

The core molecular structure of the sulfonylureas, which were introduced in the 1980s by the DuPont Co. (Beyer et al., 1982), combines the nucleus of photosynthetic-inhibiting ureas and triazines. But their primary action is inhibition of amino acid synthesis, not photosynthesis. Secondarily, they inhibit photosynthesis, respiration, and protein synthesis. Plant symptoms include chlorosis, necrosis, terminal bud death, and vein discoloration. The site of action catalyzes the first step in the biosynthesis of the three branched-chain aliphatic amino acids valine, leucine, and isoleucine. A secondary effect is cessation of plant growth (stunting) due to cessation of cell division, leading to slow plant death. Tolerance is related to a plant's ability to detoxify the herbicide.

TABLE 16.5   A Summary of Information About Some Sulfonylurea Herbicides

Common Name	Trade Name[9]	Primary Uses
Bensulfuron-methyl	Londax	Rice
Chlorimuron-ethyl	Classic	Soybean, peanut
Chlorosulfuron	Glean	Wheat, barley, oats
Halosulfuron-methyl	Permit	Corn, grain sorghum
Iodosulfuron	Autumn	Cereals, corn, rice
Metsulfuron-methyl	Ally	Wheat, barley
Nicosulfuron	Accent	Corn
Primisulfuron-methyl	Beacon	Corn
Prosulfuron	Peak	Corn, grain sorghum, wheat, barley, sugarcane
Rimsulfuron	Accent gold	Corn, potato
Sulfometuron-methyl	Oust	Noncrop, conifer plantings
Sulfosulfuron	Maverick	Wheat, triticale
Triasulfuron	Amber	Wheat, barley
Trifloxysulfuron	Enfield	Cotton, sugarcane
Triflusulfuron-methyl	Upbeet	Sugar beet

[9]Most have several trade names.

Table 16.5 shows the range of selectivity and primary use of 15 of the of 30 sulfonylurea herbicides available in the United States. Several others are available in other countries. A notable attribute of these herbicides is that they are active at rates in the range of 8–80 g (grams)/ha. This is as significant a reduction in the quantity of herbicide required per acre as that which occurred when 2,4-D was introduced and replaced the heavy metal inorganic salts. Some sulfonylureas persist in soil (e.g., bensulfuron-methyl, chlorsulfuron, sulfosulfuron, triasulfuron). Most do not persist beyond 30 days (e.g., halosulfuron-methyl, nicosulfuron, prosulfuron, triflusulfuron-methyl). Wheat is not affected by chlorsulfuron until soil concentrations approach 100 ppb. Lentil and sugar beet, on the other hand, are affected by soil concentrations of 0.1 ppb. This thousand-fold range in activity is unprecedented in herbicide chemistry. Great care is required to use these herbicides so that their activity and weed-control potential are exploited and untoward environmental problems are avoided. A significant management problem is herbicide resistance (see Chapter 19) by several weed species expressed in as little as 3 years, after annual use of one or more sulfonylureas in the same field.

3.1.2 *Imidazolinone*

The imidazolinones, also developed in the 1980s, are active at low rates. The site of action of the seven available herbicides is nearly identical to the sulfonylureas,[10] but at similar rates, their activity is lower (Shaner and O'Connor, 1991). Imazamethabenz-methyl is a selective, postemergence herbicide for control of some annual grasses and broadleaf weeds in wheat, barley, and sunflowers. Imazamox is used postemergence to control annual broadleaf and annual and some perennial grass weeds in alfalfa, edible legumes (e.g., dry beans), soybean, and crops tolerant of imidazolinones[11] (e.g., canola). Imazapic controls a wide range of annual broadleaf and annual and perennial grass weeds in peanuts. Imazapyr is not selective in crops. It does not leach vertically or laterally and is used for weed control on noncropland and in imidazolinone-tolerant corn. Imazaquin is not limited to postemergence application as most imidazolinones are. It is used to control annual grass and broadleaf weeds in soybean. Imazethapyr is used for pre- or postemergence control of annual grass and broadleaved weeds in soybean, edible legumes, alfalfa, and peanut. It has a relatively long soil persistence. Users need to be aware that some imidazolinones can persist beyond one crop season and thereby can affect rotational crops. Continued use of imidazolinones or other herbicides with the same site of action in the same field has led to weed resistance (see Chapter 19).

3.1.3 *Pyrimidinylthiobenzoate*

Chemically, pyrithiobac-sodium is a pyrimidinylthiobenzoic acid. It is the only herbicide in this chemical group available in the United States. When used pre- or postemergence it controls several annual broadleaved weeds in cotton. Although chemically distinct from other ALS inhibitors, it acts in the same way.

---

[10]Inhibition of biosynthesis of the branched chain amino acids isoleucine, leucine, and valine.

[11]Imidazolinone tolerance is present in several products all labeled Clearfield — corn, canola, lentils, rice, sunflower, and wheat. BASF Corp., which holds the patent, used a chemical mutagen or radioactivity to induce a mutation in seed. Further research led to discovery of herbicide-tolerant plants. Clearfield varieties are regarded as products of hybridization research.

3.1.4 Sulfonylamino-carbonyltriazolinone

The two herbicides in this chemical group, flucarbazone-sodium and propoxycarbazone-sodium, are both active against different annual grass weeds and some broadleaf species when applied postemergence to wheat.

3.1.5 Triazolopyrimidine – Sulfonanilide

Earlier editions of this book reported only one triazolopyrimidine, flumetsulam, that had been approved for use in the United States. It is still available for pre- or postemergence use in combination with at least one other herbicide in soybean and corn to control a range of broad-leaved weeds. It has little activity against grasses and persists in soil for 1−3 months. Four others are now available. Cloransulam-methyl is applied pre- or postemergence to control broadleaf weeds in soybean. Diclosulam is soil-applied to control broadleaf weeds and perennial nutsedge in peanuts. Florasulam is used in spring or winter cereal crops and turf. Soil life is short, so rotational use for each of these herbicides and injury to rotational crops is not a problem. Flumetsulam controls several common broadleaf species in corn and soybeans. Its soil half-life is 2 months, whereas the others in this group persist less than 1 month. Low-use rates minimize leaching in soil. The four herbicides possess a diversity of activity against weeds and selectivity in crops. They illustrate why site of action is a better system of classification than chemical group or crop of use.

3.2 Inhibition of Aromatic Amino Acid Synthesis, Specifically Inhibition of 5-Enolpyruvyl Shikimate-3-Phosphate Synthase (WSSA Class 9$^{(g)}$)

A Swiss chemist working for a pharmaceutical company, Dr. Henri Martin, discovered glyphosate in 1950 (Dill et al., 2010). Because no pharmaceutical applications were identified, the molecule was sold to a series of other companies and samples were tested for a number of possible end uses. John Franz of Monsanto Co. identified the herbicidal activity in 1970. The formulated product, Roundup, was first sold by Monsanto in 1974. It is now sold under several trade names by Monsanto and other companies. Its discovery and release was as revolutionary for weed science as the discovery of 2,4-D. The structure of the amino acid glycine is underlined in Fig. 16.1. Glyphosate, the N-phosphonomethyl derivative of glycine, is a nonselective, foliar herbicide with limited to no soil activity because of rapid, nearly complete adsorption. It controls perennial grasses and has an advantage over paraquat, which also has some activity on perennial grasses, because it translocates whereas paraquat does not. It is the only available herbicide that inhibits the enzyme EPSPS synthase, which is common in the synthetic pathways leading to the aromatic amino acids phenylalanine, tyrosine, and tryptophan. These are essential in plants as precursors for cell wall formation, defense against pathogens and insects, and production of hormones (Duke, 1990). EPSPS is not found

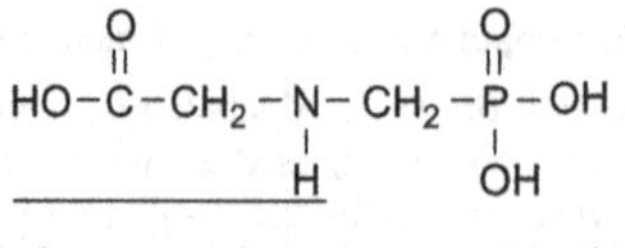

$$HO-\overset{\overset{\displaystyle O}{\|}}{C}-CH_2-\underset{\underset{\displaystyle H}{|}}{N}-CH_2-\overset{\overset{\displaystyle O}{\|}}{\underset{\underset{\displaystyle OH}{|}}{P}}-OH$$

FIGURE 16.1   Structure of glyphosate with glycine underlined.

in animals. Glyphosate has very low mammalian toxicity. Secondary effects include respiration, photosynthesis, and protein synthesis. It is active only postemergence because it is completely and rapidly adsorbed on soil colloids. Its nonselectivity means that it will affect, if not kill, most, but not all, green plants. Low-application volume is more effective than high volume, and small plants are more readily controlled than large ones. Paraquat, a photosynthetic inhibitor (see Section 7), acts quickly (1–2 days) on most plants, whereas glyphosate's effects take several days to appear. One glyphosate formulation has been approved by the US Environmental Protection Agency (EPA) for control of aquatic weeds. Transgenic crops resistant to glyphosate have been created and marketed. In January 2016, 32 weed species were reported to be resistant to glyphosate in 25 countries and in 34 US states.[12] In 2006, glyphosate resistance occurred in only 15 US states.[13] Resistant species include Palmer amaranth, common and giant ragweed, hairy fleabane, goosegrass, Italian, rigid and perennial ryegrass, horseweed, kochia, Johnsongrass, and buckhorn plantain.

4. SEEDLING GROWTH INHIBITION

All herbicides that inhibit cell division are effective on seedlings. Many herbicides affect mitosis and directly or indirectly affect microtubles (Vaughn and Lehnen, 1990). In general, they act preemergence and are absorbed by roots and shoots from soil. Many herbicides inhibit cell division as a secondary mechanism of action. The precise mitotic site of action is disruption of microtubule assembly during mitosis and cell wall formation between daughter cells (Vaughn and Lehnen).

4.1 Inhibition of Mitosis/Cell Division [WSSA Class 15$^{(K3)}$]

4.1.1 Acetamide

Two herbicides fall in this group, napropamide and flufenacet. Napropamide's mechanism of action is not well understood. It appears to inhibit growth by blocking the progression of cells through mitosis. Its primary use is for weed control in several vegetable crops and citrus orchards. It is often classified as an amide or substituted amide. The specific mechanism of flufenacet is unknown, but it appears to affect cell division in root and shoot meristems. Selectivity is obtained in several crops at several times of application. It is presently registered as a premixture with metribuzin for use in corn and soybeans. Flufenacet is often classified as an oxyacetamide.

4.1.2 Chloroacetamide (Table 16.6)

Chloroacetamides inhibit shoot growth of emerging seedlings and produce abnormal seedlings that may not emerge from soil. There is contradictory evidence of their effect on de novo (anew/from the beginning) fatty acid biosynthesis and thus on membranes. The primary site of action is presently regarded as inhibition of mitosis.

[12]International survey of herbicide resistant weeds. Weedscience.org.

[13]http://www.weedscience.org/In.asp.

TABLE 16.6   A Summary of Information About Chloroacetamide Herbicides[14]

Common Name	Trade Name[15]	Primary Uses
Acetachlor	Harness	Controls most annual grasses, yellow nutsedge, and certain small-seeded broadleaved weeds in corn and soybean.
Alachlor	Lasso	Controls many annual grasses, yellow nutsedge, and some broadleaved weeds in soybean, corn, dry bean, peanut, and grain sorghum.
Dimethenamid	Frontier	Controls yellow nutsedge, many annual grasses, and some broadleaved weeds in corn and soybean.
Propachlor	Ramrod	Controls many annual grasses and some broadleaved weeds in corn and grain sorghum.

These herbicides are important because of their widespread use in several major crops. They are relatively water soluble, soil applied, readily degraded, and not hazardous to succeeding crops. They affect germinating seedlings but do not affect seed germination. Alachlor and dimethenamid can be applied prior to planting, preemergence, or early postemergence. Propachlor is applied preemergence in corn and sorghum. It is one of the first herbicides identified as an inhibitor of long-chain fatty acids.

The range of activity in the group is illustrated by comparing butachlor, which controls many annual grasses and some broadleaved weeds selectively in rice, with the other herbicides in Table 16.6. The basic chemical nucleus of the chloroacetamides is not identical, as illustrated by dimethenamid, which has a five-member thiophene (sulfur-containing) ring instead of the six-member benzene ring common among other chloroacetamides. Crop selectivity and the weed control spectrum are similar.

The usefulness of chloroacetamides has been expanded by the development of safeners. These chemicals, also called antidotes or protectants, were developed to broaden the range of crop selectivity for particular herbicides. Acetochlor is selective in soybean but requires a prepackaged safener when used in corn. None of the safeners has phytotoxic activity.

4.1.3 Pyridine

Most pyridine herbicides act as synthetic auxins. Dithiopyr and thiazopyr disrupt cell division by inhibiting mitosis during late prometaphase. Neither binds to tubulin as the dinitroanilines do but to microtubule associated proteins associated with stabilizing tubulin molecules. Both affect miocrotubule stability. Dithiopyr is used for weed control in direct-seeded and transplanted rice and in established turf. Thiazopyr acts similarly. It is available for use in several crops.

4.1.4 Benzamide

Pronamide (aka propyzamide), a benzamide, disrupts cell division during late prometaphase as the pyridines do.

[14]See footnote 3.

[15]See footnote 3.

4.2 Inhibition of Microtubule Assembly [WSSA Class 3[(K1)]]

4.2.1 Dinitroaniline (Table 16.7)

The dinitroanilines or toluidines at one time held 8%—10% of total US herbicide sales. They are based on para-toluidine. The herbicide trifluralin (Fig. 16.2A) is related chemically, but not functionally, to the explosive TNT or trinitrotoluene (Fig. 16.2B). The dinitroaniline herbicides bind to alpha tubulin, the protein from which microtubules, required for cell division and wall formation, are composed (Duke, 1990). The binding inhibits tubulin polymerization, which prevents alignment and separation of chromosomes during mitosis and thereby prevents cell division.

TABLE 16.7 A Summary of Information About Dinitroaniline or Toluidine Herbicides

Common Name	Trade Name[16]	Primary Uses
Benefin	Balan	Controls grasses and several annual broadleaved species in lettuce, alfalfa, tobacco, and established turfgrasses.
Ethalfluralin	Sonalan	Applied preplant and incorporated it controls most annual grasses and many annual broadleaved weeds in cotton, soybean, peanut, edible bean, pea, and sunflower.
Oryzalin	Surflan	Annual grass and broadleaved species in soybean, tree fruits and nuts, and some ornamentals.
Pendimethalin	Prowl	Weed control in corn, soybean, cotton, barley, rice, sunflower, potato, pea, onion.
Prodiamine	Barricade	Many annual grasses and some broadleaved species in established turf and ornamentals.
Trifluralin	Treeflan	Most grasses and many broadleaved weeds in diverse agronomic and horticultural crops.

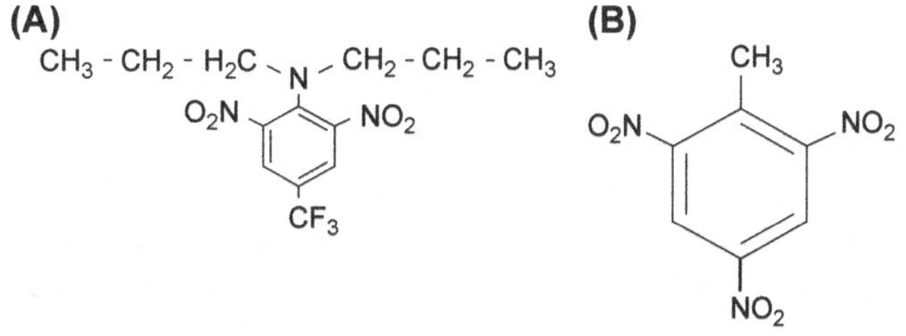

FIGURE 16.2 Structure of trinitrotoluene = TNT (A), and trifluralin (B).

[16]See footnote 3.

As a group, these herbicides control grasses and some small-seeded broadleaved weeds in cotton, soybean, dry bean, potato, canola, and many horticultural crops (Table 16.7). All are a yellow, liquid formulation, have low water solubility, little leaching, and dominantly pre-emergence activity. Soil persistence varies and soil residue problems have occurred. Dinitroanilines are usually applied prior to planting with incorporation. Some used postplanting are effective only when applied prior to seed germination. They may also be classified as root-growth inhibitors because they cause stunting and plants generally don't emerge from soil. Affected plants have short, thick lateral roots with a swollen root tip. Grass shoots are short, thick, and commonly red or purple. Broadleaved plants have swollen, cracked hypocotyls.

Volatility varies and is not a problem if recognized and controlled, usually by soil incorporation. Incorporation is essential for most dinitroanilines (not for oryzalin or pendimethalin) to prevent loss by volatilization and photodecomposition. They are poorly translocated in plants. Good incorporation places them in the weed seed-germination zone, which enhances their effectiveness.

4.2.2 DCPA

DCPA is defined in the Herbicide Handbook of the WSSA (Shaner, 2014) as a phthalic acid. It has been classified as a dibenzoic and a terephthalic acid. It is another illustration of the diversity of herbicide chemistry. Polymerization of ethylene glycol terephthalic acid yields Dacron, a textile. The primary site of action is mitotic inhibition "probably" by affecting phragmoplast microtubule arrays and cell wall formation. It is a turf herbicide, and 12—17 kg ai/ha[17] gives excellent preemergence control of crabgrass and other seedling annual grasses and some seedling broadleaved weeds. It does not interfere with seed germination, only with seedling growth (Vaughn and Lehnen, 1990). Therefore, to be effective, it must be applied preemergence—before crabgrass emerges in the spring. It is active only on germinating seedlings. It is used in horticultural crops and nurseries and in several vegetable crops.

4.2.3 Carbanilate

Only one herbicide, carbetamide, is in this group. It inhibits cell division [WSSA group [15K3]] and microtubule polymerization. Other carbanilate herbicides inhibit synthesis of long-chain fatty acids (IV-B). Some carbanilates are more properly classified as phenylcarbamates (e.g., phenmedipham) and act on photosystem II (VI-A).

5. GROWTH REGULATORS

There are at least six classes of hormones that affect plant growth: auxins, cytokinins, gibberellins, ethylene, abscisic acid, and polyamines. Plant hormones are chemicals that are produced in one location and act, in very low concentration, at another location. Auxins stimulate plant growth, particularly growth of excised coleoptile tissue. The name *auxin* generally refers to indoleacetic acid, but there are other active molecules. Gibberellins have varied

---

[17]In the United States it is common, although not universally adopted, to report herbicide and other pesticide use rates in kilograms per hectare (kg/ha), the metric standard, rather than pounds per acre (lbs/a). The metric standard will be used herein. To convert kg/ha to lbs/a, multiply kg/ha by 0.892.

effects on plant growth that differ between organs and between plants. They influence inter-node extension and thus can change dwarf to tall plants, affect cell division, induce fruit development, and can substitute for cold or light treatments required to induce sprouting or germination. There are no known herbicides whose primary mechanism of action is inter-ference with gibberellin synthesis or action, but, as noted in Section 3.2, some carbamo-thioates may interfere with gibberellin's action and mitosis as secondary actions.

Ethylene is a plant hormone involved in many aspects of growth. There are no herbicides whose primary mechanism of action is interference with ethylene action, although some non-herbicidal compounds have been developed to stimulate fruit ripening and stem growth of flowers. Auxin-like herbicides often increase ethylene production that is linked to develop-ment of injury symptoms. There are no herbicides based on cytokinin's or abscisic acid's structure, and there are no known herbicides that interfere with their action.

It is difficult to assign a specific physiological role to a compound within one of the five major hormone groups because they interact with each other and with other factors that in-fluence plant growth. In a similar way, we do not know precisely how all herbicides mimic auxin action, but we know enough about them to use them intelligently. In this text and in most classifications, herbicides that interfere with plant growth are phenoxyacetic acids, ben-zoic acids, pyridine carboxylic acids, and quinolene carboxylic acids. These growth regulator or hormone herbicides act at one or two specific auxin-binding proteins in the plasma mem-brane. They disrupt hormone balance and also affect protein sysome nthesis to yield a range of growth abnormalities.

5.1 Synthetic Auxin [WSSA Class 4$^{(O)}$]

5.1.1 Benzoic or Arylaliphatic Acid

Fig. 16.3 shows the structure of the benzoic acid aspirin and the structure of the herbicide dicamba, also a benzoic acid. These structures illustrate that herbicide chemistry is not strange or unique and that herbicides are related chemically, but not in terms of activity, to other common chemicals. Dicamba is a growth regulator, with a weed-control spectrum similar to 2,4-D, but it is more effective on many weeds at lower rates and more effective on perennial weeds, which 2,4-D does not control well. It has more foliar activity than 2,4-D and the other phenoxy herbicides, and is often used in combination with one or more of them for weed control in small grains and turf. It does not control mustards well, but is very effective on Polygonaceae species, which the phenoxy acids do not control well. This is part of the rationale for combinations. It is approved for use in cereals, corn, and sorghum and in pasture and on rangeland. It persists in soil longer than phenoxy acids.

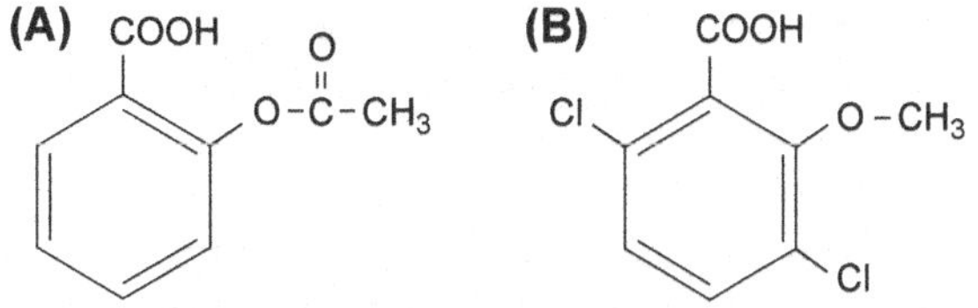

FIGURE 16.3   Structure of aspirin (A) and dicamba (B).

5.1.2 Phenoxyacetic Acid

The chloro substituted phenoxyacetic acids, 2,4-D (US) and MCPA (UK), were developed in 1940s in the United States and United Kingdom. When they were introduced widely after WWII, they revolutionized weed control because of their ability to kill selectively many annual and perennial broadleaved weeds without harming cereals and other grass crops (see Peterson et al., 2016). They were revolutionary because they were the first of many selective herbicides that made modern chemical weed control possible, and they reduced the amount required, in many cases, by nearly an order of magnitude. Grass tolerance is related to different morphology but, more importantly, to rapid, irreversible metabolism to nontoxic molecules. The susceptibility of dicotyledons is due to metabolism of the herbicides to reversible, conjugates. Farmers readily accepted the new technology because it was inexpensive and easy to apply. The inorganic salt herbicides that preceded them were not expensive but large amounts had to be applied, in large volumes of water, cost could be high, and poor weed control was common. The phenoxyacetic acids are absorbed by roots and shoots and readily translocated in plants. They have low mammalian toxicity, are nonstaining, nonflammable, and do not persist long in the environment.

Auxin-like herbicides are effective because high tissue concentrations are maintained. They affect proteins in the plasma membrane, interfere with ribonucleic acid (RNA) production, and change the properties and integrity of the plasma membrane. The rate of protein synthesis and RNA concentration increase as persistent auxin-like materials prevent normal and necessary fluctuation in auxin levels required for proper plant growth. Sugars and amino acids in reserve pools are mobilized by the action of auxin mimics. This is followed by, or occurs concurrently with, increased protein and RNA synthesis and degradation and depolymerization of cell walls.

There are chemical structural requirements that must be satisfied for an herbicide to interfere with auxin activity. These include a negative charge on a carboxyl group, which must be in a particular orientation (spatial configuration) with respect to the ring and a partial positive charge associated with the ring that is a variable distance from the negative charge. These spatial and charge requirements enable herbicide molecules to interact precisely with receptor proteins.

Growth-regulator herbicides are not metabolically stable in plants and are metabolized to a variety of different products. They are not resistant to metabolism, but plants cannot control their concentration as they can control concentration of natural plant hormones. This is an important reason for their activity. Physically, their action blocks the plants' vascular system because of excessive cell division and excessive growth with consequent crushing of the vascular transport system. External symptoms include epinastic (twisting and bending) responses, stem swelling and splitting, brittleness, short (often swollen) roots, adventitious root formation, and deformed leaves. All or a few of these symptoms may appear in particular plants, and activity is often due to two or more actions at the same time.

Translocated, auxin-like herbicides offer significant advantages, but they have limitations. Advantages include the need for only small quantities and foliar application that can kill roots deep in soil because of phloem translocation. Low doses keep residual problems to a minimum; however, limitations are just as real and important. Only roots attached to living shoots in the right growth stage are killed. A uniform stage of growth is often required and

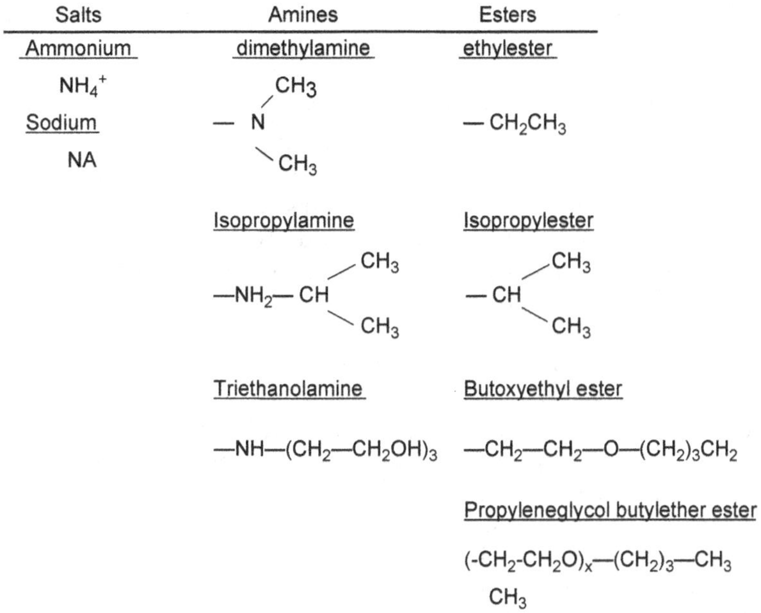

Salts	Amines	Esters
Ammonium	dimethylamine	ethylester

NH_4^+

Sodium

NA

dimethylamine:

$$-N \begin{array}{c} \diagup CH_3 \\ \diagdown CH_3 \end{array}$$

ethylester:

$$-CH_2CH_3$$

Isopropylamine

$$-NH_2-CH \begin{array}{c} \diagup CH_3 \\ \diagdown CH_3 \end{array}$$

Isopropylester

$$-CH \begin{array}{c} \diagup CH_3 \\ \diagdown CH_3 \end{array}$$

Triethanolamine

$$-NH-(CH_2-CH_2OH)_3$$

Butoxyethyl ester

$$-CH_2-CH_2-O-(CH_2)_3CH_2$$

Propyleneglycol butylether ester

$$(-CH_2-CH_2O)_x-(CH_2)_3-CH_3$$
$$CH_3$$

FIGURE 16.4 Salt, amine, and ester forms of phenoxy acids.

very difficult to achieve with a variable plant population and individuals that emerge over time and grow at different rates. Residual effects can be important if soil remains dry after application. Peterson et al. (2016) provided the most recent review of 2,4-D.

2,4-D (Fig. 16.4) is a white, crystalline solid, slightly soluble in water. Soon after it and its many relatives were developed, it became obvious that substitution for hydrogen in the carboxyl group affected activity. Therefore, a great deal of research was done on formulation to develop the ester and amine formulations (Fig. 16.4).

These forms are important because of their ability to penetrate plant cuticles and differences in volatility. In general, esters are more phytotoxic on an acid equivalent basis than the amine or salt forms. Technically, amines are also salts but have been distinguished because of their different chemical properties. Amines are, in general, soluble in water and used in aqueous concentrate formulations. Esters are oil soluble but may be applied as water emulsions with a suitable emulsifying agent. They are more toxic to plants because they are more readily absorbed by cuticle and cell membranes. The methyl, ethyl, and isopropyl esters are no longer commercially available because of high volatility. The butoxyethyl ester and propylene glycol butylether ester have low volatility and thereby reduce, but do not eliminate, volatile movement.

Symptoms often appear within hours of application and usually within a day. The most obvious symptom is an epinastic response resulting from differential growth of petioles and elongating stems. Leaf and stem thickening leading to increased brittleness often appear quickly. Color changes, cessation of growth, and sublethal responses occur. Plants often produce tumor-like proliferations and excessive adventitious roots. The effective dose varies with each weed species, its stage of growth at application, and the formulation applied. As plants mature, they can still be controlled by growth-regulator herbicides but higher rates are required.

MCPA, developed in England, differs from 2,4-D by the substitution of a methyl (CH_3) group for chlorine at the 2 position of the benzene ring. Uses are similar and performance is nearly identical. MCPA is more selective than 2,4-D in oats but less 2,4-D is required to control many annual weeds. MCPA persists 2–3 months in soil whereas 2,4-D persists about 1 month. Formulations are the same. MCPA is used more in the United Kingdom and in Europe than 2,4-D. MCPA is used in peas and flax in the United States because they are more susceptible to injury from 2,4-D.

The EPA removed 2,4,5-T from the list of approved herbicides in 1985 because of the human health effects of a dioxin contaminant found during its use for defoliation during the Vietnam war (see Chapter 21). International trade of 2,4,5-T was restricted by the Rotterdam convention of 1998. The convention promoted shared national responsibility in international trade of hazardous chemicals to protect human health and the environment.

It didn't take long after activity was found with an acetic acid derivative for researchers to examine the activity of the propionic (3 carbon), butyric (4 carbon), pentyl (5 carbon), or longer chain derivatives. Very early in the development of these compounds, it was found that a chain with an even number of carbons had herbicidal activity but a chain with an odd number did not. The even-numbered carbon chain is broken down through beta oxidation (cleavage of 2 carbon units) to produce 2,4-D, MCPA, or the appropriate analog that has a 2-carbon chain. A 3, 5, 7 carbon chain will also be broken down by beta-oxidation, but the final product is an alcohol that has no herbicidal activity. Thus, it is only the even-numbered carbon chains that are of interest as herbicides. However, as is true for many generalizations about herbicides, this one is wrong. Straight chains follow the rule, but iso- or branched chains do not. The alphaphenoxypropionic acids are widely used in Europe for weed control in small grains. Their structure has three carbons in a branched chain, which acts like a 2-carbon chain. The herbicides are dichloroprop (the analog of 2,4-D) or mecoprop (the analog of MCPA). Mecoprop was introduced in Europe as a complement to MCPA because of its ability to control catchweed bedstraw and common chickweed. Previously these weeds could only be controlled by sulfuric acid or the substituted phenols. Dichloroprop is effective against weeds in the Polygonaceae.

Another interesting part of the history of phenoxy acid herbicides are the phenoxybutyrics. MCPB and 2,4-DB [or 4-(2,4-DB)] were used widely. MCPB is selective postemergence to the crop for annual broadleaved weed control in peanut, soybean, and seedling forage legumes. Plants, through their enzyme composition, determine selectivity of 2,4-DB. Young alfalfa is less susceptible than older alfalfa because older plants have a more efficient and widespread beta-oxidation system and are able to break down 2,4-DB to 2,4-D, which is immediately toxic.

5.1.3 Pyridine Carboxylic Acid = Picolinic Acid

A group of synthetic auxin or growth-regulator herbicides is based on the pyridine ring (five carbons, one nitrogen). By adding a carboxyl group, picolinic acid is created. Moving the carboxyl group from the meta position in picolinic acid to the ortho position creates the basic molecule of nicotinic acid that is the basis of the essential B vitamin, niacin.

The development of the picolinic acid herbicides is an interesting tale. Scientists at Dow Chemical Co. were working with a pyridine-based structure to inhibit nitrification—the conversion of ammonia in soil to nitrate, the form available to plants. The general process is:

$$NH_4^+ \xrightarrow[\text{Nitrosomonas}]{} NO_2^- \xrightarrow[\text{Nitrobacter}]{} NO_3^-$$

Common form in fertilizer

Responsible microorganism

Readily available in plants and leachable

Nitrification occurs readily in many soils and is desirable. Nitrate ions are readily available to plants but leachable. Therefore, if nitrification can be slowed but not stopped, leaching will be reduced and plant availability maintained or increased. Scientists were working with the structure shown below and applying it in combination with ammonium fertilizers. The presence of the ammonia fertilizer made it possible for microorganisms to aminate the 4 position, and subsequent or simultaneous carboxylation of the trichloromethyl group was carried out by soil microorganisms to yield picloram.

The scientists saw plants dying where they weren't supposed to, and by studying their work they discovered an herbicide when they had been looking for an inhibitor of nitrification.

Picloram gives excellent control of woody plants and many annual and perennial broadleaved species. It is not effective on grasses, nor is it particularly effective on members of the Brassicaceae. It is chemically similar to, but not directly related to, other growth regulator herbicides. Picolinic acids produce epinastic and other effects typical of growth regulators and are active after absorption through foliage and roots. Picloram is effective on many perennial broadleaved plants, including field bindweed and Canada thistle. It is translocated in plants after pre- or postemergence application. Doses as low as 0.25 kg/ha are effective. Grasses, even seedling grasses, are relatively resistant. Picloram is very persistent and lasts for several months up to 1 year or longer and can affect succeeding crops. It is water soluble, not highly adsorbed, and therefore susceptible to leaching. These characteristics are undesirable, although its high activity is desirable for control of perennial weeds.

Clopyralid is less persistent and less leachable than picloram and effective for control of broadleaved species. It is selective in Christmas trees, sugar beets (a crop picloram kills) and corn, but is not effective on grasses or mustards. It is especially effective on Polygonaceae and Asteraceae in field crops and turf. A primary advantage is its high activity against Canada thistle.

Triclopyr is effective on woody plants and broadleaved weeds. It has been used for control of ash, oak, and other root-sprouting species. Most grasses are tolerant and while it is not used in many crops (it is used in rice), and in turf.

Aminopyralid (Milestone) is recommended for rangeland, permanent grass pastures, and hay crops. It can be used on noncrop areas (rights of way, campgrounds, trails, etc.). It is especially effective on several perennial weeds including Canada thistle, biennial thistles (e.g., musk thistle), spotted knapweed, yellow star-thistle, and Russian knapweed.

5.1.4 Pyrimidine Carboxylic Acid

The only herbicide is this chemical group is aminocyclopyrachlor. It is a selective, low-toxicity herbicide for pre- and postemergence control of broadleaved weeds, woody species, vines, and grasses on nonfood-use sites (e.g., rights of way, wildlife management areas, recreational areas, turf/lawns, golf courses, and sod farms). It was approved for use in 2010.

5.1.5 Quinolinecarboxylic Acid

Quinclorac is a unique chemical structure among herbicides. It controls some annual grasses, and a few annual and perennial broadleaved weeds when applied pre- or postemergence in rice. It also has good activity on some annual and perennial broadleaved weeds including field bindweed. According to the Herbicide Handbook (Shaner, 2014), its action on broadleaved weeds is as a growth regulator; however, in grasses it appears to inhibit cell wall (cellulose) biosynthesis. Exactly what it inhibits and how it does it is not clear, but its selectivity and activity are known and these enable its use.

5.2 Auxin Transport Inhibition

Naptalam, a benzamide, is a selective, preemergence herbicide for control of a wide range of annual broadleaved weeds and grasses in dicotyledonous crops, including soybean, peanut, cucumber, musk and watermelon, and established woody ornamentals. It has a unique anti-geotropic property. Because microbial breakdown is slow, it provides weed control for 3–8 weeks. Chemically it is a phthalmic or complex benzoic acid, but it can also be regarded as a substituted amide. It is not used widely but is interesting because of its ability to interfere with auxin transport.

Diflufenzopyr, a semicarbazone, is an auxin transport inhibitor that inhibits polar transport of natural auxin. It is formulated with the benzoic acid dicamba (see Section 5.1.1) and increases its activity in plants susceptible to dicamba. Many regard this compound as a growth regulator rather than as an herbicide.

6. PHOTOSYNTHESIS INHIBITORS – PHOTOSYSTEM II (TABLE 16.8)

The fundamental achievement of photosynthesis is conversion of light energy to chemical energy, a process on which all life depends. Light quanta falling on green leaves energize electrons in chlorophyll. The energy is converted to chemical energy by reducing (adding an electron) to an acceptor in the plant.

Two photosynthetic light reactions are coupled by the photosynthetic electron transport chain where photophosphorylation (production of adenosine-5′-triphosphate – ATP) occurs. ATP transports chemical energy within cells for metabolism. Photosystem I produces

TABLE 16.8 A Summary of Information About a Few Herbicides That Inhibit Photosynthesis

Common Name	Trade Name[18]	Primary Uses
Atrazine	Aatrex	Corn, conifers, macadamia nuts, sorghum sugarcane, some established turf species
Cyanazine	Bladex	Corn, cotton
Prometon	Prometon	Nonselective, noncrop, sterilant
Simazine	Princep	Almonds, apples, avocados, blueberries, established Christmas trees, grapefruit, grapes, lemons, nectarines, oranges, pears, (and several other fruits), pecans, shelterbelts, strawberries, and walnuts
Prometon	Pramitol	Noncropland
Ametryn	Evik	Banana, corn, pineapple, plantain sugarcane
Prometryn	Caparol	Cotton, celery, pigeon pea
TRIAZINONES		
Metribuzin	Sencor	Alfalfa, asparagus. potato, soybean, sugarcane, tomato
Hexazinone	Velpar	Alfalfa, pineapple, conifers
URACILS		
Bromacil	Hyvar	Citrus, pineapple, brush on noncropland alfalfa, mint, pecan, sugarcane
Terbacil	Sinbar	Alfalfa, cotton, sugarcane, pineapple, grapes, tree fruits, peppermint, spearmint, sugarcane, small fruits, deciduous tree fruits (e.g., pecans)
PHENYLUREAS		
Diuron	Karmex	Alfalfa, asparagus, birdsfoot trefoil, corn, cotton, peppermint, oats, red clover, sorghum, sugarcane, winter wheat
Fluometuron	Cotoran	Cotton
Linuron	Lorox	Asparagus, carrot, celery, corn potato, sorghum, soybean
Siduron	Tupersan	Newly seeded turf
Tebuthiuron	Spike	Pasture and rangeland, industrial sites

reduced nicotine adenine dinucleotide phosphate. Herbicides that act in photosystem I divert electrons away from photosystem I *and* generate toxic molecular species. Photosystem II, where most herbicides in this group act, begins with removal of electrons from water and production of oxygen (the Hill reaction). Most of the large and diverse group of chemicals that inhibit photosystem II block electron transport by binding to adjacent sites on the D-1 quinone protein of the photosynthetic system. It functions in the electron transport chain between the primary electron acceptor from chlorophyll A and plastoquinone.

[18]See footnote 3.

All inhibitors of photosynthesis cause gradual chlorosis in plants. The chemical groups include amide (1 herbicide), benzothiadiazole (1), nitrile (2), phenyl-carbamate (2), phenyl-pyridazine (1), pyridazinone (1), triazine (7), triazinone (3), triazolinone (1), uracil (2), and urea (7). The number of herbicides (28) that interfere with photosynthesis is among the largest of all mechanism of action groups. Table 16.8 summarizes selectivity information for some of these herbicides.

6.1 Inhibitors of Photosystem II — Site A [WSSA Class 5[(C1)]]

6.1.1 Phenyl-carbamate

The two, closely related, herbicides in this group, desmidipham and phenmedipham, may also be called bis-carbamates. They are used for postemergence weed control, especially of broadleaved weeds, in sugar beets.

6.1.2 Pyridazinone

Pyrazon, the only herbicide in the group, enters weeds by foliar and root absorption. It can be used pre- or postemergence to control a variety of broadleaved weeds in red table and sugar beets. It has low mammalian toxicity and persists from 4 to 8 weeks in soil, an advantage for use in beets.

6.1.3 Triazine

Thirteen herbicides approved for use in the United States are based on the symmetrical triazine structure. They all followed the discovery and release of simazine by Geigy Chemical Co. (now Syngenta) in 1956. Triazines were reviewed in depth by Balantine et al. (1998) and LeBaron et al. (2008).

Triazines inhibit photosynthesis following root absorption, although those with higher water solubility also have foliar activity. Translocation is apoplastic after root absorption. There are secondary mechanisms because some seedlings fail to emerge and become photosynthetic. All have relatively low mammalian toxicity. Residues of many persist in soil—an advantage where long-term, nonselective weed control is desired. Others (e.g., cyanazine and ametryn) have short soil lives. Table 16.8 includes several, but not all, triazines available in the United States.

Chlorotriazines are selective in corn. Simazine, developed for use in corn, was quickly replaced by atrazine. Both can be used as soil sterilants at doses over 20 kg/ha. The selective crop rates are between 1 and 4 kg/ha. Atrazine is more water soluble than simazine (33 vs. 6.2 ppm) and is therefore less dependent on, but not completely independent of, rainfall or irrigation for activity. Simazine's low water solubility and high soil adsorption permit use in orchards where deep-rooted trees do not come in contact with it because it leaches very slowly.

Because of its persistence and the problem of soil residues affecting a succeeding crop, some uses of atrazine have been eliminated. Atrazine has been combined in a single formulation with several herbicides, as have other chlorotriazines, to broaden the weed-control spectrum and shorten total soil life by reducing the amount of each component.

Methoxytriazines (OCH_3) are more water soluble than their chloro analogs and have more foliar activity. They are used exclusively for industrial and noncropland weed control. All are more toxic to corn than the chlorotriazines. Prometon is applied at 10–60 kg/ha. Without careful application and attention to surrounding vegetation, trees and ornamental plants adjacent to application sites can be killed.

As a result of higher vapor pressure and higher phytotoxic activity via foliar application, the methylthiotriazines (SCH_3) show high variability in selectivity. Ametryn has been used, often in combination with other herbicides, on noncropland because of its contact activity. Volatility is not a problem but persistence can be. Persistence of the methylthiotriazines is usually shorter than that of the chlorotriazines. Part of their selectivity is related to greater soil adsorption. Bananas, for example, are sensitive to atrazine because of its mobility, but not to ametryn, which is far less mobile and does not reach the banana root zone.

6.1.4 Triazinone

Triazinone or asymmetrical (*as*) triazine herbicides were first developed in 1971. Metribuzin has a shorter soil life than most symmetrical triazines. It is used for weed control in dormant alfalfa. Hexazinone can be used to control some annual and perennial broadleaved weeds and some grasses in alfalfa, pineapple, sugarcane, some coniferous species, and on noncropland.

6.1.5 Uracil

The basic uracil structure is identical to the core structure of uracil, thymine, cytosine, and guanine, the building blocks of nucleic acids and therefore of DNA. Two herbicides, bromacil and terbacil, are based on the uracil structure—an asymmetrical ring with two nitrogens. Both are absorbed primarily by roots. Bromacil is not volatile, but is moderately leachable. Its phytotoxic residues may persist up to 1 year in soil, and it can be used as a soil sterilant. Uses include: preemergence control of perennial weeds at 12–24 kg/ha, very high rates; preemergence in some perennial citrus orchards; and toxicity to a wide range of grass and broadleaved species. Terbacil is used selectively in several specialty crops (see Table 16.8). Foliar chlorosis is normal, and root and shoot inhibition and leaf necrosis are observed.

6.2 Inhibitors of Photosynthesis at Photosystem II Site A With Different Binding Behavior

6.2.1 Urea – [WSSA Class 7[(C2)]] (Table 16.8)

These herbicides, commonly called phenylureas, take their name from the organic compound urea that is the core of all urea herbicides. They are broad-spectrum soil-applied herbicides, not volatile, noncorrosive, and have low mammalian toxicity. They are absorbed by plant roots from soil and translocated to shoots in the apoplast. Leaching is variable. At doses of less than 2 kg/ha, they are selective and control seedling weeds in some crops. At higher doses most are soil sterilants. Soil persistence ranges from 2 to 6 months, but at recommended rates they usually do not affect succeeding crops. Ureas are most effective on young germinating seedlings. Because they are photosynthetic inhibitors, death occurs after emergence, as photosynthesis begins. Weeds germinating over time are controlled because ureas persist in soil. Photosynthetic inhibition is the primary mechanism of action, but they can cause chlorosis and necrosis, which are secondary effects of membrane peroxidation.

6.3 Inhibition of Photosynthesis at Photosystem II Site B. WSSA Class 6[(C3)]

6.3.1 Benzothiadiazole

Bentazon, the only herbicide in this group, inhibits photosynthesis at a different binding site than those in group A and B, above. It selectively controls broadleaved weeds in leguminous crops such as soybean, dry bean, pea, and peanut. It can be used postemergence in corn, sorghum, rice, and established spearmint and peppermint.

6.3.2 Nitrile

Hydroxybenzonitriles were introduced in the early 1960s in the United States and the United Kingdom. They are contact herbicides (no soil activity), selective in grasses, with limited translocation in shoots of some species. They are nonmobile inhibitors whose site of action is the D-1 quinone protein of the photosynthetic electron transport system. At doses of 0.21—0.56 kg ae/ha bromoxynil kills a wide range of annual weeds such as chickweed, mayweeds, and members of the Polygonaceae without injury to wheat, barley, oats, or triticale. Both bromoxynil and ioxynil can be used for weed management in onions. Bromoxynil can be used for weed control in field and popcorn, sorghum, peppermint, and spearmint. Ioxynil is effective in rice, sugarcane, flax and in pastures. Both are only effective on seedling weeds. Bromoxynil and ioxynil are often marketed in combination with a phenoxyacid herbicide to broaden the weed-control spectrum and reduce cost. Both are moderately toxic to mammals.

A third benzonitrile, dichlobenil, is structurally similar to ioxynil and bromoxynil but has little foliar activity, is volatile, primarily soil active, highly adsorbed with little leaching, and has a soil life of 2—6 months—almost the exact opposite of other substituted benzonitriles. It does not inhibit photosynthesis. Its mechanism of action is not well understood (Shaner, 2014). The primary actions are inhibition of actively dividing meristems in roots and shoots and seed germination. It also inhibits cellulose synthesis (see Section 6.2). Dichlobenil is effective on a wide range of annual and perennial weeds and is particularly effective preemergence on germinating seedlings. Primary uses are in ornamentals, turf, cranberries, and as an aquatic herbicide. Dichlobenil can be applied to soil with or without incorporation in late fall or with incorporation in spring in fruit and nut orchards, woody ornamentals, vineyards, and nursery stock containers.

6.3.3 Phenylpyridazine

Pyridate is chemically related to Pyrazon (see Section 6.1.2). It is a rapidly absorbed, postemergence, contact (foliar) herbicide that controls broadleaved weeds and some grasses in peanut and corn. It has a soil half-life of 6—7 days and does not leach. Susceptible plants turn yellow and necrotic, and a rapidly formed metabolite inhibits electron transport in photosystem II.

7. PHOTOSYNTHESIS INHIBITORS — PHOTOSYSTEM I — ELECTRON DIVERTERS. WSSA CLASS 22[(D)]

In some classifications, these herbicides are called photodynamic, a name that has not achieved wide acceptance. To call an herbicide photodynamic is to identify how it acts, but only in a general way. The bipyridiliums are photodynamic because they cause photooxidative stress by diversion of photosynthetic energy (electrons) from photosystem I. The specific action of these herbicides is reduction of molecular oxygen to a toxic superoxide radical. Rapid bleaching of photosynthetic tissue occurs. Affected plants initially appear water soaked but rapidly (several hours to a few days) become necrotic and die. Severe disruption of cell membranes is directly attributable to production of toxic oxygen species. Because of their effect on cell membranes, some systems classify them as inhibitors of cell membranes, which they are.

The bipyridiliums (diquat and paraquat) are almost completely dissociated in solution. Their action is due to the positive bipyridilium ion, reduced by drawing an electron from photosystem I (the primary site of action) to form a relatively stable free radical that continues to react and produces hydrogen peroxide, a superoxide radical O^{2-}, a hydroxyl radical (OH^-), and singlet oxygen $^1O_2$ each of which is potentially toxic to cell membranes, where the damage occurs.

They were discovered by Imperial Chemical Industries of England. Paraquat was released in 1958. They act only when absorbed by foliage and have almost no soil activity due to complete soil adsorption. Both herbicides act quickly; effects are normally seen within several hours and certainly within a few days. Translocation is poor and complete foliar coverage is essential to good weed control. They kill a wide range of annual plants and will desiccate shoots of perennials but are not translocated to roots, which is the main reason they do not provide permanent control of perennials. The addition of a nonionic surfactant or oil adjuvant improves control of many species because it aids foliar dispersal and cuticular penetration.

Paraquat is more active on grasses and diquat on broadleaves. Paraquat, a nonselective foliar herbicide, is used to control existing vegetation prior to no-till planting. It can be used for dormant-season weed control, for weed management in chemical fallow, and in more than 20 crops. Its use has diminished because of the availability of other contact, nonpersistent, less-toxic herbicides. Diquat is used to control cattails and submersed aquatic species in ponds, lakes, and drainage ditches. Both are toxic to humans from skin contact, inhalation, or ingestion. Both are nonselective and kill or affect almost any plant foliage they contact. Because of poor translocation, foliage not contacted directly is not affected. Paraquat can be used as preharvest desiccants to speed drying of some crops (e.g., potatoes).

8. CELL MEMBRANE DISRUPTORS

8.1 Inhibition of Protoporphyrinogen Oxidase (Protox = PPO) (Table 16.9)

Herbicides in this group are often called photodynamic. They inhibit the enzyme protoporphyrinogen oxidase, PPO or protox, a step in the porphyrin pathway that produces half of the chlorophyll molecule. They act independently of photosynthesis but require light for activity. In light, protox inhibition causes accumulation of the phytotoxic molecule protoporphyrin, which quickly damages lipids and proteins resulting in chlorophyll and carotenoid pigment loss and leaky membranes that lead to cell desiccation and disintegration.

TABLE 16.9   A Summary of Information About *p*-Nitrodiphenylethers

Common Name	Trade Name[19]	Primary Uses
Acifluorfen	Blazer	Controls several annual broadleaved weeds and some grasses when in peanut and soybean.
Bifenox	Modown	Used in combination with phenoxy acid or grass herbicides in rice. Not approved in United States.
Fomesafen	Reflex	Soybean.
Lactofen	Cobra	Cotton, soybean.
Oxyfluorfen	Goal	Cotton, corn, soybean, and several vegetable crops, and fruit and nut trees.

Protoporphyrinogen oxidase is responsible for the seventh step in biosynthesis of protoporphyrin IX. This porphyrin is the precursor to hemoglobin, the oxygen carrier in animals, and chlorophyll, the dye in plants. The enzyme catalyzes the dehydrogenation (removal of hydrogen atoms) of protoporphyrinogen IX to form protoporphyrin IX. Inhibition of this enzyme is a strategy used in certain herbicides.

8.1.1 Diphenylether

Ether or p-nitro substituted diphenylethers were introduced in the 1980s for postemergence broadleaved weed control in broadleaved crops. All require light for their action but are not dependent on photosynthesis; light is required to produce a substrate for their action (Duke et al., 1991). They are sometimes called photobleaching herbicides because a primary symptom is bleaching of foliage. All are active on broadleaved weeds and selective in broadleaved crops.

8.1.2 Oxadiazole

Oxadiazon has been available for some time. No others in this chemical group are available in the United States. It is of interest because of its ability to control several annual grasses and annual broadleaved species preemergence in Bermudagrass, perennial ryegrass, and fescue turf and in some ornamentals. It is strongly adsorbed by soil colloids, rarely leaches, and persists in soil.

8.1.3 Triazinone

A few triazinones (e.g., metribuzin and hexazinone, Section 6.1.4) inhibit photosynthesis. Two triazinones act quite differently. Sulfentrazone is a soil-applied preemergence herbicide used in soybean, sugarcane, sunflower, tobacco, and some turf species to control annual broadleaved weeds and some annual grasses. It is absorbed by roots and foliage.

The other member of this group, carfentrazone-ethyl, applied in the range of 4–39 g/ha, has proven to be very effective in corn, sorghum, rice, soybean, and small grains to control weeds' resistant to the imidazolinone and sulfonylurea herbicides.

[19]See footnote 3.

8.1.4 *Other*

There are four structural groups with one or two herbicides: *N*-phenylthalamides (flumiclorac and flumioxazin), phenylpyrazoles (pyraflufen), thiadiazoles (fluthiacet methyl), and trifluoromethyl uracil (saflufenacil). All are primary PPO inhibitors.

8.2 Cell Wall Synthesis Inhibition

Cell wall synthesis can be inhibited at two different sites. Action at site A disrupts production of uredine diphosphate (UDP) glucose from sucrose. UDP glucose is crucial to the biosynthesis of uronic acids, the backbone of pectic polysaccharides in the cell wall matrix, which ultimately interferes with cellulose synthesis. Site B occurs later in the same sequence when UDP glucose is prevented from being converted to cellulose (Corbett et al., 1984; Duke, 1990). Cellulose is unique to plants, but exploration of its synthesis has not resulted in other herbicides. Isoxaben, a benzamide, inhibits cellulose biosynthesis (Shaner, 2014). Indaziflam, an herbicide introduced in 2010, also inhibits cellulose biosynthesis (Sebastian et al., 2016).

9. INHIBITORS OF CAROTENOID BIOSYNTHESIS = INHIBITORS OF PIGMENT PRODUCTION

Carotenoids are essential to plant survival because they protect individual pigment-protein complexes, especially chlorophyll, and ultimately the chloroplast, against photooxidation. With high light intensity or under stressed conditions, chlorophyll molecules receive more light energy than they can transfer effectively to electron transport. The excess energy can be dissipated in several ways including production of singlet oxygen that is destructive of tissue integrity. Carotenoids protect against this by quenching excited chlorophyll molecules and singlet oxygen. Destruction of carotenoids or their biosynthesis leads to loss of the protective role (Young, 1991). A few herbicides from different chemical groups act, in slightly different ways, to disrupt carotenoid biosynthesis.

9.1 Inhibitors of 4-Hydroxyphenyl-pyruvatedioxygenase

Hydroxyphenyl-pyruvatedioxygenase (HPPD) inhibitors block 4-Hydroxyphenylpyruvate dioxygenase (4-HPPD), an enzyme that breaks down the amino acid tyrosine into components that are used to create other needed molecules. They were first marketed in 1980. They were originally used primarily in Japan in rice production, but since the late 1990s have been used in Europe and North America for weed management in corn, soybeans, and cereals. In the 21st century they have become more important as herbicide resistance increased. Genetically modified crops with resistance to HPPD inhibitors are being developed.

There are three chemical groups that interfere with 4-HPPD: triketones, isoxazoles, and pyrazolones. The triketones mesotrione and tembotrione are registered for postemergence use in corn. Isoxaflutole can be used preemergence for control of broadleaved weeds in corn and sugarcane. The pyrazolone topramezone is also registered for postemergence use in corn.

9.2 Inhibitors of Phytoene Desaturase

Norflurazon, a pyridazinone, and fluridone, a complex pyridone, block carotenoid biosynthesis by inhibiting the enzyme phytoene desaturase (PDS). This means that phytoene, a colorless carotenoid precursor, accumulates and that leads to the photodestruction of chlorophyll pigments (Bartels and Watson, 1978). Norflurazon can be used preemergence in soybeans, peanuts, cotton, tree crops (citrus, apples, apricots), nuts (almonds), vine crops, blackberries, blueberries, cherries, and hops.

Fluridone controls submerged and emerged aquatic weeds in lakes, reservoirs, and irrigation systems. It has little effect on algae and gives partial control of cattails.

9.3 Inhibition of 1-Deoxy-D-xyulose 5-Phosphate Synthetase

The chemical group isoxazolidinones has only one active herbicide: clomazone, which was released in 1986. The active (toxic) molecule is metabolized (created) in plants. It inhibits the action of the enzyme 1-deoxy-D-xyulose 5-phosphate synthetase (DOXP) synthase, a key component in plastid isoprenoid synthesis, which is in the biosynthesis of carotenoid pigments. Clomazone selectively controls many annual broadleaved and grass weeds in soybean, cotton, rice, sugarcane, tobacco, peppers, and pumpkin, It can be applied preemergence or preplanting with incorporation, but has limited postemergence activity. It rapidly turns plants white. When more than 75% of a plant's foliage is affected, death follows. There have been some important instances of drift from clomazone, made readily apparent because of its bleaching symptoms. Drift potential has been reduced by formulation.

9.4 Triazole

Amitrole, a unique five-member heterocyclic ring (a triazole), is structurally unlike any other herbicide. It is unique in that it inhibits accumulation of chlorophyll and carotenoids in the light, but exactly how this happens (the specific mechanism or site of action) is unknown. Its use in the United States was severely restricted by EPA action in 1996. In controls many annual and perennial grass and broadleaved weeds when applied postemergence on rights-of-way and noncrop land.

10. NITROGEN METABOLISM INHIBITION

10.1 Glutamine Synthetase Inhibition

Glutamine synthetase (GS) is essential for assimilation of organic nitrogen as ammonia (Duke, 1990). The enzyme plays an essential role in the metabolism of nitrogen by catalyzing condensation of glutamate in ammonia to form glutamine. The enzyme's lack leads to very high, toxic ammonia levels. Glufosinate (phosphinothricin) is the only available herbicide that inhibits GS. It is available for complete weed control in noncrop areas and as a directed spray in field- and container-grown nursery stock. It is rapidly degraded in soil (half-life 7 days). Even though it is not adsorbed tightly, it does not leach because it is degraded quickly.

Glufosinate is nearly nonselective. It has been made selective in corn because a gene coding for phosphinothricin acetyl transferase activity was isolated from the soil bacteria *Streptomyces hygroscopicus* and cloned into corn. The acetyl transferase enzyme converts glufosinate to its nonphytotoxic acetylated metabolite enabling crops to achieve resistance due to rapid glufosinate metabolism.

11. INHIBITORS OF RESPIRATION

Plants obtain energy by transforming the electromagnetic energy of the sun into stored chemical energy of carbohydrate molecules through photosynthesis. They must transform that energy to a form suitable for driving life processes. That transformation, called respiration, is analogous to the conversion of fossil fuel to electric power. Respiration is the removal of reducing power from carbohydrates, fats, or proteins and its transfer to oxygen with the concomitant trapping of released energy in ATP. Some arsenical herbicides are only of historical interest because they are no longer readily available for use. It is worthy of note that the mechanism of action of all herbicides in Section 11 is generally classified as unknown.

11.1 Uncouplers of Oxidative Phosphorylation

Uncoupling is like braking while continuing to press a car's accelerator. Energy is released as electrons pass down the electron transport chain to oxygen and is trapped by converting ADP to ATP (oxidative phosphorylation). If you uncouple—keep the accelerator down—the motor will race and overheat (Corbett, 1974).

11.1.1 Inorganic Arsenical

Arsenic has been known as a biological poison for many years. Arsenic-based insecticides were used in orchards in the late 1800s. Arsenic trioxide, an insoluble soil sterilant, was used at 400—800 kg/ha but is no longer registered for US use. Its residues remained for many years and weed control could be effective up to 5 years. Livestock were attracted to released aromatic compounds and thus could be poisoned by plants sprayed with arsenic trioxide.

The arsenicals (e.g., sodium arsenite, calcium arsenate, lead arsenate) were effective because they were translocated in plants. They cause nonspecific inhibition of sulfur-containing enzymes, precipitate proteins, and disrupt membranes. Sodium arsenite was, but is no longer, used as a preharvest desiccant in cotton. Inorganic arsenicals are poisonous to mammals and are generally regarded as nonselective, foliar-contact herbicides with soil sterilant activity because they persist in soil. Arsenic-based herbicides are no longer used except in a few combinations for soil sterilization. The precise site of action is not known. They cause rapid desiccation and disrupt cell membranes. Inorganic arsenicals are primarily of historical interest.

11.1.2 Organic Arsenical

The organic arsenicals interfere with general plant growth and may affect cell division. They are based on arsonic or cacodylic acid, both of which decompose carbonates. They have postemergence contact activity on plant foliage, are rapidly adsorbed by soil, and do not have soil activity. They are most effective at high temperatures, but rapidly lose selectivity above about 27°C. Organic arsenicals are more phytotoxic than inorganic arsenic herbicides. The most toxic form of arsenic to mammals is the AS^{+3} state, the form in inorganic arsenic compounds.[20] Organics have arsenic in the +5 state, and because it is not normally reduced to AS^{+3}, organic arsenicals are less toxic to mammals.

The principal organic arsenicals, first released in the United States in the 1950s, are monosodium methane arsenate and disodium methane arsenate (DSMA), both derivatives of arsonic acid. Cacodylic acid, an arsenic acid, is less selective. Organic arsenicals are water soluble, rapidly sorbed by soil, and do not leach, except in sandy soils, beyond 20 cm. They are much more toxic to annual than perennial grasses. DSMA has been used for selective weed control in turf and cotton. Arsenicals have been used in forest weed control. They do not persist in soil because they are rapidly and completely absorbed by soil colloids. They may affect mitosis, but the WSSA Herbicide Handbook (Shaner, 2014) lists the mechanism of action of all three as "not well understood." Mallory-Smith and Retzinger (2003) categorize them as having an unknown site of action. An excellent discussion of arsenic herbicides is in Woolson (1976).

11.1.3 Phenol

The first synthetic organic herbicide that achieved success in the field was 2-methyl-4,6-dinitrophenol, released in 1932. Several other substituted phenols followed but all are no longer approved for use in the United States and, like the arsenicals, are only of historical interest. They are intensely yellow-staining compounds, toxic to mammals, and poisonous to humans by ingestion, inhalation, or skin absorption. They were used for selective broadleaved weed control in cereals, and their activity increased directly with temperature.

2-(1-methylpropyl)-4,6-dinitrophenol (DNBP) was used in several salt forms as a selective broadleaved weed herbicide in pea, legumes, corn, and flax. It was used, with less success, in small grains and for preemergence weed control in bean and cotton. Its selectivity depended on selective retention and absorption by foliage. Good coverage was essential. The phenol derivatives have short soil persistence. They also, secondarily, inhibit other plant processes including photosynthesis and lipid, RNA, and protein synthesis. Pentachlorophenol was a widely used wood preservative. DNBP is only of historical interest as it is no longer available in the United States.

12. UNKNOWN AND MISCELLANEOUS

There are several organic herbicides with presently unknown sites/mechanisms of action. Other herbicides are described as having nonspecific action. A nonspecific or unknown site of action may mean that it is truly unknown, it has not been studied completely, or it is too new to know—it is being studied. The WSSA herbicide mechanism of action classification lists includes 30 available herbicides, seven with an unknown site of action and 23 not classified.

The site of action is unknown for most inorganic herbicides. Available studies are old and were done with far less sophisticated analytical techniques and less knowledge than more recent research. Most of these herbicides were used for many years, but their use has largely

---

[20]Arsenic in inorganic lead arsenate and calcium arsenate has a +5 valence.

disappeared as organic herbicides displaced them because they provided better weed control at lower rates and cost. A few inorganic herbicides may still be used in mixtures with organic herbicides for soil sterilization.

Nevertheless it is important to know that chemical weed control began with inorganic herbicides. Monoammonium sulfamate ($NH_2SO_3NH_4$) (AMS) was patented in 1942. It is a water-soluble contact herbicide used for brush and weed control in industrial and residential areas. It has low mammalian toxicity. Rates of 100–200 kg/ha applied in 400 L of water are required for effective brush control. A rate of 60 kg/ha in 400 L of water controls poison ivy. These rates illustrate the great change that occurred when the phenoxy acid herbicides were introduced and rates dropped to a few kilos per hectare (pounds/acre) or less.

Sodium tetraborate ($Na_2B_4O_7$) and sodium metaborate ($Na_2B_2O_4$) are nonselective, taken up by roots. Boron accumulates in reproductive structures after translocation from roots. They are used for long-term, nonselective weed control in industrial and power line areas often in combination with a triazine or urea herbicides.

Sodium chlorate ($NaClO_3$) is a nonselective soil sterilant used on noncrop land or in combination with a triazine, urea, or another organic herbicide for soil sterilization. It leaches, has foliar contact activity, and, in the past, was used widely along railroads. It is flammable when dry on foliage and many fires along railroad rights-of-way occurred when sparks from coal-fired engines landed on sprayed plants. Sodium chloride (table salt) is an example of an herbicide that desiccates and disrupts a plant's osmotic balance. It has been used for nonselective weed control for centuries. The Romans salted the fields of Carthage in 146 BCE to prevent crop growth.

Sulfuric, phosphoric, and hydrochloric acid all have burning, contact activity, but because of high toxicity to users, corrosion of equipment, and the availability of safer alternatives they are no longer used.

Among the metallic salts, copper sulfate is one of the few still used. Its toxicity is due to a nonspecific affinity for various groups in cells leading to nonspecific denaturation of protein and enzymes. It is used as an algicide.

Difenzoquat, a pyrazolium salt, is still available for selective control of wild oats in barley and wheat. Its primary site of action is unknown but includes inhibition of nucleic acid biosynthesis, photosynthesis, ATP production, potassium absorption, and phosphorus incorporation into phospholipids and DNA. It is a postemergence herbicide that has only contact foliar activity and no soil activity. It has fungicidal properties and controls powdery mildew (*Erysiphe graminis* f.sp. *Hordei*) in barley.

Pelargonic acid was introduced in 1995 as a contact, nonselective, broad-spectrum foliar herbicide. Because it is not selective, it is used only in noncrop locations and retreatment is required for plants that emerge after treatment from seed, roots, or rhizomes and other vegetative reproductive structures. It is a naturally occurring, nine-carbon fatty acid found in several plants and animals. Translocation doesn't occur so it is not effective for long-term control of biennial or perennial weeds. There is no soil residual activity. The bipyridiliums (see Section 7) are fast acting, but pelargonic acid is faster. Rate of kill is related to temperature, but even in cool conditions plants begin to exhibit damage within 15–60 min of application and die in 1–3 h. Foliage darkens and begins to look water-soaked. This is followed by rapid wilting. The site of action is unknown but it causes bleaching of chloroplasts and general ion leakage. The primary effect may be a sudden drop in intracellular pH that causes rapid membrane deterioration and leads to cell death.

13. SUMMARY

The precise site of action of most herbicides is known. Research is advancing rapidly and more precise classification will be possible with more knowledge. Any classification of a large number of chemically complex herbicides creates problems because of disagreement about:

- The best way to classify;
- The relative importance of primary and secondary sites of action; and
- Whether the classifier is a "lumper" or "splitter." Does the system integrate or separate?

Classification systems are complex, necessary, and, when adequate, will assist classification of new herbicides and integration of knowledge. This chapter has used site of action to classify many of the presently available herbicides. It does not, nor is it intended to, include all presently available herbicides. It should be possible to integrate any herbicides not mentioned herein into the classification system.

This chapter does not cover herbicide mixtures, which for good reasons are common. The reasons include broadening the spectrum of weed control, reducing the required amount of one herbicide, enhancing selectivity, or fitting a particular market niche. Nearly all mixtures will be classified in one or more of the groups mentioned previously. Therefore, determination of the site of action can be done within the classification system presented.

There are many herbicides, and a great deal on information about uses, environmental fate, and site of action is available for each one. The amount of information, even in a brief chapter, can be overwhelming. Table 16.10 lists the herbicide families included in this chapter and combines their primary site of action with the major plant function modified or irreversibly changed by the herbicide's activity. It is included to assist organization of the abundant information in this chapter.

THINGS TO THINK ABOUT

1. What is the primary site/mechanism of action of the herbicides in (site of action) group?
2. Does chemical structure always predict an herbicide's site of action? Is it a reasonably good predictor?
3. What are the major plant processes affected by herbicides?
4. Are herbicides unique chemical molecules that are unrelated to other common chemicals? Explain your answer
5. What are the major sites of herbicide action?
6. Is it likely that the next edition of this book will have a modified system of herbicide classification? Why?
7. Why do herbicides have so many different sites of action?
8. Why aren't more herbicides designed to inhibit specific plant biosynthetic processes?
9. Why don't we have a complete understanding of the precise site of action of all herbicides?

TABLE 16.10 Herbicide Site of Action

Chemical Group	Primary Site of Action	Major Function Modified or Disrupted
Phenylcarbamate		
Pyridazinone		
Triazine		
Triazinone	Photosynthesis electron	
Uracil	transport (Hill reaction)	
Phenylurea		Energy supply
Benzothiadiazole		
Nitrile		
Phenyl-pyridazine		
Organic arsenicals	Oxidative phosphorylation	
Dinitrophenols		
Photosynthesis electrons diverted from system I	Bipyrdillium	
p-nitro diphenylether		
Oxadiazole	Cell membranes - Protox	Structural organization
Triazinone	inhibitors	
Arylphenoxypropionates		
Cyclohexanediones		DEATH
Arylphenoxypropionate	Fatty acids	
Carbamothioate		
Chloroacetamide		
Amitrole		Chlorophyll production
pyridazinone/pyridone	Carotenoid synthesis	
isoxazolidinones		Low carotenoids
Dinitroaniline		
DCPA	Cell growth	
Sulfonylureas		
Imidazolinones		
Pyrimidinylthio benzoate	Amino acid synthesis	Growth
Sulfonylamino-carbonyl		
Triazolopyrimidines		
Glyphosate	EPSP synthase	Growth and reproduction
Glufosinate	Glutamine synthase	
Phenoxyacetic acid		
Arylaliphatic-Benzoicacid	Synthetic auxins	
Pyridine carboxylic acid		Growth
IAA transport	Naptalam	

Adapted from Corbett, J.R., Wright, K., Baillie, A.C., 1984. The Biochemical Mode of Action of Pesticides, second ed. Academic press, London, UK, 382 pp.

Literature Cited

Note: These sources, intentionally not cited frequently, have been used extensively.

Anonymous, 2015. Guide for Weed Management in Nebraska with Insecticide and Fungicide Information. EC-130 University of Nebraska, Lincoln, 300 pp.

Ashton, F.M., Crafts, A.S., 1981. Mode of Action of Herbicides, second ed. Wiley Interscience, NY. 525 pp.

Baker, N.R., Percival, M.P. (Eds.), 1991. Herbicides. Topics in Photosynthesis, vol. 10. Elsevier, Amsterdam, 382 pp.

Balantine, L.G., McFarland, J.E., Hackett, D.S., 1998. Triazine Herbicides: Risk Assessment. American Chemical Society, Washington, DC, 480 pp.

Bartels, P.G., Watson, C.W., 1978. Inhibition of carotenoid synthesis by fluridone and norflurazon. Weed Sci. 26, 198–203.

Beyer, E.M., Duffy, M.J., Hay, J.V., Schueter, D.D., 1982. Sulfonylurea herbicides. In: Kearney, P., Kaufmann, D.D. (Eds.), Herbicides: Chemistry, Degradation, and Mode of Action, vol. III. M. Dekker, Inc., NY, pp. 118–183.

Böger, P., Matthes, B., Schmalfuß, J., 2000. Towards the primary target of chloroacetamides — new findings pave the way. Pest Manag. Sci. 56, 497–508.

Corbett, J.R., 1974. The Biochemical Mode of Action of Pesticides, first ed. Academic Press, London, UK. 330 pp.

Corbett, J.R., Wright, K., Baillie, A.C., 1984. The Biochemical Mode of Action of Pesticides, second ed. Academic press, London, UK. 382 pp.

Devine, M., Duke, S.O., Fedtke, C., 1993. Physiology of Herbicide Action. PTR Prentice Hall, Englewood Cliffs, NJ, 441 pp.

Dill, G.M., Sammons, R.D., Feng, P.C.C., Kohn, F., Kretzmer, K., Mehrsheikh, A., et al., 2010. Glyphosate: discovery, development, applications, and properties (Chapter 1). In: Nandula, V.K. (Ed.), Glyphosate Resistance in Crops and Weeds: History, Development, and Management. Wiley, NY, pp. 1–33, 344 pp.

Duke, S.O. (Ed.), 1985. Weed Physiology. Herbicide Physiology, vol. II. CRC Press, Boca Raton, FL.

Duke, S.O., 1990. Overview of herbicide mechanisms of action. Environ. Health Perspect. 87, 263–271.

Duke, S.O., 2012. Why have no new herbicide modes of action appeared in recent years. Pest Manag. Sci. 68, 505–512.

Duke, S.O., Lydon, J., Becerril, J.M., Sherman, T.D., Lettnen Jr., L.P., Matsumoto, H., 1991. Protoporphyrinogen oxidase-inhibiting herbicides. Weed Sci. 39, 465–473.

Fedtke, C., 1982. Biochemistry and Physiology of Herbicide Action. Springer-Verlag, Berlin, 202 pp.

Gronwald, J.W., 1991a. Inhibition of carotenoid biosynthesis. In: Baker, N.R., Percival, M.P. (Eds.), Herbicides, Topics in Photosynthesis, vol. 10. Elsevier, Amsterdam, pp. 131–171.

Gronwald, J.W., 1991b. Lipid biosynthesis inhibition. Weed Sci. 39, 435–449.

Hassall, K.A., 1990. The Biochemistry and Uses of Pesticides: Structure, Metabolism, Mode of Action and Uses in Crop Protection. VCH Pub., NY, 536 pp.

Hull, H.M., 1967. Herbicide Handbook, first ed. Weed Science Society of America, Urbana, Il. 293 pp.

Kearney, P.C., Kaufman, D.D. (Eds.), 1976. Herbicides: Chemistry, Degradation, and Mode of Action, vols. 1 and 2. M. Dekker Inc., NY.

Kearney, P.C., Kaufman, D.D. (Eds.), 1988. Herbicides: Chemistry, Degradation, and Mode of Action, vol. 3. M. Dekker, Inc., NY.

LeBaron, H.M., McFarland, J.E., Burnside, O.C., 2008. The Triazine Herbicides — 50 Years of Revolutionizing Agriculture. Elsevier, Amsterdam, 584 pp.

Mallory-Smith, C.A., Retzinger Jr., E.J., 2003. Revised classification of herbicides by site of action for weed resistance management strategies. Weed Technol. 17, 605–619.

Moreland, D.E., St.John, J.B., Hess, F.D. (Eds.), 1982. Biochemical Responses Induced by Herbicides. Amer Chem. Soc. Symp. Series, vol. 181, 274 pp.

Peterson, M.A., McMaster, S.A., Riechers, D.E., Skelton, J., Stahlman, P.W., 2016. 2,4-D past, present, and future: a review. Weed Technol. 30 (2), 303–345.

Retzinger Jr., E.J., Mallory-Smith, C.A., 1997. Classification of herbicides by site of action for weed resistance management strategies. Weed Technol. 11, 384–393.

Sebastian, D.J., Sebastian, J.R., Nissen, S.J., Beck, K.G., 2016. A potential new herbicide for invasive winter annual grass control on rangeland. Rangel. Ecol. Manag. 63 (3), 195–198.

Senseman, S.A. (Ed.), 2007. Herbicide Handbook, ninth ed. Weed Science Society of America, Lawrence, KS. 458 pp.

Shaner, D.L., 2014. Herbicide Handbook, tenth ed. Weed Science Society of America, Lawrence, KS. 513 pp.

Shaner, D.L., O'Connor, S.L. (Eds.), 1991. The Imidazolinone Herbicides. CRC Press, Boca Raton, FL, 290 pp.

Vaughn, K.C., Lehnen Jr., L.P., 1990. Mitotic disruptor herbicides. Weed Sci. 39, 450–457.

Vencill, W.K. (Ed.), 2002. Herbicide Handbook, eighth ed. Weed Science Society of America, Lawrence, KS. 493 pp.

Wilkinson, R.E., 1983. Gibberellin precursor biosynthesis inhibition by EPTC and reversal by R-25788. Pestic. Biochem. Physiol. 19, 321–329.

Wilkinson, R.E., 1986. Diallate inhibition of gibberellin biosynthesis in sorghum coleoptiles. Pestic. Biochem. Physiol. 25, 93–97.

Woolson, E.A., 1976. Organoarsenicals (Chapter 15). In: Herbicides — Chemistry, Degradation, and Mode of Action, second ed., vol. 2. Marcel Dekker, Inc., New York, pp. 741–776.

Young, A.J., 1991. Inhibition of carotenoid biosynthesis. In: Baker, N.R., Percival, M.P. (Eds.), Herbicides, Topics in Photosynthesis, vol. 10. Elsevier, Amsterdam, pp. 131–171.

Zimdahl, R.L., 2010. A History of Weed Science in the United States. Elsevier Insights, London, UK, 207 pp.

17

Herbicide Formulation

FUNDAMENTAL CONCEPTS

- All herbicides are formulated.
- A formulation is a physical mixture of one herbicide or several and inert ingredients, which provides effective and economical weed control.

- The goals of formulation are to improve biological efficacy and to put the herbicide in a physical form convenient for use.
- There are several types of herbicide formulations.

LEARNING OBJECTIVES

- To know the goals of formulation.
- To understand why users can assume herbicide mixtures are homogenous.

- To know the physical characteristics of different herbicide formulations.

1. INTRODUCTION

If an herbicide (or any pesticide) has excellent biological activity and can be produced safely at reasonable cost, it may never reach the market if it cannot be formulated to retain or enhance its biological activity. All herbicides are formulated, which means they are combined with a liquid or solid carrier so they can be applied uniformly, transported, and still perform effectively. A variety of formulations have been designed for particular methods of application, to gain increased selectivity, to facilitate use, or increase effectiveness.

A formulation is a physical mixture of one herbicide or several and inert ingredients that provides effective, economical weed control. The formulation must be stable, maintain performance, and not require excessive shipping costs. It is important to note that a formulation is a physical, not a chemical, mixture. That is, chemical reactions are not the intended result of formulation. However, although no chemical reactions are involved, chemistry is. The thermodynamics of mixtures, phase equilibria, solutions, surface chemistry, colloids, emulsions, and suspensions must be considered. Each formulation has one or several biologically active chemicals. In 2016 pesticides registered by the United States Environmental Protection Agency included 1240 active ingredients, formulated in 19,290 commercial products, of which 4981 were herbicides.[1]

In addition to one or more active ingredients, formulations may include one or more inert ingredients that can enhance chemical or biological activity but have no herbicidal activity. These may be: surfactants (dispersants/wetting agents), solvents, emulsifiers, defoamers, stabilizers, antimicrobials, antifreeze, pigments/colorants, and buffers. Effectiveness (efficacy) and environmental safety are clear, objective judgments. Whether or not a formulation is economically wise is a subjective judgment of the manufacturer.

Formulation chemists have two primary goals. The first is to maintain and improve biological effectiveness by altering vapor or liquid mobility in soil, changing resistance to degradation, or improving ability to penetrate biological surfaces. The second, equally important, goal is to place the herbicide in a physical form convenient for users and appropriate for intended use(s). A formulation should also be as inexpensive as possible and have good shelf life. Exactly how formulation chemists do all of this involves art and science. Many aspects of formulation are trade secrets and there are no complete texts on the art of herbicide formulation.

Formulation can be crucial to success. The phytotoxicity of trichloroacetic acid was first discovered by a DuPont scientist and the ammonium salt was patented. The Dow Chemical Company patented the sodium salt, which is a different chemical and a different formulation. The sodium salt was the successful herbicide.

The herbicide chemical industry is dominated by a few large companies engaged in a wide range of activities in addition to herbicide production. In 1970, there were 46 separate companies in the United States engaged in synthesizing, screening, and developing herbicides (Appleby, 2005). In 2016, 11 companies remained. Three were American [Dow/DuPont (merging in 2017), FMC, and Valent]; Arysta, Isagro, Kumiai, and Nichino are Japanese; Isagro is Italian; and BASF, Bayer, and Syngenta are European. All operate internationally. Final mergers of Bayer and Monsanto and Syngenta and ChemChina are highly likely in 2017. Total sales of the top eight [Syngenta, Bayer, BASF, Dow, Monsanto, DuPont, Adama (a subsidiary of ChemChina), and FMC] global agrochemical companies were $48.5 billion in 2014 and dropped to $43.5 billion in 2015.[2] Three of the top six were European (Syngenta, Bayer, and BASF) and three American (Dow, Monsanto, and DuPont). The merger impetus has been driven by a dramatic decrease in demand for all types of chemicals and the soaring cost of finding and developing a new agrochemical. Between 1995 and 2017 the number of potential compounds that have to be tested to hopefully find a new one has risen from 50,000 to more than 140,000 (Economist, 2017).

---

[1]Personal communication, Pesticidewebcomments@epa.gov May 2016.

[2]http://news.agropages.com/News/NewsDetail—19400.htm.

A company engaged in finding and developing new herbicides may buy or synthesize (create) several hundred to several thousand new chemical structures annually. Each major group in a company examines compounds obtained or synthesized by another group to see if they have activity specific to the group's interests. Thus, the agricultural group obtains candidate chemical compounds from other groups (e.g., pharmaceutical, dye, plastics, etc.). Agricultural groups also employ organic chemists to synthesize new compounds based on known chemical groups with known or potential pesticidal activity or observed structure activity relationships.

Biologists, in the pesticide development group, receive a few to several hundred grams of a compound and preformulate the material so it can be applied to plants to determine biological activity. Chemists receive small samples and determine several physical and chemical properties such as melting point, boiling point, rate of hydrolysis, vapor pressure, specific gravity, solubility, and susceptibility to ultraviolet degradation. Inherent biological activity is determined by biologists who work with chemists.

During development, inert ingredients are selected to be formulated with a potential herbicide. The formulation chemist determines the physical properties of the formulation to appraise the compatibility of possible ingredients with the potential herbicide and container material.

When an herbicide is marketed, a farmer or custom applicator will move it from a point of distribution to a point of use or further sale where it may be stored under a variety of conditions. The formulation chemist must be concerned about stability of a herbicide formulation that may be stored in a shed where things freeze in the winter and where the temperature might be 120°F (49°C) at noon in the summertime. A formulation that spontaneously combusts or explodes above 100°F (38°C) or freezes and does not renew activity on thawing is not acceptable. Formulators must also determine the effects of temperature over time and the effects of storage conditions on stability of active ingredients, performance, and on wettability and dispersion. Chemists can modify formulations to affect solubility, volatility, and phytotoxicity.

When a user selects an herbicide and mixes it for spraying, an assumption is made. Most herbicides are added to a volume of carrier, usually water, and mixed in a spray tank. The unquestioned assumption is that the mixture is homogenous and that after agitation any volume of water taken from a spray tank contains the same amount of herbicide that any other volume of water contains. With variable water hardness (presence of divalent cations, e.g., calcium, magnesium, iron), suspended particulates and microbial activity, the formulation chemist cannot assume homogeneity. Homogeneity must be tested for and assured, which it usually is.

When a formulation is added to water in a spray tank, foaming can occur. Some foaming is inevitable if a surfactant is included, but excessive foaming is undesirable. It leads to imprecise determination of volume, and environmental contamination if foam overflows the tank. Control of foaming is a formulation problem.

Formulation chemists must be concerned about spray solution viscosity because it affects flow patterns, particle size, and weed control. The primary goal of formulation is to maintain or improve biological efficacy. A homogenous formulation that is stable at all possible temperatures to which it will be exposed, does not foam, and maintains appropriate viscosity, but during formulation loses its phytotoxicity, is a failure.

If a formulation performs satisfactorily in spray equipment and has passed all of the preceding tests, then chemists and biologists are concerned with how it interacts with plants—weeds. Among the questions asked are:

- What percent of what is applied is retained on the plant?
- What is the residual nature of the herbicide on the plant?
- Does the herbicide penetrate the plant and translocate?
- Does the herbicide form crystals on the plant surface?
- What is the formulation's biological efficacy? Does it work?
- What is the site of action?
- What is the herbicide's persistence in the environment?
- What are the possible effects of the herbicide on nontarget organisms through drift, soil residue, residue in water, and presence in other parts of a food web?

Formulation chemists are included in manufacturing decisions. Formulations must achieve the highest practical degree of efficacy at the lowest cost. Formulation chemists discuss formulations with chemical engineers to see if a formulation that works can be manufactured economically. The formulation chemist is concerned about packaging. A package that is sensitive to moisture is obviously inappropriate. One does not want a package that will be degraded by its contents. A successful formulation must maintain biological efficacy and be compatible with containers that will inevitably be stored under diverse conditions.

2. TYPES OF HERBICIDE FORMULATIONS[3]

There are two general types of herbicide formulations: liquid and dry. However, nothing as complex as herbicide formulation should be divided into two simple categories. Liquid formulations include solution concentrates, emulsifiable concentrates, and flowables. Dry formulations include wettable powders, dry flowables, and granules.

2.1 Liquid Formulations

2.1.1 Solution Concentrate

Some pesticides dissolve readily in water or a petroleum-based solvent. When mixed with the carrier they form a solution that does not settle out or separate. There are ready-to-use low-concentrate (RTU) and ultra-low volume (ULV) solutions. These formulations are not used for herbicides.

A solution concentrate is an herbicide dissolved in a solvent system designed to provide a concentrate soluble in a carrier, usually water. If an herbicide is immiscible with water, a one-phase solution containing the herbicide, one or more emulsifiers, and one or more solvents can be made to force or bridge it into solution or very fine suspension. The basic requirement for making a solution concentrate is active ingredient solubility. The herbicide must be soluble in a small enough quantity of solvent to make packaging and shipping economical. The

---

[3]I thank Randy Worthley of Loveland Products, Inc. of Loveland, CO, for his review and comments.

concentrate must be completely and rapidly soluble in water at all temperatures and concentrate—carrier ratios likely to be encountered. Usually, solution concentrates require little formulation and have a high concentration of active ingredient. Some acidic salts are formulated as solution concentrates, but few herbicides are soluble enough and capable of storage at very high or low temperatures as a water solution. The formulation is not widely used.

2.1.2 Emulsifiable Concentrate

An emulsifiable concentrate consists of an herbicide dissolved in an organic solvent with sufficient emulsifier to create an oil/water emulsion when the concentrate is added to water. Salts of acidic herbicides that are soluble in water and could be formulated as solution concentrates are commonly formulated as emulsifiable concentrates because the herbicide may react with metallic ions in water, precipitating the active ingredient and clogging spray equipment. These formulations are used when the active ingredient may not enter plant foliage readily due to high water surface tension or evaporation that leaves herbicides on foliage and results in no activity.

Other herbicides have low water solubility but can be dissolved in an organic solvent (e.g., xylene) and mixed with water to form an emulsion—one liquid suspended in another (e.g., fat globules in milk). In herbicide emulsions, water (the carrier) is the continuous phase and oil globules (solvent plus technical herbicide) are dispersed in it. These are is called oil-in-water (O/W) emulsions. Oil-soluble esters of acid herbicides and other herbicides such as the carbamothioates, dinitroanilines, and some chloroacetamides are formulated in this way. Because phases may separate, an emulsifying agent is added to keep the dispersed phase (herbicide) in suspension. This combines the two liquids without direct contact between them, and adverse reactions between the chemicals are not likely. Most agricultural emulsifiable concentrates consist of 60%—65% (by weight) of herbicide dissolved in 30%—35% organic solvent with 3%—7% of an appropriate emulsifier added to create an O/W emulsion when the concentrate is added to water. They may contain a small amount of emulsion stabilizer and a surfactant selected to permit appropriate interaction with plant surfaces. Manufacturers use the highest possible concentration, but 4 lbs/gal (approx. 0.5 kg/L) is common. Almost everyone knows that if there are 4 lbs in a gallon, there is 1 lb in a quart, and $^1/_2$ lb in a pint, etc. and pint or quart measures are easy to find in the United States.

Emulsifiable concentrates form an opaque, milky emulsion when added to water. They usually penetrate waxy foliage better than other formulations. The solvent, in some cases, can be phytotoxic and thereby aid herbicide activity. They can be applied in hard water without adverse reactions and are less apt to be washed off foliage by rain or irrigation. Herbicides formulated in this way evaporate (vaporize) slowly from plant surfaces. The formulations are easy to handle although they are easily absorbed through skin. They are inexpensive to make and easy to measure and transport. Agitation in the spray tank is not required, nozzle plugging is rare, and the active ingredient does not settle out or separate when equipment is running. They are the first method attempted for many herbicides. The formulants (solvents and other ingredients) may be toxic, and because the formulation is concentrated, overapplication is a potential problem. A challenge formulation chemists face is minimizing or, more desirably, eliminating hazards to machines and people from required toxic formulants.

2.1.3 Invert Emulsions

In an invert emulsion (W/O), oil is the continuous phase and water the dispersed phase. A common example is mayonnaise. The primary advantage of invert emulsions is drift reduction because they are more viscous and produce large drops. They have been used to formulate phenoxyacid herbicides for rangeland and industrial weed control. Inverts are nearly always applied in a large volume of diesel fuel or another low-grade petroleum product to aid plant absorption. Special emulsifiers are required. Because inverts are usually expensive and other formulations are equal or superior, they are rare.

2.1.4 Suspoemulsions

There is interest in combining several active ingredients, each of which is a different formulation type in a single formulation. The suspoemulsion combines a solid, usually formulated as emulsifiable concentrate, and liquid pesticide, usually formulated as a solution concentrate.

2.1.5 Oil Dispersions

When a pesticide is not readily soluble in a standard emulsifiable concentrate system and may not be stable in a water-based solution concentrate, an oil dispersion (OD) formulation may be used. An OD formulation does not use the toxic, flammable solvents used in emulsifiable concentrates, may have better efficacy than a solution concentrate, and is less expensive than water-dispersible granules. The carrier is typically a mineral oil or a methylated seed oil, which may improve biological efficacy. Two OD formulations of sulfonylureas were released by DuPont in 2015.

2.1.6 Flowable Concentrate

Flowable concentrates are liquid extensions of wettable powders (see Section 2.2.2). They are concentrated aqueous dispersions of an herbicide that is insoluble or nearly so in water. Not many of these formulations are available. They contain little or no organic solvent but do include clays similar to those used in wettable powders, some oil, water, an emulsifying agent, and a suspending agent. These approach other liquid formulations in ease of dispersion in water, ease of measurement, and do not require vigorous agitation. They are more difficult to make. A disadvantage is that the entire system can gel and become unusable or it can become solid with the oil portion rising to the surface.

2.1.7 Encapsulated

Encapsulated formulations enclose dry or liquid herbicide molecules in microscopic, porous polymer (plastic) capsules that are sprayed in water suspension. After application the capsule releases the herbicide slowly. Rate of release can be controlled, and timed release over a longer portion of the crop season is possible. With most formulations, maximum availability occurs at application, which may coincide with maximum crop susceptibility and the lowest weed population. These formulations are designed to change the time of maximum availability. They are applied in water, mix easily, and won't freeze under normal storage conditions.

2.2 Dry Formulations

2.2.1 Dusts and Dry Powders

Dusts are finely powdered, free-flowing, dry materials used to provide extensive surface coverage. They are relatively easy to formulate. No herbicides are formulated as dusts because of the drift potential.

2.2.2 Wettable Powders

Wettable powders are finely divided (dust-like) solids that are easily suspended in water. When an herbicide is insoluble in water or oil, the formulation chemist may turn to wettable powders. They are formulated by impregnating the active ingredient in or on an inert material such as a clay and adding wetting and dispersing agents. The wetting agent wets the active ingredient when it is mixed with water. Dispersing agents disperse the finely ground particles when mixed with water. A wettable powder with 50% active ingredient may contain 42% clay, 2% wetting agent, 2% dispersing agent, 4% inert ingredients, and 50% active herbicide. Because wettable powders form suspensions, not solutions, they will settle without continued agitation in the spray tank. They typically have less foliar activity than liquid formulations. Because they are suspended, finely divided solids, their abrasive action can wear pumps and spray nozzle tips. Frequent calibration is required. To aid dispersion and assure homogeneity, wettable powders should be mixed in a thick slurry before mixing with water in the spray tank. Most of the triazines, phenylureas, uracils, and members of several other herbicide groups have been formulated as wettable powders. A major problem with this formulation is the difficulty of measuring weight of a dry powder in the field.

Wettable powders can be an inhalation hazard to those measuring or mixing them in water. Vigorous agitation in the spray tank is required. They are the most abrasive formulation for nozzle tips and pumps, and frequent nozzle plugging can be a problem. A few decades ago they were common. This is no longer true as most formulations are now liquid.

2.2.3 Granules

Some herbicides can be formulated as granules—solid materials with 2%–10% active ingredient. The cost per unit of herbicide is high. Granules are applied as solid particles that tend to fall off plant foliage with little or no damage to plants via foliar uptake. Granular formulations are restricted to herbicides with soil activity. Equipment required for application can be inexpensive, but application can be combined with other field operations. Uniform application is difficult and granules may be moved by wind or water after application.

Granular technology combines the active ingredient with 1%–2% surfactant, and the balance is a solid carrier—the granule. Carriers must be available in a uniform size range that is free of dust and fine particles. The solid structure must not be destroyed with repeated handling. The solid carrier must have sufficient adsorptive capacity to absorb and hold the active ingredient. They are bulky to handle, costly to ship, and expensive per unit of active ingredient.

A mycoherbicide (see Chapter 12) has been formulated as a unique kind of granule (Connick et al., 1991; Daigle et al., 2002). The product (Pesta) is not commercially available, but the unique idea is interesting. Appropriate fungal propagules were entrapped in a matrix of wheat gluten. A dough prepared from wheat flour, filler, fungus, and water was rolled into a thin sheet (the process for preparation of pasta), air-dried, and ground into granules. The fungus, the active agent, grew and sporulated on the wheat granules after application to soil. Acceptable control of four broadleaved species was obtained (Connick et al., 1991). A strain of *Pseudomonas fluorescens*, a pathogen of green foxtail, has been successfully formulated with oat flour. It suppressed up to 90% emergence of green foxtail in field studies (Daigle et al., 2002). Hemp sesbania was effectively controlled (84%–88%) in soybean with microsclerotia of the bioherbicidal fungus *Colletotrichum truncatum* formulated in wheat gluten-kaolin granules called Pesta (Boyette et al., 2007).

2.2.4 Dry Flowable and Water-Dispersible Granules

Dry-flowable and water-dispersible granule formulations combine granule and wettable-powder technology and advantages. A wettable powder resembles flour, and a granule is a large particle. These formulations are dustless, small, dry particles that are measured by volume rather than weight prior to mixing in water and spray application. They offer the convenience of measurement of liquid formulations, decrease the disadvantages of liquid formulations, and retain the advantages of solid formulations. Many herbicides are formulated as dry flowables.

2.2.5 Water-Soluble Packets and Effervescent Tablets

Water-soluble packets reduce mixing and handling hazards by eliminating direct contact with the formulation. The package, containing the formulated herbicide, dissolves when placed in water. Some agitation is required to mix the formulation in the spray tank. These packets are usually small and are appropriate, although not widely available, for small land-holders in the developing world.

Effervescent tablets resemble Alka-Seltzer tablets when mixed in water. The usually palm-sized tablet can be used whole or broken into pieces. Some agitation is required to mix the formulation in the spray tank and maintain homogeneity.

2.2.6 Other

Parasitic weeds are among the most difficult challenges facing weed scientists. Among the things that make management difficult are the fact that the parasite attaches to and becomes part of its host, the crop plant. The seeds are very small, emerging plants cannot be selectively controlled, and the parasite usually does not emerge until after the crop plant has become established. Witchweeds (*Striga* spp.) have not been successfully controlled in affected crops until after some crop damage has occurred. Kannampiu et al. (2003) demonstrated season-long control of witchweed in corn (maize) in Africa by coating the crop seed with either the imidazolinone herbicide imazapyr or the benzoate herbicide pyrithiobac. The herbicides, from different chemical groups, both inhibit amino acid biosynthesis through inhibition of the acetolactate synthesis (ALS) enzyme required for biosynthesis of branched-chain amino acids. Seed coating resulted in a three- to fourfold increase in maize yield when witchweed density was high ($12/m^2$). This method is both a formulation and application challenge, in that one must be able to coat the seed with the herbicide in a durable manner or develop a way to have the seed imbibe herbicide (Kannampiu et al., 2003). The African work has shown that herbicide concentration in the immediate vicinity of the seed is quite high, dissipates before the next planting season, and does not injure legumes planted at least 15 cm from the maize row as second or companion crops. Without seed treatment there is total crop loss from witchweed.

3. SURFACTANTS AND ADJUVANTS

Surfactants, surface-active agents, do many things in formulations, including increasing wettability and spreadability, enhancing phytotoxicity, and increasing penetration. Their effects are due to their ability to increase wetting of the target surface and enhance penetration. It has been shown that one of the things surfactants do is reduce the energy required to absorb herbicides across cuticle and exterior leaf membrane barriers. Surfactants may be part of a purchased formulation (e.g., glyphosate) or added to the spray tank prior to use if recommended on the label (e.g., Gramoxone).

An adjuvant is something added that may or may not be phytotoxic. One example is addition of a surfactant to promote foliar activity, spreading, sticking, or absorption. There are safeners or protectants available for use with specific carbamothioate or chloroacetamide herbicides, which extend their range of selectivity. Spray modifiers are available in several forms to reduce drift or promote spreading and sticking. There are antifoam agents. Nitrogen has been found to enhance activity of some herbicides when added to the spray tank. Ammonium sulfate is used as an adjuvant and increases herbicidal activity in some cases. The herbicide label should always be consulted as the best and most reliable guide on use of surfactants or adjuvants.

THINGS TO THINK ABOUT

1. Why are all herbicides formulated?
2. What is an acceptable definition for each type of formulation?
3. What are the advantages and disadvantages of each type of herbicide formulation?
4. What is a surfactant, and what does it do in herbicide formulations?
5. What is an adjuvant, and what does it do in herbicide formulations?

Literature Cited

Appleby, A.P., 2005. A history of weed control in the United States and Canada — a sequel. Weed Sci. 53, 762–768.

Boyette, C.D., Jackson, M.A., Bryson, C.T., Hoaglund, R.E., Connick, W.J., Daigle, D.J., 2007. *Sesbania exaltata* biocontrol with *Colletotrichum truncatum* microsclerotia formulated in 'Pesta' granules. BioControl 52, 413–426.

Connick Jr., W.J., Boyette, C.D., McAlpine, J.R., 1991. Formulation of mycoherbicides using a pasta-like process. Biol. Control 1, 281–287.

Daigle, D.J., Connick Jr., W.J., Boyetchko, S.M., 2002. Formulating a weed-suppressive bacterium in "Pesta". Weed Technol. 16, 407–413.

Economist, May 27, 2017. Chain Reaction, p. 61.

Kannampiu, F.K., Kabambe, V., Massawe, C., Jasi, L., Friesen, D., Ransom, J.K., Gressel, J., 2003. Multi-site, multi-season field tests demonstrate that herbicide seed-coating herbicide-resistance maize controls *Striga* spp. and increases yields in several African countries. Crop Prot. 22, 697–706.

18

The Role and Future of Genetic Modification in Weed Science

FUNDAMENTAL CONCEPTS

- Genetic modification of crops has been a major scientific achievement and will be a continuing presence in agriculture.

LEARNING OBJECTIVES

- To appreciate the scientific actual and promised achievements and role of genetically modified (GM) crops.
- To know the worldwide extent of GM crops.

- To understand a little about the process of and results of genomics.
- To understand some aspects of the opposition to GM crops.

1. GENOMICS/GENETIC MODIFICATION

The process of inserting rDNA into a living cell became known as genetic engineering (Chassy, 2004). In Chassy's view, genetic engineering "catalyzed the foundation of the new industry that with great hubris called itself the biotechnology industry. The products derived from and the promises of genetic engineering have become major influences on weed management in all of agriculture.

Genomics is the study of the organization, evolution, and function of genes and other noncoding regions of a genome. There are complementary, overlapping subdisciplines (Horvath, 2010):

Functional genomics is the study of gene function and regulation.
Comparative genomics focuses on comparing the organization and sequence variation between species.
Structural genomics identifies conserved protein domains.

Population genomics is similar to comparative but focuses on evolution of gene structure and function within and between closely related species.

Genomic studies in weed science have provided insight into the physiology and evolutionary biology of developmental and environmental processes of weeds. They strive to answer fundamental questions of weed science: What makes a weed a weed? Why do they do so well? What are the traits that some plants possess that allow them to be invasive, survive environmental hazards, and often resist some of our best efforts to control or eliminate them? The genetic basis of weeds, invasive species, and their evolution is not well understood. Genomics offers the promise of greater understanding of why some plants are weeds (Stewart et al., 2009). Genetic markers have been used in horticulture and conservation biology to identify breeding lines, assess genetic diversity, and examine gene flow. Their use in weed science has not been as common because the primary focus of weed science has been on control rather than understanding of what makes a weed a weed (Bodo Slotta, 2008). Basu et al. (2004) challenged the weed science community because so little was known about the genomics of the economically important weed species. Perusal of the weed science literature affirms that their recommendation for an increase in weed genomics research has been followed. It has led to greater understanding of evolution of herbicide resistance and the basic genetic traits that make weeds successful.

2. THE PROCESS

There are three ways to make a crop resistant to one or more herbicides[1]:

- Chemical safeners—upregulate metabolic degradation of applied herbicide in the crop;
- Nontransgenically breeding or by mutation breeding (see Table 18.1);
- Transgenically—herbicide-resistant crops.

TABLE 18.1 Nontransgenic Crops Made Resistant to Herbicides by Mutation Breeding[2]

Inhibition of	Crop
AcetylCoACarbolxylase	
Fops and Dims	Corn, rice[a]
Photosynthesis at photosystem II, site A	
Triazines	Canola[a]
Acetolactate Synthase	
Sulfonylureas	Soybeans,[a] sorghum[a]
Imidazolinones	Canola, Corn, Lentils, Rice, Sunflower, Wheat

[a]*Approved and subsequently withdrawn due to low acceptance.*

[1]Personal communication, Dr. S.O. Duke, Natural Products Utilization Research Unit USDA, ARS, University, MS 38677, February 2017.

[2]See footnote 1.

There are many complex techniques in genetic engineering, but the basic principles are reasonably simple. There are five major steps in the development of a genetically modified (GM) crop:

1. DNA extraction from an organism known to have the trait of interest that will be introduced into the modified plant. Thus the first step is to discover a living organism that exhibits the desired trait. For example, the traits for Bt corn (resistance to European corn borer) were discovered at least 100 years ago. Silkworm farmers in the Orient had noticed that populations of silkworms were dying. Research discovered that a naturally occurring soil bacteria (*Bacillus thuringiensis*) produced a protein toxic to silkworms. Both insects (the silkworm and the corn borer) are members of the order Lepidoptera. Coincidentally the same protein killed the European corn borer. A similarly serendipitous discovery led to Roundup Ready glyphosate-tolerant plants being developed with a gene from bacteria found growing near a Roundup manufacturing plant.
2. Gene cloning—the gene of interest is located and isolated from the extracted DNA.
3. Modification—the gene is modified (engineered) by altering or replacing gene regions.
4. Trait insertion = transformation—the gene(s) are introduced into tissue culture cells using one of several methods. The plan is to insert the gene into the nucleus where it will become part of a chromosome. The common methods are by biolistic transformation (the equivalent of a 22-caliber gun fires a metal particle coated with DNA) and *Agrobacterium*-mediated transformation.
5. Backcross breeding and planting—the transgenic line is crossed with "elite" lines to obtain desired quality.

If everything has worked well, the genes express themselves and scientists have achieved what they set out to do.

Genome editing makes it possible to change a crop's genome down to the level of a single genetic letter. Editing imitates the process of natural mutation on which crop breeding has always depended. When a crop's DNA sequence is known, modification—that is plant breeding—can be more precise. It will no longer be necessary to grow plants to maturity to learn if they have the characteristics that the breeder wants. A quick look at the genome will suffice (Economist, 2016).

2.1 CRISPR-Cas9

Clustered Regularly Interspaced Short Palindromic Repeats (CRISPR) are segments of prokaroytic DNA. CRISPR is a naturally occurring, defense mechanism found in a wide range of bacteria. It was developed in 2013 by J. Doudna at the University of California—Berkeley; E. Charpentier at the Planck Inst, Berlin; and F. Zhang at Harvard, Cambridge. Cas9 (CRISPR-associated proteins) actually does the work of precisely snipping segments of DNA. It comes from *Streptococcus pyogenes*, better known as the bacteria that cause strep throat. It is a new technology[3] that promises to alter the future of the scientific enterprise and of weed science in predictable and unpredictable ways (Park, 2016). It is based on "a

---

[3]CRISPR's development is described well at http://wwwuser.cnb.csic.es/~montoliu/CRISPR/#4 and https://www.broadinstitute.org/what-broad/areas-focus/project-spotlight/crispr-timeline.

cluster of DNA sequences that could recognize invader viruses, deploy a special enzyme to chop them into pieces, and use the viral shards that remain to form a rudimentary immune system" (Specter, 2015). It has two components:

1. "Essentially a cellular scalpel"—Cas9, a protein—that cuts DNA.
2. A short segment of RNA, the molecule most often used to transmit biological information, which "leads the scalpel until it finds and fixes itself to the precise string of nucleotides it needs to cut" (Specter). The perhaps disturbing, but real, potential is that CRISPR will allow scientists to tinker, at will, with the genomes of plants and animals. However, the public perception against genetically modified organisms (GMOs) suggests that people will initially be hesitant to accept these plants. It heralds a new way of precise, intentional genetic engineering. It has significant advantages over conventional genetic engineering. It is precise, fast, and costs less. Genome editing with a CRISPR/Cas9 system can achieve transgene-free gene modifications and is expected to generate a wide range of new plants. The optimistic view is that when the negative and positive function of a gene sequence is known, CRISPR's search-and-replace ability will make it possible to change, delete, and replace genes in plants and animals, anywhere in the genome, including in us. CRISPR will make it possible to correct genetic errors, insert genes to improve productivity, and, one hopes, achieve the promise of plant genetic engineering— sustainable crop productivity and global food security. The other, pessimistic, more cautious view is that scientists will have the ability to rearrange the basic structure of life without knowing the result of their work. Altering the human genome will become possible and will surely be done. There is an inchoate fear of what might happen as scientists tinker with the fundamentals of life. There will be benefits and risks. The questions are scientific and moral. If CRISPR makes it easier to feed the world and reduce agriculture's environmental harm, some of the fear may lessen and may disappear.

2.2 RNAi

BioDirect is a new concept being developed by Monsanto and several other agrochemical companies. Ribonucleic acid (RNA)-interference(i) (RNAi) is a cellular process found in all higher organisms. It is a small piece of genetic code that all living things use for a specific function within a cell, including coding enzymes that plants need to survive.[4] It functions by degrading a targeted messenger RNA, thus reducing the amount of an enzyme or other protein produced by a specific gene. Most herbicides kill plants by preventing function of a specific enzyme. The hypothesis of RNAi technology is that it is possible to prevent destruction of the enzyme. Herbicide resistance has been created by inserting a gene for resistance. RNAi technology targets and prevents functioning of the resistance gene. Therefore, the herbicide will once again function to kill resistant weeds. The technology offers weed control without modifying the plant genome, without creating a GMO. A further advantage is that the interference will last only a few days or weeks. It is not permanent. If this technology can

---

[4]The function of RNAi is explained well in https://www.youtube.com/watch?v=cK-OGB1_ELE.

be perfected and is repeatable, it could solve the problem of weed resistance. Before RNAi technology can be commercialized, its safety for nontarget organisms and lack of environmental harm will need to be accepted by federal regulatory agencies.

As science and its technology alter the agricultural and, indeed the human, future in unpredictable ways, as they always have, deep forethought is required. All engaged in agriculture—farmers, scientists, technology developers, etc.—must think about whether or not what can be done is what ought to be done.

3. THE ADVENT AND GROWTH OF GENETICALLY MODIFIED CROPS

The efficacy of herbicides for weed control was evident and it was logical to search for ways to make crops resistant to successful, efficacious herbicides. It was a scientific challenge—if concern about crop selectivity could be eliminated, then an herbicide that killed most weeds could be used without fear of crop injury.

The first GM crop plant was produced in 1983, an antibiotic-resistant tobacco plant (Fraley et al., 1983). The first field trials with tobacco occurred in France and the United States in 1986. The first herbicide-resistant crop appeared in 1996 when canola resistant to atrazine was made available. It was developed as an afterthought (a good idea) following the detection of chlorotriazine-resistant broadleaved weeds in corn in Ontario, Canada (Hall et al., 1996). A breeding program was established at the University of Guelph to transfer the source of triazine resistance from birdsrape mustard to canola (rapeseed). The first atrazine-resistant canola cultivar (Ontario Agricultural College Triton) was released in 1984 (Beversdorf and Hume, 1984); others followed in 1986 and 1987. Research proceeded to identify the primary physiological mechanisms that can lead to natural or induced (created) resistance or tolerance.

Herbicide resistance is the inherited ability of a plant to survive and reproduce following exposure to normally lethal herbicide dose. Resistance may be natural or induced by genetic engineering, or selection of variants produced by tissue culture or mutagenesis. Herbicide tolerance is the inherent ability of a species to survive and reproduce after herbicide treatment. There are three physiological mechanisms for natural or induced tolerance or resistance to an herbicide:

1. Reduced sensitivity at a molecular site of action;
2. Increased metabolic degradation;
3. Avoidance of uptake or sequestration (hiding) after uptake (Duke et al., 1991).

Each of these has potential use for development of resistance in crops.

Herbicide-resistant crops and other achievements of genetic modification research have been adopted more rapidly than any other agricultural technology by farmers in the United States (Fig. 18.1) and across the world. In developing countries, weeds are the most commonly cited constraint to increasing crop production and expanding the amount of land farmed. Development of herbicide-resistant crops through biotechnology has the potential to reduce the weed control problem for farmers throughout the world. The technology has been widely promoted and adopted in developed countries. Adoption in developing countries has not

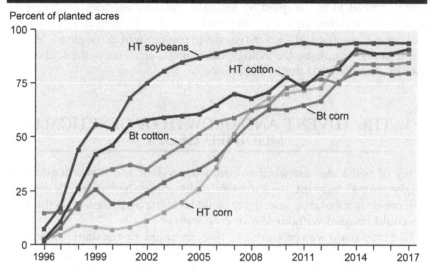

Adoption of genetically engineered crops in the United States, 1996-2017

FIGURE 18.1   Data for each crop category include varieties with both HT and Bt (stacked) traits. *Reproduced from USDA, Economic Research Service using data from Fernandez-Cornejo and McBride (2002) for the years 1996—99 and USDA, National Agricultural Statistics Service, June Agricultural Survey for the years 2000—17.*

been as rapid. Initially this was because there was little evidence of production cost reductions or increased yield (Martinez-Ghersa et al., 2003). Another reason may be that the high cost of GM seeds and herbicides has forced many small-scale farmers to assume debt beyond their ability to repay. This has led to small farmers leaving the land and consolidation of agriculture into larger farms, which, it must be admitted, many regard as desirable progress. However, debt has led to an increase in suicides among farmers in India[5] (Shiva et al., 2005; Robin, 2010).

Forty percent (175.2 million acres) of the world area of GM crops (444 million acres) were grown in the United States in 2014 (James, 2014), where use has expanded rapidly. GM crops were first grown in the United States in 1996 (4.3 million acres in the world). By 2000 the world area had grown 25 times to 109.2 million acres. In 2005 it doubled to 222 million, was 365 in 2010, 444 million acres in 2015, and 457 in 2016 (James, 2014).[6] The global area of biotech (GM) has increased at 3%—4%/year (approx. 16 million acres/year) for 19 years. In addition to the United States, the world's leading growers of GM crops in 2015 were Argentina (58.8 million acres), Brazil (72.3), India (26.7), and Canada (28.7). China, Paraguay, Pakistan, South Africa, Uruguay, and Bolivia each grew more than 1 million acres. With some exceptions in the European Union, only the Czech Republic, Portugal, Romania, Spain, and Slovakia permit GM crops.[7] James also reported that for the fourth consecutive year, developing countries planted more biotech crops (54% of total acres) than developed countries.

[5]See Bitter Seeds, a 2011 documentary film by American filmmaker and political commentator Micha Peled.

[6]Cropbiotechupdate=isaaa.org.

[7]http://www.isaaa.org/resources/publications/pocketk/16/.

Of the 28 countries that planted biotech crops, 20 were developing nations. Kalaitzandonakes et al. (2016) reported that asynchronous approval of biotech crops across the world has had mostly negative effects on global agricultural innovation, production, and trade. "As long as this situation persists, agricultural biotechnology will be prevented from delivering the full range of promised benefits of improved standard of living and food security."

4. ACTUAL AND POTENTIAL BENEFITS

Agricultural GM technology has focused primarily on corn, soybean, and cotton (Table 18.2). Table 18.3 lists the number of events[11] for 19 crops and four other plants that have been modified to be herbicide tolerant/resistant. Table 18.4 lists several of the food, feed, fiber, and other plants that have been genetically modified. Of the 11 approved for field

TABLE 18.2 Million Acres of Corn, Cotton, and Soybean Grown in the United States and the World in 2013[8]

Crop	United States		World	
	Acres	% of Total	Acres	% of Total
Corn	97.1	90	442	32
Cotton	10.1	90	84	70
Soybean	77.5	93	264	79
Total	184.7		590	

TABLE 18.3 Number of GMOs Approved for 19 Crops and Four Other Plants in 2016[9]

Crop	Countries	Events[10]
Alfalfa	9	5
Bean	1	1
Canola	13	36

(Continued)

[8]http://www.gmo-compass.org/eng/agri_biotechnology/gmo_planting/506.usa_cultivation_gm_plants_2013, http://www.gmo-compass.org/eng/agri_biotechnology/gmo_planting/257.global_gm_planting_2013.html.

[9]http://www.isaaa.org/gmapprovaldatabase/advsearch/default.asp? CropID=Any&TraitTypeID=Any&DeveloperID=Any&CountryID=Any&ApprovalTypeID=Any.

[10]When plant cells are transformed with foreign DNA, every cell that successfully incorporates the gene of interest represents a unique event. Marker genes are used to identify transformed cells, and each transgenic plant is the result of one event.

[11]A GMO and all subsequent identical clones resulting from transformation are collectively called a transformation event. If more than one gene from another organism has been transferred, the created GMO has stacked genes and is called a gene-stacked event.

TABLE 18.3   Number of GMOs Approved for 19 Crops and Four Other Plants in 2016[9]—cont'd

Crop	Countries	Events[10]
Chicory	1	3
Corn	26	147
Cotton	17	57
Eggplant	1	1
Flax	3	1
Melon	1	2
Plum	1	2
Potato	8	44
Rice	10	7
Soybean	28	32
Squash	1	2
Sugar beet	13	3
Sugarcane	1	3
Tobacco	1	1
Tomato	2	11
Wheat	5	1
Carnation	1	19
Creeping bentgrass	1	1
Poplar	1	2
Rose	1	2

release in the United States, seven included modification for herbicide resistance (NAS, 2016). Several cultivars have stacked or multiple resistance often to more than one herbicide (see Table 18.5) and may also have resistance to attack by an insect (e.g., corn rootworm). Table 18.6 lists some crops that have been modified to be resistant/tolerant to some herbicides. The dominant characteristic that has been introduced into widespread commercial use is glyphosate resistance. It is now available in eight major crops (alfalfa, corn, canola, cotton, potato, rice, soybean, sugar beet) and several others. In 2015, these crops were grown on about 12% of the world's planted cropland (NAS, 2016).

Extensive research in genetic modification has developed desirable traits (see Table 18.7 for a partial list), most of which are presently not commercially available. Although herbicide resistance/tolerance and insect resistance have dominated, several crop plants have been genetically modified: delayed browning of apples, insect resistance in eggplant, resistance

TABLE 18.4 Plants That Have Been Genetically Modified[12]

FOOD/FEED/FIBER CROPS

Alfalfa*	Apple	Canola*
Bean	Chicory	Cotton*
Eggplant	Flax	Maize-Corn (field and sweet)
Melon	Papaya	Plum
Polish canola	Potato	Rice*
Soybean*	Squash	Sugar beet*
Sugarcane	Sweet pepper	Tobacco
Tomato	Wheat*	

OTHER PLANTS

Carnation*	Creeping Bentgrass*	Eucalyptus
Petunia	Poplar	Rose

The * indicates herbicide resistance has been incorporated and approved for use in the United States.

to papaya ringspot virus, resistance to Colorado potato beetle (*Leptinotarsa decemlineata*), potato leaf roll virus X and Y, rice enriched with beta-carotene, resistance to zucchini yellow mosaic virus in watermelon, cucumber, and other cucurbits. Rao (2014) predicts several useful traits will be available over the next several years. They include:

Blight-resistant potatoes.
Enhanced photosynthetic efficiency of crop plants.
Increased nitrogen use efficiency.
Improve phosphorus efficiency and availability.
Enhanced nutritional quality of staple food crops.

TABLE 18.5 Herbicide-Resistant Gene Crops Available in the United States[13]

Crop	Resistant to	Year Approved
Corn	Glyphosate and imidazolinones	2009
	ACCase and 2,4-D	2014
	Glyphosate and glufosinate	2015
	Dicamba and glufosinate	2016

(Continued)

[12]http://www.isaaa.org/gmapprovaldatabase/cropslist/.
[13]See footnote 1.

TABLE 18.5   Herbicide-Resistant Gene Crops Available in the United States[13]—cont'd

Crop	Resistant to	Year Approved
Soybean	Glyphosate and ALS inhibitors	2008
	Glyphosate and 2,4-D	2014
	Glyphosate, glufosinate, and 2,4-D	2014
	Glufosinate and isoxaflutole (HPPD inhibitor)	2014
Cotton	Dicamba and glufosinate	2015
	Glufosinate and 2,4,D	2015

TABLE 18.6   Some Herbicides and Crops That Have Been Modified to be Resistant/Tolerant and Year Approved for a Few Herbicide—Crop Combinations[14]

Herbicide	Crop
Accase inhibitors	Corn, wheat
Bromoxynil, ioxynil	Canola (2000), cotton (1995), tobacco
Dicamba	Corn (2016), cotton (2015), soybean (2015)
Glufosinate	Alfalfa, beet, barley, canola (1995), carrot, corn (1997), cotton (2004), oats, peanut, potato, rice, sorghum, soybean (2011), sugar beet, sugarcane, tobacco, tomato
Glyphosate	Alfalfa (2005), canola (1996), corn (1998), cotton (1997), creeping bentgrass (2017), lettuce, potato, rice, soybean (1996), sugar beet (2008), tobacco, tomato, wheat
Imidazolinones	Canola, corn, cotton, lentil, peanut, potato, rice, safflower, soybean, sugar beet, tobacco, wheat
Isoxaflutole	Soybean (2013)
Sulfonylureas	Barley, canola, corn, chicory, cotton, flax, grape, lettuce, rice, soybean, sugar beet, tobacco, tomato
Triazine (e.g., Atrazine)	Broccoli, canola, rutabaga, tobacco
2,4-D	Corn (2014), cotton (2015), peanut, potato, sorghum, soybean (2014), tobacco

Development of drought tolerance.
Lengthened shelf life in major food crops.
Increased level of antioxidants in fruits.
Oilseed crops increased content of omega-3 long-chain polyunsaturated fatty acids.
Nutritional enhancement.
Bio/phytoremediation of organic soil pollutants, toxic chemicals, and explosives.
Biopharming—possible production of vaccines and pharmaceutical products from plants.

[14]Sources: Dekker and Duke (1995), Duke et al. (1991), Hopkins (1994), and National Academies (2016).

TABLE 18.7 Desirable Plant/Crop Traits That Have Been or Are Being Developed Through Research

Altered lignin production	Antibiotic resistance,
Coleopteran insect resistance	Delayed fruit softening
Drought stress tolerance	Fungal disease resistance,
Increased corn ear biomass	Lepidopteran insect resistance
Mannose metabolism	Modify alpha amylase
Modify flower color	Modify oil/fatty acid content
Modify starch/carbohydrate content	Multiple insect resistance
Nicotine reduction	Phytase production
Reduced acrylamide potential	Viral disease resistance
Visual marker	Volumetric wood increase

Most are not presently available.

The International Service for the Acquisition of Agri-biotech Applications (ISAAA) regularly publishes information on advances in biotechnology (www.isaaa.org/kc/crop biotech update). A few examples follow:

April 15, 2015—GM eucalyptus approved for commercial use in Brazil offered a 20% yield increase.
Scientists at the University of Guelph identified a gene that could block GM crops from cross-pollinating non-GM crops.
May 6, 2015—scientists at the Universities of Arizona and Illinois developed a soybean variety with reduced levels of three proteins responsible for allergenic and antinutritional effects.
June 10, 2015— Krishna et al. (2016) claimed that GM technology can help preserve the diversity of agricultural crops.
November 11, 2015—Scientists at the University of Geneva and ETH Zurich developed a variety of cassava, a dietary staple for 250 million people in Africa, that produces high levels of vitamin B6 (Li et al., 2015).
February 24, 2016—Unified support of GM technology to help meet food supply from the National Academies of Science and Engineering, the American Association for the Advancement of Science, and the American Society of Plant Biologists. Each cited a clear scientific consensus regarding the safety and efficacy of GM technology.
The Carnegie Institution for Science claims that GM crops, deployed appropriately in light of scientific knowledge and societal and environmental imperatives, can improve food and health substantially without detriment to the environment. There is considerable potential for preserving the environment through use of GMOs to reduce pesticide and fertilizer use.

March 25, 2016—US Food and Drug Administration approved Innate potatoes (less black spot bruising and reduced production of acrylamide) and Arctic apples (browning resistance caused by cuts and bruises). Canada approved sale of Arctic apples.

May 18, 2016—US National Academies find biotech crops not harmful to human health and the environment (http://nas-sites.org/ge-crops/ — released May 17, 2016).

June 8, 2016—Scientists at China's Nanjing Agricultural University and the John Innes Centre in Norwich, United Kingdom, developed rice varieties able to take up more nitrogen iron and phosphorus and increase yield up to 50%.

The ISAAA examples provide evidence that agricultural biotechnology will continue to be adopted worldwide. The technology has spread rapidly around the world because of efficacy and simplification of pest management. Optimists are sure that the anticipated developments of GM technology will lead to more sustainable agriculture, help feed the world's growing population, and meet the increasing demand for meat and a better diet as incomes increase.

5. CONCERNS AND CRITICISM OF GENETICALLY MODIFIED CROPS

In spite of the clear successes and significant promise of GM crops, genomics, and scientific research, concern and criticism dominate the media. The public's concern is derived from a general concern about genetic modification of anything. It is reasonable to claim that from the beginning television, radio, and print media coverage of GMOs/genetic engineering/genetic modification has been dominantly negative. That has slowly, perhaps imperceptibly, changed. Opposition continues. For example, the October 30, 2016, New York Times' front page article (Hakim, 2016) reported that GM crops had failed to raise yields or decrease pesticide use—thus failing to achieve the two major promises: feeding the world's growing population and requiring fewer pesticides. Since GM crops were introduced, insecticide and fungicide use in the United States have decreased by one-third. Herbicide use increased in intensity for corn, cotton, rice, and wheat, but more rapidly in nongenetically engineered (GE) crops (Kniss, 2017).

Public perceptions may be changed or, if read, at least, influenced by the National Academy's (NAS) (2016) report Genetically Engineered Crops: Experiences and Prospects, a science-based look at genetically engineered crops. The report builds on reports published between 1987 and 2010. It is an examination of the positive and adverse effects of GE crops and anticipates what emerging genetic-engineering technologies hold for the future. The report indicates where there are uncertainties about the economic, agronomic, health, safety, or other effects of GE crops and food, and makes recommendations on safety assessments, regulatory clarity, and improved access to GE technology.

Those opposed to biotechnology are not opposed to agriculture's goal of feeding the world. It is generally agreed that it is a good, ethically correct, thing to do. However, the agricultural establishments (land-grant universities, colleges of agriculture, the US Department of Agriculture, professors, research scientists, managers and employees of agriculture's supporting industries, farmers, ranchers, etc.) must consider and address the public's concerns. Ignoring or dismissing them as incorrect, nonscientific, or uninformed is wrong. The public's concern is frequently based on three issues:

1. The assumption, if not the certainty, that scientists are tampering with a boundary between humans and other creatures that should not be crossed.
2. GM foods may, for a usually unknown or unspecified reason, be harmful to those who consume them.
3. A GM food crop may have unknown, perhaps unpredictable future harmful effects on the environment and/or human health. In this regard it is worthy of mention that research in England has shown GM sugar beet to have beneficial effects: crop yield increased as much as 9%, weed seed production declined as much as 16 times, and 43% less herbicide was required (Economist, 2005). It is not this book's purpose to discuss these concerns in detail (a few appropriate sources will be cited). A brief description of some areas of public concern follow.

5.1 Public Health

People wonder, as they have since pesticides were introduced, about the safety and human health consequences of eating—what we all must do. Scientists know that it is impossible to guarantee 100% safety of anything. All foods grown from conventional crop varieties or transgenic varieties are complex mixtures of thousands of different chemical substances. Prediction of the possibility of danger from any new technology is common. There is no scientific evidence that GM food is more dangerous than non-GM food, or that it is dangerous at all (NAS, 2016; Thompson, 2015). Yet public opposition persists. Blancke et al. (2015) argue that "intuitive expectations about the world render the human mind vulnerable to particular misrepresentations of GMOs." The intuitive appeal is, in their view, a fatal attraction.

It has been proposed that GM crops that contain foreign proteins may increase the incidence of food allergies. This has not happened because tests for identifying potentially dangerous allergens are required and reliable. There is no scientific evidence that GM crops have contributed to antibiotic resistance. Unfortunately, studies that provide only one side of the story are used to justify political decisions, which often only perpetuate the controversy.

An additional aspect of GM technology is worthy of brief mention. The US Supreme Court ruled that genes can be patented, which compels legitimate concern about who owns genes. It is estimated that 90% of the world's biodiversity lies within the territories of indigenous peoples, whether in the Amazon, the Indian subcontinent, or the North Woods (La Duke, 2007). Genes have been taken from indigenous, rural communities in developing countries to the northern-based genetics industry. Patents are owned by someone or a company in the North. The question of ownership and ensuing rights are matters of continued discussion.

It is clear that the US public is divided about the health effects of organic versus GM foods (Funk and Kennedy, 2016). The PEW report notes that 48% think GM foods are no different than other food, 39% think they are worse for one's health, and 10% think they are better for health. However, the same survey revealed that 55% of US adults believe organically grown foods are healthier than conventionally grown food.

5.2 Environmental

5.2.1 Gene Flow

Resistant gene flow to sexually compatible plants is acknowledged as a potential risk of introducing any genetically engineered crop. It has been discussed for more than 20 years (Mallory-Smith and Olguin, 2011). The risk is transfer of created and desired herbicide resistance from a GM crop to a wild relative or a weed where undesirable resistance will persist by natural selection. Such transfer amounts to accepting an unknown ecological risk. It is important to recognize that for most of the world's important crops gene flow between wild and weedy relatives has always taken place (Messeguer, 2003). It is not a phenomenon that developed subsequent to introduction of transgenic plants. Conventional breeding has been used to produce more herbicide-resistant crops then genetic engineering has (Mallory-Smith and Olguin).

Corn has no wild relatives in the United States or Europe. However, it is closely related to *teosinte*. The history of modern-day corn began at the dawn of human agriculture, about 10,000 years ago. Teosinte doesn't look like corn, but its DNA is surprisingly similar. Teosinte can cross-bred with modern corn varieties and gene transfer is possible. Although corn and its progenitors are wind pollinated, "eventual introgression of transgenes from commercial hybrids into land races and wild relatives is likely if they are grown close together" (Messeguer, 2003), which is possible, but unlikely.

Transfer happened when genes from herbicide-resistant canola moved, in a short time, to a nonweedy relative in the mustard family and then to wild mustard.[15] Movement to other related wild and weedy species [turnip rape, wild cabbage, other mustard (*Brassica*) species] that are compatible breeding partners is possible (Snow et al., 1999). There can be numerous wild relatives in cultivated fields (Chévre et al., 2000). While gene flow is a real problem, the chances of success are small. Success requires sympatry with a compatible relative, flowering synchrony, successful fertilization, and viable offspring (Légère, 2005). Feral, herbicide-resistant oil rapeseed established on roadsides in Europe, but has not become a management problem (Scott and Wilkinson, 1999).

Gene flow from GM crops to near relatives has not occurred in Latin America. It is recognized as a potential problem, especially from rice to closely related weedy red rice (Gealy et al., 2003; Riches and Valverde, 2002). Dissemination of rice pollen is considered to be low because anther dehiscence occurs just before or at the same time as the lemma and palea open and most pollen falls to the ground (Messeguer, 2003). Other examples include wheat and jointed goatgrass (Hanson et al., 2005), wild and cultivated sunflower (Burke et al., 2002), canola and wild mustards (Snow et al., 1999), and wild and cultivated cucurbits (e.g., squash) (Spencer and Snow, 2001).

Messeguer (2003) discusses the possibility of gene flow from GM crop cultivars to other cultivars or wild or weedy relatives of potato, sugar beet, and cotton. There is little

---

[15]Denver Post, April 14, 1996 and New York Times, March 7, 1996

information available for asexually propagated clonal crops (e.g., potato). Potato tuber sections are planted, seed is not. Therefore, gene flow is not a concern. Transgenic and nontransgenic sugar beets can survive over winter in northern Europe and in similar cool/cold winter climates. Overwintering is normally necessary for seed production although bolting (growth of a seed stalk, flowering, and seed production) can occur in the first year. Plants that overwinter can be a source of pollen for transfer to red beets, Swiss chard, and wild beet.

If wild cotton species are present, they may be distant diploid relatives of commercial tetraploid cotton. Transgenic cotton could be dispersed into the environment, but only over distances less than 50 m. It is important to recognize that gene flow between crop cultivars and wild relatives has always occurred. It is not a new phenomenon. Messeguer (2003) correctly points out that such gene flow will modify the environment as it always has. Modification may be negative or positive, but the end result may not be known for several years. The National Academy's report (2016, p. 98) found that gene flow has occurred, but "no examples have demonstrated an adverse environmental effect…from a GE crop to a wild, related plant species."

5.2.2 Herbicide Use

A common critique of herbicide-resistant crops is that the technology will increase herbicide use, an allegation supported by The Institute of Science in Society[16] and Hakim (2016) but not by Kniss (2017). The critique includes the challenge that herbicide-resistant crops will aid continued development of, what many view as, an unsustainable, intensive, monocultural agriculture (see Duke, 2011). A survey of pesticide and herbicide use in 21 selected US crops (80% of all pesticides are used in US agriculture) from 1960 to 2008 found that in 1960, 196 million pounds of pesticides were used in the United States: 58% insecticides and 18% herbicides. In 1981, 632 million pounds were used, only 6% insecticides. Herbicides were 76% of total use, which declined to 516 million pounds in 2008 (Fernandez-Cornejo et al., 2014). From 1996 to 2008 pesticide use on US crops increased by 383 million pounds of herbicides, which some attribute to the increased acreage of GM crops. In 2008, GM crops required 26% more pounds of pesticides (not only herbicides) than conventional crops. Hellwig et al. (2003) found the best weed control and economic return in no-till, herbicide-resistant corn was obtained with an early residual herbicide followed by application of a postemergence herbicide to which corn was resistant. Research has shown that the combination of herbicides is always best, which may be evidence in support of the claim that GM crops have not affected herbicide use.

Several factors have driven changes in herbicide use: crop and input price changes, introduction of new herbicides, adoption of herbicide-tolerant crops, a shift toward conservation tillage systems, pesticide regulation, and government policies such as the incentives for ethanol producers (Fernandez-Cornejo et al., 2014). Glyphosate use increased from 112,000 pounds in 1995 to 1,646,208 in 2014; 90% was used in agriculture (Benbrook, 2016).

A related concern, noted in Latin America where herbicide-resistant technology has progressed rapidly, is that the technology may encourage expansion of agriculture to uncleared (e.g., Amazon rain forest) areas, which become economically attractive because of the economic efficiency of herbicide-resistant crops. Agricultural expansion would inevitably lead to adverse effects on nontarget organisms and ecosystems (Riches and Valverde, 2002).

[16]http://www.i-sis.org.uk/GMcropsIncreasedHerbicide.php.

It is also suggested that herbicide-resistant crops will reinforce farmers' dependence on outside, petroleum-based, potentially polluting technology. An associated concern is that there is no technical reason to prevent a company from choosing to develop a crop resistant to a profitable herbicide that has undesirable environmental qualities such as persistence, leaching, harm to nontarget species, etc. There is concern, but little evidence, that this has occurred.

It is undoubtedly true that nature's abhorrence of empty niches will mean that other weeds will move into the niches created by removal of weeds by the herbicide in the newly resistant crop. A 2015 nationwide survey (Van Wychen, 2016) identified the most common and most troublesome weeds in 26 different cropping systems. Some weeds have been with us for a long time, and others have developed more recently as important problems (e.g., Palmer amaranth). There is no evidence from the survey, nor is it implied herein, that these are result of GM crops. Herbicide-resistant crops have, as most control technologies have, solved some problems and created others. Weeds not susceptible to the herbicide to which the crop is resistant appeared. Weeds are not conscious, but they seem to be clever.

The legitimate concerns of epistasis and pleiotropy must also be recognized. Epistasis is the suppression of gene expression by one or more other genes and pleiotropy is defined as a single gene exerting simultaneous effects on more than one character. In short, one of the rules of ecology is operative: you can't do just one thing. When any genome is manipulated specific outcomes are intended. Even when these are achieved, other, unplanned (and perhaps, at least, initially, unnoticed) things may also occur. Genetic engineering, with the best intention of doing a good thing, may do unexpected things that could be good or bad.

Marketing and development of herbicide-resistant crops (HRCs) are proceeding rapidly. There are important advantages that provide good reasons for continued development. Many argue that the technology will provide lower-cost herbicides and better weed control. These powerful arguments in favor of the technology are often accompanied by the claim that HRCs will lower food costs for the consumer. It is true that HRCs are providing solutions to intractable weed problems in some crops. Glyphosate and glufosinate[17] resistance have been created in several crops. Both are environmentally favorable herbicides and therefore, many argue, it is better to use them in lieu of other herbicides that are not environmentally favorable. An important argument in favor of the technology is that it has the potential to shift herbicide development away from initial screening for activity and selectivity followed by determination of environmental acceptability to the latter occurring first. Resistance to herbicides that are environmentally favorable, but lack selectivity in any crop, could be engineered, and the herbicide's usefulness could be expanded greatly. This has important implications for minor crops (e.g., vegetables, fruits) where few herbicides are available because the market is too small to warrant the cost of development. If resistance to an herbicide already successful in a major crop could be engineered into a minor crop, manufacturers and users would benefit, especially if introduction was not followed by development of more-resistant weeds. There is a possibility that a crop genetically

---

[17]A gene coding for phosphinothricin acetyl transferase activity was isolated from *Streptomyces hygroscopicus* and cloned into several crop species. The enzyme controlled by the gene converts glufosinate to a nonphytotoxic acetylated metabolite. Crops so modified can then detoxify glufosinate (Duke, 1996).

engineered for herbicide resistance will contribute to creation of new, difficult-to-control weed hybrids.

5.3 Social

Societal concerns are real, but lack the strength of scientific evidence. Social questions are, by definition, commonly not subject to scientific or rational truth. Scientific truth can be defined mathematically, is publicly verifiable, definitive, precise, and falsifiable. Social questions are properly in the realm of personal truth where the language is vague, imprecise, symbolic, descriptive, and highly subjective.

There is a legitimate fear, but little supporting evidence, that GM/GE technology favors large farms and will inexorably lead to loss of more small farms and small-scale farmers. Evidence that small farms are disappearing from all US states is indisputable. Over the past 5 decades, the number of farmers has declined while the US population has more than doubled. By 2050, there may be 390 million people in the United States living on 2.3 billion acres compared with 152 million people in 1950. Farms occupy about half of the US land but are not protected from urban growth. Farmers are less than 1% of the US population, and many do not report farming as their primary occupation. Half of US farmland is owned by people who don't farm. US Department of Agriculture (USDA) data show that small farms account for 91% of all farms and 23% of agricultural production. There are large differences among small farms. Most production occurs on small commercial farms with annual income of at least $10,000. Most small farms are much smaller; 60% have income of less than $10,000, and 22% make less than $1000. US farm production continues to shift to larger operations, while the number of small farms and their share of farm sales continue a long-term decline.

While small farms are disappearing, there is no conclusive evidence that GM crops are *the* cause. There is no evidence that *any* specific agricultural technology has caused the decline of small farms. However, when one considers all the agricultural technology that has been introduced over the last several decades and its associated, frequently externalized costs, its role in small farm decline must be acknowledged. There is evidence that producers who have GE crops, including small-scale farms, have had favorable economic returns (NAS, 2016).

Two other matters must be considered. There is legitimate concern that the cost of food production and the cost of food to the consumer will rise as food production continues to be dominated by ever-larger farms. The decline of small farms and the associated decline of rural communities and values they embody has been the concern of many authors.[18] Secondly, a grower of a GM crop is obligated to follow some practices prescribed by those who developed and sell the technology. This can lead to deskilling (Stone, 2007). One definition is a process by which skilled labor is eliminated by the introduction of new technologies. Deskilling also refers to individuals becoming less proficient over time because someone else is doing the thinking.

Genetic modification is criticized for making workers automatons rather than artisans. To paraphrase Berry (1970), the act of thinking, "one of the happy functions of human life has

[18]An incomplete list of relevant sources (Berry, 1977, 1981, 1990; Davidson, 1990; Logsdon, 1998; Rhodes, 1989; Hanson, 1996, 2000; Kirschenmann, 2010).

been in effect abandoned." Whether or not GM technology causes deskilling has not been answered (NAS, 2016).

5.4 Labeling

Vermont's July 1, 2016 law requiring labels on food containing GM ingredients was nullified when US Senate agriculture leaders announced a deal that preempts states' labeling laws—and set a mandatory national system for GM disclosures on food products. Sen. Pat Roberts (R-KS), the chairman of the US Senate Committee on Agriculture, Nutrition, and Forestry, unveiled the plan that had been negotiated for weeks with US Sen. Debbie Stabenow (D-MI). The law (S.764, signed July 29, 2016 by President Obama) directed the Sec. of Agriculture to create a national labeling standard within 2–3 years. The law does not require producers to provide GM/GE information on products. They are only required to provide a website address where customers can obtain information. Advocates hoped the law would mandate products with genetically engineered ingredients to disclose that on packaging in one of three ways: a label, a USDA-developed symbol, or an electronic code that consumers can scan. The bill also would use a narrow definition of genetic engineering that would exempt the newest biotech methods such as gene editing from the national disclosure standards. Under the plan, food companies would be required to disclose which products contain GM ingredients. But companies would have a range of options in just how they make that disclosure: they could place text on food packaging, provide a Quick Response code, or direct consumers to a phone number or a website with more information.

News of the federal law came as many large food companies, including Campbell's Soup, Kellogg's, and General Mills, had begun labeling some of their products in anticipation of the Vermont law. Roughly 75% of processed foods on US supermarket shelves (soda, soup, crackers, and condiments) contain GM ingredients, according to generally agreed food industry estimates. Labeling has been controversial from the beginning of genetic modification. Food processors and manufacturers of products found on supermarket shelves have been opposed to labeling because of fear that widespread public distrust of GM foods would negatively affect product sales. It is an illustration of what happens when the public's right to know compels transparency on the part of manufacturers.

6. MOLECULAR BIOLOGY IN WEED MANAGEMENT

As stated before, the dominant role of molecular biology in weed management and research has been development of herbicide-resistant crops. Within weed science it is widely regarded as the most important major advance since the development of herbicides. It is important to understand as Gressel (2000) points out, "no solution in agriculture has ever been forever." Evolution is operative, and major changes are often not even observed until after they have occurred. Nature changes, weeds change, and their management must evolve. The technology offers several important benefits (Lyon et al., 2002):

- Improved weed control, particularly of difficult to control weeds.
- Reduced cost of weed control.

- Weed management is simplified. The herbicide(s) applied eliminate the need of choosing among several management techniques.
- Improved control of weeds not controlled by available herbicides or other available techniques.
  - Lengthened application time for the herbicide to which the crop has been made resistant.
- Improved crop safety because the crop is resistant to negative effects of the herbicide.
- The available evidence on yield increase is not conclusive.
  There are equally important scientific/research risks:
- The eventual need to control weeds that become resistant to the herbicide to which the crop has been made resistant.
- The introgression of transgenes into related weeds (see NAS report, 2016, p. 98).
- Overreliance on a single herbicide will lead to weed species shifts to resistant or more difficult-to-control species.
- Reliance on herbicides as the sole weed management technique, especially in large monocultures has predictably led to development of herbicide-resistant weeds (Gressel and Segal, 1978; Powles and Holtum, 1994), the presently dominant topic in the weed science literature (see Weed Science, special issue Vol. 60,2015[19]). One must wonder if GM crops have become a kind of Prozac for agriculture.[20] The inevitability of resistance and its research dominance offer a good lesson, but one that seems to be difficult to learn—overreliance on any agricultural technology tends to lead to its failure. Multiple resistance not just to the herbicide the crop has been modified to be resistant to, but to other herbicides with similar modes of action, is common. The NAS (2016) report found that herbicide-resistant crops sometimes initially decreased herbicide use but the decreases were not generally sustained.

Gressel (2000) mentions another gain from the advent of molecular biology—the positive effects on basic biological science. He suggests that in the long run it would be good to focus molecular biology research on ways to accomplish weed control without, or with fewer, chemicals. Molecular biology research in weed science has focused largely on crops of importance to developed-countries' agriculture (corn, soybean, canola, etc.). The above ISAAA reports note emphasis on crops that feed most of the world (several vegetables, pigeon pea, cassava, millets, etc.) but that may not offer as much, or perhaps any, major profit potential to corporations.

Martinez-Ghersa et al. (2003) noted that there was no evidence of production cost reductions or yield increases from molecular biological research. This is still true. Weed control is easier and perhaps more efficient, and while that is an important accomplishment, it is all that has been accomplished. Research could be conducted, and I expect it is, to make crops more competitive with weeds for light, nutrients, or water (Gressel, 2000). Allelochemical interference by crops could be enhanced. It is theoretically possible to make biocontrol agents more

---

[19]As noted in Chapter 19, from 2010 to 2016, 21.5% of the articles in Weed Science and 15.2% of those in Weed Technology dealt with herbicide resistance or tolerance.

[20]Prozac, a psychiatric medication, can help control obsessive-compulsive behavior and lead to happiness.

virulent, yet not able to spread (Gressel, 2000). Gressel also proposes that research could devise ways to make weeds less competitive through genetic modification with deleterious transposons.

7. CONCLUSIONS

Research on molecular biology applicable to agriculture and specifically to weed science will continue and new applications will be developed. Martinez-Ghersa et al. (2003) correctly note that scientific questions have dominated (What can be done?), but ethical considerations (What ought we do?) are essential to proper development of the technology, indeed to the continued success of the entire agricultural enterprise.

Thompson (2007, pp. 62–72) adverts that "biotechnology's boosters" have done "serious damage to their own case by offering several singularly bad arguments in its favor." Dealing with them at length is beyond the scope of this book, but it would be irresponsible not to mention them. The four harmful arguments he describes are:

1. The modernist fallacy appeals to an "outdated and naïve notion of technological progress." The notion is that ethical concerns are simply the price of progress and dwelling on them may deter progress. Scientific and technological progress has always created changes, most of which been beneficial; let's emphasize the positive achievements and potentials.
2. The naturalistic fallacy uses an inappropriate reference group to make comparisons of the relative risks of biotechnology. It is the claim that if something exists it must be good. Thompson offers a relevant example: "the kinds of alterations that molecular biologists are making in plants and animals are just like those that occur as a result of natural mutation. They are, therefore, an acceptable risk."
3. The argument from ignorance is an especially troubling accusation. Supporters are so convinced that there is no evidence of harm from any recombinant DNA product that it is clear to them that nothing bad will happen.
4. The argument from hunger is the claim that "agricultural biotechnology is the solution to world hunger, generally accompanied by the claim that those who oppose it are themselves ethically irresponsible in virtue of the misery from disease and starvation that their opposition is alleged to cause. That is opposing biotechnology is morally wrong. It is a bad argument because in Thompson's view it has been deployed shamelessly and cynically in a manner that promotes continued misunderstanding of the problems of global hunger and of agricultural sciences role in addressing them."

I and several others (e.g., Thompson, 2007, 2015) have dealt with ethical matters elsewhere and refer interested readers to that work for consideration of the ethics of molecular biology in agriculture (Zimdahl, 2012). It is not the purpose of this book to analyze the genomics/GMO controversy in depth. Several books (Duke, 1996; Gressel, 2002, Chapter 4; Naylor, 2012, Chapter 11) are available as are articles (e.g., Hileman, 1995) too numerous to mention. Much more work will be done and discussed, but it is important to realize that these crops are widely promoted, accepted, and used by producers.

Biotechnology was discussed by Christiansen (1991), a self-acknowledged outsider. His still-challenging view follows:

> I think it would be a pity if the power of the use of mutants and mutation to uncover and describe physiology and development were limited, in the hands of weed scientists, to the isolation and description in yet another species of yet more genes that confer resistance to yet another herbicide. To this outsider, it seems that the central issue for weed science is understanding the nature of weeds: What makes a weed a weed? How can weeds consistently come out ahead when matched up against the finest commercial varieties my plant-breeding colleagues develop? Weeds persist, they spread, and they out compete the crop plants, reducing yields when left uncontrolled. The nature of this "competitive ability" that weeds possess seems an interesting target for research and an appropriate target for analysis through generation of mutants.

Literature Cited

Basu, C., Halfhill, M.D., Mueller, T.C., Stewart Jr., C.N., 2004. Weed genomics: new tools to understand weed biology. Trends Plant Sci. 9 (8), 391—398.

Benbrook, C.M., 2016. Trends in glyphosate herbicide use in the United States and globally. Environ. Sci. Eur. 28, 3. https://doi.org/10.1186/s12302-016-0070-0.

Berry, W., 1970. A Continuous Harmony — Essays Cultural and Agricultural. Shoemaker and Hoard, Washington, DC, 176 pp.

Berry, W., 1977. The Unsettling of America: Culture and Agriculture. Avon books, NY, 228 pp.

Berry, W., 1990. What Are People for? North Point Press, San Francisco, CA, 210 pp.

Berry, W., 1981. The Gift of Good Land: Further Essays Cultural and Agricultural. North Point Press, San Francisco, CA, 281 pp.

Beversdorf, W.D., Hume, D.J., 1984. OAC Triton spring rapeseed. Can. J. Plant Sci. 64, 1007—1009.

Blancke, S., Van Breusegem, F., De Jaeger, G., Braeckman, J., Van Montagu, M., 2015. Fatal attraction: the intuitive appeal of GMO opposition. Trends Plant Sci. 20 (7), 414—418.

Bodo Slotta, T.A., 2008. What we know about weeds: insights from genetic markers. Weed Sci. 56, 322—326.

Burke, J.M., Gardner, K.A., Rieseberg, L.H., 2002. The potential for gene flow between cultivated and wild sunflower (*Helianthus annuus*) in the United States. Am. J. Bot. 89, 1550—1552.

Chassy, B.M., 2004. The history and future of GMOs in food and agriculture. Cereal Foods World 52 (4), 169—172.

Chévre, A.M., Eber, F., Darmency, H., Fleury, A., Picault, H., Letanneur, J.S., Renard, M., 2000. Assessment of inter-specific hybridization between transgenic oilseed rape and wild radish under normal agronomic conditions. Theor. Appl. Genet. 100, 1233—1239.

Christiansen, M.L., 1991. Fun with mutants: applying genetic methods to problems of weed physiology. Weed Sci. 39, 489—495.

Davidson, O.G., 1990. Broken Heartland: The Rise of America's Rural Ghetto. Anchor, NY, 206 pp.

Dekker, J.H., Duke, S.O., 1995. Herbicide Resistance in Field Crops. http://lib.dr.iastate.edu/cgi/viewcontent.cgi?article=1013&context=agron_pubs.

Duke, S.O., 2011. Comparing conventional and biotechnology-based pest management. J. Agric. Food Chem. 59 (11), 5793—5798.

Duke, S.O. (Ed.), 1996. Herbicide-Resistant Crops: Agricultural, Environmental, Economic, Regulatory, and Technical Aspects. CRC-Lewis Pub., Boca Raton, FL, 420 pp.

Duke, S.O., Christy, A.L., Hess, F.D., Holt, J.S., 1991. Herbicide Resistant Crops. Comment from CAST. No 1991-1. Council for Agric. Sci. Technol., 24 pp.

Economist, June 11, 2016. Factory Fresh, 8 pp.

Economist, January 22, 2005. Greener than You Thought, pp. 76—77.

Fernandez-Cornejo, J., Nehring, R., Osteen, C., Wechsler, S., Martin, A., Vialou, A., May 2014. Pesticide Use in U.S. Agriculture: 21 Selected Crops. EIB-124, U.S. Department of Agriculture, Economic Research Service, pp. 1960—2008, 80 pp.

Fraley, R.T., Rogers, S.G., Horsch, R.B., Sanders, P.R., Flick, J.S., Adams, S.P., et al., 1983. Expression of bacterial genes in plant cells. Proc. Natl. Acad. Sci. USA 80 (15), 4803–4807.

Funk, C., Kennedy, B., 2016. Public Opinion about Genetically Modified Foods and Trust in Scientists Connected with These Foods. http://www.pewinternet.org/2016/12/01/public-opinionabout-genetically-modified-foodsand-trustin-scientists-connected-with-these-foods.

Gealy, D.R., Mitten, D.H., Rutger, J.N., 2003. Gene flow between red rice (*Oryza sativa*) and herbicide-resistant rice (*O. sativa*): implications for weed management. Weed Technol. 17, 627–645.

Gressel, J., 2000. Molecular biology of weed control. Transgenic Res. 9, 355–382.

Gressel, J., 2002. Molecular Biology of Weed Control. Taylor and Francis, London and NY, 504 pp.

Gressel, J., Segal, L.A., 1978. The paucity of genetic adaptive resistance of plants to herbicides: possible biological reasons and implications. J. Theor. Biol. 75, 349–371.

Hakim, D., October 30, 2016. Doubts about a promised bounty. N. Y. Times 1, 22–23.

Hall, J.C., Donnelly-Vanderloo, M.J., Hume, D.J., 1996. Triazine-resistant crops: the agronomic impact and physiological consequence of chloroplast mutation. In: Duke, S.O. (Ed.), Herbicide-Resistant Crops: Agricultural, Environmental, Economic, Regulatory, and Technical Aspects. CRC-Lewis Pub., Boca Raton, FL, pp. 107–126.

Hanson, V.D., 1996. Fields Without Dreams: Defending the Agrarian Idea. The Free Press, NY, 289 pp.

Hanson, V.D., 2000. The Land Was Everything: Letters from an American Farmer. Simon and Schuster, NY, 258 pp.

Hanson, B.D., Mallory-Smith, C.A., Price, W.J., Shafii, B., Thill, D.C., Zemetra, R.S., 2005. Interspecific hybridization: potential for movement of herbicide resistance from wheat to jointed goatgrass (*Aegilops cylindrica*). Weed Technol. 19, 674–682.

Hellwig, K.B., Johnson, W.G., Massey, R.E., 2003. Weed management and economic returns in no-tillage herbicide-resistant corn (*Zea mays*). Weed Technol. 17, 239–248.

Hileman, B., August 21, 1995. Views differ sharply over benefits, risks of agricultural biotechnology. Chem. Eng. News. 8–17.

Hopkins, W.L., 1994. In: Global Herbicide Directory, first ed. Ag. Chem. Info. Services, Indianapolis, IN, pp. 157–159 http://en.wikipedia.org/wiki/genetically_modified_crops http://cera-gmc.org/index.php?action=gm_crop_database&mode=Synopsis.

Horvath, D., 2010. Genomics for weed science. Curr. Genom. 11 (1), 47–51.

James, C., 2014. Global Status of Commercialized Biotech/GM Crops: 2014. ISAAA Brief No. 49, (Int. Service for the Acquisition of Agri-biotech Applications). Ithaca, NY.

Kalaitzandonakes, N., Giddings, V., McHughen, A., Zahringer, K., 2016. The Impact of Asynchronous Approvals for Biotech Crops on Agricultural Sustainability, Trade, and Innovation. Council for Agricultural Science and Technology (CAST), Ames, IA. Commentary QTA 2016-2. 12 pp.

Kirschenmann, F.L., 2010. Cultivating an Ecological Conscience: Essays from a Farmer Philosopher. Counterpoint, Berkeley, CA, 403 pp.

Kniss, A.R., 2017. Long-term trends in the intensity and relative toxicity of herbicide use. Nat. Commun. 8, 14865. https://doi.org/10.1038/ncomms14865.

Krishna, V., Qaim, M., Zilberman, D., 2016. Transgenic crops, production risk and agrobiodiversity. Eur. Rev. Agric. Econ. 43 (1), 137–164.

La Duke, W., 2007. Rice keepers: A struggle to protect biodiversity and a native American way of life. Orion. July/August, pp. 18–23.

Légère, A., 2005. Risks and consequences of gene flow from herbicide-resistant crops: canola (*Brassica napus* L.) as a case study. Pest Manag. Sci. 61, 292–300.

Li, K., Moulin, M., Mangel, N., Albersen, M., Verhoeven-Duif, N.M., Ma, Q., et al., 2015. Increased bioavailable vitamin B6 in field-grown transgenic cassava for dietary sufficiency. Nat. Biotechnol. 33 (10), 1029–1032.

Logsdon, G., 1998. You Can Go Home Again: Adventures of a Contrary Life. Indiana University press, Bloomington, IN, 204 pp.

Lyon, D.J., Bussan, A.J., Evans, J.O., Mallory-Smith, C.A., Peeper, T.E., 2002. Pest management implications of glyphosate-resistant wheat (*Triticum aestivum*) in the western United States. Weed Technol. 16, 680–690.

Mallory-Smith, C.A., Olguin, E.S., 2011. Gene flow from herbicide-resistant crops: it's not just for transgenes. J Agric. Food Chem. C9, 5813–5818.

Martinez-Ghersa, M.A., Worster, C.A., Radosevich, S.R., 2003. Concerns a weed scientist might have about herbicide-tolerant crops: a revisitation. Weed Technol. 17, 202–210.

Messeguer, J., 2003. Gene flow assessment in transgenic plants. Plant Cell Tissue Organ Cult. 73, 201–212.

NAS — National Academy's of Sciences, Engineering, Medicine, 2016. Genetically Engineered Crops: Experiences and Prospects. The National Academies Press, Washington, DC. https://doi.org/10.17226/23395. Also see. http://nas-sites.org/ge-crops/.

Naylor, R.E.L., 2012. Weed Management Handbook. For British Crop Protection Council, ninth ed. Blackwell, London, UK. 423 pp.

Park, A., July 4, 2016. Life, the remix. TIME 42–48.

Powles, S.B., Holtum, J.A.M. (Eds.), 1994. Herbicide Resistance in Plants: Biology and Biochemistry. Lewis Pub., Boca Raton, FL, 353 pp.

Rao, V.S., 2014. Transgenic Herbicide Resistance in Plants. CRC Press, NY, 458 pp.

Rhodes, R., 1989. Farm: A Year in the Life of an American Farmer. Simon and Schuster, NY, 336 pp.

Riches, C.R., Valverde, B.E., 2002. Agricultural and biological diversity in Latin America: implications for development, testing, and commercialization of herbicide-resistant crops. Weed Technol. 16, 200–214.

Robin, M., 2010. The world according to Monsanto: pollution, corruption, and the control of the world's food supply. New Press, New York, 372 pp.

Scott, S.E., Wilkinson, M.J., 1999. Low probability of chloroplast movement from oilseed rape (*Brassica napus*) into wild *Brassica rapa*. Nat. Biotechnol. 17, 390–393.

Shiva, V., Jafri, A.H., Emani, A., Pande, M., 2005. Seeds of Suicide: The Cological and Human Costs of Globalisation of Agriculture. Navdanya Organization, Delhi, India, 298 pp.

Snow, A.A., Andersen, B., Jørgensen, R.B., 1999. Costs of transgenic herbicide resistance introgressed from *Brassica napus* into weedy *B. rapa*. Mol. Ecol. 8, 605–615.

Spencer, L.J., Snow, A.A., 2001. Fecundity of transgenic wild-crop hybrids of *Cucurbita pepo* (Cucurbitaceae): implications of crop-to-wild gene flow. Heredity 86, 694–702.

Specter, M., November 16, 2015. The gene hackers. New Yorker 52–61.

Stewart Jr., C.N., Tranel, P.J., Horvath, D.P., Anderson, J.V., Rieseberg, L.H., Westwood, J.H., et al., 2009. Evolution of weediness and invasiveness: charting the course for weed genomics. Weed Sci. 57, 451–462.

Stone, G.D., 2007. Agricultural Deskilling and the spread of genetically modified cotton in Warangal. Curr. Anthropol. 48 (1), 67–73.

Thompson, P.B., 2007. Food Biotechnology in Ethical Perspective, second ed. Springer, Dordrecht, NL. 340 pp.

Thompson, P.B., 2015. From Field to Fork: Food Ethics for Everyone. Oxford University press, UK, 343 pp.

Van Wychen, L., 2016. 2015 Survey of the Most Common and Troublesome Weeds in the United States and Canada. Weed Science Society of America National Weed Survey Dataset. Available: http://wssa.net/wp-content/uploads/2015-Weed-Survey_FINAL1.xlsx.

Zimdahl, R.L., 2012. Agriculture's Ethical Horizon. Elsevier, New York, NY, 274 pp.

The Problem and Study of Herbicide Resistance

FUNDAMENTAL CONCEPTS

- The presence and continuing occurrence of weed resistance to herbicides is a major challenge for weed management.

- Weed resistance to one or more herbicides is an inevitable consequence of repeated use of an herbicide in a field over time.

- There are many challenges to continued successful use of herbicide resistant crops.

LEARNING OBJECTIVES

- To know the definition of herbicide resistance and tolerance.

- To understand the development of herbicide resistance in cropped fields.

- To understand farmers' reasons for acceptance of resistance combined with some reluctance to make necessary management changes to delay or prevent its development.

Sir Francis Bacon introduced the idea that man should conquer nature. But the idea that science can make us "masters and possessors of nature" must be credited to the French mathematician and philosopher René Descartes, who in his Discourse on Method said that man must cease being a slave to nature and make the natural world useful by improving our knowledge of nature, but our knowledge would always be imperfect.

1. DEFINITIONS

1.1 Resistance

Herbicide resistance is a decreased response of a weed species population to an herbicide (LeBaron and Gressel, 1982). It is "survival of a segment of the population of a species following a herbicide dose lethal to the normal population" (Penner, 1994). The Weed Science

Society of America defined *resistance* (Senseman, 2007, p. 428) as the "ability to withstand exposure to a potentially harmful agent without being injured." The definition has not changed (Shaner, 2014, p. 486). It is characterized by a decreased or absence of response of a population of a weed species to an herbicide over time. Resistance, first reported in 1957, has become more widespread and obvious since the advent of genetically modified (GM) crops.

1.2 Tolerance

Resistance is often contrasted with tolerance or the natural and normal variability of response to herbicides that exists within a species and can easily and quickly occur (LeBaron and Gressel, 1982). Tolerance is characterized by survival of the normal population of a species following an herbicide dose lethal to other species (Penner, 1994). The Weed Science Society defines *tolerance* as the "ability to continue normal growth or function when exposed to a potentially harmful agent" (Shaner, 2014, p. 487). The society's definitions of resistance and tolerance point out that "there is no general agreement as to the distinction between herbicide tolerance and herbicide resistance in plants." The terms are not always clearly distinguished and often are used as synonyms. The ecological effect is the shift of the population to the resistant biotype. The weed species don't change—the ability to control them does. The topic was reviewed by Shaner (1995) who suggested that if the evident trends continued the number of herbicides effective on several weed species would decline rapidly, which has happened.

2. THE GROWTH AND EXTENT OF HERBICIDE RESISTANCE

The widespread occurrence of resistance is an example of the tragedy of the commons (see Hardin, 1968). The tragedy is found in the remorseless working out of things and the quest for a technical solution to a problem for which there may not be a technical solution. In weed science the tragedy is expressed as the failure of herbicides to control weeds that have become resistant and the consequent failure of and increased difficulty of weed management systems. Previously successful solutions no longer work. The costs (herbicide use may increase, environmental effects occur, and food production costs may increase) of the tragedy are borne by all.

For many years, entomologists knew that insects developed resistance to insecticides. More of the same insecticide did not solve the problem, and often new insecticides or new combinations did not improve control. Weed scientists assumed that weeds might become resistant to herbicides, but that it was not likely to be a major problem for several reasons. The most frequently cited reasons were (Radosevich, 1983):

1. Weeds, even annuals, have a long life cycle compared to insects.
2. Although their seeds move in several ways, weeds are not as environmentally mobile as insects.
3. There was a wide range of herbicides in use and they had several different sites of action. Insecticide resistance, it was assumed, was based on continued exposure to

chlorinated hydrocarbons or organophosphates. The two groups had different sites of action but all members of each group shared a site of action.

4. Crop rotation assured use of different herbicides in a field in different years.
5. Cultivation and other cultural techniques used in the same field where herbicides were used would kill resistant weeds. It was assumed that integration of methods was common.
6. There is and, it was assumed, always would be a large soil seed reserve.
7. Resistant species would probably be less competitive and not survive well.

Time and experience showed that although these were all logical assumptions, all were incorrect. Herbicide resistance has developed and is now an omnipresent problem that dominates weed science research.[1] The concern and action of weed scientists is illustrated by the special issue of weed science published in 2016, Human Dimensions of Herbicide Resistance. Shaw (2016) described the "wicked" nature of the herbicide-resistance problem. Wicked problems were first described by Churchman (1967) as social system problems that were ill formulated with confusing information, many clients and decision makers with conflicting values, and where the ramifications of the whole system are thoroughly confusing. His definition is appropriate for the herbicide resistant (HR) weed problem, which has become difficult or impossible to solve because it is plagued by incomplete, contradictory, and changing management requirements just as social systems are. Wicked problems, in contrast to what one might call behaved or pleasant problems, are characterized by the need for system-level thinking. Everything is interconnected. Wicked problems cannot be split into component parts because pursuit of the technical solution to one aspect of the problem may worsen other aspects. They typically have multiple ends and goals in tension with each other and often result in a proposed solution that tends to create new problems. These characteristics are an apt description of what Shaw described as the "wicked" nature of the herbicide-resistance problem.

Owen (2016) discussed diverse approaches to resistant weed management and acknowledged that the resistance problem has long been recognized as attributable to herbicide use and because acceptance of nonherbicidal tactics has been limited. Hurley and Frisvold (2016) discussed economic barriers to resistance management, and Livingston et al. (2016) discussed economic returns to resistance management. Coble and Schroeder (2016) issued a call for action on resistance management. Seven other papers are included in the 2016 special issue of Weed Science.

The wrongness of the early assumptions first became clear in 1957 when wild carrot resistance to auxin herbicides was reported in Ontario, Canada (Switzer, 1957; Whitehead and Switzer, 1957) and spreading dayflower resistance to auxin herbicides was reported in Hawaii (Hilton, 1957). These discoveries were followed by Ryan's (1970) study of the resistance of common groundsel after atrazine and simazine had been applied once or twice annually for 10 years in Washington. These studies have been cited frequently, but they were not greeted with great concern in 1957 or 1970. However, by 1986, over 50 weeds were resistant to triazines (Nat. Res. Council, 1986) and over 107 resistant biotypes had evolved

---

[1] From 2010 to 2016 the word *resistance* was included in the title of 21.5% of the papers published in Weed Science and 15.2% of the papers in Weed Technology.

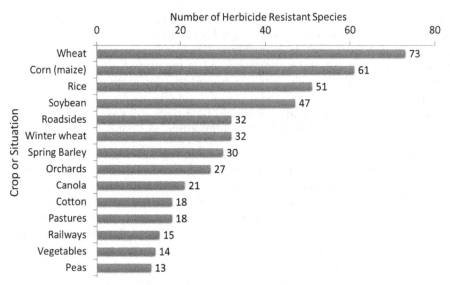

FIGURE 19.1   Number of herbicide-resistant weed species in some crops (Heap, 2016).

around the world. In 1990, 55 weeds were resistant to triazine herbicides in 31 US states, 4 Canadian provinces, and 18 other countries (LeBaron and McFarland, 1990). By the early 1990s, over 100 cases of herbicide resistance had been reported in one or more of 15 chemical families (Holt and LeBaron, 1990; LeBaron and McFarland). In 2005, 304 resistant weed biotypes from 182 species (109 dicots and 73 monocots) were known. Resistant species had been found in more than 270,000 fields in 59 countries. In view of the rapid increase, Shaner's (1995) view was prescient. In June 2017, there were 479 unique cases of resistance (species x site of action), 251 species (146 dicots and 105 monocots). Weeds have evolved resistance to 23 of the 26 known herbicide site-of-action groups and to 160 different herbicides. Resistant weeds have been reported and 91 crops in 69 countries, and the incidence of resistance continues to grow with an average yearly increase over the last 30 years of about 13 cases (Fig. 19.1).[2] For example, in 2014, 71 weed species had developed resistance to triazine herbicides although their importance (use) "had diminished with the shift to Roundup ready crops in the US and reduced herbicide use in Europe" (Heap, 2014). It is reasonable to claim that the most important HR weed problems are the weeds resistant to more than one site of action. Table 19.1 lists 15 weeds each resistant to five or more herbicide sites of action. Resistance can occur quickly. The development of glyphosate crop resistance through genetic modification revolutionized weed management in several important crops. However, in less than a decade, the utility of the "technology was threatened by the occurrence of glyphosate-tolerant/ glyphosate-resistant weed species" (Webster and Sosnoskie, 2010). In December 2016 there were 19 dicot and 17 monocot weeds resistant to glyphosate (Heap, 2016).

Several examples illustrate the problem. Wild radish collected from northern Australian wheat fields where "typical herbicide-use patterns had been practiced for the previous

---

[2]International survey of herbicide-resistant weeds: http://www.weedscience.org/in.asp.

TABLE 19.1 Fifteen Herbicide-Resistant (HR) Weeds Each of Which Exhibits Resistance to More Than One Site-of-Action Group. (See International Survey of Herbicide Resistant Weeds, http://www.weedscience.org/in.asp)

Weed Species	Resistant to Number of Sites of Action
Rigid ryegrass	11
Barnyard grass	10
Annual bluegrass	9
Blackgrass	7
Junglerice	7
Goosegrass	7
Palmer amaranth	6
Tall waterhemp	6
Perennial ryegrass	6
Smooth pigweed	5
Common ragweed	5
Wild oat	5
Horseweed	5
Kochia	5
Wild radish	5

17 seasons" exhibited multiple herbicide resistance across at least four herbicide site-of-action groups (Walsh et al., 2004). This, in the author's view (and, one must presume, in the farmer's view) presents a challenge for future wild radish management with herbicides. In a randomly collected population of rigid ryegrass from 264 fields in western Australian wheat-growing areas, 46% were resistant to diclofop-methyl (ACCase inhibition), 64% were resistant to chlorsulfuron [acetolactate synthase (ALS) inhibition], and 37% were resistant to both herbicides (Llewellyn and Powles, 2001). More rigid ryegrass populations were resistant than susceptible (28%) to the herbicides that had been most widely and successfully used. Herbicide resistance in rigid ryegrass in Australia and blackgrass in Europe has limited small grain production.

In western Nebraska, kochia is widely resistant to atrazine, which, when first introduced, quickly killed a high percentage of emerged seedlings. Because atrazine and other triazines persist in soil, they continue to kill weeds that emerge after application. Therefore, there is a long time when susceptible plants are not present to compete with resistant ones. It was believed (see point 6 above) that the soil seed reservoir, unique to plant populations, would slow the appearance of resistance because only a small percentage (2%–10%) germinates in 1 year. The large seed reserve slowed, but could not prevent expression of resistance. Kochia is also resistant to five (Table 19.1) different sites of action (synthetic auxins, inhibitors of ALS, photosystem II,[3] and EPSP synthase) (http://weedscience.org/summary/species.aspx?WeedID=101). Cao et al. (2010) claim that corn resistant to dicamba will offer enhanced weed control, lower labor and production costs, and increase environmental benefits and gains in profitability. Similar claims are made by advocates of genetic engineering to address/solve the herbicide-resistance problem. Resistance of common broadleaf weeds in wheat to sulfonylurea herbicides has been reported across the western United States. There is no question that herbicide resistance is real and widely present, however, it can be managed, albeit with required attention to the production system, not just the weeds. It is well understood that it results from repeated use of the same herbicide or herbicides with the same site of action in the same field (Owen, 2016). It is not created by the herbicides, it is selected for. Susceptible plants are killed. The resistant population survives and comes to dominate. It is evolution by chemical selection.

Monocultural cropping system that relied on glufosinate-resistant rice for weed control developed resistant weeds within 3–8 years (Madsen et al., 2002). Using tillage for supplemental weed control and increasing weedy rice seed predation delayed, but did not prevent, resistance development. Resistance to glyphosate has occurred in several populations of rigid ryegrass in Australia (Wakelin and Preston, 2006) and several other places (see Table 19.4). The resistance is encoded in the nuclear genome in the eight Australian populations studied and is inherited as a single dominant allele in four of the five resistant populations.

The time for development of resistance has proven to be short (Shaner, 1995). It took 18 years after release for resistance to 2,4-D to be reported. Compared to other herbicide sites of action the incidence of resistance among auxinic herbicides is relatively low (Mithila et al., 2011). There were 29 resistant weed species in 2011, 32 in 2016 (see footnote 2). Resistance to triallate and picloram did not appear until 25 years after introduction, for atrazine and trifluralin it took 10 years, but only 5 years for diclofop and chlorsulfuron. Sulfonylureas and imidazolinones are active at fractions of an ounce per acre, often persist in soil, and have important advantages. Resistance to some of these herbicides has developed in as little as 3 years (Gressel, 1990).

Since 1982 the number of resistant weeds has more than tripled and the land area involved has increased 10 times. Several weed species have evolved cross-resistance to more than one herbicide (Table 19.1). Multiple resistance occurs when two or more distinct resistance mechanisms appear in the same species. In general, but not always, there are enough alternative herbicides and other control measures to incorporate in an integrated weed management (IWM) program (e.g., rotation, tillage, see Chapter 10) to manage resistant weeds effectively.

If one assumes that in the first year of herbicide application there was only one resistant weed in a weed population of 100,000,000 in a large field, it would not be noticed, or if it was, the logical assumption would be that it had emerged after herbicide application or had been missed. It is likely that the resistant population would not be noticed for several years

[3]Two structural groups that differ in binding behavior inhibit Photosystem II—(1) triazines and (2) ureas and amides.

TABLE 19.2 Development of a Population of Resistant Weeds With Repeated Use of a Single Herbicide (Gressel, 1990)

Year	Susceptible Population	Resistant Population
1	100,000,000	1
2	10,000,000	4
3	1,000,000	16
4	100,000	64
5	10,000	256
6	1,000	1024
7	100	4026

(Table 19.2). It would take a person with unusual powers of observation and a keen knowledge of weeds to notice 256 weeds in a large field. If one assumed that 100,000,000 million weeds were all in a 50-acre field, there would be 0.46 weeds/ft^2 one weed/2 ft^2. That is an observable population, but it is likely that in the early 1990s, before herbicide resistance became common knowledge, one would have been delighted with the excellent weed control achieved rather than concerned about what appeared to be some escapes that were mixed with other weeds the herbicide didn't control. Another way to look at the same problem is to note that concomitant with 90% population reduction of susceptible species, the resistant species might increase by a factor of four each year. If it was a 10-acre field (435,600 square feet), with 90% control and a 4× annual increase in the resistant population, Table 19.3 shows what would happen.

TABLE 19.3 Development of a Resistant Weed Population Over 10 Years in a 10-Acre Field (Gressel, 1990)

Year	Resistant Population	Susceptible Population
1	1	4,356,000
2	4	435,600
3	16	43,560
4	64	4,356
5	256	436
6	1024	44
7	4096	4
8	16,384	1
9	65,356	0
10	130,712	0

TABLE 19.4   Number of Resistant Weeds for an Herbicide in a Few Site-of-Action Groups[4]

Inhibitor of Site-of-Action	Herbicide Example	Number of Weeds Resistant to Example	Total Number of Weeds Resistant in Each Site-of-Action Group
ALS	Chlorosulfuron	35	158
ALS	Imazamox	51	
Photosystem II	Atrazine	53	73
Photosystem II	Chlorotoluron	6	28
ACCase	Sethoxydim	14	48
EPSP	Glyphosate	37	37
Synthetic auxins	2,4-D	23	34
Electron divertors	Paraquat	33	32
Microtubule	Trifluralin	9	12
Lipid	Triallate	2	10
PPO	Oxyfluorfen	1	11
13 other groups			46

Farmers and weed scientists are now fully aware of resistance and working on management plans to anticipate, minimize, or eliminate the problem to achieve desired weed management and prevent development of resistant populations.

It may be incorrect to assume that resistance will occur with any herbicide, but the evidence that it can is persuasive. As mentioned previously it has already occurred in 479 unique biotypes from 23 of 26 known site-of-action groups. The last new site-of-action group (cellulose inhibitors) was discovered in 1998. Table 19.4 includes some examples. It is most likely to occur where some or all of the following factors are present:

- The herbicide has a high degree of control of the target species. It is very efficient.
- The weed's seed has a short life in the soil seed bank.
- The herbicide has long soil persistence.
- The herbicide is used frequently—annually for many years or more than once per year for several years.
- Annual herbicide rotation is not practiced.
- The herbicide has a single site of action.
- The herbicide's use rate is high.
- Herbicides are not mixed in a field—that is, only one is used.

[4]Www.weedscience.org/ln.asp. International Survey of Herbicide Resistant Weeds.

Weed resistance occurs more rapidly with continued use of an herbicide to which a GM crop is already resistant. The fact that not all herbicides have the same probability of selecting for resistance is illustrated well by Beckie et al. (2001). They propose that the higher the risk of an herbicide site-of-action group selecting for resistance, the less frequently herbicides from the group should be applied by a grower. The reasoning is sound, but when resistance has been identified in 23 of 26 known site-of-action groups, few herbicide alternatives may remain. Inhibitors of acetyl CoA carboxylase, ALS, and acetohydroxyacid synthase pose a high risk of rapid development of resistant weed biotypes and should not be used frequently (see Table 19.4). Inhibitors of EPSP synthase (glyphosate) and photosystem I (paraquat) electron divertors can be used preseeding to reduce the number of weeds available for selection by higher risk in-crop herbicides (Beckie et al., 2001). Fig. 19.2 (Beckie et al.) illustrates herbicide classification using the risk of resistance development. Readers are encouraged to consult local recommendations when planning a weed-management system.

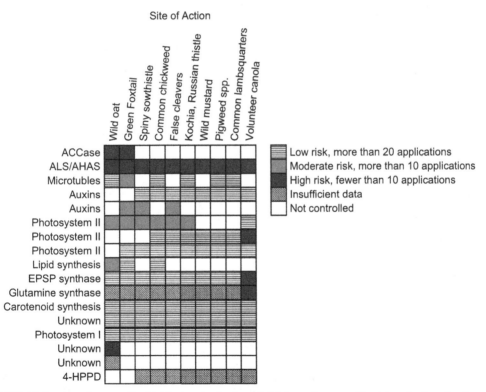

FIGURE 19.2   Classification of herbicide site of action by risk (high, moderate, and low) for selection of resistance in specific weed species in Canada (Mallory-Smith and Retzinger, 2003). *Reprinted with permission from Beckie, H.J., Hall, L.M., Tardif, F.J., 2001. Herbicide resistance in Canada — where are we today? In: Blackshaw, R.E., Hall, L.M. (Eds.), Integrated Weed Management: Exploring the Potential. Expert Comm. on Weeds. Sainte -Anne-de-Bellevue, QC, Canada, pp. 4—36, also see Retzinger Jr, E.J., Mallory-Smith, C.A., 1997. Classification of herbicides by site of action for weed resistance management strategies. Weed Technol. 11, 384—393.*

It is incorrect to assume that the phenomenon of resistance is the death knell for herbicides. Resistant weeds are not superweeds.[5] They may be less fit ecologically than their susceptible relatives. A few of the most important resistant weed species are shown in Table 19.5. It is important to recognize that resistance is not just possible, it is omnipresent.

The reasons cited at the beginning of this section explain why herbicide resistance will remain an important phenomenon. Crop rotation and herbicide rotation to combat different weed problems are important techniques and should be used in IWM systems. If the same herbicide, herbicides from the same chemical family, or those with the same site of action are used on the same land for several successive years, development of resistance is highly likely. Integration of crop rotation and other control techniques in weed management, rather than relying on herbicides to solve all problems, are important components of the solution to resistance development. Resistance is not a phenomenon

TABLE 19.5   A Few Examples of the Number of Resistant Weeds in a Few Site-of-Action Groups[6]

Site-of-action group	Resistant Weed Species	Number of Locations 2012-2016	Years[7]
ALS			
Blackgrass	10	16	1984–2015
Palmer amaranth	11	21	1993–2014
Smooth pigweed	5	10	1992–2014
Redroot pigweed	10	13	1991–2012
Common waterhemp	20	25	1993–2014
Wild oat (*Avena fatua*)	13	19	1996–2015
Kochia	23	28	1987–2015
Italian ryegrass	13	21	1995–2013
Total number of resistant species		158	
PHOTOSYSTEM II INHIBITORS - INCLUDING UREAS AND AMIDES			
Com. lambs quarters	38	40	1973–2010
Kochia	11	14	1976–2014
Annual bluegrass	13	14	1978–2013
Total number of resistant species		73	

[5]The Weed Science Society of America defines 'superweed' as a slang term used "to describe weeds that have evolved characteristics that make them more difficult to manage as a result of repeated use of the same management" methods. Superweed is commonly used incorrectly to refer to weeds that have evolved herbicide resistance and, it is incorrectly assumed, are thereby uncontrollable.

[6]See footnote 4.

[7]Years are the first and last resistance report.

TABLE 19.5 A Few Examples of the Number of Resistant Weeds in a Few Site-of-Action Groups[6]—cont'd

Site-of-action group	Resistant Weed Species	Number of Locations 2012-2016	Years[7]
ACCASE INHIBITORS			
Blackgrass	16	21	1982–2015
Wild oat (*A. fatua*)	33	35	1985–2015
Italian ryegrass	24	28	1987–2013
Johnsongrass	8	10	1991–2015
Total number of resistant species			42
SYNTHETIC AUXINS			
Total number of resistant species			32
BIPYRIDILIUMS			
Total number of resistant species			31
EPS			
Glyphosate	**Total number of resistant species**		31

limited to herbicides. It is a problem that ranges across several categories of the biological spectrum, including herbicides, insecticides, parasiticides, fungicides, and antibiotics (Table 19.6) (Rasmussen, 2016).

2.1 Soil Conservation

"There is a large and growing threat to soil conservation gains because of the clear need in many fields to manage resistant weeds by any means, including tillage" (CAST, 2012). There is no doubt that soil tillage may promote erosion, can negatively affect surface water quality, and due to soil loss can decrease productivity. With increasing herbicide resistance related to the spread of GM crops, tillage is the best available, perhaps the only way, to control HR weeds.

Duke (2011) claims that GM crops have "substantially reduced the use of environmentally and toxicologically suspect pesticides." Therefore, it is necessary that some farmers use what are regarded as older weed management methods such as cultivation, which may increase soil erosion.

3. MECHANISMS OF HERBICIDE RESISTANCE

This book will not attempt to describe the methods for study or the details of the mechanisms of herbicide resistance. The subject is vast and involves sophisticated chemical, biochemical, genetic, and physiological study beyond the scope of this book.

TABLE 19.6   Known Resistance Problems in Addition to Herbicides

Compound	Type	Targeted organisms
Bt proteins	Insecticide	European corn borer
		Colorado potato beetle
		Pink bollworm
Methoprene	Insecticide	Mosquitoes
		Stored grain insects
		Some fly species
Tolnaftate	Fungicide	Tinea root fungus
		Ring worm
Fluconazole	Fungicide	Candida yeast
		Aspergillus
Chloroquine	Parasiticide	Malaria parasite
Penicillin	Antibiotic	Staphylococcus
Fluoroquinolones	Antimicrobial	Neisseria
		Staphylococcus
Carbapenem	Antimicrobial	Enterobacteria

All herbicides must be absorbed by the target plant, moved in sufficient concentration to an active site, bound to the active site, and thereby inhibit action of a metabolic pathway. Resistance may happen when one or more of these steps does not occur. Lack of desired activity could occur for other reasons unrelated to resistance: insufficient uptake, reduced translocation to the target site, sequestration within the plant, rapid metabolism, or target-site mutation that prevents herbicide binding to the intended target. Interested readers are referred to the following papers for more specific information: Dayan et al. (2015), Délye et al. (2015), Vila-Aiub et al. (2015), Yuan et al. (2007), Mallory-Smith et al. (2015), Forouzesh et al. (2015), and Burgos (2015).

The four known mechanisms of resistance are[8]

1. Altered target site

Each herbicide has a specific site (target site of action) where it acts to disrupt a particular plant process or function (site of action). If this target site is somewhat altered, the herbicide may no longer bind to the site of action and is unable to exert its phytotoxic effect. This is the most common mechanism of herbicide resistance (Fig. 19.3).

[8]Material in this section and Figs. 19.3–19.6 are used with permission of Dr. Wayne Buhler, Department of Horticultural Science, North Carolina State University. Information can be found at: http://pesticidestewardship.org/resistance/Herbicide/Pages/Mechanisms-of-Herbicide-Resistance.aspx.

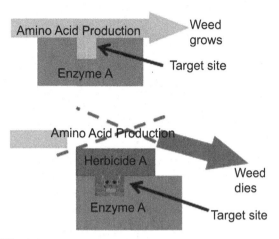

FIGURE 19.3   Mechanisms of herbicide resistance—altered target site.

2. Enhanced metabolism

Metabolism within the plant is a mechanism plants use to detoxify a foreign compound such as an herbicide. A weed with the ability to quickly degrade an herbicide can potentially inactivate it before it can reach its site of action within the plant (Fig. 19.4).

3. Compartmentalization or sequestration

Some plants are capable of restricting the movement of foreign compounds (e.g., an herbicide) within their cells or tissues to prevent the compounds from causing harmful effects. In this case, an herbicide may be inactivated either through binding (such as to a plant sugar molecule) or removed from metabolically active regions of the cell to inactive regions, the cell wall, for example, where it exerts no effect (Fig. 19.5).

4. Overexpression of the target protein

If the target protein on which the herbicide acts can be produced in large quantities by the plant, then the effect of the herbicide becomes insignificant (Fig. 19.6).

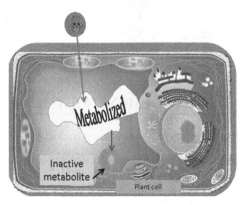

FIGURE 19.4   Mechanisms of herbicide resistance—enhanced metabolism.

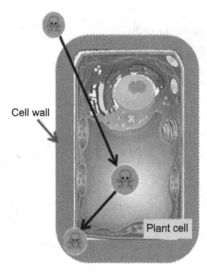

FIGURE 19.5   Mechanisms of herbicide resistance—compartmentalization or sequestration.

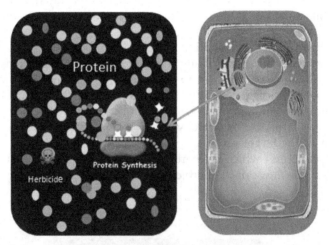

FIGURE 19.6   Mechanisms of herbicide resistance—overexpression of the target protein.

4. MANAGEMENT METHODS

Weed management practices to combat resistance include:

- Knowing what herbicide families develop resistance rapidly and minimizing or eliminating their use.
- Incorporating other weed-control techniques in a weed-management program.

Beckie and Gill (2006) list three factors that, in their view, are responsible for major herbicide-resistance weed problems:

- Multiple applications of highly efficacious herbicides with the same site of action.
- Annual weed species that occur at high densities are widely distributed, genetically variable, prolific seed producers, and have efficient gene (seed and pollen) distribution systems.
- Simple (monocultural) cropping systems that favor a few dominant weed species.

They add a fourth factor that may be the most important. Farmers have been "reluctant to proactively manage weeds to delay selection for herbicide resistance." Norsworthy et al. (2012) also cite farmer's reluctance. A farmer's reluctance is rational. It is derived from the very important desire to make a profit, the plethora of available herbicides, and the desire (need) for a quick fix.

Beckie and Gill (2006) and Vencill et al. (2012) describe several available, rational management strategies. They are not preventive strategies although if used with appropriate managerial skill they are likely to postpone appearance of the problem.

- Herbicide rotations and sequences. "Herbicide sequence is two or more applications of all appropriate herbicides to one crop in one year. Herbicide rotation is the application of herbicides to multiple crops over several growing seasons." Beckie and Gill provide several examples.
- Herbicide mixtures seem to make sense if herbicides with different sites of action that are degraded by different biochemical pathways are mixed. Beckie and Gill point out that information on the mode of degradation of herbicides in plants is not readily available to most farmers.
- Management might be aided by developing crop cultivars with alternative single or stacked herbicide-resistant traits, such as synthetic auxins, which will become available (Beckie and Tradif, 2012).
- Reducing herbicide rates is possible, but efficacy may be reduced. Therefore, this technique must be combined with other weed-management methods.
- Site-specific management is potentially useful but requires identification and monitoring of weed patches. Alternative techniques can be used to prevent seed production and reduce the potential for resistance.

A weed-control system (discussed in Chapter 3) to take advantage of resistance has been developed to control *Striga* spp. for small-scale African farmers (see Kabambe et al., 2008, Kanampiu, 2001, 2008, 2003; Ransom et al., 2012).

Beckie and Gill and Vencill et al. include discussion of nonherbicide strategies that, in the view of many, are the best way to prevent development of resistance. As mentioned before, nonherbicide strategies include many weed-management techniques that have been known and used for years (see Chapters 10 and 11): prevention, crop rotation, cover crops, intercropping, mulch, planting date, adjusting seeding rate, row spacing, soil fertility, irrigation, allelopathy, tillage, and biological control. Use of nonselective herbicides before planting the crop combined with infrequent selective herbicide use "have mitigated the evolution, spread and economic impact of herbicide-resistant wild oats and green foxtail, the most important herbicide-resistant grass weeds in the northern Great Plains of Western

Canada" (Beckie, 2007). Green and Owen (2011) acknowledge that HR crops changed the tactics that soybean and cotton growers use to manage weeds. They emphasize that it is necessary to diversify management tactics *and* to discover herbicides with new sites of action. A significant part of their solution to the problem created by herbicides is research to find new herbicides with new sites of action. That is not happening. Norsworthy et al. (2012) offer a more challenging solution. They suggest that "long-term herbicide resistance management…requires more than weed control aimed only at minimizing crop loss in any one season." Management requires:

- Reducing the number of weeds exposed to herbicide selection.
- Preventing propagation of HR weeds.
- Reducing the soil weed seed bank in future years.

To be successful these requirements must be complemented by more knowledge of weed biology, a research area that has received comparatively minor emphasis. Theirs is a management/ecological research challenge, not just a chemical challenge. It is made more difficult by the acknowledged reluctance of growers "to proactively manage weeds to delay selection for herbicide resistance" (Beckie and Gill, 2006; Norsworthy et al., 2012).

The most successful weed-management programs will also include crop rotation to prevent one species or a weed complex from dominating. The program will include consideration of the proper role of soil tillage, cultural practices to take advantage of crop competition (e.g., using narrower rows to maximize crop competitiveness), herbicides with different sites of action in successive years to slow resistance development, and herbicides with a short rather than a long soil residual life (Gaussoin et al., 2005). Management of herbicide resistance will require reducing reliance on herbicides as the primary tool for weed management and developing IWM systems that require substituting human intellect and skill for chemical technology (Shaner, 1995). Mixing herbicides with different sites of action will slow but not prevent resistance development. Long persistence has advantages and important disadvantages. Some, but not all (e.g., glyphosate) cases of herbicide resistance involve herbicides with relatively long-lasting (months to years) soil residual.

The possibility of resistant crop plants becoming hard-to-control volunteer weeds is real. This has not been shown, but Keeler (1989) urged caution and pointed out the example of wild proso millet that emerged as a weed in the 1970s after over 200 years of successful cultivation of proso millet in North America without its becoming a weed. Keeler used wild proso millet to emphasize how much we do not understand about weed evolution. Movement of glyphosate or glufosinate-resistant creeping bentgrass off-site was deemed likely but not problematic because new weed problems were not anticipated (Banks et al., 2004).

Norsworthy et al. (2012) conclude their analysis of herbicide resistance with nine weed management/HR recommendations. Their recommendations include those mentioned by others and cited previously. Knowledge of the soil weed seed bank is required. Best-management practices for crops and regions must be developed, defined, and promoted. To overcome farmers's reluctance to adopt proactive HR management, the benefits and risks of doing so must be identified and explained. Six of the nine recommendations are consistent with the view that the best solutions to HR problems will be improvement of present herbicide use practices and development of herbicides with new sites of action. Their recommendations are:

1. Include the site of action on each herbicide's label.
2. All interested parties should combine to advocate discovery of new herbicide sites of action and make all aware that the existing herbicide resource is exhaustible.
3. The chemical industry and government should develop and implement incentives to conserve critical herbicide sites of action.
4. Users should be encouraged to follow all label directions.
5. All involved should be actively engaged in the promotion of sustainable herbicide use and adoption of best management practices.
6. Funding agencies should address the knowledge gaps in best-management practices for herbicide-resistance management and prevention.

These weed-management practices have been adopted as illustrated by Cerdeira et al. (2011), Hurley et al. (2009), Ball et al. (2003) and numerous articles in Weed Science, Weed Research, Weed Technology, the online journal www.agbioforum.org, and other scientific journals cited in this chapter.

5. CHALLENGES

The herbicide chemical industry recognizes the problems of resistance and strongly advocates IWM practices that incorporate HR crops with appropriate cultural, mechanical, and biological management methods (Shaner, 1995). All major herbicide development companies and most US land-grant colleges of agriculture have research programs dealing with HR weeds and crops.

Resistance management is complicated by two other factors (Norsworthy et al., 2012) that are neither biological nor chemical. They are perfectly rational human attitudes. The first—the necessity of making a profit—is a desirable attribute of any business manager. The second is based on experience and faith in technology. Because it has happened regularly in the past several decades, it is reasonable to assume that once again technology will save us. Technology will solve the problems it has created. We are technological optimists with an enduring faith that increasing use of more-sophisticated high-energy technology is always a good thing and that any problems caused by the unintended consequences of previous technological solutions will be solved by more technology (Jensen, 2016).

Given that it takes 9–10 years from discovery to release of a new herbicide and the cost can exceed $250+ million dollars, it is not difficult to understand why agrochemical companies are not making major investments in discovery. A new herbicide may not recoup its development costs, and despite the growing resistance problem, most herbicides work well most of the time.

Harker et al. (2012) correctly note that weed scientists seem to agree that the best, perhaps the only way, to reduce selection pressure for herbicide resistance is to minimize herbicide use. They then make the well-supported (see any recent issue of Weed Science or Weed Technology) claim that the "solutions that have emerged in most recent meetings on herbicide resistance have usually involved more herbicide use." They include herbicide rotation, tank mixtures, pre- followed by postemergence herbicides, use of correct rates, and others. Harker et al. claim that weed scientists are "choosing to ignore integrated weed management"—a technique all recommend. It is an accurate, devastating challenge.

A new weed science subdiscipline has appeared: resistance management. Unfortunately, consistent with Harker et al.'s (2012) challenge, resistance management has become an accepted component of IWM. It may not be too far off the mark to say it has become the most important component. Should it be?

Fifty years ago Rachel Carson (1962) published her runaway best seller, Silent Spring. The agricultural chemical industry and the majority of weed scientists dismissed her thesis, only to have to admit much later that she was right or, at least, that she made them think. Her most-challenging claim was ecological. She asked if we have the right, if it is acceptable, to regard the creatures of the natural world simply as objects to be manipulated/engineered to meet the real or perhaps only the wants of humans? Carson's question remains as relevant today as it was 50 years ago. Can we, should we, is it best for us to use the natural world this way? Will it lead to what all seem to favor —a sustainable agriculture? Have GM crops become a kind of Prozac for agriculture and weed science (Cokinos, 2010)?

Professors of weed science have been teaching students and extension personnel have been advocating the importance of IWM for decades. It is disturbingly accurate to claim that resistance is a result of weed scientists and farmers' desire for a product that solves weed problems. Two questions dominate weed science: (1) What is that weed? (2) How can it be controlled? They are good questions. Answers to slightly different questions that are consistent with the goals of IWM are more difficult to obtain: (1) What is that weed? (2) Why is it there? Answering the second question demands knowledge about the production system and about the environment in which production occurs (Zimdahl, 1999). However, most (not all) of those engaged in agriculture have opted for a quick fix without knowing why the problem exists. IWM can use molecular, genetic, and physiological information, but its fundamental foundation is in ecology, not in chemistry or GM. The quick fix has allowed growers, advisers, and weed scientists to solve the problems herbicide resistance has created without fully understanding why the problems occurred. The quick fix is in an apparent solution that permits avoidance of research designed to understand the production system in which the problem occurred.

THINGS TO THINK ABOUT

1. What is the present and future role of HR crops?
2. Is there a potential role for HR crops in weed management in developing countries? Describe the advantages and disadvantages of HRCs for farmers in developing countries.
3. What is the future role of genetic modification in weed management?
4. What reasons can be offered for farmers' reluctance to prevent development of HR weeds?
5. Describe and discuss the several aspects of concern about genetic modification and patenting genes.
6. Are HR weeds a product of evolution?

7. Can herbicides create resistant weeds? If so, how?
8. What are the best management practices to present development of herbicide resistant weeds?
9. Is the quick fix a sustainable long-term strategy for management of HR weeds and crops?
10. Is genetic modification of crops sustainable over the long term?

Literature Cited

Ball, D.A., Rainbolt, C., Thill, D.C., Yenish, J.P., 2003. Weed Management Strategies for Clearfield Wheat Systems across PNW Precipitation Zones. pnwsteep.wsu.edu/directseed/conf2k3/dsc3ball3.htm.

Banks, P.A., Branham, B., Harrison, K., Whitson, T., Heap, I., 2004. Determination of the potential impact from release of glyphosate- and glufosinate-resistant *Agrostis stolonifera* L. In: Various Crop and Non-crop Ecosystems.

Beckie, H.J., 2007. Beneficial management practices to combat herbicide-resistant grass weeds in the northern Great Plains. Weed Technol. 21, 290–299.

Beckie, H.J., Tardif, F.J., 2012. Herbicide cross resistance in weeds. Crop Prot. 35, 15–28.

Beckie, H.J., Gill, G.S., 2006. Strategies for managing herbicide-resistant weeds. In: Singh, H.P., Batish, D.R., Kohli, R.K. (Eds.), Handbook of Sustainable Weed Management. Food Product Press a division of Haworth Reference Press, Binghamton, NY, pp. 581–625.

Beckie, H.J., Hall, L.M., Tardif, F.J., 2001. Herbicide resistance in Canada – where are we today? In: Blackshaw, R.E., Hall, L.M. (Eds.), Integrated Weed Management: Exploring the Potential. Expert Comm. On Weeds. Sainte -Anne-de-Bellevue, QC, Canada, pp. 4–36.

Burgos, N.R., 2015. Whole-plant and seed bioassays for resistance confirmation. Weed Sci. (Spec. Issue) 63, 152–165.

Cao, M., Sato, S.J., Behrens, M., Jiang, W.Z., Clemente, T.E., Weeks, D.P., 2010. Genetic engineering of Maize (*the Zea mays*) for high-level tolerance to treatment with the herbicide dicamba. J. Agric. Food Chem. 59, 5830–5834.

Carson, R., 1962. Silent Spring. Houghton Mifflin, Boston, MA, 378 pp.

CAST (Council for Agricultural Science and Technology), 2012. Herbicide-resistant weeds threaten soil conservation gains: finding a balance for soil and farm sustainability. Issue Pap. 49, 16 pp.

Cerdeira, A.L., Gazziero, D.L.P., Duke, S.O., Matallo, M.B., 2011. Agricultural impacts glyphosate-resistant soybean cultivation in South America. Agric. Food Chem. 59, 5799–5807.

Churchman, C.W., 1967. Wicked problems. Manag. Sci. – Guest Editor. 14 (4), B141–B142.

Coble, H.D., Schroeder, J., 2016. Call to action on herbicide resistance management. Weed Sci. (Spec. Issue) 64, 661–666.

Cokinos, C., Autumn 2010. Prozac for the Planet – Can Geoengineering Make the Climate Happy Again? Am. Sci. 20–33.

Dayan, F., Owens, D.K., Corniani, N., Silva, F., Watson, S.B., Howell, J., et al., 2015. Biochemical markers and enzyme assays for herbicide mode of action and resistance studies. Weed Sci. (Spec. Issue) 63, 23–63.

Délye, C., Duhoux, A., Pernin, F., Riggins, C.W., Tranel, P.J., 2015. Molecular mechanisms of herbicide resistance. Weed Sci. (Spec. Issue) 63, 91–115.

Duke, S.O., 2011. Comparing conventional and biotechnology-based pest management. J. Agric. Food Chem. 59 (11), 5793–5798.

Forouzesh, A., Zand, E., Soufizadeh, S., Foroushani, S., 2015. Classification of herbicides going to chemical family for weed resistance management strategies – an update. Weed Res. 55, 334–358.

Gaussoin, R.E., Kappler, B.F., Klein, R.N., Knezevic, S.K., Lyon, D.J., Martin, A.R., Roeth, F.W., Wicks, G.A., Wilson, R.G., 2005. Guide for weed management. In: Nebraska. Inst. of Agriculture and Nat. Resources. Univ. of Nebraska, Lincoln, NE, 168 pp.

Green, J.M., Owen, M.D.K., 2011. Herbicide-resistant crops: utilities and limitations for herbicide-resistant weed management. J. Agric. Food Chem. 59, 5819–5829.

Gressel, J., 1990. Synergizing herbicides. Rev. Weed Sci. 5, 49–82.

Hardin, G., 1968. The tragedy of the commons. Science 162 (3859), 1243–1248.

Harker, K.N., O'Donovan, J.T., Blackshaw, R.E., Beckie, H.J., C.Mallory-Smith, Maxwell, B.D., 2012. Our view. Weed Sci. 60, 143–144.

Heap, I., 2016. The International Survey of Herbicide Resistant Weeds. Online. www.weedscience.org.

Heap, I., 2014. Global perspective of herbicide-resistant weeds. Pest Manag. Sci. 70 (9), 1306–1315.

Hilton, H.W., 1957. Herbicide tolerant strains of weeds. Hawaiian Sugar Planters Assoc, p. 69. Ann. Rept.

Holt, J.S., LeBaron, H.M., 1990. Significance and distribution of herbicide resistance. Weed Technol. 4, 141–149.

Hurley, T., Frisvold, G.B., 2016. Economic barriers to herbicide resistance management. Weed Sci. (Spec. Issue) 64, 585–594.

Hurley, T.M., Mitchell, P.D., Frisvold, G.B., 2009. Weed management costs, weed best management practices, and the Roundup Ready weed management program. Agbioforum 12, 269–280.

Jensen, R., Fall 2016. What is the world? Who are we? What are we going to do about it? Land Rep. (116), p22–26.

Keeler, K.H., 1989. Can genetically engineered crops become weeds? Biotechnology 7, 1134–1139.

Kabambe, V.H., Kauwa, A.E., Nambuzi, S.C., 2008. Role of herbicide (metalachlor) and fertilizer application in integrated management of Striga asiatica in maize. Malawi Afr. J. Agric. Res. 3 (2), 140–146.

Kanampiu, F.K., 2001. Herbicide seed coating: taming the Striga witchweed. Crop Prot. 20 (10), 885–895.

Kanampiu, F.L., 2008. Herbicide seed coating: taming the Striga witchweed. In: Second Session of the Open Forum on Agricultural Biotechnology in Africa. CIMMYT, Nairobi, Kenya.

Kanampiu, F.K., Kabambe, V., Massawe, C., Jasi, L., Friesen, D., Ransom, J.K., Gressel, J., 2003. Multi-site, Multi-season Field Tests Demonstrate that Herbicide Seed-coating Herbicide-resistance Maize Controls Striga spp.

LeBaron, H.M., Gressel, J., 1982. Herbicide Resistance in Plants. Wiley Interscience, N.Y., p. xv.

LeBaron, H.M., McFarland, J., 1990. Herbicide resistance in weeds and crops: an overview and prognosis. In: Green, M.B., LeBaron, H.M., Moberg, W.K. (Eds.), Managing Resistance to Agrochemicals: From Fundamental Research to Practical Strategies, ACS Symp. Ser. 421. Amer. Chem. Soc., Washington, D.C., pp. 336–352.

Livingston, M.J., Fernandez-Cornejo, Frisvold, G.B., 2016. Economic return to herbicide resistance management in the short and long run: the role of neighborhood effects. Weed Sci. (Spec. Issue) 64, 595–608.

Llewellyn, R.S., Powles, S.B., 2001. High levels of herbicide resistance in rigid ryegrass (Lolium rigidum) in the wheat belt of Western Australia. Weed Technol. 15, 242–248.

Madsen, K.H., Valverde, B.E., Jensen, J.E., 2002. Risk assessment of herbicide-resistant crops: a Latin American perspective using rice (Oryza sativa) as a model. Weed Technol. 16, 215–223.

Mallory-Smith, C., Hall, L.M., Burgos, N.R., 2015. Experimental methods to study gene flow. Weed Sci. (Spec. Issue) 63, 12–22.

Mallory-Smith, C.A., Retzinger Jr., J., 2003. Revised classification of herbicides by site of action for weed resistance management strategies. Weed Technol. 17, 605–619.

Mithila, J., Hall, J.C., Johnson, W.G., Kelley, K.B., Riechers, D.E., 2011. Evolution of resistance to auxinic herbicides: historical perspectives, mechanisms of resistance, and implications for broadleaf weed management in agronomic crops. Weed Sci. 59, 445–457.

National Research Council, 1986. Pesticide Resistance, Strategies and Tactics for Management, National Academy press. Washington DC 11–70.

Norsworthy, J.K., Ward, S., Shaw, D., Llewellyn, R., Nichols, R., Webster, T., et al., 2012. Reducing the risks of herbicide resistance: best management practices and recommendations. Weed Sci. 60 (Special Issue), 31–62.

Owen, M.D.K., 2016. Diverse approaches to herbicide-resistant weed management. Weed Sci. (Spec. Issue) 64, 570–584.

Penner, D., 1994. Herbicide action and metabolism. In: Turf Weeds and their Control. Amer. Soc. of Agron. and Crop Sci. Soc. of Amer., Madison, WI, pp. 37–70.

Radosevich, S.R., 1983. Herbicide resistance in higher plants. In: Georghiou, G.P., Saito, T. (Eds.), Pest Resistance to Pesticides. Plenum Press, New York, NY, pp. 453–479.

Ransom, J., Kanampiu, F., Gressel, J., De Groote, H., Burnet, M., Odhiambo, G., 2012. Herbicide applied to imidazolinone-resistant-maize seed as a Striga control option for small-scale African farmers. Weed Sci. 60, 283–289.

Rasmussen, M., 2016. Resistance finds a way of life. Leopold Lett. 28 (1), 3.

Retzinger Jr., E.J., Mallory-Smith, C.A., 1997. Classification of herbicides by site of action for weed resistance management strategies. Weed Technol. 11, 384–393.

Ryan, G.F., 1970. Resistance of common groundsel to simazine and atrazine. Weed Sci. 18, 614–616.

Senseman, S.A. (Ed.), 2007. Herbicide Handbook, ninth ed. Weed Science Society of America, Lawrence, KS. 458 pp.

Shaner, D.L., 2014. In: Herbicide Handbook, tenth ed. Weed Science Society of America, Lawrence, KS. 513 pp.

Shaner, D.L., 1995. Herbicide resistance: where are we? How did we get here? Where are we going? Weed Technol. 9, 850–856.

Shaw, D.R., 2016. The "wicked" nature of the herbicide resistance problem. Weed Sci. (Spec. Issue) 64, 552–558.

Switzer, C.M., 1957. The existence of 2,4-D resistance strains of wild carrot. In: Proc. Northeastern Weed Contr. Conf., vol. 11, pp. 315–318.

Vencill, W.K., Nichols, R.L., Webster, T.M., Soteres, J.K., Mallory-Smith, C., Burgos, N.R., et al., 2012. Herbicide resistance: toward an understanding of resistance development and the impact of herbicide-resistant crops. Weed Sci. (Spec. Issue) 63, 2–30.

Vila-Aiub, M., Gundel, P.E., Preston, C., 2015. Experimental methods for estimation of plant fitness costs associated with herbicide-resistant genes. Weed Sci. 63 (Special Issue), 203–216.

Wakelin, A.M., Preston, C., 2006. Inheritance of glyphosate resistance in several populations of rigid ryegrass (*Lolium rigidum*) from Australia. Weed Sci. 54, 212–219.

Walsh, M.J., Powles, S.B., Beard, B.R., Parkin, B.T., Porter, S.A., 2004. Multiple-herbicide resistance across four modes of action in wild radish (*Raphanus raphanistrum*). Weed Sci. 52, 8–13.

Webster, T.M., Sosnoskie, L.M., 2010. Loss of glyphosate efficacy: a changing weed spectrum in Georgia cotton. Weed Sci. 58, 73–79.

Whitehead, C.W., Switzer, C.M., 1957. The differential response of strains of wild carrot to 2,4-D and related herbicides. Can. J. Plant Sci. 43, 255–262.

Yuan, J.S., Tranel, P.J., Stewart Jr., C.N., 2007. Non-target-site herbicide resistance: a family business. Trends Plant Sci. 12 (1), 6–13.

Zimdahl, R.L., 1999. My view. Weed Sci. 46, 1.

Herbicides and the Environment

FUNDAMENTAL CONCEPTS

- Herbicides are synthetic organic chemical molecules that do not occur naturally in the environment.
- All herbicides can be dangerous. Few are inherently dangerous when used properly.
- Herbicides control weeds and manage vegetation in situations where no other method is as efficient.

- Herbicide performance is measured by activity, selectivity, and soil residual behavior.
- Herbicide resistance is an important, increasingly difficult aspect of herbicide use.
- There are positive and negative interactions that occur whenever weeds are controlled.
- Science can measure risk; safety is a normative political judgement.

LEARNING OBJECTIVES

- To understand how activity, selectivity, and residual characteristics determine an herbicide's environmental interactions.
- To be aware that weed management can have positive and negative environmental effects.
- To know that herbicides and plant pathogens interact and that this may affect weed management.

- To understand some aspects of the energy relationships of herbicide use.
- To appreciate the complexity of weed management's interactions with humans and the environment.
- To understand how the LD_{50} and perception of risk affect herbicide use.
- To know rules for safe use of herbicides.

Herbicides, the weed management technique that has dominated developed country agriculture for several decades, are synthetic organic chemicals that do not occur naturally in the environment. Their dominance and chemical nature does not make them inherently evil or dangerous. They do define a need for caution and demand constant attention for possible detrimental effects that can be prevented by intelligent use. Herbicides are the most commonly used pesticide[1] in the United States in terms of the dollars spent to buy them and the number of pounds used. Agricultural chemicals consume about 4% of farm expenditures.

Glyphosate was synthesized, but not patented, in 1950 by a Swiss chemist. It was patented in 1964 as a chelating agent by Stauffer Chemical Company. Its herbicidal activity was discovered in 1970 by J.E. Franz of Monsanto during a search for chelating agents. It was released as Roundup in 1974. By 2007, it had become the most widely used herbicide in US agriculture and the second most used around homes and in gardens. More than 180 million pounds were used in 2007. It replaced atrazine as the most widely used herbicide in US agriculture. Since its introduction in 1974, 3.5 billion pounds have been used in the United States and 18.9 billion pounds in the world.

Annual US pesticide expenditures were $11.8 billion in 2006 and 12.8 in 2007, accounting for 34% of the world herbicide market and 45% of US sales.[2] The global herbicide market surpassed $14 billion in 2001, was projected to exceed $20 billion by 2016,[3] and is projected to be $31.5 billion by 2020. The Asian-Pacific region represents two-fifths of the market. Due to higher prices in North America, the revenue there is almost one-third of the global herbicide market.[4] The Economic Research Service of the US Department of Agriculture (USDA) reported that pesticide use on 21 major crops more than tripled from 1960 to 1981. Annual US pesticide use from 1960 to 2008 is shown in Fig. 20.1.

This chapter presents information on harmful and beneficial aspects of herbicides and their environmental interactions. It is not an exhaustive discussion of weed management-environment, herbicide-environment or herbicide-human interactions. That is not the purpose of this book. Sources of additional information have been cited at the end of the chapter. It is incorrect to assume that all weed management-environmental interactions or effects of herbicides are negative or harmful. Some are, most are not. Examples of both will be presented to encourage understanding of and clear thought about possible environmental interactions when herbicides are used to manage weeds.

Perhaps it will be wise to pause here to state the author's bias and thus make it clear rather than suspected. My bias, if it is that, is close to that of Berenbaum et al. (2000) and the many coauthors who wrote the US National Academy of Sciences report on the future role of pesticides in US agriculture. Their goal, expressed as a coda, follows:

> Our goal in agriculture should be the production of high-quality food and fiber at low cost and with minimal deleterious effects on humans or the environment. To make agriculture more productive and profitable in the face of rising costs and standards of human and environmental health, we will have to use the best combination of available technologies. These technologies should include chemical, as well as biological and recombinant, methods of pest control integrated into ecologically balanced programs. The effort to reach the goal must be based on sound fundamental and applied research, and decisions must be based on science.

[1]Note: *Pesticide* is a general term for all pest-control chemicals: herbicides, insecticides, fungicides, etc.

[2]http://www.epa.gov/opp00001/pestsales/07pestsales/market_estimates2007.pdf and http://epa.gov/oppfead1/cb/csb_page/updates/2011/sales-usage06-07.html.

[3]www.marketsandmarkets.com/Market-Reports/crop-protection-380.html.

[4]https://www.alliedmarketresearch.com/press-release/global-herbicides-market-is-expected-to-grow-to-31-5-billion-by-2020-allied-market-research.html.

Pesticide use in U.S. agriculture peaked in 1981 (21 selected crops, 1960 – 2008)

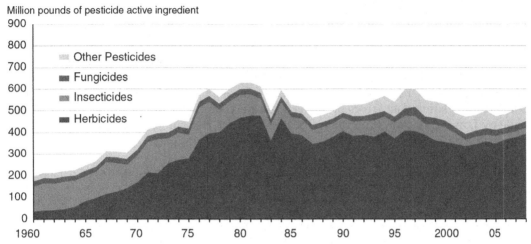

FIGURE 20.1 Annual US pesticide use 1960–2008. *Reproduced from USDA, Economic Research Service using USDA, National Agricultural Statistics Service and property data. http://www.ers.usda.gov/amber-waves/2014-june/pesticide-use-peaked-in-1981,-then-trended-downward,-driven-by-technological-innovations-and-other-factors.aspx#.VubLaPkrKhc.*

My bias[5] has also been influenced by the thoughts and challenges from others about the presumed advantages of agricultural technology and its uses.[6] My quest for a long-term view began with Rachel Carson's Silent Spring (1962) and has continued. Agricultural scientists, especially those in pest management, were, in Van den Bosch's view (1978, p. 21):

> Sucked in to the vortex, and for a couple of decades became so engrossed in developing, producing, and assessing the new pesticides that they forgot that pest control is essentially an ecological matter. Thus, virtually an entire generation of researchers and teachers came to equate pest management with chemical control.

The ecological nature of pest control was affirmed by Julian Huxley in his preface to Silent Spring and is now a consistent, but far from dominant, theme of weed science research.

For decades, agricultural scientists have been and largely remain technological optimists: new technology will solve the problems old technology created. Weed science began with the development of herbicides, in an era of forward-looking scientific optimism. The journal Weed Science first appeared in 1951, and the Weed Science Society of America was founded in December 1954 in Fargo, ND. The society's first meeting was in New York City in January 1956. The magic of herbicides appeared to be able to solve one of humankind's oldest

[5]Detecting someone's bias is difficult. It is defined as the human tendency to make systematic decisions in certain circumstances based on cognitive factors rather than evidence. It is subject to the trope that bias is what you have, I have the truth.

[6]The following few pages are an edited excerpt from Zimdahl (2012).

problems—how to reduce or eliminate weed competition in crops and reduce the need for human labor to weed crops. Early weed scientists knew they were on the cusp of a marvelous new agricultural revolution and the burden of weeding, the arduous labor of the hoe, might disappear from earth. There was no pause to consider lessons from the history of science that reveal how early claims of spectacular advances in human welfare and benefits have, with time, often been shown to demonstrate unexpected, highly negative consequences. Agricultural education emphasized chemistry and physiology while ignoring questions raised by the history of science and philosophy and those raised by the public. A few of the glaring past errors of the agricultural community are illustrated in comments made by James Davidson (Emeritus Vice President for Agriculture and Natural Resources, University of Florida):[7]

> With the publication of Rachel Carson's book entitled Silent Spring, we, in the agricultural community, loudly and in unison stated that pesticides did not contaminate the environment—we now admit that they do. When confronted with the presence of nitrates in groundwater, we responded that it was not possible for nitrates from commercial fertilizer to reach groundwater in excess of 10 parts per million under normal productive agricultural systems—we now admit they do. When questioned about the presence of pesticides in food and food quality, we reassured the public that if the pesticide was applied in compliance with the label, agricultural products would be free of pesticides—we now admit they're not.

Having clarified my thoughts and perhaps achieved some degree of transparency on the questions and debates about herbicides and the environment, I will proceed.

1. HERBICIDE PERFORMANCE

Performance— the result of weed control—is the reason herbicides are used. They do what they are designed to do—they work. Therefore, a positive aspect of herbicide-environment interactions is vegetation management and weed control. Herbicides control undesirable plants—weeds—in many places and that is an advantage to farmers and others with weed problems and to all the rest of us, consumers, who, the claim is, benefit from reduced food costs because of reduced production costs.

Herbicides used in agriculture solve one aspect of the weed problem. Weeds also cause aesthetic pollution when they interfere with enjoyment of our world. For example:

- Bicycle riders think the world would be a better place without puncturevine and sandbur. Both produce seeds with sharp, durable spines that easily puncture tires.
- Some people are very allergic to poison ivy and poison oak. They suffer bouts of itching and discomfort and don't want to tolerate either in their yard or garden. Herbicides help clear our immediate environment of these unwanted plants.
- Plants that cause allergies can be managed by herbicides. Thousands of people suffer from hay fever caused by weed pollen.

[7]Davidson's comments were made several years ago and cited by Kirschenmann (2010).

- Many weeds and other common plants are poisonous when consumed. Poisonous weeds include larkspur, monkshood, spotted water hemlock, nightshades, buttercups, poison hemlock, and jimsonweed. Other plants that are poisonous when ingested are lily of the valley, oleander, wild cherry, rhubarb (pie with lots of sugar is fine), foxglove, iris, and sadly, you can kiss or be kissed when under it but please don't ingest mistletoe.
- Most homeowners do not like crabgrass or dandelions in their lawns. Herbicides are the best way available to control these pests quickly, easily, and inexpensively. Herbicides should be combined with proper fertilization, mowing, and water management.
- Few fishermen fish where it is hard to see into the water because of aquatic vegetation. Proper weed management of aquatic sites may include herbicides because of their ability to selectively control aquatic plants without polluting water.

No one advocates herbicides in all cases where some plant bothers someone who decides it is a weed and suggests it should be controlled. The foregoing is an illustration of places where herbicides, perhaps uniquely, provide a way to control weeds. Weeds often exist in places where no other control technology is appropriate. However, their presence in such places is not a mandate or a right to control.

Herbicide performance is measured in terms of activity, selectivity, and residual characteristics. Activity is reflected in the rate used to control weeds. How much is needed is another way of asking how active the herbicide is. Selectivity (Chapter 13) determines the plants that are affected and those that are not. It determines the crops or cropping systems in which a herbicide can be used. Soil residual characteristics (Chapter 15) determine how much of the herbicide resides in soil to control weeds over time and possibly affect the next year's crop.

Each of these traits is affected by environmental factors including wind, rain, air, temperature, light, humidity, soil texture (adsorptive capacity), soil pH, and other plants. These are the givens, albeit complicated ones, of herbicide use.

2. ECOLOGICAL CHANGE

2.1 Effects of Herbicide Use

Weed control with herbicides concerns weed scientists, ecologists, other scientists, and the public because of frequently unintended, but often inevitable ecological alteration. Wheat is a major crop in the US Pacific Northwest. Phenoxy acid herbicides used on some fields for 20 years led to significant reduction in the population of annual broadleaved weeds (Table 20.1). The herbicides did what they were designed to do, but their success changed the weedy vegetation (McCurdy and Molberg, 1974). The study was completed in the early 1970s, therefore it is possible the observed poor control of redroot pigweed may actually have been the early appearance of resistance to the herbicides (see Chapter 18). Development of new herbicides helped solved the grass problem. Recent surveys[8]

---

[8]Personal communication, March 2016, L. Van Wychen, Weed Science Society of America.

TABLE 20.1   Percent Weed Reduction in Wheat Fields Treated Annually for 20 Years With a Phenoxyacid Herbicide (McCurdy and Molberg, 1974)

Weed Species	Herbicide		
	2,4-D Amine	2,4-D Ester	MCPA
Stinkweed	97	94	98
Russian thistle	88	58	35
Common lambsquarters	90	85	86
Wild buckwheat	32	54	51
Wild tomato	52	53	23
Redroot pigweed	55	15	30
Total all weeds	86	83	69

show that the most common and most troublesome weeds in winter and spring grains in the United States are nearly the same and are about evenly divided between annual broad-leaved and annual grass weeds. Failure or lack of preventive programs of field and seed sanitation contributed to development of new weeds. Without 2,4-D's success it is unlikely that annual grasses would ever have developed into dominant weed problems in small grains. It provided all the benefits of good weed control: improved yield, ease of harvest, lower production costs, etc. It also created a predictable ecological alteration—a major vegetation shift, which offered other weeds an opportunity to succeed. In weed management, as in other technological areas, a successful technique will, due to its success, often create a problem. In fact, any technology creates and solves problems at the same time. For example, computers destroyed the business of typewriter manufacture, the Model T Ford slowly eliminated buggy whip manufacturing, and one wonders what the ubiquity of cell phones will do to wristwatches. The creative destruction (Schumpeter, 1950, chap. 11, p.83) of technological advances reminds one of a fundamental ecological rule—it is not possible to do just one thing. Predicting what problems will be created is a much more difficult task than observing and recounting the benefits of the problem that was solved.

A similar situation, although not well documented, is the invasion of cornfields, particularly in Northeastern United States, by yellow nutsedge after several years of successful use of atrazine for weed control in corn. Yellow nutsedge is not affected by atrazine, and it moved into the vacant niches opened by successful control of other weeds. Atrazine's success created opportunities for invasion by crabgrass, witchgrass, fall panicum, shattercane, and wild proso millet, none of which were controlled by normal use rates of atrazine.

The phenomenon of vegetation shifts is not limited to fields with annual cropping (Table 20.2). To make other orchard maintenance activities easier and facilitate harvest, herbicides are used to control broadleaved weeds and encourage a grass ground cover. After 4 years and six applications of specific herbicides in an apple orchard there were few patches of bare ground, but the soil was not barren (Schubert, 1972). Continued application of the same herbicide or herbicides that affect the same weeds encourages unaffected weed species

TABLE 20.2 Plant Genera Encouraged After Successive Annual Applications (Brown, 1978 After Schubert, 1972)

Genera Encouraged	After Successive Annual Application of
Rumex	2,4,5-T, simazine, diuron, or terbacil
Plantago	2,4,5-T amine, diuron, or monuron
Polygonum	2,4,5-TP, simazine, or diuron
Convolvulus	Simazine, diuron, or terbacil
Rubus	Simazine, diuron, terbacil, or dichlobenil
Cerastium	Dalapon or amitrole

because those susceptible to the herbicides are controlled. It also encourages appearance of herbicide resistance (see Chapter 19).

The widely accepted lesson of these data is that herbicide rotation is a good idea. Continued use of a single herbicide for many years on one field *will* change the nature of the weedy flora. It will select for members of the species that are resistant to the herbicide, create a niche for weeds that are not susceptible to the herbicide, and complicate weed management.

On Black Mesa in western Colorado the butyl ester of 2,4-D was used to control pocket gophers (*Thomomys talpoides*) (Tietjen et al., 1967). In large doses, 2,4-D is toxic to many mammals. The amount prescribed to control weeds was too low to kill pocket gophers. The gophers were not killed, but their population was reduced because they consumed small, broadleaved forbs that were abundant on Black Mesa. Repetitive use of 2,4-D reduced the forb population from 77% (South Crystal Gulch) and 63% (Myers Gulch) to 9%, nearly eliminating the pocket gopher's food supply (Table 20.3). Use of 2,4-D closed the pocket gopher's grocery store, and they had to move to a new neighborhood or starve. Eventually the forb population recovered, but after 5 years it had not achieved pre-2,4-D levels. A detailed analysis of vegetative composition on Grand Mesa, Colorado 3 years after 2,4-D was applied showed that slender wheatgrass increased with a corresponding decrease in broadleaved species (Turner, 1969).

Santillo et al. (1989a) thought glyphosate, a contact, broad-spectrum herbicide, would affect small forest dwelling mammals by altering vegetation structure and cover and

TABLE 20.3 Pre- and Postspray Forb Composition in Two Locations (Tietjen et al., 1967)

Location	Composition of Forbs (%)		
	Prespray	Postspray	After 5 years
South crystal gulch	77	9	44
Myers gulch	63	9	10

reducing plant and insect food resources. Glyphosate applied at 4 lb ai/acre controlled 75% of nonconiferous, brushy plants. Insectivore and herbivore species were less abundant for 3 years after glyphosate application. Omnivores were equally abundant in treated and control areas. The difference in small mammal abundance paralleled herbicide-induced reductions in invertebrate species and plant cover. The total number of birds was lower on clear-cut areas treated with glyphosate (Santillo et al., 1989b).

Careful study will very likely find a negative ecological effect every time an herbicide is used, but that is not an adequate reason to ban herbicides. Eliminating herbicides because they have ecological disadvantages means their ecological advantages will also be lost. The environmental effects of unrestricted weed spread may be far more important than the negative effects of herbicide use. The disadvantages of herbicide use must be balanced against prevention of weed spread, net gains in production of useful crops, and reductions in labor required to produce crops. It is also true that although herbicides may decrease community diversity, the effect may be small and transitory.

Three herbicides effective against broadleaved species were applied to control spotted knapweed in Montana (Rice et al., 1992). Weed control was 84%–90% 2 years after application. There was a small decline in community diversity 1 year after spraying, but diversity increased relative to areas with spotted knapweed 2 years after spraying. The data suggest all herbicide-treated areas had greater diversity 3 years later. Aggressive, perennial weeds such as spotted knapweed tend to form nearly monocultural communities. The urea herbicide, tebuthiuron, enhanced rangeland diversity, increased forage production for livestock grazing, improved wildlife habitat, and protected against watershed erosion in studies in New Mexico and Wyoming (Olson et al., 1994). Controlling weedy plants with herbicides that have no other major harmful environmental effects (e.g., leaching, drift, hazard to nontarget species) is wise vegetation management.

2.2 Enhanced Soil Degradation

Because of the crop grown and the weed problem encountered some soils have been treated with the same herbicide for several years. This has led to enhanced degradation that was first reported for EPTC, a carbamothioate herbicide, in New Zealand (Rahman et al., 1979). Since then several cases of enhanced or accelerated degradation of carbamothioate or phenoxy acid herbicides have occurred. Herbicide resistance (see Chapter 18) is a reduced or total loss of efficacy with time. Enhanced degradation is the opposite problem—microorganisms responsible for herbicide degradation become more capable of degrading the herbicide and weed control decreases with time. The precise mechanism is one of four (Gressel, 1990):

1. The soil could be enriched in a population of a rare or minor microorganism that increases because of the herbicide's presence and rapidly degrades it.
2. Repeated application of the herbicide could select microorganisms from existing populations that degrade the herbicide more rapidly.
3. It is well known that substrates are capable of inducing enzymes in microorganisms. The presence of the substrate (the herbicide) could induce enzymes that rapidly degrade the herbicide *or* induce mutations in microorganisms so they are more capable of degrading the herbicide.

4. It is possible that when the herbicide is present with other soil chemicals, rapid degradation is promoted. This coinduction is related to the presence of another compound or compounds that may not be degraded.

The phenomenon of enhanced degradation has not eliminated use of susceptible herbicides. It has encouraged development of alternative control strategies and new chemicals designed to inhibit rapid degradation. Both techniques have been successful, and enhanced degradation while real is rare.

2.3 Influence of Herbicides on Soil

Most of any applied herbicide reaches the soil and soil-herbicide interactions are inevitable (see Chapter 15). An important question is, do herbicides damage soil or any of its living components? It would be tragic if an herbicide were approved for use that destroyed an important decomposer organism or affected the nitrogen cycle. This has not happened and is not likely to happen because of careful and continuing evaluation of herbicides, and all other pesticides, by the manufacturer and the US Environmental Protection Agency (EPA), before approval for use (see Chapter 21). The approval process cannot detect all possible environmental interactions because often scientists and regulators don't even know what questions to ask until after an observation has been made. However, one should not assume that pesticide use in the United States is one large experiment where no problems are anticipated or addressed until after a problem has been observed. Nature is more complex and the present level of understanding does not permit anyone to anticipate or ask every question that nature may reveal as new technology is used.

With normal use rates, the quantity of herbicide applied to soil is too small in relation to soil volume to have any detectable influence on soil's physical or chemical state. Research has shown that tilling soil has limited benefit other than weed control and preparation for planting, except the negative effects of breaking weed seed dormancy and perhaps enhancing soil erosion. Without herbicides, investigation of the effects of tillage would have been impossible because of excessive weed growth.

Part of the research on any candidate herbicide is determination of its effects on soil microorganisms. Nearly all investigations show a positive or negative effect. Reactions such as nitrification are often suppressed, but, at field use rates, suppression is not permanent. Because of large populations, short reproductive cycles, and great adaptability to environmental insult, microorganism populations are very resilient.

Metham, a dithiocarbamate, applied as a fumigant to seedbeds to control weeds and plant pathogens is a general biocide that can decimate a soil's microorganism population. But one of the most difficult things to do in the laboratory is to keep soil sterile. Microorganisms are ubiquitous and sterility, while easy to obtain, is almost impossible to maintain with exposure to air and water.

2.4 Herbicide-Disease Interactions

One of the simple rules of ecology (see Chapter 6) has become almost a cliché—in the natural world everything is connected to everything else and it is impossible to tinker with one environmental parameter without affecting others. For practical purposes, it is not possible to do just one thing. All possible effects of an environmental intervention cannot be determined in advance and food must be grown and weeds must be managed. All environmental effects of food production and weed management are not known, but we cannot stop either while all possible effects are determined. Food must be produced because we must eat.

Some herbicides may promote plant diseases and others may reduce disease incidence. Herbicides predisposed 20 hosts (crops and weeds) to higher disease levels in cases involving 20 pathogens (Altman and Campbell, 1977). One of the earliest reports was after herbicide use in peanuts. Where herbicides were used, peanuts were larger and more vigorous. The effect was most easily seen in the absence of weeds. This work involved the no longer used dinitro and other phenolic herbicides. It was proposed, and later proven, that these herbicides inhibited growth and vigor of parasitic and pathogenic fungi that affected peanuts. Sugar beets grown in nematode infested soil and treated with tillam (a carbamothioate) had a higher level of nematode infestation 6 years later than those grown in soil not treated with tillam (Altman et al., 1990). It has also been reported that the carbamothioate herbicide cycloate enhanced cyst development on sugar beet roots (Altman et al.). Soil residues of chlorsulfuron increased take-all (*Gaeumannomyces graminis* Var. *tritici*) and *Rhizoctonia*, root diseases of barley and wheat, and yield was reduced (Rovira and McDonald, 1986). The soil-applied herbicide trifluralin alters the Fusarium disease syndrome in beans.

It was not intuitively obvious that these interactions should occur. They are examples of the fact that, in nature, one cannot do just one thing. Herbicide-disease interactions are another element in the equation that must be considered in weed-management systems. The data are not available to predict if there will be an interaction and, if so, what kind it will be for all herbicide, crop, and disease combinations. The possibility exists and must be considered. A few examples are cited in Table 20.4. Norris et al. (2003) provide a detailed discussion of pest interactions.

TABLE 20.4 Examples of Herbicide-Disease Interactions (Katan and Eshel, 1973)

Organism	Disease and Crop	Herbicide
DISEASES PROMOTED		
Rhizoctonia solani	Damping off-cotton	Trifluralin
Helminthosporium sativum	Seedling disease-barley	Maleic hydrazide
Fusarium oxysporum	Wilt disease-tomato	Maleic hydrazide, Dalapon
Alternaria solani	Early blight-tomato	2,4-D
DISEASES SUPPRESSED		
Cercosporella herpotrichoides	Foot rot-wheat	Diuron
Fusarium oxysporum	Wilt disease-tomato	Propham,TCA
Alternaria solani	Early blight-tomato	Maleic hydrazide, Dalapon

3. ENVIRONMENTAL CONTAMINATION

3.1 Effects on Water

Possible pesticide contamination of ground and surface water is a matter of national importance. About 50% of the US population relies on groundwater for drinking. It is especially important for people in agricultural areas where about 95% rely on groundwater. Before the mid-1970s, the dominant scientific view was that because of its adsorptive capacity, soil would act as a slow-release filter or a binding medium that would not, or only slowly, release an adsorbed pesticide in nonconsequential amounts or it would bind it forever. We now know that adsorption occurs, binding forever is not a reasonable hypothesis, and leaching is common. Pesticides reach water-bearing aquifers below ground from applications to crop fields, seepage of contaminated surface water, accidental spills, leaks, and improper disposal.

Pesticides and degradation products are typically present throughout most of the year in watercourses that drain watersheds from large agricultural or urban areas. Pesticides were detected throughout most of the year in water from streams with agricultural (97% of the time), urban (97%), or mixed land use (94%). "Pesticides were less common in groundwater than in streams, but occurred in >50% of wells that sampled shallow groundwater beneath agricultural and urban areas. One or more pesticide compounds was detected in 33% of the deeper wells that tap major aquifers used for water supply" (Gilliom, 2007). Trends in pesticide concentration in 38 major US rivers were evaluated in relation to use trends for 11 commonly occurring pesticides (five were herbicides) (Ryberg and Gilliom, 2015). There was widespread agreement between concentration trends and use trends.

However, on a national scale less than 2% of wells sampled in multistate studies were found with pesticide concentrations above the established maximum contaminant level.

Extensive research under the National Water-Quality Assessment program confirms that streams are very vulnerable to pesticide contamination, and groundwater merits careful monitoring, especially in agricultural and urban areas (see http://water.usgs.gov/nawqa/pnsp/). Shallow groundwater in many areas is used for drinking and groundwater contamination is difficult to reverse once it occurs. The occurrence of pesticides in streams and groundwater does not necessarily cause adverse effects. Detected levels were nearly always very low. Presence is not necessarily a synonym for danger—only 1% of wells had concentrations greater than the human-health benchmark.

However, most will agree that while presence is not necessarily equal to danger, it would be best if pesticides were not there. It does not seem right or environmentally responsible when studies reveal that half of the shallow wells in agricultural areas and about one-third of shallow wells in urban areas contain two or more pesticides and their degradation products—less than 1% had 10 or more. Atrazine, deethylatrazine, simazine, metolachlor, and prometon were common in mixtures found in streams and groundwater in agricultural areas. Insecticides were common in urban streams. Pesticide degradation/metabolic products are often as common in streams and groundwater as the parent pesticide. Atrazine was found with one of its several degradates in about 75% of stream samples and about 40% of groundwater samples[9] (Ryberg and Gilliom, 2015).

The US Geological Survey (USGS) (see Footnote 9) found pesticides or their transformation products in groundwater of more than 43 states. Atrazine, simazine, alachlor, and metolachlor were among the most frequently detected herbicides (Heilprin, 2006). The proportion of sampled wells with detectable pesticide levels ranged from 4% for nationwide, rural domestic wells to 62% for postplanting sampling of wells in corn and soybean areas of the northern mid-continent. Concentrations were 1 µg or less in more than 95% of the samples.

Studies have also been done of surface waters. The herbicides detected most frequently in eight geographically dispersed US urban streams were the triazines prometon, simazine, and atrazine; the substituted urea tebuthiuron; and the chloroacetamide metolachlor (Hoffman et al., 2000). The study looked for 52 herbicides and detected 28 in one or more urban streams. Of 215 samples, only 17 detected no herbicides. The most frequently detected herbicides in streams in agricultural areas were the same as those found in groundwater: atrazine, metolachlor, and cyanazine. In 19 of every 20 streams in agricultural, urban, or mixed land-use watersheds, pesticide contamination was found at nearly all times of the year. It is important to note that the concentrations nearly always complied with the EPA's drinking-water standards, although the sample size did not reflect a person's drinking water consumption. Heilprin (2006) notes that "the large majority of pesticide detections in streams and groundwater were trace amounts, far below scientifically based minimum levels set for protecting human health and the environment."

Pesticides have been detected in surface waters in all US regions. A series of studies including 98 pesticides and 20 pesticide transformation products found 76 in one or more surface water sites.[10] The herbicides atrazine, cyanazine, simazine, metolachlor, and alachlor were detected more frequently than other pesticides. Stone et al. (2014) found over 2 decades that one or more pesticides or pesticide degradates were detected more than 90% of the time in streams across all types of land uses. During 2002−11, atrazine, deethylatrazine, metolachlor, prometon, and simazine were detected in streams more than 50% of the time. Alachlor, cyanazine, EPTC, dacthal, and tebuthiuron were detected less frequently in streams during the second decade than during the first decade. During 2002−11, only one stream had an annual mean pesticide concentration that exceeded a human health benchmark (HHB), but 17% of agriculture land-use streams and one mixed land-use stream had annual mean pesticide concentrations that exceeded HHBs between 1992 and 2001.

The National Water Quality Assessment Program has shown that pesticide contamination of streams and groundwater occurs in geographic and seasonal patterns that follow cropping patterns and associated pesticide use (see Baker and Stone, 2014). The most frequently and heavily used herbicides account for most detections. Perhaps the most important finding is that one or more pesticides were found in almost every stream sample collected. More than 95% of stream samples and nearly 50% of well samples contained at least one pesticide (Gilliom et al., 1999). "Contaminants from geologic or manmade sources were a potential

[9]The preceding data are from http://pubs.usgs.gov/fs/2006/3028/, Pesticides in the Nation's Streams and Groundwater, 1992−2001—A Summary.

[10]http://water.wr.usgs.gov/pnsp/fs-039-97/sw4.html, USGS fact sheet FS-039-97.

human-health concern in one of every five US wells sampled in the parts of aquifers used for drinking water. Differences in geology, hydrology, geochemistry, and chemical use explain how and why aquifer vulnerability and concentrations of contaminants vary across the Nation" (DeSimone et al., 2014).

3.2 Human Effects

3.2.1 General

Assessment of the effects of herbicides on people and the environment is confounded by at least four factors:

1. The changing character of the environment and our attitude toward it.
2. The changing character of the population.
3. The changing character of the problem.
4. The changing character of public health responsibility.

Everyone wants a protected and protecting environment but we must also have a productive environment. At least part of the debate about the relationship between herbicides, human health, and the environment centers on differing views of the appropriate balance among these things and how to achieve it. The discussion always includes one or more of three concepts:

1. Toxicity—the inherent capability of something to cause injury.
2. Risk or hazard—the probability that injury will occur.
3. Safety—the practical certainty that injury will not occur.

Science can measure toxicity and estimate risk, but science cannot measure or determine safety (see Section 5). Safety is a normative political judgment and its definition and determining criteria are frequently mandated by legislative acts. Safety is not primarily a scientific question; it is a social/political question. It is true as Conway and Pretty (1991, p. 576) and many others have asserted that one of the biggest problems of herbicide use is misuse. It is incorrect to claim that the pesticide industry does not seek safer herbicides. Until safer herbicides are developed, the careful and accurate application of existing herbicides is the best way to assure reduced environmental pollution and minimize harm to humans. Although herbicide misapplication and misuse are a primary cause of environmental damage, they are not the only problems. Herbicides are toxic to other forms of life and they move in the environment. These can cause major environmental effects and can affect ecological relationships. To achieve their purpose, herbicides must be used in the environment. It is wrong to assume that a herbicide will affect weeds and nothing else. That is sloppy ecological thinking and poor science. Weed management is fundamentally an ecological problem, in which the interactions between weeds, other plants, nontarget organisms, and the environment should be considered. These can be studied and, when detrimental effects are discovered, changes can and should be made in the management system.

In the past, those who worked with herbicides knew about ecological relationships but did not ask the right questions about the effect of herbicides on ecological systems. Today the right questions are being asked and environmental effects are examined, in depth, with great care. Because good questions are asked and answered, we should not assume all problems have been solved and no future environmental effects or ecological disruptions will occur because of herbicide use. The relevant questions include the following:

- What are the possible effects on public health?
- Will human food (e.g., meat, milk, fruits, or vegetables) be contaminated? If so, will the contamination be above or below established human health safety standards?
- Will domestic animals be affected?
- Will beneficial natural predators or parasites be affected?
- What is the likelihood of herbicide resistance developing?
- Will bees and other beneficial insects be affected?
- Will there be crop damage? Is the herbicide selective?
- Will ground- or surface water be contaminated?
- Will there be negative effects on fish, birds, mammals, microorganisms, or invertebrates?

The present US/EPA pesticide registration process requires that these questions be addressed and answered (see Chapter 21).

Even with the sophisticated scientific capabilities of the world's developed countries there is a limited capability to predict environmental hazards. When a limited capability to predict all possible results is coupled with the dominant attitude that, while effects may be real, they will be minimal, the result may be inattention to small but real effects and legitimate public concern. For example, the report of Heilprin (2006) cited previously notes that "the large majority of pesticide detections in streams and groundwater were trace amounts, far below scientifically based minimum levels set for protecting human health and the environment." That is, yes, they are there, but they are below the scientifically set limits for human and environmental safety. Therefore, it is not illogical to conclude they are not a problem. Stop worrying. Such data are consistent with the coda from the National Academy of Sciences report cited at the beginning of this chapter (Berenbaum et al., 2000). The conclusion is based on sound fundamental and applied research. It is a scientifically based decision, and, therefore, the argument goes, one should assume not only that it is correct but that actions or inaction based on scientific evidence are what should be done. One should have faith in the widespread applicability of decisions based on sound science.

But many people don't have faith in scientifically based decisions on pesticide safety. Why is there widespread mistrust of scientifically set limits for human and environmental safety? One reason is the arguments in favor of herbicides and other aspects of modern agricultural technology are based on what Shader-Frechette (1991) calls the realism argument. She cites Kraybill (1975, pp. 10, 16) and Furtick (1976, p. 12) both of whom make the claim that life is dangerous and while chemical pest control is risky, "it is realistic to accept the minimal degree of risk it presents, since absolute safety is unattainable in any sector of life." Bender (1994, p. 92) agrees with Shader-Frechette and notes the arguments commonly employed in defense of pesticide use include:

- Abrupt cessation of pesticide use would be calamitous for agriculture and public health.
- The key to pesticide safety is following label directions.
- Farmers have a moral obligation to feed the world.
- Life and our world are full of risks.

Bender finds problems with each of these claims, which he defines as sophistry, and denies their validity. Implicit in these claims is the claim that concerns about agricultural chemicals are all out of proportion to the actual risks (Shader-Frechette's realism argument). He claims the proponents of pesticides propose that what is needed is more knowledge and then all problems will be solved or simply disappear. More knowledge will show that many of the fears about pesticides are simply irrational (Bender, 1994, p. 111). Doering (1992, p. 239) supports Bender's argument. He claims that even if a member of the general public had the same scientific knowledge as scientists, they still might have different risk preferences. Their values about production benefits versus environmental or health concerns may be quite different. Secondly, in the view of most scientists, members of the general public may never possess adequate knowledge. Third, public perception of risk may not ever correspond with scientific facts. That is, the public may rely on other sources of information (National Public Radio, TV news, magazines, newspapers, etc.) that they regard as more trustworthy. Finally, Doering (1992) notes that scientific facts have, notably, missed some of the big trade-offs by focusing on just the facts. For example, is more yield always better than improved quality? Should more yield or more profit for someone always trump environmental improvement? What can justify harm to public health? If a pesticide is in my water is it okay if it is present below the scientifically set limits for human and environmental safety? What if, as Mackay (1988) asks, it is a chemical that has disruptive potential? What if it is a molecule that in very low concentrations has the potential to direct future events? Should we be worried? Whose values determine what should be done? These are difficult, appropriate questions without easy answers.

The realism argument correctly asserts that risks and benefits must be balanced by scientifically based decisions. What is incorrect is the claim that "the moral acceptability of a hazard, like pesticide use, is a matter only of risk magnitude or degree of physical danger" (Shader-Frechette). This view ignores consideration of whether or not the risk is distributed equitably among all that are or might be affected. It also ignores whether the risk is accepted voluntarily or imposed involuntarily. The realism argument is accepted by the agricultural community and those who favor pesticide use and scientifically based risk assessment. They accept the benefits to agriculture but have not accepted the mantle of proof to demonstrate, beyond some level of reasonable doubt, that public well-being is served. There is no doubt that there have been enormous benefits to pesticide manufacturers and some users. Corporate stockholders have benefited as manufacturers have fulfilled a primary obligation to maximize shareholder return, as they are expected to do. But corporate managers tend to filter out externalities such as consideration of public well-being, general public health, worker safety, equitable income distribution, and the well-being of natural communities, animals, plants, other sentient, and perhaps, nonsentient, creatures, soil, and the atmosphere (Nace, 2006). Culliney et al. (1992) ask a still unresolved question. Should public policy be directed toward reducing unnecessary pesticide residues in the human diet? The obvious answer is—of course it should. However, in the United States in 1993 at least 35% of food purchased by consumers contained measurable levels of pesticide residues (FDA, 1993), but only 1.1% was found to have residues above the FDA tolerance level (Pimentel, 1997). In 2005, Štepána et al. found that 59.5% of fresh apples had pesticide residues, but only 1.4% exceeded FDA standards. These and other studies are relevant to the public's concern about contamination of human food. For those interested in pursuing the topic, the USDA

Agricultural Marketing Service, Pesticide Data Program (PDP) 2014 summary shows that more than 99% of the products sampled had residues below EPA tolerances. Residues exceeding the tolerance were detected in 0.36% of the samples. If a PDP finding poses a safety risk, FDA and EPA are immediately notified. The 2014 report determined that the 99% absence and low levels of pesticide residues in a very few samples do not pose a food safety risk or safety concern. An explanatory guide is available at www.ams.usda.gov/pdp.

Many suggest concern should be even greater in other countries. For example, as much as 80% of food available in Indian markets had pesticide residues (Singh, 1993 cited in Pimentel, 1997), primarily from chlorinated hydrocarbon insecticides. Similar reports from a range of countries are available online. In view of these data, one wants to argue that public policy in some countries (Culliney et al., 1992) is inadequate.

It is often helpful when thinking about complex issues to consider related, perhaps more familiar, matters. Two will be discussed briefly.

3.2.2 The Case of Fluorides

Fluorine is poisonous and an element in a few herbicides. Its most famous and, in some circles, still controversial, use is as an intentional additive (about 1 ppm) to drinking water to prevent tooth decay. The debate has rarely focused on its efficacy. The debate is about its toxicity: people must drink water to survive and should not have to bear the risk of fluorine toxicity because they cannot avoid drinking water.

The scientific view is that the appropriate determining factor of fluoride toxicity is as it should be, for the toxicity of any chemical—dose or concentration. A 150-pound man will become ill if he ingests 0.25 mg of fluorine in 1 day. The same man will become very sick if he ingests 1 g and will die if he ingests 4—8 g of fluorine. At the prevailing level of fluorine in US drinking water, a 150-pound man would have to drink more than 42 gallons of water containing 1 ppm of fluorine to consume 0.25 mg. To ingest 1 g he would have to drink more than 3 bathtubs full. Death from water intoxication would occur long before fluorine toxicity. It is clear that the data, the scientific view, has not relieved the concern.

3.2.3 2,4,5-T

2,4,5-T controls a wide range of broadleaved and woody plants. For many years it was used selectively for weed control in crops, on home lawns, in forests, and in rice paddies. When the United States was engaged in the Vietnam war, a 2,4,5-T ester was used in combination with a 2,4-D ester as Agent Orange (the identifying color on the barrels), to eliminate unwanted vegetation. When 2,4,5-T is manufactured, temperature control is required to minimize formation of an undesirable, nonphytotoxic contaminant, 2,3,7,8-tetrachloroparadioxin. It is one member of a family of compounds know as dioxins and is a potent teratogen. A teratogen may cause terata or birth defects when pregnant women are exposed. This specific dioxin also causes chloracne, a skin condition characterized by blisters and irritation. There was never any debate about whether the dioxin contaminant in 2,4,5-T was present in Agent Orange, was a teratogen, or caused chloracne. Concern and debate ensued because of the unknown level of exposure to Agent Orange of Vietnam era service members (especially those who participated in operation Ranch Hand, the application program), Vietnamese citizens, and others. There is enduring widespread disagreement about whether the health problems experienced by many US servicemen were caused or exacerbated by Agent Orange and its dioxin contaminant. The US/EPA removed 2,4,5-T from the list of approved herbicides in 1985.

The Pesticides Monitoring Journal (publication ceased in 1981) reported surveys of pesticide levels found in the American food supply. In one report they surveyed 24,000 food samples and three contained measurable quantities of 2,4,5-T—two in milk and one in meat. Reported on a whole-milk and fresh-meat basis, the average 2,4,5-T content was 0.006 ppm. For all 24,000 samples the 2,4,5-T content was 7.5×10^{-7} ppm or roughly equal to 1 mg in 133,000 metric tons. Based on other studies, the presumed maximum nonteratogenic dose of 2,4,5-T with the dioxin contaminant for a 130-pound pregnant woman is 1.26 g/day. At the observed level of 2,4,5-T in the nation's food supply, a 130-pound pregnant woman could have consumed 170 million tons of food per day for 9 days without fear of teratogenic effects on a fetus.

Wildavsky's Chapter 3, "Dioxin, Agent Orange, and Times Beach," describes the dioxin, 2,4,5-T story in detail. He concludes that 2,3,7,8-tetrachlorodibenzo-p-dioxin is the one dioxin, among many, that has serious human health effects. It was found as a contaminant in trichlorophenoxy acetic acid products such as 2,4,5-T but was never found, for clear chemical reasons, in dichlorophenoxy acetic acid products, such as 2,4-D. These are chemical facts. Wildavsky reports that such facts paled in comparison to the political and nonscientific debates that swirled around the dioxin issue, a debate that continues (Issacson, 2007). Millions of dollars were spent to discover that dioxin has serious human health effects, but only at extremely high doses, and these occur only with unprotected, prolonged occupational exposure. There is a threshold—"below a certain level, little or no harm would occur; thus some body level might be harmless." The US government and the chemical industry paid millions of dollars to people who were not injured and equal amounts to regulate inconsequential exposures. Wildavsky concludes the discussion, as he concluded so many others in the book, with the question, "Why expend so many resources in the name of public health with so little to show for it?"

However, that sort of scientific logic does not resolve this issue or many other pesticide issues for many equally rational, thoughtful people. Many people want absolute assurance of safety and any level of risk of bearing a deformed child is too high, if the risk can be eliminated. Certainly the risk of exposure to a herbicide can be minimized if not eliminated, and even if it is infinitesimally small, many think it is too large and should be eliminated.

A Harvard scientist[11] disputed the 2,4,5-T toxicity theory and calculated the risks associated with spraying 2,4,5-T. If a person applied 2,4,5-T with a backpack sprayer 5 days a week, 4 months a year, for 30 years, the chances of developing a tumor would be 0.4 per million. The risk of developing a tumor from other activities is greater (Table 20.5).

Zimdahl (2006, 2012 – Chapter 3) includes a more complete discussion of the issues surrounding 2,4,5-T.

[11]Wilson, R. Cited in the Pesticide Pipeline. Colorado State Univ. XIV (7) July 1981.

TABLE 20.5 Risks of Developing a Tumor

Activity	Chance of Developing a Tumor (Per Million)
Smoking cigarettes	1200
Being in a room with a smoker	10
Eating 1/4 pound of charcoal-broiled steak per week	0.4
Drinking 1 can of diet soda with saccharin per day	10
Drinking milk with aflatoxin or eating 4 tablespoons of peanut butter/day	10
Drinking 1 can of beer/day	10
Sunbathing	5000

3.2.4 Summary

The data required to resolve the human and environmental questions raised herein are difficult to obtain, and the data and the solutions and actions suggested are inevitably controversial. Often the scientific data, while an essential element of the debate, are not adequate to resolve the debate and allay public concern. Yet decisions have to be made. Weeds and other agricultural pests must be controlled. Informed debate is best, but debaters should understand that such decisions, when made, will be based in part on factual information and in part on perceptions or other relevant things that may have no basis in scientific fact.

Table 20.6 shows data on the risk of death associated with certain human activities that many people do voluntarily. These data are presented not to provide a conclusion or judgment about herbicides and the environment or human welfare. Such statements can be found in several of the references cited in the supplementary literature at the end of this chapter. In most cases, thought is required. It cannot be overemphasized that the end of these debates is usually a value judgment, not a decision based solely on scientific fact. It may have been true that chances of getting cancer are increased by 1 in 1,000,000 by consuming some city's drinking water for 1 year. Residents and visitors must decide what, if anything, they propose to do about the scientific evidence. How does one judge the importance of the facts to life? This is a question that must be dealt with by those who consider the problems and advantages of herbicides in the environment.

There is another point of view that should be considered when thinking about herbicides and the environment. The United States is a rich country that can afford to ask and answer difficult environmental questions. We can afford to make decisions that favor environmental protection over productivity or the opposite. Poor countries may not choose, or be able to afford, to put productivity second. Data from the World Bank claim that 1.4 billion of the world's people survive on less than US $1/day and 2.8 billion (1.4 billion more) somehow make it on less than US $2/day. They are poor, hungry, landless, lack formal education, do not have access to adequate or, more likely, any health care, or have a hope for a brighter future. They are the least among us. They live in the shadows. If one is hungry, one has only one need—food. Obtaining or producing food is the only goal, and environmental questions,

TABLE 20.6 Acts That Increase the Risk of Death by 0.000001 (1 Chance in 1 Million)

Act	Hazard
Smoking 1.4 cigarettes/day	Cancer, heart disease
Drinking $1/2$ liter of wine/day	Cirrhosis of liver
Spending 1 h/day in a coal mine	Black lung disease
Spending 3 h/day in a coal mine	Accident
Living 2 days in New York or Boston	Air pollution
Traveling 6 min by canoe	Accident
Traveling 300 miles by car	Accident
Traveling 10 miles by bicycle	Accident
Flying 1000 miles by jet	Accident
Flying 6000 miles by jet	Cancer from cosmic radiation
Living 2 months in Denver on vacation from NY	Cancer from cosmic radiation
Living 2 months in average stone or brick home	Cancer from natural radioactivity
One chest X-ray in a good hospital	Radiation cancer
Living 2 months with a smoker	Cancer, heart disease
Eating 40 tablespoons of peanut butter	Liver cancer from aflatoxin-B
Consuming Miami drinking water for 1 year	Cancer from chloroform
Drinking 30 12-oz cans of diet soda (at one time)	Cancer from saccharin
Living 5 years in the open at the site boundary of a nuclear power plant	Radiation cancer
Living 150 years within 20 miles of a nuclear plant	Radiation cancer
Eating 100 charbroiled steaks (all at one time)	Benzpyrene-induced cancer

if thought of, are obstacles that may stand in the way of food production. One can argue that interminable discussions about environmental questions about GMOs or whether or not the risk of soil, water, personal, or environmental contamination from herbicides is acceptable, risk decreasing long-term food production and putting hungry people at further risk. Bertolt Brecht said it well in the 1928 Threepenny Opera—first comes a full stomach, then comes ethics. Attitudes toward herbicides may be very different when they are perceived, correctly or incorrectly, primarily as a way to produce food rather than primarily as environmental risks.

3.3 Global Change

It was true in the United States and is still true in most of the world's developing countries that possible or real environmental hazards of pesticides are given low priority. Many of the world's developing countries do not have good environmental policies. An important and hotly debated reason for problems with pesticides in developing countries is the lack of suitable alternatives to pesticides that offer comparable efficacy and labor efficiency.

Pimentel (1997) notes the widespread concern that as much as 40% of all food and fiber produced is lost to pests before it can be consumed by humans. This is true despite the fact that 1.3 million pounds of pesticides were applied to US crops in 2000 (200 million in California) and 5.2 billion pounds were applied in the world in 2007, 25% were herbicides, whose use has grown faster than other pesticides.[12] Most authors agree that losses would be even higher if no pesticides were used. Pest losses would be higher and crop yields lower if all pesticide use was eliminated. Although US herbicide use has increased, crop losses have not declined significantly because of changes in cropping practices including abandoning crop rotation, reduced field sanitation practices, more monocultural agriculture, and more stringent cosmetic standards for produce implemented by government regulations and encouraged by consumer demand.

Thus, we are faced with increased pesticide use and continued high losses to pests during and after production, in a world with a growing population that demands more, high-quality food. The next 50 years may be the final period of rapid agricultural expansion, which will be driven by increasing demand for food by more people who through economic development are able to afford more and better food (Tilman et al., 2001). The demand for more production will only exacerbate agriculture's already significant environmental effects. More people demand more food, and as they achieve the quite understandable goal of becoming richer, they will demand food of higher quality as well as more meat. It is forecast that as much as a billion acres of land now in natural ecosystems may be converted to agricultural use by 2050. This is a loss of natural ecosystems and the ecosystem services they provide in an area larger than the continental United States. This will be accompanied by a "2.4- to 2.7-fold increase in nitrogen- and phosphorus-driven eutrophication of terrestrial freshwater, and near-shore ecosystems and comparable increases in pesticide use" (Tilman et al.). If past trends continue, global pesticide production, which has increased for the past 40 years, will be 1.7 times greater in 2020 and 2.7 times greater than present consumption in 2050 (Tilman et al.). It is likely that herbicides will continue to dominate pesticide use. This means that humans, the environment, and other creatures will be exposed to presently unknown consequences. It also means that the social and environmental effects of agricultural technology may rival those of climate change (i.e., global warming) that the planet is now experiencing.

Agriculture is the largest and most ubiquitous human environmental interaction. It is already the greatest threat to survival of several species of birds (Green et al., 2004). Available data clearly indicate the need for what Tilman et al. (2001) call a greener revolution—an environmentally sustainable revolution. In practical terms what this means is the use of existing knowledge to reduce agriculture's inevitable environmental effects and increase productivity. Agriculture's effects on wildlife can be mitigated in one of three ways. The first is to practice wildlife-friendly farming, which is a good thing, but many believe it will almost inevitably reduce yields. The second is to use the best agricultural practices on the best land to increase yield and prevent conversion of more land to agriculture that the first solution demands (Green et al.). A third, albeit widely regarded as a more radical, option is a significant

[12]https://www.epa.gov/sites/production/files/2015-10/documents/market_estimates2007.pdf.

departure from present practices. It is development through research of new ways to practice agriculture, techniques that are not just continuation and modification of present practices. The third option is development of agricultural practices that are resilient and ultimately sustainable that enable achievement of agriculture's primary moral responsibility—feeding a growing population. Each of these solutions requires:

• Greater use of integrated weed (and other pest) management programs.
• Application specificity (e.g., herbicides and other pesticides will be applied only to areas where pests exist, not to entire fields).
• Site- and time-appropriate amounts of herbicides will be used.
• Irrigation will be site and time appropriate.
• Cover crops will be used to reduce soil erosion especially on fallow land, interrow areas, and in buffer strips between fields.
• Appropriate use of more productive cultivars to increase yield and reduce fertilizer, water, and pesticide runoff to nonagricultural areas (Tilman et al., 2001).

Yield of agricultural crops will continue to be an important measure of any program's, indeed of agriculture's, success. It cannot be the only measure if we are to have a greener revolution, indeed what Conway (1997) calls a doubly green revolution to feed all in the 21st century. If crop yields do not increase, more land will be required to feed all the people. Conway projects that just to maintain current food consumption levels will require a near doubling of cropped land by 2050 to feed the projected 9.7 billion people who will be alive. The success of a doubly green revolution must be based on all the costs (including external costs) and benefits of agricultural production, not just yield. These include "agriculture-dependent gains and losses in values of such ecosystem goods and services as potable water, biodiversity, carbon storage, pest control, pollination, fisheries, and recreation" (Tilman et al.). The balance between high-yield farming and minimal environmental effect is delicate. It is not the purpose of this book to explore the three suggested options in detail. Interested readers are directed to one or more of the references provided in the supplementary literature section at the end of this chapter.

4. ENERGY RELATIONSHIPS

In 1974 Nalewaja projected that if all the corn grown in the United States were weeded by hand in 6 weeks (a reasonable time, given the critical period for weeding corn; see Chapter 6) it would take over 17 million people working 40 h a week. That is more than four times the number of workers then employed on US farms. If weeding was not done during the critical period, yield would decrease. Therefore, there was not enough labor available to weed the US corn crop in 1974. That is still true. It is doubtful that even if people were available they would be in the right place at the right time and be willing to weed corn or any other crop. Hand weeding or hoeing are not among life's desired occupations.

Yet corn, and other crops, have to be weeded. Not all acres and all crops need the same amount of weeding, but almost all need some. In agriculture's early days, animals were substituted for hand labor as hand labor became scarce and more expensive, and as farms became larger, animals were inadequate for the task. Tractors replaced animal power. If the tractors on US farms were not used, it has been estimated that it would take more than 60 million horses and mules to replace them. The 2012 US agricultural census reported 3.6 million horses, a 10% decrease from 2007. There were 21 million horses and mules in the United States in 1900. Ergo, there aren't enough horses and mules available in the United States and the crop land and pasture required would reduce land available for food production. The shift from hand labor to animals reduced the need for human labor. The change to tractors further reduced the need for human labor, but added more petroleum energy to agriculture's input requirements. The trend has continued as agriculture has become more mechanized and chemicalized. Nitrogen fertilizer and pesticides are highly dependent on petroleum energy for manufacturing and distribution. The purpose of this section is to discuss, with reference only to the use of herbicides for weed control, US agriculture's dependence on petroleum energy.

Two relevant questions are:

1. Is energy use for herbicide manufacture and application efficient?
2. How does herbicide use efficiency compare to other methods of weed control?

The quick answer is that herbicides do not demonstrate an overdependence on energy and the efficiency of their use (energy required vs. energy out) compares well with other methods of weed control.

A study of the economic relationships for weed-control techniques in six experiments on corn showed that when weeds were controlled by hand labor there was a net economic loss (Nalewaja, 1974). When appropriate herbicides were used in corn there was a net profit (see Table 13.4). Table 13.5 shows similar data for cotton (Barrett and Witt, 1987). Of course, all costs are higher than they were in 1976, but the relationship among the techniques is still valid even though absolute costs have changed.

The data in Table 13.4 show the 1984 energy relationships for weed control in corn based on Nalewaja's (1974) study. Land was plowed, disked, and prepared in the conventional manner. The comparison is only for weed control. The data show an energy advantage for hand labor but it is not significantly better than herbicide use. Both are more advantageous than cultivation with a tractor. Corn yield with herbicide use or hand labor was nearly identical. Table 20.7 shows energy costs for several weed-control practices. Some equipment requires more energy than others, and energy costs for herbicides increase directly with rate, although application cost is constant. Hand labor is not the cheapest way to weed crops in terms of energy expended. The sulfonylurea and imidazolinone herbicides use fractions of an ounce per acre and energy costs compare favorably with any other method of weed control. The energy costs for tillage and cultivation are much higher than for herbicides.

Most US cropping systems replace human and animal energy with petroleum energy (fuel) for tractors and other machines, manufacture of nitrogen and other fertilizer, water pumping for irrigation, transportation, and for pest control. US agriculture is energy based but it is not the major energy-consuming sector of the US economy. Farm production consumes only 3% of total US energy. Petroleum energy has been substituted for hand labor and animal power. Chemical energy has replaced mechanical energy. However, despite their dominance, herbicides and the cost of application are not a significant portion of the energy cost of producing crops (Table 20.8) (Pimentel and Pimentel, 1979). The energy used for herbicides ranges from 0.1% of total energy expended to produce oranges to 27.3% for soybeans. The mean for the

TABLE 20.7 Energy Inputs Per Performance for Various Weed-Control Practices (Nalewaja, 1974)

Method	Energy Input (kcals/A)				
	Gas	Indirect Machine	Hand Labor	Herbicide	Total
Hand labor			21,770		21,770
Field cultivator	48,904	24,452	70		73,426
Tandem disk	37,674	18,837	90		56,601
Rod weeder	10,505	5,253	70		15,828
Rotary hoe	7,970	3,985	50		12,005
Row cultivator	14,852	7,426	125		22,403
Rotary tiller	106,139	53,069	375		159,583
Herbicide					
1/2 lb/A	3,260	1,630	30	5,500	10,420
1.0 lb/A	3,260	1,630	30	11,000	15,920
2.0 lb/A	3,260	1,630	30	22,000	26,920
4.5 lb/A	3,200	1,630	30	44,000	48,920

19 crops was 6.8%, which was a reasonable estimate of energy requirements for weed control in US crops. There are no current studies of comparative energy use for weed management in US agriculture.

US agriculture industry used nearly 800 trillion British thermal units (Btu) of energy in 2012, or about as much primary energy as the entire state of Utah. Agricultural energy consumption includes energy needed to grow and harvest crops and energy needed to grow livestock. Crop operations consume much more energy than livestock operations, and energy expenditures for crops account for a higher percentage of farm operating costs.[13] However, of the estimated 1.7 quadrillion Btu of total energy used by the US agricultural sector in 2002, 65% (1.1 quadrillion Btu) was consumed as direct energy (electricity, gasoline, diesel, LP gas, and natural gas), compared with 35% (0.6 quadrillion Btu) consumed as indirect energy (fertilizers and pesticides). Pesticides accounted for less than 5% of agricultural energy use.

Pesticides' share of farm production expenses has grown significantly from less than a 1% prior to 1960 to a high of nearly 5% in 1998. The oil price shocks of the late 1970s and early 1980s forced the agricultural sector to become more energy efficient. Since the late 1970s, the direct use of energy by agriculture has declined 26%, while the energy used to produce fertilizers and pesticides has declined 31%.[14]

[13]https://www.eia.gov/todayinenergy/detail.cfm?id=18431.

[14]http://nationalaglawcenter.org/wp-content/uploads/assets/crs/RL32677.pdf.

TABLE 20.8 Energy Inputs for Herbicides in Several Crops (Pimentel and Pimentel, 1979)

Crop	Location	Rate (kg/ha)	Herbicide as Percent of Total Energy for Herbicides (%)
Corn (grain)	United States	2	3.1
Corn (silage)	NY	2.5	4.0
Wheat	United States	0.5	1.8
Oats	United States	0.2	0.9
Rice	Philippines	0.6	2.4
Rice	CA	11.2	7.7
Rice	Japan	7	9.7
Sorghum	United States	4.5	8.4
Soybean	United States	5	27.3
Dry bean	United States	4	14.6
Peanuts	GA	16	14.6
Potato	NY	18	11.2
Apple	United States	2	1.1
Orange	United States	0.2	0.1
Spinach	United States	2	1.6
Tomato	CA	2	1.2
Brussels sprouts	United States	10	12.4
Alfalfa	OH	0.2	0.8
Hay	United States	1	5.8
		Average	6.8

Is the level of energy expenditure excessive? There is no clear answer. Many argue that the business of agriculture is to produce food at a reasonable cost to the consumer and profit to the grower. Agriculture's business is not to produce energy, but it must use it efficiently and responsibly. Herbicides are an efficient use of energy, in view of the energy costs and efficiency of alternative methods of weed control, which must be done.

The weed management techniques used by US agriculture could be more efficient and conserve more energy. Weed scientists are integrating weed-control techniques into new weed-management systems that use less energy. Herbicide rates are decreasing and energy used for weed control will decrease. It cannot go to zero because weed management and general crop production in conventional and organic agriculture require energy. Agriculturalists and the general society will participate in the debate over what form the energy will take and how much is needed. Given the historical trends there is no question that agriculture can become more efficient and use less of the total US energy supply. How this will be achieved is not clear. Some production systems use far less energy (Table 13.7). A rapid move to these systems would compel a sacrifice of food production for energy conservation—presently not a good trade.

5. HERBICIDE SAFETY

How safe are herbicides? It is a simple question. A definitive answer is hard to find because each answer will have a bias that should be understood. A reasonable response to the question is, compared to what? Compared to some things, herbicides are dangerous but compared to others they are relatively safe. Many people are certain that herbicides, and all other pesticides, are very dangerous. For many people pesticides and herbicides are inherently dangerous. They are regarded as poisons, things that are not safe to use or be around. If they weren't poisonous to something, they wouldn't be useful as herbicides. The suffix "icide" comes from the Latin *caedere* = to kill.

Answers to safety questions are complex. The questioner usually expects a factual response, not an opinion, but answers are nearly always composed of fact *and* opinion. Some respondents are vested, automatically, with authority and veracity by a questioner, but that does not deny the fact that most answers are part opinion, albeit, often informed opinion.

There are facts about herbicide safety; not all answers are entirely opinions. The dermal and oral LD_{50} (lethal dose at which 50% of a test population dies) for all herbicides is known. Necessary safety precautions during use and storage are well known. The agrichemical industry knows and avoids uses that create problems. The EPA and state agencies regulate herbicide use and users, to reduce if not eliminate the inherent dangers of use.

5.1 Perception of Risk

Science can measure risk and determine the probability of occurrence of a defined risk. Science cannot measure safety. It is a normative personal or political judgment. Judgment of safety is not and should not become a decision made in the scientific realm. Science plays a role in creating the data on which many judgments and decisions are based, but scientists qua scientists cannot determine what should be done about their data. Something may be described as unsafe because it is found through experiment and observation (the methods of science) to increase the risk of undesirable consequences. For example, motorcycle riding without a helmet can be fatal when an accident occurs. Scientists can measure the risk (the likelihood) of a fatal accident from riding a motorcycle without a helmet. Parents may decide not to buy a child a motorcycle, insurance companies may charge high premiums, and legislative bodies may pass laws requiring helmets because of the scientific evidence or informed opinions. Scientists may agree with these actions but science does not create them.

People perceive risk in different ways depending on where they live, how rich or poor they are, what their employment and future options are, their level of education, their friends, the scientific evidence they are aware of, what they read, etc. Perception of risk may differ from the facts as determined by scientific study. Table 20.9 (Slovik, 1982) is from a study that

TABLE 20.9 Actual Risk and the Perception of Risk by Three Groups

Rank Order of Actual Risk	Activity	Perceived Risk			
		US Deaths/ Year	League of Women Voters	College Students	Business and Professional Club Members
1	Smoking	150,000	4	3	4
2	Alcohol	100,000	6	7	5
3	Automobiles	50,000	2	5	3
4	Handguns	17,000	3	2	1
5	Electric power	14,000	18	19	19
6	Motorcycles	3,000	5	6	2
7	Swimming	3,000	19	30	17
8	Surgery	2,800	10	11	9
9	X-rays	2,300	22	17	24
10	Railroads	1,950	24	23	20
11	Private aviation	1,300	7	15	11
12	Construction	1,000	12	14	13
13	Bicycles	1,000	16	24	14
14	Hunting	800	13	18	10
15	Home appliances	200	29	27	27
16	Firefighting	195	11	10	6
17	Police work	160	8	8	7
18	Contraceptives	150	20	9	22
19	Commercial aviation	130	17	16	18
20	Nuclear power	100	1	1	8
21	Mountain climbing	30	15	22	12
22	Power mower	24	27	28	25
23	High school and college football	23	23	26	21
24	Skiing	18	21	25	16
25	Vaccinations	10	30	29	29
26	Food coloring	a	26	20	30
27	Food preservatives	a	25	12	28

TABLE 20.9   Actual Risk and the Perception of Risk by Three Groups—cont'd

Rank Order of Actual Risk	Activity	Perceived Risk			
		US Deaths/ Year	League of Women Voters	College Students	Business and Professional Club Members
28	Pesticides	a	9	4	15
29	Prescription antibiotics	a	28	21	26
30	Spray cans	a	14	13	23

*a*Unknown or not available.

Adapted from Slovik, P., 1982. Perception of risk. Science 236, 280–285.

shows how three different groups judged the risk of several common things. It is obvious that not all share the same perception of risk. In addition to reporting how people in various groups perceive risk, Table 20.9 shows the actual number of deaths from the hazard. Neither actual deaths nor perceptions of risk are an adequate way to decide what to do. The accepted standard in public health discourse and policy has become a heuristic measure of not increasing lifetime risk by more than one in a million.

It is not the purpose of this chapter to debate the question of herbicide safety but rather to frame a perspective from which the debate can proceed. The US National Safety Council estimates that 38,300 people were killed and 4.4 million injured on US roads in 2015, the largest 1-year percentage increase in half a century. There has been a steady decline. People properly conclude that automobiles are dangerous. Yet people drive too fast, without seatbelts, and often after consuming alcohol or while texting or talking on their cell phone. Because they think they are in control, many people don't think automobiles are as dangerous as they are. The danger is there, but as long as people think they are in control they believe they can control the risk. That is, the risk becomes acceptable. But the question then becomes, acceptable to whom (Starr and Whipple, 1980)? The answer may be determined legislatively or it may be determined by one's perception of the risk. Many people are much more likely to accept even a very risky activity:

Motorcycles: 4295 deaths in 2014
Hang gliding: 12 deaths in 2015
All-Terrain Vehicles: 230,666 emergency room visits in 2010.
Football: 489,676 emergency room visits in 2010; 271 concussions in 2015
Bicycle racing: 541,746 emergency room visits in 2013

If one chooses to assume a risk voluntarily, the reasons usually include the likely effects are perceived to be delayed, the risk is a known common hazard, there are no alternatives available, and the consequences are thought to be reversible. When the opposite situation prevails, risks are accepted less readily. There were 23 deaths in the United States from pesticides in 2012, most were suicides (Langley and Mort, 2012). Annually there is an average of 20,000 calls to poison control centers (Langley and Mort). It is not surprising that the general public regards pesticides as more risky and dangerous than the data show they really are. This is because they are seen as uncontrolled, involuntary risks with irreversible, severe, rapid, or lingering consequences. They are perceived as things likely to be misused and regarded as dreaded uncommon hazards.

More Americans die from bee stings (53/year), dogs (28/year),[15] prescription drug poisoning, dominantly opioids (28,000 in 2014), and falls (30,208 in 2015) than from pesticides. There are 2000–3000 cases of pesticide poisoning each year in the United States from voluntary and involuntary exposure but only a few deaths. Worldwide about 3 million people are poisoned by pesticides each year resulting in more than 250,000 deaths.[16]

There should be no debate about whether or not herbicides can be hazardous to humans. They are toxic, will poison, and may kill if not used properly. The last phrase is the key—if not used properly—but even with assured proper use, fear persists. Many prescription pharmaceuticals, household cleaning agents, aspirin, automotive fuel, and other common products are dangerous, if not used properly. Their inherent toxicity doesn't change with use but the possibility of danger increases with improper use. Stupidity doesn't increase the inherent toxicity of anything, but it increases risk.

A current, relevant example of the concern, evidence, and debate involves the carcinogenic potential of glyphosate (see New York Times, March 15, 2017). It is the world's most widely used herbicide primarily due to Monsanto's scientists pioneering efforts to genetically engineer several crops (Roundup ready) to be resistant to it. More than 200 million pounds of glyphosate were used in the United States in 2015. The crops are accepted in many countries yet the debate continues. Monsanto consistently argues that glyphosate is not a carcinogen. The federal court case is that the main ingredient of the herbicide (glyphosate) *might* cause cancer. The European Chemicals Agency has determined that Roundup herbicide is not a human carcinogen or cancer-causing substance. Resolution of the court case will depend upon the scientific evidence. Resolution of the concern in the public's mind will be much more dependent upon people's attitudes toward large corporations and their faith in scientific evidence.

5.2 Rules for Safe Use of Herbicides

There are rules for safe use of herbicides that don't change inherent toxicity but make accidents and the expression of toxicity less likely. The rules are simple, obvious, and often overlooked. A sample set of rules follows:

Before use
- Keep away from children.
- Purchase the right herbicide for the task.
- Read and be sure you understand the label use instructions.
- Follow all label instructions.
- Label equipment so cross-contamination is avoided.

[15]https://www.washingtonpost.com/news/wonk/wp/2015/06/16/chart-the-animals-that-are-most-likely-to-kill-you-this-summer/.

[16]http://www.who.int/mental_health/prevention/suicide/en/PesticidesHealth2.pdf.

Storage

- Keep in a locked storage place.
- Never store any herbicide (any pesticide) in anything other than its original container.
- Store outside the residence and away from food, feed, seed, or fertilizer.
- Protect liquids from freezing.
- Prevent access by children.

Handling

- When mixing is required do it in a well-ventilated area, preferably outside.
- Wear a protective breathing mask.
- Never smoke, eat, or chew while spraying or handling.
- Wash with soap and water and change clothing immediately if the herbicide is spilled on skin or clothing.
- Wear loose-fitting clothing.
- Wear rubber gloves and rubber boots.
- If herbicide is ingested, call a physician or the nearest hospital poison control center at once.
- If herbicide is splashed in eyes, flush with clean water immediately and call a physician.
- If symptoms of illness occur during or after handling or use, call a physician or the nearest hospital poison control center.

Application

- Prevent drift and volatility.
- Stop application if other people, especially children, appear.
- Do not contaminate wells, cisterns, other water sources, nontarget crops, or animals.
- Apply at proper time and rate with a sprayer that has been properly calibrated.
- Do not contaminate food and water containers, including those for livestock.

After use

- Dispose of empty containers properly. Puncture containers to prevent reuse.
- Wash and change to clean clothing.
- Be sure the person who washes contaminated clothing is aware of the contamination.
- Clean equipment soon after use.

The precaution concerning children is very important. Children often do unexpected things, and the best advice is keep them away. Be prepared. The other precautions for herbicide use are not difficult to understand. Most are just common sense. If poisoning occurs, treat it seriously and promptly take the victim to a physician or hospital. It is always a good idea to take along the pesticide container. Do not move victims who are in shock without treating for shock. It is often true that doing nothing except removing the victim from any possibility of further poisoning is a better thing than doing something if you are not sure what is correct.

5.3 The LD_{50} of Some Herbicides

The *Lethal Dose* at which 50% of the test population dies (LD_{50}) is a good indicator of relative toxicity and safety. It is not the only measure of safety. The LD_{50} may help understand toxicity when it is compared to the toxicity of other things. It is a measure of acute oral, not chronic toxicity. All values are expressed in milligrams (mg)/kilogram (kg) of body weight. If the LD_{50} is multiplied by 0.003 it is converted to ounces (oz)/180 pound man. The value would be different for a woman and someone of different weight.

The EPA has classified all pesticides into four groups based on their toxicity (Table 20.10). The LD_{50} of all herbicides available in the United States can be found in the Herbicide Handbook of the Weed Science Society of America (Shaner, 2014). It is wrong to assume that because two things have the same LD_{50} they are equally toxic because routes and likelihood (the chance) of exposure differ.

In one 5-ounce cup (about 150 mL) of roasted and brewed coffee there are about 85 mg of caffeine. Instant coffee has 60 mg and there are about 3 mg in decaffeinated coffee. To become poisoned from coffee, a 180-pound man would have to drink 7.6 gallons—all at once—192.5 cups, nonstop. That is impossible, and therefore it is very unlikely that anyone will die from acute caffeine intoxication.

The LD_{50} of ethyl alcohol is 4500 mg/kg and a 180-pound man would have to consume a little more than 0.8 pints to be acutely poisoned. This is not commonly done, but is possible. There are 2200 alcohol-poisoning deaths in the United States each year.

The LD_{50} as a measure of toxicity of anything is valuable, but its use must be tempered with knowledge of exposure, route of administration, rate, time, and physiological factors. It is a useful, but not a perfect, indicator of toxicity.

TABLE 20.10 Pesticide Toxicity Classes Based on LD_{50}

Toxicity Class	Signal Words[a]	LD_{50} (mg/kg)	Toxic Amount
I	Danger—poison + a skull and crossbones. Possibly followed by fatal if swallowed, poisonous if inhaled, extremely hazardous by skin contact—rapidly absorbed through skin, or corrosive—causes eye damage and severe skin burns.	56–49	Less than 5 g, about equal to a teaspoonful
II	Warning—may be fatal. Possibly followed by harmful or fatal if swallowed, harmful or fatal if absorbed through the skin, harmful or fatal if inhaled, or causes skin and eye irritation	50–499	5–30 g, about equal to 1 teaspoonful to 1 tablespoonful = 1 ounce
III	Caution—possibly followed by harmful if swallowed, may be harmful if absorbed through the skin, may be harmful if inhaled, or may irritate eyes, nose, throat, and skin.	500–4,999	More than 30 g, about equal to 1 ounce
IV	No signal word required since 2002	5,000–14,999	Practically nontoxic. Problems may result from ingestion of more than 1 pint, which is more than 455 g

[a]*Signal words must appear on pesticide label.*

THINGS TO THINK ABOUT

1. How do weeds interfere with human activities?
2. What ecological changes can occur after herbicide use?
3. Is ecological change/ecological effect inevitable?
4. How can ecological change, created by herbicides, be prevented or managed?
5. How do herbicides influence soil?
6. Is herbicide use in US agriculture energy intensive?
7. Are herbicides always harmful to people?
8. What reasons can you offer to explain why the debate about the environment, herbicides, and people is so complex?
9. What is the LD_{50} and how is it used?
10. Can herbicides be used safely? How can one maximize safety?

Literature Cited

Altman, J., Campbell, C.L., 1977. Effect of herbicides on plant disease. Ann. Rev. Phytopathol. 15, 361–385.

Altman, J., Neate, S., Rovira, A.D., 1990. Herbicide-Pathogen interactions and mycoherbicides as alternative strategies for weed control. In: Hoagland, R.E. (Ed.), Microbes and Microbial Products as Herbicides. Amer. Chem. Soc., Washington, DC, pp. 241–259.

Baker, N.T., Stone, W.W., 2014. Annual agricultural pesticide use for midwest stream-quality assessment, 2012–13. In: U.S. Geological Survey Data Series, vol. 863. https://doi.org/10.3133/ds863, 17 pp.

Barrett, M., Witt, W.W., 1987. Alternative pest management practices. In: Helsel, Z.R. (Ed.), Energy in Plant Nutrition and Pest Control. In: Stout, B.A. (Ed.), Energy in World Agriculture, vol. 2. Elsevier Press, NY, pp. 195–234.

Bender, J., 1994. Future Harvest: Pesticide-Free Farming. Univ. of Nebraska Press, Lincoln, NE, 159 pp.

Berenbaum, M., et al., 2000. The Future Role of Pesticides in US Agriculture. National Academy of Sciences, Washington, DC, 301 pp.

Brown, A.W.A., 1978. Ecology of Pesticides. J. Wiley and Sons, NY, p. 325.

Carson, R., 1962. Silent Spring (Several editions available).

Conway, G., 1997. The Doubly Green Revolution – Food for All in the 21st Century. Comstock Pub. Assoc. A division of Cornell Univ. Press, Ithaca, NY, 335 pp.

Conway, G.R., Pretty, J.N., 1991. Agriculture without pollution. In: Unwelcome Harvest – Agriculture and Pollution. Earthscan Pubs., Ltd, London, UK, pp. 555–636.

Culliney, T.W., Pimentel, D., Pimentel, M.H., 1992. Pesticides and natural toxicants in foods. Agric. Ecosyst. Environ. 41, 297–320.

DeSimone, L.A., McMahon, P.B., Rosen, M.R., 2014. The Quality of Our Nation's Waters—Water Quality in Principal Aquifers of the United States, 1991–2010. U.S. Geological Survey Circular 1360. https://doi.org/10.3133/cir1360, 151 pp.

Doering, O.C., 1992. The social and ethical context of agriculture: is it there and can we teach it?. In: Agriculture and the Undergraduate. Nat. Res. Council, Nat. Acad. Press, Washington, DC, pp. 237–244.

FDA (US Food and Drug Adminstration), 1993. Food and drug administration monitoring program. J. Assoc. Off. Anal. Chem. Int. 76 (5), 127a–141a.

Furtick, W.R., 1976. Uncontrolled pests or adequate food? In: Gunn, D.L., Stevens, J.G.R. (Eds.), Pesticides and Human Welfare. Oxford Univ. Press, Oxford, UK, pp. 228–239.

Gilliom, R.J., 2007. Pesticides in U.S. streams and groundwater. Environ. Sci. Technol. 41 (10), 3408–3414.

Gilliom, R.J., Barbash, J.E., Kolpin, D.W., Larson, S.J., 1999. Testing water quality for pesticide pollution. Environ. Sci. Technol. 33, 164A–169A.

Green, R.E., Cornell, S.J., Scharelemann, J.P.W., Balmford, A., December 23, 2004. Farming and the Fate of Wild Nature. Science Express, 10 pp. www.scienceexpress.org.

Gressel, J., 1990. Synergizing herbicides. Rev. Weed Sci. 5, 49–82.

Heilprin, J., March 4, 2006. Pesticides found in most of the nations's streams. In: Report from the Assoc. Press in the Coloradoan Newspaper, p. A-8.

Issacson, W., March 2007. The Last Battle of Vietnam. TIME.

Hoffman, R.S., Capel, P.D., Larson, S.J., 2000. Comparison of pesticides in eight U.S. urban streams. Environ. Toxicol. Chem. 19, 2249–2258.

Katan, J., Eshel, Y., 1973. Interactions between herbicides and plant pathogens. Res. Rev. 45, 145–177.

Kirschenmann, F.L., 2010. Cultivating an Ecological Conscience – Essays from a Farmer Philosopher. Counterpoint, Berkeley, CA, 403 pp.

Kraybill, H.F., 1975. Pesticide toxicity and the potential for cancer. Pest Control 43 (12), 10–16.

Langley, R.L., Mort, S.A., 2012. Human exposures to pesticides in the United States. J. Agromed. 17 (3), 300–315. https://doi.org/10.1080/1059924X.2012.688467.

Mackay, D., 1988. On low, very low, and negligible concentrations. Environ. Toxicol. Chem. 7, 1–3.

McCurdy, E.V., Molberg, E.S., 1974. Effects of the continuous use of 2,4-D and MCPA on spring wheat production and wheat populations. Can. J. Plant Sci. 54, 241–245.

Nace, T., 2006. Breadbasket of democracy. Orion 251 (3), 52–61.

Nalewaja, J.D., 1974. Energy requirements for various weed control practices. In: Proc. N. Cent. Weed Control Conf., vol. 29, pp. 19–23.

Norris, R.F., Caswell-Chen, E.P., Kogan, M., 2003. Concepts in Integrated Pest Management. Prentice Hall, Upper Saddle River, NJ, 586 pp.

Olson, R., Hansen, J., Whitson, T., Johnson, K., 1994. Tebuthiuron to enhance rangeland diversity. Rangelands 16, 197–201.

Pimentel, D., 1997. Pest management in agriculture. In: Pimentel, D. (Ed.), Techniques for Reducing Pesticide Use – Economic and Environmental Benefits. J. Wiley & Sons, New York, NY (Chapter 1).

Pimentel, D., Pimentel, M., 1979. Food, Energy and Society. E. Arnold, London, pp. 69–98.

Rahman, A., Atkinson, G.C., Douglas, J.A., 1979. Eradicane causes problems. N.Z. J. Agric. 139 (3), 47–49.

Rice, P.M., Bedunah, D.J., Carlson, C.E., 1992. Plant Community Diversity after Herbicide Control of Spotted Knapweed. USDA/Forest Service. Intermountain Res. Paper No. INT-460. 7 pp.

Rovira, A.D., McDonald, H.J., 1986. Effects of the herbicide chlorsulfuron on *Rhizoctonia* bare patch and take-all of barley and wheat. Plant Dis. 70, 879–882.

Ryberg, K., Gilliom, R.J., 2015. Trends in pesticide concentrations in use for major rivers of the United States. Sci. Total Environ. 538, 431–444.

Santillo, D.J., Leslie Jr., D.M., Brown, P.W., 1989a. Responses of small mammals and habitat to glyphosate application on clearcuts. J. Wildl. Manag. 53, 164–172.

Santillo, D.M., Brown, P.W., Leslie Jr., D.M., 1989b. Response of songbirds to glyphosate-induced habitat change on clearcuts. J. Wildl. Manag. 53, 64–71.

Schubert, O.E., 1972. Plant cover changes following herbicide applications in orchards. Weed Sci. 20, 124–127.

Schumpeter, J.A., 1950. Capitalism, Socialism and Democracy. Harper, New York, 413 pp.

Shader-Frechette, K., 1991. Pesticide policy and ethics. In: Blatz, C.V. (Ed.), Ethics and Agriculture. Univ. of Idaho Press, Moscow, ID, pp. 426–433.

Shaner, D. (Ed.), 2014. Herbicide Handbook, tenth ed. Weed Science Society of America, Lawrence, KS. 513 pp.

Slovik, P., 1982. Perception of risk. Science 236, 280–285.

Starr, C., Whipple, C., 1980. Risks of risk decisions. Science 208, 1114–1119.

Štepána, R., Ticháa, J., Hajšlováa, J., Kovalczuka, T., Kocoureka, V., 2005. Baby food production chain: pesticide residues in fresh apples and products. Food Addit. Contam. 22 (12), 1231–1242.

Stone, W.W., Gilliom, R.J., Martin, J.D., 2014. An overview comparing results from two decades of monitoring for pesticides in the nation's streams and rivers, 1992–2001 and 2002–2011. In: U.S. Geological Survey Scientific Investigations Report 2014–5154. https://doi.org/10.3133/sir20145154, 23 pp.

Tietjen, H.P., Halvorsen, C.H., Hegdal, P.L., Johnson, A.M., 1967. 2,4-D herbicide, vegetation and pocket gopher relationships on Black Mesa, Colorado. Ecology 48, 635–643.

Tilman, D., Fargione, J., Wolff, B., D'Antonio, C., Dobson, A., Howarth, R., et al., 2001. Forecasting agriculturally driven global environmental change. Science 292, 281–292.

Turner, G.T., 1969. Responses of mountain grassland vegetation to gopher control, reduced grazing, and herbicide. J. Range Manag. 22, 377–383.

Van den Bosch, R., 1978. The Pesticide Conspiracy. Univ. of California Press, Berkeley, CA, 226 pp.

Zimdahl, R.L., 2006. Agriculture's Ethical Horizon. Elsevier, Inc., San Diego, CA, 235 pp.

Zimdahl, R.L., 2012. Agriculture's Ethical Horizon, second ed. Elsevier Insights, London, UK. 274 pp.

Recommended Supplementary Literature

Interested readers are encouraged to explore the work of Wendell Berry and Wes Jackson.

Anonymous, 1995. Agriculture and Spirituality — Essays from the Crossroads Conference at Wageningen Agricultural University. International Books, Utrecht, The Netherlands, 140 pp.

Bailey, L.H., 1915. The Holy Earth. Sowers Printing Company, Lebanon, PA, 117 pp.

Barrons, K.C., 1981. Are Pesticides Really Necessary? Regnery Gateway, Inc., Chicago, IL, 245 pp.

Berry, T., 1988. The Dream of the Earth. Sierra Club, San Francisco, CA, 247 pp.

Bird, E.A.R., Bultena, G.L., Gardner, J.C., 1995. Planting the Future — Developing an Agriculture that Sustains Land and Community. Iowa State Univ. Press, Ames, IA, 276 pp.

Brown, L.R., 2003. Plan B — Rescuing a Planet under Stress and a Civilization in Trouble. W.W. Norton & Co, New York, NY, 285 pp.

Busch, L., Lacy, W.B., Burkhardt, J., Hemken, D., Moraga-Rojel, J., Koponen, T., Silva, J.S., 1995. Making Nature, Shaping Culture — Plant Biodiversity in Global Context. University of Nebraska, Lincoln, NE, 231 pp.

Colburn, T., Dumanoski, D., Myers, J.P., 1996. Our Stolen Future: Are We Threatening Our Fertility, Intelligence, and Survival?: A Scientific Detective Story. Penguin Books USA, Inc., NY, 306 pp.

Cummings, C.H., 2008. Uncertain Peril — Genetic Engineering and the Future of Seeds. Beacon Press, Boston, MA, 232 pp.

Davis, E.F., 2009. Scripture, Culture and Agriculture — An Agrarian Reading of the Bible. Cambridge University Press, Cambridge, UK, 234 pp.

Evans, L.T., 1998. Feeding the Ten Billion — Plants and Population Growth. Cambridge Univ. Press, Cambridge, UK, 247 pp.

Giron, L., Glover, S., Smith, D., September 17, 2011. Drug Deaths Now Outnumber Traffic Fatalities in U.S. Los Angeles Times.

Graham Jr., F., 1970. Since Silent Spring. Houghton Mifflin Co, Boston, MA, 333 pp.

Gunn, D.L., Stevens, J.G.R., 1976. Pesticides and Human Welfare. Oxford Univ. Press, Oxford, U.K., 278 pp.

Hanson, V.D., 1996. Fields Without Dreams — Defending the Agrarian Idea. The Free Press, New York, NY, 289 pp.

Hanson, V.D., 2000. The Land Was Everything — Letters from an American Farmer. The Free Press, New York, NY, 258 pp.

Kimbrell, A. (Ed.), 2002. Fatal Harvest — The Tragedy of Industrial Agriculture. Island Press, Washington, DC, 384 pp.

Limerick, P.N., 2000. Something in the Soil — Legacies and Reckonings in the New West. W.W. Norton & Company, NY, 384 pp.

Logan, W.B., 1995. Dirt — The Ecstatic Skin of the Earth. The Berkley Publishing Group, New York, NY, 202 pp.

Logsdon, G., 1994. At Nature's Pace — Farming and the American Dream. Pantheon Books, New York, NY, 208 pp.

Lomborg, B., 2001. The Skeptical Environmentalist — Measuring the Real State of the World. Cambridge University press, Cambridge, UK, 515 pp.

Manning, R., 2004. Against the Grain — How Agriculture Has Hijacked Civilization. North Point Press, NY, 232 pp.

Marco, G.J., Hollingworth, R.M., Durham, W. (Eds.), 1987. Silent Spring Revisited. Amer. Chem Soc., Washington, DC, 214 pp.

Montgomery, D.R., 2007. Dirt — The Erosion of Civilizations. The University of California Press, Berkeley, CA, 285 pp.

Orr, D.W., 1994. Earth in Mind — On Education, Environment, and the Human Prospect. Island Press, Washington, DC, 213 pp.

Patel, R., 2012. Stuffed and Starved — The Hidden Battle for the World Food System. Melville House, Brooklyn, NY, 418 pp.

Pimentel, D., Zehman, H. (Eds.), 1993. The Pesticide Question: Environment, Economic, and Ethics. Chapman and Hall, NY, NY, 441 pp.

Pollan, M., 2006. The Omnivore's Dilemma — A Natural History of Four Meals. The Penguin Press, NY, 450 pp.

Pyle, G., 2005. Raising Less Corn, More Hell — The Case for the Independent Farm and against Industrial Food. Public Affairs, NY, 229 pp.

Ragsdale, N.N., Kuhr, R.J. (Eds.), 1987. Pesticides — Minimizing the Risks. ACS Symposium Series 336. American Chemical Society, Washington, DC, 183 pp.

Ridley, M., 2010. The Rational Optimist — How Prosperity Evolves. Harper Collins, New York, 438 pp.

Rieff, D., 2015. The Reproach of Hunger — Food, Justice, and Money in the Twenty-First Century. Simon & Schuster, NY, 402 pp.

Roberts, P., 2008. The End of Food. Houghton Mifflin Co, Boston, MA, 390 pp.

Runge, C.F., Senauer, B., Pardey, P.G., Rosegrant, M.W., 2003. Ending Hunger in Our Lifetime — Food Security and Globalization. The Johns Hopkins Univ. Press, Baltimore, MD, 288 pp.

Sheets, T.J., Pimentel, D. (Eds.), 1979. Pesticides: Contemporary Roles in Agriculture, Health, and the Environment. Humana Press, Clifton, NJ, 186 pp.

Smil, V., 2000. Feeding the World — A Challenge for the Twenty-First Century. The MIT Press, Cambridge, MA, 360 pp.

Thompson, P.B., 2010. The Agrarian Vision — Sustainability and Environmental Ethics. The University Press of Kentucky, Lexington, KY, 323 pp.

Whorton, J., 1974. Before Silent Spring: Pesticides and Public Health in Pre-DDT America. Princeton Univ. Press., Princeton, NJ, 288 pp.

Whitten, J.L., 1966. That We May Live. D. Van Nostrand Co, Inc., Princeton, NJ, 251 pp.

Wright, A., 1990. The Death of Ramon Gonzalez: The Modern Agricultural Dilemma. U. of Texas Press, Austin, 337 pp.

Pesticide Legislation and Registration

FUNDAMENTAL CONCEPTS

- The pesticide registration process is complex, mandatory, and based on state and federal legislation.

- The US Environmental Protection Agency is the federal agency that administers federal pesticide laws and is responsible for pesticide registration.

LEARNING OBJECTIVES

- To understand the purpose and complexity of federal pesticide laws.

- To be aware of the protection the regulatory process provides the US public.

- To become familiar with the basic steps of pesticide registration under US law.

- To understand the role of pesticide registration and regulation.

1. THE PRINCIPLES OF PESTICIDE REGISTRATION

Most of the world's nation states have some sort of pesticide[1] registration procedure. In some countries procedural and data requirements are few to nonexistent, possibly due to fiscal constraints and unawareness of the need. It is also true that the laws in some countries are not fully implemented. Countries that regulate pesticides share the goal of providing protection from real and potential adverse effects of pesticides and gaining the benefits of pesticide use (Snelson, 1978). These objectives are achieved through the registration process and subsequent control of the pesticide label. Registration enables the regulatory agency to

[1]US federal and state pesticide regulations apply equally to all classes of pesticides. The general term *pesticide* will be used in most places in this chapter instead of the specific term *herbicide*.

exercise control over use, performance claims, label directions and precautions, packaging, and advertising to ensure proper use and environmental and human protection. In general, the process protects the public's interest and the manufacturer's rights (Snelson).

It is apparent from the questions regularly raised in the news media that the public does not know the intricacies of pesticide registration and the laws that govern the process. This chapter describes some general aspects of US pesticide registration.

In the world's developed countries pesticides have been subject to governmental regulation for about 100 years. The public is aware of pesticides and fearful because of mistakes that have occurred. Nearly everyone knows something negative about DDT. The fact that more than 1 billion pounds of pesticides are applied in the United States every year is disturbing to many. The widespread use and public concern, if not fear, justifies the need for governmental regulation. This causes the public to ask, as Wildavsky (1997) did, Is it true? His answer, in most cases, where the public became concerned was, No, it is not true. Although concern was real and legitimate, irrational fear was and continues to be unwarranted. There are other agricultural examples:

- The mid-1960s and continuing controversy over the real and suspected hazards of 2,4,5-T, a component of Agent Orange used in Operation Ranch Hand, a vegetation control program during the Vietnam war. It was the first major public debate that challenged the intellectual foundation of weed science's dependence on herbicides.
- There were few immediate, reasonable solutions when weed resistance first appeared in 1957 (Switzer). Now resistance to herbicides has appeared in 479 unique cases (species x site of action) in 146 dicots and 105 monocot weeds for 163 different herbicides in 91 crops in 69 countries.
- Pesticide poisoning: The World Health Organization estimates 3 million cases of pesticide poisoning occur every year, resulting in 250,000 deaths, many by suicide.[2] Developing countries use 25% of pesticides but experience 99% of the deaths (Goldman, 2004). The US Environmental Protection Agency (EPA) estimates that 10,000–20,000 physician-diagnosed pesticide poisonings occur each year among the approximately 2 million US agricultural workers. Groundskeepers, pet groomers, fumigators, and workers in a variety of other occupations are at risk for exposure to pesticides.[3]
- In the mid-1980s, Bhopal was a city of nearly 1 million people in India's Madhya Pradesh region between New Delhi and Bombay. On December 2, 1984, workers at Union Carbide's pesticide plant were making the pesticide Sevin. An explosion led to what many consider to be the worst industrial accident in history. Reports claim 3000–4000 deaths and 50,000 treated for illnesses related to the explosion.
- In October 1997 a warehouse in Birmingham, Alabama, United States, housing 18,000 gallons of the termite killer Dursban caught fire. Firefighters put out the fire, but created a solution of pesticide and water that flowed into Village Creek and then the Warrior River and caused a 10-mile fish kill.
- The debate within and outside the agricultural community over the risks and ultimate benefits of genetic modification of crops has raised legitimate economic, social, and

[2]http://www.who.int/mental_health/prevention/suicide/en/PesticidesHealth2.pdf.

[3]http://www.cdc.gov/niosh/topics/pesticides/.

biological concern. The primary concern for weed science is the widespread adoption of herbicide-resistant technology.

- Air and water pollution and animal suffering from Confined Animal Feeding Operations.
- Mad cow disease, swine flu, bird flu, meat recalls, and antibiotic resistance are all of concern or have been of major societal concern in the past.
- The ecological dead/hypoxic zone in the Gulf of Mexico extends over a large area from the terminus of the Mississippi river. There are more than 400 similar hypoxic zones in the world.

Many Americans are aware of these and other problems created by agricultural technology. However, most people are not aware of the plethora of laws and the intricate and continually reviewed procedures for registration of a pesticide prior to use. These regulations are intended to prevent disaster. Registration, a complex process, is not as simple as registering a pet or a car. It is not simply recording ownership and paying a nominal fee. Registration requires compliance with legal requirements that establish a regulatory process that demands proof of safety. It does not require proof of efficacy because it is assumed that no manufacturer will attempt to market a pesticide that does not work as advertised. Different nation states establish registration processes that conform to their needs. The system in the United States is among the most complex and successful. It is not perfect, and there are many complaints about it from those who argue that protection is not sufficient and, in contrast, from manufacturers who find the process slow, expensive, and unnecessarily cautious. The United Kingdom used to work with a voluntary approval scheme wherein a consensus was reached between the manufacturer, government, and users about appropriate regulation. The procedure was abandoned in the mid-1980s because of pressure from the European Economic Community for uniform standards. The UK scheme and the European Union process now resemble the procedures followed in the United States, including provisions that regulate advertising, storage, application, and crop use. Many nations follow the standards put forth by the UN Food and Agricultural Organization Codex Committee on Pesticide Residues that establishes maximum residue limits for pesticides in food. The CODEX also guides countries on safety regulations for use, storage, and analysis of pesticides.

This chapter provides a brief description of the history of pesticide legislation and registration in the United States and general procedures that must be followed.

2. FEDERAL LAWS

2.1 Food and Drug Act of 1906

The purpose of the first law administered by the US Department of Agriculture (USDA) was "to halt the exposure of the general public to filthy rotten food, adulteration, substitution, and misleading claims." Several cases of arsenic poisoning in England from imported US apples stimulated passage of the law.

The first US federal law that directly involved pesticides was the Insecticide Act of 1910, which set standards for chemical quality and provided consumer protection. It did not address the growing issue of environmental damage and biological health risks. It was passed to stop unethical persons from selling ineffective or adulterated products and was specifically aimed at Paris green, lead arsenate, and other insecticides and fungicides. There were no effective herbicides in 1910. The law, administered by the USDA, introduced a labeling requirement that mandated an ingredient statement and the manufacturer's name. The late 1960s and 1970s were a period of slow development of pesticides, and there was little public concern about them because they were generally considered useful. In contrast to present regulations, the law emphasized proof of efficacy. It neither demanded proof of safety nor evaluation of possible hazards of the pesticides it regulated. Chemical analysis for crop residue was the most important enforcement procedure. The Insecticide Act protected the public against the possible loss of crops or damage to property from pesticide use, but there was no assurance that pesticides were not human health hazards.

2.2 Food, Drug, and Cosmetic Act of 1938

The US Federal Food, Drug, and Cosmetic Act (FFDCA) was passed in 1938 to gain more control of adulteration; misbranding; substitution of improper ingredients in food, drugs, and cosmetics; and to ensure the integrity and safety of food moving in interstate commerce. Beginning as the Division of Chemistry and after July 1901 the Bureau of Chemistry, the modern era of the US Food and Drug Administration (FDA) dates to 1906 with the passage of the Federal Food and Drug Act; this added regulatory functions to the agency's scientific mission. The Bureau of Chemistry's name changed to the Food, Drug, and Insecticide Administration in July 1927, when the nonregulatory research functions of the bureau were transferred elsewhere in the department. In July 1930 the name was shortened to the present version. FDA remained under the Department of Agriculture until June 1940, when the agency was moved to the new Federal Security Agency. In April 1953 the agency again was transferred, to the Department of Health, Education, and Welfare (HEW). Fifteen years later FDA became part of the Public Health Service within HEW, and in May 1980 the education function was removed from HEW to create the Department of Health and Human Services. The agency has grown from a single chemist in the US Department of Agriculture in 1862 to a staff of approximately 15,000 employees and a budget of $4.4 billion in 2014.

The 1938 law gave authority to the FDA to oversee the safety of food, drugs, and cosmetics. At the present time, the law is administered by the Department of Health and Human Services and the EPA. The need for the law was justified by the increasing use of potentially adulterating chemicals in food, drugs, and cosmetics. Manufacturers were required to prove safety and usefulness. The Federal Security Agency established safe levels of residues, a responsibility transferred to HEW in 1953. Thus, a health agency made safety decisions and was required to make agricultural decisions (on usefulness). This arrangement was cumbersome and objectionable to users and manufacturers.

The law requires tolerances be established for all pesticides used in or on food or in a manner that will result in a residue in or on food or animal feed. A tolerance (now set by EPA) is the maximum permissible level for pesticide residues allowed in or on human food and animal feed. This law includes strong provisions for protecting infants and children, as well as other sensitive subpopulations.

2.3 Federal Insecticide, Fungicide, and Rodenticide Act of 1947

Pesticide development and increasing use during and after World War II created a need for a stronger law. USDA, supported by the pesticide industry, developed the Federal Insecticide, Fungicide, and Rodenticide Act (FIFRA), which became law in 1947. The law retained the key portions of the Insecticide Act of 1910 and extended the principle that a pesticide formulation should meet proper standards. No other federal law had authority over the pesticide and its labeling. The FIFRA added two new ideas to pesticide regulation. The first was that all pesticides intended for shipment in interstate commerce must be registered with the US Secretary of Agriculture before shipment. The second stipulation was that the USDA was given control over all precautionary statements on the pesticide label. USDA was empowered to review the public presentation of safety procedures so important to proper use. The law also placed the burden of proving efficacy and safety on the manufacturer. These provisions stopped shipment of untested or improperly labeled products in interstate commerce by requiring that labeling be adequate and that all labels be approved (registered) by the USDA. Withholding registration was an effective way to stop use and transport of pesticides that were faulty due to contamination, mislabeling or lack of efficacy. The USDA could withhold registration until data were provided to prove that the pesticide would give the degree of pest control claimed or implied on the label. Labels could also be withheld pending submission of adequate evidence of human safety based on appropriate scientific data. The act included several specific, new items:

- Protection of the user from physical injury or economic loss.
- Protection of the public from injury. Previous laws only protected the users.
- The manufacturer had to provide evidence that the pesticide was effective for its intended use.
- A pesticide was defined and limited to economic poisons defined as "any substance or mixture of substances intended for preventing, destroying, repelling, or mitigating any insects, rodents, fungi, weeds, and other forms of plant or animal life or viruses except viruses on or in living man or animals, which the Secretary of Agriculture shall declare to be a pest."

The major public protection came from strict control over every feature of labeling. FIFRA has no control over the user of the product. Users bear the responsibility to read and heed the label and avoid misuse and environmental contamination.

Questions about coverage of economic poisons not specifically defined under FIFRA were raised. Therefore, the law was amended in 1959 to include regulation of residues of nematicides, plant regulators, defoliants, and desiccants as economic poisons in or on food or feed crops. This broadened scope did not include adequate protection for fish and wildlife, and the law was further amended to include pesticides sold for control of moles, birds, predatory animals, and other nonrodent pests. It also included certain plants and viruses when they are injurious to plants, domestic animals, or to man. Thus, it was possible to regulate pesticides designed to control specific things. The regulations were expanded to include:

- Mammals, including but not limited to dogs, cats, moles, bats, wild carnivores, armadillos, and deer;
- Birds, including but not limited to starlings, English sparrows, crows, and blackbirds;

- Fish, including but not limited to jawless fish (e.g., sea lamprey), cartilaginous fish (e.g., sharks), and bony fish (e.g., carp);
- Amphibians and reptiles, including but not limited to poisonous snakes;
- Aquatic and terrestrial invertebrates, including but not limited to slugs, snails, and crayfish;
- Roots or other plant parts growing where they are not wanted; and
- Viruses other than those in or on living man or other animals.

One might ask who was enamored of armadillos or English sparrows and managed to have them included. To ask the question misses the point that the law was being expanded in scope consistent with Congressional interpretation of the public's desire for environmental safety.

2.3.1 *Federal Insecticide, Fungicide, and Rodenticide Act Amendments*

The Miller Pesticide Amendment, or PL-518, passed in 1954 amended the FFDCA of 1938, corrected cumbersome enforcement procedures, and defined procedures for setting safety limits for pesticide residues on raw agricultural commodities. It was formulated by a Committee of the House of Representatives chaired by Representative James Delaney (D, NY) that was formed to investigate chemicals in food and cosmetics. The committee decided that a better way to establish tolerances on food crops was required and assumed the initiative to formulate a way. Congressman Arthur Miller (R, NE) formulated the recommendations in a bill known as the Pesticide Chemicals Amendment to the FFDCA. A pesticide chemical was defined as "any substance which alone, in chemical combination, or in formulation with one or more other substances, is an economic poison as defined by the FIFRA of 1947, as now or as hereinafter amended and which is used in the production, the storage, or the transportation of raw agricultural commodities." This definition through use of the term *economic poison* related the Miller amendment to FIFRA.

The amendment established new procedures for obtaining tolerances. The HEW Secretary was charged with establishing tolerances or maximum allowable limits of pesticides on raw agricultural commodities moving in interstate commerce. A raw agricultural commodity was defined as "any food in its raw or natural state including all fruits in a washed, colored, or otherwise treated state in their unspoiled form prior to marketing." This formalized the establishment of tolerances in the Federal Food and Drug Administration (FFDA) of HEW. The USDA had to certify to the FFDA that the chemical for which a petition for tolerance had been filed would be useful for the purposes described and express an opinion as to whether the tolerance requested reasonably reflected the residues likely to remain on the crop when the pesticide was used as directed. This change assigned agricultural functions to the USDA, an agricultural agency, and health functions to the FFDA under HEW, a health agency. Tolerances were obtained from FFDA and use clearance from USDA. Prior to filing a tolerance petition, the chemical must have been registered under the FIFRA. All registration functions are now handled by the EPA.

The 1968 Color Additive Amendment subjected all color additives to the provisions for food additives. The 1964 Seed Coloring Amendment subjected all seed colorings to the provisions for food additives. The most controversial amendment to the Federal Food and Drug Control Act was the widely and hotly debated Delaney Cancer Clause that was included with the 1958 food additive amendment. It said, "no additive is deemed safe if found to induce cancer when ingested by man or animal or if it is found after tests which are found appropriate for the evaluation of the safety of food additives to induce cancer in man or animals." Thereafter, much of the controversy concerning use and misuse of pesticide chemicals focused on the restrictions of the Delaney amendment. It prohibited setting of a tolerance in a processed food, although tolerances could be set in raw agricultural commodities. It is important to note that it said nothing about dose, nor the length of time within which cancer must be induced. It applied only to processed food, not raw agricultural commodities. After years of debate, many (but not all) were pleased when the 38-year-old Delaney clause was removed in the 1996 Food Quality Protection Act (FQPA) signed into law by President Clinton on August 3, 1996. Since the Delaney clause was enacted (1958) chemical analytical technology progressed so it is possible to routinely detect parts per billion (ppb) or trillion (ppt) concentrations that are well below what could be detected in 1958 and, based on scientific data, pose no known human health hazard. The standard of reasonable certainty is now defined, in part, as "no more than a one-in-one-million chance of getting cancer after a lifetime of exposure." Replacement of Delaney standards with new health-based standards that do not distinguish between raw and processed agricultural commodities has not eliminated concern about health issues. The new standards may be just as tough or tougher than the old, widely discussed standard.

The FQPA included major changes to the FFDCA. EPA was required to consider children's special sensitivity and exposure to pesticides, group compounds for tolerance with a common site of action, consider cumulative exposure through contact with air, food, water, and other routes of exposure, and to reevaluate all existing tolerances within 10 years. The FQPA mandated testing for endocrine disruptors—compounds that block or mimic the effects of human hormones, e.g., estrogen (See Colburn et al., 1996).

2.4 Federal Environmental Pesticide Control Act of 1972

On October 21, 1972, President Nixon signed the Federal Environmental Pesticide Control Act (FEPCA). This law amended the FIFRA of 1947, in several ways, but FIFRA remained as the primary law. FEPCA was designed to protect man and the environment and extended federal regulation to all pesticides including those manufactured and used within a state.

Responsibility for use and misuse was now lodged with the pesticide applicator. In addition, no pesticide could be registered or sold unless it and its labeling were designed to prevent injury to man and any unreasonable effects on the environment. Future label evaluation by EPA had to consider the public interest, including benefits from pesticide use. However, under the FIFRA ultimate responsibility for pesticide use and misuse was born by the manufacturer who prepared the label. Under FEPCA manufacturers still had to establish safety, but responsibility rests with the user if failure to follow label directions results in human or environmental harm. Violators could be prosecuted under civil or criminal law.

FIFRA required the registration of pesticides moving in interstate commerce, whereas FEPCA requires registration of all pesticides regardless of their point of manufacture or use. All registered pesticides are classified for restricted or general use. General use pesticides can be purchased and used by anyone who, it is assumed, will follow label directions. The restricted category includes all pesticides that demonstrate the potential for harm to human health or the environment even when used according to the label. The restricted classification must appear on the front label of the pesticide package. The FEPCA requires certification of commercial applicators that involves demonstration of a minimum level of knowledge and competence about pesticides. A commercial applicator is one who may use or supervise the use of restricted-use pesticides. Private applicators are recertified by participation in approved training programs. Most states require a written test. A private applicator is a certified applicator who uses or supervises use of restricted-use pesticides for purposes of producing any agricultural commodity on property owned or rented by the individual or an employer or on property of another person (if such application is without compensation other than trading personal services).

The FEPCA strengthened EPA enforcement procedures. All pesticide-producing establishments had to be registered and regularly report sales and production. A pesticide's registration can also be revoked by EPA. Under the amended FIFRA, nonessentiality is not a sufficient reason to deny registration. This means that if one pesticide is already available for a specific use, registration cannot be denied to a new product. States may impose more stringent pesticide regulations than the amended FIFRA. In the past, some states had no pesticide laws, but the FEPCA required all states to have them or federal regulations automatically applied.

EPA can cancel registration after 5 years even when continued use is requested by the manufacturer. EPA also has the power to reclassify, suspend, or cancel a pesticide if it causes unreasonable adverse effects on the environment when used as directed. Unreasonable adverse effects include any unreasonable risk to humans or the environment. The decision must be based on the economic, social, and environmental costs and benefits of pesticide use. If a use (or all uses) of a pesticide is suspended (it is taken off the market) and later canceled, the law provides indemnities to the manufacturer and other owners. Such indemnities are designed to protect the manufacturer who has met all legal requirements and may suffer large monetary losses when new knowledge demonstrates that continued use of the product may be hazardous.

The most recent change in federal legislation is the Pesticide Registration Improvement Act (PRIA) of 2004. The consolidated appropriations act of 2004 established a new pesticide registration system. PRIA is now Section 33 of FIFRA. It created a registration service fee for applicants and established new tolerances for maximum residue levels in food and feed.

2.5 Endangered Species Act of 1973

The Endangered Species Act, passed in 1973, requires federal agencies to ensure that any action they authorize, fund, or carry out will not likely jeopardize the continued existence of any listed species, destroy, or adversely modify any critical habitat for those species. Part of EPA's determination is an assessment of whether listed endangered or threatened species or their designated critical habitat may be affected by use of the product.

2.6 Food Quality Protection Act of 1996

The 1996 law amended FIFRA and FFDCA and required EPA to find that a pesticide poses a "reasonable certainty of no harm" before it can be registered for use on food or feed. Every registered pesticide must be reviewed at least once every 15 years. Factors must be addressed before a tolerance can be established, including:

- Aggregate, nonoccupational exposure from the pesticide (exposure through diet and drinking water and from using pesticides in and around the home).
- Cumulative effects from exposure to pesticides that have a common mechanism of toxicity, that is, two or more pesticide chemicals or other substances that cause a common toxic effect(s) by the same, or essentially the same, sequence of major biochemical events (i.e., interpreted as site of action).
- Whether there is increased susceptibility to infants and children, or other sensitive subpopulations, from exposure to the pesticide.
- Whether the pesticide produces an effect in humans similar to an effect produced by a naturally occurring estrogen or produces other endocrine-disruption effects (See Colburn et al., 1996).

2.7 Pesticide Registration Improvement Act of 2003

PRIA amended FIFRA and FFDCA. It was reauthorized by the Pesticide Registration Improvement Renewal Act of 2007 and the Pesticide Registration Improvement Extension Act of 2012. Under PRIA:

- Companies must pay service fees according to the category of the registration action.
- EPA must meet decision review time periods, which result in a more predictable evaluation process for companies.
- Shorter decision review periods are provided for reduced-risk registration applications.

3. THE ENVIRONMENTAL PROTECTION AGENCY

In December 1970 the EPA was created. The entire pesticide regulations division of the USDA and, somewhat later, the pesticide office of the FDA were transferred to the EPA Office of Pesticides Program, which included the pesticide registration and enforcement divisions. The sections of the pesticide regulation division include: Invertebrate-Vertebrate —three branches, Fungicide, Herbicide, Fungicide-Herbicide, Minor Use and Emergency Response, and Chemistry, Inerts, and Toxicology Assessment.

The EPA office of enforcement and compliance "goes after pollution problems that impact American communities through vigorous civil and criminal enforcement that targets the most serious water, air and chemical hazards." EPA pledged to Congress that it would do all it could to significantly reduce pesticide use in the United States and has worked to fulfill that pledge.

4. PROCEDURAL SUMMARY

The 1988 FIFRA amendments focused on ensuring that previously registered pesticides met current scientific and regulatory standards. The procedure for pesticide registration is complex and the details are beyond the scope of this book. A summary of the process can be found at https://www.epa.gov/pesticide-registration/about-pesticide-registration.

Petitions must be supported by abundant prescribed data including the identity and composition of the pesticide chemical, appropriate methods of analysis, complete information on proposed uses, full reports of investigations made on residues produced, and toxicity information. During preregistration, other involved federal agencies can express an opinion regarding use and registration of a pesticide. These agencies include the Forest Service, Department of the Interior, Bureau of Land Management, and other conservation and wildlife interests (including private interests). These organizations cannot accept or reject a pesticide, but their opinions are of great value to decision makers. Other federal agencies are involved in pesticide use but not in registration. They include the National Research Council that promulgates information on safe pesticide use, the Occupational Safety and Health Administration that protects workers who handle pesticides, and the Federal Aviation Administration that regulates aspects of safety for aerial pesticide application.

A certificate of usefulness may or may not be issued. If it is denied, the pesticide will not be registered and approved for use. If it is issued, the Tolerance Division has 90 days to act on a petition for tolerance and publish residue regulations in the Federal Register. If a tolerance is not established, the pesticide may fail to be registered. It is possible to obtain either without the other but both are necessary for registration.

Under existing federal law, the EPA will register a pesticide only when the following criteria are met:

- The composition must warrant the claims proposed by the registrant.
- The proposed label must include all relevant federal regulatory requirements.
- It must perform its intended function without unreasonable adverse environmental effects.
- It must not cause unreasonable adverse environmental effects when used in accordance with accepted practices.

In each case the burden of proof is on the manufacturer (registrant). The EPA may waive the requirement to prove efficacy on the assumption that manufacturers will not be so foolish as to risk their reputation by marketing a product that does not do what they claim it will do.

However, there are foolish people and irresponsible manufacturers.

A manufacturer may apply to EPA for an experimental use permit (EUP) before a pesticide is granted full registration. Such permits are usually granted for 1 year and crops may be used or destroyed as determined by EPA when the permit is granted. EUPs permit the manufacturer to sell the product while gathering performance information under field conditions to support full registration.

Section 5 of FIFRA permits EPA to authorize use of a pesticide before full registration if an emergency condition can be established. Permits are granted only when the weed (or other pest) problem is urgent and nonroutine and no other registered pesticides will provide effective control and no other control measures are economically or environmentally feasible.

A final procedural matter relates to the ability of states to regulate sale or use of federally registered pesticides under section 24c of FIFRA. State regulatory agencies may register a federally approved active ingredient or product for a special local need that is not part of the EPA-approved label.

5. TOLERANCE CLASSES

Federal law requires that a tolerance be set, before registration, for any pesticide used on food or feed products. Tolerances vary from crop to crop depending on the many safety factors involved. All pesticides are in one of four tolerance classes.

5.1 Exempt

Some pesticides are exempt because there is no known human or animal health concern. They are generally recognized as safe (GRAS). Castor oil, cinnamon, corn gluten meal, corn oil, cottonseed oil, lemongrass oil, linseed oil, mint oil, peppermint oil, potassium sorbate, sodium chloride, soybean oil, and white pepper are included. The 600+ things on the GRAS list include more than 80 spices and natural seasonings and more than 150 essential oils, oleoresins, and natural extractives.

5.2 Zero

If, because of toxicological characteristics, the Pesticide Regulation Division (PRD) of EPA decides it is not in the public interest to accept any detectable residue of a given chemical, it can establish a zero tolerance. That means none of the chemical is permitted in any crop. It is not possible to register a pesticide with a zero tolerance for use on food crops. This is a regulatory position and applies to pesticides even when no manufacturer has applied for a tolerance. Zero tolerance used to mean, when used according to label directions, no detectable residue would remain. However, more sensitive methods of detection invalidated the concept. Today, no one knows what level will be detectable tomorrow. Parts per trillion are not uncommon and smaller amounts can be found. As of 1966, a finite tolerance must be established but zero can still be used. EPA may apply the zero tolerance to a pesticide and refuse to register it for use on food crops.

5.3 Finite

A finite tolerance is used when chemical residues are known to exist. It is the tolerance under which most pesticides are registered. A finite tolerance is usually larger than a negligible residue. However, it is still well below possible toxic levels. For a pesticide used on a food product a tolerance must be set, unless it is exempt from tolerance. Any raw agricultural commodity moving in interstate commerce and found to have pesticide residues over the stated amount is subject to seizure by the FDA. Studies on test animals, including fish, birds, and mammals, are done to determine the acute and chronic toxicity of the chemical. The length of time the pesticide remains in the environment is measured. Possible long-term effects such as buildup in animals or in the environment are studied. All these factors (and others) are taken into account before setting a tolerance. The tolerance is usually set at least 100 times smaller than the highest dose that has no effect in test animals.

5.4 Negligible Residue Tolerance

A residue may occur from indirect contact even though no pesticide was ever directly applied. For example, livestock may eat sprayed forage. Therefore, edible meat may contain a residue and a negligible residue tolerance is required. A preemergence herbicide may create a soil residue, which the crop may absorb. If the residue is in the crop at harvest, a negligible residue tolerance must be set. A negligible residue is usually one-tenth (0.1) of a part per million or less, far below a toxic level.

No residue registration was eliminated for pesticides in 1970. The concept was that if no residue could be detected by the best analytical method available at the time, a compound could be registered under the no residue provision. The problem, as mentioned earlier, is that detection methods have improved so much that a compound that originally could not be detected by methods sensitive to a part per million can now be detected in parts per trillion, or less. In 1967, no residue registrations were gradually converted, on petition of the manufacturer, to finite tolerances. If a manufacturer did not request a finite tolerance, the pesticide's uses were canceled.

6. THE PROCEDURE FOR PESTICIDE REGISTRATION

A complete registration petition contains a great deal of information necessary for full consideration of benefits and risks. At a minimum, the petition must contain:

1. A statement of active and inert ingredient, chemical and physical properties of the formulation, complete quantitative formula of the product, its environmental stability, and known impurities in the formulation.
2. Multiple copies of the proposed label, including:
    a. brand name
    b. complete chemical name and physical and chemical properties
    c. ingredients including samples of the chemical and its formulation(s)
    d. directions for use, specifying crops or sites intended for treatment
    e. amount(s) to be used, that is, rate of application per unit area
    g. timing of application
    h. any precautions or limitations on use
    i. warning statement for protection of nontarget species
    j. the antidote in case of human consumption
    k. warning to keep out of reach of children
    l. manufacturing details
    m. net weight statement
    n. restricted versus general use statement
3. Full description including the data from scientific tests used to determine efficacy and safety.
4. Complete toxicity report of tests on lab animals, and the methods for obtaining the data. At a minimum, such tests must include items a to k that follow. Studies are often expanded to include data on oncogenicity, spermatogenicity, aspects of mutagenicity, and other risk-related factors that research may identify.
    a. two-year rate feeding study to determine reproductive and carcinogenic effects
    b. eighteen-month mouse feeding study to determine reproductive and carcinogenic effects
    c. two-year dog feeding study

    d. dominant lethal mutagenic possibilities in mice

    e. teratogenic study in rabbits

    f. three-generation rat reproduction study and reproduction studies in chickens

    g. meat residue in cows, chickens, and swine; milk residue in cows; egg residue

    h. ninety-day rate feeding study to determine mammalian metabolites

    i. twenty-one-day subacute toxicity in rabbits, and oral and dermal LD_{50} in rats

    j. eye and skin irritation

    k. tests to determine effects on two species of fish and quail

5. Results of tests on the amount of residue remaining and the description of analytical methods. This is critical for the tolerance requested. Tolerances are set on the amount of residue remaining and not on the highest figure permissible from a health standpoint. EPA is interested in data that show the amount of residue on the crops and animals on which the pesticide will be used and in the soil (or other portion of the environment) that will be treated with the pesticide.

6. Practical methods for removing residues exceeding any proposed tolerance, including a description of the method.

7. Proposed tolerance for the pesticide and reasons for the level requested.

8. Reasonable grounds in support of the petition, including a summary of data in the entire petition and a summary of benefits when used in agriculture.

If there are no food residues and if no other residue exists, and if EPA's PRD concludes that the pesticide is safe and conforms to the manufacturer's claims, then the Fish and Wildlife Service of the Department of the Interior and the Hazard Evaluation Division of EPA are notified of the intent to register. These agencies and others can concur or reject the petition. If they reject, a reevaluation must occur. When use will result in residue at harvest of crops or slaughter of animals, registration is subject to the requirements of the Miller Amendment. In seeking to register such a compound, a petition proposing a tolerance or exemption must be submitted to EPA. It must provide information on the pesticide, its use, and reports of safety tests. The safety information must include results of animal susceptibility experiments, tests on residues, the analytical methods employed, and practical methods for removing residues that exceed proposed tolerances. Table 21.1 summarizes the process.

The FIFRA, through its registration and enforcement features, provides the primary public protection against improperly labeled or adulterated pesticides. The law is also the primary effort to protect the health of users and consumers against potential adverse effects of pesticides.

About 30 pesticides were registered for use in the United States in 1936, and fewer than a million pounds were applied. Thirty-five years later (1971) more than 900 were registered and more than a billion pounds were applied (McEwen, 1978). In 2012, if proprietary labels (a trade name owned by a manufacturer) are considered, more than 1000 chemical and biological compounds are used for pest control in the world (Hopkins, 1994). The EPA reports 17,026 pesticides registered under Section 3 of FIFRA and 2264 under section 24c. In 2016 EPA registration included 1240 active ingredients, formulated in 19,290 commercial products, of which 4981 were herbicides. There were 2077 inert ingredients.[4]

---

[4]Personal communication — EPA PesticideWebComments@epa.gov. May 2016.

TABLE 21.1   A Summary of the U.S. Pesticide Registration Procedure

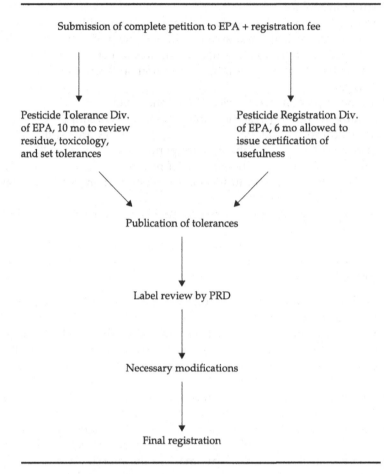

It is accepted that herbicides for weed control are an important and growing part of world agriculture as well as a major part of the agrochemical industry. Their development as primary tools for weed control in the United States is shown by the fact that the number of herbicides included in the Herbicide Handbook of the Weed Science of America has grown from 97 representing 27 site-of-action groups in the first edition (1967) to 222 from 20 groups in the ninth edition (2007) to 232 from 24 groups in the 10th edition (2014).

The Pesticides Registration Division of EPA:

- Certifies the chemical is useful for the use for which the label is requested, and
- Expresses an opinion as to whether the tolerance requested reasonably reflects the residues likely to remain on the treated crop.

The burden of proof is always on the applicant. EPA's Pesticide Tolerances Division establishes tolerances or maximum allowable limits of pesticide residues in or on raw agricultural commodities in interstate commerce. EPA's role does not necessarily stop at this point. EPA and state officials collect unregistered pesticides, look for misbranded or adulterated products, and can take legal action against offenders. The FDA monitors the nation's food supply for pesticide residues. The program was established under the FFDCA of 1938. The FDA regularly samples US and imported food products for pesticide residues. In 2008, 473 pesticides were included in the analytical system, and residues of 161 were found. In 2008 the program analyzed 5053 food samples: 1398 domestic (from 41 states) and 3655 imported.

In 2012[5] the import violation rate was 11.1% and the domestic violation rate was 2.8%, consistent with those in recent years, which have ranged from 0.7% to 2.4%; however, the import sample violation rate is up from 2.6% to 7.6% from previous years. Under regulatory monitoring, 5523 samples were analyzed. Of these, 1158 were domestic foods and 4365 were imported foods. In 2012, 97.2% of all domestic foods analyzed by FDA were in compliance, i.e., no residues were found or residues found were not at violative[6] levels. The compliance rate for domestic foods from 1996 to 2011 was between 97.6% and 99.3%. As in earlier years, fruits and vegetables accounted for the largest proportion of the domestic commodities analyzed in 2012, comprising 75.0% of the total number of domestic samples. Of the 1158 domestic samples, 57% had no detectable residues and 2.8% had violative residues. The logical conclusion is that the US food supply is monitored regularly and our food is safe with respect to pesticide residues.

States, through their Department of Agriculture, Environmental Agency, or other designated body, have the power to register and regulate the intrastate use of pesticides. States are allowed to register additional uses of federally registered pesticides for special local needs (FIFRA, Section 24-c registrations). States cannot invoke any regulations that are less stringent than the amended FIFRA, but can impose additional requirements on the registrant. Pesticides formulated and distributed within a state must be registered by the federal EPA. State registrations cannot be obtained if EPA has already denied FIFRA registration for a particular use.

7. A FINAL COMMENT

It is patently obvious that the public is concerned about pesticides and that the concern has increased with time. Most people no longer believe that chemistry always leads to better living. Average citizens are not involved in agriculture and know little, if anything, about farming, how food is produced, or the pesticide safety procedures described herein. All are concerned about the safety of the food they must eat. Most people cannot cite specific problems or explain the reasons for their concern. However, the public's ignorance of agriculture and the apparent inability of many to provide clear reasons does not diminish the responsibility and should encourage weed scientists and others (manufacturers, scientists, farmers) who work with pesticides to be aware of, listen to, and understand the public's

[5]www.fda.gov/Food/FoodSafety/FoodContaminantsAdulteration/Pesticides/ResidueMonitoringReports/ucm228867.htm#Figure_1.

[6]A violative residue is one that exceeded a tolerance or was a residue of regulatory significance for which no tolerance had been established in the sampled food.

concerns. A few examples of agricultural mistakes that have occurred and why people are fearful are shown herein. Nearly everyone knows something negative about pesticides, and those things form their view and explain their fear. A 2010 article (Leslie) illustrates the point. California's central valley is one of the world's most productive agricultural regions. In early 2010 11 local babies were born with severe defects (cleft palate, Down syndrome, missing brain parts, cleft lips and palates), one was stillborn, and three died in infancy. This could have been just bad luck, a statistical anomaly. Local residents didn't believe it was. Pesticides are among the causes cited. No one knows for sure, but in spite of the rigid EPA requirements, mistakes happen and people blame environmental factors. Most can't or don't elaborate, but some (perhaps many) would agree that the modern highly productive agricultural system that successfully dominates nature cannot be sustained, especially when it may (there is no scientific proof) harm people.

What one believes is what matters. We say, When I see it I will believe it. However, it is also true that when one believes something it will be seen—it will be real. While the evidence of adverse health consequences of agricultural practice is far from conclusive, policy makers and regulators must consider their obligation to create a better system for decision-making about pesticides and their safety that recognizes and incorporates the public's concern. Absolute safety cannot be guaranteed. Food must be produced. Food's producers deserve a just reward for their talent and labor. There is no guarantee that decision makers will always be right, but protecting those at greatest risk should not be deferred (Groopman, 2010).

THINGS TO THINK ABOUT

1. Why does a nation bother to register and thereby regulate pesticide use?
2. What federal agency governs US pesticide registration?
3. What federal acts govern pesticide registration in the United States?
4. What does it mean when a pesticide is registered?
5. Why is the pesticide label important?
6. What is the significance of the Miller pesticide amendment to the FIFRA?
7. What things must a manufacturer prove to register a pesticide?
8. What are the tolerance classes under which a pesticide may be registered.
9. What information must be included with a petition for registration and who bears the responsibility for preparing the information?

Literature Cited

Colburn, T., Dumanoski, D., Myers, J.P., 1996. Our Stolen Future: Are We Threatening Our Fertility, Intelligence, and Survival? A Scientific Detective Story. The Penguin Group, New York, NY, 306 pp.

Goldman, L., 2004. Childhood Pesticide Poisoning: Information for Advocacy and Action. United Nations Environment Programme, Châtelaine, Switzerland, 37 pp.

Groopman, J., May 31, 2010. The Plastic Panic: How Worried Should We Be About Everyday Chemicals. New Yorker, pp. 26–31.

Hopkins, W.L., 1994. Global Herbicide Directory. Ag Chem Information Services, Indianapolis, IN, 181 pp.

Leslie, J., July 22, 2010. How gross is my valley: America's toxic agricultural capital. New Repub. 10–11.

McEwen, F.L., 1978. Food production — the challenge for pesticides. Bioscience 28 (12), 773–777.

Snelson, J.T., 1978. The need for and principles of pesticide registration. FAO Plant Prot. Bul. 26, 93–100 (NOTE. This issue contains several articles on pesticide registration).

Switzer, C.M., 1957. The existence of 2,4-D resistant strains of wild carrot,. Proc. Northeast. Weed Control Conf. 11, 315–318.

Wildavsky, A., 1997. But Is it True? A Citizens Guide to Environmental Health and Safety Issues. Harvard Univ. Press, Cambridge, MA, 574 pp.

22

Weed-Management Systems

FUNDAMENTAL CONCEPTS

- Weed management systems neither stand alone nor should they be imposed in isolation of other aspects of crop production and plant growth.

- Good weed management systems reduce but do not eliminate weed problems.

- The goal of integrated weed management (IWM) is population stabilization at a low level with techniques that are economically and environmentally sound and sustainable.

- Six logical steps are fundamental to all successful IWM systems.

LEARNING OBJECTIVES

- To understand the logical steps that are part of successful weed-management systems.

- To understand the complexity of combining available weed control techniques into a weed management system.

- To understand the design and implementation of weed-management systems for a few crops and cropping situations.

- To understand the complexity and role of computer-based decision-aid models in weed management.

1. INTRODUCTION

Weed control, an old practice, is a process of reducing weed growth to an acceptable level. Weed management includes the dictionary sense of management—"taking charge of and directing" the growth of weeds and a complementary definition "handling carefully" that includes the concept of husbandry—to manage economically and conserve.

Weed management has been defined as, "An environmentally sound system of farming using all available knowledge and tools to produce crops free of economically damaging, competitive vegetation" (Fischer et al., 1985). This definition lacks the specificity of an economic threshold, and inclusion of "free of" implicitly advocates no yield loss (i.e., no economic damage). It could be interpreted as advocating a high level (100%) of weed control because it mentions crops free of economically damaging, competitive vegetation. Fryer (1985) defined weed management as the "rational deployment of all available technology to provide systematic management of weed problems in all situations." Unfortunately, there is no agreement on what is rational or systematic. The definition implies that everyone already knows.

Ideally weed management is a systematic approach to minimize weed effects and optimize land use. It combines prevention and control (Aldrich, 1984) and emphasizes minimizing the effects of weeds, but not necessarily eliminating all in a field. Weeds will be accepted as a normal, manageable part of agriculture, albeit a part one must learn how to manage and live with. The objective will be manipulation of the crop-weed relationship so crop growth is favored over weeds.

It is worthy of note that in the 1990s weed management was a new concept in weed science. The first appearances of the term *weed management* may have been in Buhler et al. (1996) and Forcella et al. (1996). Buhler's book Expanding the Concept of Weed Management was published in 1998.

The idea of an integrated weed-management system (IWMS) was first mentioned by Shaw in 1982, and Smith's book appeared in 1995. However, the concept of, indeed the term IWM first appeared in the weed science literature in Liebman and Gallandt's (1997) paper, followed by Blackshaw et al. (2000), Mortensen et al. (2000), Bussan and Boerboom (2001), Buhler (2001, 2002) and Liebman and Davis (2009). The current common use of weed management and integration has shifted emphasis away from control. Buhler (1999) posited that weed management "should be viewed as an integrated science," which would move attention from control to "prevention of propagule production, reduction of weed emergence, and minimizing weed competition with the crop." He claimed that weed science had lagged behind other pest-management disciplines in developing integrated management systems.

IWM has evolved to mean "decision support systems for the selection and use of pest (weed) control tactics singly or harmoniously coordinated into a management strategy, based on cost-benefit analyses that takes into account the interests of and impacts on producers, society, and the environment" (Norris et al., 2003, p. 11). Young (2012) described IWM as "putting components (of weed control) together, not taking them apart." Young challenged weed scientists by asserting that "weed science has stopped at the 'field edge' in assembling the components into a truly integrated approach." Prior to that, Doohan et al. (2010) suggested the necessity of understanding why farmers often make decisions contrary to science-based recommendations. Their paper explores the human dimension of weed management and strongly suggests that most weed scientists "have ignored such questions or considered them beyond their domain and expertise." Weed scientists focused on fundamental weed science and technology—how to control weeds. Doohan et al. found that "farmers expressed an overwhelming preference for weed control with herbicides. They had "a typical inverse relationship between perceived risk and benefit." Farmers' preferences reflect people's choices in many human activities. If an activity is perceived as beneficial, it is in our nature to automatically perceive the risk is low. In general, farmers saw great risk and little benefit in preventive measures. They applauded the efforts of weed scientists to develop effective control techniques to avoid weed resistance. They challenged weed scientists to look beyond control to what they called "the crucible around which social, biological and environmental scientists" could/should cooperate. Similarly, Bastianns et al. (2008) proposed that the general reliability and efficacy of cultural compared to chemical control is only moderate. Cultural weed management relies on a combination of measures. It is a more complex system that requires more managerial skill. Therefore, farmers, understandably, choose the simpler, chemical-based system as opposed to the more complex ecological weed management.

When research has provided an adequate base for IWMSs, they will include five components:

- Incorporation of ecological principles.
- Use of plant interference through knowledge of the quantitative effects of crop-weed competition.
- Incorporation of economic and damage thresholds.
- Integration of several weed-control techniques.
- Supervised weed management frequently by a professional weed manager employed to develop an IWM system for each crop-weed situation.

Systems will be designed to prevent or reduce the inevitable probability that new weeds will appear. Managers will try to anticipate and plan for the future. Systems will be designed to manage/control weeds that, if ignored, will reduce yield. Ecological considerations will include natural weed mortality, inter- and intraspecific competition, crop plant density, and the benefits and risks of genetically modified crop varieties that may be more susceptible to control techniques such as tillage and herbicides, and could be more competitive, or allelopathic. Successful weed management programs will include precise timing of cultural practices such as tillage to maximize benefit and careful selection of rate and application time of herbicides.

IWM is not new; Shaw urged weed scientists in 1982 to develop IWMSs for crops *and* specific weeds or weed complexes in crops. These demanded integration of the whole agricultural system, not just its parts, and resulted in three choices (Weiner, 1990) to achieve:

1. maximum short-term yield
2. maximum sustainable yield
3. maximum yield sustainability (minimum risk)

Maximum short-term yield dominates developed-country agriculture. Maximum yield sustainability characterizes third-world agricultural systems that are dominated by low, stable (barring environmental disasters such as floods or drought), long-term production. Jackson (1984) and Weiner (1990) suggest that choosing maximum short-term yield "requires high input costs, high environmental costs, and high nutrient and capital fluxes." They advocate low input, low environmental cost, low nutrient, and low capital flux agricultural systems. Therefore, in their view, option 2 is the most desirable. It is most likely to produce systems that integrate the fewest inputs to achieve the desired result. Organic-production systems come close to meeting the criteria of option 2 (see Chapter 11).

Results will always be important. One measure of results of a weed-management system is how well it manages the problem in the year it is first used—the farmer's view (see Doohan et al., 2010). Another essential measure of success is whether or not it reduces the likelihood of future problems. A primary question is, will there be fewer seeds or vegetative propagules after the management system is imposed than before? If the answer is, yes, it's probably a good system that will reduce, but not eliminate, weed problems. The need for effective, annual weed management programs will continue.

This chapter illustrates some principles and available components of weed-management systems. Each system is incomplete, partially because the research base is still developing. It is also incomplete because few management systems will ever be fully complete and fixed for all time. There are principles, but no enduring formula. Weeds in a place will change as they always have and management systems must be dynamic. There is no attempt herein to discuss every weed problem in every crop or cropping situation. The weed problem in eight production systems and the techniques that can be integrated will be described. This is not a weed-control guide or how-to manual. Discussion principles are available (Smith, 1995). Weed-management systems neither stand alone nor are imposed in isolation. They are part of agriculture and landscape management. Each must consider soil conditions, tillage practices, economic and political realities, and the social and aspects of plant culture. The principles developed in this chapter should and can be applied and adapted to weed-management situations important in a region.

2. A METAPHOR FOR WEED MANAGEMENT

Managing weeds is necessary when a place, a field, is selected for planting. The history of the place is important. Past cropping sequences and weed-control methods reveal the kinds of weeds to be expected. The way in which the soil and seedbed have been prepared will be important. Plowing the field exposes a different population of weed seed than disking or chiseling. The kind and timing of irrigation influence weed species. If the land has been observed carefully and edges, ditches, and fences have been kept clean of new sources of weed infestation, the weeds present will be different than if field sanitation has never been practiced. Past and present insect and disease management must be integrated with weed management because they affect each other. Many crop growers do all or some of these things but few pay attention to their effects on the weed problem or the weed management program. Research is needed to determine specific effects of each management practice on specific weed species.

Weed-management systems can be compared to a carpenter's toolbox. A good carpenter's toolbox contains a large assortment of tools many of which the noncarpenter doesn't know the purpose of. Almost everybody recognizes and knows something about a hammer, a screwdriver, or a tape measure. But a good carpenter's toolbox contains an assortment of tools, the uses of which are unknown to most of us. Their purpose is a mystery. Many people would be pleased to have the toolbox of a good carpenter and would quickly use several tools. Other tools would be used later as knowledge developed or after someone explained their purpose and how to use them. Still other tools might remain interesting and unused. The purposes and uses of some tools will be obvious. Others will look familiar but their uses won't be obvious. There will probably be many with no clear purpose and one may even wonder why they are in the box.

Similar to the carpenter's toolbox, there are some features common to all weed-management systems—all the toolboxes—which all know how to use. Each weed management toolbox should have three compartments: weed prevention, weed control, and weed eradication. The prevention compartment will have the tools used to keep weeds from occurring in a previously clean area. The control compartment will be the largest, and many tools that belong there will also be found in the other compartments. Control tools are things used to reduce the weed population. The eradication compartment is the smallest one, or, if not the smallest, the least used. The tools in it are not more complex than others but they require great persistence and often just don't seem to work as well as others. Their purpose is complete removal of a weed species and its propagules from an area—a difficult task.

Dewey et al. (1995) proposed that noxious weed management could be regarded as forest managers think of wildfire management. They regarded weeds as a raging biological wildfire. All aspects of wildfire and weed management are similar except two. The first dissimilarity is that wildfires spread more rapidly than weeds even though their patterns of spread are similar. The second, more important, difference is that forest managers never fail to see a wildfire. Fires cannot be ignored, weeds can be. Weeds don't obviously destroy things, and because the occupied area increases slowly, they can be and often are, ignored until they dominate large areas. Then people say, "Why didn't someone do something?" It is too late then for other than expensive, time consuming, and potentially environmentally harmful attempts to control. Action needs to be taken when fires and weeds begin. The next section describes the logical steps for developing a weed management system to take the necessary action.

3. THE LOGICAL STEPS OF WEED MANAGEMENT

Most toolboxes and good tools come with a set of instructions on purpose and proper use. Weed management tools also come with instructions. Presently, instructions are general but they will improve and become more specific as knowledge expands. At a minimum the instructions for weed-management systems for nearly all areas and all cropping systems will include seven logical steps:[1]

3.1 Prevention

The first, most obvious, and perhaps the most frequently omitted step is prevention. Weeds that don't appear because clean seed is planted, machines are cleaned, and new cattle are separated (see Chapter 10, Section 1.1) don't have to be controlled. Early detection is part of weed prevention. Detecting a new weed does not prevent its arrival, but quick action can prevent its spread. Preventive action can be as simple as bending over and pulling the weed and removing it from the site.

3.2 Mapping

An accurate map of weed infestations should be made before a good management program can commence. Typically people think of a map as a drawing on a piece of paper.

---

[1]I am indebted to Dr. K. G. Beck, Professor Emeritus of Weed Science, Dept. of Bioagricultural Sciences and Pest Management, Colorado State Univ., for the concept of these management steps.

A mental map may be adequate. Weeds must be defined by species and located in the field or the area for special attention before solutions are proposed. No one wants their physician to treat an unspecified, unknown illnesses. An accurate diagnosis is expected before treatment. Weed managers must be as careful and know the problem before prescribing a solution. The best weed-management systems will be designed to control specific weeds in specific places. One must diagnose the problem before attempting to control it.

It is presently not and probably never will be technically feasible to map every weed. Major weed species and those likely to become problems (e.g., parasitic weeds, perennials, hard-to-control annuals) should be located and the size of the infestation defined. Early detection of new weeds is part of mapping and prevention.

Weed scientists agree that weeds rarely, if ever, exist in a uniform density over an entire field or area. They frequently exist in patches (see Fig. 6.11). In other words, weed populations have a spatial dimension, which must be defined prior to development of a site-specific integrated management plan. The fact of variable spatial distribution has been neglected because available weed-control technology prescribes, if not requires, uniform control techniques for entire fields. Farmers and land managers know that fields and weed populations are heterogeneous. Heterogeneity has increased as fields have grown larger and monoculture has increased (Mortensen et al., 1998). There is adequate justification for what Maxwell and Luschei (2005) call site-specific weed management: "there is no need to attempt to control weeds where they are not present in crop production fields." Limiting control to places where weeds are reduces the environmental effects of control techniques, saves money for weed managers, but may increase risk (escapes, missed areas) for growers who choose to manage conservatively (Mortensen et al.).

3.3 Prioritization

Money, time, technology, or labor are often lacking. It is not possible to do everything. The best weed managers will know as much as possible about the weeds present and select those to be managed. The species that pose the greatest threat to present or future land use should receive highest priority.

3.4 Development of an Integrated Weed-Management System

When management begins it is necessary to look closely at the array of tools in the box to see what is available and if a particular tool or set of tools will be most effective for the weeds to be managed. The best weed-management systems will not rely on a single technique. All appropriate tools will be examined, and an integrated approach including two or more tools or strategies will be selected. The ideal IWM system will employ what Liebman and Gallandt (1997) called "many little hammers" rather than a single hammer, no matter how effective it may be. Integration will consider cultural methods such as grazing management, fertility, irrigation, seeding rate, and use of competitive cultivars. Mechanical methods include tillage before and during the crop's growth, mowing, burning, flooding, and mulching. Biological and chemical control will also be considered, but their use will shift from a "yes, provided that" to a "no, unless" decision (Mortensen et al., 2000). All of these methods will be in the weed-management toolbox. Some tools will be more numerous, more apparent, or easier to grab and will be used more frequently. For several years, herbicides have been the method of choice in many situations. Other tools always seem to wind up on the bottom of the box and aren't selected often. This could be because they are not as easy to use, the knowledge of how to use them is not available, or others (usually herbicides) have worked well and are easy to use (see Doohan et al., 2010). For example, soil fertility influences weeds and should be considered in weed management. The knowledge of how to manipulate fertility to complement weed management is not abundant. Fig. 22.1 is a conceptual model of weed-management systems. It is a glimpse into the weed-management toolbox. After completing the preliminary steps shown in Fig. 22.1, control options are selected to develop an integrated system. When the methods are selected, weed managers must ask what can be done (a scientific question), what should be done (a social/moral question), and what farmers will or are likely to do. Some things that are scientifically possible may not be socially, culturally, politically, or environmentally desirable. For example, intensive tillage might increase soil erosion and intensive herbicide use might pollute water or harm nontarget species. The best weed-management systems will be integrated with other aspects of crop management

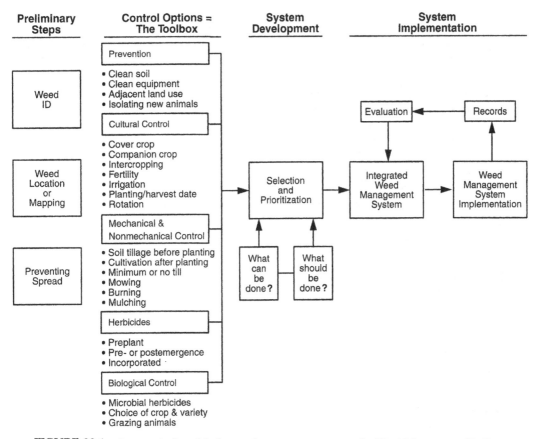

FIGURE 22.1 A conceptual model of a weed-management system, the Weed-Management Toolbox.

(e.g., insect or disease control, fertility of the crop), the environment, and the area or field in which weeds are to be managed.

IWMSs supported by extensive research are now available for many cropping systems but are not widely used by growers (Norris, 1992). Perhaps the most important reason for lack of use is the difference between the goals and needs of agricultural researchers and those of producers (Norris, 1992; Doohan et al., 2010). Researchers are concerned about developing effective control techniques, the ecological effects of control techniques, and long-term effects. Producers want as much certainty of yield and profit as possible at reasonable cost in the current year. A producer, quite understandably, needs to know what will solve the weed problem that exists or is anticipated in a year. What techniques can be combined that will provide reasonable assurance of solving the problem, optimizing yield, and maximizing profit this year, is the grower's crucial question.

Weed control research, in recognition of the grower's need, has emphasized herbicide development and considered combinations with mechanical control (tillage), weed-crop competition, competitive cultivars, biological control, and, on occasion, allelopathic cover crops (Wyse, 1992). These aspects of control science and technology are necessary parts of IWMS but they are not a sufficient base for system development. They must be combined with what Wyse calls principles-based research. Development of these principles demands a shift from weed control in a crop to a total systems approach to crop protection (an integrated approach). Such an approach will begin to solve the "escalating economic and environmental consequences of combating agricultural pests" (Lewis et al., 1997). Weed control has been dominated by the search for "silver bullet" products to control weeds (Lewis et al., Mortensen et al., 2000). Such therapeutic interventions have been effective short-term (a crop year) techniques. Long-term management will be achieved by changing the approach from annual weed control to long-term management. It will take advantage of natural preventive strengths (Lewis et al., 1997). Short-term therapies are not abandoned. They may become supplements rather than a primary defense. The attitude is the "no, unless view" expressed before. Some tools are not used unless they are the only alternative (Mortensen et al., 2000). The natural preventive strategies for weed management emphasize development of diverse systems that are biologically robust (Van Acker, 2001) and "inherently less susceptible to weed invasion, proliferation and interference." The required data on "weed demographics and competition" include studies of (Buhler, 2001; modified from Wyse, 1992):

Seed and bud dormancy mechanisms
Seed germination, development, and production
Seed banks and emergence dynamics
Population genetics
Population shifts
Spatial distributions
Modeling of weed-crop systems
Weed-crop interaction studies

These studies will serve as and are the necessary foundation for development of weed management technologies that will reduce soil erosion, surface and groundwater chemical contamination, and maintain an acceptable level of weed control (Wyse, 1992). The essence of these recommendations is that understanding the organism's biology and ecology is required before it can be controlled well with minimal or no undesirable environmental or social effects. IWMS cannot be separate components of crop production. They must be regarded as part of the design of integrated-cropping systems. Prevention rather than control becomes the key objective (Mortensen et al., 2000).

3.5 Implementation of Systematic Management

With a map that shows where the weeds are and an integrated crop-management plan, managers can design and implement an IWM program. Not everything has to be accomplished in one season or with one technique. Such plans will necessarily be crop, place, and possibly weed specific. The program will be systematic in that timely, planned management will be done over several seasons. The manager may decide to begin with control of perimeter weed infestations to prevent their spread and eliminate sources of future problems, with the caveat that some weeds may be hosts of beneficial insects that prey on harmful insects and therefore should not be eliminated.

Another approach could be spot treatment of a patch of parasitic or perennial weeds. Based on available techniques, their cost, environmental acceptability, and adaptability to the situation, the manager can choose among several courses of action (Dewey et al., 1995). If an infestation is small but the weed is very aggressive and likely to spread quickly, attempts to eradicate may be the best choice. A weed could be a serious threat, but not easy to control. In this case it could be contained or confined with other tools to be employed when the weed threatens a crop. In some cases managers might opt to do nothing except monitor the weed regularly and evaluate management options.

3.6 Records and Evaluation

Records of what was done and its success must be kept (Fig. 22.1). Records allow the manager to repeat successes and learn from mistakes. Evaluation should be continuous, not just a week or month after control was done. Evaluation over 2 or 3 years is required to measure success and to observe what did not work well. For example, students are required to write final examinations at the end of classes. A better measure of learning (although one that would not be popular with students) might be an examination 3–5 years after completing a class. Such an examination would measure knowledge retention. It is the same with a weed-management program. Success over time is more important than success in the short run. It is a bad mistake to develop a system, assume it works, and avoid doing regular evaluation (go out and look) to verify the system's success.

3.7 Persistence

Modern weed-control techniques have been used for several decades, but in spite of the successes that have been achieved, weeds are still a predictable threat in cropped fields and in many other places (Buhler, 2001). One should not assume the weeds have won, but that they have adapted rapidly. Resistance to herbicides (see Chapter 19) and weed population shifts are continuing challenges. Successful management is rarely achieved after application of one control technique in one season. Integration of techniques based on basic biological and ecological knowledge is required. The soil seed bank and new sources of infestation demand continued attention. Vigilance is the price of success. Weed management is not a one-step (one-control technique) process. Methods must be integrated over time to reduce or eradicate some weeds (e.g., parasitic weeds) and manage other populations to reduce yield loss and crop injury. Successful agriculture includes a well-designed site-specific weed management program. It is rarely thought if as such, but it is as complex as rocket science.

4. WEED-MANAGEMENT PRINCIPLES IN EIGHT SYSTEMS

There are as many weed-management systems as there are cropping systems and weed complexes. The number may equal the number of weed managers. Each manager puts unique touches on any system, but each system shares some general characteristics. The feature that all systems share is integration, which has led weed scientists to develop and improve IWMS. Not all systems include all possible methods of weed control but all include or at least recommend consideration of all methods appropriate to the cropping system and the environment in which weeds are to be managed. This review includes weed-management systems for small grain crops (Donald and Eastin, 1995; Donald, 1990), corn and row crops, turf (Bingham et al., 1995), pastures and rangeland (Smith and Martin, 1995; Bovey, 1995), perennial crops, aquatic, woody plants (Bovey, 2001, 1995; McNabb et al., 1995), and organic agriculture.

Most weed-management systems are based on herbicides as the primary control technique. This is primarily because herbicides work well, which is especially true if externalities are not considered.[2] Herbicides are generally reliable in that they do what they are advertised to do, are selective, and, relative to other methods, are not expensive. Recent thought about IWMS emphasizes two approaches (Mortensen et al., 2000). The first, often called the curative approach, emphasizes use of better application technology, improved application timing, site-selective application (apply only where weeds are), lower doses, and use of herbicides with reduced or minimal environmental effects. However, herbicides are the primary tool. The second approach reduces dependence on herbicides as the primary technique. The goal shifts from controlling weeds to preventing their appearance through development of an integrated-management system. These systems emphasize techniques that reduce the fitness of weeds and kill or reduce their numbers through crop competition, rotation, planting time, companion cropping, mulches, etc. When a weed's biology and ecology are understood, that knowledge can be used to manipulate crop-weed interactions to the benefit of the crop. Altieri and Liebman (1998) take the latter approach in their review of ecological approaches to weed management.

---

[2]An externality is a cost not reflected in price, or more technically, a cost or benefit for which no market mechanism exists. It is a cost that a decision maker does not have to bear, or a benefit that cannot be captured. From a self-interested view, an externality is a secondary cost or benefit that does not affect the decision maker.

The systems described herein should be thought of as generic management systems that are generally, but not specifically, for place, time, etc. complete and correct. It is assumed that weed identification, mapping, preventive measures, record keeping, and evaluation are part of each system. How these things are done differs with each system. The following examples focus on identifying components of IWMS. The examples are not complete, readily adaptable, prescriptive systems. They are intended to be a basis for discussion of such systems and their use and further development. Weed-management systems that are not included herein are covered well in Smith (1995), including those for oil-seed crops (Wilcut et al., 1995), pasture and hay crops (Smith and Martin, 1995), and horticultural crops (Smeda and Weston, 1995).

Weed scientists, extension agents, and practitioners across 49 US states, Puerto Rico, and eight Canadian provinces participated in a 2015 Weed Science Society of America (WSSA) survey, which identified the most common weeds (those most frequently seen) and the most troublesome weeds (those most difficult to control) in 26 different cropping systems and natural areas (Van Wychen, 2016). The data available at http://wssa.net/wssa/weed/ surveys illustrate the ubiquity of species across large areas and crops and may serve as a guide for weed managers.

4.1 Small-Grain Crops

Weed-management systems for small grains, including winter and spring wheat, barley, oat, sorghum, and rice, were reviewed by Donald (1990) and by Donald and Eastin (1995).

4.1.1 Prevention

Preventive strategies, the first phase of weed management in small grain crops, are not complex; they are the basis of good farming practices. The first preventive step is identical for all crops and cropping systems—plant weed-free seed.

Custom combines and other itinerant machines are sources of weed seed and should be cleaned before leaving a farm and before moving from an area of known contamination to a weed-free area. Competitive weeds on field edges and roadsides should be managed because they are sources of new field infestations. Trucks or wagons used to transport grain should be covered to prevent wind dissemination of weed seed from uncleaned grain. The Harrington seed destructor (see Chapter 10) is clearly a preventive technique.

An ecological (integrated) approach to weed management in winter wheat developed by Anderson (2005) emphasizes prevention and aims to reduce herbicide use. A second approach suggested by Mortensen et al. (2000) includes three primary goals: enhancing natural seed loss through leaving weed seeds on the soil surface where death is more likely, reducing weed seedling establishment, and minimizing seed production by established weeds. Combining these goals means growers may have to change the way crops are grown. Crop rotation, crop sequencing (fall and summer planting), crop residue management, and competitive cultivars will have to be included (Anderson, 1994, 2003).

4.1.2 Mechanical Methods

When preparing land for small grains, there is a wide choice of techniques. Traditionally soil has been plowed, but plowing is dependent on prevailing weather, implements available, and grower preference. Preplant tillage ranges from plowing 8–16 in. deep, disking or field cultivating up to 6 in. deep, surface tillage 1–3 in. deep, or no tillage and direct planting. Each of these practices and their timing affects the type, presence, and abundance of weeds. No cultivation and a shallow, noninverting cultivation increases the incidence of perennial weeds and decreases annuals (especially broadleaved species). It is important to understand that soil tillage controls some weeds and creates an environment in which others flourish. It is also important to acknowledge that tillage, especially in areas subject to flooding or wind, may lead to erosion of the earth's most valuable resource—soil.

No tillage tends to increase the incidence of annual grass weeds such as wild oats, annual brome grasses, annual bluegrass, common rye, and jointed goatgrass. Plowing and disking may prevent or facilitate spread of perennials, but neither alone will control Canada thistle or field bindweed effectively. Plowing is 10%–20% more effective than shallow cultivation or disking for control of perennials, but it returns previously buried roots and rhizomes to the surface where, if they do not desiccate rapidly, they will produce new plants. After plowing or without plowing, early cultivation of land to be planted to small grains stimulates germination of seeds of annual weeds. The seedlings can be controlled by subsequent cultivation.

Shallow cultivation of stubble in fields from which a small grain has been harvested can aid control of perennial grasses and prevent some annual weeds from producing seed. If stubble cultivation is done at the wrong time or with weeds that survive cultivation, the weed problem can worsen. For example, stubble cultivation soon after harvest could bury wild oat seeds and reduce loss through natural causes (e.g., cold weather). Seedlings of winter annuals such as downy brome easily survive shallow, noninverting tillage and partial burial.

Fallow (no crop), or fallow combined with tillage is an effective weed management technique. Seedlings can be eliminated with cultivation. More than one cultivation may be required to control most emerged seedlings. No-till systems have enabled wheat producers in the semiarid US central Great Plains to change the rotation from endless wheat-fallow to one that includes a spring-planted or warm-season crop (Anderson, 2005).

4.1.3 Cultural

Farmers, for many reasons, want to plant early and date of sowing determines what weeds will be present and which ones will dominate. The earlier a crop is planted, the less time is available for weeding of any kind before planting, which may increase the chance that weeds will germinate and grow with the crop. Seeding winter wheat at a higher rate reduces competitiveness of blackgrass and wild oats. Increasing crop density by using a higher seeding rate (O'Donovan et al., 2006) or narrowing row width tends to increase competitiveness of wheat and other cereals against spring-emerging grass weeds.

Delaying winter wheat planting until emerged weeds can be killed by light tillage is an effective, inexpensive weed-management technique. On the other hand, a quick-emerging, vigorous, dense crop stand is also an important technique. For example, early planting of spring grains may allow crop development before foxtails germinate. Small grains are normally planted in 7-inch rows with adequate rain or irrigation or in about 14-inch rows on dry land. This accepted agronomic practice is usually not changed for weed-management reasons even though row width and subsequent crop density affects weeds and their control.

Correct seedbed preparation for the soil and cultural system, cultivar selection; use of high quality, clean seed; and careful attention to optimum fertility to produce rapid emergence of vigorous crop plants contribute to weed management. Many farmers save seed of small grains from year to year to save money. With no or poor cleaning of saved seed, weed seed can be planted (see Chapter 5).

As Italian ryegrass density increased, wheat yield decreased and semidwarf cultivars had lower yield than tall cultivars with the same density of Italian ryegrass (Appleby et al., 1976; Table 22.1). In Canada, green foxtail was more competitive in semidwarf than in tall spring-wheat cultivars (Blackshaw et al., 1981). These data point out the importance of cultivar selection, crop canopy development, and crop competitiveness. Semidwarf cultivars have a more open crop canopy, permit more light to reach weeds, and allow Italian ryegrass to be more competitive. Cultivars are not often chosen for weed management but their influence should not be ignored. Unfortunately, the basis for cultivar competitiveness is too poorly understood. It is a tool that is not used in most weed-management systems.

Crop rotation breaks a weed's life cycle by altering the crop it must associate with. It demands use of weed-management techniques adapted to different crops. Rotation effectively manages winter-annual grasses in winter grains or summer-annual grasses in spring grains. Each crop has its own set of cultural practices that create habitats for certain weeds. Changing crops changes available habitats and weeds.

TABLE 22.1 Yield of Four Wheat Cultivars Grown With Three Densities of Italian Ryegrass (Appleby et al., 1976)

Cultivar Height	Ryegrass Plants (Plants/sq yd)	Wheat Yield (lb/A)
Tall	0	3096
	33.4	2520
	82.8	2232
Tall	0	3924
	33.4	2925
	89.4	2709
Semidwarf	0	3042
	32.6	2214
	80.3	1908
Semidwarf	0	3465
	36.8	2565
	82.8	2115
LSD @ $P = .05$		423

Weed management must be done in winter wheat-fallow systems. Weeds such as downy brome, jointed goatgrass, and rye use moisture during fallow, and the seed produced easily infests the next wheat crop. Their life cycles are similar to winter wheat so inclusion of a spring crop (e.g., barley or corn) is a useful weed-management technique. The spring crop may permit use of cultivation and herbicides that cannot be used in winter wheat. Adding summer-annual crops to a winter wheat-fallow rotation lengthens the time before the next wheat crop, reduces the annual weed problem, and increases weed seed mortality. To reduce the need for herbicides to control weeds during fallow, some producers grow legumes such as peas or lentils for 6–8 weeks during fallow to suppress weed growth in spring (Anderson, 2005). The legumes are killed by application of glyphosate or by tillage, which incorporates plant organic matter in soil. The goal is to gain nitrogen from the legume, reduce the need for another herbicide application, and reduce the noncrop interval before winter wheat planting and thus the period for weed growth (Anderson, 2005).

Precise fertilizer placement is a regular practice in row crops but not in small grains, where it may have potential for weed management. Placing nitrogen fertilizer in the crop seed row, away from weeds, achieved a small reduction in seed production of rye and jointed goatgrass. Greater reduction was obtained when fertilizer placement was integrated with an increased wheat seeding rate and a taller, more competitive cultivar (Anderson, 1994, 2003). In a barley-field pea-barley-field pea rotation in Alberta, Canada, fertilizer application timing had little effect on weed competition in barley, but spring- compared to fall-fertilizer application increased yield of field peas in 2 of 4 years (Blackshaw et al., 2005).

4.1.4 Biological

There are few biological weed-control techniques available for use in small grains. Use of an endemic anthracnose disease to control northern jointvetchjoint vetch in rice is one example (see Chapter 12). If developed, biological control agents will have to be integrated with other weed-management techniques.

4.1.5 Chemical

Herbicides used in small grains are generally safe, efficient, and profitable; but, like other methods, when used alone, they do not solve all weed problems. Herbicides must be regarded as one of the little hammers (Liebman and Gallandt, 1997) in an IWM system. For maximum effectiveness, herbicides should be applied when weeds are young and have not yet affected crop yield.

Information on proper herbicide application is critical to successful use. Always read and follow label directions. If herbicides are mixed, follow label directions. Combining herbicides with different actions and activity can improve the weed-control spectrum. When mixed, the rate of one or both herbicides may be reduced. The presence of resistant weeds and the possibility of creating them must be considered.

For the many weed problems in small grains, there are herbicides or herbicide combinations that provide good control and crop safety if they are properly applied at the right time and rate. Manufacturers determine rates that work across many environments and climatic conditions. These may not apply to all fields. Reduced rates may work well in some conditions but local recommendations should always be followed.

Most postemergence, foliar-absorbed herbicides require actively growing weeds for maximum effectiveness. Weed growth is reduced by cool temperatures and drought. Herbicides with soil activity are less affected by temperature but their activity may be reduced in dry soil. A key to successful weed control with herbicides is early use when weeds are most susceptible.

Managers should know what weeds are to be controlled before selecting an herbicide. The herbicide should be applied uniformly or to selected patches with a properly calibrated sprayer. Reading the part of the herbicide label specific to each crop-weed situation is essential. Local recommendations summarize the attributes of several herbicides that can be used for weed control in small grains in a state or region. These recommendations change as efficacy changes, resistance develops, or herbicides are removed from the market. This book is not intended to and does not recommend herbicides.

4.2 Corn and Row Crops

One of every four US crop acres grows corn. It has been selected as representative of the many row crops grown in the United States: soybeans, dry beans, sugar beets, cotton, tobacco, sorghum, peanuts, vegetables from broccoli to zucchini squash, and potatoes. Each crop has its own, unique weeds (Van Wychen, 2016; WSSA, 2015 survey cited earlier), weed management requirements, and solutions. Examples from studies with a few of these crops are included. Because they are all annual row crops, they share some weed-management principles.

Corn, which most of the rest of the world calls maize (mais), will be used to emphasize the shared principles of weed management. The United States produces 37% and China 21% of the world's 15,148 million bushels. Less than 3% of the US corn crop is eaten by humans directly. Animals consume approximately 43% (decreasing), 30% is used to produce ethanol (increasing), 15% is exported, and 7.8% is used for high-fructose corn syrup. Corn has over 3000 uses in more than 1200 food items ranging from corn syrup to margarine. Other uses include paper production, plastics, cleaning agents, cosmetics, additives for pesticides, and ethanol for fuel.

Corn and several other crops are called row crops because they are planted in rows from 20 to 30 or more inches apart. Small-grain crops are also planted in rows but the rows are narrow and mechanical; interrow cultivation is impossible. Rows were invented because of the necessity of cultivation for weed control. Rows also make planting and harvesting easier and modern equipment demands straight rows. Manufacturing facilities concentrate human power, talent, and resources in factories for mass production. Agriculture requires a different spatial geometry, and the advantages of concentration are limited. Spacing in rows is required for optimum yield per unit area for all row crops, and yield is not increased if plants become too crowded. The results of the WSSA (2015) nationwide survey (Van Wychen, 2016), which identified the most common and most troublesome weeds in corn, are shown in Table 22.2. The 2015 nationwide survey complemented the surveys done for the Southern US states (Webster, 2005, 2008, and 2009). Those surveys summarized by Webster and Nichols (2012) showed that common and troublesome weeds in corn, cotton, soybeans, and wheat had changed primarily due to the influence of herbicide-resistant weeds.

TABLE 22.2 The Most Common and Most Troublesome Weeds in Corn (Van Wychen, 2016)

Most Common	Most Troublesome
Common lambs quarters	Common waterhemp
Giant foxtail	Palmer amaranth
Common waterhemp	Giant ragweed
Common ragweed	Common lambs quarters
Velvetleaf	Morning-glory spp.
Palmer amaranth	Johnsongrass
Giant ragweed	Giant foxtail
Large crabgrass	Kochia
Redroot pigweed	Velvetleaf
Morning-glory spp.	Yellow nutsedge

4.2.1 Prevention

Weed-prevention strategies are similar for most crops. There is nothing sophisticated or mysterious about them. Most practices are just common sense and should be incorporated in all good weed-management systems. See Chapter 10, Section 2 for a brief discussion of preventive practices.

4.2.2 Mechanical

Not too many years ago it was standard practice to moldboard plow the preceding fall or in the spring before planting corn. Plowing has not been abandoned as a weed-management/ seedbed preparation technique, but its use is diminishing. Plowing controls emerged weeds and buries weed seeds but brings other seeds to the surface where they can germinate. It, like many practices that affect seeds, controls and encourages weeds at the same time. Plowing is usually followed by other tillage operations to prepare a seedbed. Disking and harrowing break down clods and make crop planting easier with seed drills and corn planters. They also create ideal conditions for germination of weed seeds with, or just before, the crop. Plowing is a soil-inverting operation, chiseling is not. It brings fewer weed seeds to the surface. Tillage operations, subsequent to plowing, are shallower and encourage germination of shallowly buried seeds but not those deeper in the profile. The effects of the two kinds of tillage on weed populations were discussed in Chapter 10.

The use of conservation or reduced tillage has expanded greatly in the last decade and interest in adopting some form of reduced tillage has expanded even faster. These systems range from surface disking to breaking up the residue of the preceding crop to no tillage at all with planting directly into the preceding crop's residue using specially designed no-till planters. In the first year or two after no tillage is begun, weed problems decrease dramatically, and without careful management weeds can increase in subsequent years. For example, in one experiment with monoculture corn and conservation tillage all plots were weed free the first year. In the second and third years, fall panicum dominated and smooth pigweed dominated in the fourth and fifth years, reaching densities of 85% of total plot area (Coffman and Frank, 1992). The authors related the change of weed flora to continuous use of certain herbicides. Fall panicum dominated in plots treated with atrazine and a carbamothioate herbicide. A triazine-resistant biotype of smooth pigweed dominated in plots treated with atrazine plus cyanazine (Coffman and Frank, 1992).

Ridge-till systems are used to reduce soil erosion and the need for herbicides in some corn-soybean rotations. A ridge-till cultivator makes ridges over the crop row during the final, summer cultivation of either crop. The ridges, which are disturbed at harvest and during spring planting, are leveled by moving some soil into the furrows. Immediately after smoothing or "knocking off" the ridges, the crop is sown on the remainder of the ridge and the ridges are gradually rebuilt, during cultivation, as the crop matures. The system was most effective in Minnesota when corn and soybeans were rotated (Forcella and Lindstrom, 1988a). Ridges crack, and weeds emerge when corn is grown continuously.

Ridge tillage as used in corn-soybean rotations is not without problems. Knocking off ridges controls many weeds and ridging soil during the summer encourages germination of numerous weed seeds that produce seeds to infest the next crop. Conventionally tilled corn had about two-thirds fewer weed seeds than ridge-tilled corn because of the large seed production by weeds that germinated when ridges were rebuilt (Forcella and Lindstrom, 1988b). Ridge tillage is not successful unless herbicides are used to control late-emerging weeds.

Studies in Indiana evaluated no-tillage, moldboard plow, and chisel plow systems in three rotation systems, each of which included corn (Martin et al., 1991). Net incomes for no-till systems on all farms were lower than incomes for moldboard or chisel plow systems due to slightly lower yields and higher herbicide costs. In general, farm incomes were higher with moldboard as opposed to chisel plow systems. It is important to note that these studies were done on highly productive, flat, well-drained soils that were not highly eroded. A different situation for another system will probably yield different results. Agriculture and agricultural research are definitely site specific as are herbicide recommendations (Treadway-Ducar et al., 2003).

After land preparation, corn can be cultivated mechanically, one or more times, with various types of implements ranging from a straight shank to several different, duckfoot-shaped tools. A rotary hoe can also be used to kill weeds between rows. Cultivation can move soil into corn rows and cover emerging seedling weeds, but it cannot till in the row. Some cultivation implements operate close to the crop row and make herbicide banding (application in a narrow band just over the crop row) attractive. If all weeds between rows are controlled mechanically, then herbicide quantity and cost can be reduced by applying the herbicide only in a band over the crop row. This requires more application skill and accuracy than broadcast application. Postemergence flaming has been used selectively in corn and cotton.

4.2.3 *Cultural*

Crop rotation can be a profitable and useful weed-management technique. Corn is often grown in monoculture or in a limited rotation with soybeans or another row crop. Rotating to a small-grain or hay crop, or both, in succession often results in reasonable yields that may be lower than those in a corn-soybean rotation (Helmers, 1986). Rotational possibilities are limited by land, climate, market opportunities, and the availability of suitable rotational crops. Rotating corn with other crops is critical to managing weeds in systems attempting to reduce herbicide use. Introduction of crops with different life cycles and cultural practices deters growth of summer-annual weeds with life cycles similar to corn, whose growth is encouraged by continuous or frequent corn crops. Rotation reduced annual-grass weeds in corn in the Central Great Plains of the United States (Wicks and Smika, 1990). Many annual grasses germinate in May and set seed by late August before corn is harvested. As mentioned earlier, rotating to winter wheat changes the times of tillage and crop presence and disrupts the life cycle of annual weeds.

It has also been shown that planting corn at higher densities (100,000 plants/ha = about 40,000/A) in a dry year in Ontario, Canada, or in narrower rows [38 versus 76 cm (15–30 in.)] in a wet year provided greater weed suppression (Shrestha et al., 2001). Higher densities reduced early weed competition and narrow rows reduced late-emerging weeds.

In contrast to Helmers's (1986) study, Liebman et al. (2008) showed that low-external-input cropping systems were as or more productive and profitable than conventional 2-year-rotation corn systems in the central US corn belt. Three rotations were compared:

Rotation	Years	Nitrogen Inputs Compared to 2 year rotation	Herbicide Use Compared to 3 year rotation	Corn and Soybean Yield Compared to 2 year rotation	Net Return
Corn/soybean/	Two				intermediate
Corn/soybean/red clover	Three	59% lower	76% lower	equal or higher	lowest
Corn/soybean/small grain + alfalfa/alfalfa	Four	74% lower	82% lower	equal or higher	Highest

When government subsidies were included, differences among the three systems were smaller because subsidies favored the 2-year system, but the rank order of net return of the three systems remained the same.

4.2.4 Biological

Currently there are no biological-control agents used routinely in corn or other row crops. Because row crops usually have several weeds rather than just one, the specificity of a biological-control organism may not fit the weed-control need. Integrated weed control in soybean with a combination of the phytopathogenic bacterium (*Pseudomonas syringae* pv. *tagetis*), which attacks Canada thistle, a highly competitive soybean cultivar, and the herbicide bentazon was investigated (Hoeft et al., 2001). The bacterium reduced Canada thistle growth but was less effective than the herbicide bentazon. The competitive soybean cultivar was not beneficial and there was no synergy (added benefit) between the two control techniques.

4.2.5 Chemical

The main weed control technique that dominates most weed-management systems in corn is herbicides. Some, but not all, of the herbicides available for use in field corn may also be used in sweet or popcorn. Label and local recommendations must always be consulted before using any herbicide. The dominant herbicide families used for weed control in corn are triazine (atrazine), chloroacetamide (alachlor, metolachlor), phenoxy and benzoic acid (2,4-D and dicamba), and sulfonylurea, and EPSP synthase inhibitors (glyphosate). Herbicides in these families may be used alone but are most often used in combinations. It is common to find state recommendations that include more than 40 preplanting (soil-applied), 30 or more preemergence, and 30 or more postemergence herbicide combinations for weed control in corn. About one-third of the soil-applied treatments combine atrazine with one or more other herbicides. The greatest variety of herbicides is found in the postemergence group. There is a wide array of postemergence herbicides dominated by glyphosate. Many are directed at postemergence annual-grass-control problems. Chapter 19 addresses the seemingly omnipresent problem in corn and many other crops of herbicide resistance.

4.2.6 Integrated Strategies

Simulation models for weed management in corn (King et al., 1986; Lybecker et al., 1991) suggest that flexible weed-management strategies, based on control variables, outperform fixed or prescriptive weed-management programs. The variable used for deciding what and how much herbicide to use is the number of weed seeds in soil. The models require knowledge of losses due to specific weed densities, percent emergence of weed seed from the soil seed bank, and the efficacy of each herbicide against each weed. The models don't consider the effect of weed escapes on the next year's crop, which, in one study (Swanton et al., 2002), was not important. A flexible strategy lowered total herbicide cost and the quantity of herbicide used, increased postemergence herbicide use, decreased preemergence herbicide use, and increased the farmer's gross profit margin (Lybecker et al., 1991, 1992). The models were developed for irrigated corn. Most models have not incorporated mechanical methods of weed control. They herald a new era of weed management when decisions will be informed by knowledge of weed seed in soil and the efficacy of different control measures. Weed-management decisions have often been made on the basis of what someone thought the problem was going to be and have, therefore, been prophylactic rather than directed at a specific problem.

Integrated management of itchgrass in corn was studied in Costa Rica employing velvetbean as a leguminous cover crop, a preemergence herbicide (pendimethalin) and classical biological control with the head smut [*Sporisorium ophiuri* (P.Henn) Vanky], which is host specific for itchgrass (Smith et al., 2001). The head smut affects itchgrass seedlings as they emerge and leads to seed sterility rather than plant death. Thus, it is a preventive measure. Velvetbean planted at either of two densities between corn rows was very effective in reducing itchgrass populations from 54 to 17 plants/m^2. Pendimethalin's contribution to weed control was modest. When the cover crop was effective and the head smut achieved 50% infection, significant income benefits were obtained by corn growers because the biological control was less expensive than the herbicide (Smith et al., 2001). Similar results were achieved in control of cogongrass in corn in Nigeria (Chikoye et al., 2001). Twelve months after planting corn with velvetbean as a cover crop, corn averaged 65% less cogongrass biomass at three locations than weedy control plots without velvetbean. Corn grain yields were up 25%–50% over the three locations. Similar results were obtained for cassava yield with velvetbean (Chikoye et al., 2001).

So far, weed-management systems in cotton are dominantly curative (Mortensen et al., 2000) and emphasize herbicides. Burke et al. (2005) studied soil-applied herbicides for use with glyphosate-resistant cotton. As herbicide inputs increased, cotton yield increased. Environmental concerns (soil erosion and pesticide use) have led to study of conservation tillage (reduced tillage) for cotton production. The curative approach demonstrated that optimum cotton yield was achieved with reduced tillage only when it was combined with broadcast application of preemergence or early postemergence herbicides (Toler et al., 2002).

Potato research has followed a similar curative approach. Systems that include a rye cover crop, reservoir tillage,[3] and herbicide banded over the crop row reduced preemergence herbicide use up to two-thirds and maintained tuber yield (Boydston and Vaughn, 2002).

4.3 Turf

Turfgrass is what most Americans grow, love, and struggle to maintain. It is a benefit and a puzzle.

> God and St. Francis were discussing gardens and nature on earth. God asked Francis -
> What is going on down there? What happened to the dandelions, violets, milkweeds
> and stuff I started? I had a perfect no-maintenance garden plan. Those plants grow
> in any type of soil, withstand drought and multiply with abandon. The nectar from
> the long-lasting blossoms attracts butterflies, honey bees and flocks of songbirds.
> I expected to see a vast garden of colors by now. But, all I see are these green rectangles.
> Francis said - It's the tribes that settled there, Lord. The Suburbanites call your
> flowers weeds and go to great lengths to kill them and replace them with grass.
> God said - Grass? But, it's so boring. It's not colorful, doesn't attract butterflies,
> birds and bees; only grubs and sod worms, and it's sensitive to temperature. Are you
> sure these suburbanites really want all that grass? Apparently so, Lord. They go to
> great pains to grow it and keep it green. Each spring they fertilize grass and poison any
> other plant that appears in what they call the lawn. But Francis, the spring rains and
> warm weather probably make grass grow really fast. That must make them happy.
> Apparently not, Lord. As soon as it grows a little, they cut it—sometimes twice a week.
> They cut it? Do they bale it like hay? Not exactly, Lord. Most of them rake it up and
> put it in bags. They bag it? Why? Is it a cash crop? Do they sell it?
> No, Sir, just the opposite. They pay to throw it away. Now, let me get this straight.
> They fertilize grass so it will grow. And, when it does grow, they cut it off and pay to
> throw it away? These suburbanites must be relieved in the summer when we cut back on
> the rain and turn up the heat. That surely slows growth and saves them a lot of work. You
> aren't going to believe this, but when the grass stops growing so fast, they drag out
> hoses and pay money to water it, so they can continue to mow it and pay to get rid of it.

Kolbert notes that we no longer choose to keep lawns, we just keep on keeping them. Our efforts (yes, I have a lawn) gain benefits. Lawns are aesthetically pleasing, improve home value, cool the environment, and enhance the beauty of parks and public areas. They absorb

---

[3]Reservoir tillage, also called dammer tillage, is used in potatoes and carrots in the US Pacific Northwest. A rotating paddle wheel creates depressions about 10 inches deep (small water reservoirs) in the furrow between crop rows.

TABLE 22.3 The Most Common and Most Troublesome Weeds in Turf in the United States (Van Wychen, 2016)

Most Common	Most Troublesome
Crabgrass	Annual bluegrass
Dandelion	Common blue violet
Annual bluegrass	Dallisgrass
White Clover	Yellow nutsedge
Smooth crabgrass	Ground Ivy
Plantain spp.	Bermuda grass
Goosegrass	Goosegrass
Ground Ivy	Large crabgrass
Yellow nutsedge	Virginia buttonweed
Common chickweed	Green kyllinga

carbon dioxide and trap dust and dirt. They also supply employment for many and recreation and enjoyment (maybe a bit of frustration) for the nation's more than 26 million - golfers on more than 15,000 golf courses. Sports fields with turf are safer for players. Turf purifies water and releases oxygen.

Desirable turfgrass, usually divided into cool- and warm-season species, varies with climate, rainfall, and intended use. The United States has about 25 species that can be used for turf (Vengris and Torello, 1982). They are usually perennials that do well with continuous close mowing. Cool-season species grow best during cool (60–75°F), wet conditions in the spring and fall and may become partially dormant during hot summer months. Warm-season species grow vigorously during hot, dry times when temperatures are above 80°F. Vengris and Torello list 108 weeds that invade turf. Those that occur most frequently in the United States and Canada are crabgrass, dandelion, annual bluegrass, common chickweed, plantains, and prostrate knotweed; there are many others. Of the 108 common turf weeds 17 are perennial monocots, 11 annual monocots, 44 perennial dicots, 29 annual dicots, four winter annual dicots, and three biennials. A nationwide survey conducted in 2015 (Van Wychen, 2016) documented the most common and most troublesome weeds in turf (Table 22.3). Weed control on many turf sites is principally elimination of broadleaved species, and because of the variety of herbicides available, the task is not difficult. Weed-management systems for turf were reviewed by Bingham et al. (1995).

4.3.1 Prevention

Turf is no different from any other crop in the sense that prevention is an essential component of weed management. Preventive practices are used by turf managers when turf is established and during its life. Turfgrasses are particularly poor competitors during establishment and elimination of weeds by thorough tillage prior to planting or use of preplanting herbicides is important. Imported topsoil is almost always contaminated with weed seeds and vegetative propagules. Delaying planting of turfgrass until some of these germinate and can be controlled is wise. Planting the correct turfgrass or grass mixture is an obvious preventive strategy. Vigorous emergence and growth reduces weed growth. Grass seed must be of high quality (high percent germination), free of weed seed, and sown at the right time for the climate. Cool-season grasses do best when planted in the fall or early spring and warm-season grasses do best when planted in spring or early summer.

Weeds in established turf can often be traced to windblown seed sources and poor management practices. Weeds most easily invade cool-season grasses such as Kentucky bluegrass, ryegrass, or fescue when they are mowed less than 1.5 to 2 in. high. On the other hand, weeds most easily invade stoloniferous and rhizomatous Bermuda grass turf when it is mowed too high (above 1 in.). Too little water may stress turfgrasses and drought-tolerant broadleaved weeds will invade. Too much water will create ideal conditions for establishment of annual and winter annual grasses. Turf is commonly fertilized with nitrogen or a mix of nitrogen, phosphorus, and potassium to keep it vigorous and maintain desired color. Fertilization appropriate to the climate and turf species helps prevent weed invasion. Prevention can also be practiced by controlling turf wear from constant use of certain spots. Change of traffic or play patterns help maintain a vigorous turf and prevent weed invasion.

4.3.2 Hand Weeding

The oldest method of weed control, hand pulling, is still appropriate and common in turf. For a home lawn, it is efficient, even if not pleasant—but pulling or attempting to pull dandelions and bindweed requires more persistence than most homeowners have. Eventually it will work for dandelions and bindweed, but eventually will be a very long time. For golf courses and public lawns it is inefficient and expensive. Tillage is not appropriate in established turf unless extensive renovation is undertaken. Scarification or vigorous raking is used to thin stoloniferous grasses and control some broadleaved species. If overdone it thins turf and allows weed invasion. Aeration is used in many climates to reduce compaction and stimulate vigorous turf growth, thereby reducing weed competition.

4.3.3 Chemical

Because turf is valuable, fumigation, which requires professional expertise and is expensive, may be a desirable preplanting control strategy. Herbicides are commonly used in turf but are most appropriate after all preventive-management techniques have been employed. There are at least 30 different herbicides that can be used in turf but not all herbicides can be used with all turf species. For example, several herbicides (dithiopyr, oxadiazon, pendimethalin, and prodiamine) control crabgrass in cool-season, perennial turfgrasses. All of them can be used on Bermuda grass turf, but some hybrids and fine-leaved varieties can be injured (Elmore, 1985). Local recommendations should be consulted before any herbicide is applied to bentgrass turf. There are no selective herbicides for removal of coarse-leaved, perennial-grass weeds from perennial-turf species; this is a major unsolved weed-control problem. It can be accomplished by spot application of a translocated, nonresidual, nonselective herbicide such as glyphosate, which will kill nearly all other plants it contacts. The most common herbicide used in turf is 2,4-D alone or in combination with other growth-regulator herbicides to control a range of annual- and perennial-broadleaved species without injuring most turfgrasses. Injury of seedling grasses less than 4–6 weeks old is likely, Because of the nature of plant growth and the translocation pattern in perennials, fall application is often the most effective. As in other crops, application when the crop is growing vigorously and the weeds are young is best.

Warm-season grasses (e.g., St. Augustine, Bermuda, Zoysia) in warm climates present a unique weed-control situation. These grasses commonly are dormant during the winter and herbicides such as paraquat that would desiccate Bermuda grass foliage if applied when it is actively growing can be used to control cool-season grasses and annual-broadleaved weeds that grow when the grass is dormant. Postemergence application of the organic arsenical MSMA selectively controls some annual grasses in warm-season turfgrasses. More than 20 herbicides are available for weed control in turf. Local recommendations and turf managers should always be consulted prior to herbicide choice and application.

4.3.4 A Question

An interesting, important aspect of weed control in turf is the practical question of how long we will choose to pay the cost of maintaining it. US turf occupies at least 40 million acres, a little smaller than the state of Washington (71,300 mi^2 = 42 million acres). The United States has more than 15,000 golf courses on 2.2 million acres = 3507 mi^2 (roughly the size of Delaware and Rhode Island). More acres are devoted to turf of all kinds in the United States than to any other crop. Homes, golf courses, and parks occupy more land than corn and wheat. Turf is the largest irrigated crop in the United States and the area is increasing. Lawns occupy three times more land than irrigated corn.[4] Nearly one-third of all residential water in the United States is used for landscaping. Not all turf is irrigated, and it consumes only a small percentage of total water used in the United States, but competition for water will be under increasing pressure as populations grow. It is a finite resource.

When digging them out was the only option, weed control was essentially hopeless. All homeowners seem to want a lawn that is weed free. That can be achieved with regular watering, fertilizer, and herbicides. Insecticides and fungicides may also be used.

Ultimately, the turf question will become an economic and perhaps a moral issue. As the area devoted to turf increases and more water, pesticides, and fertilizer are used to maintain it, we must ask if this is what we ought to do? We know what must be done to create a desirable lawn. Is it what we ought to do? Are we being careless with our resources? It is not the purpose of this book to answer the questions, but they should not be ignored.

4.4 Pastures and Rangeland

Pastures and rangeland cover more than 40% of the world's agricultural acres. These diverse habitats exist in all topographies, climates, and over most soil types. The desirable vegetation is equally diverse, ranging from short-grass prairies of the mid-United States to the oak/pine associations found in western states to the designed, planted, irrigated pastures

---

[4]Earthobservatory.nasa.gov/IOTD/view.php?id=6019.

of irrigated and rain-fed areas. With the exception of intensively managed, planted pastures, rangeland and pasture may have over 100 species/m^2. These areas are often very large and are located on hilly or mountainous terrain that makes access for weed management difficult, if not impossible. Weed-management systems for pastures and hay crops were reviewed by Smith and Martin (1995) and Sellers and Ferrell (2016). Several state publications are available.

4.4.1 Burning

Controlled burning has been a common management technique to reduce competition from woody species and competition for water. It has some obvious environmental drawbacks (smoke pollution, potential erosion of bare soil, and of most importance—escape). Its intentional use has been reduced due to its environmental effects and the increasing number of homes in natural areas. Intentional and accidental burning is often followed, on large areas, by reseeding, commonly by air.

4.4.2 Mechanical

Mechanical removal with bulldozers and chaining is used on rangeland, but both are expensive and results are temporary. Mowing is a good technique for control of annual weeds in pastures but not on large areas of rangeland. Growth-regulator herbicides can be used to control woody species such as big sagebrush or greasewood on rangeland.

4.4.3 Biological

Biological control with grazing animals, managed grazing, is a desirable weed-management technique. Goats are particularly good browse animals, but have to be carefully managed so they don't overgraze or compete with cattle. Some of the major successes in biological control of weeds with insects have been achieved on rangeland such as the use of *Cactoblastis cactorum* to control prickly pear and *Chrysolina quadrigemina* to control St. John's wort (see Chapter 12).

A major problem in the arid western states is the perennial, herbaceous weed leafy spurge (see Chapter 12). Successful management has been accomplished when techniques have been integrated. Sheep or goats (biological control) will graze the weed early in the growing season to release desirable grasses from leafy spurge competition and make the leafy spurge more susceptible to fall-applied herbicides. By 2006, 12 insects had been released in the United States for biological control of leafy spurge, including six species of *Aphthona* beetles (Bourchier et al., 2006). The leafy spurge hawkmoth (*Hyles euphorbiae*) eats leafy spurge leaves and bracts during its larval stage (Harris et al., 1985), but alone the hawkmoth is not an effective biological-control agent (Coombs et al., 2004).

A root- and stem-boring beetle (*Oberea erythrocephala*) imported from Italy was established in Montana and North Dakota (Lehninger, 1988). Adult beetles feed on leaves and stems, which does not kills plants (Coombs et al., 2004). Stem girdling by the adults, with subsequent egg laying, usually results in shoot death. Larvae bore into stems and move to roots where they mature and exist on carbohydrate root reserves. Boring allows other pathogens to enter. A good example is the adult and larval *Aphthona* spp. Beetles; two of six species—*A. lacertosa and A.* czwalinae—have been released. They feed on leaves and flowers, and the larvae bore into roots and feed on root hairs and young roots. They destroy vascular structure while feeding. Grass infested with leafy spurge will be favored by use of the cultural controls, fertilization and irrigation, if either is economically feasible. These strategies reduce competition and permit efficient grazing by animals such as cattle that do not eat leafy spurge. In the fall, leafy spurge can be sprayed with selective herbicides (chemical control). Neither herbicides nor grazing animals have greatly affected vitality or future performance of biological-control insects. It is a certainty that integration of methods will not eliminate leafy spurge in one season, but it will keep the population at a level that permits efficient land use. Persistence, defined as continued use of several techniques and continued evaluation, is required.

4.4.4 Chemical

Several herbicides are available for pasture and rangeland weed management of diverse annual and perennial species. Perennial weeds such as Canada thistle are controlled better when herbicides and mowing are combined. Milestone (Dow Chemical Co.) at 5–7 oz/A is very effective for Canada thistle control. Mowing improves pastures and stresses perennial weeds that may then be more susceptible to herbicides. The value of combining herbicides and mowing is illustrated well by control of the exotic invasive weed, perennial, tropical soda apple in Florida (Mullahey et al., 1996). It was first found in 1981 but not identified as an important invader until 1990 when it was estimated to occur on about 25,000 acres in south Florida. By 1993 it occupied 388,000 acres. In 1995 the area had grown to almost 750,000 acres, and in 2010 it had invaded almost 1,000,000 acres of Florida pastureland where it flowers and sets seed throughout the year (Westbrooks, 1998). It has been found in eight southern states and California and Pennsylvania. Control has been best when plants are mowed or chopped 60 days prior to spraying with the growth-regulator herbicide triclopyr. Mowing three times 60 days apart gave 83% control after 180 days. Mowing or chopping 60 days prior to triclopyr application was 93%–100% effective, 180 days after herbicide application. Further mowing is not required, but spraying escaped plants is recommended. Cattle ranchers are urged to isolate new cattle and monitor cattle movement between pastures because cattle eat the fruit and seed easily passes through their digestive system to reinfest pastures. Cattle isolation is a part of an integrated weed-management system.

4.5 Perennial Crops

Perennial crops (e.g., alfalfa) grow for several years and then the land may be rotated to an annual or another perennial crop. Other perennial crops are grown for many years (e.g., apples, almonds), after which the trees are removed and another orchard may be established on the same site. A diverse group of weeds succeeds in perennial crops including annuals, biennials, and perennials favored by perennial culture. Some perennial crops such as alfalfa, peppermint, asparagus, or strawberries are not commonly cultivated mechanically and without good crop competition and weed management, perennials can invade and succeed. Cultivation can be a part of a weed-management program in tree fruits and nuts that have wide rows and low crop density. Weed-management options are limited because the crop's longevity precludes use of rotation, and in some crops, mechanical tillage. Cover crops and mulching are feasible in perennial crops and can be incorporated in weed-management plans. Biological control must be chosen based on the weeds present and cannot be prescribed for all perennials.

4.5.1 Prevention

Vigilance is a prerequisite for a good weed-management program. The manager must be aware of sources of weed infestation and take appropriate action to prevent invasion. In perennial cultures actions could include: screening of irrigation water to prevent import of weed seed, careful selection of clean mulch material, composting of manure to kill weed seed before spreading, mowing to prevent seed production, careful selection of adapted crop cultivars to maximize competitiveness, and planting weed-free seed or seeding stock. The last two can only be done at planting, an opportunity that should not be lost.

Site selection is a weed-management technique. Perennial weeds are favored in perennial crops. Insofar as possible, sites without them should be selected for initial planting. Annual weeds will be present on almost any site, and some control can be achieved by preplant tillage, just as it can be before annual crops are planted. Perennials are not controlled easily by tillage, and preventing them is always good planning.

4.5.2 Cultural

Timing of planting is a cultural control and preventive technique. Planting should be done with a quick-emerging seed (or established transplant), which is likely to assure a vigorous crop. For example, alfalfa planted in the fall in southern California becomes established and is a better competitor with weedy-spring grasses than spring- or summer-planted alfalfa (Mitich, 1991). Planting time varies with climate and environmental conditions, but its role in weed management should not be ignored.

Irrigation timing is an important cultural practice that influences weed presence and growth. Barnyardgrass and yellow foxtail establish readily when alfalfa is stressed before or during harvest and water is applied when there is little alfalfa growth to shade soil. When water is applied near alfalfa cutting, weed invasion is reduced. Grazing animals on perennial cropland contributes weeds in manure, and overgrazing always encourages weed invasion. Grazing animals control some annual weeds.

For some short duration (3–5 years) perennials (e.g., alfalfa), planting with a nurse or cover crop is a useful weed-management technique. Crop yield may be reduced in the first year but subsequent crops have fewer weeds. Cover crops, ground covers, or grassed, mowed alleyways are part of good orchard management in perennial row crops and fruit orchards. Ground cover species adapted to local environments should be selected based on local recommendations. Cover crops and ground covers compete directly with weeds but should not compete with the crop. They may also have allelopathic effects. Regular mowing of grassed interrow areas changes ecological relationships and affects weed populations. Rows may be treated with herbicides.

4.5.3 Mechanical

For alfalfa, peppermint, and similar crops that are not planted in wide rows and eventually cover the soil surface, cultivation is not possible. In tree crops, clean cultivation is a widely practiced weed-management technique but precludes grassed interrow areas or ground covers. Clean cultivation is common in many nut orchards and is usually combined with chemical methods of weed control. It is a desirable management technique but increases the risk of soil erosion from water or wind.

4.5.4 *Chemical*

Herbicides for perennial crops are as diverse as the crops. Because this is a book of principles rather than recommendations, the several herbicides available for perennial crops are not listed. Local recommendations should always be consulted for each crop. Many persistent herbicides including dinitroanilines (used in peppermint, spearmint, sugarcane), triazines (used in asparagus, alfalfa, citrus fruits, nuts, pineapple, sugarcane), and uracils (used in peppermint, spearmint, pineapple) are approved for use in perennial crops. Decisions on herbicide use must be based on the weeds to be controlled and how the herbicide affects other weed-management strategies (especially incorporation of a permanent ground cover). Herbicides are valuable tools in these crops and should be regarded as part of the overall weed-management program but not as a complete management technique.

Herbicides were essential parts of a program to manage leafy spurge on native rangeland (Masters and Nissen, 1998). Invasion of native range by leafy spurge was directly linked to past management practices that reduced native species diversity and opened niches for leafy spurge. Leafy spurge biomass was lowest in areas where tall grass (e.g., big bluestem, switchgrass) yield was highest. This was accomplished best by combining vegetation suppression with fall-applied herbicides, burning standing dead plant residue, and no-till planting of desirable native tall grasses in the spring. The same conclusion was reached for control of Russian knapweed on rangeland (Laufenberg et al., 2005). Herbicides were most effective when combined with revegetation in areas that lacked a diverse mixture of desired species.

4.6 Aquatic Weed Sites

It is seldom necessary or desirable to remove all vegetation from water. The aquatic weed manager must decide if complete eradication is desirable or if a reduced level of control (selective management) is more appropriate. It may be possible to control an especially troublesome or dominant species and leave others undisturbed. Control can be infrequent by mechanical or chemical means or it can be just removal of excessive growth for part of a season.

4.6.1 *Classification of Weeds*

A brief introduction to aquatic weeds was given in Chapter 3, Section 3.2.4. It uses the simple classification of free-floating, submersed, and emersed weeds. They are useful divisions, but the aquatic world is more complicated. An explanation of the complexity is offered by McNabb and Anderson (1985) who subdivide the categories to provide more information about habit of growth and plant type (Table 22.4). The aquatic weed manager must know exactly what weed is to be controlled or managed and its method of reproduction. Algae reproduce asexually by cell division. Completely submerged plants reproduce by fragmentation, vegetatively by rhizomes and runners, and by specialized submerged buds (turions) and tubers. Submersed plants that have some floating leaves such as American pondweed reproduce by seed as do emerged plants. Free-floating plants (e.g., water hyacinth) reproduce by seed and asexually or clonally by fragmentation and some reproduce by spores and clonally (e.g., salvinia). With dual modes of reproduction, some weeds cannot

TABLE 22.4   Classification of Aquatic Weeds (McNabb and Anderson, 1985)

Type of Plant	Growth Habit	Examples
Algae	Unicellular or microscopic colony	
	Free-floating attached to substrate	Phytoplankton, diatoms
	Filamentous-green,	Cladophora
	Colonial— attached or floating	Spirogyra
	Blue-green	Nostoc
Vascular plants	Completely submerged	Sago pondweed, hydrilla
		Eurasian watermilfoil
		Some mosses
	Submersed with floating leaves	Water lilies
		American pondweed,
		Arrowhead
	Emergent	Cattail, Bulrush
		Several grasses
	Free-floating	Water hyacinth
		Duckweed, Azolla
		Salvinia (a fern)

be managed by preventing seed production as is the case with terrestrial annuals. If the manager doesn't know the plant's growth habits and method(s) of reproduction, poor or no control may result from improper choice of methods.

4.6.2 Prevention

Preventive strategies depend on knowledge of the factors affecting growth of aquatic vegetation. These include light, nutrients, water depth, water flow rate, the rooting medium (water or soil) and its nutritional status, dissolved gases, and temperature. While the latter two are important, there is little that can be done to change them. Light can be managed by control of water depth. Water management to control weeds by reducing depth through intentional drawdown manages some weeds effectively. It has little effect on floating vascular plants or algae but will aid control of rooted species. It is not permanent because other weeds adapt to new water levels, but it can help manage current weed problems. Eurasian watermilfoil, arrowhead, and water lilies can be managed by a drawdown of water (decreasing water level), but most pondweeds, cattails, and rushes are not affected. Drawdown can be done at any time, but most irrigation structures were not designed to facilitate the technique and it is not used widely. If the manager understands the biology of the weed to be controlled, drawdown can be timed to stop production of reproductive structures. The opposite—ponding or deepening water—can be used to manage some aquatic species.

Apparent water depth for plants is also affected by turbidity. Turbid, or more nearly opaque, water provides less light to submerged species. If turbidity can be tolerated, it can be created by stirring or intentional incorporation of silt or other soil particles. Turbidity can also be created by fertilization to promote algal blooms (i.e., abundance). This technique can cause other problems because algae may be toxic or otherwise undesirable. Fertilizer stimulates algal growth that shades plants that root underwater and do not emerge. A bloom must be maintained through a growing season to achieve control. Careful monitoring is required. More commonly, nutrients from surrounding fields or other sources encourage growth of weedy plants and worsen the weed problem. Dredging or reshaping a pond to remove shallow areas reduces light on the edges and reduces growth of submerged and emersed weeds.

Irrigation water inevitably brings weeds with it (see Chapter 5, Section 2.3). Prevention of water movement to ponds and lakes is nearly impossible. Animals, birds, and humans transport seeds and vegetative reproductive organs to water and, with the exception of human transport, cannot be prevented easily, if at all.

4.6.3 Mechanical

Mechanical methods of weed control adapted for use on aquatic sites frequently require large, specially adapted machinery. Aquatic weeds can be mowed with floating mowers but these are expensive and there is a problem of disposal of the mowed, inevitably smelly vegetation. Repeated mowing is required and the method is not adapted to large areas. As is true for terrestrial perennial weeds, mowing may release dormant, vegetative buds and actually worsen the weed problem unless it is done frequently or integrated with another method. It is only appropriate for rooted plants. Floating plants can be gathered by a large machine that collects what it removes (Fig. 22.2). Control would be an incidental similar to

FIGURE 22.2 A weed mower for aquatic weeds.

physical removal of terrestrial vegetation. Removal is a good technique for floating plants. The biomass of aquatic plants is often quite high and large equipment is needed to collect it. The method is only appropriate for small to medium lakes or straight waterways. Vegetative reproducers will quickly repopulate an area and disposal of collected biomass is a problem. Physical disruption of rooted plants by chaining or dredging is a good technique for straight irrigation ditches. It immediately reduces clogging by weeds, but the relief is only temporary for plants that grow back from severed roots. Dredging or reshaping a ditch or pond can be more effective if roots and vegetative reproductive organs are removed, but it is expensive.

Burning in spring is used in much of the western United States to remove plant residue from irrigation ditches before water flows. It is more for sanitation and good housekeeping than weed control. Temperatures are not high enough to kill buried seeds or vegetative organs. If young seedlings are emerging they will be controlled, but the main benefit is sanitation.

4.6.4 Biological

The same criteria for success of a biological control organism apply to aquatic and terrestrial environments (see Chapter 12). *Agasicles hygrophila*, a South American flea beetle, has been used to clear southern US waterways of alligatorweed, a free-floating plant. *Cercospora rodmanii* has shown promise for control of water hyacinth in tropical and semitropical waterways (Strobel, 1991). Although *Agasicles* has been very successful, there are no other examples of widely adapted, successful insects or pathogens for control of aquatic plants.

Other organisms have been used for control of aquatic plants. They include the sea manatee, a large aquatic mammal, and two fish, the grass carp or white amur (*Ctenopharyngodon idella*) and tilapia (*Tilapia melanopleura*). They eat aquatic vegetation but are generally nonselective. However, that is not regarded as a serious disadvantage. Survival and reproduction are problems with any biological-control organism. The manatee (*Trichechus* spp.) has no known enemies except man and a pathogen that reduced their population in 1996. Manatees survive well in warm freshwater but do not reproduce well.

Fish provide an alternative crop in many aquatic environments. Tilapia (*Tilapia* spp.) are intolerant of cold water and are, therefore, adapted only to subtropical and tropical climates. The triploid grass carp is a sterile version of the white amur. They can eat 2 to 3 times their body weight daily and can grow to marketable size. They were not, but are now, approved for use in California (McNabb and Anderson, 1985). Failure to reproduce keeps a population under control and prevents escape of an otherwise desirable control organism. It also requires restocking in 3–5 years.

4.6.5 Chemical

Aquatic plants are susceptible to a wide range of herbicides. Fewer than 10 are approved for use in water. However, herbicides are the most common method of weed control in the United States. These apparently contradictory statements are easily explained. Water has multiple uses and it moves. Herbicides offer the same advantages in the aquatic environment that they do in field crops. They are selective, easy to apply, act quickly, are relatively inexpensive, and can be used where other methods don't work well. Nevertheless, their use is limited in water. Aquatic herbicides are applied directly to emerged plants or to water. Some herbicides can be applied to exposed soil after water is drawn down or removed. In all cases, water's multiple uses must be considered. Label and local recommendations must be consulted prior to use. As noted, there are very few herbicides for aquatic use and restrictions limit use of approved herbicides.

McNabb and Anderson (1985) developed a decision-making scheme for integration of methods. The first question in all cases is, what will the water be used for? The answer determines control options, and other questions can only be considered after the first has been answered. They divide water into three categories: industrial, potable or recreational, and irrigation. Their scheme integrates all control methods through a series of logical questions regarding the possibility of nutrient control, drawdown of water level, and control of downstream flow through ponding or temporarily holding water. Their scheme is presented for one situation in Table 22.3.

Some aquatic weeds can be managed by skillful combination of preventive and control techniques in an integrated system.

4.7 Woody Plants

Woody plants are perennials that produce secondary growth in the form of wood. Some are useful for their wood. Others provide browse grazing for animals, shade, shelter, medicinal extracts, and aesthetic benefits. Many become undesirable (weedy) when they dominate an area to the exclusion of desirable vegetation. Woody plants and their management have been dealt with in detail in a major book by Bovey (2001) and in reviews by Bovey (1995) and McNabb et al. (1995). Interested readers are referred to those works for complete coverage of woody-plant management.

TABLE 22.5   Decision Scheme for Control of Attached Submersed Weeds Where Water is Potable

Can Nutrients Be Reduced?		
Yes		No
Can water be held temporarily?		Is drawdown possible?
Yes	No	Excavate
Compatible herbicide	Herbicide with potable use	Dredge
Excavate	Mechanical removal	Foliar herbicide for plants with floating leaves
Dredge	Herbivorous fish	
Line pond bottom	Reduce light	Yes
Foliar herbicide	Plant beneficial competitive plants	Soil-incorporated herbicide
Soil-incorporated herbicide		Line bottom

4.8 Organic Agriculture

There are six components of an integrated weed-management system that could be included when designing a system for any crop/range/land/aquatic/etc. site. They are: prevention, hand weeding, mechanical methods, cultural, biological, and chemical components. Each can be and most are used in organic-crop production. The primary difference between an IWMS for organic agriculture and all other production systems is the fact that synthetic organic and inorganic chemicals (pesticides and fertilizers) cannot be used in organic systems and more management skill and decisions are required (Sørensen et al., 2005). However, all other weed-management/control methods are used. Another difference is that weed management is considered to be part of all crop-production activities rather than a short-term once-a-year activity. Weed management in organic agriculture is a systematic, year-round process that minimizes weed effects and optimizes land use. It combines prevention and control (Aldrich, 1984) to minimize, but not necessarily eliminate, all weeds. Weeds are accepted as a normal, manageable part of the agricultural community and may even be welcomed because of their contribution of organic matter to the soil during tillage. Sørensen et al. included discussion of the use of a weeding robot, and an integrated system for band steaming both increased the capital investment required and reduced labor demand 83%–85% for sugar beet and 60% for carrots and thereby improve profitability 72%–85%. The benefit from robotic weeding was dependent on weed population and initial price of the equipment. The robot was profitable when its weeding efficiency was greater than 25%.

5. WEED-MANAGEMENT DECISION AID MODELING[5]

Weed-crop competition was the first process that weed scientists modeled. Modeling has proven to be good research but has yielded little of practical value. Decision aid models, if present, are not obvious components of the weed-management toolbox. Accurate yield loss estimates are needed to create weed-management decision aid models and to evaluate economic thresholds. Weed-management models have been of two broad types (Lundkvist, 1997; Swinton and King, 1994): research models (to develop an understanding of processes) and practical models (decision aid or weed-management tools). Research models are designed to quantify the effects of the density of one species, usually a crop, on its own yield or biomass production and on the yield or biomass production of a competing weed species (Lundkvist, 1997; Radosevich, 1987). Practical models incorporate scouting and economic thresholds and purport to be decision aids for weed management (Wilkerson et al., 2002).

Lundkvist (1997) concluded that although research models had clarified principles, practical models were still only potential tools—a situation that still prevails. They remain potential primarily because of the regional nature of crops and the weeds that infest them. Models and integrated weed-management systems must reflect regional crop and weed diversity and the diversity of ecosystems in which crops are grown. This is a daunting task (see Gunsolus et al., 2000). Models of crop-weed competition have been categorized

---

[5]Much of the material in this section has been excerpted with permission from Zimdahl, R.L., 2004. Weed-Crop Competition: A Review, second ed. Blackwell Publishing. Ames. IA. 220 pp.

as conceptual, simulation (generally used synonymously with analytical), and empirical (generally used synonymously with mechanistic or ecophysiological). Conceptual models are research tools, developed to provide insight into the competitive process. Most practical models are empirical, and much can be and has been learned from empirical modeling of weed-crop competition (Cousens, 1985b), but modeling has remained a research exercise with limited to no practical application.

Coble and Mortensen (1992) reported the four most common definitions of threshold used in weed science. The threshold to be determined depended on the response measured; it is not a fixed definition. The most common adjectives were damage, economic, period, and action. Damage is used to define the weed population that caused a yield reduction. The economic threshold (Jordan, 1992) is the weed population at which the cost of control is equal to the increase in crop value from control. It is further complicated because it may be used for single- or multiple-season effects. A period threshold implies that there are times in the growth of a crop when weeds are more damaging. The action threshold is often related to the period threshold and is the point at which control is (should be) initiated. Action is usually based on cost but may include risk aversion, desire for clean fields (the neighbor's opinion effect) (Wilkerson et al., 2002), or other considerations. Models, independent of the threshold used, have given primary emphasis to aiding the decision to use or not use an herbicide or another control method—the putative aim of all models. Some models also incorporate mechanical and cultural methods (Wilkerson et al., 2002). Weed science models and modeling emphasize decisions about herbicide use. Attempts to determine the economic threshold have been most common.

Zanin and Sattin (1988) conducted four tests to determine the economic threshold for velvetleaf in corn and velvetleaf seed production with different levels of infestation with and without a corn crop. The economic threshold was calculated with Cousens (1985a) model and varied from 0.3 to 2.4 plants/m^2. Corn reduced velvetleaf seed production by 50%. But even when only four or five velvetleaf/m^2 competed with corn, velvetleaf still produced 8000 to 10,000 seeds/m^2. Zanin and Sattin questioned the value of a threshold density for weed management when one must consider velvetleaf's ecological characteristic that permits great seed production at low density. Cardina et al. (1995) found the single-year economic threshold for velvetleaf in corn ranged from 0.4 to 14 velvetleaf/m^2 in conventional tillage and 0.13 to 3.13/m^2 with no tillage. Cardina et al. also questioned the value of the economic threshold because of seasonal variation and high seed production from subthreshold velvetleaf populations. Economic thresholds that were predicted using yield goal information deviated from the actual threshold values by -43 to $+30\%$.

Roberts and Hayes (1989) proposed a decision criteria model for Johnsongrass control in soybean, based on actual data, which describes the relationship between Johnsongrass density and soybean yield loss. When these data are combined with the cost of control and the expected soybean price, the result can be used to show the weed density threshold at which Johnsongrass control becomes profitable. Toler et al. (1996) tested an additive-response model and a product-response model to predict yield reductions due to Johnson-grass and smooth pigweed interference in soybean. Both models predicted higher soybean yield losses than were observed. When growing conditions were favorable and the competitive effects of weeds were low, both models adequately predicted soybean yield decline. If the weather was dry, the product response model was superior. Smooth pigweed was 80% of the biomass when species were grown together (Toler et al.). The modeling was complicated by the fact that as Johnsongrass density increased the reduction in soybean yield was linear, whereas an exponential response characterized the decrease in soybean yield due to smooth pigweed.

Practical application of single-season economic thresholds for postemergence weed-control decisions have been frustrated by the variable effect of differences in climate between growing seasons, different soils, and variable crop-weed interactions (McDonald and Riha, 1999)—the same factors that plague developers of quantitative models. Simulations showed that when weeds do not emerge before corn, corn will only suffer a yield reduction in two of 10 years (McDonald and Riha, 1999). Therefore, economic thresholds based solely on the level of weed infestation (the density) are inherently flawed. McDonald and Riha advocated shifting the focus from measuring weed density to assessing the competitive status of the crop indirectly with climate information, which would alleviate many of the problems of inaccuracy associated with present threshold-management strategies. This view is supported by the work of McGiffen et al. (1997) who found that economic thresholds for foxtail interference in corn are not constant but vary with weather, cropping system, and soil type. McGiffen and Reicosky offered the pessimistic view that widespread management of weeds with economic thresholds is an unrealistic goal until the stability (i.e., accuracy across years and regions) of interference models improves. Jasieniuk et al. (1999) expressed the same view based on a multistate, multiyear study of crop yield loss—weed density relationships between wheat and jointed goatgrass. Site-to-site and year-to-year variation in winter wheat and jointed goatgrass yield loss parameter estimates demonstrated that management recommendations made by a bioeconomic model cannot be based on a single yield loss function with the same parameter values for different winter wheat-producing regions. Jasieniuk et al. advocated that models would be improved when yield loss functions incorporating time of emergence and crop density are built into the model's structure. Subsequently, Jasieniuk et al. (2001) evaluated three models that empirically predict crop yield from crop and weed density for their fit to 30 data sets from a multistate, multiyear winter wheat—jointed goatgrass study. They used seven criteria to evaluate the models to determine which one best fit the objectives of a bioeconomic model that seeks to identify economic optimum weed management recommendations. The earlier paper (Jasieniuk et al., 1999) used the rectangular hyperbolic model proposed by Cousens (1985a). The later paper (Jasieniuk et al., 2001) compared three modifications of Cousens' model. The first involved the use of two linked hyperbolic equations derived from Cousens' hyperbolic model. The second modification was first proposed for aboveground biomass by Baeumer and de Wit (1968) and, as mentioned by Jasieniuk et al. (2001), was applied to marketable yield by Weiner (1982). It was the best compared to six other models for predicting barley and winter wheat yield (Cousens, 1985b). The third modification involved the use of a model derived from "a crop density—yield loss model" proposed by Martin et al. (1987) who modified Cousens' hyperbolic model. The conclusion of this very detailed manuscript is that no one model was superior unless one selected and defined the criteria of evaluation, that is, they defined what made the model superior. The common choices are the proportion of regressions that converge on a solution and more readily exhibit asymptotic behavior or statistical significance and a linear relationship between yield and crop density under the constraint of limited data. Some, albeit limited, research continues to develop models that combine reliability across years and locations with statistical reliability and conformity to biological reality.

Three computer-based models were developed for corn, cotton, peanut, and soybean (Bennett et al., 2003). The Herbicide Application Decision Support System (HADSS) was designed for office, desktop use; Pocket HERB is designed for onsite, field use; and WebHADSS uses field-specific information to estimate crop yield loss if no weed control is done, to compare and eliminate inappropriate herbicides, and to estimate yield results after recommended treatments. Each model is a curative, herbicide-based system. Each has been modified for use in several southern states and in Canada (Weaver et al., 1999) for site- and weed-specific conditions.

Most recently Lacoste and Powles (2015) evaluated their ryegrass integrated management model. They examined the biological, agronomic, and economic components of the model with particular emphasis on the biological characteristics of ryegrass that are critical for effective management. They acknowledged "extreme variability" of the defined parameters and "subsequent limits" of the model. It required compromises that were achieved by emphasizing the primary end use of the program—supporting, but not making, a grower's or advisor's decision. Their conclusion is similar to that of Bajwa et al. (2015) who noted that because "no single strategy is perfect, an integrated approach may provide better results."

Even with the significant research effort that has been devoted to decision aid models, they are not widely used and are often not superior to a farmer's decision. Swinton et al. (2002) compared three models (Michigan WEEDSIM/GWM, CORNHERB, and SOYHERB) and found no model was statistically superior to weed-management decisions made by farmers unaided by decision aid models. Development of decision aid models "has been and will continue to be, an evolutionary process" (Bennett et al., 2003). Models and modelers will evolve and models will improve. Whether the time and trouble will be cost effective for growers remains to be determined. As Masin et al. (2005) point out for WeedTurf, there is a "possibility of developing interactive computer software to determine the critical timing of weed removal and provide improved recommendations for herbicide application timing." Whether models will ever be superior to the knowledge and experience of growers remains to be determined.

Norris's (1999) extensive survey concluded that in spite of the abundant literature on the effects of weed density and duration on competition (see Zimdahl, 2006), improved computer technology, and decision aid models, the information on weed crop competition has had almost no effect on weed-management practice. Norris argued persuasively for greater emphasis on research on weed biology and the mechanisms of competition. His plea has not resulted in a significant change in the type of research weed scientists do. Norris's view is supported by arguments presented by Wilkerson et al. (2002) who note that models may *not* be necessary because farmers want a weed-free crop, and herbicide-resistant crops, it was assumed, would eliminate the need for models. In addition, an expert can usually make a good and much quicker recommendation without collecting the data models require. Even with the required data, the model may not change the accuracy of a recommendation derived from experience and expertise. Norris also advocated a no-seed production threshold. That is, if no weeds are allowed to produce seed, future problems will be reduced and gradually may be eliminated. A model is not needed to justify a no-seed production threshold. Zero is a difficult goal and achieving it is a decision that may not be aided by today's models. Hill et al. (2016) found that when growth of common lambs quarters, common ragweed, and giant foxtail was terminated prior to flowering, additions to the soil weed seed bank were reduced, thus supporting management to eliminate seed production. Work by Maxwell and Ghersa (1992) with a theoretical model to assess the relative importance of weed competition and seed dispersal on long-term crop yield losses also supported the no-seed threshold concept. Simulations using extant data of green foxtail competition in spring wheat showed that seed dispersal from the invading weed might have more influence on yield than the relative competitive ability of the weed. Maxwell and Ghersa (1992) also suggested that if the weed was uniformly distributed at a high density, seed dispersal was less important relative to competitive ability.

Jones and Medd (2000) support Norris's (1999) concept of no-seed production as the proper goal. They suggest that although economic thresholds are strongly embedded in weed management, perhaps because profit is a primary goal, they may not be the best approach. Jones and Medd suggest a population management approach that includes the "intertemporal effects" of management decisions. The proper focus, in their view, is to manage weed populations over time rather than to minimize the effect of weeds in a crop in a year, which is what most economic thresholds and the associated models advocate. The goal, consistent with Norris's recommendation, is to deplete the seed bank over time. Jones and Medd (2000) tested this approach using wild oat invasion of spring wheat in Australia and found the economic benefits from the population-management approach were significantly greater than the typical economic threshold approach. Sattin et al. (1992) found that the economic threshold for velvetleaf in corn varied between 0.3 and 1.7 plants/m^2. Their findings agree with those of Jones and Medd who suggested the proper focus is one that includes measurements over time. Therefore, one must conclude that a single-season economic threshold is not the best model or management strategy.

6. SUMMARY

There are few, if any, fully integrated, widely applicable weed-management systems. Each developing system must be adapted to local environmental, economic, and farming realities, and therefore no single system will be appropriate for a crop everywhere it is grown. For the foreseeable future, herbicides will continue to be important components of most weed-management systems, with the exception of organic systems. Their use may be reduced as integrated systems become more common and effective. Some research has been done to develop effective, integrated weed-management systems that minimize cost, optimize weed control, and are sustainable with changing economic conditions.[6]

These systems will not solve all weed-management problems. Integrated systems will stabilize weed populations at a low level by employing an array of control techniques. Systems will evolve because of failure to prevent invasion by new weeds, development of resistance to one or more control techniques, and development of an uncontrolled population

[6]A search of Weed Science, Weed Technology and Weed Research using the key words decision aid, decision aids, decision support system, and models found only 19 papers published between 2002 and 2016.

that is favored by a given management system. Weeds will never be eliminated. They can be managed.

The emphasis in modeling has shifted away from weed management to studies of the effect of climate change on invasive species and prediction distribution of invasive species.

THINGS TO THINK ABOUT

1. What are the basic weed-management techniques that should be considered for all weed-management systems?
2. Describe the components of a good weed-management system.
3. Design a weed-management system that includes several techniques for a crop or weed-management situation of your choice.
4. What things other than weeds must be considered in the design of a weed-management system?
5. What information is essential to create a good weed-management system?
6. Explain the reasons for the limited role of biological control in present weed-management systems.
7. Explain the reasons for the dominant role of herbicides in present weed-management systems.
8. Explain reasons for the role of mechanical methods in weed-management systems.
9. How can cultural control techniques be incorporated in weed-management systems?
10. How and why does weed management in organic systems differ from other production systems?
11. Discuss the role of models and modeling in weed management.

Literature Cited

Aldrich, R.J., 1984. Weed-Crop Ecology: Principles in Weed Management. Breton Publishers, North Scituate, MA.

Altieri, M.A., Liebman, M., 1988. Weed Management in Agroecosystems: Ecological Approaches. CRC Press, Boca Raton, FL, p. 354.

Anderson, R.L., 1994. Management strategies for winter annual grasses in winter wheat. In: Murphy, L.S. (Ed.), Proc. Maximum Economic Yield Conference. Wheat Management Conf. Sponsored by the Phosphate and Potash Inst and the Foundation for Agronomic Research, Denver, CO, pp. 114–122.

Anderson, R.L., 2003. An ecological approach to strengthen weed management in the semiarid Great plains. Adv. Agron. 80, 33–62. Academic Press. New York, NY.

Anderson, R.L., 2005. A multi-tactic approach to manage weed population dynamics in crop rotations. Agron. J. 97, 1579–1583.

Appleby, A.P., Olsen, P.D., Colbert, D.R., 1976. Winter wheat yield reduction from interference by Italian ryegrass. Agron. J. 68, 463–466.

Baeumer, K., de Wit, C.T., 1968. Competitive interference of plant species in monocultures and mixed stands. Neth. J. Agric. Sci. 16, 103–122.

Bajwa, A.A., Mahajan, G., Chauhan, B.S., 2015. Nonconventional weed management strategies for modern agriculture. Weed Sci. 63, 723–747.

Bastianns, L., Paolini, P., Baumann, D.T., 2008. Focus on ecological weed management: what is hindering adoption? Weed Res. 48 (6), 481–491.

Bennett, A.C., Price, A.J., Sturgill, M.C., Buol, G.S., Wilkerson, G.G., 2003. HADSS™, Pocket HERB™, and WebHADSS™: decision aids for field crops. Weed Technol. 17, 412–420.

Bingham, S.W., Chism, W.J., Bhowmik, P.C., 1995. Weed management systems for turfgrass. In: Smith, A.E. (Ed.), Handbook of Weed Management Systems. M. Dekker, Inc., New York, NY, pp. 603–665.

Blackshaw, R.E., Stobbe, E.H., Sturko, A.R.W., 1981. Effect of seeding dates and densities of green foxtail (*Setaria viridis*) on the growth and productivity of spring wheat (*Triticum aestivum*). Weed Sci. 29, 212–214.

Blackshaw, R.E., Molinar, L.J., Muendel, H.H., Saindon, G., Li, X., 2000. Weed Technol. 14 (2), 327–336.

Blackshaw, R.E., Moyer, J.A., Harker, K.N., Clayton, G.W., 2005. Integration of agronomic practices and herbicides for sustainable weed management in a zero-till barley field pea rotation. Weed Technol. 19, 190–196.

Bourchier, R., Hansen, H., Lym, R., Norton, A., Olsen, D., Bell, C., et al., 2006. Biology and Biological Control of Leafy Spurge, p. 138.

Bovey, R.W., 1995. Weed management systems for rangeland. In: Smith, A.E. (Ed.), Handbook of Weed Management Systems. M. Dekker, Inc., New York, NY, pp. 519–662.

Bovey, R.W., 2001. Woody Plants and Woody Plant Management: Ecology, Safety and Environmental Impact. M. Dekker, Inc., New York, NY, p. 564.

Boydston, R.A., Vaughn, S.F., 2002. Alternative weed management systems control weeds in potato (*Solanum tuberosum*). Weed Technol. 16, 23–28.

Buhler, D.D. (Ed.), 1999. Expanding the Context of Weed Management. Food Products Press, Binghamton, NY, p. 289.

Buhler, D.D., 2001. Developing integrated weed management systems. In: Blackshaw, R.E., Hall, L.M. (Eds.), Integrated Weed Management: Explore the Potential. Expert Comm. on Weeds. Sainte-Anne-de-Bellevue, Canada, pp. 37–46.

Buhler, D.D., 2002. Challenges and opportunities for integrated weed management. Weed Sci. 50 (3), 273–280.

Buhler, D.D., King, R.P., Swinton, S.M., Gunsolus, J.L., Forcella, F., 1996. Field evaluation of a bioeconomic model for weed management in corn (*Zea mays*). Weed Sci. 44, 915–923.

Burke, I.C., Troxler, S.C., Askew, S.D., Wilcut, J.W., Smith, W.D., 2005. Weed management systems in glyphosate-resistant cotton. Weed Technol. 19, 422–429.

Bussan, A.J., Boerboom, C.M., 2001. Modeling the integrated management of velvetleaf in a corn-soybean rotation. Weed Sci. 49 (1), 31–41.

Cardina, J., Regnier, E., Sparrow, D., 1995. Velvetleaf (*Abutilon theophrasti*) competition and economic thresholds in conventional and no-tillage corn (*Zea mays*). Weed Sci. 43, 81–87.

Chikoye, D., Ekleleme, F., Udensi, U.E., 2001. Cogongrass suppression by intercropping cover crops in corn/cassava systems. Weed Sci. 49, 658–667.

Coble, H.D., Mortensen, D.A., 1992. The threshold concept and its application to weed science. Weed Technol. 6, 191–195.

Coffmann, C.B., Frank, J.R., 1992. Corn-weed interactions with long-term conservation tillage management. Agron. J. 84, 17–21.

Coombs, E.M., Clark, J.K., Piper, G.L., Cofrancesco Jr., A.F., 2004. Biological Control of Invasive Plants in the United States. Oregon State Univ. Press, Corvallis, OR, p. 467.

Cousens, R., 1985a. A simple model relating yield loss to weed density. Ann. Appl. Biol. 107, 239–252.

Cousens, R., 1985b. An empirical model relating crop yield to weed and crop density and a statistical comparison with other models. J. Agric. Sci. (Cambr.) 105, 513–521.

Dewey, S.A., Jenkins, M.J., Tonioli, R.C., 1995. Wildfire suppression – a paradigm for noxious weed management. Weed Technol. 9, 621–627.

Donald, W.W., 1990. Systems of weed control in wheat in North America. In: Weed Sci. Soc. of America. Champaign, IL, p. 488.

Donald, W.W., Eastin, E.F., 1995. Weed management systems for grain crops. In: Smith, A.E. (Ed.), Handbook of Weed Management Systems. M. Dekker, Inc., New York, NY, pp. 401–476.

Doohan, D., Wilson, R., Canales, E., Parker, J., 2010. Investigating the human dimension of weed management: new tools of the trade. Weed Sci. 58, 503–510.

Elmore, C.L., 1985. Ornamentals and turf. In: Principles of Weed Control in California. Thomson Pub, Fresno, CA, pp. 387–397.

Fischer, B.B., Yeary, E.A., Marcroft, J.E., 1985. Vegetation management systems. In: Principles of Weed Control in California. Thomson pub., Fresno, CA, pp. 213–228.

Forcella, F., Lindstrom, M.J., 1988a. Movement and germination of weed seeds in ridge-till crop production systems. Weed Sci. 36, 56–59.

Forcella, F., Lindstrom, M.J., 1988b. Weed seed populations in ridge and conventional tillage. Weed Sci. 36, 500–503.

Forcella, F., King, R.P., Swinton, S.M., Buhler, D.D., Gunsolus, J.L., 1996. Multi-year validation of a decision aid for integrated weed management. Weed Sci. 44, 650–661.

Fryer, J.D., 1985. Recent research on weed management: new light on an old practice. In: Fletcher, W.W. (Ed.), Recent Advances in Weed Research. The Gresham Press, Old Working, Surrey, UK (Chapter 9).

Gunsolus, J.L., Hoverstad, T.R., Potter, B.D., Johnson, G.A., 2000. Assessing integrated weed management in terms of risk management and biological time constraints. In: Kennedy, G.C., Sutton, T.B. (Eds.), Emerging Technologies for Integrated Pest Management: Concepts, Research, and Implementation. APS Press, St. Paul. MN, pp. 373–383.

Harris, P., Dunn, P.H., Schroeder, D., Vormos, R., 1985. Biological control of leafy spurge in North America. In: Watson, A.K. (Ed.), Leafy Spurge Monograph No. 3. Weed Sci. Soc. Am., Champaign, IL, pp. 79–82.

Helmers, G.A., 1986. An economic analysis of alternative cropping systems for East Central Nebraska. Am. J. Altern. Agric. 1, 153–158.

Hill, E.C., Renner, K.A., VanGessel, M.J., Bellinder, R.R., Scott, B.A., 2016. Late-season weed management to stop viable weed seed production. Weed Sci. 64, 112–118.

Hoeft, E.V., Jordan, N., Zhang, J., Wyse, D.L., 2001. Integrated cultural and biological control of Canada thistle in conversation tillage soybean. Weed Sci. 49, 642–646.

Jackson, W., 1984. Toward a unifying concept for an ecological agriculture. In: Lowrance, R., Stinner, B.R., House, G.J. (Eds.), Agricultural Ecosystems. J. Wiley, NY, pp. 209–221.

Jasieniuk, M., Maxwell, B., Anderson, R.L., Evans, J.O., Lyon, D.J., Miller, S.D., Morishita, D.W., et al., 1999. Site-to-site and year-to year variation in *Triticum aestivum-Aegilops cylindrica* interference relationships. Weed Sci. 47, 529–537.

Jasieniuk, M., Maxwell, B., Anderson, R.L., Evans, J.O., Lyon, D.J., Miller, S.D., Morishita, D.W., et al., 2001. Evaluation of models predicting winter wheat and jointed goatgrass densities. Weed Sci. 49, 48–60.

Jones, R.E., Medd, R.W., 2000. Economic thresholds and the case for longer-term approaches to population management of weeds. Weed Technol. 14, 350–378.

Jordan, N., 1992. Weed demography and population dynamics: implications for threshold management. Weed Technol. 6, 184–190.

King, R.P., Lybecker, D.W., Schweizer, E.E., Zimdahl, R.L., 1986. Bioeconomic modeling to simulate weed control strategies for continuous corn (*Zea mays*). Weed Sci. 34, 972–979.

Kolbert, E., July 21, 2008. Turf War — Americans Can't Live without Their Lawns—but How Long Can They Live with Them? The New Yorker, pp. 80–86.

Lacoste, M., Powles, S., 2015. RIM: anatomy of a weed management decision support system for adaptation and wider application. Weed Sci. 63 (3), 676–689.

Laufenberg, S.M., Sheley, R.L., Jacobs, J.S., Borkowski, J., 2005. Herbicide effects on density and biomass of Russian knapweed (*Acroptilon repens*) and associated plant species. Weed Technol. 19, 62–72.

Leininger, W.C., 1988. Non-chemical Alternatives for Managing Selected Plant Species in the Western United States. Colo. State Univ. Coop. Ext. Ser. Pub. No. XCM-118, p. 40.

Lewis, W.J., van Lenteren, J.C., Phatak, S.C., Tumlinson III, J.H., 1997. A total system approach to sustainable pest management. Proc. Nat. Acad. Sci. USA 94, 12243–12248.

Liebman, M., Davis, A.S., 2009. Managing weeds in organic farming systems: an ecological approach. In: Francis, C.F. (Ed.), Organic Farming: The Ecological System. Monograph 54. American Society of Agronomy, Madison, WI, pp. 173–196.

Liebman, M., Gallandt, E.R., 1997. Many little hammers: ecological approaches for management of crop-weed interactions. In: Jackson, L.E. (Ed.), Ecology in Agriculture. Academic Press, San Diego, CA, pp. 291–343.

Liebman, M., Gibson, L.R., Sundberg, D.N., Hegenstaller, A.H., Westerman, P.R., chase, C.A., et al., 2008. Agronomic and economic performance characteristics of conventional and low-external-input cropping systems in the central corn belt. Agron. J. 100 (3), 600–610.

Lundkvist, A., 1997. Weed management models: a literature review. Swedish J. Agric. Res. 27, 155–166.

Lybecker, D.W., Schweizer, E.E., Westra, P., 1991. Computer Aided Decisions for Weed Management in Corn. West. Agri. Econ. Assoc, Portland, OR, pp. 234–239.

Lybecker, D.W., Schweizer, E.E., Westra, P., 1992. Reducing Herbicide Loading in Corn with Weed Management Models. Abst., Div. Environ. Chem., Amer. Chem. Soc., San Fran., CA.

Martin, R.J., Cullis, B.R., McNamara, D.W., 1987. Prediction of wheat yield loss due to competition by wild oats (*Avena* spp.). Aust. J. Agric. Res. 38, 487–499.

Martin, M.A., Schreiber, M.M., Riepe, J.R., Bahr, J.R., 1991. The economics of alternative tillage systems, crop rotations, and herbicide use on three representative East-Central corn belt farms. Weed Sci. 39, 299–307.

Masin, R., Zuin, M.C., Archer, D.W., Forcella, F., Zanin, G., 2005. WeedTurf: a predictive model to aid control on annual summer weeds in turf. Weed Sci. 53, 193–201.

Masters, R.A., Nissen, S.J., 1998. Revegetating leafy spurge (*Euphorbia esula*)-infested rangeland with native tallgrasses. Weed Technol. 12, 381–390.

Maxwell, B.D., Ghersa, C., 1992. The influence of weed seed dispersion versus the effect of competition on crop yield. Weed Technol. 6, 196–204.

Maxwell, B.D., Luschei, E.C., 2005. Justification for site-specific weed management based on ecology and economics. Weed Sci. 53, 221–227.

McDonald, A.J., Riha, S.J., 1999. Model of crop-weed competition applied to maize:*Abutilon theophrasti* interactions. II. Assessing the impact of climate: implications for economic thresholds. Weed Res. 39, 371–382.

McGiffen Jr., M.E., Forcella, M., Lindstrom, J., Reicosky, D.C., 1997. Covariance of cropping systems and foxtail density as predictors of weed interference. Weed Sci. 45, 388–396.

McNabb, T., Anderson, L.W.J., 1985. Aquatic weed control. In: Principles of Weed Control in California. Thomson, Fresno, CA, pp. 440–455.

McNabb, T., Smith, D.B., Mitchell, R.J., 1995. Weed management systems for forest nurseries and woodland. In: Smith, A.E. (Ed.), Handbook of Weed Management Systems. M. Dekker, Inc., New York, NY, pp. 667–711.

Mitich, L.W., 1991. Alfalfa. In: Principles of Weed Control in California. Thomson Pub. Inc., Fresno, CA, pp. 232–237.

Mortensen, D.A., Dieleman, J.A., Johnson, G.A., 1998. Weed spatial variation and weed management. In: Hatfield, J.L., Buhler, D.D., Stewart, B.A. (Eds.), Integrated Weed and Soil Management. Ann Arbor Press, Chelsea, MI, pp. 293–309.

Mortensen, D.A., Bastianns, L., Sattin, M., 2000. The role of ecology in the development of weed management systems: an outlook. Weed Res. 40, 49–62.

Mullahey, J.J., Mislevy, P., Brown, W.F., Kline, W.N., 1996. Tropical soda apple, an exotic weed threatening agriculture and natural systems. Down Earth 51 (1), 10–17.

Norris, R.F., 1992. Have ecological and biological studies improved weed control strategies?. In: Proc. Int. Weed Cont. Cong., Melbourne, Australia, pp. 7–29.

Norris, R.F., 1999. Ecological implications of using thresholds for weed management. J. Crop Prod. 2, 31–58.

Norris, R.F., Caswell-Chen, E.P., Kogan, M., 2003. Concepts in Integrated Pest Management. Prentice Hall, Upper Saddle River, NJ, p. 586.

O'Donovan, J.T., Blackshaw, R.E., Harker, K.N., Clayton, G.W., 2006. Wheat seeding rate influences herbicide performance in wild oat (*Avena fatua* L.). Agron. J. 98, 815–822.

Radosevich, S.R., 1987. Methods to study interactions among crops and weeds. Weed Technol. 1, 190–198.

Roberts, R.K., Hayes, R.M., 1989. Decision criterion for profitable johnsongrass (*Sorghum halepense*) management in soybeans (*Glycine max*). Weed Technol. 3, 44–47.

Sattin, M., Zanin, G., Berti, A., 1992. Case history for weed competition/population ecology: velvetleaf (*Abutilon theophrasti*) in corn (*Zea mays*). Weed Technol. 6, 213–219.

Sellers, B.A., Ferrell, J.A., 2015. Weed Management in Pastures and Rangeland—20161. The Institute of Food and Agricultural Sciences, Univ. Of Florida, p. 16.

Shaw, W., 1982. Integrated weed management systems. Weed Sci. 30 (Suppl. 2), 2–12.

Shrestha, A., Rajcan, I., Chandler, K., Swanton, C.J., 2001. An integrated weed management strategy for glufosinate-resistant corn (*Zea mays*). Weed Technol. 15, 517–522.

Smeda, R.J., Weston, L.A., 1995. Weed management systems for horticultural crops. In: Smith, A.E. (Ed.), Handbook of Weed Management Systems. M. Dekker, Inc., New York, NY, pp. 553–601.

Smith, A.E., 1995. Handbook of Weed Management Systems. M. Dekker, Inc., N.Y., p. 741

Smith, A.E., Martin, L.D., 1995. Weed management systems for pastures and hay crops. In: Smith, A.E. (Ed.), Handbook of Weed Management Systems. M. Dekker, Inc., New York, NY, pp. 477–517.

Smith, M.C., Valverde, B.E., Merayo, A., Fonseca, J.F., 2001. Integrated management of itchgrass in a corn cropping system: modeling the effect of control tactics. Weed Sci. 49, 123–134.

Sørensen, C.G., Madsen, N.A., Jacobsen, B.H., 2005. Organic farming scenarios: operational analysis and costs of implementing innovative technologies. Biosyst. Eng. 91, 127–137.

Strobel, G.A., July 1991. Biological control of weeds. Sci. Am. 72–78.

Swanton, C.J., Shrestha, A., Clements, D.R., Booth, B.D., Chandler, K., 2002. Evaluation of alternative weed management systems in a modified no-tillage corn-soybean-winter wheat rotation: weed densities, crop yield and economics. Weed Sci. 50, 504–511.

Swinton, S.M., King, R.P., 1994. A bioeconomic model for weed management in corn and soybean. Agric. Syst. 44, 313–335.

Swinton, S.M., Renner, K.A., Kells, J.J., 2002. On-farm comparison of three postemergence weed management decision aids in Michigan. Weed Technol. 16, 691–698.

Toler, J.E., Guice, J.B., Murdock, E.C., 1996. Interference between johnsongrass (*Sorghum halepense*), smooth pigweed (*Amaranthus hybridus*), and soybean (*Glycine max*). Weed Sci. 44, 331–338.

Toler, J.E., Murdock, E.C., Keeton, A., 2002. Weed management systems for cotton (*Gossypium hirsutum*) with reduced tillage. Weed Technol. 16, 773–780.

Treadway-Ducar, J., Morgan, G.D., Wilkerson, J.B., Hart, W.E., Hayes, R.M., Mueller, T.C., 2003. Site-specific weed management in corn (*Zea mays*). Weed Technol. 17, 711–717.

Van Acker, R., 2001. Pesticide-free production: a reason to implement integrated weed management. In: Blackshaw, R.E., Hall, L.M. (Eds.), Integrated Weed Management: Explore the Potential. Expert Comm. on Weeds, Sainte-Anne-de-Bellevue, Canada, pp. 61–73.

Van Wychen, L., 2016. 2015 Survey of the Most Common and Troublesome Weeds in the United States and Canada. Weed Science Society of America National Weed Survey Dataset. Available. http://wssa.net/wp-content/uploads/2015-Weed-Survey_final.xlsx.

Vengris, J., Torello, W.A., 1982. Lawns — Basic Factors, Construction and Maintenance of Fine Turf Areas, third ed. Thomson Pub, Inc., Fresno., CA, pp. 133–167.

Weaver, S., Sturgill, M.C., Wilkerson, G.G., Coble, H.D., Buol, G.C., 1999. HADSS User's Manual. Ontario Version 2.0. Harrow, Ontario, Canada Agriculture and Agri-Food, p. 24.

Webster, T.M., 2005. Weed survey — southern states: broadleaf crops subsection. In: Vencill, W.K. (Ed.), Proceedings Southern Weed Science Society, pp. 291–306.

Webster, T.M., 2008. Weed survey — southern states: grass crop subsection. In: Vencill, W.K. (Ed.), Proceedings Southern Weed Science Society, pp. 224–243.

Webster, T.M., 2009. Weed survey — southern states: broadleaf crops subsection. In: Webster, T.M. (Ed.), Proceedings Southern Weed Science Society, pp. 509–524.

Webster, T.M., Nichols, R.L., 2012. Changes in the prevalence of weed species in the major agronomic crops of the southern United States: 1994/1995 to 2008/2009. Weed Sci. 60, 145–157.

Weiner, J., 1982. A neighbourhood model of annual plant interference. Ecology 63, 1237–1241.

Weiner, J., 1990. Plant population ecology in agriculture. In: Carroll, C.R., Vandermeer, J.H., Rosset, P.M. (Eds.), Agroecology. McGraw-Hill, NY, pp. 235–262.

Westbrooks, R.G. (Ed.), 1998. Invasive Plants, Changing the Landscape of America: Fact Book. Federal Interagency Committee for the Management of Noxious and Exotic Weeds, Washington, DC, p. 109.

Wicks, G.A., Smika, D.E., 1990. Central great plains. In: Donald, W.W. (Ed.), Systems of Weed Control in Wheat in North America. Weed Sci. Soc. Am., Champaign, IL, pp. 127–157.

Wilcut, J.W., York, A.C., Jordan, D.L., 1995. Weed management systems for oil seed crops. In: Smith, A.E. (Ed.), Handbook of Weed Management Systems. M. Dekker, Inc., New York, NY, pp. 343–400.

Wilkerson, G.G., Wiles, L.J., Bennett, A.C., 2002. Weed management decision models: pitfalls, perceptions, and possibilities of the economic threshold approach. Weed Sci. 50, 411–424.

Wyse, D.L., 1992. Future of weed science research. Weed Technol. 6, 162–165.

Young, S.L., 2012. True integrated weed management. Weed Res. 52, 107–111.

Zanin, G., Sattin, M., 1988. Threshold level and seed production of velvetleaf (*Abutilon theophrasti*) in maize. Weed Res. 28, 347–352.

Zimdahl, R.L., 2006. Agriculture's Ethical Horizon. Academic Press, San Diego, CA, p. 235.

Weed Science: The Future

FUNDAMENTAL CONCEPTS

- Weed science has a rich, productive history and weed management will continue to contribute to agriculture and other disciplines where weeds occur.

- There are many new research areas in weed science that create a challenging future.

LEARNING OBJECTIVES

- To know the promising areas for weed science research.

- To understand that weed science is an evolving discipline.

- To understand that research opportunities in weed science and its potential contributions

to food production and alleviation of world hunger are great.

- To understand the opportunities and problems related to use of biotechnology in weed science.

But you who seek to give and merit fame,
And justly bear a critic's noble name,
Be sure yourself and your own reach to know,

How far your genius, taste, and learning go;
Launch not beyond your depth, but be discreet,
And mark the point where sense and dullness meet. *Alexander Pope, An Essay on Criticism, Part I*

Those who have read this far know the end is near, but the end of the book is not the end of the story. This book has presented a brief, accurate history of weed science and described the present situation. Now it is time to think a bit about the future. Sir Winston Churchill said he always "avoided prophesying before hand, because it is a much better policy to prophesy after the event has already taken place." Available Google evidence suggests that the Danish physicist Niels Bohr actually was the first to say, "Prediction is difficult, especially about the future," or something very like that. It has been repeated often and its first use has been incorrectly attributed to Yogi Berra and Mark Twain, who in his autobiography said "Prophecies which promise valuable things, desirable things, good things, worthy things, never come true." However, one should not fear expressing what might be regarded as an eccentric or wild, potentially wrong opinion. Bertrand Russell said, "do not fear to be eccentric in opinion, for every opinion now accepted was once eccentric."

When the past is known and the present is understood it is tempting to try to glimpse the future. What follows will be a glimpse because as Bohr reminds us the future cannot be predicted with absolute confidence, but some trends can be seen. This will be, one hopes, not beyond my depth and not dull. I will not promise, only explore what might be. To begin, I assert that weed science, although young among the agricultural sciences, has an enviable record of achievement and a good future.

The practice of weed control was a recognition of necessity by farmers who had been controlling weeds long before herbicides were invented. The advent of herbicides changed the way weed control was done but didn't change its fundamental purpose—to improve yield of desirable species. Herbicides replaced human, animal, and mechanical energy with chemical energy. No other method of weed control, before herbicides, was as efficient at reducing the need for labor or as selective. People with hoes could discriminate between weeds and crops and weed selectively. Mechanical and cultural methods, while effective, were not selective enough. Herbicides enabled prevention, reduced the weed population, and selectively removed weeds from crops. Other methods could do these things but not as well, as easily, or as cheaply. Weed control in the world's developed countries now depends on herbicides as the primary technology. This situation will prevail well into the 21st century.

1. RESEARCH NEEDS

There are at least three important problems that have and may continue to hinder progress in Weed Science. First is the assumption that anyone can control weeds. Those who make this assumption understand neither the complexity of weed problems nor their solutions. They do not know how much specialized knowledge is required to control weeds correctly. Those who study weeds and their control know how wrong this assumption is. The second problem is that weeds have been and will continue to be components of agriculture and the environment. They lack the appeal and urgency of sudden, serious infestations of other pests. Other pests are serious but that does not mean they are more serious than weeds or deserve more attention. If a problem is always there, it doesn't receive the attention or funding that new, obvious, but perhaps no more serious problems receive. A reasonable analogy can be drawn with thoughts about world hunger. Famines, that is food production, receive the greatest attention. But persistent, widespread hunger and malnutrition, the most serious aspects of world hunger, are not just production problems. They are also, and more importantly, social, political, and moral problems. The third problem is lack of people and research funds. A great deal of research is done by large herbicide development and sales corporations. It is good work but it is inevitably and understandably oriented toward sales. Research on weed biology, ecology, seed dormancy, and other problems that leads to basic understanding rather than immediate control is done by too few scientists. Careful observers will note that most of the literature cited in Chapters 4 and 5 of this book is old. Research is neither being done nor planned. Public funding of agricultural research and specifically weed research is not growing. There continues to be a rapid increase in the research capability of private sector corporations, but publicly funded agricultural research and development is and, it seems, will remain, a minor component. In 1985, Buttel observed that private sector research and funding was driving the nature of public sector research rather than the opposite. In agriculture this is still true. Weed science has few scientists and too few of them work on long-term, sustainable approaches to weed management. Research in weed science now emphasizes invasive species and herbicide resistance.

In 2000, Hall et al. offered their thoughts on future directions for weed science. They offered nine advantageous research areas. In short, they wanted weed scientists to do a little bit of everything.

In 2008, Moss charged that the overall balance direction to much of weed research was wrong. There was too much emphasis on scientific effect at the expense of practical application. He suggested that his colleagues lacked an awareness of the complexities and resources needed to translate research results into actions for farmers. In 2014, Ward et al. charged that two broad aims were driving weed science research: improved weed management and improved understanding of weed biology and ecology. They granted that "excellent work is being done." However, they also charged that "agricultural weed research has developed a very high level of repetitiveness, a preponderance of purely descriptive studies, and has failed to clearly articulate novel hypotheses linked to established bodies of ecological and evolutionary theory." Although they accepted that studies of weed management remain important, they urged weed scientists to recognize the benefits of deeper theoretical justification, a broader vision, and increased collaboration across diverse disciplines. One must conclude that perhaps weed science research is not as good as many colleagues think it should be.

This chapter will deal with the future of weed science in terms of research possibilities rather than in terms of what will be accomplished. It is conjecture, not prophecy. It might be best conceived as a proposal of what ought to be done. It may not be what will be done because research does not always follow a straight path and other developments may change what ought to be done. For example, environmental legislation that mandated reduced herbicide use could rapidly change the way agriculture is practiced. A description of research needs is a safer prophetic stance. It describes what could be done rather than describing what the situation will be several years hence. This approach, of course, reduces the possibility that the prophet will be wrong.

1.1 Weed Biology

1.1.1 Weed Biology and Seed Dormancy

Weed scientists know that dependence on herbicides is equivalent to treating the symptoms of a disease without actually curing the disease, but there has been little choice. Agriculture would be far better served if weed scientists learned how to control weed seed dormancy and seed germination so weeds could be prevented, rather than controlled after they appear. No one knows enough about weed seed dormancy and much research remains to be done to reach the prevention goal.

Entomology and plant pathology began as scientific disciplines long before weed science (Zimdahl, 2010). Both began before insects and plant diseases could be controlled. Both were begun by men (all were men) who studied and described the organisms. Their goal was not control, which was beyond their capability. The goal was understanding. In contrast, weed science began when weeds could be controlled. Empirical herbicide testing made it easy to control weeds without studying their biology. Those who attempt to control must know what weeds are to be controlled and where they are growing. That is, control is not blind. There is an object to be controlled and it is known. But, with herbicides, while it has been necessary to know something (what is that weed?), very little more was required. Whereas when science made it possible to control insects and diseases, entomologists and plant pathologists knew a great deal about life history, stages of growth, anatomy, habitat, etc. of the organisms to be controlled.

In general, herbicide development has neither exploited weak points in a plant's life cycle nor used specific physiological knowledge for control purposes. The safest approach has been to aim for, although it is rarely achieved, complete control of weeds in a crop. As knowledge grows, scientists find that a few weeds are not injurious in some crops and control is not necessary. Some plants now considered weeds, may be beneficial and should not be controlled. The required knowledge includes the entire life cycle of a weed and knowledge of the dormancy of its seeds and vegetative reproductive organs. A marvelous series of projects could be developed on the biology of perennial weeds. These projects should include the objectives mentioned by Wyse (1992):

Regulation of weed seed dormancy
Regulation of bud dormancy of perennials
The development and life of reproductive propagules of perennials
Weed population genetics
Modeling of crop-weed systems

Several research programs are oriented toward modeling crop-weed systems. Models take several forms and several are available (see Chapter 6, Section 11 and Chapter 22, Section 5). Models attempt to combine effective use of several weed management techniques aided by computer technology to find strategies that minimize control efforts and herbicide use. Models routinely include knowledge of the size of the soil seed bank, rate of seed emergence, and seedling survival. It is only logical to assume control and management methods will improve as knowledge of weed biology and seed dormancy improves.

2. WEED-CROP COMPETITION AND WEED ECOLOGY

Much of the basic information required to develop computer-based models of weed-crop systems that lead to best use of available weed control techniques has come and will continue to be derived from weed biology and ecology research. What plants compete for and when competition is most severe between crops and weeds is known in sufficient detail to be useful in development of weed-management systems. The old, but still used, period threshold concept of weed competition (Dawson, 1965) affirms that weed competition is nearly always time dependent. Seedling weeds in a crop very early in the season (i.e., at emergence) are less detrimental than those that compete with the crop later on. This principle led to timely use of herbicides and other techniques for weed management. Some crop cultivars are more competitive than others, and this needs to be considered in developing weed-management systems and is a basis for cooperative work with plant breeders, very little of which has been done.

Weed populations change with time, and the reasons are beginning to be understood. A major reason that presently dominates weed research is the development of weeds resistant to some herbicides, often after only a few years use in one field (see Chapter 19). Active research on why resistance occurs is coupled with development of techniques to combat it. The chances of selection for resistance are increased when a persistent herbicide with a single site of action is used for several years. When resistance has developed it has not led to totally unmanageable weed populations because other weed-control techniques (e.g., cultivation or crop rotation) are available. Other reasons that weed populations change with time are found in the study of weed ecology, a relatively new area of emphasis in weed science. Understanding why populations change and the weed-management implications of population shifts is important to development of successful, sustainable weed-management systems. As mentioned in Chapter 19, a weed science subdiscipline has appeared—resistance management. It is an accepted component of integrated weed management (IWM) and may have become the most important. There is no doubt about its importance. However, as Harker et al. (2012) note, the best way to reduce selection pressure for herbicide resistance is to reduce herbicide use, but the dominant weed-management programs all seem to advocate more herbicide use. The best way to manage resistance is a question that has not been answered.

Even the casual observer of the world of weeds will recognize that weed problems change (see Van Wychen, 2016). Some of the most difficult weeds to control in most crops today are not those that were important 10 or 20 years ago. That can be interpreted as evidence that weed scientists have developed successful solutions to some weed problems. However, it is also true that many common weeds (e.g., pigweeds, lambs quarters, velvetleaf, Canada thistle, cheatgrass) have been targets of control programs for many years. Thus, we have simultaneous evidence of success and continuing problems. It is also evidence that nature abhors empty niches. When successful control efforts have reduced the population of, a species they have inevitably left space unoccupied and resources unused. Other species move into empty niches created by successful weed control.

Solutions to this dilemma take two forms. The first solution is to reduce the attractiveness of the niche. Farmers typically overprovide for crops. Fertilizer placement and precise rate recommendations have reduced surplus nutrients, but nitrogen runoff due to excessive application is a significant problem with notable externalities.[1] Whole fields are irrigated and light cannot be controlled. If water could be placed (e.g., drip irrigation) as precisely as fertilizer

[1]The main cause of the Gulf of Mexico hypoxic (<2 ppm dissolved O_2) or dead zone is excess nutrients, especially nitrogen, washed into the Gulf from the Mississippi River. The river drains 40% of the United States and carries 1.6 million metric tons of nitrogen, more than half of which enters the river in the form of fertilizer runoff. The rest comes from animal waste, industrial and sewage treatment wastewater, and atmospheric pollution. In 1988, the Gulf hypoxic zone was 15 mi^2, in 2002 it was 8400 mi, in 2015 it was 6474 mi^2, larger than Connecticut. It is the world's second largest hypoxic area.

and only as much of each that was needed was provided, the attractiveness of the niche and the success of potential invaders could be reduced. This is a preventive approach to weed management. The second approach has an element of prevention. Few look carefully, but some of the important problem weeds of the next decade are already in fields or lurking on the edges. If they were identified and their weedy potential determined, weed scientists could try to predict which ones would be successful invaders and they could be controlled or managed before they invaded. Invasive plant management is now a major area of research for weed scientists as indicated by the 2008 launch of the journal Invasive Plant Science and Management (see Chapter 8).

More basic biological-ecological knowledge is essential to either approach. Without this knowledge, weed science may be doomed to endure the Red Queen effect (a character in Lewis Carroll's classic book Through the Looking Glass). The Red Queen tells Alice, "In this place, it takes all the running you can do to keep in the same place." Weeds and our ability to control them, especially with herbicides, seem to be evolving at about the same rate. In trying, and often succeeding, to eliminate weeds from fields, weed scientists have created, in a sense, better, more ecologically successful weeds as well as negative environmental effects of dominant control technologies.

3. ALLELOPATHY

Allelopathy could be discussed with herbicide technology (the following section). It is a well-known plant phenomenon that has been studied for a long time but has not been commercially exploited (see Chapter 9). It is intriguing that plants have natural chemical defenses that could be discovered and might be exploited as herbicides. It is one of those things that seems to be forever just beyond our grasp. Maybe there has been too little research. Maybe the responsible chemicals are very common in nature, not selective enough, or not active enough for commercial use. Maybe some observed effects are not allelopathy at all. The lure remains. It will take time and good research to discover if a natural herbicide (an allelochemical) can be found and developed for commercial use and to ask if they are, in any way, environmentally superior to synthetic organic herbicides.

3.1 Biological Control

What if the agrichemical industry had never developed? Just suppose that someone had discovered an organism in 1944 that selectively controlled annual broadleaved weeds in small grains. What would weed science be like? Would we recommend as many herbicides as we now do? Would our environmental problem be what to do with mutating organisms rather than polluting chemicals? Hypothetical questions lead to interesting discussions, but they don't always solve problems. Biological control is still in its infancy compared to other control methods (see Chapter 12). Its theoretical potential, unrealized in all but a few cases, has had little effect on IWM systems for agronomic or horticultural crops. Future research may discover specific biological control organisms and combinations of organisms that are effective and safe and can be integrated with other methods of weed management in crops. There aren't many now.

Those who understand the techniques of biological weed management readily acknowledge that it is not devoid of environmental concerns. However, the "massive accumulation of environmental problems—air pollution, acid rain, the greenhouse effect, ozone depletion, climate change—suggests that few of our citizens and virtually none of our politicians have seriously considered that the very meaning of progress in the future must be different" (Bellah et al., 1991). What the public may regard as the thoughtless exploitation of the earth at the expense of future generations will not be tolerated, and weed management may have to rethink and redefine what proper weed control is.

3.2 Weed Control

Rotation of crops in a field is an effective way to reduce weed competition and will be a more important part of future weed science research. Future recommendations may include use of more smother and green manure crops to keep land covered and protected from wind and water erosion more of the year. Research may demonstrate that in addition to their environmental value, such crops may have economic value. It is well known that rotations reduce soil erosion and manage weeds, but the research has not been done so that one can say exactly what weeds will be controlled effectively and how rotations can be used to greatest advantage. Rotational research will have to include research on the relative competitiveness of crop cultivars and weeds.

Cultivation, from plowing to interrow tillage, has been part of growing crops and weed management for years, but it has been employed without knowing its full potential in integrated-management systems. Cultivation will continue to be part of many crop and weed-management systems, but its potential to encourage soil erosion must be recognized. Worldwide, 2 billion hectares of soil (greater than the area of the United States and Canada combined) has been degraded (Cummings, 2006). Sir Albert Howard's 1940 book An Agricultural testament criticized the rise of what he called "scientific agriculture." Following the work of Leibig in the mid-19th century, many thought that all plants needed from soil was the correct quantities of nutrients. Without Liebig's work and the development of nitrogen fertilizer, it is highly probable that 40% of humanity would not be here.

Howard was concerned with the success of farmers and with feeding people. He was confident that both were more dependent on the health of the soil upon which, in his view, the health of a nation depended. Dr. Rick Haney, a US Department of Agriculture (USDA) soil scientist, agrees with Howard (Ohlson, 2016). Haney claims that "Our entire agricultural industry is based on chemical inputs, but soil is not a chemistry set, it's a biological system. We've have treated it like a chemistry set because chemistry is easier to measure and study than the soil's biology." Not applying chemicals is the key to sustainable food production (Ohlson). To reduce soil health to a few chemical inputs was the worst of reductionist science (Shapin, 2006). Nature has always had the science right (Haney in Ohlson).

In Africa, where many of the world's persistently hungry people and difficult weed problems are, three-quarters of arable land is severely degraded (Cummings, 2006). The vast majority—99.7% of human food comes from cropland—is shrinking by more than 10 million hectares (almost 37,000 square miles) a year due to soil erosion. Worldwide, erosion removes about 75 billion tons of soil a year. The economic effect of soil erosion in the United States costs about \$37.6 billion each year in productivity losses. In Africa, if current trends of soil degradation continue the continent may be able to feed only one-quarter of its population by 2025. The cost of soil erosion in the United States may be as high as \$44 billion and \$45.5 billion in the 24 European Union countries. As a result of erosion over the past 40 years, 30% of the world's arable land has become unproductive.[2] Weed management has a role in solving this problem. It may be possible, as unlikely as it may seem, to farm land without creating a weed problem (Faulkner, 1943). Soil tillage routinely buries weed seeds for future recovery and growth. That does not seem to be good weed management.

Research will have to be done to determine the effect of cultivation timing with different types of implements on specific weeds. Much of the work will emphasize modeling of the crop-weed system to determine optimum timing for different weed-management techniques during the crop's life (Wyse, 1992).

3.3 Bioeconomic Models

Models and modeling were discussed in Chapter 6. In the early 1980s, two decision aid, computer-based models were available. By the early 1990s, 21 mostly crop-based, weed-control-management, decision aid computer-based software models were available to researchers and farmers (Mortensen and Coble, 1991).

Available models will be refined and new models will be developed. Some weed scientists suggest the ideal is a mechanistic weed-crop competition model that considers and responds to changing environments. The essential knowledge of weed biology to construct such models is not yet available (Schweizer et al., 1998; Weaver, 1993). In most models, weed density is the sole variable used to estimate crop yield loss. Future models will incorporate a variable for relative time of crop and weed(s) emergence, crop density, climate variation, method and amount of fertilizer applied, and weed density (O'Donovan, 1996). Rather than just predicting crop loss, as important as that is, future models will also enable realistic monitoring of weed population development and long-term implications of failure to control, or, to say it another way, the long-term implication of seed production by uncontrolled weeds (O'Donovan, 1996).

As more biological knowledge of weeds becomes available, as weed seed–sampling techniques improve, and as models improve, they will be used by decision makers. Weed-management decisions guided by models will lead away from prophylactic control methods. Such methods have accounted for the clear and repeatable success of ridding a field of weeds for a season. They have not provided long-term control of weeds in several crops in economically efficient, environmentally sound ways. Models and modeling have emphasized herbicides that have been profitable and efficacious. These will continue to be important criteria, but long-term weed-management success will become a more important criterion of success. As new knowledge is incorporated, computer-based, weed-management decision aid models will provide greater assurance of achieving profitability and appropriate long-term weed management. Some of the knowledge required to develop better models is shown in Table 23.1.

---

[2]http://www.news.cornell.edu/stories/march06/soil.erosion.threat.ssl.html.

TABLE 23.1   Examples of Knowledge Required to Develop Improved Weed-Management Systems and Decision Aid Models.

Management Goal	Research Needed
Management decision aid models	Relationship of the size of the weed seed bank to final weed population
	Emergence rate of individual weed species
	Determination of economic optimum thresholds for control
	Interaction of management practice and weed seed production
	Effect of weed density on control
Prediction of seedling emergence	Mechanisms of dormancy
	Determination of interaction of environmental conditions, seed
	germination, and dormancy
Effect of management	Effect of crop rotation of weed seed bank size on weed seed
	Effect of living and dead mulches
	Rate of seed predation and decay
	Rate of seedling mortality
	Light requirement for seed germination
	Role of tillage and cultural practices

Adapted from Buhler, D.D., Hartzler, R.G., Forcella, F., 1997. Implications of weed seed bank dynamics to weed management. Weed Sci. 45, 329–336.

3.4 Herbicide Technology

Herbicides are the most successful weed-control technology ever developed. They are selective, not too expensive compared to other techniques, fairly easy to apply, have persistence that can be managed, and many formulations and chemical types are available. In spite of their many advantages, herbicides are far from perfect even in the eyes of their staunchest advocates. Ideal herbicides are not available. The ideal herbicide would be:

- Effective on a spectrum of weeds, or able to control a single species selectively.
- Very selective in at least one major crop and some minor crops.
- Not toxic to nontarget species or humans.
- Persistent in soil but not beyond the period of intense weed competition.
- Easily and quickly degraded to innocuous breakdown products by soil microorganisms or nonenzymatic soil processes.
- Applied postemergence to avoid the prophylactic nature of preemergence use.
- Active at very low rates (mg or g/ha = ounces or fractions of an ounce/A).
- Not leachable and not volatile.

There are some that meet nearly all of these criteria but none are perfect and much remains to be done to develop better herbicides. There is no adequate justification for abandoning herbicides for weed management. There is adequate justification for reducing their use.

3.4.1 *Study of Plant Biochemistry and Physiology*

It is known that dichlobenil and isoxaben affect a plant's ability to synthesize cellulose. Neither was developed to do so. It is serendipitous that they interfere with a specific plant process. There are four ways to discover new herbicides (Beyer, 1991; Evans, 1992):

Method 1. Random selection of chemicals submitted to targeted biological screening.
   This relies on carefully developed, targeted biological-screening techniques to detect chemicals with activity and selectivity. Candidate chemicals are obtained from a company's several divisions or purchased. This is often referred to as the "blue sky" approach (Evans, 1992).
Method 2. Screening of chemical derivatives of herbicides with known activity.
   Once activity is found in method 1, derivative or chemical analog development ensues and further screening attempts to gain more activity or greater or different selectivity.
Method 3. Development of leads taken from natural products with biological activity. Nature can be viewed as an intense arena of complex chemical activity. Allelopathy is one. Biologists and chemists can find clues in this chemistry that could lead to successful herbicides. There have been few successes (Evans, 1992). Natural products are inherently complex and the chemicals frequently have insufficient potency or the wrong activity or selectivity to provide strong leads for chemical synthesis (Evans, 1992).
Method 4. Rational design based on biochemical principles and knowledge of plant physiology and biochemistry.

The most intellectually attractive concept is biorational design. The organism to be controlled is considered to be target and an enzyme, essential to its survival, is chosen for direct attack. Candidate chemicals are developed to inhibit an essential plant function, based on biochemical knowledge of the target site. To date there has been little success but the technique is advancing (Evans, 1992).

Scientists know how to use all four techniques. In practice the first two have provided almost all presently available herbicides. Method 1, in which a large number of chemical compounds is screened for possible activity, has been the most common and successful method of herbicide discovery. The screening process includes structural relatives of compounds with known activity and chemical structures with unknown activity. Method 4, biorational design, has become increasingly feasible and useful. Combination of biological performance, toxicological properties, and environmental behavior with computer analysis make this a powerful technique for herbicide development. The herbicide industry will likely base future screening programs on greater understanding of plant biochemistry and physiology (Method 4). It is a certainty that future screens will more precisely target specific plant processes as the biochemistry of those processes becomes known. However, as pointed out in Chapter 16, discovery and release of new herbicides has continued, but no new sites of action have been discovered in more than 25 years (Duke, 2012). Reasons include company consolidation, rational management decisions, availability of generic formulations, the dominance of genetically modified (GM) crops, and the possibility that the best target sites are known (Duke, 2012), although the latter is debatable.

3.4.2 Rate Reduction and Precise Application

The advent of sulfonylurea and imidazolinone herbicides (see Chapter 16, Section 3.11 and 3.12) reduced herbicide rates by an order of magnitude, from kilograms to grams/ha (from pounds to ounces/A). That is a desirable direction from an economic and environmental perspective. Further reductions are possible when there is more information on threshold levels for phytotoxic activity. Most herbicides are now broadcast over the entire target area, some may be applied in a band over the plant row. Advances in application technology combined with GPS systems will allow herbicide application only to weeds rather than an entire field. Present methods require enough herbicide to satisfy the soil's adsorptive requirements and must account for dilution in soil. When knowledge of precise thresholds is combined with precise application, significant reductions in the amount applied per acre will be achievable. An essential question is, what is the minimum amount that can be applied to achieve the desired control?

3.4.3 Soil Persistence and Controlled Soil Life

It is at once a significant advantage and a major problem that herbicides persist in soil. Excessive persistence may affect succeeding crops, lead to contamination of ground or surface water, or cause undesirable residues in crops. Some persistence is good because it gives weed control over time, avoids repeated herbicide applications, and eliminates or reduces the necessity of using other weed-control methods in the crop. Regulating or controlling soil life is a desirable goal for future herbicide development. If nonpersistent herbicides could be given a few weeks of persistence, and the soil life of herbicides with long persistence but desirable activity and selectivity could be shortened, it would be good. If these things are achieved it will probably be through controlled-release formulations that have already achieved some success (e.g., encapsulation). The technology for these formulations is a special challenge in the complex soil environment.

3.4.4 Formulation Research

A few decades ago only a few formulations were available (see Chapter 17). Most herbicides were emulsifiable concentrates, solution concentrates, or wettable powders. A few granular formulations were available but application technology limited their use. Today formulation chemists have reduced dust, foaming, and storage problems, made handling easier and safer, and improved efficacy. Users can choose from all the previous formulations and several improved ones. Further improvement will occur to make formulations safer and easier to use. Controlled-release formulations could reduce volatility, leaching, and use rates, and increase selectivity. They could give nonpersistent herbicides a desirable soil life without fear of residual carryover to the next crop.

3.4.5 Perennial Weeds

There are some herbicides that control some perennial broadleaved or grass species. Perennial members of the *Cyperus, Cynodon, Sorghum,* and *Convolvulus* genera are not generally controlled selectively by available herbicides. Few herbicides that are sufficiently active on perennial weeds are also selective enough to be used in most crops and several persist in soil. It is relatively easy to control the emerged shoots of perennial weeds but far more difficult to assure translocation of an adequate amount of the herbicide throughout the extensive root, root runner, rhizome, or stolon system of a perennial weed. Selectivity is a greater problem than activity. A continuing problem for turf managers is selective control of coarse-leaved perennial grasses (e.g., quackgrass, tall fescue) that infest fine-leaved turf (e.g., Kentucky bluegrass). The advent and selective activity of aminopyralid (Milestone) has been an important contribution to control of Canada thistle.

3.4.6 Aquatic Weeds

The difficulty of controlling aquatic weeds is related to their habitat not their life cycle. Seventy percent of the earth's surface is covered with an interconnected water system. Ultimately, all water flows to the sea and all contaminants can be carried along as dissolved solutes or adsorbed to eroded soil. Almost all water contamination is, in the minds of most people, unacceptable. Because of heightened and enlightened concern about environmental quality, the unacceptability of contaminating water has resulted in legislation to prevent, control, and, if necessary, punish water pollution. Only a few herbicides can be used in aquatic systems. Present and future herbicides must be compatible with all other actual and potential uses for water. This is a very difficult requirement, and extensive, expensive research will be required if acceptable weed control with herbicides is to be developed. In the aquatic environment, it is likely that proof of safety will be demanded by an anxious public. Reasonable assurance or reasonable doubt will not suffice. It is equally true that expensive research will be required if nonchemical techniques for aquatic weed management are to be developed. Because of these appropriate concerns, the aquatic environment is a likely site for development of nonchemical-control techniques.

3.4.7 Parasitic Weeds

Parasitic weeds are present in the United States, but unless one has them they are not regarded as major problems in most states. They are major problems in several of the world's developing countries (Gressel et al., 2004). Few selective herbicides or other control techniques are available (see Chapter 3, Section 3.4). Research has not focused on them because they are very difficult problems and because herbicide development and developers have concentrated on large acreage crops of the developed world. It takes only a little experience with parasitic weeds to recognize how devastating they can be and how large crop yield losses can be. Farmers in developing countries where parasitic weeds are common stop growing susceptible crops, abandon fields, and are often defeated by them. Developing reliable, affordable management techniques for parasitic weeds would be a significant scientific achievement and a major contribution to development of agriculture and feeding people in the world's poor countries. Imaginative, scientifically sound solutions are required to solve parasitic weed problems. One approach to the *Striga* problem in Africa was mentioned in Chapter 3. It involves applying very low doses of an appropriate herbicide to maize seed after the maize plant has been rendered resistant to the herbicide through now standard nontransgenic techniques (Gressel et al., 2004; De Groote et al., 2007). ImiR (nontransgenic maize) was developed by CIMMYT.[3] It is available for short season corn (95 days) in Kenya and

[3]CIMMYT, Centro Internacional Mejoramiento Maiz y Trigo, International Maize and Wheat Improvement Center, located in Mexico.

neighboring countries from a variety of seed companies. Resistance will take a long time to evolve as the high local rate (in the rhizosphere — not/ha) requires homozygosity. This will not be so in long season corn (115 + days)—as toward the end of the season enough herbicide dissipates to allow heterozygous R individuals to survive. The strategy is seed treatment with an imidazolinone herbicide to create a weed-free zone in the vicinity of the corn and up to 12 weeks free of *Striga*, followed by a mid-season application of glyphosate, which requires that the corn be glyphosate/imidazolinone resistant. Each herbicide would kill surviving *Striga* and be appropriate for long-season corn[4] (Gressel, 2009; Ransom et al., 2012). To date, there is little evidence that the herbicide industry, governments, or other research agencies have adequately funded and thereby encouraged the work to solve a major third-world (particularly African) weed problem.

3.4.8 Packaging and Labeling

Major herbicide manufacturers have developed safe packaging and are concerned about personal safety of users. Systems that minimize human exposure are available and will become more common. Manufacturers and users are working together to minimize the hazards of herbicide handling and container disposal. Safe, efficient herbicide delivery systems will be needed and will demand adaptations of formulations and application methods.

Herbicide labels and use instructions in the world's developed countries are explicit, readily available, and contain adequate instructions for all approved uses. Sadly, the same is not true in much of the rest of the world. Manufacturers are well aware of the problem. Labels have to be developed that are clear, simple, adequate to the task, but not too complex. When potential users may be illiterate, clear instructions (often pictorial) are imperative. It is a difficult challenge to create instructions that combine the need for clarity and accuracy given the growing complexity of herbicide chemistry and use.

3.4.9 The Agricultural Chemical Industry

Innovation and progress in herbicide development depend on the agrichemical industry, where consolidation has led to domination by a few large, world-oriented European and US companies. There is market saturation in the developed countries and limited prospects for quick expansion in the large, diffuse market in the world's developing countries. The trend toward no- and minimum-tillage, evolving soil tillage practices, genetic modification to create herbicide resistance in crops, and the changing weed spectrum in crops will affect the herbicide market.

Mergers among agrichemical companies began in 1986, when DuPont bought Shell's pesticide division and the agricultural groups of Dow and Eli Lilly merged their agricultural chemical divisions into a new company called DowElanco. The French firm Rhone-Poulenc bought Union Carbide's agrochemical business, which had previously absorbed AmChem (that previously had purchased the herbicide business begun by American Paint Co.). These units all merged with AgrEvo, which had been formed from NOR-AM and Hoechst-AG, to form Aventis in 1999, which merged with Bayer Crop Science in 2002. The American division of Britain's Imperial Chemicals Co. (ICI) bought the Stauffer Chemical group from Cheesebrough-Pond, and the Swiss company Sandoz bought VS Crop Protection (Velsicol).

---

[4]Personal communication Professor Jonathan Gressel, Weizmann Institute of Science Rehovot, Israel 76,100.

Sandoz Agrochemical and Ciba combined to form Novartis, which merged in 2000 with AstraZeneca to form Syngenta. In the same year, BASF bought the agricultural division of American Cyanamid. In 1999, five European and four US companies dominated the world herbicide market. Mergers have continued. In 2016, ChemChina, a state-owned company, offered to buy Syngenta (Swiss), and Bayer (Germany) reached an agreement in September to buy Monsanto (United States). For a summary of herbicide company genealogy prepared and maintained by A. P. Appleby, Professor Emeritus, Oregon State University, see: cropand-soil.oregonstate.edu/herbgnl.

To the outsider it seemed that boards of directors must have said we must get big or get out. There has been no decrease in the availability or number of herbicides (see Chapter 13). The global agrochemicals market in 2013 was US $41.12 billion, 43% growth compared to $28.8 billion in 2007. The cost of development for a single agrochemical (that must pay for all failures) was estimated to be $286 million in 2014 (McDougall, 2016). The agricultural chemicals business is not one for the timid. The cost of herbicide development has increased and new, profitable, herbicides are rare (see Duke, 2012).

It remains for historians to ascertain the effect of these mergers on agriculture and farmers. There is a downward price pressure due to patent expirations, market saturation, and competition. Biorational approaches that consider biological efficacy, toxicology, and environmental interactions, and, presumably save money, will dominate design of new herbicides. These approaches will be combined with more specialized chemical, biological, and safety testing to determine site of action, use, economic benefit, and environmental acceptability.

The green or environmental movement and widespread public concern have helped create ever more restrictive herbicide legislation and regulation, which the public thinks is justified. Some claim there is more emotion and law than science in decisions that affect herbicide use and development and govern environmental decisions. Regulations that are too restrictive, it is claimed, suppress herbicide development and therefore agricultural production.

These brief comments on the agrichemical industry are included to give the reader insight into the business side of agriculture that is often not noticed by those who control weeds. Business decisions in corporate executive offices may have a much greater effect on future weed-management programs and the direction of weed science than anything that occurs in a research laboratory.

3.5 Biotechnology and Herbicide-Tolerant Crops[5]

All major herbicide companies have research programs to incorporate herbicide resistance into crops (transgenic crops, GM). Success has been achieved with several herbicides and several major crops: corn, soybean, wheat, rice, cotton, and tobacco (Duke et al., 1991). From its introduction in the mid-1980s to 1994, more than 1500 approvals for field testing of a wide range of transgenic organisms were granted and 40% were for herbicide tolerance (Hopkins, 1994). Between 1996 and 1999, the world's commercial area (albeit dominantly in the United States) planted with transgenic crops increased from 1.7 to 39.9 million hectares (James, 1999) and to 81 million hectares (200 million acres in 2004), a steady double-digit growth rate. By 2005, 8.3 million farmers (http://www.isaaa.org) grew transgenic varieties

---

[5]Chapter 18 discusses herbicide resistance and GM crops.

on more than 90 million hectares (222 million acres) acres in 21 countries. Transgenic crops (GMs) have been the most rapidly adopted technology in agricultural history. Soybeans are the most common GM crop. Corn is second with more than 21 million hectares (51.8 million acres) of corn modified to resist an herbicide or tolerate insects in 2005, up from 300,000 ha (741,000 acres) in 1996 (Halweil, 2006).

In 2005, more than 7 million of the 8.5 million adopters were small-holder farmers in the world's developing countries (Chassy et al., 2005). For example, transgenic cotton was first planted in China in 1998 and quickly grew to 50% of the cotton acreage. Yields improved 10%–30%, pesticide use declined 50%–80%, and farmer profit increased (Chassy et al.). In fact, insecticide use in agriculture has declined due to transgenic crops (Chassy et al., www.ers.usda.gov/publications/aer810/). But Lang (2006) warns that all is not rosy. Cotton resistant to bollworm was planted widely, worked well, and failed. After 7 years, bollworms resistant to the Bt cotton have not appeared, but secondary pests have appeared in abundance. Farmers are now spraying cotton up to 20 times a season to control the secondary pests (e.g., mirids) that formerly were controlled by the insecticides used to control bollworm.

Because transgenic crops have been available for some time, it is not unreasonable to ask why they are discussed in a chapter on the future of weed science and in Chapter 18. The reason is that although we know what has been done with agricultural biotechnology and herbicide resistant crops, the technology is so new and changing so rapidly that we do not know, perhaps we cannot know, what might be done. That is, the direction of research is clear, the final destination is not. Modernity cannot be identified with any particular technological or social breakthrough (Kirsch, 2016). It is a subjective condition, a feeling or intuition that our science of weed management is in some profound sense different We cannot be sure what new possibilities will be discovered as the technology continues to develop. Adoption of molecular-based methods in weed science research will bring a new dimension to the science and can have "far-reaching benefits in agriculture and biotechnology" (Marshall, 2001). One potential, presently unachieved, benefit of genomics research is the discovery of new targets for herbicide action (Hess et al., 2001). Other benefits may include identification and use of genes that contribute to a crop's competitive ability (e.g., early shoot emergence, rapid early growth, fast canopy closure, production of allelochemicals). Genomics may also discover genes that contribute to weediness (see Chapter 2), a plant's perennial growth habit, seed dormancy, and allelopathy (Weller et al., 2001).

3.5.1 Criticism and Risk

Criticism of herbicide resistant crops is common and is usually related to all or some of four perceived risks: public health, environmental, social, and weed control (see Chapter 18 and Zimdahl, 2006, 2012 for a discussion of ethical concerns).

3.5.1.1 PUBLIC HEALTH

Some concern focuses on the possibility of water or food contamination from increased herbicide use. Additional concern centers on use of an herbicide(s) in crops that do not metabolize the herbicide. Therefore, the unaltered herbicide could be consumed by people and harm, while not assured, is assumed. As biotechnology and herbicide-resistant crops develop, it is important to remember that no technology is ever proved to be perfectly safe. Science deals with probability, which, effectively prohibits absolute verification of safety. In addition, scientists cannot prove a negative—that something won't happen. Indeed, it is impossible to prove something will not happen. Scientists look for evidence of harm, and if none is found, logically conclude there is none, or that it must be looked for in a different way (Anonymous, 2005). However, it is also logical to conclude that the absence of evidence is not evidence of absence of possible harm. Secondly, this technology has, as all technologies have (e.g., herbicides, cell phones, computers[6]), good and bad uses. We must be cautious about demonizing the potential, but unknown, bad effects of legitimate uses by good people and weigh them carefully against illegitimate uses by bad people.

3.5.1.2 ENVIRONMENTAL

Hubbell and Welsh (1998) suggest that transgenic crops have the potential to create a more sustainable and environmentally favorable agricultural system than chemically based systems but will fail "in enabling a fully sustainable agriculture." Hubbell and Welsh divide these genetic traits into three categories: transitional, compatible, and sustainable. Transitional traits reduce environmental damage (e.g., glyphosate tolerance); compatible traits are those with limited future value but during their useful lives they enhance use of sustainable practices (cropping systems that reduce or eliminate use of toxic chemicals). Sustainable traits are those that encourage use of sustainable practices (there are presently no transgenic crops in this category). Examples include plants that make more efficient use of natural fertilizers, conversion of annuals to perennials, or enhancing nitrogen-fixing ability to reduce fertilizer use. They claim that genetic traits that have a higher potential of enabling truly sustainable agricultural systems have not been developed for three main reasons:

1. The lack of US Environmental Protection Agency and USDA regulatory policies that specifically promote sustainable traits.
2. An agricultural biotechnology industry that is dominated by agricultural chemical companies.
3. Patent law and industry policies that prevent farmers from saving transgenic seed and thus tailoring transgenic crops to their local ecological conditions.

3.5.1.3 SOCIAL

Social concern is related to a fear that the technology favors large farms and will lead to loss of small farms, small-scale farmers, and rural communities. It is estimated that as many as 1.4 billion small-scale farmers grow 15%–20% of the world's food from seed saved from the previous year's harvest. Much of this food, grown by women, is not sold. It is eaten or bartered locally and not included in national production figures. Loss of the ability to save seed may drive these farmers out of farming (Sexton, 1998). A secondary concern is that the cost of food production and food cost to the consumer will rise.

3.5.1.4 WEED CONTROL

There are three concerns related to weed control:

1. More weeds will develop resistance to one or more herbicides because of continued use of an herbicide to which a crop is resistant (see Sandermann, 2006 and Chapter 18).

---

[6]See Carr, N., 2011. The Shallows—What the Internet is Doing to Our Brains. W.W. Norton, NY. 280 pp.

2. Resistant gene flow to sexually compatible plants is acknowledged as a potential risk of introducing any genetically engineered (GM) crop. The risk is transfer of desired herbicide resistance from the crop to a weed where undesirable resistance persists by natural selection. It is worth noting that this has happened when genes from herbicide resistant canola moved to a nonweedy relative in the mustard family and then to wild mustard in a short time.[7] The risk is especially great when a crop and weed are closely related and can interbreed. For example, red rice and rice or Johnsongrass and grain sorghum.

Resistant crop plants may become hard-to-control volunteer weeds. This has not been shown, but Keeler (1989) urged caution and pointed out the example of wild proso millet that emerged as a weed in the 1970s after over 200 years of successful cultivation of proso millet in North America without its becoming a weed. Keeler used wild proso millet to emphasize how much we do not understand about weed evolution.

3.5.1.5 EPISTASIS AND PLEIOTROPY

The quite legitimate concerns of epistasis and pleiotropy (see Chapter 7) must also be recognized. Epistasis is suppression of gene expression by one or more other genes and pleiotropy is when a single gene exerts simultaneous effects on more than one character. In short, an ecological rule applies: It is impossible to do just one thing. When science manipulates a genome, any genome, specific outcomes are intended and, even when these are achieved, other, unplanned, things may also occur. Genetic engineering, with the best intention to do a good thing, may do unexpected things that could be good or bad.

3.5.1.6 PROMOTE HERBICIDE USE

Another common critique of herbicide-resistant crops is that the technology will promote, not decrease, herbicide use, while promoting what many view as an unsustainable, chemical, energy, and capital intensive monocultural agriculture. It is also suggested that herbicide-resistant crops will reinforce farmers' dependence on outside, petroleum-based, potentially polluting technology. An associated concern is that there is no technical reason to prevent a company from choosing to develop a crop resistant to a profitable herbicide that has undesirable environmental qualities such as persistence, leachability, harm to nontarget species, etc. This has not been done. It is undoubtedly true that nature's abhorrence of empty niches will mean that other weeds will move into the niches created by removal of weeds by the herbicide used in the newly resistant crop. In other words, herbicide resistance will solve some but not all weed problems. Weeds not susceptible to the herbicide to which the crop is resistant will appear. Weeds are not conscious, but they seem to be clever.

3.5.2 *Further Development*

Development of herbicide-resistant crops has and,is proceeding rapidly. There are important advantages that justify their development. Many argue that the technology will provide lower cost herbicides and better weed control. These are powerful arguments in favor of the technology because both can, but may not, lead to lower food costs for the consumer. It is also true that herbicide-resistant crops are providing solutions to intractable weed problems in

[7]Denver Post April 14, 1996 and New York Times March 7, 1996.

some crops. Glyphosate resistance has been created in several crops. It is an environmentally favorable herbicide, and, therefore, many (e.g., Hubbell and Welsh, 1998) argue that it is better to use it in lieu of other herbicides that are not environmentally favorable. An important argument in favor of the technology is that it has the potential to shift herbicide development away from initial screening for activity and selectivity and later determination of environmental acceptability to the latter occurring first. Resistance to herbicides that are environmentally favorable but lack adequate selectivity in any crops or in a major crop so their development will be profitable could be engineered and the herbicide's usefulness could be expanded greatly. This has important implications for minor crops (e.g., vegetables, fruits) where few herbicides are available because the market is too small to warrant the cost of development. If resistance to an herbicide already successful in a major crop (e.g., cotton) could be engineered into a minor crop, manufacturers and users would benefit.

Biotechnology was discussed by Christianson (1991), a self-acknowledged outsider, and his view is quoted here:

> I think it would be a pity if the power of the use of mutants and mutation to uncover and describe physiology and development were limited, in the hands of weed scientists, to the isolation and description in yet another species of yet more genes that confer resistance to yet another herbicide.

The central issue for weed science is understanding the nature of weeds: What makes a weed a weed (see Chapter 2)? How can weeds consistently come out ahead when matched up against the finest commercial varieties plant breeders have developed? Weeds persist, they spread, and they outcompete crop plants, reducing yields when left uncontrolled. The nature of the competitive ability weeds possess seems an interesting target for research and an appropriate target for analysis through generation of mutants.

Goethe's "The Sorcerer's Apprentice" and "Faust," Mary Shelley's "Frankenstein," and, more recently, Michael Crichton's "Jurassic Park" reinforce the often-inchoate fear of intelligent, rational people that some powerful form of life manufactured by man with good intentions, but excessive hubris, might one day slip out of control (Specter, 2016). The 1950s gave us catchy phrases that still resonate—Better Living Through Chemistry and Atoms For Peace. We don't hear similar things now. Chernobyl/Fukushima nuclear reactors, agent orange, space shuttle crashes, ozone destruction, pesticides in food, and climate change dominate the public's thoughts about science that while it clearly solves it simultaneously causes problems. These disasters combined with human drug disasters have soured the public and made people suspicious of the efficacy and trustworthiness of science and scientists (Lemonick, 2006). It is this context that public doubts about genetic modification of anything are raised and must be addressed. Weed scientists and others involved with GM technology often think that if they could educate people about what scientists do (William et al., 2001) they would understand the value of science's results. Education is important but a conversation among equals may be a better course, especially in a time when science has made mistakes and is regarded with some well-founded suspicion. Weed scientists should not regard themselves as the only acceptable arbiters of the way new developments in their science should be used. Through the public's perception of greed and arrogance on the part of developers, and misunderstanding, many view genetic modification as a hazard, not a salvation, and reject it (Specter, 2016).

As mentioned before, transgenic crops have developed more rapidly than any preceding agricultural technology (see Hileman, 1995; Zimdahl, 2006, 2012). It is not the purpose of this text to analyze the controversy in depth. Others have done this (Duke, 1996, 2006; Zimdahl, 2006, 2010). Scientific journal and popular press articles too numerous to mention are available. Much more work will be done and discussed, but it is important to realize that GM technology is widely promoted, accepted, and used. Specter (2016) reminds us that the real question is not whether genetic modification will occur but how it will be used.

3.6 Organic Agriculture

A paraphrase of G. K. Chesterton's comment on Christianity makes my point—organic agriculture, within weed science, with some exceptions, hasn't been tried and found wanting, it has been found difficult and left untried. Evans (2002, p. 14) said it differently but gave the same message:

> Because weeds are inextricably both products of psychology and ecology, weed problems are best addressed by considering not only the agroecosystems that produce them but also the culture that informs how we think and farm. Recognition of this point is potentially threatening and subversive for it challenges the very social structure of farming upon which employment of weed scientists depends. They are, in effect, servants of large-scale, single-crop, commercial agriculture and if that were to disappear so too would a large proportion of their jobs.

Weed scientists have been conditioned to take pride in, achieve production agriculture where no weeds exist, only the crop (Harker, 2004). That may be the wrong goal. The goals of organic agriculture encourage us to think about what Evans (2002, p. 51) calls a "blindly oppositional attitude" that led scientists to "frantically search for immediate solutions to an ever-worsening" weed problem. That search often obscured the need for different approaches to weeds and to needed changes in North American farming systems. Herbicides helped weed scientists define weeds as the enemy, and only in the late 1990s did this begin to change. Dependence on herbicides led weed scientists and farmers to neglect other weed-management strategies such as crop rotation, tillage, fertility management, elimination of herbicides, and a change of attitude about what a weed is and does (Harker, 2004), which organic agriculture has encouraged thought about. The demands and popularity of organic agriculture may compel weed scientists to broaden their horizons and consider other weed-management tools more carefully.

3.6.1 Engineering Research

No matter how selective or active herbicides are, they have to be applied in the environment to fulfill their purpose. The herbicide in the package is interesting, but not functional. The Weed Science Society of America monograph on applying herbicides (McWhorter and Gebhardt, 1987) points out that over 90% of all herbicides applied annually in the United States were sprayed using sprayers with the same four basic components—tank, pressure regulator, pump, and nozzles—that have not changed since herbicides were first sprayed. The technology has evolved and application equipment is more precise, more durable, and more flexible than older pieces of equipment.

Must herbicides be sprayed? Is there a better way? Some of all sprayed herbicides never hit the target. Can spray that doesn't hit the target be recovered and reused or at least handled so it doesn't reside in the environment without fulfilling its intended purpose? Low-volume and ultralow-volume application techniques are available but not widely used. There is potential for decreasing the volume of spray required.

Sprayers are being developed that use remote-sensing technology to sense weed presence and just spray where weeds are. Other techniques sense variations in soil type and adjust herbicide concentration to account for differences. Automated detection and identification of weeds are the greatest obstacles to development of practical, affordable site-specific weed-management systems (Brown and Noble, 2005). However, before site-specific systems can be perfected, accurate, affordable techniques to sample weed populations must be developed (Wiles, 2005). Sampling to determine the location and identity of the weed problem is a fundamental principle of IWM. Broadcast herbicide application reduces or eliminates the need to know what the problem is or where it may be. Weeds are assumed to be everywhere, which is usually not true. Sampling is a decision-making tool that has not been fully exploited by weed science primarily because there are no simple, affordable methods for weed seed sampling. Machine vision technology with appropriate data processors are now available and enable recognition of crop growth patterns and use of automated systems that can remove intrarow weeds as well as thin crops (Fennimore et al., 2016). These are particularly applicable in specialty crops such as flowers and vegetables where appropriate herbicides are often not available. "Automated weed removal equipment continues to improve and become more effective" (Fennimore et al., 2016).

In addition to affordable sampling techniques, other research necessary to perfect remote sensing technology includes: the need for artificial lighting, definition of spectral band requirements, techniques (required to recognize different weeds and distinguish them from the crop), rapid image processing, multiple spatial resolution systems, and techniques for dealing with multiperspective images (Brown and Noble, 2005). For example, leafy spurge can become, and in many places is, a major problem in the western United States. It can be detected easily by extensive, expensive ground surveys. Use of Space Imaging's 4-m multispectral Ikonos imagery was effective for detecting leafy spurge patches under some circumstances, but in areas with a high forb component, identification of leafy spurge was not as effective (Casady et al., 2005). Shaw (2005) evaluated use of remote-sensing data for weed management and saw it as an area of "tremendous opportunity." He predicted that proper use of remote sensing would result in reductions in inputs, reduced environmental liability from applying herbicides to entire areas rather than just to weed patches, increased crop yield due to better management and early detection, and more effective management of invading species. A less sanguine appraisal of the profitability and economic feasibility of site-specific weed management was based on the advantages of reduced herbicide use, the additional costs of scouting for weeds, preparing treatment maps, and applying herbicides only to identified patches (Swinton, 2005).

Ecologically based, site-specific weed management may offer revenue gains to farmers compared to broadcast-herbicide applications because while herbicides are very effective for weed control, further yield gains from their use are unlikely, therefore revenue gains, if any, will come from higher crop prices or environmental stewardship payments from government programs, which presently do not exist (Swinton, 2005). Policy makers and consumers may be willing to pay for increased freedom from the perceived and real environmental risks of herbicides. No one knows if this is true or if such policies can be developed and implemented. Remote sensing aided by global-positioning technology will permit precise herbicide application on weed patches. The goal is to improve application accuracy and operator safety while maintaining environmental quality and protecting crop yield.

The Lettuce Bot (Economist, 2012), a robot developed by engineers, identifies a weed or a lettuce plant that grows too close to another one. A nozzle at the back of the machine squirts a concentrated dose of fertilizer and/or an effective herbicide. It is selective, effective weed control and supplies fertilizer to the remaining lettuce. Lettuce is California's main vegetable crop ($2.25 billion in 2015—71% of US lettuce). The inventors of the Lettuce Bot are developing another robot that controls only identified weeds mechanically with a rotating blade (essentially a mower). It will be especially attractive to organic growers.

Herbicide application is not the only area in which engineers can contribute to improved weed management. The effects of tillage over time on recurrent weed populations is not known. There is room for improvement in methods of mechanical control of annual and perennial weeds. Cultivators are available to weed the entire area between crop rows and leave only a narrow band over the crop row. This is a vast improvement over a person with a hoe or an animal-drawn cultivator, but the person with a hoe is selective. Further research will reveal even more specialized tillage and cultivation methods for weeds in crops. Organic farmers use cultivation and several tillage implements effectively (see Chapter 11).

Tillage research has shown that much is unnecessary and, in fact, complicates weed management. Tillage exposes weed seeds to sunlight and encourages germination (see Chapter 10). No tillage leaves seeds buried and prevents or inhibits germination. The best tillage for weed management may be none. That does not mean that weeds will not be present. No tillage creates a niche for weeds that do well without tillage. It will change, but not eliminate, the need for weed management.

3.7 Integrated Weed Management

In science as in most human activities, movements occur, directions change, and progress may result (Kuhn, 1970). Some movements are called bandwagons. Each has associated words and phrases that define and identify it, often called buzzwords. Each movement makes its contribution to the parade of ideas and contributes to the general cacophony of competing ideas, which, one hopes will yield a harmonious new paradigm. Some ideas assume a position at the head of the line. IWM has assumed a position of centrality and leadership. The concept of integrated pest management (IPM), particularly in entomology, can be traced to the late 1800s when ecology became the foundation for scientific plant protection (Kogan, 1998). The term IPM did not appear in the literature until 1967 (Smith and van den Bosch, 1967). The conceptual origin of IPM can be traced to the term integrated control, which was introduced by entomologists at the University of California—Riverside and Berkeley (Stern et al., 1959). IPM was embraced by biocontrol advocates and challenged by advocates of aggressive, well-funded insecticide control programs. IPM has endured as the more desirable alternative to previous pest-management systems, especially for insect control. It has been challenged because from 1968 to 1992, when US interest in IPM grew steadily, pesticide use in US crops increased 125% (Gardner, 1996). Many see IPM as a buzzword that hasn't changed pest control. If IWM systems are to endure, change will be required. The direction and scope of change may determine the enduring success of IWM. Historians tell us, after the fact, what things have endured and why, and why other ideas were only temporary. Judgments from the present are often flawed because one is so close that subjectivity dominates. The perspective of time is often a prerequisite to objectivity.

There is a risk that weed management may be another buzzword like Andy Warhol's (1968) remark about people, "In the future, everyone will be world-famous for 15 min[8]." But IWM makes so much sense that it is likely to endure. It is not perfect but it is better than anything else we have. The evidence in this book is sufficient to demonstrate that weed-management systems for crops are incomplete. They are developing, but the research gaps identified herein and summarized in Table 23.1 preclude defining complete systems. It is true that weed management has become dependent on herbicides and, within the last decade, herbicide-resistant crops. If truly IWM systems that are sustainable over time are to be developed, they will have to include, as Young (2012) points out, components of weed control that must be put together rather than taking them apart. Weed science, in his view, "has stopped at the field edge in assembling the components into a truly integrated approach." Integration of other weed-control techniques in addition to herbicides and GM crops has not occurred. Even casual observation of capital, energy, and chemical-dependent agriculture demands answers to questions about the system's sustainability. In view of the rapid appearance of herbicide-resistant weeds, increased herbicide use and emphasis on short-term solutions, sustainability does not appear to be the goal. It should be (Mortensen et al., 2012). "The monocrop factory mode that dominates US agriculture today certainly differs from what has characterized farming for most of history and in most of the world" (Peterson, 2000). "Instead of natural ecosystems with their haphazard mix of species..., farms (now) are taut, disciplined communities conceived and dedicated to the maintenance of a single species: us" (Mann, 2012). Many argue that this kind of agriculture cannot achieve what all regard as a proper goal—sustainability. The foundation of weed science has been control technology rather than understanding weeds, why and how they compete so well, and their role in the agroecosystem (Buhler, 2006). That must change.

Several research questions remain:

- What seeds emerge first from complex soil seed banks? Why?
- What percentage of seeds, from each species, emerges each year?
- What is the precise percent control for different weed control techniques and for different weeds at different growth stages?
- If soil seed bank composition is known, can the weed complex that will appear in a crop be predicted?
- How do weed-management techniques affect other pests and other pest management techniques?
- How can the advances in automated machinery for real-time detection and control of weeds developed by engineers and computer scientists be integrated into a weed-management system (Young, 2012).

---

[8]Although Warhol accepted the credit for the remark, although he acknowledged that he never said it.

- Although IWM is widely promoted within the weed-science community as the best way to manage weeds, it is not widely practiced because current systems failed to meet performance expectations of growers (Young et al., 2017). Why is that true, and how can it be corrected?

It is likely that successful IWM systems will have to be developed rapidly within the opportunities and constraints of agricultural industrialization. Industrialization is a process whereby agricultural production is structured under the pressure of increasing levels of capital and technology in a way that allows management systems to integrate each step in the economic process to maximize efficiency of capital, labor, and technology (Urban, 1991; Keeney, 1995). A primary question is whether this process is compatible with, and capable of, achieving a sustainable agriculture. Keeney thinks that if agriculturalists, environmentalists, and government work together, a sustainable rural landscape can be achieved. If they do not, "today's haphazard and divisive times" will continue. Urban believes that the industrializing forces of consumer desires and demand, prescription agricultural products, molecular biology, and the changing nature of farming combine to make agricultural industrialization inevitable. Many argue that modern industrial agriculture built on and dependent on scientific knowledge is the only way to feed the 9 billion people expected on the planet by 2050. Others, with equal passion and commitment (see Wendell Berry, Wes Jackson, Fred Kirschenmann) argue that if the dominance of energy-, chemical-, and capital-dependent agriculture is not drastically changed, the 9 billion will not be fed, because the system is not sustainable (Mann, 2012). Without change, the demise of the Jeffersonian agricultural heritage of the farmer is inevitable. The result of the debate will affect weed science.

IWM should not limit its focus to weed control. To be successful, the focus must be the total vegetation complex or better habitat management rather than weed control in a year in a crop. Perhaps it is most correct to say that industrialization should, although it may not, change the scale of concern. Sustainable IWM systems should extend concern to environmental quality and future generations. These are large-scale concerns. Small-scale concerns such as how to control weeds in a crop in a year have dominated, but future agricultural systems will require major changes. Environmental concerns demand large scale thought. Small-scale thought suffices for individual concerns. Large thoughts are needed for large systems. Everything needs to be integrated to have a complete crop-management system. It won't be easy to do. It is necessary.

3.8 Other Challenges

3.8.1 Scientific

There are several other scientific research areas that should be considered when planning weed science's future. They are less developed than those that preceded but may be just as important. They include:

- What is the value and what are the advantages and disadvantages of monocultural agriculture?
- What is the role of companion cropping and regular inclusion of cover crops in weed management? Can weeds be cover crops?

- What are the long-term effects of soil erosion after regular plowing and cultivation? One effect is all too apparent in the brown color of many country's rivers (see Logan, 1995; Montgomery, 2007). Weed scientists were not too concerned with long-term effects when the science was developing. Weeds decreased crop yield—a detrimental long-term effect. The vision didn't extend much farther. Solving the weed problem was a sufficient challenge. Any technology, used for enough time, has demonstrable environmental and social effects. A longer-term view will help reveal these effects and compel their consideration before widespread use is achieved.
- Weed scientists must begin to work more closely with economists who ask, what does it cost and what is it worth? What is it worth to do the work to develop a more competitive cultivar, to deplete the soil seed bank, to have assurance of 80% or 100% weed control? What will it be worth to be able to predict weed problems? No one knows, but the answers are important to IWM systems.
- How will nanotechnology affect weed science? Nano integrates biological material with synthetic materials to build new molecular structures. Synthetic biology goes beyond moving existing genes to creating new ones that are programmed to perform specific tasks. Nanobiotechnology operates at the nano scale: one billionth of a meter $= 10^{-9}$ m of living and nonliving parts. It has enormous potential for good and for harm (Shand and Wetter, 2006). Syngenta sells two pesticide products that contain nanoscale active ingredients. They prevent filter clogging and are readily absorbed by plants (Shand and Wetter, 2006, p. 82).

3.8.2 Social and Moral

Weed scientists are aware of the scientific research opportunities (Table 23.1) and challenges. There are equally important, although less discussed, social and moral challenges. The primary goal of agricultural scientists has been to develop technologies that enable achieving the maximum yield of a few crops in the world's developed (rich) countries. It is a good goal, but one must ask if it is the right goal (see Kirschenmann, 2012; Zimdahl, 2012). Is it more important than enabling the poor of the world to feed themselves? Can the seemingly unending task of discovering new technologies to maximize yields lead to a sustainable agricultural system that will feed 9 billion of more people? Is maintaining rural communities a proper goal for agricultural science or is that someone else's task? Should achieving maximum yield and profit always take precedence over preserving the environment? Liu et al. (2015) found cultural practices with negative effects on global food production. "It is crucial for agricultural sustainability to increase crop yields **and** simultaneously decrease environmental impacts of agricultural intensification" (Liu et al., 2015).

3.8.2.1 SUSTAINABILITY

Achieving a sustainable agriculture is a goal all agricultural scientists share. Even a cursory review of current writing on agriculture reveals that achieving sustainability has obtained the generally revered status of motherhood with one important difference. There is little debate about what motherhood is or about its worth, its goodness. The difference is that in spite of the nearly universal adulation of agricultural sustainability, there is little agreement on its nature, on what is to be sustained, or on how it is to be accomplished (Zimdahl, 2012, p. 121; Gliessman, 2001). Production is and always will be important, but it is not possible to create a sustainable agriculture without a sustainable culture. The reverse is also true (LeVasseur, 2010). It is impossible to have a serious, comprehensive discussion of sustainable agriculture without including community and culture (Holthaus, 2009). Within the agricultural community achieving sustainability is viewed as mainly or wholly technical in nature. All it requires is different farming methods and adoption of alternative technologies (Morgan and Peters, 2006), which will be significantly aided by advances in biotechnology. This view ignores the moral, educational, and political tasks that must be considered. In Morgan and Peters' view it requires a commitment to "philosophical principles that depart from the utilitarian premises of industrial agriculture." It is a demanding task that requires new thinking and a change in attitude toward the earth. It requires us to cease attempting to achieve dominion over the earth, but to achieve humility and reverence before the world (Berry, 2002).

Liu et al. (2015) found that current cultural practices will have negative effects on global food production. "It is crucial for agricultural sustainability to increase crop yields **and** simultaneously decrease environmental impacts of agricultural intensification" (Liu et al.). The majority of the mainstream agricultural community does not agree with Liu et al. The dominant view is that supporting crop intensification is the best route to feed 9 billion people and protect the environment. But there is no room for complacency, especially of the kind invoked by some advocates of biotechnology (Fischer et al., 2014).

Finally, a caution. Those engaged in agriculture and its subdiscipline weed science possess a definite but unexamined moral confidence or certainty about the correctness of what they do. The basis of the moral confidence is not obvious to those who have it, or to the public. In fact, the moral confidence that pervades agriculture is potentially harmful because it is unexamined. It is necessary that all those engaged in agriculture and weed science analyze what it is about their science and their society that inhibits or limits their science. All should strive to nourish and strengthen the aspects that are beneficial and change those that are not. To do this agricultural people must be confident to study themselves, their institutions, and be dedicated to the task of modifying the goals of both (Zimdahl, 2012).

3.8.2.2 THE HUMAN DIMENSION OF AGRICULTURE

Doohan et al. (2010) claim "the human dimension of weed management is most evident when farmers make decisions contrary to science-based recommendations." Weed scientists may be aware that their recommendations are often ignored but usually do not ask why because such questions are beyond their area of expertise. Scientists do science that leads to science-based recommendations. When recommendations are ignored, it is assumed, the reasons could be: economic (too expensive), stubbornness, lack of trust, and different perceptions of risk and benefit. Doohan et al. argue that farmers exhibit an inverse relationship between perceived risk and benefit. If any technology is regarded as beneficial it is automatically perceived as low risk. That, of course, is not true. Ignoring farmers' reasons is perilous for weed science.

3.8.2.3 GOALS

Weed science has made major contributions to increasing crop production over several decades. Herbicides have been the primary control technique. Because of their efficacy and ease of use there has been an overreliance on them at the expense of other weed-control methods (Blackshaw et al., 2008). If the only or primary goal is to increase production, then the quest for better herbicides must continue. If the goal is sustainable weed management in a sustainable environment and society, then other control techniques must be investigated. Farmers must be shown that adopting nonherbicidal weed management will not increase the risk of crop failure or reduce profit. The new goal should be development of successful weed-management systems that have minimal or no effect on the flora and fauna of soil, water, or air and have no adverse effects on people.

4. POLITICAL CONSIDERATIONS

Weed scientists and most people engaged in agriculture are not, by nature or choice, good politicians. Many agriculturalists consider a career as a politician to be more noble than being a Mafia Don or shoplifter but perhaps about on a par with those folks who call you, just at supper time, to persuade you your carpets need cleaning. Failure or inability to consider the fact that we live in a political world and are affected by it is a prescription for disappointment or disaster. Political considerations affect our daily life. A major political accomplishment in many countries is cheap food, especially in urban areas. Most enjoy the benefit of this, often unstated, government policy. It affects the way we practice agriculture and manage weeds. If the government removed itself from agricultural policy making and the market, cheap food might disappear. Given agricultural and environmental history, concern about environmental pollution from agriculture is a fairly recent political development. It wasn't too long ago that pesticide use in agriculture meant prosperity and progress rather than environmental pollution and corporate irresponsibility. For example, a study commissioned by the American Farm Bureau, an organization noted for its defense of agriculture (King, 1991), showed that only 15% of the American public was in favor of abolishing pesticide use in agriculture. However, 66% of the people surveyed thought pesticide use should be limited in the future and 38% thought farmers were using more pesticides than they had in the past. Such information and concern has political meaning and consequences. The findings are ignored or dismissed only by those who willfully ignore the effects of political action. Political acts change many things and agriculture has to recognize and work in a political milieu or suffer the consequences of regulation by those who do.

5. CONCLUSION

The American author and farmer Wendell Berry (1981b) has written often and eloquently about problems facing American agriculture and about their solutions. He advocates solving for pattern. Berry says that "to the problems of farming, then, as to other problems of our time, there appear to be three kinds of solutions." The first kind causes a ramifying series of new problems. The only limitation on the new problems is that they "arise beyond the purview of the expertise that produced the solution." That is those who are encumbered by the new problems are not those who devised solutions for the old problem. This kind of solution shifts the burden away from those who created the problem.

The second kind of solution is one that immediately worsens the problem it is intended to solve. These are often quick-fix solutions that, within weed science, take the form of questions such as, what herbicide will kill the weed? Adopting this kind of problem solving leads to the need for more quick-fix solutions. Everyone who has tried to fix something is familiar with this kind of solution. What was tried first didn't work and some study (but perhaps little knowledge) revealed that loosening another bolt or screw would do it. Alas, loosening that screw was the wrong thing to do because it loosened other things and suddenly parts were everywhere and neither the source of each part nor a way to fit them back together was known. I am quite familiar with these solutions.

The third most desirable solution creates a ramifying series of solutions. Parts don't fly off in all directions; they fit together. These solutions make, and keep, things whole. For Berry (1981b) a good solution is one that acts constructively on the larger pattern of which it is a part. It is not destructive of the immediate pattern or the whole. People who devise the best solutions recognize the pattern in which they must fit and work to create a set of solutions that maintains the essential pattern. Good solutions solve for the whole system, not for a single goal or purpose.

Those who will create the next generation of weed-management systems for simple and complex weed problems will do well to remember Berry's admonition as they search for the best solutions. One must know the whole system and devise solutions that create more solutions that maintain the pattern or make it better. Weed scientists should view the agricultural system as a good family physician views patients—in family, not just individual terms. It is the entire system not just the weeds that must be managed.

M. S. Swaminathan, the creator of India's green revolution, former Director General of the Int. Rice Research Institute, and recipient of the first World Food prize delivered the B. Klepper endowed lectureship at the 2005 meeting of the Tri-Societies (Agronomy, Soil, and Crop Science). He said, "The most cruel form of inequity is malnutrition." This is a large-scale concern. Contributing to the elimination of hunger in the world is a proper goal for weed science. It is a goal consistent with the Millennium goals of the UN (Sachs, 2005, pp. 211–212). Two of the goals are relevant to agriculture and worthy of the attention of weed scientists. These large-scale goals include:

- Eradicating extreme poverty and reducing hunger by half by 2015;
- Ensuring environmental sustainability.

In his Recollected Essays, Berry (1981a, p. 98) writes eloquently about a vision of the future that is shared by those who want to create alternative futures including alternative, improved, sustainable agricultural systems. His words are a good place to end thoughts about the future of weed science. I leave it to readers to determine if I have reached beyond my knowledge and ability.

> We have lived by the assumption that what was good for us would be good for the world. We have been wrong. We must change our lives, so that it will be possible to live by the contrary assumption that what is good for the world will be good for us. And that requires that we make the effort to know the world and to learn what is good for it. We must learn to cooperate in its processes, and to yield to its limits. But even more important, we must learn to acknowledge that the creation is full of mystery; we will never clearly understand it. We must abandon arrogance and stand in awe. We must recover the sense of the majesty of the creation, and the ability to be worshipful in its presence. For it is only on the condition of humility and reverence before the world that our species will be able to remain in it.

Job offers a similar challenge:

> Gird up now thy loins like a man; for I will demand of thee, and answer thou me. Where wast thou when I laid the foundation of the earth? Declare if thou hast understanding. *Job 38:3—4*

THINGS TO THINK ABOUT

1. Why are herbicides the primary weed control technique in the world's developed countries?
2. What are the major problems that impede progress in weed science?
3. What will be the characteristics of progress in weed science?
4. Identify and explain the nature of important, future research problems?
5. Why is understanding weed-crop competition crucial to the future of weed science and weed management?
6. Why are studies of seed dormancy and seed germination important to weed management?
7. What will be the future role for bioeconomic models of weed-crop competition and what parameters should new models incorporate?
8. What are the characteristics of an ideal herbicide?
9. Why are perennial weeds so hard to control?
10. Why are parasitic weeds so hard to control?
11. What lessons, if any, does weed management in organic agriculture offer?
12. What are the major problems with herbicide use for aquatic weed management?
13. Why is there concern about packaging and labeling of herbicides?
14. What has been the recent evolutionary trend in the agrichemical industry?
15. Why are herbicides sprayed?
16. In what areas can engineering research contribute to weed management?
17. Why are political considerations important to weed science?
18. What is solving for pattern and how does it relate to weed science and weed management?
19. Is Wendell Berry right or wrong? State your position and defend it.

Literature Cited

Anonymous, March 26, 2005. The shock of the new. Economist 13.

Bellah, R.N., Madsen, R., Sullivan, W.M., Swidler, A., Tipton, S.M., 1991. The Good Society. A.A. Knopf, New York, NY, p. 97.

Berry, W., 1981a. A native hill. In: Recollected Essays. 1975—1980. North Point Press, San Francisco, CA, pp. 73—113.

Berry, W., 1981b. Solving for pattern. In: The Gift of Good Land. North Point Press, San Francisco, CA, pp. 134—145.

Berry, W., 2002. The Art of the Commonplace. Shoemaker and Hoard, Washington, DC, p. 330.

Beyer Jr., E.M., 1991. Crop Protection — Meeting the Challenge. In: Proc. British Crop Prot. Conf. — Weeds, vol. 1, pp. 3—22.

Blackshaw, R.E., Harker, K.N., O'Donovan, J.T., Beckie, H.J., Smith, E.G., 2008. Ongoing development of integrated weed management systems on the Canadian Praires. Weed Sci. 56, 146—150.

Brown, R.B., Noble, S.D., 2005. Site-specific weed management: sensing requirements—what do we need to see? Weed Sci. 53, 252–258.

Buhler, D.D., 2006. Approaches to integrated weed management. In: Singh, H.P., Batish, D.R., Kohli, R.K. (Eds.), Handbook of Sustainable Weed Management. Food Products Press, Binghamton, NY, pp. 813–824.

Buhler, D.D., Hartzler, R.G., Forcella, F., 1997. Implications of weed seed bank dynamics to weed management. Weed Sci. 45, 329–336.

Buttel, F.H., 1985. The land-grant system: a sociological perspective on value conflicts and ethical issues. Agric. Hum. Values 93–94 .

Casady, G.M., Hanley, R.S., Seelan, S.K., 2005. Detection of leafy spurge (*Euphorbia esula*) using multidate high-resolution satellite imagery. Weed Technol. 19, 462–467.

Chassy, B.M., Parrott, W.A., Roush, R., October 2005. Crop Biotechnology and the Future of Food: A Scientific Assessment. CAST Commentary, p. 6. QTA 2005-2.

Christianson, M.L., 1991. Fun with mutants: applying genetic methods to problems of weed physiology. Weed Sci. 39, 489–495.

Cummings, C.H., 2006. Ripe for change: agriculture's tipping point. World Watch. 19 (4), 38–39.

Dawson, J.H., 1965. Competition between irrigated sugar beets and annual weeds. Weeds 13, 245–249.

De Groote, H., Wangare, L., Kanampiu, F., 2007. Evaluating the use of herbicide-coated imidazolinone-resistant (IR) maize seeds to control Striga in farmers' fields in Kenya. Crop Prot. 26, 1496–1506.

Doohan, D., Wilson, R., Canales, E., Parker, J., 2010. Investigating the human dimension of weed management: new tools of the trade. Weed Sci. 58, 503–510.

Duke, S.O. (Ed.), 1996. Herbicide-resistant Crops: Agricultural, Environmental, Economic, Regulatory, and Technical Aspects. CRC-Lewis Pub., Boca Raton, FL, p. 420.

Duke, S.O., 2006. Taking stock of herbicide-resistant crops ten years after introduction. Pest Manag. Sci. 61, 211–218.

Duke, S.O., 2012. Why have no new herbicide modes of action appeared in recent years. Pest Manag. Sci. 68, 505–512.

Duke, S.O., Christy, A.L., Hess, F.D., Holt, J.S., 1991. Herbicide resistant crops. Comment from CAST. No 1991-1. Counc. Agric. Sci. Technol. 24.

Economist, December 1, 2012. March of the lettuce bot. Technol. Q. 5.

Evans, D.A., 1992. Designing More Efficient Herbicides. In: Proc. First Int. Weed Control Congress, vol. 1, pp. 34–41.

Evans, C.L., 2002. The War on Weeds in the Prairie West: An Environmental History. Univ. Of Calgary press, Calgary, Alberta, Canada, p. 309.

Faulkner, E.H., 1943. Plowman's Folly. Univ. of Oklahoma Press, Norman, OK, pp. 150–151.

Fenimore, S.A., Slaughter, D.C., Siemens, M.C., Leon, R.G., Saber, M.N., 2016. Technology for automation of weed control in specialty crops. Weed Technol. 30, 823–837.

Fischer, T., Byerlee, D., Edmeades, G., 2014. Crop Yields and Global Food Security – Will Yield Increase Continue to Feed the World. Australian Centre for International Agricultural Research, p. 634. Monograph #158.

Gardner, G., March/April 1996. IPM and the war against pests. World Watch. 21–27.

Gliessman, S.R., 2001. Agroecosystem Sustainability – Developing Practical Strategies. CRC Press, Boca Raton, FL, p. 210.

Gressel, J., 2009. Crops with target-site herbicide resistance for *Orobanche* and *Striga* control. Pest Manag. Sci. 65, 560–565.

Gressel, J., Hanafi, A., Head, G., Marasas, W., Obilana, A.B., Ochanda, J., Souissi, T., Tzotzos, G., 2004. Major heretofore intractable biotic constraints to African food security that may be amenable to novel biotechnological solutions. Crop Prot. 23, 661–689.

Hall, J.C., Van Eerd, L.L., Miller, S.D., Owen, M.D.K., Prather, T.S., Shaner, D.L., et al., 2000. Future directions for weed science. Weed Technol. 14, 647–658.

Halweil, B., 2006. Grain harvest flat. In: Assadourian, E. (Ed.), Vital Signs 2006–2007. W.W. Norton & Co, New York, NY, pp. 22–23.

Harker, K.N., 2004. My view. Weed Sci. 52, 183–184.

Harker, K.N., O'Donovan, J.T., Blackshaw, R.E., Beckie, H.J., Mallory-Smith, C., Maxwell, B.D., 2012. Our view. Weed Sci. 60, 143–144.

Hess, F.D., Anderson, R.J., Reagan, J.D., 2001. High throughput synthesis and screening: the partner of genomics for discovery of new chemicals for agriculture. Weed Sci. 49, 249–256.

Hileman, B., August 21, 1995. Views differ sharply over benefits, risks of agricultural biotechnology. Chem. Eng. News 8–17.

Holthaus, G., 2009. From Farm to the Table: What All Americans Need to Know about Agriculture. University Press of Kentucky, 363 pp.

Hopkins, W.L., 1994. In: Global Herbicide Directory, first ed. Ag. Chem. Info. Services, Indianapolis, IN, pp. 157–159.

Howard, A.S., 1940. An Agricultural Testament. Oxford Univ. Press, London, UK, 253 pp.

Hubbell, B.J., Welsh, R., 1998. Transgenic crops: engineering a more sustainable agriculture. Agric. Hum. Values 15, 43–56.

James, C., 1999. Global Review of Commercialized Transgenic Crops 1999. Int. Service for the Acquisition of Agri-biotechnology Applications. Briefs No. 12 Preview Int. Serv. for the Acquisition of Agri-biotechnology, Ithaca, NY.

Keeler, K.H., 1989. Can genetically engineered crops become weeds? Biotechnology 7, 1134–1139.

Keeney, D., 1995. Can sustainable agriculture landscapes accommodate corporate agriculture? In: Environmental Enhancement through Agriculture. Proc. of a Conference. Tufts Univ., Boston, MA, pp. 173–181.

King, J., 1991. A matter of public confidence. Agric. Eng. 72 (4), 16–18.

Kirsch, A., September 5, 2016. What makes you so sure? The New Yorker 71–74.

Kirschenmann, F., 2012. From commodities to communities. Leopold Lett. 24 (1), 5.

Kogan, M., 1998. Integrated pest management: historical perspectives and contemporary developments. Annu. Rev. Entomol. 43, 243–270.

Kuhn, T.S., 1970. The Structure of Scientific Revolutions, second ed. University of Chicago Press. 210 pp.

Lang, S., July 25, 2006. Seven-year glitch: Cornell warns that Chinese GM cotton farmers are losing money due to secondary pests. Chron. Online. http://www.newscornell.edu/stories/july06.Bt.cotton.China.ssl.html.

Lemonick, M.D., 2006. Are we losing our edge? TIME 167, pp. 22–30, 33.

LeVasseur, T.J., 2010. Gary Holthaus: from farm to the table: what all Americans need to know about agriculture. J. Agric. Environ. Ethics 23, 301–302.

Liu, Y., Pan, X., Li, J., 2015. Current agricultural practices threaten future global food production. J. Agric. Environ. Ethics 28, 203–216.

Logan, W.B., 1995. Dirt — The Ecstatic Skin of the Earth. Riverhead Books, NY, 202 pp.

Mann, C.C., November/December 2012. State of the species. Does success spell doom for *Homo sapiens*? Orion 16–27.

Marshall, G., 2001. A perspective on molecular-based research: integration and utility in weed science. Weed Sci. 49, 273–275.

McDougall, P., 2016. The cost of new agrochemical product discovery, development and registration in 1995, 2000, 2005–2008 and 2010 to 2014. R&D expenditure in 2014 and expectations for 2019. Crop Life Int. http://www.croplifeamerica.org/cost-of-crop-protection-innovation-increases-to-286-million-per-product/.

McWhorter, C.G., Gebhardt, M.R. (Eds.), 1987. Methods of Applying Herbicides. Weed Sci. Soc. of America., Champaign, IL. Monograph 4. 358 p.

Montgomery, D.R., 2007. Dirt — The Erosion of Civilizations. University of California Press, Berkeley, CA, 285 pp.

Morgan, P.A., Peters, S.J., 2006. The foundations of planetary agrarianism. Thomas Berry and Liberty Hyde Bailey. J. Agric. Environ. Ethics 19, 443–468.

Mortensen, D.A., Coble, H.D., 1991. Two approaches to weed control decision-aid software. Weed Technol. 5, 445–452.

Mortensen, D.A., Egan, J.F., Maxwell, B.D., Ryan, M.R., Smith, R.G., 2012. Navigating a critical juncture for sustainable weed management. Bioscience 62 (1), 75–84.

Moss, S.R., 2008. Weed research: is at delivering what it should? Weed Res. 48 (5), 389–393.

O'Donovan, J.T., 1996. Computerized decision support systems: aids to rational and sustainable weed management. J. Plant Sci. 76, 3–7.

Ohlson, K., March/April 2016. Dirt first — a renegade soil scientist is transforming American agriculture. Orion 14–21.

Peterson, A., 2000. Alternatives, traditions, and diversity in agriculture. Agric. Hum. Values 17, 95–106.

Ransom, J., Kanampiu, F., Gressel, J., De Groote, H., Burnet, M., Odhiambo, G., 2012. Herbicide applied to imidazolinone resistant-maize seed as a *Striga* control option for small-scale African farmers. Weed Sci. 60, 283–289.

Sachs, J.D., 2005. The End of Poverty: Economic Possibilities for Our Time. The Penguin Press, New York, NY, 396 pp.

Sandermann, H., 2006. Plant biotechnology: ecological case studies on herbicide resistance. Trends Plant Sci. 11, 324–328.

Schweizer, E.E., Lybecker, D.W., Wiles, L.J., 1998. Important biological information needed for bioeconomic weed management models. In: Hatfield, J.L., Buhler, D.D., Stewart, B.A. (Eds.), Integrated Weed and Soil Management. Ann Arbor press, Chelsea, MI, pp. 1–24.

Sexton, S., Hildyard, N., Lehmann, L., 1998. Food? Health? Hope? Genetic engineering and world hunger. Cornerhouse Briefing No. 10.

Shand, H., Wetter, K.J., 2006. Shrinking science: an introduction to nano technology. In: Starke, L. (Ed.), State of the World – 2006. W.W. Norton & Co., New York, NY, pp. 78–95.

Shapin, S., May 15, 2006. Paradise sold. The New Yorker 84–88.

Shaw, D.R., 2005. Translation of remote sensing data into weed management decisions. Weed Sci. 53, 264–273.

Smith, R.F., van den Bosch, R., 1967. Integrated Control. In: Apple, J.L., Smith, R.F. (Eds.), Pest Control: Biological, Physical, and Selected Chemical Methods. Plenum Press, New York, NY, pp. 295–340, 457 pp.

Specter, M., January 2, 2016. Rewriting the code of life. The New Yorker 34–43.

Stern, B.M., Smith, R.F., van den Bosch, R., Hagen, K.S., 1959. The Integrated Control Concept. Hilgardia 29 (2), 81–101.

Swinton, S.M., 2005. Economics of site-specific weed management. Weed Sci. 53, 259–263.

Urban, T.N., 1991. Agricultural industrialization: its inevitable. Choices, 4th Quarter 6 (4), 4–6.

Van Wychen, L., 2016. 2015 Survey of the Most Common and Troublesome Weeds in the United States and Canada. Weed Science Society of America National Weed Survey Dataset. Available: http://wssa.net/wp-content/uploads/2015-Weed-Survey_final.xlsx.

Ward, S.M., Cousens, R.D., Bagavathiannan, M.V., Barney, J.N., Beckie, H.J., Busi, R., et al., 2014. Agricultural weed research: a critique and two proposals. Weed Sci. 62, 672–678.

Weaver, S., 1993. Simulation of crop-weed competition: models and their application. In: Symposium on Weed Ecology, Expert Comm. on Weeds, pp. 3–11. Edmonton, Canada.

Weller, S.C., Bressan, R.A., Goldsbrough, P.B., Fredenburg, T.B., Hasegawa, P.M., 2001. The effect of genomics on weed management in the 21st century. Weed Sci. 49, 282–289.

Wiles, L.J., 2005. Sampling to make maps for site-specific weed management. Weed Sci. 53, 228–235.

William, R.D., Ogg, A., Rabb, C., 2001. My view. Weed Sci. 49, 149.

Wyse, D.L., 1992. Future of weed science research. Weed Sci. 6, 162–165.

Young, S.L., Pitla, S.K., Van Evert, F.K., Schueller, J.K., Pierce, F.J., 2017. Moving integrated weed management from low level to a truly integrated and highly specific weed management system using advanced technologies. Weed Res. 57, 1–5.

Young, S.L., 2012. True integrated weed management. Weed Res. 52, 107–111.

Zimdahl, R.L., 2006. Agriculture's Ethical Horizon, first ed. Elsevier, Inc., San Diego, CA. 235 pp.

Zimdahl, R.L., 2012. Agriculture's Ethical Horizon, second ed. Elsevier, Inc., San Diego, CA. 274 pp.

Zimdahl, R.L., 2010. A History of Weed Science in the United States. Elsevier, Inc., San Diego, CA, 207 pp.

Appendix 1
List of Crop and Other Nonweedy Plants Cited in Text, Alphabetized by Common Name[1]

Common Name	Scientific Name
Alfalfa	*Medicago sativa* L.
Almonds	*Prunus dulcis* (Mill.) D.A. Webb
American cranberry	*Vaccinium macrocarpon* Ait.
Apple	*Malus* spp.
Apricot	*Prunus armeniaca* L.
Argentine canola	*Brassica napus* L.
Artichoke	*Cynara scolymus* L.
Ash	*Fraxinus* spp.
Asparagus	*Asparagus officinalis*
Avocado	*Persea americana* Mill
Azalea	*Rhododendron* spp.
Bahiagrass	*Paspalum notatum* Flueggé
Bamboo	*Bambusa vulgaris* Schrad. ex J.C. Wendl
Banana	*Musa paradisiaca* L.
Barley	*Hordeum vulgare* L.
Bean	*Phaseolus* spp.
Begonia	*Begonia* spp.
Beet	*Beta vulgaris* L.
Bellpepper	*Capsicum annuum* L.
Bermudagrass	*Cynodon dactylon* L.

[1]With some exceptions (e.g., clovers), plants are listed alphabetically the way the name is used in the text. That is, white ash is listed with those plants beginning with "w" rather than as "ash, white".

Common Name	Scientific Name
Berseem clover	*Trifolium alexandrinum* L.
Big bluestem	*Andropogon gerardii* Vitm.
Birdsfoot trefoil	*Lotus corniculatus* L.
Blackberries	*Rubus allegheniensis* Porter
Black currant	*Ribes nigrum* L.
Black walnut	*Juglans nigra* L.
Blueberries	*Vaccinium angustifolium* Ait.
Bluebunch wheatgrass	*Agropyron spicatum* (Pursh) Scribn. & Sm.
Blue grama	*Bouteloua gracilis* (Willd. ex Kunth) Lag ex Griffiths
Bluegrass	*Poa pratensis* L.
Broadbean	*Vicia faba* L.
Broccoli	*Brassica oleracea* L.—Botrytis group
Brussel sprouts	*Brassica oleracea*—Gemmifera group
Buckwheat	*Fagopyrum esculentum* Moench.
Buffalo grass	*Buchloe dactyloides*
Bunchgrass	A general category with several genera
Cabbage	*Brassica oleracea* L.
Calliandra	*Calliandra calothyrsus* Meissn
Camellia	*Camellia* spp.
Canola = rapeseed	*Brassica napus* L.
Cantaloupe	*Cucumis melo* var. *cantalupensis* Naudin
Caper spurge	*Euphorbia lathyris* L.
Carnation	*Dianthus caryophyllus* L.
Carrot	*Daucus carota* L.
Cassava = manioc	*Manihot esculenta* Crantz
Catnip	*Nepeta cataria* L.
Celery	*Apium graveolens* L.
Centro	*Centrosema pubescens* Benth
Cherry	*Prunus* spp.
Chicory	*Chicorum intybus* L.
Chestnut oak	*Quercus montana* Willd.
Chickpea	*Cicer arietinum* L.

Common Name	Scientific Name
Chinese cabbage	*Brassica rapa* L.
Clove	*Syzygium aromaticom* (L.) Merr. & L.M. Perry
Clover	*Trifolium* spp.
Coconut	*Cocos nucifera* L.
Coffee	*Coffea arabica* L.
Collards	*Brassica oleracea* L.
Common rye	*Secale cereale* L.
Corn = maize	*Zea mays* L.
Corallita	*Antigonon leptopus* Hook & Arn.
Cotton	*Gossypium hirsutum* L.
Cottonwood	*Populus deltoides* Marsh
Cowpea	*Vigna unguiculata* (L.) Walp.
Crambe	*Crambe abyssinica* Hochst.
Cranberry	*Viburnum opulus* L.
Crested wheatgrass	*Agropyron desertorum* (Fisch. ex Link) Schult.
Crimson clover	*Trifolium incarnatum* L.
Crotalaria	*Crotalaria grahamiama* Wight and Arn
Crownvetch	*Coronilla varia* L.
Cucumber	*Cucumis sativus* L.
Cypress	*Taxodium distichum* L.
Date, desert	*Balanites aegyptiaca* (L.) Del.
Date palm	*Phoenix dactylifera* L.
Delphinium	*Delphinium* spp.
Douglas fir	*Pseudotsuga menziesii* (Mirbel) Franco
Dry bean	*Phaseolus* spp.
Durum wheat	*Triticum durum* Desf.
Egusi melon	*Citrullus lanatus* L. subsp. *mucospermus*
Eggplant	*Solanum melongena* L.
English ivy	*Hedera helix* L.
Fava/faba bean	*Vicia faba* L.
Fescue	*Festuca* spp.

Common Name	Scientific Name
Field pea	*Pisum sativum* L.
Fig	*Ficus* spp.
Flax	*Linum usitatissimum* L.
Foxtail millet	*Setaria italicum* (L.) Beauv.
Foxglove	*Digitalis purpurea* L. (Several species)
Garbanzo bean	*Cicer arietinum* see Chickpea
Globe artichoke	*Cynara scolymus*
Grapes	*Vitis* spp.
Grapefruit	*Citrus paradisi*
Groundnut = peanut	*Arachis hypogaea* L.
Guayule	*Parthenium argentatum* A. Gray
Hairy vetch	*Vicia villosa* Roth
Hops	*Humulus lupulus* L.
Hyacinth bean = lablab	*Lablab purpureus* L.
Idaho fescue	*Festuca idahoensis* Elmer
Intermediate wheatgrass	*Thinopyrum intermedium* (Host) Barkworth and D.R. Dewey
Iris	*Iris pseudacorus* L.
Jack bean	*Canavalia ensiformis* L.
Jatropha	*Jatropha curcas* L.
Jojoba	*Simmondsia chinensis* L.
Juniper	*Juniperus* spp.
Jute	*Corchorus olitorius* L.
Kentucky bluegrass	*Poa pratensis* L.
Kernza = intermediate wheatgrass	*Thinopyrum intermedium* (Host) Barkworth and D.R. Dewey
Lacquer tree	*Toxicodendron vernicifluum* (Stokes) F. Barkley
Lemon	*Citrus limon* L.
Lentil	*Lens culinaris* Medik. = *L. culinaris* L.
Lettuce	*Lactuca sativa* L.
Lily-of-the-valley	*Convallaria majalis* L.
Lima bean	*Phaseolus limensis* L.

Common Name	Scientific Name
Lupine	*Lupinus* spp.
Macadamia	*Macadamia* spp.
Maize = corn	*Zea mays* L.
Maple	*Acer* spp.
Medic	*Medicago lupulina* L.
Melon	*Cucumis melo* L.
Mint	*Mentha piperita* L.
Millet	*Pennisetum* spp.
Mistletoe	*Arceuthobium* spp.
Mung bean	*Vigna radiata* (L.) Wilczek var. radiata
Muskmelon	*Cucumis melo* L.
Narcissus	*Narcissus poeticus* L
Nectarine	*Prunus persica* L.
Norway maple	*Acer platanoides* L.
Oak	*Quercus* spp.
Oats	*Avena sativa* L.
Oleander	*Nerium oleander* L.
Olive	*Oleo europaea* L.
Onion	*Allium cepa* L.
Orange	*Citrus sinensis* (L.) Osb.
Orchard grass	*Dactylis glomerata* L.
Pangola grass	*Digitaria decumbens* Stent
Papaya	*Carica papaya* L.
Pea	*Pisum sativum* L.
Peanut	*Arachis hypogaea* L.
Pear	*Pyrus* spp.
Pearl millet	*Pennisetum glaucum* L.
Pecan	*Carya illinoinensis* L.
Peppermint	*Mentha piperita* L.
Peppers	*Capsicum* spp.
Petunia	*Petunia* × *hybrida*

Common Name	Scientific Name
Phacelia	*Phacelia tanacetifolia* Benth.
Pigeon pea	*Cajanus cajan* (L.) Millsp.
Pine	*Pinus* spp.
Pineapple	*Ananas comosus* L.
Plantain	*Musa paradisiaca* L.
Plum	*Prunus domestica* Stanley
Polish canola	*Brassica rapa* L.
Ponderosa pine	*Pinus ponderosa* Dougl. ex P. & C. Laws.
Poplar	*Populus* spp.
Potato	*Solanum tuberosum* L.
Proso millet	*Panicum miliaceum* L.
Psoralea	*Psoralea* spp.
Pubescent wheatgrass	*Thinopyrum intermedium* subsp. *barbulatum*
Pumpkin	*Cucurbita pepo* L.
Radicchio	*Cichorium intybus* var. *foliosum* Radicchio Group
Radish	*Raphanus sativus* L.
Rapeseed = canola	*Brassica napus* L.
Raspberry	*Rubus* spp.
Rattan	*Calamus*—largest genus with 370 species
Red alder	*Alnus rubra* L.
Red clover	*Trifolium pratense* L.
Red currant	*Ribes sativum* Syme
Red maple	*Acer rubrum* L.
Red table beets	*Beta vulgaris* L.
Rhododendron	*Rhododendron* sp.
Rhubarb	*Rheum rhaponticum* L.
Rice	*Oryza sativa* L.
Rose	*Rosa* spp.
Rutabaga	*Brassica napobrassica* L.
Rye	*Secale cereale* L.
Ryegrass	*Lolium* spp.

Common Name	Scientific Name
Sacramento (mountain) thistle	*Cirsium vinaceum* Woot. & Standl.
Safflower	*Carthamus tinctorius* L.
Salvia	*Salvia officinalis* L.
Sand dropseed	*Sporobolus cryptandrus* (Torr.) A. Gray
Sainfoin	*Onobrychis viciifolia*
Satsuma orange	*Citrus nobilis* Lour.
Sava medic	*Medicago scutella* L.
Seashore paspalum	*Paspalum vaginatum*
Sesbania	*Sesbania sesban* (L.) Merr.
Siratro	*Macroptilium purpureum* (DC) Urb.
Side-oats grama	*Bouteloua curtipendula* (Michx.) Torr.
Silk-cotton tree	*Ceiba pentandra* (L.) Gaertner poll.
Slender wheatgrass	*Agropyron desertorum* (Fisch. ex Link) Schult.
Smooth bromegrass	*Bromus inermis* Leyss
Snapbean	*Phaseolus vulgaris* L.
Snapdragon	*Antirrhinum majus* L.
Snowberry	*Symphoricarpos* spp.
Sorghum	*Sorghum bicolor* (L.) Moench
Soybean	*Glycine max* (L.) Merr.
Spearmint	*Mentha spicata* L.
Spinach	*Spinacia oleracea* L.
Squash	*Cucurbita pepo* L.
St. Augustine grass	*Stenotaphrum secundatum* Hitchcock, A.S. (rev. A. Chase)
St. Johnwort	*Hypericum perforatum* L.
Strawberry	*Fragaria vesca* L.
Subterranean clover	*Trifolium subterraneum* L.
Sudangrass	*Sorghum sudanense* (Piper) Stapf
Sudax	*Sorghum—sudangrass* cross
Sugarbeet	*Beta vulgaris* L.
Sugarcane	*Saccharum officinarum* L.
Sunflower	*Helianthus annua* L.

Common Name	Scientific Name
Sweet potato	*Ipomoea batatas* Lam
Sweet wormwood	*Artemisia annua* L.
Swiss chard	*Beta vulgaris* L. subsp. *vulgaris*
Switchgrass	*Panicum virgatum* L.
Teosinte	*Zea perennis* (Hitchc.) Reeves & Manglesdorf
Timothy	*Phleum pratense* L.
Tobacco	*Nicotiana tabacum* L.
Tomato	*Lycopersicon esculentum* L.
Triticale	*Triticosecale*
Tropical kudzu	*Pueraria phaseoloides* (Roxb.) Benth.
Velvet bean	*Mucuna cochinchinensis* (Lour.) A. Chev.
Violet	*Viola* spp.
Watermelon	*Citrullus lanatus* (Thunb.) Matsum. & Nakai
Walnut	*Juglans* spp.
Wheat	*Triticum aestivum* L.
White ash	*Fraxinus americana* L.
White oak	*Quercus alba* L.
White clover	*Trifolium repens* L.
Wild winged bean	*Psophocarpus palustris* Desv.
Willow	*Salix* spp.
Wild cherry	*Prunus* spp.
Winter camelina	*Camelina sativa* (L.) Crantz
Yellow Indian grass	*Sorghastrum nutans* (L.) Nash
Zoysiagrass	*Zoysia japonica* Steud.
Zucchini squash	*Cucurbita maxima* L.

Appendix 2
Weeds Cited in Text[1]
Alphabetized by Common Name

Common Name	Scientific Name[2]
Alder	*Alnus* spp.
Alkaliweed	*Cressa truxillensis* H.B.K.
Alligatorweed	*Alternanthera philoxeroides* (Mart.) Griseb.
Amaranth	*Amaranthus* spp.
American black nightshade	*Solanum americanum* Mill.
American dragonhead	*Dracoephalum parviflorum* Nutt.
American pondweed	*Potamogeton nodosus* Poir.
Annual bluegrass	*Poa annua* L.
Annual bromes	*Bromus tectorum* L. and *Bromus japonicus* Thunb. ex Murr. = Japanese brome
Annual ryegrass	Probably—*Lolium temulentum* L.
Annual sowthistle	*Sonchus oleraceus* L.
Arrow grass	*Triglochin maritima* L.
Arrowhead	*Sagittaria sagittifolia* L.
Artichoke thistle	*Cynara cordonculus* L.
Arundo	*Arundo donax* L.
Aspen = quaking aspen	*Populus tremuloides* Michx.

[1]With a few exceptions, weeds are listed alphabetically the way the name is used in the original source. That is, field bindweed is listed with those plants beginning with "f" rather than as bindweed, "field."

[2]The authority for the scientific name is given when it could be found after a reasonable search.

Common Name	Scientific Name[2]
Autumn olive	*Elaeagnus umbellata* Thunb.
Azolla = pinnate mosquitofern	*Azolla pinnata* R. Brown
Barnyardgrass	*Echinochloa crus-galli* (L.) Beauv.
Bearded sprangletop	*Leptochloa fascicularis* (Lam.) Gray
Beggarstick	*Bidens* spp.
Bengal dayflower	*Commelina benghalensis* L.
Bentgrass	*Agrostis* spp.
Bermudagrass	*Cynodon dactylon* (L.) Pers.
Bigleaf maple	*Acer macrophyllum* Pursh
Big sagebrush	*Artemisia tridentata* Nutt.
Bindweed	*Convolvulus* spp.
Birdsrape mustard	*Brassica rapa* L.
Bitterbush	*Eupatorium odoratum* L. = *Chromolaena odorata* (L.) R.M. King & M. Robinson
Bittercress	*Cardamine* sp.
Blackgrass	*Alopecurus myosuroides* Huds.
Black henbane	*Hyoscyamus niger* L.
Black medic	*Medicago lupilina* L.
Black mustard	*Brassica nigra* (L.)W.J.D. Koch
Black nightshade	*Solanum nigrum* L.
Blackseed needlegrass	*Stipa* (species unknown)
Bluebunch wheatgrass	*Pseudoroegneria spicata* (Pursh) Á. Löve
Blue pimpernel	*Anagallis coerulea* Nathh.
Bohemian knotweed—hybrid of Japanese and giant knotweed	*Polygonum* × *bohemicum* (J. Chertek & Chartkova) Zirka and Jacobsen
Bouncing bet	*Saponaria officinalis* L.
Brackenfern	*Pteridium aquilinum* (L.)Kuhn
Branched broomrape	*Orobanche ramosa* L.
Brazilian peppertree	*Schinus terebinthifolius* Raddi.
Brazilian satintail	*Imperata brasiliensis* Trin.
Bristly starbur	*Acanthospermum hispidum* DC.
Broadleaf filaree	*Erodium botrys* (Cav.) Bertol.
Broadleaf plantain	*Plantago major* L.
Brome	*Bromus* spp.

Common Name	Scientific Name[2]
Broomrape	*Orobanche* spp.
Broom snakeweed	*Gutierrezia sarothrae* (Pursh) Britt & Rusby
Buckhorn plantain	*Plantago lancolata* L.
Bufflegrass = arrow-arum	*Pennisetum ciliare* (L.) Link
Buffalobur	*Solanum rostratum* Dun
Bulbous buttercup	*Ranunculus bulbosus* L.
Bull mallow	*Malva nicaeensis* All.
Bull thistle	*Cirsium vulgare* (Savi) Tenore
Bulrush	*Scirpus* spp.
Burcucumber	*Sicyos angulatus* L.
Burdock	*Arctium* spp.
Bursage	*Ambrosia* spp.
Buttercup	*Ranunculus* spp.
Cacti	*Opuntia* spp.
California sagebrush	*Artemisia californica* Less.
California chaparral = whiteleaf sage	*Salvia leucophylla* Greene
California peppertree	*Schinus molle* L.
Camel thorn	*Alhagi pseudalhagi* (Bieb.) Desv.
Canada thistle	*Cirsium arvense* (L.) Scop.
Canary grass	*Phalaris* spp.
Carpet grass	*Axonopus affinis* Chase
Catchweed bedstraw	*Galium aparine* L.
Catclaw mimosa	*Mimosa pigra* L.
Catnip	*Nepeta cataria* L.
Cattail	See common cattail
Celosia	*Celosia argentea* L.
Charlock	*Brassica* spp.
Chicory	*Cichorium intybus* L.
Chinese sprangletop	*Leptochloa chinensis* (L.) Nees
Chinese tallow	*Sapium sebiferum* (L.) Roxb
Chokecherry	*Prunus virginiana* L.
Coat buttons	*Tridax procumbens* L.
Cocklebur	*Xanthium* spp.

Common Name	Scientific Name[2]
Coffee senna	*Cassia occidentalis* L.
Cogongrass	*Imperata cylindrica* (L.) Beauv.
Coltsfoot	*Tussilago farfara* L.
Common burdock	*Arctium minus* (Hill) Bernh.
Common cattail	*Typha latifolia* L.
Common chickweed	*Stellaria media* (L.) Vill.
Common cocklebur	*Xanthium strumarium* L.
Common crupina	*Crupina vulgaris* Cass.
Common elodea	*Elodea canadensis* L.C.Rich
Common foxglove	*Digitalis purpurea* L.
Common groundsel	*Senecio vulgaris* L.
Common hempnettle	*Galeopsis tetrahit* L.
Common hogweed	*Heracleum sphondylium* L.
Common lambsquarters	*Chenopodium album* L.
Common mallow	*Malva neglecta* Wallr.
Common milkweed	*Asclepias syriaca* L.
Common mullein	*Verbascum thapsus* L.
Common nettle	*Urtica* spp.
Common plantain	*Plantago* spp.
Common pokeweed	*Phytolacca americana* L.
Common purslane	*Portulaca oleracea* L.
Common ragweed	*Ambrosia artemisiifolia* L.
Common reed	*Phragmites australis* (Cav.) Trin. ex Steud.
Common rye	*Secale cereale* L.
Common St. Johnswort	*Hypericum perforatum* L.
Common sunflower	*Helianthus annuus* L.
Common tansy = tansy ragwort	*Tanacetum vulgare* L.
Common teasel	*Dipsacus fullonum* L.
Common velvetgrass	*Holcus lanatus* L.
Common vetch	*Vicia sativa* L.

Common Name	Scientific Name[2]
Common waterhemp	*Amaranthus rudis* Sauer
Corallita	*Antigonon leptopus*
Coontail	*Ceratophyllum demersum* L.
Corn cockle	*Agrostemma githago* L.
Corn marigold	*Chrysanthemum segetum* L.
Corn poppy	*Papaver rhoeas* L.
Corn speedwell	*Veronica arvensis* L.
Cottonwood	*Populus* spp.
Cowcockle	*Vaccaria pyramidata* Medicus
Cowpea witchweed	*Striga gesnerioides* (Willd.) Vatke
Crazyweed	*Oxytropis* spp.
Crowfoot grass	*Dactyloctenium aegypticum*
Crabgrass	*Digitaria* spp.
Creeping bentgrass	*Agrostis stolonifera* L.
Creeping buttercup	*Ranunculus repens* L.
Creeping daisy	*Wedelia trilobata* (L.) Hitchc.
Creeping woodsorrel	*Oxalis corniculata* L.
Cress	*Lepidium sativum* L.
Croftonweed	*Ageratina adenophora* (Sprent.) King and H.E. Robins
Curly dock	*Rumex crispus* L.
Curlycup gumweed	*Grindelia squarrosa* (Pursh) Dunal
Cut-leaf groundcherry	*Physalis angulata* L.
Daisy fleabane = annual fleabane	*Erigeron annuus* (L.) Pers.
Dalmatian toadflax	*Linaria dalmatica* (L.) Mill.
Dandelion	*Taraxacum officinale* Weber in Wiggers
Darnel = poison ryegrass	*Lolium temulentum* L.
Dayflower	*Commelina* spp.
Deadly nightshade	*Atropa belladonna* L.
Devil's-claw	*Proboscidea louisianica* (Mill.) Thellung
Diffuse knapweed	*Centaurea diffusa* Lam.
Dock	*Rumex* spp.

Common Name	Scientific Name[2]
Dodder	*Cuscuta* spp.
Downy brome—synonyms Cheatgrass, drooping brome	*Bromus tectorum* L.—synonym *Anisanta tectorum* (L.) Nevski
Duckweed	*Lemna* spp.
Dwarf mistletoe	*Arceuthobium vaginatum* M. Bieb.
Dyer's woad	*Isatis tinctoria* L.
Eastern black nightshade	*Solanum ptycanthum* Dun
Eclipta	*Eclipta prostrata* L.
English ivy	*Hedera helix* L.
Elodea	*Eledoa* spp.
Eurasian watermilfoil	*Myriophyllum spicatum* L.
European barberry	*Berberis thunbergii* DC.
European blackberry	*Rubus fruticosus* L.
European buckthorn	*Rhamnus cathartica* L.
Evening primrose	*Oenothera biennis* L.
False brome = slender false brome	*Brachypodium sylvaticum* (Huds.) Beauv
False cleavers	*Galium spurium* L.
False flax	*Camelina* spp.
Fall panicum	*Panicum dichotomiflorum* Michx.
Fescue	*Festuca* spp.
Field dodder	*Cuscuta campestris* Yuncker
Field bindweed	*Convolvulus arvensis* L.
Field horsetail	*Equisetum arvense* L.
Field pennycress	*Thlaspi arvense* L.
Field pepperweed	*Lepidium campestre* (L.)R.Br.
Field poppy	*Papaver dubium* L.
Field violet	*Viola arvensis* Murr.
Fireweed	*Epilobium angustifolium* L.
Flaxweed = flatseed false flax	*Camelina alyssum* (Mill.) Thell
Flixweed	*Descurainia sophia* (L.) Webb. ex Prantl
Florida beggarweed	*Desmodium tortuosum* (Sw.) DC.

Common Name	Scientific Name[2]
Florida pusley	*Richardia scabra* L.
Foxtails	*Setaria* spp.
Foxtail barley	*Hordeum jubatum* L.
Foxtail millet	*Setaria italica* (L.) Beauv.
Fringed sagebrush	*Artemisia frigida* Willd.
Fumitory	*Fumaria officinalis* L.
Galinsoga	*Galinsoga* spp.
Gambel oak	*Quercus gambelii* Nutt.
Garden cress	*Lepidium sativum* L.
Garlic mustard	*Alliaria petiolata* (Bieb.) Cavara & Grande
Giant duckweed = greater duckweed	*Spirodela polyrrhiza* (L.) Schleid
Giant foxtail	*Setaria faberi* Herrm.
Giant hogweed	*Heracleum mantegazzianum* Sommler & Levier
Giant knotweed	*Polygonum sachalinense* F. Schmidt ex Maxim
Giant ragweed	*Ambrosia trifida* L.
Giant reed	*Arundo donax* L.
Giant Salvinia	*Salvinia auriculata* Aubl.
Giant smutgrass	*Sporobolus indicus* (L.) R. Br. var. *pyramidalis*
Giant witchweed = purple witchweed	*Striga hermonthica* Benth.
Gill-over-the-ground = creeping Charlie	*Glechoma hederacea* L.
Globe fringe rush	*Fimbristylis miliacea* (L.)Vahl
Goldenrod	*Solidago* sp.
Goosegrass	*Eleusine indica* (L.) Gaertn.
Gray rabbitbrush	*Chrysothamnus nauseosus* (Pallas)Britt.
Greasewood	*Sarcobatus vermiculatus* (Hook.) Torr.
Green bristlegrass	*Setaria magna* Griseb
Green foxtail	*Setaria viridis* (L.) Beauv.
Green sorrel	*Rumex acetosa* L.
Groundcherry	*Physalis* spp.
Ground ivy	*Glechoma hederacea* L.
Groundsel	*Senecio* spp.

Common Name	Scientific Name[2]
Guayule	*Parthenium argentatum* A. Gray
Guineagrass	*Panicum maximum* Jacq.
Hair fescue	*Festuca filiformis* Pourret
Hairy beggarticks	*Bidens pilosa* L.
Hairy fleabane	*Conyza bonariensis* (L.) Cronq.
Hairy nightshade	*Solanum sarrachoides* Sendtner
Hairy vetch	*Vicia villosa* Roth
Hairy willowherb	*Epilobium hirsutum* L.
Halberdleaf orach	*Atriplex patula* var. *hastata* (L.) Gray
Halogeton	*Halogeton glomeratus* (Stephen ex. Bieb.) C. A. Mey
Hayfield tarweed	*Hemizonia congesta* DC.
Heath	*Erica scoparia* L.
Hedge bindweed	*Calystegia sepium* (L.) R. Br.
Heliotrope	*Heliotropium* spp.
Hemp dogbane	*Apocynum cannabinum* L.
Hemp sesbania	*Sesbania exaltata* (Raf.) Rtdb. Ex A.W. Hill
Henbane	*Hyoscyamus* spp.
Henbit	*Lamium amplexicaule* L.
Hoary alyssum	*Berteroa incana* (L.) DC.
Hoary cress = whitetop	*Cardaria draba* (L.) Desv.
Horsenettle	*Solanum carolinense* L.
Horsetail	*Equisetum* spp.
Horseweed	*Conyza canadensis* (L.) Cronq.
Houndstongue	*Cynoglossum officinale* L.
Hydrilla	*Hydrilla verticillata* (L.f.) Royle
India crabgrass	*Digitaria longiflora* (Retz.) Pers.
Indian balsam	*Impatiens glandulifera* Royle
Indian tobacco	*Lobelia inflata* L.
Italian ryegrass	*Lolium multiflorum* Lam.
Itchgrass	*Rottboellia cochinchinensis* (Lour.) W.D. Clayton
Ivyleaf morningglory	*Ipomoea hederacea* (L.) Jacq.

Common Name	Scientific Name[2]
Japanese brome	*Bromus japonicus* Thunb. ex Murr.
Japanese honeysuckle	*Lonicera japonica* Thunb.
Japanese knotweed	*Polygonum cuspidatum* Seib. & Zucc. = *Fallopia japonica* (Houtt.) Ronse. Ducr.
Japanese stiltgrass	*Microstegium vimineum* (Trin.) A. Camus var. *Imberbe* (Ness) Honda
Jaraguagrass	*Hyparrhenia rufa* (Nees) Stapf.
Java plum	*Syzygium cumini* (L.) Skeels
Jerusalem artichoke	*Helianthus tuberosus* L.
Jimsonweed	*Datura stramonium* L.
Johnsongrass	*Sorghum halepense* (L.) Pers.
Jointed goatgrass	*Aegilops cylindrica* Host
Junglerice	*Echinochloa colona* (L.) Link
Karibaweed	*Salvinia molesta* Mitch.
Kikuyugrass	*Pennisetum clandestinum* Hochst. ex Chiov.
Klamath weed	See St. Johnswort
Knapweed	*Centaurea* spp.
Kochia	*Kochia scoparia* (L.)Schrad.
Kudzu	*Pueraria montana* var. *lobata* (Willd.) Maesen & S. Almeida
Ladysthumb	*Polygonum persicaria* L.
Lambsquarter	*Chenopodium album* L.
Lantana	*Lantana* spp.
Largeleaf lantana	*Lantana camara* L.
Large crabgrass	*Digitaria sanguinalis* (L.)Scop.
Larkspur	*Delphinium* spp.
Leafy spurge	*Euphorbia esula* L.
Little mallow	*Malva parviflora* L.
Locoweed	*Astragalus* spp.
Longspine sandbur	*Cenchrus longispinus* (Hack.) Fern.
Madagascar periwinkle	*Catharanthus roseus* (L.) G.Don
Mallow	*Malva* spp.

Common Name	Scientific Name[2]
Mandrake	*Mandragora officinarum*
Manzanita	*Arctostaphylos mendocinoensis*
Marestail	*Hippuris vulgaris* L.
Marijuana	*Cannabis sativa* L.
Marshpepper smartweed	*Polygonum hydropiper* L.
Mayweed = dog fennel	*Anthemis cotula* L.
Meadow barley	*Hordeum brachyantherum* Nevski
Meadow foxtail	*Alopecurus pratensis* L.
Meadow salsify	*Tragopogon pratensis* L.
Meadowsweet	*Spirea latifolia* (Ait.) Borkh.
Mediterranean sage	*Salvia aethiopis* L.
Medusahead	*Taeniatherum caput-medusae* (L.) Nevski
Melaleuca	*Melaleuca quinquenervia* (Cav.) Blake
Mesquite	*Prosopis* spp.
Miconia	*Miconia* spp.
Mile-a-minute weed	*Polygonum perfoliatum* L.
Mimosa vine = sensitive vine	*Mimosa strigillosa* Torr. & Gray
Mistletoe	*Viscum* spp.
Mock bishopsweed	*Ptilimnium capillaceum* (Michx.) Raf.
Molassesgrass	*Melinus minitiflora* Beauv.
Monkshood = Aconite	*Aconitum napellus* L.
Monochoria	*Monochoria vaginalis* (Burm.f.)Kunth
Morningglory	*Ipomoea* spp.
Moth mullein	*Verbascum blattaria* L.
Mouse-ear cress	*Arabidopsis thaliana* (L.) Heynh.
Mugwort	*Artemisia vulgaris* L.
Mulberry weed	*Fatousa villosa* (Thunb.) Nakai.
Mullein	*Verbascum blattaria* L. or *Verbascum thapsus* L.
Multiflora rose	*Rosa multiflora* Thunb. ex Murr.
Musk thistle	*Carduus nutans* L.
Mustard	*Brassica* spp.

Common Name	Scientific Name[2]
Myrtle spurge	*Euphorbia myrsinites* L.
Needle-and-thread grass	*Stipa comata* Trin. & Rupr.
Nepalese browntop	*Microstegium vimineum* (Trin) A. Camus
Netseed lambsquarter	*Chenopodium berlandieri* Moq.
Nettle	*Urtica* spp.
Nightshade	*Solanum* spp.
Nodeweed	*Synedrella nodiflora* (L.) Gaertn
Northern jointvetch	*Aeschynomene virginica* (L.)B.S.P.
Nostoc	*Nostoc* spp.
Nutsedge	*Cyperus* spp.
Oak	*Quercus* spp.
Oakleaf goosefoot	*Chenopodium glaucum* L.
Ohio buckeye	*Aesculus glabra* Willd.
Oldfield cinquefoil	*Potentilla simplex* Michx.
Oldfield toadflax	*Linaria canadensis* (L.) Dumont
Orange hawkweed	*Hieracium aurantiacum* L.
Orchard grass	*Dactylis glomerata* L.
Orobanche	*Orobanche* spp.
Oxalis	*Oxalis pes-caprae* L.
Palmer amaranth	*Amaranthus palmeri* S. Wats.
Pangola grass	*Digitaria decumbens* Stentx
Paragrass	*Brachiaria mutica* (Forsk.)Stapf
Parthenium ragweed	*Parthenium hysterophorus* L.
Passionflower	*Passiflora* spp.
Pennsylvania smartweed	*Polygonum pensylvanicum* L.
Pennycress—see field pennycress	*Thlaspi arvense* L.
Peppergrass = greenflower pepperweed	*Lepidium densiflorum* Schrad.
Perennial pepperweed	*Lepidium latifolium* L.
Perennial ryegrass	*Lolium perenne* L.
Perennial sowthistle	*Sonchus arvensis* L.
Persian darnel	*Lamium persicum* Boiss & Hohen ex Boiss

Common Name	Scientific Name[2]
Persian speedwell	*Veronica persica* Poir.
Phacelia	*Phacelia tanacetifolia* Benth. var. Boratus
Phragmites	See common reed
Pigweed	*Amaranthus* spp.
Pineapple weed	*Matricaria matricarioides* (Less.) C.L.Porter
Pinnate tansymustard	*Descurainia pinnata* (Walt.) Britt
Pitted morningglory	*Ipomoea lacunosa* L.
Plains pricklypear	*Opuntia polyacantha* Haw.
Plantain	*Plantago* spp.
Plumeless thistle	*Carduus acanthoides* L.
Poison hemlock	*Conium maculatum* L.
Poison-ivy	*Toxicodendron radicans* (L.)Ktze
Poison-oak	*Toxicodendron toxicarium* (Salisb.) Gillis
Poison ryegrass = darnel = tares	*Lolium temulentum* L.
Pondweed	*Potamogeton* spp.
Poppy	*Papaver* spp.
Prickly lettuce	*Lactuca serriola* L.
Prickly pear cactus	*Opuntia littoralis* (Engelmann) Cockerell
Prickly sida	*Sida spinosa* L.
Prince's feather	*Polygonum orientale* L.
Prostrate knotweed	*Polygonum aviculare* L.
Prostrate pigweed	*Amaranthus graecizans* auctt., non L.
Prostrate spurge	*Euphorbia supina* Raf.
Puncturevine	*Tribulus terrestris* L.
Purple moonflower	*Ipomoea alba* L.
Purple loosestrife	*Lythrum salicaria* L.
Purple nutsedge	*Cyperus rotundus* L.
Purple witchweed	*Striga hermonthica* Benth.
Purslane	*Portulaca* spp.
Quackgrass	*Elytrigia repens* (L.)Nevski
Queen-of-the-meadow	*Filipendula ulmaria* (L.) Maxim.

Common Name	Scientific Name[2]
Rabbitbrush	*Chrysothamnus* spp.
Ragweed	*Ambrosia* spp.
Rattail fescue	*Vulpia myuros* (L.) K.C.Gmel.
Red alder	*Alnus rubra* Bong.
Red = ivyleaf morningglory	*Ipomoea coccinea* L.
Red rice	*Oryza sativa* L.
Red sepal evening primrose	*Oenothera glazioviana* Micheli
Red sorrel	*Rumex acetosella* L.
Redroot pigweed	*Amaranthus retroflexus* L.
Red sprangletop	*Leptochloa filiformis* (Lam.)Beauv.
Redtop	*Agrostis gigantea* Roth
Reed canarygrass	*Phalaris arundinacea* L.
Rice flatsedge	*Cyperus iria* L.
Rigid ryegrass	*Lolium rigidum* Gaudin
Rocky mountain beeplant	*Cleome serrulata* Pursh
Rosy periwinkle	See Madagascar periwinkle
Roughstalk bluegrass	*Poa trivialis* L.
Round-leaved mallow	*Malva rotundifolia* Am. auctt., non L. = *Malva pusila* Sm.
Rubber vine	*Cryptostegia grandiflora* R. Br.
Rush	*Scirpus* spp.
Rush skeleton weed	*Chondrilla juncea* L.
Russian knapweed	*Acroptilon repens* (L.) DC.
Russian olive	*Elaeagnus angustifolia* L.
Russian thistle	*Salsola iberica* Sennen & Pau
Rye	*Secale montanum* Guss. origin of *S. cereale* L.
Ryegrass	*Lolium* spp.
Sage	*Artemisia* spp.
Sagebrush	*Artemisia* spp.
Sago pondweed	*Potamogeton pectinatus* L.
Sakhalin knotweed = S. knotgrass	*Polygonum sachalinense* F. Schmidt ex Maxim.
Saltbush	*Atriplex* spp.

Common Name	Scientific Name[2]
Salt cedar = tamarisk	*Tamarix ramosissima* Ledeb.
Saltgrass	*Distichlis spicata* (L.) Greene
Salvinia	*Salvinia molesta* Mitch.
Sandbur	*Cenchrus* spp.
Sand dropseed	*Sporobulus cryptandrus* (Torr.) Gray
Sand sagebrush	*Artemisia filifolia* Torr.
Saw grass	*Cladium mariscus* (L.) Pohl ssp. *jamaicense* (Crantz) Kukenth.
Seagrass	There are four families, and 12 genera with 60 species.
Scotch broom	*Cytisus scoparius* (L.) Link
Scotch thistle	*Onopordum acanthium* L.
Seashore paspalum	*Paspalum vaginatum* Sw.
Sensitiveplant	*Mimosa pudica* L.
Sessile joyweed	*Alternanthera sessilis* (L.) R.Br. ex DC.
Shattercane	*Sorghum bicolor* (L.) Moench
Shepherd's purse	*Capsella bursa-pastoris* L. Medicus
Showy crotalaria	*Crotalaria spectabilis* Roth
Showy milkweed	*Asclepias speciosa* Torr.
Siam weed	*Chromolaeum odorata* (L.) R.M. King & H. Robinson
Sicklepod	*Senna obtusifolia* (L.)
Signal grass	*Brachiaria decumbens* Stapf. = *Urochloa decumbens* (Stapf) R.D. Webster
Silver grass	*Miscanthus sinensis* Anderss.
Silverleaf nightshade	*Solanum elaeagnifolium* Cav.
Skeletonleaf bursage	*Ambrosia tomentosa* Nutt.
Skeleton weed	*Lygodesmia juncea* (Pursh) D. Don
Slender amaranth	*Amaranthus viridis* L.
Slender speedwell	*Veronica filiformis* Sm.
Slimleaf lambsquarters	*Chenopodium leptophyllum* (Moq.)Nutt. ex S. Wats.
Small smutgrass	*Sporobolus indicus* (L.) R. Br. var. *indicus*
Smallflower galinsoga	*Gallinsoga parviflora* Cav.
Smallflower morningglory	*Jacquemontia tamnifolia* (L.) Griseb.

Common Name	Scientific Name[2]
Smallflower umbrella sedge	*Cyperus difformis* L.
Smooth brome	*Bromus inermis* Leyss
Smooth sumac	*Rhus glabra* L.
Smartweed	*Polygonum* spp.
Smooth barley	*Hordeum murinum* L. spp. *glaucom*
Smooth dock	*Rumex* (species unknown)
Smooth pigweed	*Amaranthus hybridus* L.
Smooth sumac	*Rhus glabra* L.
Soft brome	*Bromus mollis* L.
Soft rush	*Juncus effusus* L.
Sorghum-almum	*Sorghum almum* Parod.
Sour paspalum	*Paspalum conjugatum* Bergius
Southern crabgrass	*Digitaria ciliaris* (Retz.)Koel
Southern sandbur	*Cenchrus echinatus* L.
Sow thistle	*Sonchus* spp.
Spear grass—see cogon grass	There are several plants with this common name—this is probably *I. cylindrica* L.
Speedwell	*Veronica* spp.
Spiny amaranth	*Amaranthus spinosus* L.
Spiny sowthistle	*Sonchus asper* (L.) Hill
Spotted cat's ear	*Hypochaeris radicata* L.
Spotted geranium	*Geranium maculatum* L.
Spotted knapweed	*Centaurea maculosa* Lam. = *Centaurea stobe* L.
Spotted waterhemlock	*Cicuta maculata* L.
Sprangletop	*Leptochloa* spp.
Spreading dayflower	*Commelina diffusa* Burm. f.
Spreading dogbane	*Apocynum androsaemifolium* L.
Spurge	*Euphorbia* spp.
Spurred anoda	*Anoda cristata* (L.) Schlecht.
St. Augustine grass	*Stenotaphrum secundatum* (Walt.) Ktze.
St. Johnswort/Tipton's weed/ Klamath weed	*Hypericum perforatum* L.

Common Name	Scientific Name[2]
Star chickweed	*Stellaria pubera* Michx.
Stink grass	*Eragrostis cilianensis* (All.) E. Mosher
Stinkweed	*Pluchea camphorata* (L.) DC.
Strangler vine	*Morrenia odorata*
Sudangrass	*Sorghum bicolor* subsp. *drummondii*
Sulfur cinquefoil	*Potentilla recta* L.
Sumac	*Rhus* spp.
Sunflower broomrape	*Orobanche cumana* Wallr.
Sweet clover	*Melilotus* spp.
Syrian sage	*Salvia syriaca* L.
Tall fescue	*Festuca arundinacea* Schreb.
Tall morningglory	*Ipomoea purpurea* (L.) Roth
Tall oatgrass	*Arrhenatherum elatius* (L.) Beauv. ex J. & C. Presl
Tamarisk = salt cedar	*Tamarix ramosissima* Ledeb.
Tansy	*Senecio* spp
Tansy ragwort	*Senecio jacobaea* L.
Tares = common vetch	*Vicia sativa* L.
Darnel	*Lolium temulentum* L.
Texas panicum	*Panicum texanum* Buckl.
Texasweed	*Caperonia palustris* (L.) St. Hil
Toadflax	*Linaria* spp.
Toothed spurge	*Euphorbia dentata* Michx
Torpedograss	*Panicum repens* L.
Trailing crownvetch	*Coronilla varia* L.
Travelers joy/clematis/old man's beard	*Clematis vitalba* L.
Tree cactus	*Opuntia megacantha* Salm-Dyck
Tree-of-heaven	*Ailanthus altissima* (Mill.) Swingle
Tropic ageratum	*Ageratum conyzoides* L.
Tropical soda apple	*Solanum viarum* Dunal
Tropical spiderwort	*Commelina benghalensis* L.
Tumble mustard	*Sisymbrium altissimum* L.

Common Name	Scientific Name[2]
Tumble pigweed	*Amaranthus albus* L.
Turnip rape	*Brassica rapa* L.
Velvet grass	*Holcus lanatus* L.
Velvetleaf	*Abutilon theophrasti* Medikus
Venice mallow	*Hibiscus trionum* L.
Vetch	*Vicia* spp.
Virginia pepperweed	*Lepidium virginicum* L.
Watergrass	*Echinochloa* spp.
Waterreed	*Phragmites communis trin*
Waterhemlock	*Cicuta* spp.
Waterhemp	*Amaranthus* spp.
Waterhyacinth	*Eichhornia crassipes* (Mart.) Solms
Waterlettuce	*Pistia stratiotes* L.
Water lily	The term is used to describe several plants from the *Nymphaeaceae* and the *Nelumbonaceae*
Wavyleaf thistle	*Cirsium undulatum* (Nutt.) Spreng.
Western elodea	*Elodea nuttallii* (Planch.) H. St. John
Western ragweed	*Ambrosia psilostachya* D.C.
White campion	*Silene latifolia*
White mustard	*Sinapis alba* L.
Whitetop	*Cardaria* spp.
Whorled milkweed	*Asclepias verticillata* L.
Wild beet	*Beta vulgaris* L.
Wild blackberry	*Rubus* spp.
Wild cabbage	*Brassica oleracea* L.
Wild cane	*Gynerium sagittatum* (Aubl.) Beauv.
Wild carrot	*Daucus carota* L.
Wild buckwheat	*Polygonum convolvulus* L.
Wild garlic	*Allium vineale* L.
Wild marigold	*Tagetes erecta* L.
Wild melon	*Cucumis melo* L.

Common Name	Scientific Name[2]
Wild mustard	*Sinapis arvensis* L.
Wild oats	*Avena fatua* L.
Wild onion	*Allium canadense* L.
Wild proso millet	*Panicum miliaceum* L.
Wild radish	*Raphanus raphanistrum* L.
Wild rice	*Zizania* spp.
Wild tomato	*Solanum triflorum* L.
Winged water primrose	*Ludwigia decurrens* Walt.
Winter bent grass	*Agrostis hyemalis* (Walt.)
Winter wild oats	*Avena ludoviciana* Durieu
Witchgrass	*Panicum capillare* L.
Witchweed	*Striga asiatica* (L.) Ktze.
Wood sorrel	*Rumex* spp.
Wooly cupgrass	*Eriochloa villosa* (Thunb.) Kunth
Yarrow	*Achillea millefolium* L.
Yellow bristlegrass	*Setaria parviflora*
Yellow charlock	*Brassica arvensis* Ktze = *Sinapis arvensis* L.
Yellow devil hawkweed	*Hieracium caespitosum* Dumort
Yellowflag iris	*Iris pseudacorus* L.
Yellow foxtail	*Setaria pumila* (Poir) Roem & Schultes—formerly *Setaria glauca* (L.) Beauv.
Yellow hawkweed	*Hieracium pratense* Tausche
Yellow mustard	*Brassica hirta* Moench
Yellow nutsedge	*Cyperus esculentus* L.
Yellow rocket	*Barbarea vulgaris* R. Br.
Yellow star thistle	*Centaurea solstitialis* L.
Yellow sweetclover	*Melilotus officinalis* (L.) Lam.
Yellow toadflax	*Linaria vulgaris* Mill.
Yucca	*Yucca* spp.

Glossary of Terms Used in Weed Science[1]

Absorption The process by which herbicides are taken into plants, by roots or foliage (stomata, cuticle, etc.). (*see* Adsorption).

Achene A small, dry, thin-walled fruit that does not split open when ripe, e.g., dandelion.

Acid equivalent (ae) The theoretical yield of parent acid from an active ingredient in acid-based herbicides.

Acid soil Soil with a pH less than 7.0.

Acre A common unit of land measure, equal to 43,560 square feet or 0.405 hectares.

Acropetal Toward the apex; generally upward in shoots and downward in roots. (*see* Basipetal).

Active ingredient (ai) Chemical(s) responsible for herbicidal effects.

Acute toxicity Toxicity that reaches a crisis rapidly.

Adjuvant An ingredient that facilitates or modifies the action of the principle ingredient; an additive.

ADP (adenosine diphosphate) An adenosine-derived ester formed in cells, converted to adenosine triphosphate for energy storage.

Adsorption Chemical or physical attraction of a substance to a surface; can refer to gases, dissolved substances, or liquids on the surface of solids or liquids. (*see* Absorption).

Adventitious Having plant organs such as shoots or roots that are produced in an abnormal position or at an unusual time of development.

Alien A plant native to one region but naturalized in another. An alien plant may also be one that is native to a region and has invaded another region but has not yet become naturalized.

Aliphatic Derived from straight chain hydrocarbons.

Alkaline soil Soil with pH above 7.0.

Allelopathy Adverse effect of one plant or microorganism on another caused by release of a chemical from living or decaying organisms.

Allelopathy, functional Toxins resulting from transformation after release by the plant or during decomposition of plant residues.

Allelopathy, true Allelochemicals that enter the environment through volatilization or root exudation and move through soil by leaching.

Allopatry Occurring in separate, widely differing geographic areas. (*see* Sympatry).

Aneuploid Having an extra chromosome = trisomic.

Angiosperm A plant that has seeds borne within a pericarp.

Annual A plant that completes its life cycle in 1 year (i.e., germinates from seed, grows, flowers, produces seed, and dies in the same season); examples are redroot pigweed, common ragweed, mustards, foxtails, and large crabgrass.

Anticlinal Inclining in opposite directions from a central axis. Used with reference to cell walls that are perpendicular to the surface of an organ.

Apoplast The continuous, nonliving cell wall phase that surrounds and contains the symplast.

Aquatic weed A weed that grows in water. There are three kinds: (1) submerged—grow beneath the surface (e.g., sago pondweed, elodea, water milfoil); (2) emerged—grow above the water (e.g., cattails and water lilies); and (3) floaters—float on the surface (e.g., water hyacinth).

Aromatics Compounds derived from the hydrocarbon benzene.

ATP (adenosine triphosphate) An adenosine-derived nucleotide. The primary source of energy through conversion to adenosine diphosphate.

[1]A glossary of terms also appears on pages 457–462 of Vencill, W.K. (Ed.), 2002. Herbicide Handbook, eighth ed. Weed Science Society of America, Lawrence, K.S. and on pages 424–430 of Senseman, S.A. (Ed.), 2007. Herbicide Handbook, eighth ed. Weed Science Society of America. Lawrence, KS, 458 pp.

Available water The portion of water in soil that can be absorbed by plant roots.

Band application Application to a continuous restricted band such as in or along a crop row, rather than over the entire field.

Basal treatment Herbicide applied to the stems of woody plants at and just above the ground.

Basipetal Toward the base; generally downward in shoots and upward in roots. (*see* Acropetal).

Bed A narrow flat-topped ridge on which crops are grown with a furrow on each side for drainage of excess water, or an area in which seedlings or sprouts are grown before transplanting.

Biennial A plant that completes its growth in 2 years. The first year, it produces leaves and stores food; the second year, it produces fruits and seeds (e.g., wild carrot, bull thistle).

Biological control Controlling a pest with natural or introduced enemies.

Biotype A population within a species that has distinct genetic variation.

Blind cultivation Cultivating before plant emergence.

Broadcast application Application over an entire area rather than only on rows or beds, or between rows.

Broad-leaved plants In general, used as an antonym for grass plants.

Brush control Control of woody plants.

Bulb A subterranean leaf bud with fleshy scales or coats.

Calibration A series of operations to determine the amount of solution (volume) applied per unit area of land and the amount of pesticide to add to a known volume of diluent.

Capitulum A dense head-like cluster of stalkless flowers.

Carcinogenic Capable of causing cancer in mammals.

Carrier Liquid or solid material added to a chemical compound to facilitate its application (usually water, but diesel oil has been used with water for brush control).

Cas9 CRISPR-associated proteins.

Cation exchange capacity The total exchangeable cations a soil can adsorb; expressed as moles or millimoles of negative charge per kilogram of soil (or other exchange material, e.g., clay).

Chemical name The accepted name of a chemical compound according to the rules of the International Union of Pure and Applied Chemistry.

Chlorosis The loss of green foliage color.

Chronic toxicity Prolonged or lingering toxicity.

Clay (1) Soil consisting of particles <0.002 mm in diameter, (2) soil textural class; soil containing >40% clay, <45% sand, and <40% silt.

Cleistogamous Having small, unopened, self-pollinated flowers.

Common name The generic name for a chemical compound or a plant. For example, the common name "dandelion" is used instead of the scientific name (*Taraxacum officinale* L.) and the common name "2,4-D" is used instead of the chemical name, "2,4-dichlorophenoxy acetic acid."

Compatibility The quality of two compounds that permits them to be mixed without effect on the properties of either.

Compensation point The light intensity at which the rate of photosynthesis and the rate of respiration in a leaf are equal.

Competition The active acquisition of limited resources by an organism that results in a reduced supply and consequently reduced growth of other organisms.

Concentration The amount of active ingredient in a given volume of diluent. Recommendations for concentration of herbicides are normally based on the weight or volume of active ingredient or product per-unit volume of diluent.

Contact herbicide An herbicide that kills primarily by contact with plant tissue rather than after translocation.

Corm The enlarged fleshy base of a stem, bulb-like but solid.

Cotyledon First leaf or pair of leaves of the embryo of seed plants.

CRISPR Clustered regularly interspersed short palindromic repeats.

Culm The stem of sedges and grasses.

Cuticle The layer of cutin, a wax-like, water-repellent material that covers the epidermis of plants.

Defoliator or defoliant Causes foliage to fall from plants.

Desiccant A desiccant promotes dehydration of plant tissue and may lower moisture content of seeds to facilitate harvest.

Dicot Abbreviation of dicotyledon. A member of the Dicotyledoneae. One of two classes of angiosperms (see Monocotyledoneae) usually characterized by having two seed leaves (cotyledons), net leaf venation, and a root system with a taproot.

Diluent A liquid or solid to dilute an active ingredient in the preparation of a formulation.

Dioecious Having male and female reproductive organs on separate plants; literally = two houses.

Diploid Having two copies of each chromosome.

Directed spray Precise application to a specific area or plant organ such as a row or bed, or just plant leaves or the base of stems.

Dormant = dormancy A condition in which seeds or other living plant organs are not dead but do not grow. A state of suspended animation.

Dormant spray Chemical application in winter or very early spring before plants have begun active growth.

Drift Movement as a liquid.

Ecosystem Ecological entity composed of the biotic community and nonliving environmental phases functioning together in an interacting system.

Ecotype A population within a species that develops distinct morphological or physiological characteristics in response to a specific environment. The changes persist when species move to a different environment.

Edaphic Of or pertaining to soil.

Emergence Appearance of a plant above the soil.

Emersed plant Rooted or anchored aquatic plant that grows with most of its stem tissue above the water surface.

Emulsifiable concentrate (EC) Single-phase, liquid formulation that forms an emulsion when added to water.

Emulsifier A material that facilitates suspension of one liquid in another.

Emulsion A mixture in which one liquid is suspended in minute globules in another liquid without either losing its identity (e.g., oil in water).

Encapsulated formulation An herbicide enclosed in capsules (or beads) to control the rate of release of the active ingredient and thereby extend the period of activity.

Epidermis The outer protective, nonvascular layer of cells or protective covering of a plant or plant part.

Epinasty Increased growth on one surface of plant organ or part, causing it to bend.

Euploid A plant with a full chromosome set.

Eutrophication A term used to designate a body of water in which the increase in mineral and organic nutrients has reduced the dissolved oxygen, thus creating an environment that favors plant over animal life.

Exchange capacity The total ionic charge of the soil adsorption complex.

Exotic From another part of the world; foreign.

Field capacity The percent water remaining in soil after free drainage has ceased.

Floating plant Free-floating or anchored aquatic plant adapted to grow with most vegetative tissue above the water surface; plants rise and fall with water level.

Flowable Two-phase formulation containing solid herbicide suspended in liquid that forms a suspension when added to water.

Formulation A pesticide preparation supplied by a manufacturer. The process of preparing pesticides for commercial use.

Fumigant A volatile liquid or gas used to kill insects, nematodes, fungi, bacteria, seeds, roots, rhizomes, or entire plants. They are usually applied in an enclosure of some kind or to covered soil.

Gamete Male and female reproductive cells.

Genotype The genetic constitution of an organism.

Germination The activation of metabolic sequences culminating in renewed growth of the seed embryo, which is morphologically observable as radicle protrusion through the seed coat.

Granular A dry formulation consisting of discrete particles usually less than 10 mm^3, designed to be applied without water.

Green manure Green plant material incorporated into soil.

Growth regulator A substance effective in minute amounts for controlling or modifying plant processes.

Growth stages of cereal crops (1) Tillering: when additional shoots are developing from the crown; (2) jointing: when stem internodes begin elongating; (3) boot: when leaf sheath swells owing to the growth of developing spike or panicle; (4) heading: when seed head emerges from the sheath.

Hard water Water that contains minerals, usually calcium and magnesium sulfates, chlorides, or carbonates, in solution, to the extent of causing a curd or precipitate, rather than a lather, when soap is added.

Harvest index The amount (weight) of grain versus total plant foliar dry weight.

Hectare An area of land equal to 10,000 square meters or 2.47 acres.

Herbaceous plant A vascular plant without woody tissues.

Herbicide A chemical used for killing or inhibiting the growth of plants; phytotoxic chemical (from Latin *Herba*, plant and *caedere*, kill).

Herbicide resistance The inherited ability of a plant to survive and reproduce after exposure to a dose of herbicide normally lethal to the wild type. Resistance occurs through natural selection. Weed Technology 12 (4), 789, 1998.

Herbicide tolerance The inherent ability of a species to survive and reproduce after herbicide treatment. Weed Technology 12 (4), 789, 1998.

Heterozygous Having identical pairs of genes for any given pair of hereditary characteristics.

Homozygous Having dissimilar pairs of genes for any hereditary characteristic.

Hormone A growth-regulating substance occurring naturally in plants or animals; refers to certain man-made or synthetic chemicals with growth-regulating activity; more correctly called synthetic regulators, not hormones

Hypocotyl The portion of the stem of a plant embryo or seedling between the cotyledons.

Imbibition To imbibe is to absorb or take in water or any liquid. Imbibition is the act of imbibing.

Incorporate To mix or blend herbicides in soil.

Interference Total adverse effect that plants exert on each other when growing in a common ecosystem; includes competition and allelopathy.

Invasive species An alien species that becomes established in natural or seminatural ecosystems or habitats is an agent of change and threatens native biological diversity.

Invert emulsion An emulsion in which water is dispersed in oil; oil forms the continuous phase with water dispersed therein; usually a thick, mayonnaise-like mixture results.

Involucre A circle or collection of bracts surrounding a flower cluster or head, or a single flower.

Kairomone An allelochemical of favorable adaptive value to the organism receiving it.

K_d The ratio of sorbed to dissolved pesticide at equilibrium in a water/soil slurry.

K_{oc} Soil organic carbon sorption coefficient; K_d divided by the weight fraction of organic carbon in soil.

Label Directions for herbicide use created by the manufacturer and approved by federal or state regulatory agencies.

Lay-by Refers to the stage of crop development (or the time) when the last regular cultivation is done.

LD_{50} The dose (quantity) of a substance that causes 50% of test organisms to die; usually expressed in weight (milligrams) chemical per unit of body weight (kilograms).

Leaching Usually refers to movement of water through soil, which may move soluble plant foods or other chemicals.

Lodge (lodging) To beat down plants from action of rain and wind; often encouraged by high rates of nitrogen fertilizer.

Mass screening Use of selected plant species to ascertain whether a candidate herbicide has activity (kills or affects plant growth) and selectivity (affects some, but not all species).

Mechanism of action The precise biochemical or biophysical reaction or series of reactions that create an herbicide's final or ultimate effect; many herbicides have primary and secondary mechanisms of action.

Meristem The growing point or area of rapidly dividing cells at the tip of the stem root or branch.

Mesocotyl The elongated portion of the axis between the cotyledon and coleoptile of a grass seedling.

Miscible liquids Two or more liquids capable of being mixed, which will remain mixed under normal conditions.

Mode of action The sequence of events that occur from an herbicide's first contact with a plant until its final effect (often plant death) is expressed.

Monocot = monocotyledon One of two classes of angiosperms, usually characterized by having one seed leaf (cotyledon), parallel leaf venation, and root systems arising adventitiously that are usually diffuse and fibrous. (See Dicot).

Mulch Material (grass, straw, plastic, and plant residue) spread on soil to cover or protect soil.

Naturalized Adaptation or acclimation of a plant (or any species) to a new environment. To conform to nature.

Necrosis = necrotic Localized death of tissue, usually characterized by browning and desiccation.

Niche The area within a habitat occupied by an organism. It is also the set of functional relationships of an organism or a population to the environment it occupies. The term is used to describe a species place in the community, including when it is present, what place (space) it occupies, and what function(s) it fulfills in the community. The ecological concept of niche includes a species specialization: its special or unique function in the community.

Nonselective herbicide Herbicide used to kill plants generally without regard to species.

No-till = no-tillage Planting without prior soil disturbance.

Noxious weed Plant defined by law as being especially undesirable, troublesome, and difficult to control.

Pappus The modified calyx in some members of the Compositae that forms a crown or a tuft of bristles at the summit of the achene.

Perennial A plant that lives from year to year; in most cases, in cold climates, stems and foliage die but roots persist, (e.g., field bindweed, dandelion, Canada thistle, johnsongrass, leafy spurge).

Pesticide Any substance or mixture of substances intended for controlling insects, rodents, fungi, weeds, and other forms of plant or animal life considered to be pests.

Phenology Naturally occurring phenomena that recur periodically, e.g., flowering and time of seed germination.

Phloem Living plant tissue that transports metabolic compounds from site of synthesis to storage or site of use.

Phytotoxic Something that is poisonous or inhibitory to growth of plants (from Greek *phyton*, plant and *toxikon*, poison).

Plagiotropic A term used primarily for roots, stems, or branches to describe growth at an oblique or horizontal angle.

Plasmalemma A membrane that surrounds a cell, especially one immediately within the wall of a plant cell.

Polyphagous Feeding on a using a variety of foods.

Postemergence (post) After emergence of a specified weed or crop.

Preemergence (pre) After a crop is planted but before it emerges.

Preplant Application of an herbicide (or anything) before planting.

Preplant incorporated (PPI) Herbicide applied and blended into soil before planting.

Radicle The part of the plant embryo that develops into the primary root.

Rate and dosage These are synonyms. Rate is preferred and usually refers to the amount of active ingredient applied to a unit area regardless of the percentage of chemical in the carrier.

Registration The process of gaining approval (registration) from the US Environmental Protection Agency to sell an herbicide or other pesticide as governed by the amended Federal Insecticide, Fungicide, and Rodenticide Act.

Residual Applied to an herbicide, the sense is to have a continued effect over a period of time.

Resistant The decreased response of a population to an herbicide (*see* Tolerance).

Rhizome An underground stem capable of sending out roots and leafy shoots.

Ribonucleic acid (RNA) A polymeric constituent of all living cells, which consists of a single strand of alternating phosphate and ribose units with the bases adenine, guanine, cytosine, and uracil bonded to the ribose. The structure and base sequence determine the proteins synthesized.

RNAi Ribonucleic acid (RNA)-interference(i).

Runner A plagiotropic shoot that may root when in contact with soil.

Safener A substance that reduces an herbicide's phytotoxicity.

Selectivity The property of differential tolerance' some plants are affected whereas others are not. It is an essential attribute of most herbicides.

Sequester To remove or set apart.

Sink A plant site with a high rate of metabolic activity where food resources are used.

Soil incorporation The mechanical mixing of herbicides in soil.

Soil injection Mechanical placement of an herbicide beneath the soil surface.

Soil persistence The length of time an herbicide remains in soil. It may refer to effective life (i.e., the time during which plants are killed) or to total soil residence time.

Soil sterilant An herbicide that prevents growth of all plants. The effects may be temporary (a few months) or long-term (years).

Solution A homogenous mixture of two or more substances.

Solution concentrate A liquid formulation that forms a solution when added to water.

Soluble powder A dry formulation that forms a solution when added to water.

Spike stage The early emergence stage of corn in which leaves are tightly rolled to form a spike, usually before corn is more than 2 inches tall.

Spot treatment Application of herbicide to localized or restricted areas, as opposed to overall, broadcast, or complete coverage.

Spray drift Movement of airborne liquid spray particles.

Stolon An aboveground creeping stem that can root and develop new shoots (e.g., Bermuda grass).

Stunting Retardation of growth and development.

Submersed plant An aquatic plant that grows with all or most vegetative tissue under water.

Surfactant Material added to pesticide formulations to impart spreading, wetting, dispersibility, or other properties that modify surface interactions.

Suspension A liquid or gas in which very fine solid particles are dispersed but not dissolved.

Sympatry Occurring in one area (*see* Allopatry).

Symplast A functionally integrated unit consisting of all living cells of a multicellular plant.

Synergism The action of two or more substances that creates a total effect greater than the sum of independent effects; achievement of an effect by two substances that neither is capable of achieving alone.

Systemic herbicide An herbicide that is translocated in plants to produce an effect throughout the entire plant system.

Taxa The plural of taxon.

Taxon A group of organisms constituting one of the categories or formal units in taxonomic classification, such as a phylum, order, family, genus, or species, and characterized by common characteristics in varying degrees of distinction.

Teratogenic Something that produces birth defects.

Tetraploid Having four copies of each chromosome. Also referred to as polyploid.

Tolerance Amount of pesticide chemical allowed by law to be in or on the plant or animal product sold for human consumption (legal definition). Natural and normal variation that exists within a species and can evolve quickly (*see* Resistance).

Tolerant Capable of withstanding effects (*see* Resistance).

Toxicity The potential to cause injury, illness, or undesirable effects.

Trade name A trademark or other designation of a commercial product.

Translocation Transfer of photosynthate or other materials such as herbicides from one plant part to another.

Transposons Transposable genetic elements.

Tuber A much-enlarged portion of a subterranean branch (stolon) that has buds on the sides and at the tip.

Turion Scaly shoot developed from a bud on a subterranean or submerged rootstock.

Volatility A measure of the tendency to change state from liquid to gas.

Weed Any plant that is objectionable or interferes with the activities and welfare of humans (definition accepted by the Weed Science Society of America).

Weed control (1) The process of limiting weed infestations so crops can be grown profitably or other operations can be conducted efficiently; (2) the process of reducing weed growth or weed infestation to an acceptable level.

Weed eradication Complete elimination of all live plants, plant parts, and weed seeds from an area.

Weed management A relatively new term in the lexicon of weed science that has not obtained an official definition (Vencill, 2002). A synthesis of definitions follows rational deployment of appropriate technology to minimize weed effects, provide systematic management of weed problems, and optimize intended land use. (It is likely that the evolving definition will incorporate determination of an economic threshold.)

Weed prevention The process of stopping weeds from invading an area.

Wettable powder A powder that forms a suspension (not a true solution) in water.

Wetting agent A chemical that, when added to a spray solution, causes the solution to contact (wet) plant surfaces more thoroughly.

Winter annual A usually temperate climate plant that starts germination in the fall, lives over winter, and completes its growth, including seed production, the following season (e.g., downy brome grass); many summer annuals can behave as winter annuals if they germinate in fall and live over the winter.

Xylem The nonliving plant tissue that conducts water and solutes from roots to shoots.

Index

'Note: Page numbers followed by "f" indicate figures and "t" indicate tables.'

Printed in the United States
By Bookmasters